**The Geology of North America**
**Volume O-2**

# *Hydrogeology*

Edited by

William Back
U.S. Geological Survey
431 National Center
Reston, Virginia 22092

Joseph S. Rosenshein
U.S. Geological Survey
414 National Center
Reston, Virginia 22092

Paul R. Seaber
Illinois Geological Survey
Natural Resources Building
615 East Peabody Drive
Champaign, Illinois 61820

**1988**

**Acknowledgment**          CEN    SCT

Publication of this volume, one of the synthesis volumes of *The Decade of North American Geology Project* series, has been made possible by members and friends of The Geological Society of America, corporations, and government agencies through contributions to the Decade of North American Geology fund of the Geological Society of America Foundation.

Following is a list of individuals, corporations, and government agencies giving and/or pledging more than $50,000 in support of the DNAG Project:

Amoco Production Company
ARCO Exploration Company
Chevron Corporation
Cities Service Oil and Gas Company
Diamond Shamrock Exploration
 Corporation
Exxon Production Research Company
Getty Oil Company
Gulf Oil Exploration and Production
 Company
Paul V. Hoovler
Kennecott Minerals Company
Kerr McGee Corporation
Marathon Oil Company
Maxus Energy Corporation
McMoRan Oil and Gas Company
Mobil Oil Corporation
Occidental Petroleum Corporation

Pennzoil Exploration and Production
 Company
Phillips Petroleum Company
Shell Oil Company
Caswell Silver
Standard Oil Production Company
Sun Exploration and Production Company
Superior Oil Company
Tenneco Oil Company
Texaco, Inc.
Union Oil Company of California
Union Pacific Corporation and
 its operating companies:
  Union Pacific Resources Company
  Union Pacific Railroad Company
  Upland Industries Corporation
U.S. Department of Energy

Published by The Geological Society of America, Inc.
3300 Penrose Place • P.O. Box 9140, Boulder, Colorado 80301

Printed in U.S.A.

**Library of Congress Cataloging-in-Publication Data**

Hydrogeology / edited by William Back, Joseph S. Rosenshein, Paul R.
 Seaber.
     p.    cm.—(The Geology of North America ; v. O-2)
  Bibliography: p.
  Includes index.
  ISBN 0-8137-5206-X
  1. Water, Underground—North America.   I. Back, William, 1925–
. II. Rosenshein, Joseph S.   III. Seaber, Paul R.   IV. Decade of
North American Geology Project.  V. Series.
QE71.G48      1986 vol. O-2
[GB1012]
557 s—dc19
[551.4'9'097]                                    88-24732
                                                    CIP

10  9  8  7  6  5  4  3

**Front Cover:** Thunder Spring, perennial source of Thunder River, a tributary of Tapeats Creek in the Grand Canyon of the Colorado River, Arizona. The spring issues from above a relatively impermeable layer in the Cambrian Muav Limestone. Photo by John E. Warme and Lewis C. Kleinhans, Colorado School of Mines, July 1988.

# Contents

## Contents

### Central Cratonic Sector

### Appalachian Sector

**Coastal Plain Sector**

**Island Sector**

**Permafrost**

# III. COMPARATIVE HYDROGEOLOGY

## IV. GROUNDWATER AND GEOLOGIC PROCESSES

## V. OUTLINE FOR THE FUTURE

### Plates

Plate 1. Hydrogeologic map of North America showing the major rock units that underlie the surfical layer
R. Heath

Plate 2. Hydrogeologic map of North America showing the major units that comprise the surfical layer
R. Heath

Plate 3. Ground-water flow systems in the Great Basin
J. R. Harrill, J. S. Gates, and J. M. Thomas, with additions by M. Mifflin

# *Preface*

*The Geology of North America* series has been prepared to mark the Centennial of The Geological Society of America. It represents the cooperative efforts of more than 1,000 individuals from academia, state and federal agencies of many countries, and industry to prepare syntheses that are as current and authoritative as possible about the geology of the North American continent and adjacent oceanic regions.

This series is part of the Decade of North American Geology (DNAG) Project which also includes eight wall maps at a scale of 1:5,000,000 that summarize the geology, tectonics, magnetic and gravity anomaly patterns, regional stress fields, thermal aspects, seismicity, and neotectonics of North America and its surroundings. Together, the synthesis volumes and maps are the first coordinated effort to integrate all available knowledge about the geology and geophysics of a crustal plate on a regional scale.

The products of the DNAG Project present the state of knowledge of the geology and geophysics of North America in the 1980s, and they point the way toward work to be done in the decades ahead.

In addition to the contributions from organizations and individuals acknowledged at the front of this book, major support has been provided to the editors of this volume by the Regional Aquifer Systems Analysis Program of the U.S. Geological Survey.

A. R. Palmer
General Editor for the volumes
published by The Geological Society
of America

J. O. Wheeler
General Editor for the volumes
published by the Geological
Survey of Canada

# *Foreword*

This volume is about hydrogeology—with emphasis on the geologic aspects. It is a synthesis of current hydrogeologic understanding merged with principles and processes from other sub-disciplines of geology.

Although some general studies of ground water in the United States and Canada were made prior to 1870, it was not until the 1890s and early 1900s that American and Canadian geologists directed serious attention to ground water and began to publish comprehensive areal reports on the subject; investigations of ground water began in Mexico about a decade later. The year 1888 marks not only the founding of the Geological Society of America but also the first appropriation in the United States for funding of water research that developed into the Water Resources Division of the U.S. Geological Survey. This volume is dedicated to our North American predecessors of the last hundred years who set the foundation and published the science that made this book possible.

In many ways, this volume can be considered a modern (expanded to include Canada and Mexico) version of Meinzer's U.S. Geological Survey Water Supply Paper 489, published in 1923 as the first of a planned series of six papers on ground water in the United States. Because of the progress that has been made in understanding ground-water science in North America since 1923 and the diversity of topics now encompassed by hydrogeology, this volume could not have been written by a single author.

The volume demonstrates hydrogeologic principles, concepts, and processes that control the occurrence, movement, storage, and chemical character of ground water. One goal of this volume is to identify, clarify, and describe systematically the basic relation of hydrogeology to other sub-disciplines of geology such as geomorphology, stratigraphy, structure, and historical geology.

Thus, one prime purpose of this volume is communication and stimulation. It is expected that the authors' contributions will provide the basis for significant geologically oriented research, creative exploration and management of ground-water resources, and advances in knowledge of regional and comparative hydrogeology in North America.

This volume is not an atlas that evaluates ground water as a resource, nor is it a discussion of anthropogenic effects. Discussions of man's effects during the past several centuries are limited to those necessary to demonstrate principles, concepts, and processes. Excellent reports and journals published by many water-resource organizations and societies throughout North America are abundantly available, and readers wishing to know about such subjects as water resources or contamination problems can turn to those.

Although many excellent hydrogeology textbooks exist today, a modern, comprehensive one-volume documentation of the relation of ground water to geology does not exist. The editors and authors of the individual chapters are extremely grateful to the Geological Society of America (GSA), and the Decade of North American Geology (DNAG) Project, for the opportunity to prepare such a book.

This book is the collective effort of members of the Hydrogeology Division of GSA and had its origin in the summer of 1980 when Bruce Hanshaw, of the U.S. Geological Survey, proposed two volumes, one each on surface water and ground water, to be the Division's Centennial contribution. These topics were added to the planned set of volumes on *The Geology of North America* being developed by the DNAG Project. At the 1981 Annual Meeting of GSA in Cincinnati, Bill Back and Paul Seaber agreed to serve as editors, and an annotated list of topics deemed suitable for the volume was prepared for the 1982 Annual Meeting in New Orleans. Thirty-five hydrogeologists who attended the Hydrogeology Division meeting that year expressed an interest in helping to write this volume, and 26 volunteered to attend a planning workshop to be held in Denver, Colorado, June 18–19, 1983.

In January, prior to that meeting, the attendees were given general guidelines for the planning workshop and identified chapters they would like to prepare. The attendees divided into four groups, each with a temporary leader, who later became the coordinator for a section of this volume: Gerry Meyer (Introduction), Joe Rosenshein (Regional Synthesis), Stan Davis (Comparative Hydrogeology), and Pat Domenico (Ground Water and Geologic Processes). The editors and authors are grateful to these Section Coordinators for their leadership and editing efforts.

A second workshop, attended by 38 of the chapters' authors, was held in Denver June 8–9, 1985. That meeting reviewed the rough drafts of many chapters and discussed progress and problems. During that meeting, Joe Rosenshein, as coordinator for more than half of this volume, agreed to become a co-editor. Although the other section coordinators could justifiably be listed as co-editors, they chose to remain in their coordinating roles. Informal meetings to discuss progress and problems in the development of this volume were held by editors and authors in attendance at every Hydrogeology Division annual meeting from 1983 to 1987.

This volume is a well-diversified North American effort, involving hydrogeologists from Canada, the United States, and Mexico. Federal, Provincial, and State agencies, consulting firms, industry, and universities, as well as private individuals, some of whom are officially retired, contributed to the success of this volume. Many thanks go to the organizations who were employers of the editors and authors. The U.S. Geological Survey is especially thanked for the support given, particularly by the Regional Aquifer System Analysis Program, to the three editors and many of the authors of the volume. Throughout the entire process, we had the firm and experienced guiding hand of A. R. (Pete) Palmer, to whom we express our sincere appreciation. Above all, we express our appreciation to the authors who not only completed their own chapters but also so willingly agreed to review the chapters of their colleagues. Their enthusiasm and consistent encouragement made the preparation of this book a most pleasant and memorable experience.

The Editors
July 1988

The Geology of North America
Vol. O-2, Hydrogeology
The Geological Society of America, 1988

# Chapter 1

# *Historical perspective*

**Gerald Meyer**
*U.S Geological Survey, 409 National Center, Reston, Virginia 22092*
*With contributions by*
**George Davis**
*Woodward-Clyde, Rockville, Maryland*
**P. E. LaMoreaux**
*P. E. LaMoreaux and Associates, Tusacaloosa, Alabama*

## INTRODUCTION

In any field of science, soundness of conclusions is dependent upon the validity of basic concepts and principles, accuracy of collected facts, and the level of understanding of processes at work. This chapter, therefore, summarizes evolvement of the primary concepts and principles on which modern hydrogeology is based and the basic processes perceived to be operating. Methods of investigation and problem-solving developed along with the science; the origins of modern methods of regional hydrogeologic investigation are an important element of the growth of hydrogeology.

## REGIONAL HYDROGEOLOGIC PRINCIPLES AND PROCESSES

Harold E. Thomas (1952) noted that "the science of hydrology would be relatively simple if water were unable to penetrate below the earth's surface." In the absence of infiltration, natural water flow would be reduced to precipitation, overland runoff, and evapotranspiration. Subsurface geology would have little influence on hydrology and water resources, and the science of hydrogeology would not have arisen.

However, water does indeed enter, collect in, migrate through, react chemically with, and discharge from the earth's rocks; subsurface water is a large and important segment of the hydrologic cycle. Evolvement of the primary hydrogeological concepts, principles, and processes pertaining to ground-water occurrence in North America is only summarized here, but these fundamentals are addressed more fully in succeeding chapters in their relations to the regional (Part II) and comparative (Part III) hydrogeological descriptions and in regard to the role of ground water in geologic processes (Part IV).

Hydrogeology is a young science resting on a foundation of physico-chemical principles and processes, identified for the most part during only the past 100 years. Most hydrogeological concepts originating in the latter part of the nineteenth century and in the early decades of this century, though modified to some extent in subsequent years, remain accepted and actively applied today. Modern contributions have added significant advances, of course, but the basic foundation of fundamental concepts, principles, and processes still determines the nature of the science and its applications.

Hydrogeology is an interdisciplinary science that is dependent upon many branches of the physical, chemical, and biological sciences. It is an offspring of geology and hydrology, and hydrogeology could develop as an organized science only after maturation of those two mother sciences in the eighteenth and nineteenth centuries. Advances in the physical (Freeze and Back, 1983) and geochemical (Back and Freeze, 1983) branches of hydrogeology were identified in two volumes of benchmark papers selected from thousands of hydrogeological papers. These two scientific anthologies concentrate on North American ground-water literature and therefore are especially germane to the purposes of this volume.

A number of other books and papers contribute to documentation of the history of advances in hydrogeology. Meinzer (1942) described the founding and rise of hydrology from antiquity to the 1930s, including the late emergence of the ground-water science. Bredehoeft (1976) assessed the status of quantitative ground-water hydrology, and Rouse and Ince (1957) concentrated on hydraulics. Landa and Ince (1987), the most recent publication on history of hydrology, contains invited papers that were presented at symposia sponsored by the American Geophysical Union. Biswas (1970) traced the history of hydrology, and Adams (1954) provided a parallel comprehensive his-

Meyer, G., 1988, Historical perspective, in Back, W., Rosenshein, J. S., and Seaber, P. R., eds., Hydrogeology: Boulder, Colorado, Geological Society of America, The Geology of North America, v. O-2.

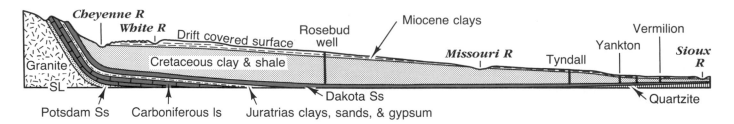

Figure 1. Vertical section from eastern portion of the Black Hills across South Dakota, showing the attitude and relations of the water-bearing Dakota sandstone; looking north; 483 km (300 mi) long; vertical scale considerably exaggerated (from Darton, 1897).

tory of geology. Ferris and Sayre (1955) identified the origins and rise of quantitative ground-water hydrology. Current trends in scientific hydrogeology are characterized in a recent publication (Narasimhan, 1982) containing invited papers that cover concepts, methods of measurement, mathematics, and—especially relevant to the topic of this volume—regional ground-water flow and geochemistry (Bredehoeft and others, 1982, p. 297–316).

The most significant advances in physical and chemical hydrogeology with respect to regional ground-water systems are summarized in the following two subsections.

### Physical hydrogeology

Early North American hydrogeology arose in company with pragmatic ground-water exploration, mainly for irrigation water supply in mid-western and western United States. The enlarging and widening water requirements for agricultural development of subhumid and semiarid regions spurred interest in regional ground-water systems underlying the Great Plains and western physiographic provinces. In the absence of ready access to steam or electrical power to operate pumps, ground-water systems that produced flowing artesian wells were especially desired, and much of the early hydrogeologic research in the United States and Canada dealt with regional geology and artesian phenomena in the central and western parts of the continent. Investigations in industrial and populated areas of the East, on the other hand, focused on localized water development, including well-field and well-construction design, water-treatment technology, and other engineering aspects of hydrology. Hydrogeological and water-resources literature of the late nineteenth and early twentieth centuries reflects these regional differences in ground-water development and their impacts on the science.

A number of distinguished early North American geologists turned their talents to hydrogeology in efforts to identify the fundamental controls exercised by lithology and structure on the continuity of aquifers and flow of ground water. Chamberlin's report to the Director of the Geological Survey in 1885 outlined the geologic and hydrologic conditions necessary for distant migration and artesian flow of ground water. His clear explanation of regional ground water occurrence and the functioning of artesian systems may be said to mark the formal beginning of hydrogeology as an organized science (Freeze and Back, 1983, p. 291). Chamberlin described water-bearing and confining beds and their controlling influences on ground water, and particularly on artesian flow, noting importantly that totally impermeable confining beds do not exist. Darton (1897) added additional fundamental concepts in his report on classical initial field studies of the Dakota sandstone, an extensive artesian aquifer system of the north-central plains region of the United States. Darton identified recharge to the system in the topographically high Black Hills of southwestern South Dakota, eastward flow, and discharge in areas 300 to 500 km to the east (Fig. 1). In a more rigorous analysis of a much smaller system, King (1898) related hydraulic-head gradients to flow of ground water in a shallow, unconfined aquifer near Madison, Wisconsin. In perhaps the first such usage, he employed a flow net to depict the recharge, flow-through, and discharge of ground water.

Those and many additional important hydrogeological studies were recorded in the Annual Reports of the U.S. Geological Survey just before and after the turn of the century. The need for a publication series devoted to water supply became apparent, and the Geological Survey's Water-Supply and Irrigation Paper series was initiated in 1896. (In 1907, the title was contracted to the now-familiar Water-Supply Paper.) During the first dozen years of the existence of the series, a large number of substantive reports on hydrogeological field studies in many varied geologic terranes were published, providing general ground-water understanding for most of the principal rock types underlying the continent. Leverett (1899), for example, prepared hydrogeologic

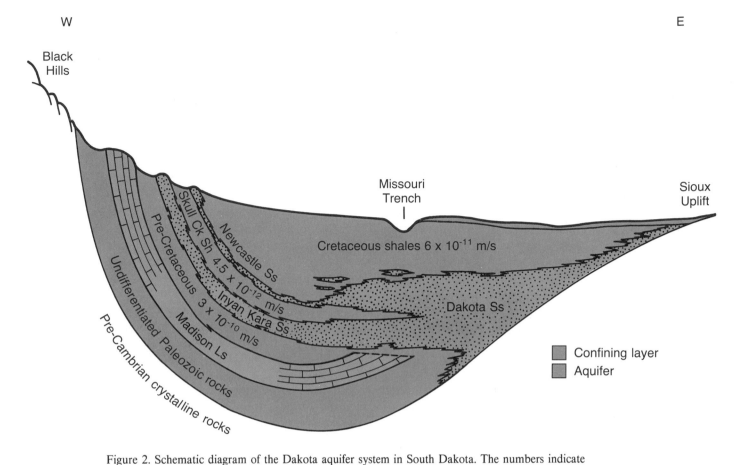

Figure 2. Schematic diagram of the Dakota aquifer system in South Dakota. The numbers indicate vertical hydraulic conductivity of the confining layers (after Neuzil, 1980).

maps of Indiana showing the distribution of glacial deposits and their well-yield potential. Mendenhall (1905) investigated the hydrogeology of alluvial deposits of the San Bernardino Valley, California, and Ellis (1906) examined the storage and flow controls exercised by fracture systems in the crystalline rocks of Connecticut.

Geologists, geochemists, and hydrologists at state geological surveys and universities contributed significantly to the hydrogeological literature and understanding of those early years (the late 1800s and early 1900s). These agencies and institutions continue to be important contributors today. Studies range from systematic well and spring inventories to complex hydrologic and geochemical investigations.

Examples of important early work at the state level include a report on the chemistry of ground water in Kansas (Bailey, 1902), on the artesian systems of the Gulf Coastal Plain (McCallie, 1898), on the nature of occurrence and availability of ground water in Alabama (Smith, 1907), and on the geology and ground-water resources of Mississippi (Hilgard, 1860).

Both scientific and economic interest in artesian systems carried into the twentieth century. As noted by Freeze and Back (1983, p. 292), the Dakota sandstone continued as a model for

study and debate long after publication of Darton's early explanation of artesian flow, which rested on the relative differences between the slope of the land surface and the slope of the potentiometric surface. Russell, in 1928, proposed a different concept of the origin of artesian pressure based on the hydrodynamic loading of lenticular sandstone layers, but Meinzer's (1928) explanation, published about the same time and based on the concept of compressibility and elasticity of artesian aquifers, received greater acceptance. Consideration of possible hydraulic connection between the Dakota sandstone and the deeper-lying Madison limestone (Swenson, 1968) has expanded perception of the possible magnitude of the Dakota flow system and sources of recharge to it. The general principles of ground-water occurrence in the Dakota aquifer system identified by Chamberlin and Darton are still in acceptance today. Geological and hydrological knowledge of this and similar systems has increased, particularly with respect to the hydrologic interaction between aquifers and confining layers. For instance, compare the present-day understanding of the hydrogeology of the Dakota aquifer system, expressed by Neuzil (1980) as reflected in Figure 2, with Figure 1.

Papers of Meinzer (1928), Russell (1928), and Terzaghi (1929) led to the realization that much of the water drawn from artesian aquifer systems could be accounted for by compaction of the sediments and its consequence, subsidence of the land surface. Tolman (1937) described land subsidence of more than 1 m in the Santa Clara Valley, California, and attributed it to compaction within the confined aquifer system. Smith (1934, cited in Tolman, 1937) was perhaps the first worker to compute the volume of ground water released by compaction over a broad area—the Livermore Valley, California.

Following World War II, pronounced land subsidence, on the scale of meters, occurred in several areas in California. This phenomenon, which was sufficient to cause severe damage to large structures that depended upon gentle gradients, such as irrigation canals, drainage ditches, and levee systems, led to the establishment in 1954 of an Inter-Agency Committee on Land Subsidence to study the problem. This committee, under the leadership of J. F. Poland of the U.S. Geological Survey, carried out studies over a period of three decades that elucidated the mechanics of aquifer systems as related to compaction and suggested effective methods for predicting and ameliorating the land subsidence problem (Bull, 1975; Bull and Poland, 1975; Helm, 1982; Meade, 1968; Miller, 1961; Poland and Davis, 1969; Poland and others, 1975; Poland, 1984).

Expanded pumping of ground water in many developing parts of the continent in the 1920s and succeeding decades accommodated growth of population, industry, and irrigation agriculture. Localized or more extensive lowering of water levels followed, occurring most notably within heavily pumped semi-arid regions where natural replenishment rates are small. The increasing pumping lift required and the accompanying rises in power costs further stimulated interest in the systemwide behavior of ground-water systems under pumping stress. Many now-classic papers appeared; literature of the 1930s and 1940s reflects the increasing scientific attention devoted to well and aquifer hydraulics (Theis, 1935; Hubbert, 1940; Wenzel, 1936, 1942), to the mechanics and significance of "leaky" aquifer systems (Jacob, 1946), and to systemwide ground-water flow (Stringfield, 1936; Sayre and Bennett, 1942).

Cooperation among local, state, and federal water agencies in the United States, and among equivalent jurisdictions in Canada and Mexico, accelerated description of the ground-water systems of the continent. In 1934, V. T. Stringfield of the U.S. Geological Survey, and Herman Gunter, Director of the Florida Geological Survey, together identified the major water problems in Florida and devised a unique broad-scale cooperative federal-state investigative program of the state's water resources. Manpower, funds, and equipment were pooled in a joint venture of mutual benefit to the state and federal governments. Similar statewide programs followed in other states. A piezometric map of the Floridan aquifer, Florida's principal source of ground water, was one early and valuable product of the cooperative studies in the state; it revealed the recharge and discharge areas of the system, hydraulic gradients, and directions of lateral movement of the artesian ground water. The map was a forerunner of the synoptic potentiometric maps now released annually.

Thus, the performance of the aquifers, confining beds, ground-water basins, and regional systems was fairly well understood by the mid-twentieth century, both in theory and in applications to hydrological problems. The good progress enabled geological and water-resources agencies of the Canadian, Mexican, and U.S. governments to expand programs of systematic areal investigation, yielding many thousands of authoritative hydrogeological reports collectively covering large portions of the continent. (See, for example, *Publications of the Geological Survey, 1879–1961,* and subsequent catalogs in that series; *Index of Publications of the Geological Survey of Canada [1845–1958],* and subsequent indexes; and the Mexico (City) National University, Institute of Geology, *Publications Issued from the Beginning [1895] to the Middle of July 1957.)*

This comprehensive DNAG volume was feasible only because of the wealth of information assembled through national, state, and provincial programs of systematic areal coverage. Although most of those studies sought to describe the ground-water resources as an aid to development, management, and regulation, many contributed to the illumination of principles and processes of occurrence as well. These dual accomplishments reflect the scientific orientation characteristic of most ground-water investigators. Pioneer principles of ground-water law and regulation of extractions in New Mexico grew from investigations of shallow and artesian ground water in the Roswell Basin of southeastern New Mexico (Fiedler and Nye, 1933). Jacob's (1946) investigations of ground water on Long Island, New York, advanced the theory of flow in elastic artesian aquifers in the course of developing the hydrological understanding needed to design sound programs for management of increasing pumpage and threats to chemical quality on the island.

Most methodologies for ground-water measurement or analysis evolved in company with field investigations. Until about 1960, regional studies dealt mainly with laterally extensive and comparatively thin ground-water systems, and areal potentiometric contour maps were employed to depict and analyze flow. As Freeze and Back (1983, p. 294) note, Canadian sedimentary and glacial environments under investigation in the Great Plains region, where flow is more localized and vertical components of flow more significant, led to development of methods for depicting and evaluating these systems by use of vertical cross sections (Toth, 1962; Freeze and Witherspoon, 1966). Similarly, numerical computer simulation of ground-water systems and associated predictive methodologies (Prickett, 1975)—now standard techniques in ground-water studies—evolved in association with pragmatic hydrogeological studies.

### Chemical Hydrogeology

Chemical hydrogeology is a younger arm of the ground-water science than its physical counterpart, arising in the twentieth century after the basic geologic and hydrodynamic elements were in place. The advances and present status of chemical hydrogeology are charted in some detail by Back and Freeze (1983), covering the development of chemical hydrogeology in

North America, and by Bredehoeft and others (1982) with respect to the geochemistry of regional flow systems, shale membrane phenomena, and heat flow (p. 306–312). Those publications and the citations they contain constitute a comprehensive history and analysis of chemical phases of hydrogeology, which is addressed only in an abbreviated form here with emphasis on regional aspects in North America.

Systematic measurement of the chemistry of many streams in the United States had been instituted by the U.S. Department of the Interior's Geological Survey and Reclamation Service (now Bureau of Reclamation) by the first decade of the twentieth century (Dole, 1909). Ground-water chemistry, on the other hand, from the beginning commonly was measured and studied integrally with the physical hydrogeology in a manner quite similar to modern practice. The report by Meinzer and Kelton (1913) on the geology and water resources of Sulphur Spring Valley, Arizona, demonstrates the early integrated, joint investigation of chemical and physical hydrogeology. Periodic measurement of changes in ground-water chemistry in a manner comparable to the regular sampling of streamflow was, as today, limited to specific cases requiring time-related chemical data, reflecting early awareness of imperceptibly slow natural changes in the chemical character of ground water. Ground-water samples collected and analyzed in decades past by and large remain representative of ground-water chemistry at the sampled sites today.

The early rooting of chemistry in ground-water science set the stage for further advances in the next several decades toward fuller integration of hydrochemistry with rapidly advancing understanding of regional and basin ground-water flow and geologic controls. See, as an example, the 1916 investigation of ground water in the San Joaquin Valley, California, by Mendenhall, Dole, and Stabler—geologist, chemist, and engineer, respectively. Investigation of saltwater encroachment into fresh coastal ground water during the 1920s and subsequent decades provided a stimulus to integrated study of chemical and hydrodynamic facets of ground-water hydrology (Brown, 1925; Brown and Parker, 1945; Piper and others, 1953).

Many of the chemical principles and processes of fundamental importance to the understanding of ground-water chemistry had been identified by the 1950s. They include the primary chemical reactions and their controls on the chemical nature of ground water, the concept of mineral equilibria in natural water environments, and oxidation-reduction reactions pertaining to the ground-water environment (Back and Freeze, 1983, p. 2). By the end of the decade, the important role of mass balance, mass transfer, mineral equilibria, and kinetics in ground-water chemical systems were identified and in use. Schoeller's (1955) treatise on the chemistry and thermodynamics of ground water captured the status of understanding of principles and processes at the midcentury period. Foster (1950) explained the ion-exchange origin of bicarbonate water in regional aquifers of the Atlantic Coastal Plain through laboratory experimentation and interpretation of field chemical and hydrologic data, paving the way for further integration of chemical and physical hydrogeology.

The marked advances in the principles of water geochemistry and applications to ground water that followed in the 1960s were attributable predominantly to growing appreciation of the important, fundamental role of chemical thermodynamics. Garrels (1960) and coworkers laid the conceptual foundation in the laboratory for modern hydrogeochemical understanding resting on the principles of thermodynamics. Their investigations of the chemistry of low-temperature aqueous systems led to the concept of transfer of mass in flowing water. Back (1961a) demonstrated the application of the law of mass action to equilibrium principles and to the chemistry of ground water, and he (1961b, 1966) also explained and lucidly illustrated the role of hydrochemical facies in the interpretation of hydrogeologic systems. Figure 3, a fence diagram of the middle Atlantic Coastal Plain in the vicinity of Washington, D.C., depicts the cross-sectional geology, on which are superimposed the cationic chemical facies of the ground water and (in the central panel) the pattern of flow of ground water from highland to lowland areas as interpreted from the progressive spatial changes in ground-water chemistry.

With the advent of the nuclear age, hydrogeologists were presented with several new and unique tools, in the form of environmental isotopes, for investigating the occurrence and movement of ground water. The term applies to the isotopes (forms of the same element of different atomic weight) introduced to the hydrologic cycle naturally or by man's actions on a worldwide scale. The isotopes of greatest application in hydrogeology are those of the light elements (hydrogen, oxygen, carbon, nitrogen, chloride, silicon, and sulphur) that form the principal inorganic components and solutes of water.

Radioactive isotopes ($^3$H, $^{14}$C, and $^{36}$Cl), formed in the upper atmosphere by cosmic ray reactions and also through nuclear weapons explosions, decay at uniform, known rates. Thus, the time of recharge of ground water removed from atmospheric additions of radioactive isotopes may be determined by the rate of loss of its radioactive components.

Nonradioactive, or stable isotopes ($^2$H, $^{18}$O, $^{13}$C, $^{15}$N, and $^{34}$S) provide valuable information on environmental conditions at the time of ground-water recharge, and on the origin of dissolved matter and the post-recharge chemical history of the ground water. For example, the ratios of $^2$H/$^1$H and $^{18}$O/$^{16}$O, being temperature sensitive, give information on the temperature of recharge as well as altitude. The $^{13}$C/$^{12}$C ratio is subject to fractionation effects in living organisms; thus this ratio indicates not only whether dissolved carbon is of inorganic or organic origin but also the relative contributions of each type to mixtures.

Much of the impetus for the development of isotope hydrology has come through the efforts of the International Atomic Energy Agency (IAEA), which has the responsibility for promoting peaceful uses of nuclear energy. The IAEA, through support of research, conferences, and publications, has brought isotope hydrology to a mature state; it is now accepted as routine practice in ground-water investigations.

Computer technology improved contemporaneously with advances in understanding of the principles and processes of

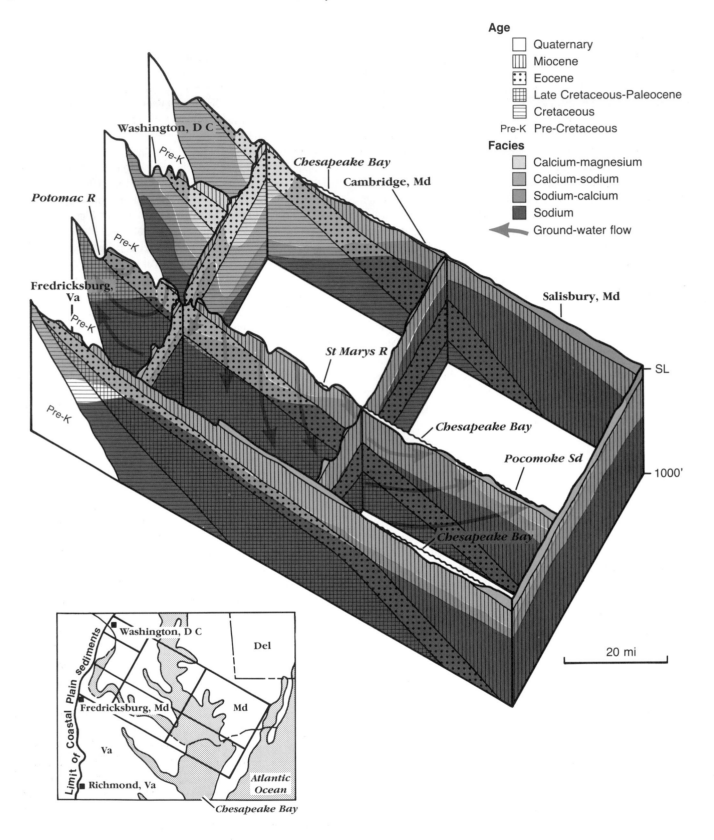

Figure 3. Fence diagram of part of the middle Atlantic Coastal Plain showing the geology, hydrochemical facies, and ground-water flow pattern (middle panel) interpreted from the chemical data (Back, 1961b, p. D–382).

ground-water chemistry and their applications to ground-water systems study. Concepts of mass (solute) and heat transport have continued to develop during the 1970s and the 1980s, aided by better simulation methods and numerical techniques. The joint progress has enabled the modeling of many chemical processes acting in ground-water systems and a good start on the integration of chemical models with flow models to permit combined systems analysis and improved prediction capability. Schwartz and Domenico (1973) explained the fundamental chemical controls on simulation of regional geochemical patterns, and Plummer (1977) and Plummer and Back (1980) employed mass-balance relationships and mass-transfer calculations to define chemical reaction paths, thereby simulating observed water-chemistry changes in conjunction with flow in regional carbonate-rock aquifers.

According to the solute transport equation, movement of solutes in ground-water systems is governed by ground-water velocity, hydrodynamic dispersion, and the net mass transfer resulting from all chemical reactions. At the present time, applica-tions to field problems necessitate many physical and chemical assumptions primarily because of the difficulties in obtaining accurate, relevant field measurements of the controlling parameters. The present status of modeling to simulate the movement of solutes, including contaminants, is described by Anderson (1979). As she notes (p. 142), improvement in field measurements and development of new field techniques, along with improvement in the theoretical framework, are necessary before geochemical models can be routinely applied to contaminant transport problems. These are two primary thrusts of chemical-hydrogeology research today.

Increasing concern about pollution of ground water by industrial chemicals, mainly petroleum products and chlorinated hydrocarbon solvents, has led to burgeoning research and many important findings in the organic chemistry of ground water. An overview of much of what is known of chemical processes in ground water (Cherry and others, 1984) indicates that most existing knowledge of the chemistry of contaminants relates to inorganics.

## REFERENCES CITED

Adams, F. D., 1954, The birth and development of the geological sciences: New York, Dover Publications, 506 p.

Anderson, M. P., 1979, Using models to simulate the movement of contaminants through groundwater flow systems: CRC Critical Reviews in Environmental Control, v. 9, p. 97–156.

Back, W., 1961a, Calcium carbonate saturation in ground water, from routine analyses: U.S. Geological Survey Water-Supply Paper 1535-D, p. D1–D14.

—— , 1961b, Techniques for mapping of hydrochemical facies, *in* Short papers in the geologic and hydrologic sciences: U.S. Geological Survey Professional Paper 424-D, p. D380–D382.

—— , 1966, Hydrochemical facies and ground-water flow patterns in northern part of Atlantic Coast Plain: U.S. Geological Survey Professional Paper 498-A, p. A1–A42.

Back, W., and Freeze, R. A., eds., 1983, Chemical hydrogeology: Stroudsburg, Pennsylvania, Hutchinson Ross Publishing Co., Benchmark Papers in Geology, v. 73, 416 p.

Bailey, E.H.S., 1902, Special report on mineral waters: Kansas Geological Survey, v. 7, 343 p.

Biswas, A. K., 1970, History of hydrology: Amsterdam, North-Holland Publishing Co.; New York, Elsevier Publishing Co., 336 p.

Bredehoeft, J. D., 1976, Status of quantitative groundwater hydrology, *in* Saleem, Z. A., ed., Advances in groundwater hydrology: Minneapolis, American Water Resources Association, p. 8–14.

Bredehoeft, J. D., Back, W., and Hanshaw, B. B., 1982, Regional ground-water flow concepts in the United States; Historical perspective, *in* Narasimhan, T. N., ed., Recent trends in hydrogeology: Geological Society of America Special Paper 189, p. 297–316.

Brown, J. S., 1925, A study of coastal ground water, with special reference to Connecticut: U.S. Geological Survey Water-Supply Paper 537, 101 p.

Brown, R. H., and Park, G. G., 1945, Salt water encroachment in limestone at Silver Bluff, Miami, Florida: Economic Geology, v. 40, p. 235–262.

Bull, W. B., 1975, Land subsidence due to ground-water withdrawal in the los Banos–Kettleman City area, California; Part 2, Subsidence and compaction of deposits: U.S. Geological Survey Professional Paper 437-F, 90 p.

Bull, W. B., and Poland, J. F., 1975, Land subsidence due to ground-water withdrawal in the Los Banos–Kettleman City area, California; Part 3, Interrelations of water-level change, change in aquifer system thickness, and subsidence: U.S. Geological Survey Professional Paper 437-G, 67 p.

Chamberlin, T. C., 1885, Requisite and qualifying conditions of Artesian Wells: U.S. Geological Survey 5th Annual Report, p. 125–173.

Cherry, J. A., Gillham, R. W., and Barke, J. R., 1984, Contaminants in ground water; Chemical processes, *in* Groundwater Contamination: Washington, D.C., National Academy Press, p. 46–64.

Darton, N. H., 1897, Preliminary report on artesian waters of a portion of the Dakotas: U.S. Geological Survey 17th Annual Report, part 2, p. 609–694.

Dole, R. B., 1909, The quality of surface waters of the U.S.; Part 1, Analyses of water east of the One-Hundredth Meridian: U.S. Geological Survey Water-Supply Paper 236, 123 p.

Ellis, E. E., 1906, Occurrence of water in crystalline rocks, *in* M. L. Fuller, Underground-water papers: U.S. Geological Survey Water-Supply Paper 160, p. 19–28.

Ferris, J. G., and Sayre, A. N., 1955, The quantitative approach to ground-water investigations: Economic Geology, 50th Anniversary issue, p. 714–747.

Fiedler, A. G., and Nye, S. S., 1933, Geology and ground-water resources of the Roswell artesian basin, New Mexico: U.S. Geological Survey Water-Supply Paper 639, 372 p.

Foster, M. D., 1950, The origin of high sodium bicarbonate waters in the Atlantic and Gulf Coastal Plains: Geochimica et Cosmochimica Acta, v. 1, p. 33–48.

Freeze, R. A., and Back, W., eds., 1983, Physical hydrogeology: Stroudsburg, Pennsylvania, Hutchinson Ross Publishing Co., Benchmark Papers in Geology, v. 72, 448 p.

Freeze, R. A., and Witherspoon, P. A., 1966, Theoretical analysis of regional groundwater flow. 1. Analytical and numerical solutions to the mathematical model: Water Resources Research, v. 2, p. 641–656.

Garrels, R. M., 1960, Mineral equilibria at low temperature and pressure: New York, Harper and Brothers, 254 p.

Helm, D. C., 1982, Conceptual aspects of subsidence due to fluid withdrawal, *in* Narasimhan, T. N., ed., Recent trends in hydrogeology: Geological Society of America Special Paper 189, p. 103–142.

Hilgard, E. W., 1860, Report on the geology and agriculture of the State of Mississippi: Jackson, Mississippi, Geological Survey of Mississippi, 391 p.

Hubbert, M. K., 1940, Theory of ground-water motion: Journal of Geology, v. 48, p. 785–944.

Jacob, C. E., 1946, Radial flow in a leaky artesian aquifer: EOS Transactions of the American Geophysical Union, part 2, p. 582.

King, F. H., 1898, Principles and conditions of the movements of ground water: U.S. Geological Survey 19th Annual Report, part 2, p. 207–245.

Landa, E. R., and Ince, S., 1987, The history of hydrology; v. 3, History of geophysics: Washington, D.C., American Geophysical Union, 122 p.

Leverett, F., 1899, Wells of northern Indiana: U.S. Geological Survey Water-Supply and Irrigation Paper no. 21, 82 p.

McCallie, S. W., 1898, A preliminary report on the artesian well system of Georgia: Geological Survey of Georgia Bulletin no. 7, 214 p.

Meade, R. H., 1968, Compaction of sediments underlying areas of land subsidence in central California: U.S. Geological Survey Professional paper 497–D, 39 p.

Meinzer, O. E., 1928, Compressibility and elasticity of artesian aquifers: Economic Geology, v. 23, p. 263–291.

—— , 1942, Physics of the earth; Part 9, Hydrology: New York, McGraw Book Co., 712 p.

Meinzer, O. E., and Kelton, F. C., 1913, Geology and water resources of Sulphur Spring Valley, Arizona, with a section on agriculture by R. H. Forbes: U.S. Geological Survey Water-Supply Paper 320, 231 p.

Mendenhall, W. C., 1905, The hydrology of San Bernardino Valley, California: U.S. Geological Survey Water-Supply and Irrigation Paper 142, 124 p.

Mendenhall, W. C., Dole, R. B., and Stabler, H., 1916, Ground water in San Joaquin Valley, California: U.S. Geological Survey Water-Supply Paper 398, 310 p.

Miller, R. E., 1961, Compaction of an aquifer system computed from consolidation tests and decline in artesian head: U.S. Geological Survey Professional Paper 424–D, p. B54–58.

Narasimhan, T. N., ed., 1982, Recent trends in hydrogeology: Geological Society of America Special Paper 189, 448 p.

Neuzil, C. E., 1980, Fracture leakage in the Cretaceous Pierre Shale and its significance for underground waste disposal [Ph.D. thesis]: Baltimore, Johns Hopkins University, 150 p.

Piper, A. M., and others, 1953, Native and contaminated water in the Long Beach–Santa Ana area, California: U.S. Geological Survey Water-Supply Paper 1136, 320 p.

Plummer, L. N., 1977, Defining reactions and mass transfer in part of the Floridan aquifer: Water Resources Research, v. 13, p. 801–812.

Plummer, L. N., and Back, W., 1980, The mass balance approach; Application to interpreting the chemical evolution of hydrologic systems: American Journal of Science, v. 280, p. 130–142.

Poland, J. F., 1984, Case history No. 9.14, Santa Clara Valley, California, USA, in Poland, J. F., ed., Guidebook to studies of land subsidence due to groundwater withdrawal: Paris, United Nations Educational Scientific and Cultural Organization, p. 279–290.

Poland, J. F., and Davis, G. H., 1969, Land subsidence due to withdrawal of fluids: Geological Society of America Reviews in Engineering Geology II, p. 187–269.

Poland, J. F., Lofgren, B. E., Ireland, R. L., and Pugh, R. G., 1975, Land subsidence in the San Joaquin Valley as of 1972: U.S. Geological Survey Professional Paper 437–H, 78 p.

Prickett, T. A., 1975, Modeling techniques for ground water evaluation, in V. T. Chow, ed., Advances in Hydroscience, v. 10: New York, Academic Press, p. 1–143.

Rouse, H., and Ince, S., 1957, History of hydraulics: Ames, Iowa State University, Iowa Institute of Hydraulic Research, 269 p.

Russell, W. L., 1928, The origin of artesian pressure: Economic Geology, v. 23, p. 132–157.

Sayre, A. N., and Bennett, R. R., 1942, Recharge, movement, and discharge in the Edwards Limestone reservoir, Texas: EOS Transactions of the American Geophysical Union, pt. 1, p. 19–27.

Schoeller, H., 1955, Geochimie des eaux souterraines: Revue de l'Institut Francais du Pétrole, v. 10(3), p. 181–213; v. 10(4), p. 219–246.

Schwartz, F. W., and Domenico, P. A., 1973, Simulation of hydrochemical patterns in regional ground-water flow: Water Resources Research, v. 9, no. 3, p. 707–720.

Smith, E. A., 1907, The underground water resources of Alabama: Alabama Geological Survey Monograph 6, 388 p.

Smith, M. B., 1934, Ground water in the Livermore Valley, California [M.A. thesis]: Stanford, California, Stanford University, 83 p.

Stringfield, V. T., 1936, Artesian water in the Florida peninsula: U.S. Geological Survey Water-Supply Paper 773–C, p. 115–195.

Swenson, F. A., 1968, New theory of recharge to the artesian basin of the Dakotas: Geological Society of America Bulletin, v. 79, p. 163–182.

Terzaghi, C., 1929, Origin of artesian pressure: Economic Geology, v. 24, no. 1, p. 96–97.

Theis, C. V., 1935, The relation between the lowering of the piezometric surface and the rate and duration of discharge of a well using ground-water storage: EOS Transactions of the American Geophysical Union, v. 16, p. 519–524.

Thomas, H. E., 1952, Ground-water regions of the United States; Their storage facilities: U.S. 83d Congress, House Interior and Insular Affairs Committee, The physical and economic foundation of natural resources, v. 3, 78 p.

Tolman, C. F., 1937, Ground Water: New York, McGraw-Hill, 593 p.

Toth, J., 1962, A theory of ground water motion in small drainage basins in central Alberta, Canada: Journal of Geophysical Research, v. 67, p. 4375–4387.

Wenzel, L. K., 1936, The Thiem method for determining permeability of water-bearing materials and its application to the determination of specific yield— Results of investigations in the Platte River valley, Nebraska: U.S. Geological Survey Water-Supply Paper 679–A, p. 1–57.

Wenzel, L. K., 1942, Methods of determining permeability of water-bearing materials, with special reference to discharging-well methods: U.S. Geological Survey Water-Supply Paper 887, 192 p.

MANUSCRIPT ACCEPTED BY THE SOCIETY APRIL 12, 1988

The Geology of North America
Vol. O-2, Hydrogeology
The Geological Society of America, 1988

# Chapter 2

# *Hydrostratigraphic units*

**Paul R. Seaber**
*Illinois State Geological Survey, Natural Resources Building, 615 Peabody Drive, Champaign, Illinois 61820*

## INTRODUCTION

Geologists have long agreed upon the need for consistent, explicit methods of dividing local rock sequences into separate units. The stratigraphic procedures and principles that have evolved define a stratigraphic unit as a naturally occurring body of rock distinguished from adjoining rock on the basis of some stated property or properties (North American Commission on Stratigraphic Nomenclature, 1983, p. 847). Codes of stratigraphic nomenclature prepared by the North American Commission on Stratigraphic Nomenclature (NACSN, 1983) and the International Subcommission on Stratigraphic Classification (ISSC, 1976) are now widely used for stratigraphic terminology and classification. A stratigraphic code or guide is a formulation of current views on stratigraphic principles and procedures and is designed to promote standardized classification and formal nomenclature of rock materials.

In stratigraphic work, two basic categories of units are now recognized: (1) material units, based on actual bodies of rock; (2) nonmaterial units, based on the abstract concept of time (Owen, 1987, p. 363). A combination of these two categories, a chronostratigraphic unit, is also commonly recognized. These are equivalent to the older "holy trinity" of rock units, time units, and time-rock units.

A standard method of classification facilitates the systematic arrangement and partitioning of rock or unconsolidated materials of the Earth's crust into units based on their inherent properties and attributes, and in turn, the geologic mapping and regional correlation of such units. This standardization promotes the systematic and rigorous study of the composition, geometry, sequence, history, and genesis of rocks and unconsolidated materials. A number of different standardized bases exist for dividing rock units now in use in North America. Commonly used properties include composition, texture, included fossils, magnetic signature, radioactivity, seismic velocity, and age (NACSN, 1983, p. 847).

Units based on one property commonly do not coincide with those based on another, and therefore, geologists have found that distinctive terms and criteria are needed to identify and describe the property used in defining each type of material unit (NACSN, 1983, p. 847). Stratigraphic units used by hydrogeologists, herein termed hydrostratigraphic units, are an example of a type of unit lacking distinctive criteria that permits formal definition and classification. The adjective stratigraphic is used here in the broad sense to refer to stratigraphic procedures and principles which are now applied to all classes of earth materials.

Hydrostratigraphic units were originally defined as bodies of rock with considerable lateral extent that compose a geologic framework for a reasonably distinct hydrologic system (Maxey, 1964, p. 124). A hydrostratigraphic unit was intended by Maxey to become a specialized category within the framework of a code of stratigraphic nomenclature that would allow geologists to express clearly and precisely the similarities, parallels, and contrasts between ground-water entities and other units recognized by the code. A hydrostratigraphic unit was also intended to be the practical mappable unit for ground-water studies, equivalent in rank to formation in rock-stratigraphic classification. It was intended to serve as a fundamental unit for describing hydrogeologic systems in the field based on the properties of the rock that affect ground-water conditions and would be of tested mappability. However, Maxey (1964) included the dynamics of the hydrological regime in his concept and definition of the hydrostratigraphic unit. This condition, because it is not a material property of any rock unit, is a factor that does not allow formal adoption of this concept of hydrostratigraphic units by the North American Commission on Stratigraphic Nomenclature.

Rules of the U.S. Geological Survey (USGS, 1890, 1903) and the codes of stratigraphic nomenclature by the predecessors of NACSN, the Committee on Stratigraphic Nomenclature (CSN, 1933) and the American Commission on Stratigraphic Nomenclature (ACSN, 1961, 1970), as well as the *International Stratigraphic Guide* (ISSC, 1976), designate the formation (Lyell, 1858) as the basic building block of all geologic mapping. Originally, a formation could mean any assemblage of rocks that had some character in common, whether of age, origin, or composition. Only later (ACSN, 1961) was emphasis placed on the lithic or solid rock characteristics. Hydrogeologists have related their hydrogeologic units to a classical geologic framework of lithologic character, stratigraphic association, and sometimes, to fossils contained in the rocks. This practice has created problems and

Seaber, P. R., 1988, Hydrostratigraphic units, *in* Back, W., Rosenshein, J. S., and Seaber, P. R., eds., Hydrogeology: Boulder, Colorado, Geological Society of America, The Geology of North America, v. O-2.

difficulties for field hydrogeologists who have mapped and defined hydrogeologic units based not only on the character of the solid rock material (lithic), but also on that of the intersticies.

Hydrogeologists have recognized the need to develop an appropriate set of principles and procedures for classifying and naming hydrostratigraphic units that would allow consistent mapping of these units. Problems with usage of stratigraphic terminology are discussed by Owen (1987). Jorgensen and Rosenshein (1987) review the confusion resulting from naming aquifers, which probably occurs because of inexact or inaccurate definitions of an "aquifer." Much of the confusion in classifying and naming hydrostratigraphic units occurs because the nature and boundaries of the unit have not been defined before mapping the unit. An additional disagreement among hydrogeologists seems to be whether to map and name the flow system and the rock body separately, or to find a means to combine the two concepts into one system of mapping and nomenclature.

Current stratigraphic and hydrogeologic concepts and principles must be used to solve this dilemma, and the classification procedures and principles adopted must be flexible in order to provide for both change and additions that may be needed to improve their relevance to new scientific problems. To be effective, the procedures must be widely used and accepted. This chapter discusses the historical background and presently used concepts in the mapping, classification, and nomenclature of water-bearing and non-water-bearing deposits in North America.

## EVOLUTION OF A STRATIGRAPHIC CODE

The scientific classification and nomenclature of hydrogeologic units closely parallels, but has lagged behind, the conceptual evolution of stratigraphic procedures and practices applied to other types of rock units. Scientific classification of rocks deals with defining the nature and boundaries of a unit. Nomenclature deals with establishing the rank and hierarchy of that unit and assigning a unique word or words to give it a distinctive designation. Arguments over rank and nomenclature are futile whenever the criteria used to determine the nature and boundaries of a unit are not clear, which is the present case for hydrostratigraphic units.

This chapter focuses on the relation of stratigraphy and hydrogeology. This section of the chapter deals with evolving stratigraphic concepts, particularly of the term formation, which is entrenched in hydrogeologic literature in relation to the term aquifer. In practice, at present, an aquifer is an abstraction that is by no means equivalent to a single geologic, lithologic, or stratigraphic unit (Marsily, 1986, p. 115). Many hydrogeologists have extensive geologic backgrounds and they use stratigraphy to name aquifers (Jorgensen and Rosenshein, 1987, p. 210). Conversely, many other geohydrologists, without this extensive geologic background, have used hydrologic terminology and concepts to define aquifers. Therein lies a major source of confu-

sion in the classification and nomenclature of hydrogeologic units.

The attempts of geologists to separate, correlate, and map rocks have focused on: (1) cartographic representation, (2) intelligible description, (3) historical interpretation, and (4) regional correlation. Geologists have always made "must be mappable" a significant and critical criterion in rock classification and nomenclature. Historically, geologists have used the formation as the fundamental mapping unit of all rocks from the time of Lyell (1858) to the publication of the "North American Stratigraphic Code" (NACSN, 1983). Lyell's basic building block of geology was defined as "any assemblage of rocks which have some character in common, whether of origin, age or composition." It was mappable. The common character could be lithology, genesis, or age.

One of the earliest attempts at standardization of representation of mapping units is given in the Tenth Annual Report of the U.S. Geological Survey (USGS, 1890, p. 63–79), which for cartographic purposes divided all rocks into four great classes: (a) fossiliferous clastic rocks, (b) surficial deposits, (c) ancient crystalline rocks, and (d) volcanic rocks. For clastic rocks the two classes or divisions were the structural and time divisions. The time divisions were the 11 periods of geologic time. The structural divisions were the units of cartography used for mapping and were designated formations. Formation essentially meant Lyell's definition but, "In every case the definition should be that which best meets the practical requirements of the geologist in the field and the prospective user of the map: that is to say, each formation should possess such characteristics that it may be recognized on the ground alike by the geologist and by the layman" (USGS, 1890, p. 64). Hydrogeologic units were not recognized in this report.

The Twenty-Fourth Annual Report of the U.S. Geological Survey (1903) extensively revised the earlier rules and recognized for cartographic purposes three great classes of rock: (a) sedimentary, (b) igneous, and (c) metamorphic. All cartographic (mappable) units were called "formations." For sedimentary rocks, "The formation should . . . be traced and identified by means of its lithologic character, its stratigraphic association, and its contained fossils" (USGS, 1903, p. 23). Igneous formations were discriminated on the basis of mode of occurrence, chemical composition, and mineral and textural characteristics (USGS, 1903, p. 24). Metamorphic formations were discriminated on the basis of their petrographic characters (USGS, 1903, p. 25).

This system prevailed until the first code of stratigraphic nomenclature was published (Committee on Stratigraphic Nomenclature, 1933). The 1933 code and the 1903 U.S. Geological Survey "rules" are very similar, and most state geological surveys adopted the 1933 code. In the 1933 code, a sedimentary rock formation was defined as a genetic unit formed under uniform depositional conditions, was of limited horizontal extent, and of supposedly uniform lithology. Hydrogeologists of the day related ground water to this classical framework of lithologic character, stratigraphic association, and contained fossils, where present in sedimentary rocks, or to the igneous and metamorphic formation characteristics. Neither aquifers nor any other hydrogeologic units are mentioned in the 1933 code.

A new Code of Stratigraphic Nomenclature was developed between 1933 and 1961 (ACSN, 1961). Categories of stratigraphic units were named rock-stratigraphic, soil-stratigraphic, biostratigraphic, and time-stratigraphic, and the nonmaterial units were termed geochronologic and geologic-climate units. The formation was still the basic rock-stratigraphic (lithostratigraphic) formal mapping unit for all rock types. The major change from the 1933 code was that time, genesis, and fossil content were no longer basic criteria used in defining a formation. Formal units, essentially those requiring stability of nomenclature, as well as informal units, were used. Aquifers were mentioned as informal units in the 1961 code, and were lumped with other economic and utilitarian units such as coal beds, quarry layers, oil-bearing reefs, and oil sands.

The 1970 American Code of Stratigraphic Nomenclature (ACSN, 1970) and the 1976 International Stratigraphic Guide (ISSC, 1976) mention aquifers, which could, if stratigraphically significant, be recognized as beds, members, and formations. The term formation is applied in all three of these codes (1961, 1970, and 1976) to all rock types, based strictly on observable physical features. A formation had to possess some degree of internal lithologic homogeneity or distinctive lithologic features. Accordingly, formations were lithic units and remained the basic practical unit of geologic work. The terms rock-stratigraphic and lithostratigraphic were considered to be synonymous.

The 1983 Code of the North American Commission on Stratigraphic Nomenclature contains six major categories of material units and several units based on age. This code has material categories of units based not only on lithic (used as synonymous with "lithologic") characteristics, but also on content, inherent attributes, or physical limits. Rock units, as used in earlier rules and codes, is a term that could be used as a synonym for material categories of units. The material categories include lithostratigraphic, lithodemic, magnetopolarity, biostratigraphic, pedostratigraphic, and allostratigraphic units. The basic building blocks for most geologic work are rock bodies defined on the basis of composition and related lithic characteristics, or on their physical, chemical, or biologic content or properties. Emphasis is placed on the relative objectivity and reproducibility of data used in defining units within each category (NACSN, 1983, p. 848).

In the 1983 code, the term "formation" is restricted to lithostratigraphic units that are generally stratified, tabular rock bodies conforming to the Law of Superposition. It is not used with lithodemic units, which are mostly igneous or metamorphic rocks, or with allostratigraphic units, which are recognized by bounding disconformities. Neither does it apply to the other material categories. For nonstratiform rocks (of plutonic or tectonic origin, for example), the parallel term lithodeme is used. The term rock-stratigraphic is no longer used formally in the 1983 Code nor in the International Stratigraphic Guide, since stratigraphic in a restricted sense in limited to stratiform rocks.

The interstices (voids) in rocks, which create the porosity and permeability mapped by hydrogeologists, are not mentioned in the 1983 code as a rock property. Aquifers are mentioned as informal economic units or as possible formal lithostratigraphic units, but not as lithodemic or allostratigraphic units even though the latter two types of units may be water bearing. Definitions of hydrogeologic units, such as aquifer, that rely on the formation as a basic part of their definition are thus outdated if the 1983 code is adhered to for hydrogeologic work.

## HISTORY OF HYDROGEOLOGIC MAPPING

Perhaps the best publication dealing with the historical treatment of hydrostratigraphic units (i.e, ground-water units) is entitled *Hydrological Maps* (UNESCO, 1977, p. 135–190), wherein four stages of development of ground-water mapping are discussed. Paraphrasing freely, these are: (1) relating ground water to the classical geologic (lithostratigraphic) framework without detailed regard for hydraulic continuity, (2) relating ground-water occurrence to the hydraulic properties within the classical geologic framework, (3) showing ground water as a resource, and (4) showing the relationship of ground water to the water-bearing characteristics of the rocks and to the dynamics of the hydrological regime. The kind of ground-water mapping done was strongly influenced by the existing rules or stratigraphic code of the day.

Before 1923, geologists doing hydrogeological mapping, such as Chamberlain (1897), Darton (1896), Ellis (1916), King (1898), Leverett (1896, 1899), Mendenhall (1905), Mendenhall and others (1916), Norton and others (1912), and Slichter (1899), simply spoke of a formation as a water-bearing or non-water-bearing unit. This early stage of ground-water mapping had the advantage of showing the extent to which the geology of an area could provide clues to the occurrence, distribution, and movement of ground water without regard to hydraulic continuity.

Meinzer (1923b, p. 30) states, "An aquifer is a formation, group of formations, or part of a formation that is water-bearing." This definition essentially classifies and maps rocks and formations according to their water-bearing properties and relates ground-water occurrence to the hydraulic properties within the classical geological framework. Meinzer (1923a, 1923b) clearly recognized in his definition of an aquifer that formations were not everywhere exactly equivalent to either water-bearing or non-water-bearing units. He would have had no need to invoke the concept of an aquifer if they were, and could have easily referred to them simply as water-bearing or non-water-bearing formations as previous geologists did. His definition of an aquifer was utilitarian and practical, and he used the terms ground-water reservoir, water-bearing bed, water-bearing stratum, and water-bearing deposit as synonyms of aquifer. Meinzer (1923a) also classified and mapped water-bearing units by geologic age. He coined no word or term for non-water-bearing rocks except confining beds, which, because of their position and their impermeability or low permeability relative to that of the aquifer, give the water in the aquifer either artesian or subnormal head (Meinzer, 1923b, p. 40). Other workers used terms such as aquitard, aquiclude, and aquifuge to classify rocks that were not aquifers but were equivalent, in part, to Meinzer's "confining bed" concept.

Maps that show ground water as a resource began appearing in the 1950s, but initially these maps showed no direct relation with either the rock column or the flow system. About the same time, maps that related the water-bearing characteristics of the rocks to the dynamics of the flow system appeared. An early example of this type of map is that of the Baltimore, Maryland, area (Bennett and Myer, 1952), which built on the earlier work of Darton (1896, p. 142–148). The important differences between this type of map and earlier maps is that aquifers and confining beds are mapped as distinctive, independent units whereas earlier hydrogeologic maps related the water-bearing characteristics to specific geologic units. This new concept in mapping turned out to have many practical and utilitarian purposes, particularly in modeling efforts.

Maxey (1964) first proposed the term hydrostratigraphic unit in order to map ground-water occurrence and relate it to the water-bearing characteristics of the rocks and to the dynamics of the hydrologic regime. The terms aquifer and confining bed were redefined (Lohman, 1972) to show their relationship to the rock column and to the dynamics of the hydrological system, as did Poland and others (1972), who also defined the term aquifer system. Many aquifers and confining beds have been mapped. Most frequently these are stratiform water-bearing deposits (aquifers) and, if present, their overlying and/or underlying confining beds.

Another stage of development of ground-water mapping is to relate water-bearing and non-water-bearing characteristics to the rock column without trying to make these physical rock attributes (porosity and permeability) conform to other mappable material (rock) units, or to the dynamics of the hydrologic regime (Seaber, 1982, 1986). Such a technique forces a redefinition of Maxey's hydrostratigraphic unit and defines, as well as maps, a hydrogeologic unit as a body of rock distinguished and characterized by its porosity and permeability. This redefinition of a hydrostratigraphic unit ties such a unit to the rock column, but not necessarily to the dynamics of the flow system, which can be ephermal in terms of geologic time and is subject to changes by the work of man. This type of mapping allows hydrostratigraphic units to be considered for formal inclusion in the North American Stratigraphic Code as a unique type of material unit, which thus could be mapped independently of, but related to, other rock units.

Many types of available geologic maps can be utilized to aid in hydrogeologic mapping. King (1969, Fig. 5, p. 12–13) presents three ways of depicting the geology of an area. The area shown on his maps represents an area in eastern Tennessee and western North Carolina, including portions of the Valley and Ridge and Appalachian Plateau physiographic provinces. The maps show: (a) time-stratigraphic units (geologic systems), (b) rock-stratigraphic systems (structural stages), and (c) structure showing folds and faults. A hydrogeologic map would most clearly resemble the rock-stratigraphic map, although it would not everywhere be a map of permeability and porosity (i.e., hydraulic properties). However, all three types of geologic mapping are important aids to some aspect of hydrogeologic mapping.

## HISTORICAL DEVELOPMENT OF HYDROGEOLOGICAL CLASSIFICATION AND NOMENCLATURE

Conventional geologic mapping has tended to concentrate only on the solid rock material. The character of the interstices in the rock might be noted (e.g., whether they are primary or secondary), but interstices are never used as a primary feature for mapping. Geologists, however, recognize that intrinsic permeability and porosity are fundamental attributes of all rocks, whether consolidated or unconsolidated, stratiform or nonstratiform. Hydrogeologists use combinations of the character of the solid rock material, the type and variations of the interstices, and the flow patterns of the water in the rock as mapping criteria. Field work has demonstrated that hydrostratigraphic units can be defined and mapped on the basis of physical properties and attributes, and also that flow systems in rocks can be mapped using water level and other data. But hydrogeologists do this mapping without uniform practices and procedures; they are lacking a "Hydrogeologic Code" similar to the present 1983 North American Stratigraphic Code.

Without formally established and universally accepted hydrostratigraphic procedures and principles, it is difficult for hydrogeologists to bring scientifically ordered terminology to their field studies, including mapping of hydrogeologic units. This problem is compounded because hydrogeologists themselves have not agreed upon a scheme for hydrostratigraphic classification and nomenclature based on mappable hydraulic properties or other attributes important to hydrogeology and ground-water investigations.

Meinzer (1923a) coined the words aquifer and confining bed. Tolman (1937, p. 36–37) coined the word aquiclude to categorize a "formation," that absorbs and transmits water slowly, and gave credit to Davis (1930, p. 488) for the word aquifuge, which is used to describe impervious rocks. Davis originally used "aquifuge" to apply to rocks with the characteristics of both "aquiclude" and "aquifuge" as used by Tolman, but later agreed to Tolman's more restricted usage of the word.

John Ferris coined the word "aquitard" (as better Latin) to replace the term aquiclude in courses given to U.S. Geological Survey ground-water personnel in 1958. Aquiclude means confinement of, rather than the exclusion of, water. Aquiclude, in this sense, may be thought of as a synonym for the hydraulic term confining bed. Todd (1980, p. 26) discusses the Latin origin of some hydrostratigraphic terms: "The word aquifer can be traced to its Latin origin. Aqui- is a combining form of aqua ("water") and -fer comes from ferre ("to bear"). Hence, an aquifer is a water-bearer. The suffix -clude of aquiclude is derived from the Latin claudere ("to shut or close"). Similarly, the suffix -fuge of aquifuge comes from fugere ("to drive away"), while the suffix -tard of aquitard follows from the Latin tardus ("slow")." Hydrogeologists tend to use interchangeably the prefixes aqui- (Latin) and hydro- (Greek) to mean water, although these terms also mean fluid (Spilman, 1955, p. 25, 108).

The U.S. Geological Survey, since 1972 (Lohman), has used only two terms—aquifer and confining bed—for "hydrostratigraphic units." The term "confining bed" is used to collectively supplant aquiclude, aquitard, and aquifuge. The U.S. Geological Survey is presently preparing guidelines for its authors in the naming and classification of aquifers (Laney and Davidson, 1986).

A hydrostratigraphic unit, as redefined here, is a body of rock distinguished and characterized by its porosity and permeability. A hydrostratigraphic unit may occur in one or more lithostratigraphic, allostratigraphic, pedostratigraphic, or lithodemic units and is unified and delimited on the basis of its observable hydrologic characteristics that relate to its interstices. Hydrostratigraphic units are defined by the number, size, shape, arrangement, and intercollection of their interstices, and are recognized on the basis of the nature, extent, and magnitude of the interstices in any body of sedimentary, metamorphic, or igneous rock (Seaber, 1982, 1986).

Hydrogeologists have described many hydrostratigraphic units that are of broad areal extent, particularly as the results of the Regional Aquifer System Analysis (RASA) investigations of the U.S. Geological Survey (Bennett, 1979; Sun, 1986). Most of these regionally mappable units require consistency and stability of nomenclature, because they are likely to extend far beyond the locality in which they were first recognized and mapped.

In delineating hydrostratigraphic units, much discretion must be left to the field hydrogeologists, as is the case for geologists dealing with all other types of geologic units within the 1983 Code. Hydrostratigraphic units must, of course, be practical and utilitarian. Whereas establishing exact unit boundaries is up to the practicing field hydrogeologist, each unit must be mappable and should preferably be recognizable to geologist, engineer, and layman alike. Hydrostratigraphic units may vary from area to area depending on the relative permeability or porosity of the rocks, but must be reproducible and traceable.

Uncertainty presently exists in defining the classification, nature, boundaries, rank, hierarchy, and nomenclature of rocks in terms of their interstices (Seaber, 1982, 1986). This situation applies to both water-bearing and non-water-bearing deposits. One difficulty has been that the water-bearing nature of the rocks is essentially an economic and, in places, somewhat ephemeral characteristic, but is the principal focus of most ground-water investigations. The interstices are considered to be an attribute or characteristic of the rock or geologic unit that has supposedly been defined, classified, named, and mapped using other stratigraphic criteria. Hydrogeologists have, in practice, been mapping both the water and the rock for years, but the need for formalization and standardization of hydrogeologic mapping units is not yet widely accepted.

In 1890 the U.S. Geological Survey issued the dictum, reiterated in 1903 (p. 23), that: "The selection of formations shall be such that they will best meet the practical and scientific needs of the users of the map. In every case the definition of a formation in the folio text should include a statement of the important facts that led to its discrimination and of the characteristics by which it may be identified in the field, whether by geologist or layman." This should apply today to the classification, nomenclature, and mapping of hydrostratigraphic units. The need for consistent and uniform usage in classification and terminology, and therefore the promotion of unambiguous communication, is becoming greater today because of the many laws and regulations being written by nonhydrogeologists that deal with hydrogeology. One logical approach would seem to be the inclusion of hydrostratigraphic units in the North American Stratigraphic Code in order to achieve uniform standards and common procedures for these units recognized by hydrogeologists and, as importantly, their peers in the world of earth science and the general public.

Deliberations along these lines by the Committee on Hydrostratigraphic Units, Hydrogeology Division, Geological Society of America, have led to recommendations for amendments to the North American Stratigraphic Code to provide formal procedures for classifying and naming hydrostratigraphic units (Seaber, 1986). The proposed addition of hydrostratigraphic units to the 1983 code is an attempt to bring hydrogeology into harmony with the changing concepts of scientific classification and nomenclature for all types of rock units. The purpose of the addition is to promote uniform and unambiguous methods to be used in partitioning any type of body of rock into hydrostratigraphic units, based on their inherent, mappable porosity and permeability attributes.

## SUMMARY

The scientific classification and nomenclature of hydrostratigraphic units closely parallels that of other material rock units. Presently, there are no available categories in any major codes of stratigraphic nomenclature that can adequately accommodate hydrostratigraphic units. The question remains open with many hydrogeologists as to whether interstices need to be a formally recognized mappable criterion, and more important, whether hydrostratigraphic mapping criteria and nomenclature can or need to be included in the North American Stratigraphic Code or in the revised International Stratigraphic Guide.

Hydrogeologic terminology is presently not uniform or consistent. Any terminology set forth should be applicable to all earth materials and all disciplines. This would assure consistent and uniform usage in classification, mapping, and terminology of hydrostratigraphic units, at least among geologists and hydrogeologists. The remaining chapters in this book illustrate the present diversity of hydrostratigraphic classification and nomenclature. Perhaps Meinzer's (1923b, p. 2) advice that "It is very useful to have definite terms to denote important scientific concepts, but it is of less consequence what these terms are, provided there is general agreement as to them" is very appropriate.

# REFERENCES

American Commission on Stratigraphic Nomenclature, 1961, Code of Stratigraphic Nomenclature: American Association of Petroleum Geologists Bulletin, v. 45, no. 5, p. 645–665.

——, 1970, Code of Stratigraphic Nomenclature (2nd edition): American Association of Petroleum Geologists, 45 p.

Bennett, G. D., 1979, Regional ground-water systems analysis: Water Spectrum, v. 11, no. 4, p. 36–42.

Bennett, R. R., and Meyer, R. R., 1952, Geology and ground-water resources of the Baltimore area: Maryland Department of Geology, Mines, and Water Resources Bulletin 4, 573 p.

Chamberlain, T. C., 1897, Requisite and qualifying conditions of artesian wells: U.S. Geological Survey 5th Annual Report, p. 125–173.

Committee on Stratigraphic Nomenclature, 1933, Classification and nomenclature of rock units: Geological Society of America Bulletin, v. 44, no. 2, p. 423–459, and American Association of Petroleum Geologists Bulletin, v. 17, no. 7, p. 843–868.

Darton, N. H., 1986, Preliminary report on artesian waters of a portion of the Dakotas: U.S. Geological Survey 17th Annual Report, part 2, p. 609–694.

Davis, W. M., 1930, Origin of limestone caverns: Geological Society of America Bulletin, v. 41, p. 475–628.

Ellis, A. J., 1916, Ground water in the Waterboro area, Connecticut: U.S. Geological Survey Water-Supply Paper 397, 73 p.

International Subcommission on Stratigraphic Classification (ISSC), 1976, *in* Hedberg, H. D., ed., International Stratigraphic Guide: New York, John Wiley and Sons, 200 p.

Jorgensen, D. G., and Rosenshein, J. S., 1987, Naming aquifers: EOS Transactions of the American Geophysical Union, v. 68, no. 15, p. 210–211.

King, F. H., 1898, Principles and conditions of the movements of ground water: U.S. Geological Survey 19th Annual Report, part 2, p. 207–245.

King, P. B., 1969, The tectonics of North America; A discussion to accompany the tectonic map of North America: U.S. Geological Survey Professional Paper 628, 94 p., scale 1:5,000,000.

Leverett, F., 1896, The water resources of Illinois: U.S. Geological Survey 17th Annual Report, part 2, p. 695–849.

——, 1899, Wells in northern Indiana: U.S. Geological Survey Water-Supply and Irrigation Paper no. 21, 82 p.

Laney, R. L., and Davidson, C. R., 1986, Aquifer-nomenclature guidelines: U.S. Geological Survey Open-File Report 86–534, 46 p.

Lohman, S. W., ed., 1972, Definitions of selected ground-water terms; Revisions and conceptual refinements: U.S. Geological Survey Water-Supply Paper 1988, 21 p.

Lyell, C., 1858, Manual of Geology (6th edition): London, England, p. 2.

Marsily, G. de, 1986, Quantitative Hydrogeology: New York, Academic Press, Harcourt Brace Jovanovich, Publishers, 440 p.

Maxey, G. B., 1964, Hydrostratigraphic Units: Journal of Hydrology, v. 2, p. 124–129.

Meinzer, O. E., 1923a, The occurrence of ground water in the United States, with a discussion of principles: U.S. Geological Survey Water-Supply Paper 489, 321 p.

——, 1923b, Outline of ground-water hydrology, with definitions: U.S. Geological Survey Water-Supply Paper 494, 71 p.

Mendenhall, W. C., 1905, The hydrology of San Bernadino Valley, California: U.S. Geological Survey Water-Supply and Irrigation Paper 142, 124 p.

Mendenhall, W. C., Dole, R. B., and Stabler, H., 1916, Ground water in San Joaquin Valley, California: U.S. Geological Survey Water-Supply Paper 398, 310 p.

North American Commission on Stratigraphic Nomenclature, 1983, North American Stratigraphic Code: American Association of Petroleum Geologists Bulletin, v. 67, no. 5, p. 841–875.

Norton, W. H., Hendrixson, W. S., Simpson, H. E., Meinzer, O. E., and others, 1912, Underground water resources of Iowa: U.S. Geological Survey Water-Supply Paper 293, 994 p.

Owen, D. E., 1987, Commentary; Usage of stratigraphic terminology in papers, illustrations, and talks: Journal of Sedimentary Petrology, v. 7, no. 2, p. 363–372.

Poland, J. F., Lofgren, E. E., and Riley, F. S., 1972, Glossary of selected terms useful in studies of the mechanics of aquifer systems and land subsidence due to fluid withdrawal: U.S. Geological Survey Water-Supply Paper 2025, 9 p.

Seaber, P. R., 1982, Definition of hydrostratigraphic units: 2nd Annual Symposium on Florida Hydrogeology, Northwest Florida Water Management District Public Information Bulletin 82–3, p. 25–26.

——, 1986, Evaluation of classification and nomenclature of hydrostratigraphic units: EOS Transactions of the American Geophysical Union, v. 67, no. 16, p. 281.

Slichter, C. S., 1899, Theoretical investigation of the motion of ground water: U.S. Geological Survey 19th Annual Report, part 2, p. 305–328.

Spilman, M., 1955, Medical Greek and Latin: Salt Lake City, Utah, 137 p.

Sun, R. J., ed., 1986, Regional aquifer-system analyses program of the U.S. Geological Survey, Summary of projects, 1978–84: U.S. Geological Survey Circular 1002, 264 p.

Todd, D. K., 1980, Groundwater Hydrology: New York, John Wiley and Sons, Inc., 535 p.

Tolman, C. F., 1937, Ground Water: New York and London, McGraw-Hill Book Company, Incorporated, 593 p.

United Nations Educational, Scientific and Cultural Organization; and World Meteorological Organization, 1977, Preparation of ground-water maps, *in* Hydrological maps: Louvain, International Association of Hydrological Sciences, Studies and Reports in Hydrology no. 20, p. 135–192.

U.S. Geological Survey, 1890, Nomenclature: Tenth Annual Report of the Director 1888–1889, part I–Geology, p. 63–79.

——, 1903, Nomenclature and classification for the geologic atlas of the United States: Twenty-fourth Annual Report 1902–1903, p. 21–27.

MANUSCRIPT ACCEPTED BY THE SOCIETY MARCH 15, 1988

# Chapter 3

# *Hydrogeologic setting of regions*

**Ralph C. Heath**
*U.S. Geological Survey, 4821 Kilkenny Place, Raleigh, North Carolina 27612*

## HYDROGEOLOGY OF NORTH AMERICA

Ground water occurs in the openings in the rocks that form the North American continent. The volume of the openings and their hydraulic characteristics depend on the structure and the texture of the rocks and on their mineral composition. It is useful in hydrogeology to classify openings in rocks on the basis of origin. Openings formed at the same time as the rock are referred to as primary openings, and include pores in sedimentary deposits, and vesicles, lava tubes, and cooling fractures in basalt. Openings that develop after a rock is formed are referred to as secondary openings and include joints and faults in igneous, metamorphic, and consolidated sedimentary rocks, and solution-enlarged openings in carbonate and other relatively soluble rocks (Fig. 1).

Pores in sedimentary deposits range from the microscopic openings in clays to openings a centimeter or more in width in very coarse grained, well-sorted gravel. The volume of the pores in sedimentary deposits depends primarily on sorting and the shape of the grains. Finer-grained sediments such as clay and silt tend to be very well sorted and consequently have the largest porosity.

Igneous, metamorphic, and consolidated sedimentary rocks commonly are collectively referred to as bedrock. Interconnected, water-bearing openings in these rocks consist of fractures along which the rocks have been broken by tectonic forces (Fig. 1). Joints, the most common type of fracture, generally occur in groups of similar strike and dip, which are referred to as joint sets. Relatively massive rocks tend to be broken by two steeply dipping ("vertical") sets of joints and a third ("horizontal") set more or less parallel to the land surface. Similar joint sets exist in unfolded consolidated sedimentary rocks. In such rocks, the horizontal set is developed along bedding planes, and this set is usually most important in the movement of ground water through the rocks. In folded consolidated sedimentary rocks, one of the "vertical" joint sets develops at many places along bedding planes, and the horizontal set is either absent or poorly developed. Because of these structural differences, relatively flat-lying sedimentary rocks generally form more productive aquifers than folded sedimentary rocks.

Plate 1 is a map of North America showing the major rock units of the ground-water system, differentiated primarily on the basis of their water-bearing characteristics—that is, on the nature and other aspects of their openings. The rock units shown on the plate are those that underlie the surficial cover. Thus, in most of the continent they are the units that commonly would be referred to as bedrock. The principal exceptions are the coastal areas, such as the Atlantic and Gulf Coastal Plains, the alluvial basins in the southwestern United States and Mexico, and the High Plains area of the U.S., which are underlain by thick sections of unconsolidated sedimentary deposits older than the Pleistocene. Because the emphasis in Plate 1 is on the nature of the water-bearing openings, extrusive igneous rocks, which contain both primary and secondary openings, are differentiated from intrusive igneous and metamorphic rocks that contain, for practical purposes, only secondary openings. Consolidated sedimentary rocks are also divided into two units; those that are flat lying to gently dipping, and those that are folded and relatively intensively faulted. In most of the continent, the rocks shown on Plate 1 are the principal sources of ground water and the dominant units of the ground-water system.

As noted in the preceding paragraph, Plate 1 deals only with rock units below the surficial layer—that is, the rock units that are generally older than the Pleistocene. Plate 2 is a map of North America showing the major units that compose the surficial layer. This layer, where it is relatively thin and mantles bedrock, is commonly referred to as regolith. In coastal plains, intermontane basins, most glaciated regions, and other areas where the depth to bedrock may range upward from about a hundred to a thousand meters or more, the term regolith is not normally applied to the surficial layer. In the remaining areas, where a surficial layer of unconsolidated material either never formed or has been removed by glacial or other erosion, the land surface is underlain by bedrock.

The surficial layer is clearly of Pleistocene age in coastal plains and in alluvial basins and in the parts of the continent covered by ice sheets and glaciers during the Pleistocene. In the remainder of the continent, the surficial layer consists of residuum—that is, of material formed essentially in place by the

Heath, R. C., 1988, Hydrogeologic setting of regions, *in* Back, W., Rosenshein, J. S., and Seaber, P. R., eds., Hydrogeology: Boulder, Colorado, Geological Society of America, The Geology of North America, v. O-2.

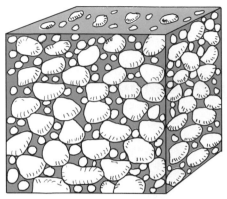

A. Pores in unconsolidated
sedimentary deposits

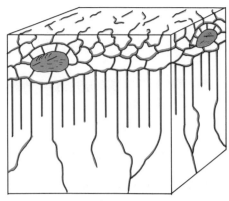

B. Lava tubes and cooling
fractures in extrusive
igneous rock

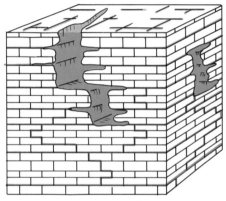

C. Caverns and other solution-
enlarged openings in
limestone

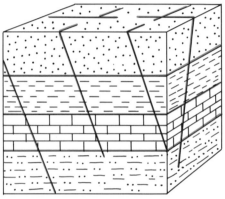

D. Joints in flat-lying
consolidated sedimentary
rock

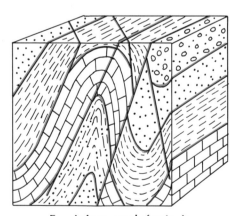

E. Joints and fault in
folded consolidated
sedimentary rock

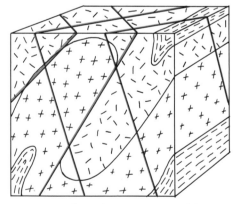

F. Joints in metamorphic
and intrusive igneous
rocks

Figure 1. Types of openings in selected water-bearing rocks. Block A is a few millimeters to a few tens of centimeters wide, depending on the nature of the deposit, and the remaining blocks are a few tens of meters wide. Openings in block A and B are of primary origin, and those in the remaining blocks are of secondary origin.

mechanical and chemical breakdown (weathering) of the underlying bedrock. In these areas, new material forms at the base of the layer while the processes of erosion remove material from the surface (from the top of the soil zone). Therefore, depending on the rates of weathering and erosion, at least the upper parts of the surficial layer may predate the Pleistocene in parts of these areas.

The surficial layer contains the upper part of the saturated zone in many areas and is the layer through which most ground-water recharge moves downward to the saturated zone. It is also the layer most directly involved in pollution of ground water by wastes placed on or near the land surface. Thus, the surficial layer is a very important part of hydrogeology.

The two features of the surficial layer that are most important to hydrogeology are its thickness and its composition. Both of these features are to some extent related to the origin of the layer, as noted in the following discussion. Its thickness ranges from less than a meter near rock outcrops to 50 m or more in some areas of active deposition such as the delta of the Mississippi River, and to 100 m or more in some of the valleys that contain glacial deposits.

The composition of the surficial layer also ranges between wide extremes. In coastal plains, alluvial basins, flood plains, and other areas of active deposition, the surficial layer consists largely of interbedded (stratified) clay, silt, and sand with minor admixtures of gravel. In the areas formerly occupied by ice sheets and glaciers, it consists mostly of two relatively distinct types of material. One of these is till, an unsorted mixture of particles ranging in size from clay to boulders, deposited directly by the ice. The other is sorted material deposited both in the channels of meltwater streams and in lakes that formed around the margins of the melting ice. The channel deposits include sand interbedded with gravel that, in most places, is overlain by interbedded sand, silt, and clay. The lacustrine deposits consist principally of interbedded fine-grained sand, silt, and clay.

A major part of the surficial layer in a broad arcuate band in the central part of the United States consists of an unconsolidated, nonstratified, calcareous deposit of fine sand, silt, and clay referred to as loess, which was at least partly deposited by wind. This deposit is more than 50 m thick in central Nebraska.

In most of the remainder of the continent, the surficial layer consists of nonstratified material derived from weathering of the underlying bedrock. Its composition depends on the mineral composition of the bedrock. Where the bedrock consists of igneous and metamorphic rocks, carbonates, and shale, the surficial layer tends to be clay rich. Where the bedrock consists of quartzite and sandstone and interbedded sandstone and shale, the surficial layer is mostly sand-size particles of quartz. In the areas where the surficial layer is derived from weathering of the bedrock, its thickness ranges from a meter or less, where it occurs on carbonate rocks that contain relatively little arenaceous material, to 25 m or more in areas where the bedrock consists of igneous and metamorphic rocks or shale.

An important factor, in addition to geologic structure and texture, that affects the volume and hydraulic characteristics of openings is mineral composition. Mineral composition is primarily of importance because of its effect on rock solubility. Thus, for hydrogeologic purposes, rocks can be classified either as relatively soluble or virtually insoluble. Ground water moving through relatively soluble rock slowly dissolves the rock and thus increases the size and the hydraulic efficiency of the openings. The more important rocks that are relatively soluble in water include, in decreasing order of solubility, gypsum, limestone, marble, dolomite, and diorite and related calcium-rich igneous rocks. Of these, the most widespread and most important to hydrogeology are limestone and dolomite, which commonly occur together and are collectively referred to as carbonate rocks. Figure 2 is a map of North America showing the major areas in which carbonate rocks and (or) gypsum are either exposed at the land surface or are in the zone of active ground-water circulation—that is, within a thousand meters or so of the land surface. It was not feasible to differentiate between carbonate rocks and gypsum on the map, but it should be noted that gypsum occurs mainly in western Texas and eastern New Mexico.

## CHARACTERISTICS USED IN THE DELINEATION OF GROUND-WATER REGIONS

In order to deal effectively with the principles of occurrence and movement of ground water in a large, geologically complex area like North America, subareas that have similar ground-water conditions must first be delineated. Such subareas are commonly referred to as ground-water regions or provinces (Fuller, 1905; Meinzer, 1923; Thomas, 1952; Heath, 1984). Because geology controls the occurrence of ground water, ground-water regions are areas in which the structure, arrangement, and composition of the rock units comprising the ground-water system are similar (Heath, 1982). For the purpose of delineating the ground-water regions, geologic features need to be restated in terms of hydrogeologic features.

Five hydrogeologic characteristics are of prime importance in the delineation of ground-water regions. These are (1) the presence and character of aquifers and confining beds, (2) the nature of the water-bearing openings of the dominant aquifer or aquifers with respect to whether they are of primary or secondary origin, (3) the mineral composition of the rock matrix of the dominant aquifer or aquifers with respect to whether the material is soluble or relatively insoluble, (4) the water storage and transmission characteristics of the dominant aquifer or aquifers, and (5) the nature and location of natural recharge and discharge areas. These five hydrogeologic characteristics are listed in Table 1, together with explanatory information.

## GROUND-WATER REGIONS OF NORTH AMERICA

Using the characteristics listed in Table 1, North America is divided into the 28 regions shown on Figure 3. The close relation between the ground-water regions and hydrogeology can be observed by comparing Figure 3 with the hydrogeologic maps,

**TABLE 1. CHARACTERISTICS OF GROUND-WATER SYSTEMS USED IN THE DELINEATION OF GROUND-WATER REGIONS**
(Heath, 1984, Table 4)

| Characteristic | Aspect | Range in Conditions | Significance of Characteristic |
|---|---|---|---|
| Presence and character of aquifers and confining beds | Unconfined aquifers | Thin, discontinuous, hydrologically insignificant.<br>Minor aquifer, serves primarily as a storage reservoir and/or as a conduit for recharge of an underlying aquifer or for discharge into an adjoining surface-water body.<br>The dominant aquifer. | Affects response of the ground-water system to pumping and other stresses.<br>Determines location and distribution of recharge and discharge areas and affects recharge and discharge rates<br>Determines susceptibility of the system to pollution. |
| | Confining beds | Not present, or hydrologically insignificant.<br>Thin, markedly discontinuous, or very leaky.<br>Thick, extensive, and of low permeability.<br>Complexly interbedded with aquifers or productive zones. | |
| | Confined aquifers | Not present, or hydrologically insignificant.<br>Thin or not highly productive.<br>Multiple thin aquifers interbedded with nonproductive zones.<br>The dominant aquifer—thick and productive. | |
| | Presence and arrangement of components | A single, unconfined aquifer.<br>Two interconnected aquifers with different characteristics but essentially of equal hydrologic importance.<br>A three-unit system consisting of an unconfined aquifer, a confining bed, and confined aquifer.<br>A complexly interbedded sequence of aquifers and confining beds. | |
| Water-bearing openings of dominant aquifer or aquifers | Primary openings | Pores in unconsolidated deposits.<br>Pores in semi-consolidated rocks.<br>Vesicles, tubes, and cooling fractures in volcanic (extrusive-igneous) rocks. | Controls water-storage and transmission characteristics.<br>Affects dispersion and dilution of wastes. |
| | Secondary openings | Fractures and faults in crystalline and consolidated sedimentary rocks.<br>Solution-enlarged openings in limestone and other soluble rocks. | |
| Composition of rock matrix of dominant aquifer | Rock solubility | Essentially insoluble.<br>Both relatively insoluble and soluble constituents.<br>Relatively soluble. | Affects size of openings and thus affects water-storage and transmission characteristics.<br>Has major influence on water quality. |
| Storage and transmission characteristics of dominant aquifer | Effective porosity | Large, as in well-sorted, unconsolidated deposits.<br>Moderate, as in poorly sorted unconsolidated deposits and semiconsolidated rocks.<br>Small, as in fractured crystalline and consolidated sedimentary rocks. | Controls response of aquifers to pumping and other stresses.<br>Determines yield of wells.<br>Affects long-term yield of the ground-water system.<br>Affects rate at which pollutants move. |
| | Hydraulic conductivity | Large, as in cavernous limestone, some lave flows, and clean gravels.<br>Moderate, as in well-sorted, coarse-grained sands, and semiconsolidated limestones.<br>Small, as in poorly sorted, fine-grained deposits and most fractured rocks.<br>Very small, as in confining beds. | |
| Nature and location of natural recharge and discharge areas | Recharge | In areas between streams, particularly in humid regions.<br>Through channels of losing streams. | Affects response to and recovery from stress.<br>Determines susceptibility to pollution<br>Affects water quality. |
| | Discharge | Through springs or by seepage to stream channels, lakes, estuaries, or the ocean.<br>By evaporation on flood plains and in basin "sinks." | |

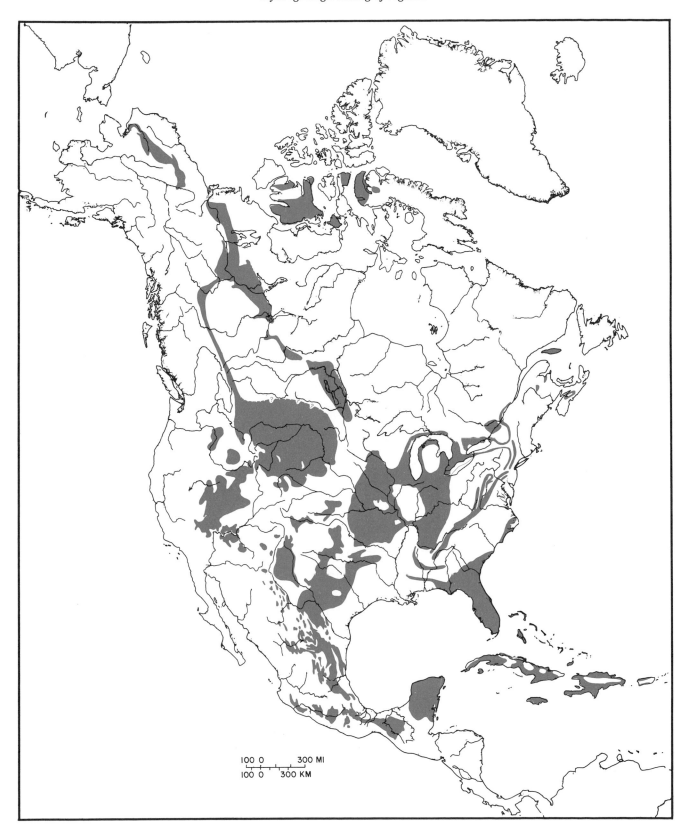

Figure 2. Areas of North America in which carbonate rocks and/or gypsum are either exposed at the land surface or are in the zone of active ground-water circulation. The gypsum occurs mainly in western Texas and eastern New Mexico. Most other areas are limestone and/or dolomite.

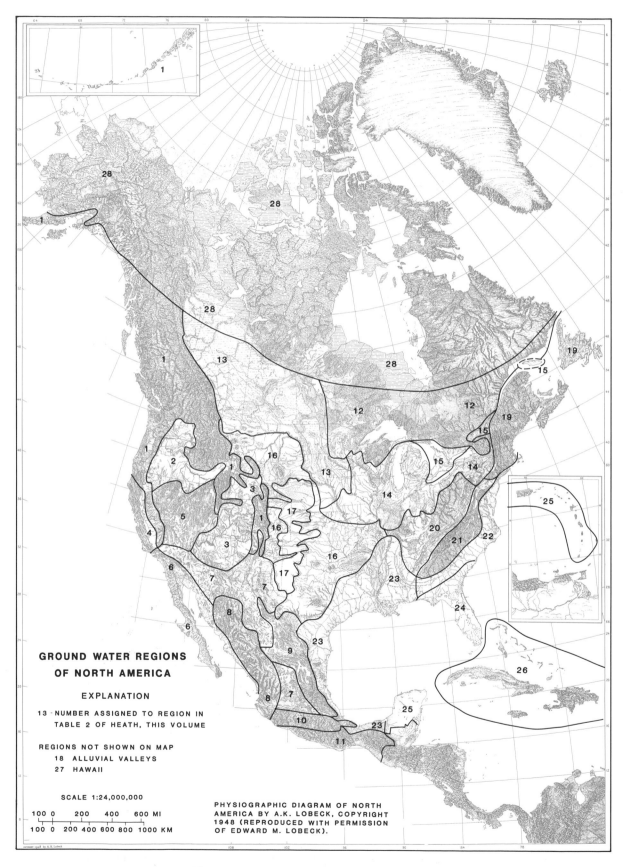

Figure 3. Ground water regions of North America.

**TABLE 2. HYDROGEOLOGIC SITUATION AND COMMON RANGES OF SELECTED HYDRAULIC CHARACTERISTICS
OF THE GROUND-WATER REGIONS OF NORTH AMERICA**

| Region | Hydrogeologic Situation* | Common Ranges in Hydraulic Characteristics of the Dominant Aquifers[†] | | | |
|---|---|---|---|---|---|
| | | Transmissivity ($m^2$ day$^{-1}$) | Hydraulic Conductivity (m day$^{-1}$) | Recharge Rate (mm yr$^{-1}$) | Well Yield ($m^3$ min$^{-1}$) |
| 1. Western Mountain Ranges | Mountains with thin soils over fractured rocks of Precambrian to Cenozoic age alternating with valleys underlain by alluvial and glacial deposits of Pleistocene age. Includes the Puget-Willamette Trough of Washington and Oregon. | 0.5-500 | 0.01-20 | 3-50 | 0.05-1 |
| 2. Columbia Lava Plateau | Low mountains to undissected plains underlain by discontinuous alluvial deposits and thin soils over a thick, complex sequence of lava flows irregularly interbedded with unconsolidated deposits. The lava flows and sediments range in age from Miocene to Holocene. | 2,000-500,000 | 1-3,000 | 5-250 | 0.9-180 |
| 3. Colorado Plateau and Wyoming Basin | A region of canyons, cliffs, and plains underlain by thin residuum and colluvium over horizontal to gently dipping consolidated sedimentary rocks of Paleozoic to Cenozoic age. | 0.5-100 | 0.01-5 | 3-50 | 0.05-2 |
| 4. Central Valley and Pacific Coast Ranges | Broad, relatively flat valleys underlain by thick alluvial deposits bordered along the coast by low mountains composed of semiconsolidated sedimentary rocks and volcanic deposits of Mesozoic and Cenozoic age. | 10-10,000 | 30-600 | 10-100 | 0.5-20 |
| 5. Great Basin | Alternating wide, relatively flat-floored basins underlain by thick alluvial deposits and short, subparallel mountain ranges composed, in part, of crystalline and sedimentary rocks of Paleozoic and Mesozoic age and, in part, of volcanic rocks of Cenozoic age. | 10-10,000 | 10-500 | 5-50 | 0.5-10 |
| 6. Coastal Alluvial Basins | Relatively flat valleys underlain by thick alluvial deposits separated by mountain ranges composed of metamorphic and sedimentary rocks of Mesozoic age and volcanic rocks of Cenozoic Age. | 10-1,000 | 5-100 | 5-50 | 0.5-10 |
| 7. Central Alluvial Basins | Relatively flat valleys underlain by thick alluvial deposits separated by elongated, discontinuous mountain ranges composed, in part, of sedimentary rocks of Paleozoic and Mesozoic age and, in part, of volcanic rocks of Cenozoic age. | 10-2,000 | 10-200 | 5-50 | 0.5-10 |
| 8. Sierra Madre Occidental | A relatively high, dissected region of thin regolith over a complex sequence of volcanic rocks of Cenozoic age. | 100-10,000 | 10-500 | 10-100 | 0.5-20 |
| 9. Sierra Madre Oriental | A relatively high area of anticlinal mountain ranges and synclinal valleys underlain by thin regolith over sedimentary rocks of Mesozoic age. | 10-1,000 | 1-50 | 5-50 | 0.5-5 |
| 10. Faja Volcanica Transmexicana | A high, mountainous area underlain by thin regolith over a complex sequence of volcanic rocks of Cenozoic age. | 100-10,000 | 10-500 | 10-100 | 0.5-20 |
| 11. Sierra Madre Del Sur | A highly dissected mountainous area underlain by thin regolith over metamorphic rocks of Precambrian and Paleozoic age, sedimentary rocks of Mesozoic age, and volcanic rocks of Mesozoic and Cenozoic age. | 5-5,000 | 1-300 | 10-100 | 0.1-10 |
| 12. Precambrian Shield | A hilly terrane underlain by glacial deposits over complexly folded to flat-lying metamorphic rocks of Precambrian age. Along the southwest side of James Bay, the bedrock consists of relatively flat-lying consolidated sedimentary rocks mostly of Paleozoic age, which overlap the metamorphic rocks of the Precambrian Shield. | 10-500 | 0.1-30 | 10-300 | 0.1-1 |

TABLE 2. (CONTINUED)

| Region | Hydrogeologic Situation* | Common Ranges in Hydraulic Characteristics of the Dominant Aquifers[†] | | | |
|---|---|---|---|---|---|
| | | Transmissivity (m² day⁻¹) | Hydraulic Conductivity (m day⁻¹) | Recharge Rate (mm yr⁻¹) | Well Yield (m³ min⁻¹) |
| 13. Western Glaciated Plains | Hills and relatively undissected plains underlain by glacial deposits over relatively flat-lying consolidated sedimentary rocks of Mesozoic and Cenozoic age. | 25-2,500 | 0.2-100 | 5-200 | 0.2-2 |
| 14. Central Glaciated Plains | An area of diverse topography, ranging from the plains of Iowa to the Catskill Mountains of New York, underlain by glacial deposits over flat-lying consolidated sedimentary rocks of Paleozoic age. | 100-2,000 | 2-300 | 5-300 | 0.2-2 |
| 15. St. Lawrence Lowlands | A hilly area underlain by glacial deposits over flat-lying consolidated sedimentary rocks of Paleozoic age. | 100-2,000 | 2-300 | 5-300 | 0.2-2 |
| 16. Central Non-glaciated Plains | Relatively undissected plains underlain by thin regolith over flat-lying consolidated sedimentary rocks of Paleozoic to Cenozoic age. Includes the Black Hills, where metamorphic rocks of Precambrian age are exposed in a dome structure surrounded by the upturned truncated edges of the sedimentary rock layers. | 300-10,000 | 3-300 | 5-300 | 0.4-10 |
| 17. High Plains | A relatively undissected, eastward-sloping tableland underlain by thick alluvial deposits over consolidated sedimentary rocks of Paleozoic to Cenozoic age. | 1,000-10,000 | 2-100 | 1-150 | 0.4-10 |
| 18. Alluvial Valleys | Thick deposits of sand and gravel, in places, interbedded with silt and clay of Pleistocene and Holocene age, underlying flood plains and terraces of streams. | 200-50,000 | 30-2,000 | 50-500 | 0.4-20 |
| 19. Northeastern Appalachians | Hilly to mountainous area underlain by glacial deposits over folded metamorphic rocks of Paleozoic age complexly intruded by igneous rocks. | 50-500 | 0.3-50 | 30-300 | 0.1-2 |
| 20. Appalachian Plateaus and Valley and Ridge | Hilly to mountainous area underlain by thin regolith over flat-lying to folded consolidated sedimentary rocks of Paleozoic age. The folded rocks of the Valley and Ridge province differ somewhat in their water-bearing characteristics from the flat-lying rocks of the Appalachian Plateaus but not enough to warrant including them in a separate region. | 300-10,000 | 3-300 | 5-400 | 0.4-10 |
| 21. Piedmont and Blue Ridge | Hilly to mountainous area underlain by thick regolith over folded metamorphic rocks of Paleozoic age complexly intruded by igneous rocks. Includes downfaulted basins containing rocks of early Mesozoic age. | 10-200 | 0.1-2 | 30-300 | 0.1-1 |
| 22. Atlantic and Eastern Gulf Coastal Plain | A relatively undissected low-lying plain underlain by complexly interbedded sand, silt, and clay that thicken progressivley toward the east and south. The coastal plain deposits are of Mesozoic and Cenozoic age and overlie truncated crystalline rocks similar in composition and structure to the bedrock in region 20. | 500-10,000 | 5-100 | 50-500 | 0.5-20 |
| 23. Gulf of Mexico Coastal Plain | A relatively undissected low-lying plain underlain by complexly interbedded sand, silt, and clay of Mesozoic and Cenozoic age that thicken progressively toward the coast and the center of the Mississippi Embayment. | 500-10,000 | 5-100 | 50-500 | 0.5-20 |
| 24. Southeastern Coastal Plain | A relatively low-lying area underlain by thick layers of sand and clay over semiconsolidated carbonate rocks of Cenozoic age. | 1,000-100,000 | 30-3,000 | 30-500 | 5-50 |

TABLE 2. (CONTINUED)

| Region | Hydrogeologic Situation* | Common Ranges in Hydraulic Characteristics of the Dominant Aquifers[†] | | | |
|---|---|---|---|---|---|
| | | Transmissivity (m² day⁻¹) | Hydraulic Conductivity (m day⁻¹) | Recharge Rate (mm yr⁻¹) | Well Yield (m³ min⁻¹) |
| 25. Yucatan Peninsula | A flat, low-lying area underlain by thin regolith over semiconsolidated carbonate rocks of Cenozoic age. | 500-5,000 | 10-500 | 20-200 | 0.5-5 |
| 26. West Indies | Mostly hilly and mountainous islands underlain by intrusive igneous and volcanic rocks of Mesozoic and Cenozoic age, which are overlain by thin regolith and bordered, in part, by semiconsolidated carbonates and unconsolidated alluvial deposits of Cenozoic age. | 100-10,000 | 3-300 | 5-300 | 0.1-10 |
| 27. Hawaiian Islands | Mountainous islands underlain by thin regolith and discontinuous alluvial deposits over a complex sequence of lava flows of Cenozoic age. | 10,000-100,000 | 200-3,000 | 30-1,000 | 0.4-20 |
| 28. Permafrost Region | A topographically diverse area ranging from the highest mountains in North America eastward across the tundra plains of northern Canada. Commonly underlain by unconsolidated deposits, partly of glacial origin, overlying fractured igneous, metamorphic, and consolidated sedimentary rocks of Precambrian to Cenozoic age. Hydrogeologic conditions are dominated by continuous permafrost in the northern part of the region and discontinuous permafrost in the southern part. | 10-2,000 | 10-200 | 5-100 | 0.01-2 |

*An average thickness of about 5 m was used as the break point between thick and thin.
[†]All values rounded to one significant figure.

Plates 1 and 2, and Figure 2. This comparison will, however, reveal two exceptions that should be noted. The first relates to regions that are similar as far as the first four hydrogeologic characteristics are concerned but differ with regard to recharge and discharge conditions. The most notable example is provided by the Great Basin and High Plains ground-water regions (regions 5 and 17, Fig. 3 and Table 2), both of which are underlain by thick alluvial deposits. In the Great Basin, recharge mainly is from losing streams that enter the basins from the bordering mountains and discharge is by evaporation in basin "sinks." In the High Plains, on the other hand, recharge mainly is in interstream areas and discharge is through springs and seepage to streams.

The second exception relates to the need to divide some areas into two or more regions in order to have manageable-sized units, although from the standpoint of hydrogeology and recharge and discharge conditions, these could have been treated as single regions. By doing this, it was also possible to obtain maximum benefit from the specialized knowledge of different groups of ground-water hydrologists. For example, the Western Glaciated Plains (region 13), Central Glaciated Plains (region 14), and St. Lawrence Lowlands (region 15) are all underlain by glacial deposits over flat-lying consolidated sedimentary rocks and, thus, are hydrogeologically similar. Dividing this area into three regions facilitated preparation of the text by hydrogeologsts familiar with different parts of the area. Similarly, regions 16 and 20 are hydrogeologically indistinguishable, as are regions 22 and 23.

A list of the regions together with a brief description of the hydrogeologic situation in each region and ranges in the hydraulic characteristics of the dominant aquifers are shown in Table 2.

## REFERENCES CITED

Fuller, M. L., 1905, Underground waters of eastern United States: U.S. Geological Survey Water-Supply Paper 114, 285 p.

Heath, R. C., 1982, Classification of ground-water systems of the United States: Ground Water, v. 20, no. 4, p. 393–401.

——— , 1984, Ground-water regions of the United States: U.S. Geological Survey Water-Supply Paper 2242, 78 p.

Meinzer, O. E., 1923, The occurrence of ground water in the United States, with a discussion of principles: U.S. Geological Survey Water-Supply Paper 489, 321 p.

Thomas, H. E., 1952, Ground-water regions of the United States; Their storage facilities, *in* The physical and economic foundations of natural resources: U.S. 83rd Congress, House Committee on Interior and Insular Affairs, v. 3, p. 3–78.

Manuscript Accepted by the Society June 3, 1987

The Geology of North America
Vol. O-2, Hydrogeology
The Geological Society of America, 1988

# Chapter 4

# *Region 1, Western mountain ranges*

**Bruce L. Foxworthy**
*U.S. Geological Survey, 601 N. Baker Avenue, Apartment 102, East Wenatchee, Washington 98802*
**Debra L. Hanneman**
*Department of Geology, Montana College of Mineral Science and Technology, Butte, Montana 59701*
**Donald L. Coffin**
*U.S. Geological Survey, MS 406, Box 25046, Denver Federal Center, Denver, Colorado 80225*
**E. Carl Halstead**
*National Hydrologic Research Institute, 1001 West Pender Street, Vancouver, British Columbia V6B 1R8, Canada*

## INTRODUCTION

The Western Mountain Ranges region (Fig. 3, Table 2, Heath, this volume) owing to its areal extent (more than 1.5 million km$^2$) and large range in altitude (5,000 m) and latitude (35° to 62° N), includes perhaps the greatest diversity of geologic conditions and processes of all the regions of North America. Consequently, ground-water conditions also are highly varied.

The region includes 15 subregions. Of the subregions, 12 are mountainous subregions and 3 are intermontane and lowland subregions. Included with the mountainous subregions are intermountain basins in Montana and Colorado and individual coastal lowlands along the entire Pacific Coast. The generalized boundaries and subregions of the Western Mountain Ranges region, as described in this chapter, are shown in Figure 1.

Although the region boundaries shown in Figure 1 are based mainly on topographic breaks in slope, at places they represent compromises between lithologic and physiographic boundaries. The northern boundaries also are compromises between major topographic features and the areas of permafrost (subregions 4 through 7). Most of the mountain ranges trend generally north–south or northwest–southeast. Notable exceptions are ranges in Alaska and the Uinta Mountains in Utah, which trend nearly east–west.

The mountain ranges are mostly steep and spectacular, and include some of the world's most rugged terrain. The mountains tend to be separated by relatively narrow valleys. Most ranges owe their origin to uplift resulting from plate convergence, thrusting, and/or block faulting, although many of the higher peaks are of volcanic origin. The peaks range in altitude to 4,418 m (Mount Whitney, California), and they rise above adjacent lowlands by heights ranging to about 4,000 m (Mount Rainier, Washington).

Knowledge about ground-water conditions in many mountainous areas is meager because of remoteness, harsh environment, complex geology (Fig. 2), and limited resource utilization.

These factors result in many gaps in detailed information about the geologic framework, and in a paucity of wells that would provide subsurface information. In vast areas, the only available information about the hydrogeology consists of inferences that can be made from springflow, drainage characteristics of mines and tunnels, low-flow characteristics of streams, and the general geologic and climatic setting. Available information about hydrogeologic conditions in the various subregions of the Western Mountain Ranges is summarized in Table 1.

## GEOLOGIC CONTROLS

Geologic conditions and processes, in the broadest sense, control virtually all aspects of the occurrence of ground water in the region. The most fundamental controls are provided by the lithology and structure, which largely determine the existence and characteristics of openings in which ground water occurs. Lithology, including types and solubility of the minerals, constitutes the greatest controlling factor of the natural quality of the ground water. Geologic processes, past and present, that significantly affect ground water in the region include faulting, folding, volcanism, glaciation, weathering, and erosion.

The Western Mountain Ranges region consists largely of accretionary terranes (Sullivan, 1984, p. 57–65) comprising consolidated igneous, metamorphic, and sedimentary rocks. Virtually the entire region has been subjected to intense folding and faulting (Fig. 2) during more than one episode of orogeny. For many parts of the region the mountain-building processes are still active.

The consolidated rocks generally have little primary porosity and storage capacity for ground water. Significant primary openings exist in some consolidated and semiconsolidated sandstone units that constitute a relative small part of the bedrock.

Foxworthy, B. L., Hanneman, D. L., Coffin, D. L., and Halstead, E. C., 1988, Region 1, Western mountain ranges, *in* Back, W., Rosenshein, J. S., and Seaber, P. R., eds., Hydrogeology: Boulder, Colorado, Geological Society of America, The Geology of North America, v. O-2.

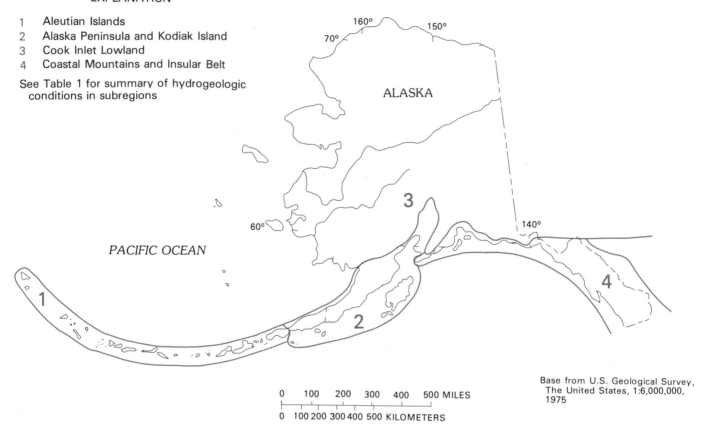

EXPLANATION

1    Aleutian Islands
2    Alaska Peninsula and Kodiak Island
3    Cook Inlet Lowland
4    Coastal Mountains and Insular Belt

See Table 1 for summary of hydrogeologic
   conditions in subregions

Figure 1. Subregions of the Western Mountain Ranges. A. Northwestern part; B (facing page) Southeastern part.

These pore spaces commonly are not well connected, and the permeability is only moderate. Of all the bedrock types, the greatest primary porosity and hydraulic conductivity occur in mafic lava and associated pyroclastic rocks of Tertiary and Quaternary age. Flow breccia, vesicular zones, lava tubes, tephra beds, cooling joints, and incomplete molding of molten lava into the irregular surface of preceding flow layers (interflow zones) all contribute to primary porosity and permeability of the volcanic units. In general, the volcanic rocks of Quaternary age are more permeable than those of Tertiary age.

Secondary openings resulting from jointing, fracturing, dolomitization and solution of calcareous rocks, and weathering have increased permeability of bedrock in many areas. However, these processes do not necessarily increase permeability significantly.

The region is seismically active, particularly in the western and northern parts along the Pacific Rim, and faults of many ages are common. The faults range from those apparently sealed completely by their own gouge to highly permeable breccia zones that serve as extensive ground-water conduits. In general, the more indurated and less weathered the bedrock, the greater the likelihood that jointing and fracturing will increase permeability. The mapping of lineaments, joint sets, and fracture trends, often by remote sensing, is a common way of selecting sites for test-drilling for water in bedrock.

The permeability imparted by jointing, fracturing, and weathering generally decreases with depth, because the increasing lithostatic pressure reduces openings in the rock materials. This aspect is borne out by many unsuccessful well-drilling attempts in bedrock as well as by mining and other tunneling. Seldom does the draining of a mine or tunnel in the bedrock present much of an engineering problem, and some mine works (e.g., the Omineca–Columbia Mountains subregion in northeast Washington) have been extended at relatively shallow depths beneath rivers without appreciable leakage from the river to the mine. The main exceptions are in areas of highly permeable volcanic rocks or cavernous limestone.

Calcareous rocks are not abundant in many parts of the region, but where they have interconnected solution openings they can be productive sources of spring and well water. Dolomitization has increased porosity and permeability but decreased

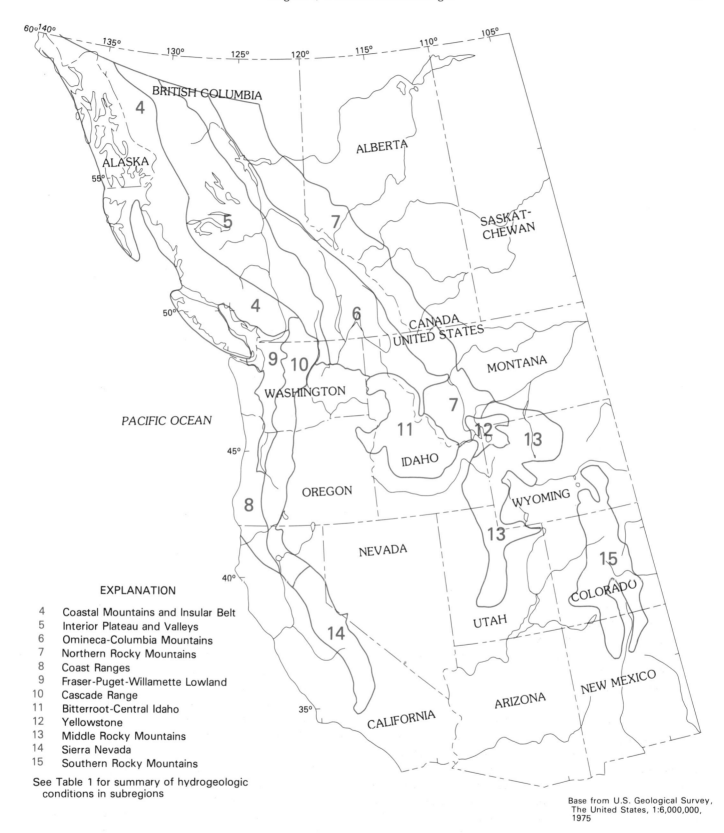

EXPLANATION

4    Coastal Mountains and Insular Belt
5    Interior Plateau and Valleys
6    Omineca-Columbia Mountains
7    Northern Rocky Mountains
8    Coast Ranges
9    Fraser-Puget-Willamette Lowland
10   Cascade Range
11   Bitterroot-Central Idaho
12   Yellowstone
13   Middle Rocky Mountains
14   Sierra Nevada
15   Southern Rocky Mountains

See Table 1 for summary of hydrogeologic
conditions in subregions

Base from U.S. Geological Survey,
The United States, 1:6,000,000,
1975

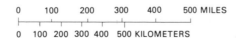

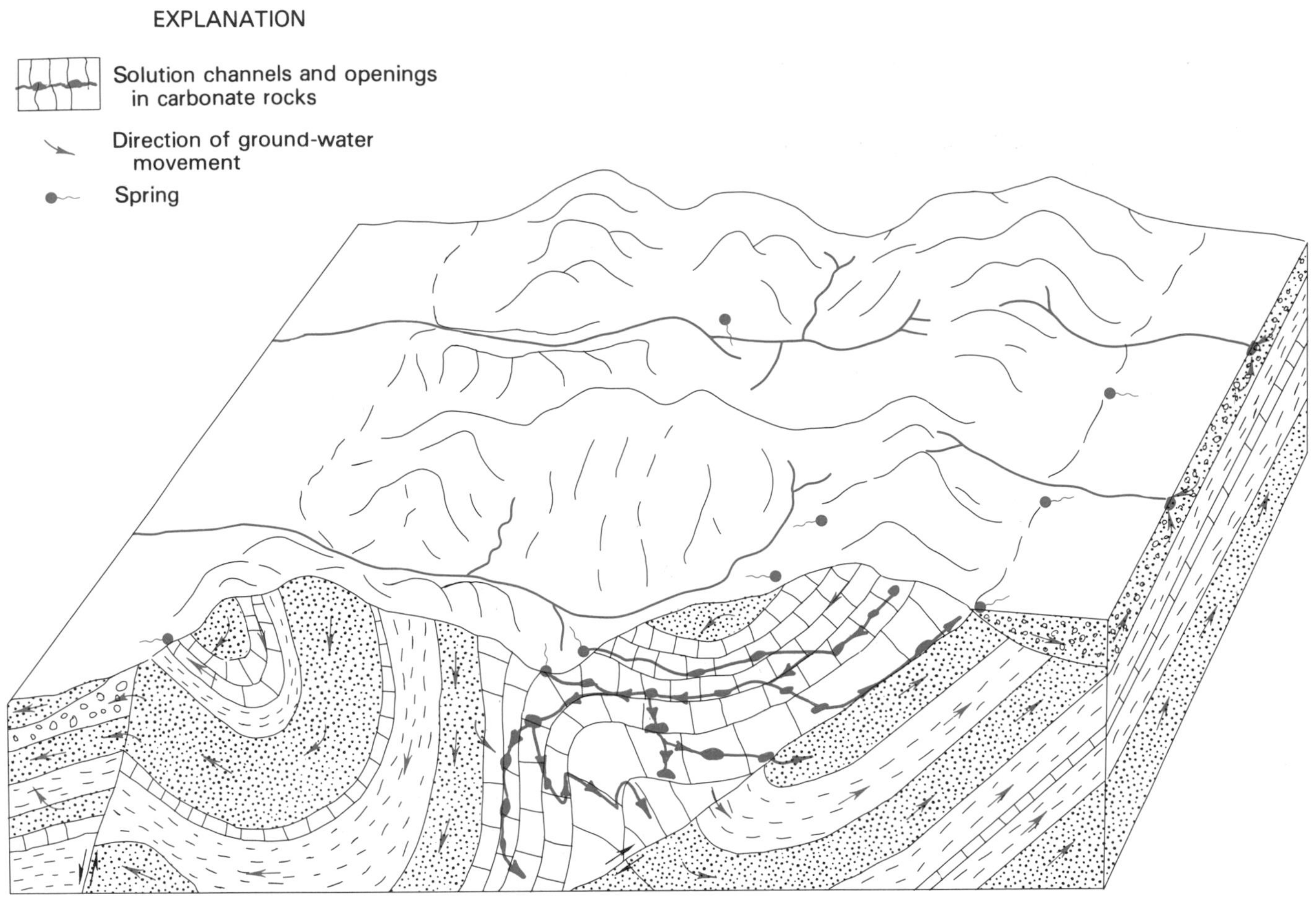

EXPLANATION

Figure 2. Diagram showing complexities of ground-water movement in folded and faulted rocks of the Middle Rocky Mountains subregion (Modified from Lines and Glass, 1975).

solubility of some limestone. However, some of the calcareous rocks (mainly dolomite) apparently were never affected by solution activity, or the solution openings filled.

Folding and faulting enhance ground-water recharge in parts of the region. In bedded rocks containing permeable zones, the opportunity for recharge to the beds has been greatly increased where uptilting has exposed the permeable zones at the land surface.

Weathering increases water-storage capacity in crystalline rock by creating a porous regolith over the intact bedrock. In high, steep parts of the mountains, however, the regolith may be thin because erosion can keep pace with weathering. Such may hold true even in the areas of rapid weathering; for example, the perennial freeze-thaw zones and the more humid areas. In the valleys and along the lower mountain fronts, the detritus from the peaks accumulates to provide significant storage space for ground water.

The most accessible productive ground-water reservoirs in the region are unconsolidated clastic deposits that underlie floors

of valleys and other lowlands. Outside of areas of continental glacier deposits, valley-fill deposits consist mainly of products of weathering and erosion on adjacent slopes. Therefore, the permeability of the deposits in those valleys depends on the type of adjacent bedrock and the way it weathers. Stream valleys in areas of granite or aphanitic basalt in the Cascade Range, for example, tend to have coarse, permeable beds of gravel in valley-fill deposits. In contrast, streams draining marine sedimentary rocks in the Coast Range of Oregon have deposited mostly fine-grained material that often is about as poorly permeable as the parent marine beds.

Glaciers in the past have exerted a strong influence on the hydrogeology of the region and continue to do so now. Virtually all the region north of the 47th parallel was covered by the Cordilleran Ice Sheet or by associated alpine or valley glaciers. Contemporaneous alpine glaciers also existed in higher mountains as far south as Colorado in the eastern part of the region and southern California in the west. In the broader valleys and lowlands formerly occupied by the ice sheet, such as the Fraser–

Puget Sound lowland and the interior valleys of British Columbia and Montana, the most productive aquifers commonly are proglacial and recessional outwash deposits, which generally are incompletely separated by one or more till aquitards.

Alpine and valley glaciers persist in parts of Alaska, British Columbia, Montana, and Colorado, and alpine glaciers occupy the high peaks as far south as the southern part of the Sierra Nevada. The present-day glaciers continue to erode and transport detritus from the higher peaks and, along with discharging ground water, maintain the dry-season flow of streams. In the Rocky Mountains from Montana southward, however, the mountain glaciers are retreating, and currently have little effect on erosion or transport.

## CLIMATIC EFFECTS

Although the geologic framework imposes the fundamental controls on the region's ground-water systems, geohydrologic conditions in the region cannot be adequately addressed without consideration of climatic differences and their effects. Climatic differences in the region result from the range in latitude, in altitude and relief of the land, and in orographic differences.

The most obvious effect of climate on ground water is the availability of recharge, which is determined by the amount of precipitation and the freeze-and-thaw cycles. However, the climatic differences within the region also have greatly affected weathering, erosion, and deposition, which in turn, are important influences on ground-water occurrence and chemical character.

## CLIMATE

The region includes some of the wettest parts of North America as well as semi-arid areas. In the western and northern parts of the region near the Pacific Ocean, annual precipitation commonly exceeds 2,000 mm, and amounts as great as 4,900 mm have been reported for the northern part of the Coast Ranges subregion (Fig. 1). In both drier and wetter parts, a strong direct correlation exists between altitude and annual precipitation. For some ranges, the difference in precipitation between upper and lower slopes exceeds 1,000 mm per year. Because many storms move east, the west sides of the mountains receive the most precipitation, and some eastern foothills and adjacent lowlands are in the rain shadows of the taller ranges and are dramatically drier. The mountain ranges generally receive progressively less precipitation from west to east. Inasmuch as the Western Mountain Ranges collect precipitation on their west sides but reduce precipitation on their east sides, this region has a large control on recharge of ground water in western and midcontinent regions of the United States.

In many of the mountains, the annual precipitation is fully adequate for, or exceeds, the intake capacity of the exposed soil and rocks. The highest mountains receive more than enough precipitation to fill the openings in the local rocks, and the adjacent lower areas receive runoff that recharges valley-fill aquifers by infiltrating along streams or the edges of floodplains, through talus slopes or, in some fault-block valleys, through the bordering fault zones.

The seasonal distribution of precipitation in much of the region is favorable for recharge of ground water, inasmuch as a large part of the precipitation occurs during fall and winter when it is not subject to much evapotranspiration. Above altitudes of about 1,000 m, most of the precipitation is snow, which together with melt from alpine glaciers, provides streamflow from spring through winter. In the southeastern part of the region, more of the annual precipitation falls during the growing season, in response to the influences of continental, rather than oceanic, air masses.

In a few areas, high precipitation on highly permeable rocks results in exceptional amounts of ground-water recharge, storage, and discharge. In Oregon, for example, some of the higher parts and eastern slopes of the Cascade Range are underlain by permeable volcanic rocks, which in places are able to absorb the 700 to 3,000 mm of annual precipitation. Some rivers that drain the area have remarkably constant flows that are little affected by normal storm runoff. Elsewhere in that area, the ground water travels for tens of kilometers eastward before discharging to streams outside the region (Foxworthy, 1979, p. 17–19).

## CONDITIONS IN MOUNTAIN ZONES

The influences of geologic processes on ground-water conditions differ throughout the region, but equally important are climatic conditions. These climatic conditions may be grouped as three different mountain zones.

### Upper mountain zone

This zone is characterized by perennial snow and ice and local permafrost (see Sloan and van Everdingen, Chapter 31.). The upper mountain zone also includes much of the steepest land in the region and is present in the highest mountain ranges and/or the highest latitudes. From the steep slopes the rates of erosion and transport by streams, glaciers, and snowslides is usually greater than the rate of weathering. Consequently, detritus within or from the upper zone tends to be coarse and unweathered. Exceptions occur in areas of rocks that are readily susceptible to weathering or produce only fine-grained detritus—tuff, for example—but such rock types are in the minority in the upper zone.

Meltwater from the alpine and valley glaciers in the upper mountain zone not only adds to the runoff available for recharge of aquifers in the middle and lower mountain zones, but also can directly recharge any permeable rock materials underlying and adjacent to the glaciers. Also, in the northern Cascade Range and probably elsewhere, some alpine glaciers have englacial liquid water, akin to perched ground water, in porous rock material and subglacial water (Tangborn and others, 1975).

TABLE 1. SUMMARY OF GROUND-WATER CONDITIONS IN SUBREGIONS OF THE WESTERN MOUNTAIN RANGES
(Subregion numbers correspond to map delineations in Figure 1.)

| Subregion | Dominant Geologic Terrane | Significant Aquifers | Ground-water Occurrence | Remarks and Selected References |
|---|---|---|---|---|
| 1. Aleutian Islands | Mafic volcanic rocks and beach deposits; glaciated volcanic peaks (many active) reach altitudes as high as 1,857 m. | Volcanic rocks and unconsolidated beach deposits. | Joints and pores of volcanic rocks (largely confined conditions); primary openings in beach and glacial deposits (mainly unconfined conditions). | Includes about 150 small islands, mostly uninhabited, extending for about 1930 km. Ground-water conditions are largely undefined, but general rock types and ample recharge are favorable. Obvious potential for geothermal ground water, both fresh and saline. Nearshore freshwater aquifers are vulnerable to salt-water intrusion (Coats, 1956; Gates and others, 1971; Feulner and others, 1976). |
| 2. Alaska Peninsula and Kodiak Island | Rugged, glaciated volcanic mountains reaching altitudes as high as 2,714 m, and lowlands of alluvial and glacial deposits with numerous lakes. Very active seismically and volcanically. | Alluvial and glacial outwash deposits. | Primary openings, confined to unconfined. | Largely unpopulated except in nearshore areas. Aquifer capabilities of lava rocks largely undefined; geothermal potential near active volcanoes. Dissolved iron is commonly present in concentrations greater than desirable (≥ 0.05 mg/L) (Jones and others, 1978; Zezone and Anderson, 1978.) |
| 3. Cook Inlet Lowland (Alaska) | Structurally complex bedrock of consolidated sedimentary, igneous, and metamorphic rocks, overlain by alluvial and glacial outwash deposits. Seismically very active. | Thick alluvium and glacial outwash deposits.<br><br>*Consolidated and semiconsolidated bedrock. | Primary openings; confined to unconfined.<br><br>*Secondary and primary openings; generally unconfined. | Ground water provides large public supplies for Anchorage and industrial supply for Kenai Peninsula; no evidence of overpumping. Unconsolidated deposits, where sufficiently thick, have great potential for recharge from streams. Dissolved iron commonly is present in greater than desirable concentrations. Man-caused contamination occurs locally.<br>*Bedrock aquifers are the principal supply for most private wells in upland areas (Cederstrom and others, 1964; Freethey and Scully, 1980). |
| 4. Coastal Mountains and Insular Belt (British Columbia and southeastern Alaska) | Steep, rugged, structurally complex mountains with adjacent islands and narrow shorelines indented by long fjords. Mountains consist mainly of intrusive igneous rocks, predominantly granitic. Hills and emerged shorelands that constitute the Insular Belt are underlain mainly by unmetamorphosed and metamorphosed sedimentary and volcanic rocks. Seismically very active. Cordilleran Ice Sheet extended throughout subregion. | Sand and gravel in alluvial and glaciofluvial deposits beneath valley floors and terraces.<br><br>*Massive crystalline rocks having well-developed joint systems (mainly in the mountains).<br><br>*Folded and faulted sedimentary rocks (mainly in the Insular Belt). | Primary openings; confined to unconfined.<br><br>*Joint systems; confined to unconfined. | Virtually all population is in island and nearshore areas, so ground-water conditions in mountains are largely unexplored. Glaciers, perennial snow fields, and permafrost at higher altitudes and latitudes. Ground-water availability is limited by storage capacity of rocks rather than by recharge, which is abundant. Water quality is generally excellent. Mineralized water (≥ 1000 mg/L) from bedrock is fairly common; however, concentration depends mainly on depth and type of source rocks. Potential for sea-water intrusion limits pumping withdrawals in some nearshore areas.<br>*Bedrock aquifers are the only sources of fresh water for mainly island, peninsular, and upland areas. |

TABLE 1. (continued)

| Subregion | Dominant Geologic Terrane | Significant Aquifers | Ground-water Occurrence | Remarks and Selected References |
|---|---|---|---|---|
| 5. Interior Plateau and Valleys (British Columbia and north-central Washington) | Unmetamorphosed sedimentary and volcanic rocks, glaciated and overlain by younger lava and glaciofluvial deposits. | Gravel and sand along streams and courses of former (glacial meltwater) rivers. | Primary pore spaces; mainly unconfined. | Population and ground-water development are greatest in southern part. Wells commonly produce a few tens to more than 100 L/s. Water quality is generally excellent. |
|  |  | *Flat-lying to gently dipping lava. | *Joints and basal and interflow zones. | *Yields ≥ 10 L/s rarely have been reported for wells that tap lava aquifers, possibly because prospecting has been limited. Lava is extensive and could be a source of large water supplies. |
| 6. Omineca-Columbia Mountains (British Columbia and northern Washington, Idaho, and Montana) | Glaciated mountains and valleys underlain mostly by metamorphic, granitic, and consolidated sedimentary rocks. Structurally controlled valleys such as the Rocky Mountain Trench contain unconsolidated to semiconsolidated deposits. | Fluvial and glaciofluvial gravel and sand in valleys; sand and gravel lenses in semiconsolidated deposits. | Primary pore spaces; unconfined to confined. | Comprises the Omineca Belt of Monger (1978) and similar glaciated mountains and valleys of northern U.S. The eastern margin comprises the Rocky Mountain Trench and similar structural valleys. Yields of 10 L/s or more to wells are common. Natural water quality is good, but unconsolidated aquifers are vulnerable to pollution from land surface. |
|  |  | *Consolidated sedimentary, igneous, and metamorphic rocks. | *Joints and weathered zones; mostly unconfined. | *Bedrock aquifers are important as sources for springs, thousands of domestic and stock wells, and base flow of upland streams (Coffin and others, 1971; Konizeski and others, 1968; Lawson, 1968; Pluhowski and Thomas, 1968). |
| 7. Northern Rocky Mountains (British Columbia, Alberta, and Montana) | Folded and faulted sedimentary strata, forming mountains that reach altitudes generally above 3,000 m. In southwestern Montana, intermountain valleys several km wide contain thick basin-fill deposits. | Sheetlike sedimentary beds, generally gently dipping but repeated by thrust faulting. Basin-fill deposits yield water from lenses of sand and gravel. | Primary pores and secondary joints; confined and unconfined. | Extends from about lat 45N (in U.S.) NNW to the generalized boundary of the zone of discontinuous permafrost. Ground-water system functions as a number of separate hydrologic basins, with leakage between them depending on relative permeabilities. Dissolved-solids concentrations greater (as much as 800 mg/L) in eastern foothills areas than elsewhere. Some thermal springs yield water rich in sulfate (Cherry and others, 1972; Hackett and others, 1960). |
| 8. Coast Ranges (Washington, Oregon, and northern California) | Coastal mountains and seaward lowlands underlain by sedimentary, metasedimentary, and lesser amounts of volcanic rocks. Elevations generally below 1,000 m except for north and south parts, where peaks reach well above 2,000 m. Seismically active. | Gravel and sand in valley-fill and terrace deposits. Sand and gravel of dune and beach deposits. | Primary pore spaces; unconfined to confined. | Significant aquifers are very limited in extent; permeable gravel and sand in mostly thin, narrow deposits in valleys in or adjacent to resistant consolidated rocks; dune and beach deposits in discontinuous, narrow reaches of coastal plains. Quality of water from these aquifers is generally good; saline water in places from adjacent marine sediments or sea-water intrusion. |
|  |  | *Weathered and fractured zones in igneous, sedimentary, and metamorphic rocks. | *Secondary openings; mostly unconfined. | *Preponderance of fine-grained sedimentary bedrocks, along with deep saprolite, makes yields of wells in most places small and base flow of many streams low. Quality of shallow ground water commonly good; salinity often increases with depth (Hampton, 1963). |

TABLE 1. (continued)

| Subregion | Dominant Geologic Terrane | Significant Aquifers | Ground-water Occurrence | Remarks and Selected References |
|---|---|---|---|---|
| 9. Fraser-Puget-Willamette Lowland (British Columbia, Washington, Oregon) | Structural trough underlain by bedrock consisting mainly of fine-grained sedimentary rocks with basaltic lava and metamorphic rocks in places. Seismically active. North one-half was occupied by Cordilleran Ice Sheet, which resulted in thick outwash and till deposits; east side received outwash from glaciers in the Cascade Range. | Sand and gravel facies of glacial and alluvial deposits.<br><br>*Consolidated and semi-consolidated sandstone and metamorphic rocks; basaltic lava. | Primary pore spaces; confined to unconfined; semiperched in some coastal and valley terraces.<br><br>*Primary pore spaces and secondary joints in sedimentary rocks; solution openings in rare carbonate rocks; joints and interflow zones in lava. | Most populous subregion. Surficial deposits in N part and along Columbia River are highly permeable in many places; could support large additional withdrawal. Natural quality of water from surficial deposits generally is good to excellent; excessive iron is a common problem; some salinity from relict sea water or from sea-water intrusion. Pollution of shallow aquifers is a growing problem.<br>*Low-yield bedrock aquifers are important in many island, peninsula, and upland areas as sources of base flow of streams and of water for domestic supplies. Water quality generally ranges from good to acceptable; tends to degrade with increasing depth of producing zone (Armstrong, 1983; Brown, 1967; Columbia-North Pacific Technical Staff, 1970). |
| 10. Cascade Range (British Columbia, Washington, Oregon, northern California) | Steep, rugged mountains composed predominantly of crystalline intrusive and metamorphic rocks in north part, and volcanic rocks in south part. Very active seismically and volcanically; about 15 major volcanic peaks and many smaller cones. Crest commonly reaches 2,500 m elevation (Mt. Rainier is highest peak at 4392 m). | Valley-fill gravel derived from igneous and other consolidated rocks throughout subregion. Basaltic and andesitic lava flows, breccia, and scoria in the central and southern parts of the subregion. | Secondary openings; mostly unconfined, in clastic deposits. Primary openings; unconfined to confined, locally perched, in some lava of the high Cascades in south part of subregion. | Ground water is little utilized because the population is sparse, but ground water is important for maintaining strong base flow of streams, especially in southern part. Abundant rain and snowmelt provide ample recharge. Lava aquifers of high Cascades in southern part are among the most permeable in the region. Water quality is generally excellent, except for excessive iron (common) and other metals (local). Thermal springs and geothermal energy potential are associated with active volcanoes. Glaciers and perennial snowfields are common in higher parts. (Foxworthy, 1979; Godwin and others, 1971). |
| 11. Bitterroot-Central Idaho (Idaho, eastern Oregon, western Montana) | Metamorphic and intrusive igneous rocks, limestone and other sedimentary rocks, form mountains with peaks reaching above 3,000 m. Seismically active. | Gravel and sand, mostly in narrow floodplain and terrace deposits in many small valleys.<br><br>*Fractured and weathered crystalline rocks. | Primary pore spaces; unconfined to confined.<br><br>*Secondary openings; mostly unconfined. | Includes extensive uninhabited areas where ground-water conditions can only be inferred from geology and streams. Water quality is generally good (dissolved solids < 250 mg/L).<br>*Bedrock aquifers are the only sources of ground water in extensive areas (Columbia-North Pacific Technical Staff, 1970). |
| 12. Yellowstone (northwestern Wyoming, southwestern Montana, eastern Idaho) | Volcanic, igneous, metamorphic, and sedimentary rocks forming plateaus lying above 3,000 m and mountains reaching 1,000 m higher. | Sand and gravel in alluvial, terrace, and outwash deposits.<br><br>*Fractured and weathered volcanic rocks. | Primary pore spaces; mostly unconfined.<br><br>*Secondary openings; confined and unconfined. | Includes the geothermal area in and near Yellowstone National Park where hot springs and geysers are common. In the non-geothermal areas, water quality is generally good (dissolved solids < 500 mg/L). In the geothermal areas, dissolved solids may be several thousand mg/L, and the water may be high in silica, chloride, and fluoride (Stauffer and others, 1980). |

TABLE 1. (continued)

| Subregion | Dominant Geologic Terrane | Significant Aquifers | Ground-water Occurrence | Remarks and Selected References |
|---|---|---|---|---|
| 13. Middle Rocky Mountains (western Wyoming, eastern Idaho, south-central Montana, central and eastern Utah) | Mostly sedimentary rocks, but includes igneous and metamorphic rocks. Relief exceeds 2,500 m; valley walls are steep, and stream gradients are high. | Consolidated sedimentary rocks, basin-fill deposits, sand and gravel along streams and beneath terraces. | Primary pore spaces, joints and fractures, solution openings (in carbonate rocks); unconfined to confined. | Includes the Bighorn Basin in north-central Wyoming, where well yields are generally less than 10 L/s; however, around the edge of the basin, springs and wells yield more than 150 L/s from fractures and solution openings in carbonate rocks (Cox, 1976; Lowry and others, 1976). |
| 14. Sierra Nevada (California, Nevada) | Mostly granitic rocks, except in northern and western parts, where metasedimentary and mafic volcanic rocks predominate. Comprises a great block, faulted on the east side and tilted westward. Seismically very active. A few peaks higher than 4,000 m; some glaciers and perennial snowfields. | Gravel and sand in valley-fill deposits; colluvium and fan deposits on lower slopes. *Fractured and weathered crystalline rocks. | Primary pore spaces; mostly unconfined. *Secondary openings; mostly unconfined. | Some withdrawals in isolated areas, but main ground-water significance is as a source of runoff that recharges aquifers in adjacent foothills and valleys. Water quality generally good (Bateman and others, 1963). |
| 15. Southern Rocky Mountains (southern Wyoming, central Colorado, and northern New Mexico) | High-grade metamorphic and intrusive igneous rocks form the core of the mountain ranges, which are flanked by sedimentary rocks. Relief is as much as 3,000 m. | Sand and gravel along streams and beneath terraces; weathered and fractured igneous and metamorphic rocks. | Primary pore spaces, joints and fractures; mostly unconfined. | Includes intermontane valleys that are structurally controlled, underlain by consolidated sediments that are buried beneath fine-grained basin-fill deposits. In the mountainous areas, water generally contains < 300 mg/L dissolved solids; sedimentary rocks and basin-fill deposits generally yield water considerably higher in dissolved solids (Cherry and others, 1972). |

*Bedrock aquifers rarely yield as much as 10 L/s to wells, but have special significance as sources of supply in areas where more productive aquifers are absent.

## Middle mountain zone

This zone comprises areas where the dominant form of weathering is freeze-thaw cycles. The middle mountain zone includes the intermediate slopes and valleys of the higher mountain ranges, is the most areally extensive zone, and contains the steepest land. Rates of erosion and transport, by seasonal snowslides and all other forms of gravity transport, commonly exceed the rate of weathering. At any place, however, the relation between weathering and erosion depends largely on rock types and the amount and pattern of precipitation. These same factors determine permeability of clastic materials derived from higher lands and deposited along valleys or in other depressions.

Colluvial deposits along bases of valley slopes commonly are among the most permeable deposits in the middle zone. These deposits are recharged by streamflow from valley sides, as well as direct infiltration of precipitation. However, they commonly contain no permanent body of ground water because they underlie steep slopes and are drained except during wet seasons. Deposits underlying valley terraces also tend to be largely drained. The most important hydrologic function of colluvial aprons, alluvial fans, or terrace deposits along valley sides is in conveying the infiltrating water from valley sides to valley-floor deposits, which in turn discharge into valley streams.

Valley-floor deposits in the middle zone include some of the most permeable aquifers in the region—glaciofluvial gravels along streams that drain or formerly drained glaciers. However, the associated till and morainal deposits are poorly permeable. Unlike the deposits along the valley sides, which depend mainly on the geologic conditions and processes on adjacent slopes, the valley-floor deposits are influenced by conditions and processes everywhere upstream within the drainage basin.

Assessment of bedrock aquifers in the middle zone generally includes evaluation of springs and low flow of streams. For indurated bedrock, the study of joint patterns may be a beneficial guide for test-drilling. Some of the most permeable bedrock in the middle zone is volcanic rock of Quaternary age. However, many of the permeable volcanic rock masses are in areas of high relief, so that, even in the more humid areas, sizable zones of saturation in the volcanic rock do not extend much above the levels of adjacent major streams.

## Lower mountain zone

This zone comprises the lower slopes and mountain valleys where the freeze-thaw cycle is not a dominant mode of weathering. The lower mountain zone, as defined here, may not exist in all the mountain ranges and is most common and extensive in ranges near the Pacific Ocean and at lower latitudes.

Differences in some of the characteristic features between the middle and lower zones are gradational and often subtle. Slopes and stream gradients in the lower zone generally are gentler, and the valley floors wider than in the adjacent middle zone. Terraces along the valley sides are more common, and the alluvial deposits tend to include finer-grained but better-sorted sediments. Terrace deposits in the lower zone may be extensive enough to contain sizable zones of saturation, but tend to have a lower average permeability than in the middle zone.

In the lower zone, the weathering processes depend mainly on the rock types and the amount of precipitation. An example is provided by basalt of Miocene age in the lower eastern slopes of the Cascade Range, where annual precipitation averages about 300 mm, and the extension of the basalt on the lower western slopes of the range, where annual precipitation is about 1,000 mm or more. The basalt in both areas is generally aphanitic, and the individual flow layers are commonly 10 to 30 m thick. In its western occurrence, the upper part of the basalt often has been weathered to a reddish, amorphous saprolite as thick as 10 m. Even below the saprolite, weathering of the basalt has obliterated many of the cross-bed cooling joints that otherwise would be avenues for downward-migrating recharge. In comparable occurrences on the eastern side of the Cascade Range, the basalt also is noticeably weathered, commonly about a meter inward from cooling joints and flow surfaces. However, the minerals are not so hydrated as to close the joints, and the weathered rock usually has not lost structural integrity. Consequently, residual soil on this eastern basalt is very thin, and this basalt is better suited to accept whatever water is available for infiltration.

## UTILIZATION OF GROUND WATER

The significance of any resource, and of the natural processes that control it, depends mainly on its usefulness and uses. A few general comments about ground-water utilization in the region, including limitations and future opportunities, help to put the ground-water situation into the perspective of human concerns.

In general, the pattern of ground-water development and use closely reflects the population distribution. Consequently, lowland areas near population centers are centers for intensive ground-water withdrawal. The only major exceptions are some semiarid lowlands where irrigation is common. Ground-water problems in the areas of intensive development include potential overdraft in relatively small areas, but most reported problems involve deteriorating water quality. In virtually all parts of the region, ground-water utilization has a strong inverse relation to the availability of good-quality surface water. Much of the available information on ground-water conditions and utilization in the lowlands is contained in the references.

Future uses of ground water will be mostly unchanged from the present. However, a significant new use of ground water, in the mountains as well as lowlands, may be for geothermal energy development. Geothermal areas have been identified in several parts of the region, mostly associated with Holocene volcanic activity, and thermal ground water at several places in and adjacent to the Western Mountain Ranges has been used succesfully for space heating and moderate-temperature commercial applications. However, no geothermal reservoirs having the rare combination of both ground-water yields and water temperatures high enough for efficient and economically competitive electric power generation have been identified in this region. Reviewed interest in geothermal energy has been sparked by recent and currently advancing research on energy development using deep, hot, "dry" rocks as the heat source, and circulating water from the surface through them by means of paired wells. In North America to date, the most promising dry-rock, geothermal, experimental results were obtained in May–June 1986 by a team from the Los Alamos National Laboratory at a site in the Jemez Mountains of New Mexico (western limb of the Southern Rocky Mountains subregion, Fig. 1). In that experiment, water from a surface reservoir was circulated via two wells through artificially fractured granite at a depth of about 4,000 m (Franke and others, 1986) and returned at a temperature as great as 190°C. At the end of the experiment, the system was producing enough usable heat to supply electrical power for a town of 2,000 people. Geothermal gradients that are considerably higher than the average (about 30°C per kilometer of depth), and therefore favorable for dry-rock energy development, have been found in several parts of the Western Mountain Ranges (Smith, 1982). Water for such a system could be supplied by shallow aquifers at the injection sites.

## REFERENCES CITED

Armstrong, J. E., 1983, Environmental and engineering applications of the surficial geology of the Fraser Lowland, British Columbia: Geological Survey of Canada Paper 83-23, 54 p.

Bateman, P. C., Clark, L. D., Huber, N. K., Moore, J. G., and Rinehart, C. D., 1963, The Sierra Nevada batholith; A synthesis of recent work along the central part: U.S. Geological Survey Professional Paper 414–D, p. D1–D46.

Brown, I. C., ed., 1967, Groundwater in Canada: Geological Survey of Canada, Economic Geology Report No. 24, 228 p.

Cederstrom, D. J., Trainer, F. W., and Waller, R. M., 1964, Geology and groundwater resources of the Anchorage area, Alaska: U.S. Geological Survey Water-Supply Paper 1773, 101 p.

Cherry, J. A., van Everdingen, R. O., Meneley, W. A., and Toth, J., 1972, Hydrogeology of Rocky Mountains and Interior Plains: International Geological Congress, Field Guide A26, 42 p.

Coats, R. R., 1956, Reconnaissance geology of some western Aleutian Islands, Alaska: U.S. Geological Survey Bulletin 1028–E, p. 83–100.

Coffin, D. L., Brietkrietz, A., and McMurtrey, R. G., 1971, Surficial geology and water resources of the Tobacco and upper Stillwater River Valleys, northwestern Montana: Montana Bureau of Mines and Geology Bulletin 81, 48 p.

Columbia-North Pacific Technical Staff, 1970, Water resources *in* Columbia-North Pacific comprehensive framework study of water and related lands: Vancouver, Washington, Pacific Northwest River Basins Commission, app. 5, 1022 p.

Cox, E. R., 1976, Water resources of northwestern Wyoming: U.S. Geological Survey Hydrologic Investigations Atlas HA-558, 3 sheets.

Feulner, A. J., Zenone, C., and Reed, K. M., 1976, Geohydrology and water supply, Shemya Island, Alaska: U.S. Geological Survey Open-File Map Report 76–82, 1 sheet.

Foxworthy, B. L., 1979, Summary appraisals of the nation's ground-water resources; Pacific Northwest Region: U.S. Geological Survey Professional Paper 813–S, 39 p.

Franke, P. R., Brown, D. W., Smith, M. C., and Mathews, K. L., 1986, Hot dry rock geothermal energy program: Los Alamos National Laboratory Report LA–10661–HDR, 31 p.

Freethey, G. W., and Scully, D. R., 1980, Water resources of the Cook Inlet Basin, Alaska: U.S. Geological Survey Hydrologic Investigations Atlas HA–620, 4 sheets.

Gates, O., Powers, H. A., and Wilcox, R. E., 1971, Geology of the Near Islands, Alaska, with a section on surficial geology, by J. P. Schafer: U.S. Geological Survey Bulletin 1028–U, p. 709–822.

Godwin, L. H., Haigler, L. B., Rioux, R. L., White, D. E., Muffler, L.J.P., and Wayland, R. G., 1971, Classification of public lands valuable for geothermal steam and associated geothermal resources: U.S. Geological Survey Circular 647, 18 p.

Hackett, O. M., Visher, F. N., McMurtrey, R. G., and Steinhilber, W. L., 1960, Geology and ground-water resources of the Gallatin Valley, Gallatin County, Montana, with sections on the surface-water resources, by F. Stermitz and F. C. Boner, and chemical quality of the water, by R. A. Krieger: U.S. Geological Survey Water-Supply Paper 1482, 282 p.

Hampton, E. R., 1963, Ground water in the coastal dune area near Florence, Oregon: U.S. Geological Survey Water-Supply Paper 1539–K, p. K1–K36.

Jones, S. H., Madison, R. J., and Zenone, C., 1978, Water resources of the Kodiak-Shelikoff subregion, south-central Alaska: U.S. Geological Survey Hydrologic Investigations Atlas HA–612, 2 sheets.

Konizeski, R. L., Brietkrietz, A., and McMurtrey, R. G., 1968, Geology and ground-water resources of the Kalispell Valley, northwestern Montana: Montana Bureau of Mines and Geology Bulletin 168, 42 p.

Lawson, D. W., 1968, Groundwater flow systems in the crystalline rocks of the Okanogan Highland, British Columbia: Canadian Journal of Earth Sciences, v. 5, p. 813–824.

Lines, G. C., and Glass, W. R., 1975, Water resources of the Thrust Belt of western Wyoming: U.S. Geological Survey Hydrologic Investigations Atlas HA–539, 3 sheets.

Lowry, M. E., Lowham, H. W., and Lines, G. C., 1976, Water resources of the Bighorn Basin, northwestern Wyoming: U.S. Geological Survey Hydrologic Investigations Atlas HA–512, 2 sheets.

Monger, J.W.H., 1978, Evolution of the Cordillera: Geos, Fall 1978, p. 5–8.

Pluhowski, E. J., and Thomas, C. A., 1968, A water-balance equation for the Rathdrum Prairie ground-water reservoir, near Spokane, Washington: U.S. Geological Survey Professional Paper 600–D, p. D75–D78.

Smith, M. C., 1982, The hot dry rock geothermal energy program: Los Alamos National Laboratory Mini-review LALP–81–45, 4 p.

Stauffer, R. E., Jenne, E. A., and Ball, J. W., 1980, Chemical species of selected trace elements in hot-springs drainage of Yellowstone National Park: U.S. Geological Survey Professional Paper 1044–F, p. F1–F20.

Sullivan, W., 1984, Landprints; On the magnificent American landscape: New York, The New York Times Book Co., 384 p.

Tangborn, W. V., Krimmel, R. M., and Meier, M. F., 1975, A comparison of glacial mass balance by glaciological, hydrological, and mapping methods, South Cascade Glacier, Washington: International Association of the Hydrologic Sciences, Publication 104, p. 185–196.

Zenone, C., and Anderson, G. S., 1978, Summary appraisals of the nation's ground-water resources; Alaska: U.S. Geological Survey Professional Paper 813–P, 28 p.

MANUSCRIPT ACCEPTED BY THE SOCIETY SEPTEMBER 19, 1987

Printed in U.S.A.

The Geology of North America
Vol. O-2, Hydrogeology
The Geological Society of America, 1988

# Chapter 5

# *Region 2, Columbia Lava Plateau*

**G. F. Lindholm**
*U.S. Geological Survey, 230 Collins Road, Boise, Idaho 83702*
**J. J. Vaccaro**
*U.S. Geological Survey, 1201 Pacific Avenue, Suite 600, Tacoma, Washington 98402*

## INTRODUCTION

The Columbia Lava Plateau (Fig. 3, Table 2, Heath, this volume) is defined by Heath (1984, p. 28) as "an area of 366,000 km$^2$ in northern California, eastern Washington and Oregon, southern Idaho, and northern Nevada" (Fig. 1). Parts of the plateau are flat relative to bordering mountains, parts are gently to moderately rolling, and the remainder is highly dissected and rugged. The plateau ranges in altitude from 500 to 3,000 m above sea level, most being less than 1,800 m. Bordering mountains on the east, north, and west are in the Western Mountain Ranges ground-water region (Heath, 1984, p. 17). The Basin and Range Province on the south is in the Alluvial Basins ground-water region.

Most of the plateau is drained by the Columbia River and its tributaries. The largest tributary is the 1,700-km long Snake River, which drains most of southern Idaho and parts of Wyoming, Nevada, Oregon, and Washington.

Much of the Columbia Lava Plateau has an arid to semiarid climate. Annual precipitation ranges from about 200 to 1,200 mm; it is least immediately east of the Cascade Range and on the Snake River Plain (Heath, 1984, p. 31). Bordering mountains, particularly the northern Cascades, receive substantially greater amounts. Most of the Cascades receive about 1,500 mm of annual precipitation and, in places, more than 4,000 mm. On the east, most of the bordering Rocky Mountains receive more than 1,000 mm per year.

The Columbia Lava Plateau is dominated by Tertiary and Quaternary volcanic rocks, mainly basalt. On the basis of predominant rock type and differences in the basalt, the plateau has been divided into areas underlain by: (1) the Columbia River Basalt Group, (2) the Snake River Group, and (3) other rock units. This chapter summarizes the geology, hydrology, and geochemistry of each area. The discussion of "other rock units" is generalized and brief. Much of the other rock unit area is more appropriately part of the Great Basin ground-water region (Mifflin, this volume).

Foxworthy (1979) made a summary appraisal of ground-water resources in the Pacific Northwest that includes most of the Columbia Lava Plateau. Brown (1981) compared hydrologic properties of layered volcanics in Idaho, Oregon, and Washington.

## HYDROGEOLOGIC PROBLEMS

Volcanic rocks that compose the Columbia Lava Plateau are characterized by their diversity. They differ chemically, mineralogically, structurally, and in mode of emplacement. Consequently, their hydraulic properties differ greatly, and collectively they constitute a complex, heterogeneous, and anisotropic ground-water system. The complexity and diversity of rocks that compose the Columbia River Basalt Group and the Snake River Group preclude detailed understanding of their hydraulic properties. Only recently have regional stratigraphic relations of the Columbia River Basalt Group been defined.

High hydraulic conductivities of basalt of the Snake River Group and, to a lesser degree, other basalt underlying the Columbia Lava Plateau limit the depth to which basalt aquifers have been investigated. Most wells completed in the Snake River basalt penetrate less than 60 m of the aquifer, many less than 15 m, before desired yields are obtained. As a result, deep drill-hole information is sparse, and little is known about basalt aquifers at depth. Test drilling in some basalt presents problems because of high porosity and high hydraulic conductivity. Lost-circulation problems are common, and return of drill cuttings is generally poor. Core drilling and geophysical logging help maximize information return. Subsurface stratigraphic correlation of flows or groups of flows that compose the Snake River Group is usually difficult. The degree of difficulty is increased because of the large numbers of relatively thin flows of small areal extent. Major sedimentary interbeds are usually the best stratigraphic markers. Sedimentary interbeds in the Columbia River Basalt Group are relatively thick and areally extensive, aiding separation of the basalt sequence into several major units. Interbeds in the Snake

Lindholm, G. F. and Vaccaro, J. J., 1988, Region 2, Columbia Lava Plateau, *in* Back, W., Rosenshein, J. S. and Seaber, P. R., eds., Hydrogeology: Boulder, Colorado, Geological Society of America, The Geology of North America, v. O-2.

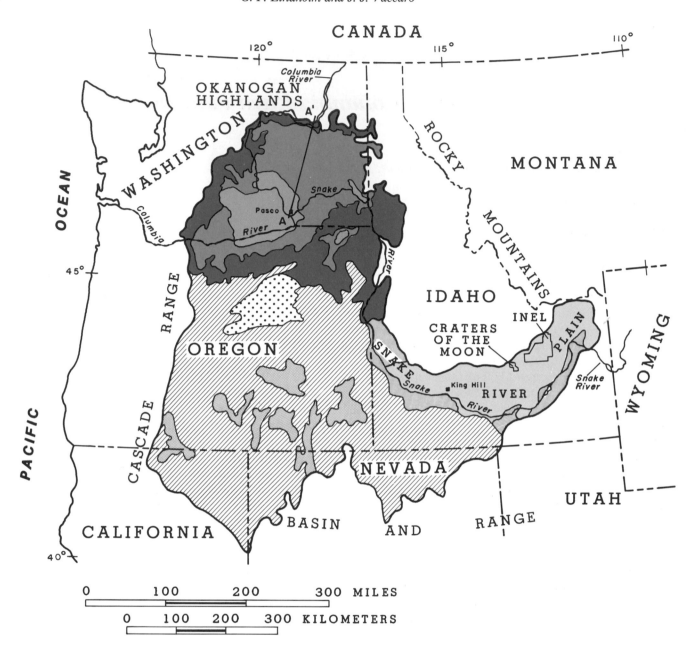

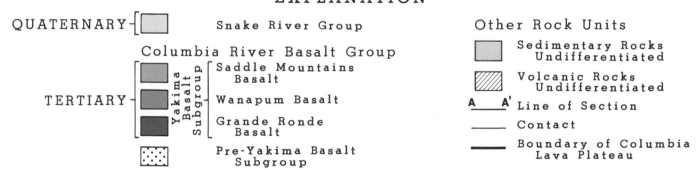

Figure 1. Generalized geology of the Columbia Lava Plateau.

River Group are thinner and less extensive, making regional stratigraphic separation extremely difficult. Petrography, mineralogy, chemical analysis, age dating, and paleomagnetism are also used to make stratigraphic correlations in basalt.

The use of geophysical techniques to define subsurface geology is complicated in areas of volcanic rock. Magnetic and gravity surveys have been used with moderate success to define the distribution of major rock types and basement configuration. Depth soundings by direct-current electrical resistivity methods have been used on the Snake River Plain with moderate success to define basalt-sediment contacts. Geophysical references for the Snake River Plain were compiled by Whitehead (1986a).

The diversity of volcanic rocks makes it difficult to determine their hydraulic properties in the field. Although it is known that multiple interflow zones result in high total horizontal transmissivity, the transmissivity of individual zones within the basalt sequence is generally unknown. The isolation of an interflow zone for testing is commonly difficult and is complicated by irregular fracturing that provides hydraulic connection between zones. Little is known about the extent and physical and hydraulic properties of sedimentary interbeds. They impede water movement in many areas, but the degree to which they do so has as yet been determined largely by theoretical methods with virtually no verification in the field.

## COLUMBIA RIVER BASALT GROUP

### *Geology*

Miocene rocks of the Columbia River Basalt Group underlie about 200,000 km$^2$ of the Columbia Lava Plateau (Fig. 1). Basalt flows were extruded from a system of northwest-trending linear vents in southeastern Washington, northeastern Oregon, and western Idaho (Swanson and others, 1975). Eruptions occurred from about 16 to 6 Ma, mostly over a short time period centered around 15 Ma (Swanson and Wright, 1978, p. 37, 41). From 14 to 6 Ma, eruptions were less frequent, generally allowing time for erosion and structural deformation between subsequent eruptions. Numerous successive lava flows and lesser amounts of interbedded sedimentary rocks compose the Columbia River Basalt Group. Flows were voluminous and advanced as sheetfloods over large areas. Total thickness locally may be as much as 4,500 m (Reidel and others, 1982). Subsidence was concurrent with volcanism, forming both a structural and topographic basin in which the lowest point is near Pasco, Washington (Fig. 1). Along the borders of the basin, rocks of the Columbia River Basalt Group are underlain by a complex of Precambrian to lower Tertiary volcanic and metamorphic rocks. The nature of rocks underlying the Columbia River Basalt Group in the center of the basin is largely unknown.

In eastern Washington, rocks of the Columbia River Basalt Group are intensely folded and faulted. The area is characterized by long, narrow anticlines and broad synclines that trend in an easterly to southeasterly direction. Folds are generally asymmet-

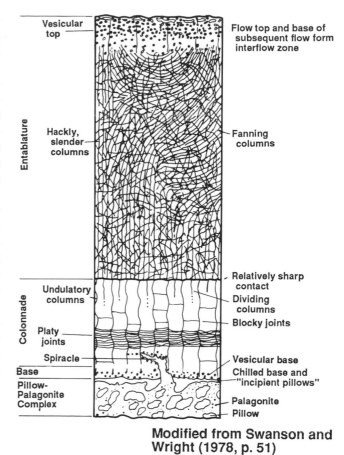

Modified from Swanson and Wright (1978, p. 51)

Figure 2. Intraflow structure in the Columbia River Basalt Group.

rical with the steeper limb to the north. Most faults are thrust or reverse faults whose strikes are the same as anticlinal fold axes and which probably are contemporaneous with the folding. Northwest-to-north–trending shear zones and minor folds commonly transect the major folds (Barrash and others, 1983). In eastern Washington, basalt flows have a regional dip of less than 5° to the southwest.

Following extrusion, flows cooled rapidly, gases were expelled, and vesicles and cooling joints formed. Upper surfaces of flows were broken by subsequent internal lava movement, which produced brecciated flow tops. Individual flows range from a few tens of centimeters to more than 100 m in thickness and average 30 to 40 m (Swanson and Wright, 1978, p. 49). From bottom to top, individual flows generally consist of a flow base, colonnade, and entablature (Fig. 2). Thickness of each depends on total flow thickness. The flow base is typically vesicular basalt and constitutes less than 5 percent of total flow thickness. If flows entered water, a pillow-palagonite complex of variable thickness is present at the base. The colonnade, which averages about 30 percent of total flow thickness, consists of nearly vertical three- to eight–sided columns of basalt, less vesicular than the base. Individual columns are commonly about 1 m in diameter and 7.5 m in length (Swanson and Wright, 1978, p. 50–51). Columns may be crosscut by systems of joints. The denser entablature, which averages about 70 percent of total flow thickness, consists of small-diameter (average less than 0.5 m) basalt columns, sometimes in fan-shaped arrangements. Cross joints in the entablature are less consistently oriented and interconnected than in the colonnade; hackly joints are common. The upper part of the entablature is commonly vesicular and may be rubbly and clinkery. The vesicular upper part of the entablature, in combination with the superposed flow base, is called the interflow zone.

Element chemistry, magnetic polarity, sedimentary interbeds, and members belonging to the Miocene Ellensburg Formation make it possible to subdivide the Columbia River Basalt Group. Swanson and others (1979) divided the Columbia River Basalt Group into major basalt units, the top three of which compose the Yakima Basalt Subgroup. Pre-Yakima basalt units are present at only scattered locations in the basin and little is known about them. Therefore, they will not be discussed further.

The distribution and stratigraphic relations of major basalt units composing the Yakima Basalt Subgroup are shown below and in Figures 1 and 3.

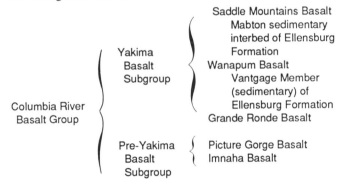

The Grande Ronde Basalt of the Yakima Basalt Subgroup underlies essentially the entire area shown as Columbia River Basalt Group in Figure 1. It constitutes nearly 85 percent of the total volume of the Columbia River Basalt Group (Reidel, 1982, p. 78). The Grande Ronde Basalt ranges in thickness from a few meters along the northern margin of the basin, where it pinches out against basement rocks, to more than 1,200 m in the central and southwestern parts (Drost and Whiteman, 1985). At least 30 and perhaps as many as several hundred individual flows compose the Grande Ronde Basalt. Most of the basalt is fine grained with scattered plagioclase microphenocrysts and plagioclase-clinopyroxene microphyric clots (Swanson and Wright, 1978, p. 42). Sedimentary interbeds are extremely rare and generally only a few meters thick. A sedimentary interbed as much as 30 m thick, the Vantage Member of the Ellensburg Formation, separates the Grande Ronde Basalt from the overlying Wanapum Basalt in much of the area.

Wanapum Basalt crops out or is covered by a thin veneer of sediment throughout most of the basin. In the central part, it is generally covered by a thick sequence of sediment or by the

Saddle Mountains Basalt. The Wanapum Basalt averages about 120 m thick and generally includes 10 flows (Drost and Whiteman, 1985). Thickness varies from a few meters along its lateral extent where it pinches out against exposures of the Grande Ronde Basalt to more than 300 m in the southwestern part of the basin (Drost and Whiteman, 1985). Most Wanapum Basalt is medium grained and slightly to moderately plagioclase-phyric (Swanson and Wright, 1978, p. 45). Sedimentary interbeds are more common in the Wanapum Basalt than in the Grande Ronde Basalt but are still relatively rare and at most only a few meters thick. A sedimentary interbed as much as 45 m thick, informally known as the Mabton interbed of the Ellensburg Formation, separates the Wanapum Basalt and the overlying Saddle Mountains Basalt in the southwestern part of the basin.

Saddle Mountains Basalt underlies only the southwestern part of the basin. It has a maximum thickness of more than 240 m and averages about 120 m (Drost and Whiteman, 1985). Individual flows vary greatly in texture and composition. Sedimentary interbeds are common and relatively thick; many are 15 m or more.

Interbed materials were derived from older rocks surrounding the basin, from contemporaneous volcanic rocks in the Cascade Range, and from the basalt itself. The areal extent, thickness, and lithology of interbeds are dependent on (1) duration of exposure of a flow top to atmospheric weathering before subsequent eruptions and burial, (2) duration of a Cascade eruption, (3) structural position, (4) proximity to an older rock source, and (5) proximity to a stream. Most interbeds exhibit lateral facies changes (Swanson and Wright, 1978, p. 56); grain size ranges from clay to gravel.

## Hydrology

The Columbia River Basalt Group is a complex, regional multiaquifer system in which permeable parts of basalt flows constitute numerous small aquifers, unconfined to confined. Basalt, by nature of formational processes, is extremely heterogeneous with respect to its hydraulic properties. Within a single flow, basalt exhibits a large range in vertical and horizontal hydraulic conductivity. Interflow zones (about 5 to 10 percent of total flow thickness) generally have the highest conductivities and form a series of superposed aquifers. Fracture assemblages in the entablature and colonnade are better connected vertically than horizontally, allowing movement of water between interflow zones. Vertical water movement is typically slow, owing to the general tightness of joints. Lateral water movement predominates in interflow zones, and vertical movement predominates in the central parts of flows. On a small scale, the distribution of hydraulic conductivity, fracture geometry, and position in a local flow system enable water to move three dimensionally through all parts of a basalt flow. On a regional scale, water movement is also three dimensional, as shown in Figure 3. Position in the regional flow system and variations in hydraulic conductivity create further head differences with depth that are larger than those in an iso-

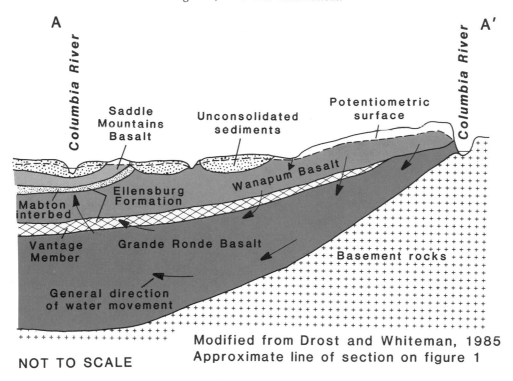

Figure 3. Stratigraphic sequence and general direction of water movement in the Columbia River Basalt Group.

tropic aquifer. Head differences are such that net flow is downward in much of the area and upward near topographic lows, usually in the vicinity of stream valleys.

Water in the Columbia River Basalt Group originates mainly as precipitation on the plateau. The part of precipitation that does not run off, evaporate, or transpire is available to recharge the ground-water system. Recharge from precipitation generally is greatest at higher altitudes on the plateau where precipitation is greatest, although there are exceptions in some places owing to topography, soil type, and vegetation. Minute amounts of water are added to the system from underlying and surrounding basement rocks. The basalt is artificially recharged by irrigation water diverted from the Columbia, Yakima, Umatilla, Snake, and Walla Walla rivers.

Natural ground-water recharge in the area underlain by the Columbia River Basalt Group ranges from 0 to 250 mm/yr, but is generally less than 50 mm/yr (H. H. Bauer, U.S. Geological Survey, personal communication, 1985). Artificial recharge in parts of the area irrigated by surface water is as much as 370 mm/yr (Tanaka and others, 1974, p. 46; Bolke and Skrivan, 1981, p. 28; Prych, 1983, p. 53).

After percolating to the saturated zone, water moves as part of local and regional flow systems. Lateral water movement is from areas of higher altitude, along the borders of the plateau, toward the center of the structural basin where altitude is lowest

(Fig. 3). Direction of regional flow is governed by physiographic and structural controls, by distribution of hydraulic conductivity, and by differences in head due to natural and artificial recharge and to pumping. Water movement in the uppermost basalt is further controlled by small streams, which constitute numerous local drains. Major streams, such as the Columbia and Snake rivers, are the major drains for the entire basalt sequence.

Horizontal conductivity of the Columbia River Basalt Group has been estimated by aquifer tests and from specific capacity data. Aquifer tests of cased wells in basalt generally yield site-specific values that are representative of a productive zone. To obtain an understanding of the range in values for the entire basalt sequence and of the effects of geologic controls on transmissivity, many areally distributed tests of vertical hydraulic conductivity are needed. Sparsity of aquifer-test data and inconsistency among those available necessitate the use of specific capacity data to estimate hydraulic conductivity. Estimates thus made represent the integrated effects of the entire open interval of a well, including all parts of a lava flow or flows and any included sediment interbeds.

Most wells finished in the Columbia River Basalt Group are uncased. Horizontal hydraulic conductivity values for each major basalt unit were estimated from specific capacity data by H. H. Bauer and J. J. Vaccaro (U.S. Geological Survey, personal communication, 1985). The Grande Ronde and Saddle Mountains basalts have a median conductivity of about 0.65 meters per day (m/d), whereas the Wanapum Basalt has a median of about 1.6 m/d. Values of horizontal hydraulic conductivity for all three units range from 0.002 to 1,600 m/d and average about 18 m/d with a median of 1 m/d. Hydraulic conductivity values follow a skewed bimodal distribution with 75 percent of the values being less than 8 m/d and 15 percent being in the range from 70 to 265 m/d. The lower average hydraulic conductivity of Saddle Mountains Basalt is attributed to numerous sedimentary interbeds. The large range in hydraulic conductivity is indicative of the inherent heterogeneity of basalts. Recent modeling studies (A. Davies-Smith, U.S. Geological Survey, personal communication, 1984; F. A. Packard, U.S. Geological Survey, personal communication, 1985) suggest that, within a basalt unit, average formational horizontal hydraulic conductivity on some geologic structures is as low as $10^{-3}$ m/d.

Horizontal conductivity may be much larger than estimated from specific capacity data because when wells are pumped water moves across the dense, central parts of flows that have lower hydraulic conductivities than do interflow zones. Cascading water in many uncased wells is attributed to the relative ease with which water moves horizontally along interflow zones, compared with vertical movement through the central parts of flows.

Vertical hydraulic conductivity of the Columbia River Basalt Group is largely unknown. Estimates made during recent modeling studies (MacNish and Barker, 1976, p. 5; Prych, 1983, p. 35; F. A. Packard, U.S. Geological Survey, personal communication, 1985) are rough and account for the integrated effects of all basalt flows and sedimentary rock interbeds. Vertical hydraulic conductivity of the Columbia River Basalt Group as determined from modeling studies ranges from $2 \times 10^{-3}$ to 1.5 m/d. True vertical hydraulic conductivity probably ranges from $10^{-8}$ to 10 m/d, depending on the degree of fracturing, presence of pillow-palagonite complexes (Swanson, 1967, p. 1083), and geologic structure.

Transmissivity of the Columbia River Basalt Group was estimated by J. J. Vaccaro on the basis of work by Drost and Whiteman (1985) using the above-described conductivity values. Transmissivity commonly ranges from about 2,000 to 100,000 $m^2/d$ with extreme values of 0.15 and 200,000 $m^2/d$ and an average of about 7,600 $m^2/d$. Averages for basalt units in the Yakima Basalt Subgroup are, from oldest to youngest, 8,600, 3,200, and 2,800 $m^2/d$. The Grande Ronde Basalt has the highest transmissivity owing to its great thickness.

The centers of some flows are locally semiconfining but are not considered to be regionally confining. Thus, changes in hydraulic head with depth are highly variable, and aquifers range from effectively unconfined to confined. Ground water follows paths of highest conductivity, which are generally parallel to the direction of dip of individual basalt flows. As such, water moves most readily down flow dip to a point or area of discharge, usually a structural low. Dip of basalt flows results in confined water in some structurally low areas. Wells completed in the Columbia River Basalt Group yield as much as 370 liters per second (l/s) but commonly yield 0.85 to 170 l/s with 1 to 60 m of drawdown. High yields are generally obtained from deep wells (more aquifer zones penetrated), wells some distance from a steeply sloping land surface, and wells in structural lows. Wells drilled in areas of sloping topography, such as on an anticline, generally yield less water owing to the combination of drawing water across the central parts of flows and declining heads with depth.

## Geochemistry

Rocks composing the Columbia River Basalt Group consist primarily of labradorite (plagioclase feldspar), augite (pyroxene), and opaque metal oxides (commonly titanomagnetite). The most common accessory minerals are apatite, olivine, and metallic sulfides, in varying and relatively minor amounts. Accessory minerals are present as intergrown and isolated crystals in a glassy to cryptocrystalline matrix. The most abundant secondary alteration products are a nontronitic smectite, quartz, and clinoptilolite (Ames, 1980; Benson and Teague, 1982; Hearn and others, 1985).

The Grande Ronde, Wanapum, and Saddle Mountains basalts are distinct compositionally, and individual flows within the three units also may be compositionally different. On the basis of bulk compositions, Swanson and Wright (1978, p. 48) described eight chemically different types of basalt in the northern part of the plateau. The calculated average composition is similar to that of other tholeiites but composition differs from other basalts in the Columbia Lava Plateau, such as the Snake River Group (Hamilton, 1963).

The chemical composition of water in the Columbia River Basalt Group results from the interaction of water with crystalline and amorphous minerals in the rocks. (See also discussion on geochemistry in Wood and Fernandez, this volume.) The suite of solutes at any point along a flow path is dependent upon (1) initial composition of the recharge water, (2) composition and relative solubility of the rock through which the water has passed, and (3) residence time of the water.

The capacity of recharge water to react with rock components is a function of water composition when it enters the aquifer. A soil zone in the recharge area augments the dissolution capacity, inasmuch as carbonic acid is a recognized weathering agent. Following the dissolution of minerals, solution chemistry is changed and new minerals are precipitated. These processes act in concert and result in water of the composition present in the geohydrologic system.

Water in the Grande Ronde, Wanapum, and Saddle Mountains basalt units contains dissolved-solids concentrations generally less than 500 mg/l. Concentrations of most individual dissolved species generally increase downgradient, with depth, and with residence time of water in the aquifer. As a result, the chemical type of water in the Columbia River Basalt Group changes from calcium-magnesium bicarbonate to sodium-potassium bicarbonate (Bortleson and Cox, 1985; Hearn and others, 1985). Deviations from this scheme are due to local structural features, ground-water withdrawals, recharge from irrigation, rock-water interactions, deposition of secondary minerals, and man-caused contamination.

Areal variations in ground-water chemistry can be determined from water quality and aquifer mineralogy. Conceptual models can be developed and used to track the chemical evolution of ground water. One such model was discussed by Hearn and others (1985) and is essentially as follows:

Mildly acidic water enters the basalt aquifer and reacts with the rock framework. Glassy and cryptocrystalline phases, pyroxene, and plagioclase begin to dissolve, owing to the corrosiveness of the water and hydrolysis of the vitric and less crystalline silicate phases. Dissolved oxygen is consumed in the oxidation of ferrous to ferric iron, which precipitates as an amorphous oxyhydroxide. The solubility limits of amorphous compounds of aluminum and silicon are quickly exceeded, essentially limiting the solutional accumulation of these species. Amorphous phases become structurally more ordered with time and form smectite and iron minerals.

Hydroxyl ions are a product of hydrolysis, and their accumulation raises ground-water pH. The rising pH eventually exceeds the stability limit for carbonate minerals, which then begin to precipitate. Calcite is the first to precipitate, subsequently lowering calcium concentrations. Dissolved carbon is also removed and tends to buffer the rising pH. The process is likely a function of the availability of dissolved carbon dioxide, which is either atmospheric or is derived from the oxidation of organic material in soil and sedimentary interbeds. Concentrations of sodium and potassium increase significantly with residence time. The increases, however, may be limited by (1) the solubility of clinoptilolite, which is essentially the exclusive zeolitic phase in the upper kilometer of basalt, and (2) removal from solution by ion exchange.

## SNAKE RIVER GROUP

### *Geology*

Much of the 40,400 km$^2$ Snake River Plain in southern Idaho is underlain by Quaternary basalt of the Snake River Group. Basalt predominates in the eastern part of the plain, east of King Hill (Fig. 1), where aggregate thickness is as much as 1,500 m (Whitehead, 1986b). In the western part, Tertiary and Quaternary sedimentary rocks predominate. Miocene basalt and silicic volcanic rocks, chiefly Idavada Volcanics, underlie basalt of the Snake River Group and the Tertiary and Quaternary sedimentary rocks.

The eastern part of the plain is bounded by Cenozoic, Mesozoic, and Paleozoic rocks, chiefly limestone, sandstone, and shale of the Basin and Range Physiographic Province. Tertiary rhyolitic and basaltic rocks bound the western part of the plain on the south; Cretaceous granitic rocks of the Idaho batholith bound the western part on the north.

Basalt of the Snake River Group was extruded from a series of northwesterly trending volcanic vents, most of which are aligned along volcanic rift zones (Kuntz and Dalrymple, 1979, p. 5). Rift zones are in line with normal faults that flank basin-and-range type structures on both sides of the eastern plain.

Basalt is thickest near the center of the eastern plain and thins toward the borders where basalt is intercalated with sedimentary rocks. Snake River basalt is younger than the Columbia River Basalt Group, ranging in age from about 500 to 2 ka; the youngest basalt is at Craters of the Moon National Monument. Kuntz and others (1980, p. 10) used potassium-argon dating to identify major eruptive episodes about 450, 225, and 75 ka at the INEL (Idaho National Engineering Laboratory) site (Fig. 1). Eruptions were separated by relatively long periods of quiescence during which fluvial, lacustrine, and eolian sediments were deposited, much as happened during late phases of extrusion of the older Columbia River Basalt Group.

Individual lava flows generally range from 3 to 15 m in thickness and average about 6 to 7 m (Mundorff and others, 1964, p. 143). They commonly cover 130 to 300 km$^2$ and are as much as 30 km in length. A typical flow consists of (1) a basal layer, less than 1 m thick, of oxidized, fine-grained scoriacious basalt; (2) a massive central layer of coarser-grained basalt whose thickness varies, depending on total thickness of the flow, but is commonly several meters to 20 m thick; and (3) a top layer, less than 2 m thick, of fine-grained, vertically and horizontally jointed, vesicular, clinkery basalt (Kuntz and others, 1980, p. 11).

Layering is prominent in the thick sequence of successive lava flows that were erupted from numerous centers. The result is a complex overlapping and interlocking of flows (Nace and others, 1975, p. 15). Although basalt of the Snake River Group exhibits various types of fractures, tension joints formed by cooling and contraction are characteristic. Joints in the massive central parts of flows are typically almost vertical and form polygonal (commonly hexagonal) columns of basalt. Other less regular fractures developed along the leading edges of flows and as collapse features. Joints, open fissures, and minor displacement faults are concentrated in rift zones. Depths of fractures visible at the surface are unknown, but the presence of cinder cones, lava cones, fissure flows, and other volcanic landforms along rift zones suggests that the fractures may have been healed at depth by movement of magma subsequent to fracturing. Open fissures are common on pressure ridges formed by more recent lava flows. Voids in basalt differ greatly in size and degree of interconnection. Macroporosity consists of lava tubes as much as 5 m or more in diameter and 1,000 m or more in length (Nace and others, 1975, p. 17). Vesicles may be minute to several tens of millimeters in diameter and, if elongate, as much as several centimeters in length. Vesicles may exceed 25 percent of total rock volume, and 10 to 20 percent is common in the upper parts of flows. Laboratory tests on cores of basalt of the Snake River Group at the INEL site indicated that total porosity ranged from about 6 to 37 percent, and effective (interconnected) porosity ranged from 4 to

22 percent (Johnson, 1965, p. 59–61). Interconnection of vesicles and effective porosity are greatly enhanced by fracturing.

Separation of the Snake River Group into mappable basalt units in the subsurface was moderately successful on a local scale at the INEL site. There, the basalt sequence was divided into seven groups of flows on the basis of stratigraphic, radiometric, and paleomagnetic information (Kuntz and others, 1980). Flow groups typically are separated by sedimentary interbeds. Surface electrical-resistivity soundings made as part of a regional aquifer study (Whitehead, 1986b) suggested, and data from a test hole (Whitehead and Lindholm, 1985) verified, that Tertiary and Quaternary sedimentary rocks several tens of meters thick in the Idaho Group separate basalt of the Snake River Group from Tertiary basalt of the Idaho Group at the test-hole site about 50 km east of King Hill, Idaho (Fig. 1). Idaho basalt is mineralogically similar to Snake River basalt but is typically highly weathered. Many vesicles, vugs, and other openings are partially or entirely filled with clay minerals and calcite (Whitehead and Lindholm, 1985). However, at the U.S. Geological Survey test-hole site, the upper 30 m of Idaho basalt is highly vesicular with little secondary mineralization. A lithologic sequence similar to that in the U.S. Geological Survey test hole was penetrated in a deep test hole at the INEL site (Doherty and others, 1979). Stratigraphic correlation between the two widely separated holes is highly speculative. No regional stratigraphic subdivision has been made of volcanic rocks underlying the eastern Snake River Plain.

Tertiary and Quaternary sedimentary rocks as much as 1,500 m thick predominate in the western plain. They are largely clay and silt deposits that contain diatomite layers and scattered, thin ash beds. They also include some sand and gravel lenses, which are of small areal extent and are most common along the margins of the plain and along river courses. Lateral facies changes are numerous and, in many places, abrupt.

Compared with the variety and complexity of large-scale geologic structures in mountainous areas bordering the Snake River Plain and in the Columbia River Basalt Group, basalt of the Snake River Group exhibits relatively few large structural features. Deep-seated structures may be buried by the thick sequence of Quaternary basalt flows. However, the flows themselves exhibit a variety of small-scale internal structures, including layering, fractures, open fissures, and various types of primary voids.

## Hydrology

The Snake Plain aquifer, as defined by Mundorff and others (1964, p. 142), is "the series of basalt flows and intercalated pyroclastic and sedimentary materials that underlie the Snake River Plain east of Bliss (near King Hill)." The aquifer, which is herein referred to as the Snake River Plain aquifer, is composed largely of Quaternary basalt of the Snake River Group. The aquifer stores vast quantities of water under generally unconfined conditions.

Natural recharge to the Snake River Plain aquifer is generally less than 25 mm/yr. Prior to irrigation, about two-thirds of all recharge was derived from streamflow and underflow from tributary drainage basins. Most northern tributary streams lose all their flow to ground water within a short distance after flowing onto the plain. Losing streams, mainly upper reaches of the Snake River, contribute about 20 percent of total recharge, and direct precipitation contributes the remainder.

Natural discharge from the Snake River Plain aquifer is primarily spring flow to the Snake River. Discharge is greatest from several groups of springs that issue from the north wall of the Snake River canyon upstream from King Hill (Fig. 1). Included along a 145-km reach of the Snake River above King Hill are 11 of the 65 springs in the United States that discharge an average of more than 2.8 $m^3$/s (Meinzer, 1927, p. 42–51). After nearly 100 years of irrigation, the hydrologic system has changed significantly, owing to increased recharge from surface-water irrigation. Irrigation water now supplies about two-thirds of the total recharge. Initially, ground-water levels rose several tens of meters in large areas and spring discharges increased about 60 percent. Since the 1950s, these trends have reversed as the result of increased use of ground water for irrigation, decreased diversions of surface water, increased efficiency in water use, and climatological changes.

Basalt devoid of soil and vegetation cover composes the land surface in about 10 percent of the Snake River Plain (Lindholm and Goodell, 1986). In much of the rest of the plain, soil and sedimentary rocks, one to several hundred meters thick, overlie or are stratigraphically equivalent to the basalt. Along the northern margins of the plain, coarse-grained, largely unconsolidated sedimentary rocks as much as several hundred meters thick predominate. They consist of materials derived from the bordering mountains. Lacustrine sedimentary deposits are more common along the southern margins of the plain than elsewhere.

The irregular, broken upper parts of flows and fractures (mainly vertical) in the central parts of flows provide conduits for water to infiltrate to the zone of saturation. Where basalt is covered by unconsolidated sedimentary rocks, deep percolation to the basalt is controlled by the texture and thickness of the overlying sediment. Fine-grained sediment intercalated with basalt greatly impedes the vertical movement of water. In recharge areas, percolating water may be temporarily perched above the regional water table by sedimentary interbeds. Sedimentary interbeds are semiconfining in discharge areas, as shown in Figure 4.

Intraflow structure and hydraulic conductivity vary considerably within individual flows (Fig. 4). The fractured, rubbly top of an individual flow is typically very porous and has a high horizontal hydraulic conductivity. The central part of a flow may have moderately high porosity depending on the degree of vesicularity and jointing, but vertical and horizontal hydraulic conductivities are considerably lower than in interflow zones. The base of a flow is typically scoriaceous and, in places, has high hydraulic conductivity.

Horizontal hydraulic conductivity of basalt in the Snake

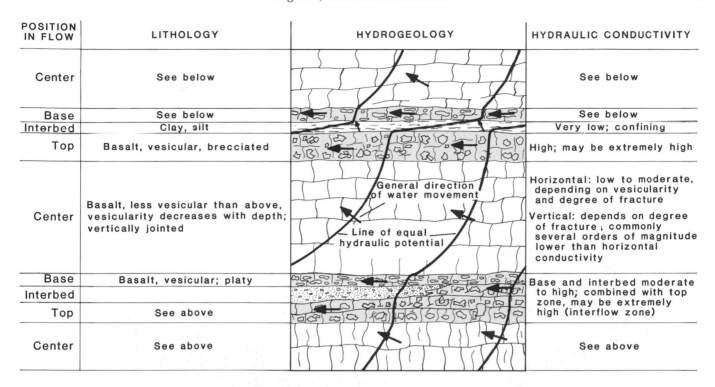

| POSITION IN FLOW | LITHOLOGY | HYDROGEOLOGY | HYDRAULIC CONDUCTIVITY |
|---|---|---|---|
| Center | See below | | See below |
| Base | See below | | See below |
| Interbed | Clay, silt | | Very low; confining |
| Top | Basalt, vesicular, brecciated | | High; may be extremely high |
| Center | Basalt, less vesicular than above, vesicularity decreases with depth; vertically jointed | General direction of water movement — Line of equal hydraulic potential | Horizontal: low to moderate, depending on vesicularity and degree of fracture — Vertical: depends on degree of fracture ; commonly several orders of magnitude lower than horizontal conductivity |
| Base | Basalt, vesicular; platy | | Base and interbed moderate to high; combined with top zone, may be extremely high (interflow zone) |
| Interbed | | | |
| Top | See above | | |
| Center | See above | | See above |

Figure 4. Intraflow structural controls on water movement in a discharge area of the Snake River Group.

River Group is as high as 3,000 m/d and most commonly ranges from 150 to 1,500 m/d (S. P. Garabedian, U.S. Geological Survey, personal communication, 1984). Barraclough and others (1976, p. 49) reported that one flow at the INEL site has a ratio of horizontal to vertical hydraulic conductivity of 3.7 to 1; vertical hydraulic conductivity is lower owing to the massive central part of the flow. Robertson (1977, p. 26) estimated that the vertical hydraulic conductivity of sedimentary interbeds above the regional water table at one location on the INEL site ranges from $3 \times 10^{-2}$ to $3 \times 10^{-6}$ m/d. It follows that fine-grained sedimentary interbeds are generally the greatest impedence to vertical flow. Although water moves vertically through fractures in basalt, the preferred direction of water movement is lateral, along interflow zones. Lateral hydraulic gradients in the basalt generally range from 2 to 12 m/km.

Although actual flow velocities vary greatly owing to aquifer heterogeneity, horizontal water movement through basalt of the Snake River Group is fast, relative to movement through sedimentary rock aquifers. Robertson and others (1974, p. 13) reported that average water velocity in the basalt ranges from 1.5 to 6.0 m/d.

Saturated Quaternary basalt of the Snake River Group in the eastern plain may be as much as 1,200 m thick (Whitehead, 1984a), but it is thought that porosity and hydraulic conductivity decrease with depth and that most water moves through the upper 60 to 150 m of the aquifer. As determined by model studies, transmissivity of the Quaternary basalt aquifer in the eastern plain commonly ranges from 28,000 to 121,000 m²/d, with a maximum of about 450,000 m²/d (Garabedian, 1986), and transmissivity of the upper 30 m ranges from 19,000 to 93,000 m²/d (S. P. Garabedian, U.S. Geological Survey, personal communication, 1985). The number and thickness of highly transmissive interflow zones are important factors in total aquifer transmissivity. The most transmissive basalts are pillow lavas that were deposited in ancestral channels of the Snake River. The largest springs issue from such pillow lavas to the present Snake River between Milner and King Hill (Fig. 5).

Wells completed in basalt of the Snake River Group yield as much as 450 l/s and commonly yield 35 to 190 l/s with less than one to several meters of drawdown.

Hydraulic head in the basalt sequence decreases with depth in recharge areas and increases with depth in discharge areas, as shown in Figure 4. Head differences are such that, in the vicinity of the springs, ground water moves upward as well as laterally to the Snake River. Within the discharge area, sedimentary interbeds and massive crystalline basalts act as semiconfining beds and impede upward movement of water (Figs. 4 and 5). Increases of head with depth were verified in a U.S. Geological Survey test hole near the springs and the Snake River (Whitehead and Lindholm, 1985).

### Geochemistry

The mineral and chemical compositions of basalt of the

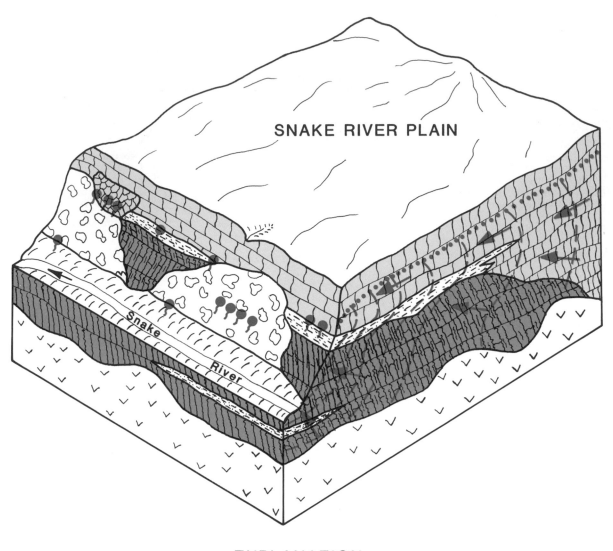

SNAKE RIVER PLAIN

EXPLANATION

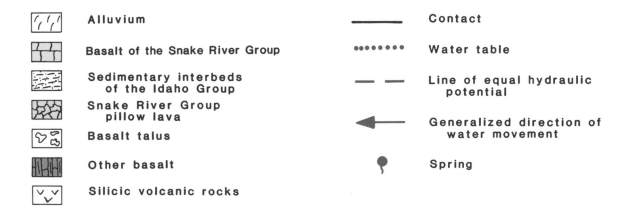

| | Alluvium | | —————— | Contact |
| | Basalt of the Snake River Group | | •••••••• | Water table |
| | Sedimentary interbeds of the Idaho Group | | — — — | Line of equal hydraulic potential |
| | Snake River Group pillow lava | | | |
| | Basalt talus | | ◀————— | Generalized direction of water movement |
| | Other basalt | | | |
| | Silicic volcanic rocks | | 🌢 | Spring |

Figure 5. Geologic controls on ground-water movement to the Snake River canyon in the vicinity of major springs.

Snake River Group are remarkably uniform, though texture and structure vary widely. Basalt at the INEL contains about 35 to 60 percent calcic plagioclase, 25 to 50 percent pyroxene, 5 to 10 percent olivine, and small amounts of accessory magnetite and ilmenite (Nace and others, 1975, p. 13). The rocks are typically porphyritic and contain up to 20 percent by volume phenocrysts of olivine and plagioclase (labradorite) with feldspar predominant (Leeman, 1982, p. 182). The groundmass consists of olivine, labradorite, clinopyroxene, ilmenite, magnetite, apatite, and glass (W. W. Wood and W. H. Low, U.S. Geological Survey, personal communication, 1985).

Secondary alteration products are most commonly iron oxides, ilmenite, and clay minerals. Calcite and silica precipitates line some vesicles and other rock openings. Clay minerals, common in openings near the tops of flows, are mostly of eolian origin. Sedimentary interbeds are of eolian, fluvial, and lacustrine origin and are extremely variable in grain size, thickness, and areal extent. Sediment was derived by weathering of local basalt and a variety of other rock types in highlands bordering the plain. Barraclough and others (1976, p. 71) determined that the cation exchange capacity of 56 bulk samples of sedimentary interbed materials ranged from 1.1 to 45 milli-equivalents (meq)/100 g; cation exchange capacity is generally proportional to the amount of expandable clays present.

Geochemical processes controlling solute concentrations in basalt of the Snake River Group are similar to those defined for the Columbia River Basalt Group. (See also discussion on geochemistry in Wood and Fernandez, this volume.) Water in Snake River basalt contains dissolved-solids concentrations generally less than 400 mg/l. On the basis of predominant cations and anions, water in the Snake River Group is largely of the calcium-sodium bicarbonate type (W. W. Wood and W. H. Low, U.S. Geological Survey, personal communication, 1984). Wood and Low studied the mechanisms of water-rock reactions in four lithologically representative drainage basins tributary to the plain. They estimated that about 80 percent of solutes in ground water in the eastern plain are imported as components of tributary basin drainage and 20 percent are from dissolution of the rock framework within the plain. Included in the tributary drainage is an unknown but assumed small amount of solutes whose origin can be attributed to human activities. Water in streams draining areas underlain by granitic rocks is typically least mineralized; water in streams draining areas of carbonate rocks is most mineralized. Locally on the plain, excess water from irrigation, municipal, industrial, and domestic uses recharges and contributes additional solutes to the ground-water system. Solute input from human activities is indicated by an increase in sodium and chloride in and downgradient from irrigated areas.

Major ions constitute about 95 percent of the solute load, but minor and trace elements are also present. Most minor and trace elements are derived from weathering of the rock framework; however, some have been introduced by human activities. Lithium, strontium, and nitrogen generally increase as total dissolved solids increase.

## OTHER ROCK UNITS

### *Geology*

Underlying the remainder of the Columbia Lava Plateau are Paleozoic to Cenozoic sedimentary rocks and Cenozoic (Tertiary and Quaternary) volcanic rocks (Fig. 1). In northeastern California and central Oregon, Quaternary basalt similar to basalt of the Snake River Group predominates. Basalt forms lava plains that on the south merge with rocks of the Basin and Range Physiographic Province. Some of the youngest basalt is less than 1,000 years old (Heath, 1984, p. 28).

In the rest of the area, Tertiary (Pliocene) rhyolitic rocks are interbedded with lesser amounts of basaltic and sedimentary rocks. Many of the rhyolitic rocks are volcaniclastic (ash, tuff, and agglomerate). Overlying the Tertiary rocks are Quaternary basalt and unconsolidated sedimentary rocks. Sedimentary rocks fill basins between block-faulted mountains in the plateau. A large area in southeastern Oregon, southwestern Idaho, and northern Nevada consists largely of Tertiary rhyolitic lava flows and volcaniclastic rocks. This area is a highly dissected plateau that includes the Owyhee Mountain Range.

### *Hydrology*

The only regional definition of aquifer units in this part of the plateau is for those in eastern Oregon (Gonthier, 1985). Much of the area is remote and sparsely settled and, except in local areas, little is known about hydraulic properties of rock units and about ground-water flow systems.

Although valley-fill deposits and other sedimentary rocks yield sizable amounts of water to wells in places, the best and most developed source of water in this part of the Columbia Lava Plateau is Quaternary basalt. The basalt consists of many thin flows and, therefore, includes numerous interflow zones with generally good hydraulic connection between zones. In some areas, volcanic and sedimentary rock aquifers are hydraulically connected.

Wells ranging from 180 to 1,100 m in depth produce as much as 190 l/s of thermal water (to more than 80°C) from confined silicic and basaltic rock aquifers in southwestern Idaho and north-central Nevada (Young and Lewis, 1982, p. 4). Most of these wells penetrate or are in the vicinity of faults that apparently provide avenues for water movement. Hydraulic heads in the geothermal aquifers are high and many wells flow.

### *Geochemistry*

No reports describe geochemical studies on this part of the plateau. Sparse water-quality data are reported for a few areas.

Geochemical reactions in areas dominated by rhyolitic volcaniclastic rocks should be similar to those in the Amargosa Desert, Nevada, as described by Claassen and White (1979) and summarized by Claassen (1983, p. 20–21). Water quality in vitric

TABLE 1. SUMMARY OF GEOLOGIC, HYDROLOGIC, AND GEOCHEMICAL PROPERTIES OF MAJOR ROCK UNITS UNDERLYING THE COLUMBIA LAVA PLATEAU

| Rock Unit | Geology Physical Characteristics and area extent | Hydrology Hydraulic Conductivity (m/d) | Transmissivity (m²/d) | Storage Coefficient | Well Yield (L/s) | Geochemistry | Problems |
|---|---|---|---|---|---|---|---|
| Columbia River Basalt Group | Tholeiitic basalt, dense to vesicular; flood type, individual flows to several thousand km²; commonly 15-30 m thick, maximum total thickness as much as 4,500 m, in places folded and faulted; interflow zones vesicular and brecciated, flow centers vertically jointed; sediment interbeds commonly 10-15 m thick, thicker near margins of the plateau. | *Horizontal* Saddle Mountains Basalt: median, 0.65 Wanapum Basalt: median, 1.6 Grande Ronde Basalt: median, 0.65 Range, 0.002-1600 *Vertical* Basalt, may be many orders of magnitude lower than horizontal Range $10^{-8}$ to 10 | 2,000-100,000 (as high as 200,000) | 0.0001-0.01 | 0.85-170, with drawdowns of 1-60 m; as much as 370 with increased drawdowns | Alteration products: nontronite, quartz Secondary precipitates: clinoptilolite, calcite Calcium-magnesium bicarbonate to sodium-potassium bicarbonate type water | Lack of information on deeply buried basalts. Few data on vertical hydraulic connection between aquifers. Aquifer discharge difficult to estimate. Numerous uncased wells limit areal data by aquifer |
| Snake River Group | Olivine basalt, dense to vesicular; extruded from numerous vents, individual flows commonly 130-300 km², commonly 3-15 m thick, maximum total thickness exceeds 1,500 m; interflow zones vesicular and brecciated, flow centers vertically jointed; sediment interbeds commonly one to several m thick, thicker near margins of plain. | *Horizontal* Basalt: range, 150-1,500; maximum, 3,000; decreases with depth *Vertical* Basalt: <1 to several orders of magnitude lower than horizontal sediment interbeds: range, $3\times10^{-6}$ to $3\times10^{-2}$ | 28,000-121,000 (as high as 450,000) | 0.01-0.07 | 35-180, with drawdowns of 0.3-3 m; as much as 450 with similar drawdowns | Alteration products: iron oxides, ilmenite, and other clay minerals Secondary precipitates: calcite, silica Calcium-sodium bicarbonate type | Paucity of deep drillholes. May be extremely heterogeneous, local hydraulic characteristics difficult to determine in field and may not be representative |
| Other rock units | Rhyolitic rocks including many volcaniclastics; basalt similar to Snake River Group; locally abundant sediment interbeds; sedimentary rocks predominate in some areas | | Generally unknown but highly variable | | As much as 180 from geothermal wells in southwestern Idaho | Undefined | Paucity of data owing to scattered development |

tuffs changes with time, owing to the reaction of dissolved carbon dioxide ($CO_2$) in recharge waters with the tuff by both ion-exchange and ion-diffusion processes. As the reaction takes place, montmorillonite and clinoptilolite are precipitated. The greater the percentage of sodium in ground water, the greater the ratio of clinoptilolite to montmorillonite. Claassen (1983) used the concentrations of dissolved sodium, calcium, and magnesium in ground water to help determine the source of recharge. He concluded that residence time of ground water in that area is less important than the initial concentration of $CO_2$ in determining the concentration of dissolved solids.

## SUMMARY

Two major basalt units, the Columbia River Basalt Group and the Snake River Group, crop out or are at shallow depth in nearly half of the Columbia Lava Plateau. Various basaltic, rhyolitic, and sedimentary rock units underlie the remainder of the plateau. Volcanic rocks, particularly basalt, store and yield large quantities of good-quality water. A summary of geologic, hydrologic, and geochemical properties of major rock units underlying the Columbia Lava Plateau is given in Table 1.

## REFERENCES CITED

Ames, L. L., 1980, Hanford basalt flow mineralogy: Richland, Washington, Pacific Northwest Laboratory, INL-2847, 447 p.

Barraclough, J. T., Robertson, J. B., and Janzer, V. J., 1976, Hydrogeology of the solid waste burial ground, as related to the potential migration of radionuclides, *in* Saindon, L. G., ed., Idaho National Engineering Laboratory, with a section on drilling and sample analyses: U.S. Geological Survey Open-File Report 76-471, 183 p.

Barrash, W., Bird, J., and Venkatakrishnan, R., 1983, Structural evolution of the Columbia Plateau in Washington and Oregon: American Journal of Science, v. 238, p. 897–933.

Benson, L. V., and Teague, L. S., 1982, Diagenesis of basalts from the Pasco Basin, Washington; I., Distribution and composition of secondary mineral phases: Journal of Sedimentary Petrology, v. 52, no. 2, p. 0595–0613.

Bolke, E. L., and Skrivan, J. A., 1981, Digital-model simulation of the Toppenish Aquifer, Yakima Indian Reservation, Washington: U.S. Geological Survey Open-File Report 81-425, 34 p.

Bortleson, G. C., and Cox, S. E., 1985, Occurrence of dissolved sodium in ground waters of basalts underlying the Columbia Plateau, Washington: U.S. Geological Survey Water-Resources Investigations Report 85-4005, scale 1:500,000, 6 sheets.

——, 1981, Regional management considerations of the ground-water resources in layered volcanics of Idaho, Oregon, and Washington: Pullman, Washington, Washington State University Department of Ecology Research Report 81-068, 103 p.

Claassen, H. C., 1983, Sources and mechanisms of recharge for ground water in the west-central Amargosa Desert, Nevada; A geochemical interpretation: U.S. Geological Survey Open-File Report 83-452, 61 p.

Claassen, H. C., and White, A. F., 1979, Application of geochemical kinetic data in ground-water systems; A tuffaceous-rock system in southern Nevada, *in* Jenne, E. A., ed., Chemical modeling in aqueous systems: American Chemical Society Symposium, ser. 93, p. 771–793.

Doherty, D. J., McBroome, L. A., and Kuntz, M. A., 1979, Preliminary geological interpretation and lithologic log of the exploratory geothermal test well (INEL-1), Idaho National Engineering Laboratory, eastern Snake River Plain, Idaho: U.S. Geological Survey Open-File Report 79-1248, 7 p.

Drost, B. W., and Whiteman, K. J., 1985, Surficial geology, structure, and thickness of selected geohydrologic units in the Columbia Plateau, Washington: U.S. Geological Survey Water-Resources Investigations Report 84-4326, scale 1:500,000, 10 sheets.

Foxworthy, B. L., 1979, Summary appraisals of the Nation's ground-water resources; Pacific Northwest Region: U.S. Geological Survey Professional Paper 813-S, 39 p.

Garabedian, S. P., 1986, Application of a parameter-estimation technique to modeling the regional aquifer underlying the eastern Snake River Plain, Idaho: U.S. Geological Survey Water-Supply Paper 2278, 60 p.

Gonthier, J. B., 1985, A description of aquifer units in eastern Oregon: U.S.

Geological Survey Water-Resources Investigations Report 84-4095, 39 p.

Hamilton, W., 1963, Columbia River basalt in the Riggins Quadrangle, western Idaho: U.S. Geological Survey Bulletin 1141-L, 137 p.

Hearn, P. O., Steinkampf, W. C., Bortleson, G. C., and Drost, B. W., 1985, Geochemical controls on dissolved sodium in ground water in Columbia Plateau basalts, Washington: U.S. Geological Survey Water-Resources Investigations Report 85-4048, 26 p.

Heath, R. C., 1984, Ground-water regions of the United States: U.S. Geological Survey Water-Supply Paper 2242, 78 p.

Johnson, A. I., 1965, Determination of hydrologic and physical properties of volcanic rocks by laboratory methods, *in* Wadia, D. N., ed., Commemorative volume: Mining and Metallurgical Institute of India, p. 78 and 50–63.

Kuntz, M. A., and Dalrymple, G. B., 1979, Geology, geochronology, and potential volcanic hazards in the Lava Ridge–Hells Half Acre area, eastern Snake River Plain, Idaho: U.S. Geological Survey Open-File Report 79-1657, 66 p.

Kuntz, M. A., Dalrymple, G. B., Champion, D. E., and Doherty, D. J., 1980, Petrography, age, and paleomagnetism of volcanic rocks at the Radioactive Waste Management Complex, Idaho National Engineering Laboratory, Idaho, with an evaluation of potential volcanic hazards: U.S. Geological Survey Open-File Report 80-388, 63 p.

Leeman, W. P., 1982, Olivine tholeiitic basalts of the Snake River Plain, Idaho, *in* Bonnichsen, B., and Breckenridge, R. M., eds., Cenozoic geology of Idaho: Idaho Bureau of Mines and Geology Bulletin 26, p. 181–191.

Lindholm, G. F., and Goodell, S. A., 1986, Irrigated acreage and other land uses on the Snake River Plain, Idaho and eastern Oregon: U.S. Geological Survey Hydrologic Investigations Atlas 691, scale 1:500,000.

MacNish, R. D., and Barker, R. A., 1976, Digital simulation of a basalt aquifer system, Walla Walla River Basin, Washington and Oregon: Washington State Department of Ecology Water-Supply Bulletin 44, 51 p.

Meinzer, O. E., 1927, Large springs in the United States: U.S. Geological Survey Water-Supply Paper 557, 93 p.

Mundorff, M. J., Crosthwaite, E. G., and Kilburn, C., 1964, Ground water for irrigation in the Snake River basin in Idaho: U.S. Geological Survey Water-Supply Paper 1654, 224 p.

Nace, R. L., Voegeli, P. T., Jones, J. R., and Deutsch, M., 1975, Generalized geologic framework of the National Reactor Testing Station, Idaho: U.S. Geological Survey Professional Paper 725-B, 49 p.

Prych, E. A., 1983, Numerical simulation of ground-water flow in Lower Satus Creek basin, Yakima Indian Reservation: U.S. Geological Survey Water-Resources Investigations Report 82-4065, 66 p.

Reidel, S. P., 1982, Stratigraphy of the Grande Ronde Basalt, Columbia River Basalt Group, from the lower Salmon River and northern Hells Canyon area, Idaho, Oregon, and Washington, *in* Bonnichsen, B., and Breckenridge, R. M., eds., Cenozoic geology of Idaho: Idaho Bureau of Mines and Geology Bulletin 26, p. 77–101.

Reidel, S. P., Long, P. E., Myers, C. W., and Mase, J., 1982, New evidence for greater than 3.2 km of Columbia River basalt beneath the central Columbia Plateau: American Geophysical Union Transactions (EOS), v. 63, no. 8, p. 173.

Robertson, J. B., 1977, Numerical modeling of subsurface radioactive solute transport from waste-seepage ponds at the Idaho National Engineering Laboratory: U.S. Geological Survey Open-File Report 76-717, 68 p.

Robertson, J. B., Schoen, R., and Barraclough, J. T., 1974, The influence of liquid waste disposal on the geochemistry of water at the National Reactor Testing Station, Idaho, 1952–1970: U.S. Geological Survey Open-File Report OF-73-0238, 345 p.

Swanson, D. A., 1967, Yakima Basalt of the Tieton River area, south-central Washington: Geological Society of America Bulletin, v. 78, p. 1077–1110.

Swanson, D. A., and Wright, T. L., 1978, Bedrock geology of the southern Columbia Plateau and adjacent areas, *in* Baker, V. R., and Nummedal, D., eds., The Channeled Scabland: Washington, D.C., National Aeronautical and Space Administration Planetary Geology Program, p. 37–57.

Swanson, D. A., Wright, T. L., and Helz, R. T., 1975, Linear vent systems and estimated ranges of magma production and eruption for the Yakima Basalt on the Columbia Plateau: American Journal of Science, v. 275, p. 877–905.

Swanson, D. A., Wright, T. L., Hooper, P. R., and Bentley, R. D., 1979, Revisions in stratigraphic nomenclature of the Columbia River Basalt Group: U.S. Geological Survey Bulletin 1457-G, 59 p.

Tanaka, H. H., Hansen, A. J., and Skrivan, J. K., 1974, Digital-model study of ground-water hydrology, Columbia Basin Irrigation Project area, Washington: Washington State Department of Ecology Water-Supply Bulletin 40, 60 p.

Whitehead, R. L., 1986a, Compilation of selected geophysical references for the Snake River Plain, Idaho and eastern Oregon: U.S. Geological Survey Geophysical Investigations Map 869, scale 1:1,000,000.

——, 1986b, Geohydrologic framework of the Snake River Plain, Idaho and eastern Oregon: U.S. Geological Survey Hydrologic Investigations Atlas 681 scale 1:1,000,000, 3 sheets.

Whitehead, R. L., and Lindholm, G. F., 1985, Results of geohydrologic test drilling in the eastern Snake River Plain, Gooding County, Idaho: U.S. Geological Survey Water-Resources Investigations Report 84-4294, 30 p.

Young, H. W., and Lewis, R. E., 1982, Hydrology and geochemistry of thermal ground water in southwestern Idaho and north-central Nevada: U.S. Geological Survey Professional Paper 1044-J, 20 p.

MANUSCRIPT ACCEPTED BY THE SOCIETY MARCH 6, 1987

# Chapter 6

# *Region 3, Colorado Plateau and Wyoming Basin*

**O. James Taylor**
*Wright Water Engineers Inc., 2490 W. 26th Ave., Suite 55A, Denver, Colorado 80211*
**J. W. Hood**
*1209 Princeton Ave., Salt Lake City, Utah 84105*

## INTRODUCTION

The Colorado Plateau and Wyoming Basin (Fig 3; Table 2, Heath, this volume) include parts of Wyoming, Utah, Colorado, New Mexico, and Arizona (Heath, 1984). This region, which has an area of 414,000 km², is bounded by mountain ranges on the north, west, and east. Landforms within the region include broad plains, deeply incised canyons, relatively flat flood plains, and many scenic erosional features.

Rock types and their hydrologic characteristics are diverse in the Colorado Plateau and Wyoming Basin. Consolidated sedimentary rocks of Paleozoic, Mesozoic, and Cenozoic age attain a maximum thickness of more than 6,000 m. Limestone, dolomite, sandstone, and shale beds include major aquifers in some areas; limestone, dolomite, shale, and evaporite deposits are common confining layers (Price and Arnow, 1974). Minor rock types include alluvium and volcanic rocks. Alluvial deposits are aquifers only in relatively short reaches of several major stream valleys. Volcanic rocks may include aquifers of local importance.

Broad features of most ground-water flow systems are controlled by geologic structure, climate, and erosion. Repeated tectonic activity has deformed the region into numerous basins and uplifts (Fig. 1); structural relief exceeds 6,000 m in parts of the region. Because of these structures, aquifers that are stratigraphically low in the geologic section and at great depth in basins are exposed in or near uplifted areas. Exposed aquifers are recharged by snow and rain in mountainous uplifted areas. Some aquifers that are stratigraphically high in the geologic section have broad exposures at lower altitudes in basins where they also are recharged. In general, recharge is less at lower altitudes because of less precipitation and greater evaporation than in mountainous uplifted areas. Annual precipitation ranges from about 150 to 1,000 mm. Most aquifers discharge to stream valleys, where erosion has exposed the aquifers at relatively low altitudes or has decreased their depth below land surface. In some areas, uplift and erosion have resulted in removal of entire aquifers.

## HYDROGEOLOGIC CHARACTERISTICS

The geologic history of sedimentary rocks is reviewed briefly here to describe the origin and sequence of the hydrogeologic framework of aquifers and confining layers.

### *Paleozoic rocks*

In early Paleozoic time, clastic and carbonate sediments were deposited in transgressive and regressive seas; however, sedimentation was not continuous. The resulting laterally continuous beds of sandstone and quartzite of Cambrian age probably include aquifers, but their hydrologic characteristics generally are unknown (Fig. 2). A Cambrian and Ordovician sequence of mostly limestone, dolomite, and shale, including confining layers and aquifers, is stratigraphically above the sandstone and quartzite rocks. Solution openings in zones within the carbonate rocks have resulted in aquifers, as evidenced by springs, but the permeability of most of the carbonate sequence is low.

From Late Devonian through Mississippian time, a thick sequence of carbonate rocks was deposited under stable tectonic conditions; this sequence includes the Ouray, Madison, Redwall, and Leadville Limestones. These rocks are a major regional aquifer because of the development of solutional channels, including karst topography, and because of fracturing that resulted from later tectonic activity.

Beginning in Pennsylvanian time, orogenic activity uplifted areas from which large thicknesses of sandstone and shale were eroded and deposited elsewhere. Carbonate rocks, evaporite minerals, and additional clastic rocks were deposited in Permian time. A hiatus occurred at the end of the Permian. In general, the sandstone beds are aquifers, especially where fractured; shale and carbonate beds normally are not aquifers, but they may be permeable if fracturing and solution activity are extensive. The permeability of most evaporite beds is low.

Taylor, O. J., and Hood, J. W., 1988, Region 3, Colorado Plateau and Wyoming Basins, *in* Back, W., Rosenshein, J. S. and Seaber, P. R., eds., Hydrogeology: Boulder, Colorado, Geological Society of America, The Geology of North America, v. O-2.

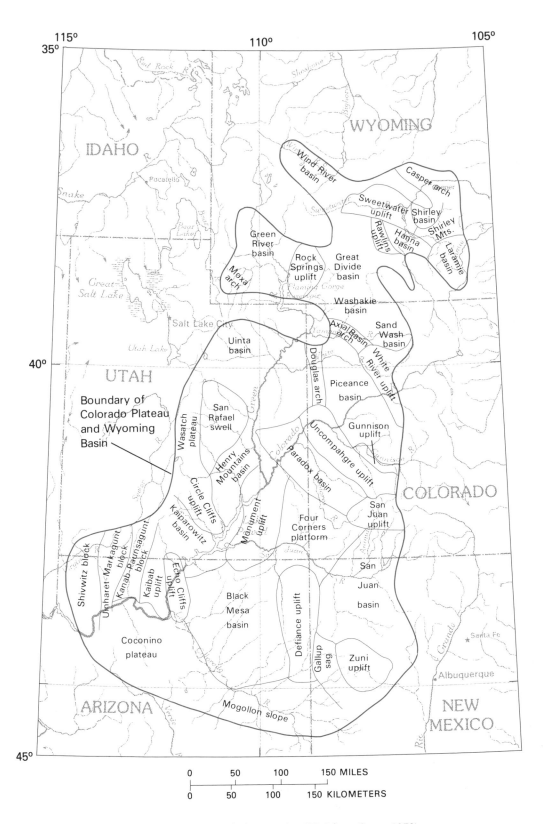

Figure 1. Principal tectonic features (modified from Grose, 1972).

| AGE | FORMATION | HYDROGEOLOGIC UNIT |
|---|---|---|
| PERMIAN | State Bridge Formation (part) | Aquifer |
| PENNSYLVANIAN | Maroon Formation | |
| PENNSYLVANIAN | Minturn Formation | Confining layer |
| PENNSYLVANIAN | Belden Formation | Confining layer |
| PENNSYLVANIAN | Molas Formation | Confining layer |
| MISSISSIPPIAN | Leadville Limestone | Aquifer |
| DEVONIAN | Chaffee Formation | Aquifer |
| ORDOVICIAN | Fremont Limestone | |
| ORDOVICIAN | Harding Sandstone | Confining layer |
| ORDOVICIAN | Manitou Dolomite | Confining layer |
| CAMBRIAN | Dotsero Formation | |
| CAMBRIAN | Peerless Formation | Aquifer (?) |
| CAMBRIAN | Sawatch Sandstone | Aquifer (?) |

Figure 2. Representative Paleozoic section and hydrogeologic units of Piceance basin, Colorado (modified from MacLachlan, 1987).

| AGE | FORMATION | | HYDROGEOLOGIC UNIT |
|---|---|---|---|
| JURASSIC | Morrison Formation | | Confining layers and aquifers |
| JURASSIC | San Rafael Group | Summerville Formation | Confining layers and aquifers |
| JURASSIC | San Rafael Group | Curtis Formation | Confining layer |
| JURASSIC | San Rafael Group | Entrada Sandstone | Aquifer |
| JURASSIC | San Rafael Group | Carmel Formation | Confining layer |
| TRIASSIC | Glen Canyon Group | Page Sandstone | Confining layer |
| TRIASSIC | Glen Canyon Group | Navajo Sandstone | Aquifers |
| TRIASSIC | Glen Canyon Group | Kayenta Formation | Aquifers |
| TRIASSIC | Glen Canyon Group | Wingate Sandstone | Aquifers |
| TRIASSIC | Chinle Formation | | Confining layers |
| TRIASSIC | Moenkopi Formation | | Confining layers |

Figure 3. Stratigraphic units of Triassic and Jurassic age near Henry Mountains basin, Utah (modified from Hintze, 1973).

## Mesozoic rocks

The hiatus at the end of the Permian was ended by the encroachment of Mesozoic seas; the resulting limestones, shales, and sandstones subsequently were partly removed and those remaining were covered with continental and shallow marine deposits (Fig. 3). Most of the Triassic rocks are a composite-confining bed, but they locally contain minor aquifers.

The Glen Canyon Group of Late Triassic and Early Jurassic age and the overlying Jurassic San Rafael Group are continental deposits of mostly clastic origin and are aquifers. The Glen Canyon Group is a major regional aquifer.

Most Middle and Upper Jurassic rocks contain aquifers of local importance; none of these aquifers is as important as the Glen Canyon Group. Parts of the Morrison Formation, of Latest

Jurassic age, together with the Dakota Sandstone of Early or Late Cretaceous age, constitute extensive but minor aquifers.

With the major change in depositional patterns that occurred in Cretaceous time, seas encroaching westward from the central seaway left a regional blanket of black-to-gray marine shale. The Mancos Shale and equivalent shale beds in the Wyoming Basin, in contrast to the Glen Canyon Group, are major confining beds, even though locally they may contain tongues of sandstone aquifers. The Mancos Shale thickness is 1,500 m or more.

Uppermost rocks of Late Cretaceous age, together with those that extend across the Cretaceous-Tertiary time boundary, contain regionally extensive sandstone aquifers. These youngest Mesozoic aquifers are important aquifers, mainly in parts of the Wyoming Basin.

### Cenozoic rocks

In Paleocene time, complexes of terrigenous deposits resulted from erosion related to the Laramide orogeny. Sandstone beds are discontinuous aquifers in many areas. Swamps and lakes formed; the area of one large lake continued to expand into Eocene time. Resulting thick lake deposits include shale, mudstone, sandstone, arkose, and conglomerate. Extensively fractured shale beds and sandstone beds are aquifers in the Piceance Basin; coarse-grained fluvial deposits are aquifers in the Green River Basin (Fig. 1).

In late Eocene and early Oligocene time, fluvial deposition was dominant. Near the mountains, swift streams deposited sand, gravel, and boulders that were consolidated into conglomerate. Silt and clay were deposited in basins distant from the mountains. In Miocene time, rapid erosion began in response to regional uplift. Many of the resulting deposits are permeable; however, they are discontinuous or partly drained because of erosion.

## REPRESENTATIVE AQUIFERS

Details of the hydrogeologic characteristics of the areally extensive and thick aquifer systems of the Colorado Plateau and Wyoming Basin cannot be discussed in a brief report. Therefore, general characteristics of several representative aquifer systems are discussed, including a Paleozoic carbonate aquifer, a Mesozoic sandstone aquifer, and a Cenozoic shale aquifer.

### Carbonate rocks of Devonian and Mississippian age

Carbonate rocks of Devonian and Mississippian age at various sites consist of one to four of the following formations: the Darby Formation, Chaffee Formation, Ouray Limestone, Elbert Formation, Temple Butte Limestone, Madison Limestone, Deseret Limestone, Redwall Limestone, and Leadville Limestone. These formations extend throughout much of the study area, but they are absent in the Uncompahgre uplift, Mogollon slope, Gallup sag, and part of the San Juan Basin (Fig. 1). They are exposed in the Green River basin and in the San Juan and Zuni uplifts. The combined thickness of these formations ranges from about 90 to 500 m, increasing toward the west, except in the southeastern part of the Paradox Basin.

These carbonate rocks constitute a major aquifer throughout most of the region. Movement of water in the aquifer in part of the Colorado Plateau and Wyoming Basin has been studied by plotting shut-in heads from drill-stem tests and by contouring the potentiometric surface (Lindner-Lunsford and others, 1987). The resulting regional flow pattern is disrupted by

the Uncompahgre uplift but not by the numerous other uplift and basin structures (Fig. 4). The aquifer is recharged through outcrops by precipitation, and water movement is mostly through solution channels and fractures. The direction of movement generally is toward or along major streams. Water discharges from the aquifer into the valleys of streams, where it contributes to stream flow. In recharge areas, dissolved-solids concentration is less than 1,000 mg/L. At depth in the Uinta, Piceance, and Paradox Basins, the dissolved-solids concentration commonly ranges from 100,000 to 300,000 mg/L. Ground water with this range of salinity and associated density may not be moving as part of the regional flow system. The dense water may be trapped below less saline and less dense water.

The Uncompahgre uplift is a major structure feature in which the aquifer is absent. Emergence of the incipient uplift began in Devonian time (Baars, 1972). Major uplifting probably began during Mississippian time, allowing erosion to remove all Mississippian and older strata from the uplift. The Uncompahgre uplift is fault bounded on the southwestern margin and upwarped on the northeastern margin (Tweto, 1980).

Regional flow in the aquifer is disrupted by the Uncompahgre uplift. Water in the carbonate aquifer appears to remain within the carbonate rocks rather than moving into and through other aquifers that are structurally higher on the uplift. On the north side of the uplift, the aquifer is overlain by the Molas Formation, which consists of claystone, shale, siltstone, and conglomerate. These beds confine the aquifer from the overlying Triassic and Jurassic strata on the uplift that unconformably overlie Precambrian rocks. On the south side of the uplift, displacement along the normal faults has offset the carbonate aquifer, causing it to butt against the granite core of the uplift (Tweto, 1983), isolating the aquifer from overlying strata.

### Glen Canyon Group

The Glen Canyon Group is a principal aquifer in the Mesozoic rocks of the Colorado Plateau and Wyoming Basin (Freethey and others, 1984). The group includes (in ascending order) the Wingate Sandstone, the Kayenta Formation, and the Navajo Sandstone. The Page Sandstone at the base of the overlying San Rafael Group (Fig. 3) is hydrologically part of the Navajo Sandstone aquifer system. Stratigraphically equivalent or partially equivalent formations are the Moenave Formation in north-central Arizona and south-central Utah, Glen Canyon Sandstone in northeastern Utah and northwestern Colorado, and the Nugget Sandstone of north-central Utah and southwestern Wyoming. The group is missing east of a north-trending line that bisects the western part of Colorado. The group thickens westward to about 400 m in east-central Utah and attains a maximum thickness of about 800 m in south-central Utah.

The sandstone aquifers of the Glen Canyon Group are recharged by precipitation and stream flow on outcrops and by leakage from other aquifers. Natural discharge generally is to the nearest deeply incised stream valley. Where major stream valleys

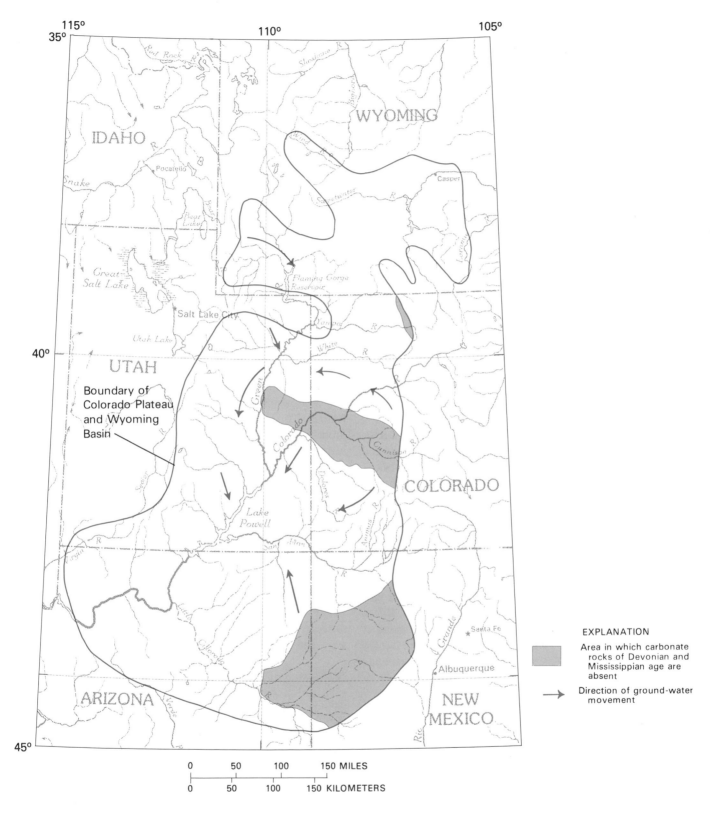

Figure 4. Ground-water movement in carbonate rocks of Devonian and Mississippian age.

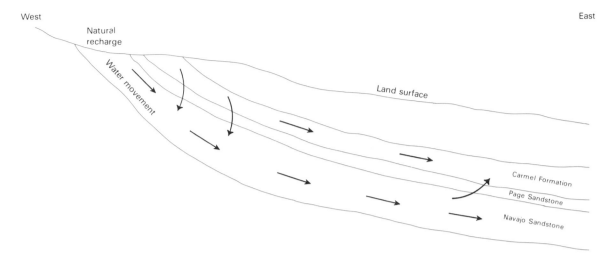

Figure 5. Schematic diagram showing interchange of water between Carmel Formation and Navajo Sandstone near San Rafael swell, Utah.

are at great distances from the recharge areas, it is probable that the sandstone aquifers also discharge to the overlying aquifers.

The primary permeability of the sandstone is minimal, as might be expected from the fine grain size, but the permeability is further decreased by the diagenetic addition of siliceous cement, which decreases pore size and interconnection of pores. The sandstone contains small quantities of feldspar, part of which has been altered to clay. Even small quantities of included clays that swell cause the sandstone to be less permeable to fresh water than to saline water or brine. The sandstone near outcrops is well weathered and has a greater permeability than sandstone at a depth of about 10 m. At least locally, however, the discharge of water at the outcrop appears to have left a zone partly sealed by minerals precipitated from the ground water.

Laboratory tests of the effects of overburden loading indicate that the permeability of the Navajo Sandstone also decreases with depth of burial (Hood and Patterson, 1984). This decreased permeability would result in slow circulation at great depths. In the Uinta and Green River basins, the Navajo Sandstone contains water with a dissolved-solids concentration in excess of 100,000 mg/L, mainly the result of slow circulation or near-stagnant conditions in the depths of the basins. Near sandstone outcrops that are recharged by precipitation, dissolved-solids concentration in the water from the Navajo Sandstone in some areas is only several hundred milligrams per liter.

In much of the region, the sandstone is slightly to extensively fractured, resulting in secondary permeability. However, many fractures are found to be filled with silica, calcium carbonate, iron compounds, or secondary gypsum; at depth, fractures tend to close. The resulting net effect of fracturing is only a small increase in permeability.

In the southern part of the San Rafael swell of Utah, the water quality of the Navajo Sandstone is affected by structural deformation and the related changes in the flow system (Fig. 5). The water movement is from west to east in the Navajo and Page Sandstone aquifers and in local permeable zones in the overlying Carmel Formation. A basal siltstone in the Carmel Formation probable slowed the interchange of water between the Carmel and underlying sandstone aquifers prior to structural deformation. As fracturing increased the permeability and allowed greater natural recharge, water that dissolved gypsum and salt deposits from the Carmel Formation migrated downward into the Navajo Sandstone. This migration is evidenced by saline water in the upper part of the Navajo Sandstone. The migration probably was augmented as the solution activity removed salts in quantities sufficient to increase the permeability of the Carmel Formation even more (Hood and Danielson, 1981). Farther to the east, the migration is reversed; water migrates upward from the Navajo and into the Carmel. Similar exchanges of water can be inferred for other regions of south-central Utah where the Navajo Sandstone and Carmel Formation have been folded.

### Oil-shale aquifers of Piceance Basin

The largest known resource of oil shale occurs within the Green River and Uinta Formations of Eocene age in the Piceance basin of Colorado. These formations are fine-grained lake beds that were deposed in Eocene time. The evaporite minerals nahcolite and dawsonite also occur in the northern part of the basin. Interest in development of oil-shale reserves and other minerals has resulted in exploratory drilling in the basin and various investigations of the geology and water resources.

Normally, shale and other fine-grained rock do not transmit water readily, but many parts of the formations that consist of oil shale are extensively fractured. Fractures with apertures greater than several centimeters were encountered during shaft construction at one mine. In contrast, fractures are virtually nonexistent in some outcrops. The network of fractures allowed water to enter evaporite deposits, dissolve minerals, and create openings. Because of the fractures in the oil shale and the solution openings within the evaporite deposits, the Green River and Uinta Formations contain large volumes of water.

Two groups of major aquifers and one confining layer generally are recognized in the basin (Fig. 6). The lower bedrock aquifers have a total thickness of about 210 m; they extend from the top of the Garden Gulch Member of the Green River Formation to the base of the Mahogany zone in the Parachute Creek Member of the same formation. The Mahogany zone is a rich oil-shale zone that averages 50 m thick and is a confining layer. From the top of the Mahogany zone to the land surface are the upper bedrock aquifers with an average thickness of about 270 m.

The effects of the fracture system on suspected directional permeability have been studied by Robson and Saulnier (1981) and Taylor (1982). In parts of the aquifer system, lateral permeability varies with direction, probably because of some combination of fracture aperture, continuity, and spacing. Vertical permeability is less than lateral permeability, especially at depth, either because vertical fractures are less abundant or because the overburden pressure has decreased the aperture of fractures at depth.

The north-south section illustrated in Figure 7 shows natural recharge of the aquifers from snowmelt, at altitudes as much as 2,800 m. Part of the recharged water discharges to the south through springs along deeply incised valleys. These springs contribute to the flow of the Colorado River and its tributaries. Because of the depth of the incised valleys, the streams cannot contribute water to the bedrock aquifers. Recharged water also moves to the north, where it discharges to the valleys of other creeks in the Colorado River system at altitudes as low as 1,700 m. These shallow stream valleys result in stream-aquifer systems in which the streams and bedrock aquifers may interchange water.

The lower aquifers are recharged and discharged through the Mahogany zone, a confining layer. This flow system illustrates that the true definition of a confining layer is one that may transmit water slowly, not a layer that does not transmit water at all. Because of the large areal extent of the Mahogany zone, it is able to transmit water under steady conditions and to maintain the recharge and discharge of the upper and lower aquifer series.

Aquifer testing of the upper and lower bedrock aquifers indicates that they are confined, or artesian, aquifers. Although the Mahogany zone caps the lower aquifers, the upper aquifers are not capped by any confining beds. Apparently the minimal vertical permeability of the upper aquifer impedes vertical drainage and causes the aquifer to respond like an artesian aquifer.

<table>
<tr><th>AGE</th><th>FORMATION</th><th>MEMBER</th><th>HYDROGEOLOGIC UNIT</th></tr>
<tr><td rowspan="5">TERTIARY</td><td>Uinta Formation</td><td></td><td rowspan="2">Upper aquifers</td></tr>
<tr><td rowspan="4">Green River Formation</td><td rowspan="3">Parachute Creek Member</td></tr>
<tr><td>Confining layer (Mahogany zone)</td></tr>
<tr><td>Lower aquifers</td></tr>
<tr><td>Garden Gulch Member</td><td>Confining layer</td></tr>
</table>

Figure 6. Generalized correlation of Eocene stratigraphic and hydrogeologic units, Piceance basin, Colorado (modified from Taylor, 1982).

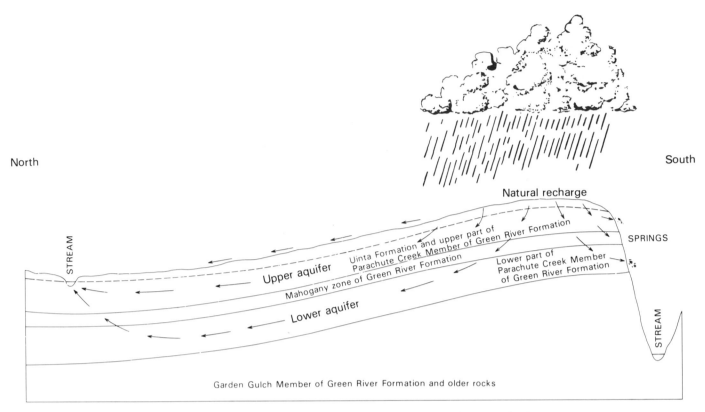

Figure 7. Diagrammatic north-south section of hydrologic system in Piceance basin, Colorado.

## REFERENCES

Baars, D. L., 1972, Devonian System, *in* Geologic atlas of the Rocky Mountain region: Denver, Rocky Mountain Association of Geologists, p. 90–99.

Freethey, G. W., Kimball, B. A., Wilberg, D. E., and Hood, J. W., 1984, General hydrogeology of the aquifers of Mesozoic age, Upper Colorado River Basin, excluding the San Juan basin, Colorado, Utah, Wyoming, and Arizona: U.S. Geological Survey Open-File Report 84–716, 11 p.

Grose, L. T., 1972, Tectonics, *in* Geologic atlas of the Rocky Mountain region: Denver, Rocky Mountain Association of Geologists, p. 35–44.

Heath, R. C., 1984, Ground-water regions of the United States: U.S. Geological Survey Water-Supply Paper 2242, 78 p.

Hintze, L., 1973, Geologic history of Utah: Provo, Utah, Brigham Young University Geology Studies, v. 20, pt. 3, 181 p.

Hood, J. W., and Danielson, T. W., 1981, Bedrock aquifers in the lower Dirty Devil River basin area, Utah: Utah Department of Natural Resources Technical Publication 68, 143 p.

Hood, J. W., and Patterson, D. J., 1984, Bedrock aquifers in the northern San Rafael swell area, Utah, with special emphasis on the Navajo Sandstone: Utah Department of Natural Resources Technical Publication 78, 128 p.

Lindner-Lunsford, J. B., Kimball, B. A., Chafin, D. T., and Bryant, C. G., 1987, Hydrology of aquifers of Paleozoic age, Upper Colorado River Basin, excluding the San Juan basin, in Colorado, Utah, Wyoming, and Arizona: U.S.

Geological Survey Hydrologic Investigations Atlas (in press).

MacLachlan, M. E., 1987, General geology of the Piceance basin, *in* Taylor, O. J., compiler, Oil shale, water resources, and valuable minerals of the Piceance basin, Colorado; The challenge and choices of development: U.S. Geological Survey Professional Paper 1310 (in press).

Price, D., and Arnow, T., 1974, Summary appraisals of the nation's ground-water resources; Upper Colorado region: U.S. Geological Survey Professional Paper 813–C, 40 p.

Robson, S. G., and Saulnier, G. J., Jr., 1981, Hydrogeochemistry and simulated solute transport, Piceance basin, northwestern Colorado: U.S. Geological Survey Professional Paper 1196, 65 p.

Taylor, O. J., 1982, Three-dimensional mathematical model for simulating the hydrologic system in the Piceance basin, Colorado: U.S. Geological Survey Water-Resources Investigations Open-File Report 82–637, 35 p.

Tweto, O., 1980, Tectonic history of Colorado, *in* Kent, H. C., and Porter, N. W., eds., Colorado geology: Denver, Rocky Mountain Association of Geologists, p. 5–9.

—— , 1983, Geologic sections across Colorado: U.S. Geological Survey Miscellaneous Investigations Map I–1416, scale 1:500,000.

MANUSCRIPT ACCEPTED BY THE SOCIETY AUGUST 15, 1987

## Chapter 7

# *Region 4, Central Valley and Pacific Coast Ranges*

**C. D. Farrar**
*U.S. Geological Survey, P.O. Box 1298, Santa Rosa, California 94502*
**G. L. Bertoldi**
*U.S. Geological Survey (retired), Route 1, Box 448, Wilton, California 95693*

## INTRODUCTION

Region 4 (Fig. 3; Table 2, Heath, this volume) includes two distinctly different geologically and hydrologically separated subregions—the Central Valley and Coast Ranges of California. The hydrogeology of both subregions is discussed together here because they evolved geologically from the same large-scale processes acting on the continental margin of western North America during middle Mesozoic through Cenozoic time.

The area of Region 4, as defined in this paper, is bounded on the north by the Klamath Mountains and Cascade Range, on the east by the Sierra Nevada, on the south by the Transverse Ranges, and on the west by the Pacific Ocean (Plate 1). Total land area is about 160,000 km$^2$; of that area the Central Valley, also known as the Great Valley of California, contains about 52,000 km$^2$. The line of demarcation between the Central Valley and the surrounding mountains is the 150-m contour line. Generally, below the 150-m altitude the land surface has very low relief, the result of millions of years of fluvial deposition of sediments derived from the bordering mountain ranges. In contrast to the topographic monotony of the valley, the Coast Ranges rise abruptly from the Pacific Ocean to form a northwest–southeast-trending belt of alternating shallow alluvium-filled valleys and moderately dissected mountain ranges. Altitudes of range crests are moderate, generally less than 1500 m, but relief is great. Peaks as high as 1500 m are within 6 km of the coast line. The northern Coast Ranges are higher than the southern Coast Ranges; maximum altitudes are 2312 m in the north and 2083 m in the south.

## GEOLOGIC SETTING

During the past two decades the concept of plate tectonics has become a unifying principle in the interpretation of various types of geologic and geophysical data (Blake and Jones, 1981). Plate tectonics is especially important as a framework for understanding the geologic evolution and hydrologic setting of Region 4. Region 4 lies adjacent to the western continental margin of North America. During much of the latter half of the Mesozoic

and early part of the Cenozoic eras, this part of the Earth's crust was the site of converging plates along which the oceanic plates in relative easterly movement were overridden and consumed beneath the North American plate.

Although the Coast Ranges and Central Valley are adjacent subregions and evolved from the same large-scale geologic processes, they are very different physiographically, geologically, and hydrologically. The two subregions form adjoining elongate bands that are subparallel to the California coast. The geologic differences are attributed to the differing geologic processes along a subducting plate margin moving from active trench to forearc basin (Dickinson, 1981).

The sedimentary environment and tectonic processes active in the subduction-zone complex produced the lithologic framework of the Coast Ranges. Cenozoic faulting, erosion, and mass wasting have overprinted on this framework. Thrust faulting and tectonic mixing produced mélange and shear zones and introduced exotic blocks of rock from different surrounding terrains into the aggregation of sedimentary rocks. Translational movement associated with the San Andreas fault system displaced the plutonic and metamorphic rocks of the Salinian block hundreds of kilometers northwestward from its original position (Page, 1981). The Salinian block is now flanked on either side by rocks of the Franciscan Complex.

In contrast, sedimentation during Jurassic through early Cenozoic time in what is now the Central Valley took place in the forearc basin with relatively little subsequent deformation, resulting in typical deep marine, shallow marine, and deltaic deposition of clastic material. More recent geologic events have modified both subregions, but the fundamental differences, both geologically and hydrologically, were determined by the positions of both these subregions in a region of active plate tectonics along a convergent plate margin.

### *Central Valley*

The Central Valley lies east of the Coast Ranges and averages about 80 km in width. The valley extends about 650 km

---

Farrar, C. D., and Bertoldi, G. L., 1988, Region 4, Central Valley and Pacific Coast Ranges, *in* Back, W., Rosenshein, J. S., and Seaber, P. R., eds., Hydrogeology: Boulder, Colorado, Geological Society of America, The Geology of North America, v. O-2.

Figure 1. Location of geologic and geographic features.

The western parts of both the San Joaquin and the Sacramento Valleys contain the thickest sedimentary sections in the Central Valley. In addition to this east–west asymmetry, the basin also has a regional southward tilt. The southerly tilt is interrupted where the basement rocks are bowed downward several thousand meters under the Sacramento Valley and also in the southern San Joaquin Valley. The basin is transected by two cross-valley faults, the Stockton fault and White Wolf fault (Hackel, 1966). The Stockton fault and associated Stockton arch together form a geologic divide between the Sacramento and the San Joaquin Valleys. The only other prominent structure in the entire Central Valley is the Sutter Buttes, a Pliocene and Pleistocene volcanic plug that rises abruptly to an altitude of 600 m above the flat valley floor. Sutter Buttes is about 15 km in diameter, and associated intrusions have been injected locally into sedimentary valley fill.

The Jurassic, Cretaceous, and Tertiary sedimentary rocks and deposits, in order of prevalence, consist of siltstone, claystone, sandstone, and conglomerate that are predominately of marine origin (Hackel, 1966). The Cretaceous deposits are most prevalent in the Sacramento Valley; the thickest Tertiary sections occur in the San Joaquin Valley. Throughout the Jurassic and Cretaceous, the source area for sediments was most probably the Klamath Mountains and the Sierra Nevada (Bailey and others, 1964). Sedimentation during these periods produced thick deposits with little variation in lithology over large areas.

The Tertiary Period in California was marked by episodes of tectonism in both the Coast Ranges and the Sierra Nevada that included uplift, folding, faulting, and volcanism. Concomitant erosion in these two provinces provided massive volumes of sediments to the intervening Central Valley. Rapid changes in base level and sediment source areas produced an incomplete Tertiary section at any one place, but taken as a whole, sedimentation in the Central Valley was continuous throughout the Tertiary. Deposition varied with time and location between marine and continental environments. As a result, the Tertiary strata are more variable in lithology than the Mesozoic strata.

Sedimentation during Pleistocene and Holocene time has left fluvial and lacustrine deposits throughout the valley. These continental deposits are generally unconsolidated and range in thickness from less than 30 m to more than 1000 m (Poland and Evenson, 1966).

### Coast Ranges

The Coast Ranges extend for about 950 km along the California coast from the Santa Ynez River to north of Humboldt Bay (Fig. 1). The San Francisco Bay Region divides this province into the northern Coast Ranges and the southern Coast Ranges. Geologic differences between the northern and southern parts may be related to the relative position of the San Andreas fault system (Norris and Webb, 1976). The San Andreas fault transects the ranges obliquely from an eastern position in the south to a western position near San Francisco Bay and lies offshore farther to the north.

northwest from the Tehachapi Mountains to near Redding (Fig. 1). For convenience in discussion, the valley can be divided into two parts: the northern one third is known as the Sacramento Valley and the southern two thirds as the San Joaquin Valley.

The Central Valley is an asymmetric synclinal trough that contains Jurassic to Holocene sedimentary rocks and sediments as much as 15,000 m thick in the Sacramento Valley and as much as 10,000 m in the San Joaquin Valley (Repenning, 1960). On the eastern side, the valley fill overlies a gently westward-sloping surface of basement rocks that are composed primarily of granitic and metamorphic rocks of pre-Tertiary age that are the subsurface continuation of the Sierra Nevada. On the western side, the sedimentary rocks and deposits of the Central Valley are steeply dipping, having been overthrust on rocks of the Coast Ranges. Mafic and ultramafic rocks demark the sole of the overthrust sheet (Suppe, 1979; Etter and others, 1981). These mafic and ultramafic rocks probably form the basement under the western part of the Central Valley.

Basement rocks are of two types: the assemblage of rocks composing the Franciscan Complex and the plutonic and metamorphic rocks of the Salinian block. The Franciscan Complex is a structural complex consisting of an aggregation of rocks that have been faulted and tectonically mixed. Relatively intact blocks and fault slices of bedded rocks are imbricated and interleaved with shear zones, mélanges, and broken formations (Fox, 1983). The parental rocks were predominantly graywacke interbedded with siltstone, conglomerate, chert, and less commonly limestone (Bailey and others, 1964). Deposition most likely involved turbidity currents and mudflows; during and subsequent to deposition, tectonic mixing disrupted and deformed the original bedding. In addition, during subduction of the oceanic crust, compressional forces caused thrust faulting and squeezed up blocks of material both from the subducted crust and from the mantle (Dickinson, 1981). These blocks range from less than a meter to several kilometers in size and consist of serpentine, greenstone, and ophiolitic rocks locally zeolitized and metamorphosed to low-grade blueschist, greenschist, or amphibolite. Exotic blocks of high-grade blueschist and eclogite also are prevalent in the mélanges.

The second type of Coast Range basement rocks are the crystalline rocks of the Salinian block. The basement rocks of this tectonic sliver lying between the San Andreas and Nacimiento faults consist of granite plutons and metamorphic rocks. The metamorphic rocks include gneiss, schist, marble, and quartzite and are intruded by granitic rocks of Cretaceous age.

In places, both types of basement are overlain by the marine sedimentary rocks of the Great Valley sequence. Although of the same age as the parent rocks of the Franciscan Complex, the Great Valley sequence consists of sandstone, siltstone, and conglomerate beds that have not undergone the tectonic disruption evident in the Franciscan Complex. The Great Valley sediments were derived from the Sierra Nevada and deposited in the marine forearc basin during Late Jurassic to Late Cretaceous time. These sediments were then thrust over the basement rocks to the west.

Cenozoic time is represented by thick local accumulations of marine sediments, most of continental shelf or deep basin origin. Repeated and localized tectonism during the Cenozoic Era has resulted in incomplete sedimentary records at any one place, although viewed as a whole the entire Cenozoic is represented in the Coast Ranges. Miocene deposition resulted in widespread and thick accumulations of siliceous-phosphatic sediment in a marine environment. During the Pliocene the sea began to withdraw, and fluvial and lacustrine deposits became more widespread. Volcanic activity began in the Miocene in the San Francisco Bay area and shifted northward to the Clear Lake region by Pleistocene time.

The dominant northwesterly geologic structure prevalent throughout the Coast Ranges is graphically displayed in the topographic expression of the entire province. Linear mountain ranges with intervening valleys are closely determined by geologic structures.

Major northwest-trending faults include the San Andreas, Nacimiento, and Maacama faults. Each of these faults is hundreds of kilometers in length and actually consists of several subparallel en echelon fault branches. Displacements on some of these faults have produced earthquakes prior to and during historic time. North of San Francisco Bay, several small alluvium-filled valleys were formed as pull-apart basins that were produced from fault displacements along en echelon branches of the Maacama fault zone during Pliocene and Pleistocene time (McLaughlin and Nilsen, 1981). Folds, although not as prominent as the fault systems in the Coast Ranges, also are important. Some of the folds have hydrologic significance, having provided basins for sediment accumulation, as in the Salinas Valley (Fig. 1).

## HYDROLOGIC SETTING

The pronounced geologic and physiographic differences between the Coast Ranges and the Central Valley are paralleled by significant differences in hydrologic characteristics. The Central Valley, as its name implies, is virtually one large alluvium-filled valley occupying a central position in California between the Coast Ranges and the Sierra Nevada. The Coast Ranges are predominately a grouping of rugged mountain ranges that include many small hydrologically separate alluvium-filled valleys.

The ultimate source of water for both subregions is from the moist airmasses that are swept inland from the Pacific on the prevailing westerlies. Mean annual precipitation exceeds 2500 mm in parts of the northern Coast Ranges but rarely exceeds 550 mm in the northern part of the Central Valley. Although topographic altitude is the major influence on precipitation, both subregions show distinct variations in precipitation from north to south. In the southern Coast Ranges, mean annual precipitation generally ranges from 250 to 500 mm, and in the north it ranges from 750 to 2500 mm. In the Central Valley, precipitation ranges from about 150 mm in the south to 550 mm in the north.

The total quantity of surface water discharged from the region, a mean of about $5.4 \times 10^{10}$ m$^3$ annually, accounts for about 60 percent of the total annual discharge for the State of California (Kahrl, 1978). Natural surface-water drainage systems vary considerably in both geometry and source between the Central Valley and the Coast Ranges.

### Central Valley

The Central Valley is drained by two large river systems: the Sacramento River drains the northern part and the San Joaquin River drains most of the southern part. The two rivers join in a deltaic area about 50 km east of San Francisco and drain to the Pacific Ocean through the San Francisco Bay estuary. The southern part of the San Joaquin Valley is a closed basin with the Kings and Kern Rivers carrying surface runoff to closed depressions that in the recent past contained Tulare Lake and Buena Vista Lake (Fig. 1).

Prior to the development of agriculture in the San Joaquin Valley, flow in the San Joaquin River was sustained entirely from runoff of precipitation that falls on the western slope of the Sierra Nevada. Subsequent to agricultural development, flow in the river is sustained by a combination of Sierran runoff and agricultural wastewater derived from drained fields that use irrigation water imported from outside the San Joaquin drainage area.

The Sacramento River has its headwaters in the Trinity Mountains west of Mount Shasta (Fig. 1); however, a large quantity of flow is contributed to the Sacramento River from the Pit River, which drains about 13,000 km$^2$ of the southern Cascade Range and the Modoc Plateau of northeastern California. Several large streams with steep gradients flow westward from the Sierra Nevada and join these rivers nearly at right angles close to the axis of the valley. Flows in the Sierran streams are seasonally variable, with high flows coinciding with the snowmelt that comes in late spring. Flows are generally intermittent in the streams that drain eastward from the Coast Ranges to the Central Valley.

### Coast Ranges

In contrast to the Central Valley, the rugged mountains of the Coast Ranges divide the area into numerous smaller surface-water drainage basins. The flow in each basin is derived from precipitation that falls within a more localized area. Periods of high flow coincide with or shortly follow precipitation events throughout the wet season. Near the coast, streams are short, with steep gradients that carry runoff directly to the Pacific. Drainage from inland areas is influenced by the northwesterly trending geologic structure. Many of the larger Coast Range rivers, such as the Salinas and the Eel (Fig. 1), flow for long distances in structurally controlled interior valleys that run nearly parallel to the coastline.

The importance of ground water to the region is apparent when it is recognized that the major areas of population and most intensive agricultural development are centered in water-deficient areas. Ground water supplements surface-water supplies for agricultural, municipal, domestic, and industrial uses throughout the region. Locally, and in times of drought, ground water is the major source of water supply. Ground water supplies about half of the required $2.7 \times 10^{10}$ m$^3$ of water used annually for irrigation in the Central Valley. In the Salinas Basin, within the Coast Ranges, ground water accounts for more than 95 percent of the $5.6 \times 10^8$ m$^3$ of water used annually (Durbin and others, 1978).

## HYDROGEOLOGY

The ground-water systems within the Central Valley and Coast Ranges can be contrasted in several ways. The aquifer system of the Central Valley is essentially hydraulically continuous throughout its 52,000-km$^2$ area, whereas the principal aquifers of the Coast Ranges consist of numerous small hydraulically separate alluvial basins scattered throughout less permeable rocks composing the mountainous terrain. In the Central Valley, geologic structures are few in number and have little impact on the flow of ground water. But in the Coast Ranges, folds, faults, and associated fracture zones are ubiquitous and greatly influence the movement of ground water. The diverse lithologies present in the Coast Ranges are the primary cause of the extreme variability of ground-water availability; variability is much less in the Central Valley.

The relation of stratigraphy to the occurrence and movement of ground water is more obvious in the Coast Ranges than in the Central Valley. In the Coast Ranges, unconsolidated alluvial deposits are the most permeable geologic units and store the greatest quantity of ground water. However, some lithified formations, such as the Purisima Formation, Butano Sandstone, and Vaqueros Sandstone, in the Santa Cruz area (Fig. 1) contain permeable zones that are saturated with ground water. The variable lithology within these formations restricts the occurrence of ground water to specific zones that are separated by intervening impermeable rock. In the Central Valley, ground-water occurrence is not closely related to mappable stratigraphic units. Many of the stratigraphic units recognized in outcrops by physiographic, pedologic, or weathering criteria, cannot be identified in the subsurface because there are no apparent differences in hydrologic or lithologic properties (Page, 1986).

### Central Valley

The chief source of fresh ground water (dissolved solids less than 2,000 mg/L) throughout the Central Valley is the upper 300 m of unconsolidated continental deposits and sedimentary rocks of post-Eocene age (Page, 1986). Fresh ground water occurs to depths of more than 1,000 m in the sedimentary units that fill structural down-warps along the west side of Sacramento Valley and in the southern part of the San Joaquin Valley. Below the freshwater zone is saline water, primarily connate, contained in the thick, marine sedimentary rocks of Cretaceous to Eocene age. In places, such as around Sutter Buttes and near Sacramento, saline water is contained at shallow depth in continental deposits. Such occurrences of saline water may result from ground-water upflow, evaporation, or estuarine water trapped during sedimentation. Underlying the thick pile of saturated marine sedimentary rocks and continental rocks and deposits is the relatively impermeable basement. The basement consists of Sierran crystalline rocks under the eastern part and lithified rocks, which are extensions of the Coast Ranges, under the western part of the valley.

In the Central Valley, considerable variation exists in the hydraulic properties of aquifers from place to place. The variability is related to differences in lithology and depositional processes. Sedimentary rocks and deposits compose the bulk of the fill overlying the basement, but volcanic rocks are also present. Basalt flows, dikes, and sills occur in the Sacramento Valley, and volcanic rocks of rhyolitic to andesitic composition make up the Sutter Buttes. As a group, the hydraulic properties of the volcanic rocks are extremely variable. Fine-grained, compact tuff beds and cemented mudflow units are practically impermeable; rubble zones between flows or fractured flow rocks are highly permeable.

The most productive aquifers are within the coarse-grained deposits of fluvial origin. Wells completed in these deposits commonly provide 30–60 $\ell$/sec of water. Alluvial fans have formed on all sides of the Central Valley. In this depositional setting, the fine-grained detritus is carried farther toward the valley axis, leaving the coarse-grained materials closer to the valley margins. Over time, shifting stream channels cause the fans to coalesce, forming broad sheets of interfingering wedge-shaped lenses of gravel, sand, and finer detritus. Near the valley axis, deposition in lacustrine and flood plain environments produced thick beds of clay and silt. During the Pliocene, as much as 1,900 km² of the San Joaquin Valley was covered by lakes that accumulated up to 50 m of diatomaceous clay.

The lithologic differences observed around the valley are partly related to the different source rocks that provide detritus, and partly to the frequency and intensity of local precipitation that, in part, determines the mode of deposition. The crystalline rocks of the Sierra Nevada are generally more resistant to erosion than the Coast Range rocks and provide a greater percentage of coarse-grained detritus to the Central Valley. Where precipitation and runoff are greatest, in the northern Coast Ranges and the Sierra Nevada, fluvial processes are most effective in sorting the sediment and bedload carried by streams flowing into the Central Valley.

Fresh ground water occurs under confined and unconfined conditions in the Central Valley. Large differences in hydraulic head have been recorded in many parts of the valley. The direction of ground-water flow is determined by the variations in head. Under predevelopment conditions, ground-water flow in both the confined and unconfined aquifers was generally from the valley sides toward the axis and from the north and south toward the Sacramento–San Joaquin Delta (Davis and others, 1959; Olmsted and Davis, 1961). Large-scale ground-water development has modified the natural flow pattern by creating cones of depression in major pumping centers. More recently, importation of surface water has, to some extent, decreased the magnitude of these man-induced changes by decreasing the need for pumping and also by increasing recharge in these areas.

### Coast Ranges

Geohydrologically, the rocks and sediments of the Coast Ranges can be divided into two broad groups: consolidated rocks and poorly consolidated to unconsolidated deposits. The consolidated rocks include the sedimentary and low-grade metamorphic rocks of the Franciscan Complex, crystalline rocks of the Salinian block, and Cenozoic marine sedimentary rocks. The consolidated rocks are exposed over about 80 percent of the Coast Ranges and make up the mountainous terrain. The poorly consolidated to unconsolidated deposits are restricted to narrow coastal terraces and valley floors and margins. These deposits consist primarily of uncemented Pleistocene and Holocene alluvial deposits and loosely cemented Pliocene and Pleistocene sediments.

The differences in hydrologic properties between the two groups approach the extremes found in nature. In most areas, the low porosity and permeability of the consolidated rocks so limit their capacity to store or transmit ground water that, in a hydrologic sense, they are more important as barriers to the movement of ground water and for providing boundaries for sediment-filled basins within them. However, because of the diverse lithologies included in this group and the common occurrence of fractured zones associated with faulting, locally the consolidated rocks do contain ground water that can be extracted from carefully placed wells.

The poorly consolidated and unconsolidated deposits are porous and contain saturated sections below generally shallow water tables. The abundance of these permeable deposits is greatest in the southern Coast Ranges, where valley fill is as much as 600 m thick in the Salinas Basin (Durbin and others, 1978) and 700 m thick in Santa Maria Valley (Miller and Evenson, 1966). Along the southern part of San Francisco Bay, porous materials in alluvial fans, derived from the mountains around Santa Clara Valley, contribute significant storage capacity for ground water. No large alluvium-filled valleys occur in the northern Coast Ranges. Accumulations of more than 100 m thickness occur in only a few small structural basins along the Russian and Eel River drainages (Fig. 1). The Santa Rosa Plain is the largest ground-water basin in the northern Coast Ranges (250 km²) and contains an average thickness of about 120 m of permeable, saturated alluvial fill.

The great lithologic diversity of the valley-fill deposits, consisting of varying mixtures of clay, silt, sand, gravel, and boulders, causes correspondingly large variations in permeability. Sorting rather than grain size is commonly the major factor in determining the degree of permeability. Ubiquitous clay and silt, especially prevalent away from the present-day stream courses, inhibit the flow of ground water by filling the interstices of the coarser materials.

Discontinuous marine terraces underlain by unconsolidated or loosely consolidated sand and gravel deposits are distributed along many parts of the California coast. These deposits, although porous, are only partly saturated seasonally and provide limited storage capacity for ground water because they are generally less than 30 m thick. In places, windblown sand has accumulated as dunes along the coast. Although the sand is highly permeable and provides a surface for ground-water recharge, these deposits are generally unsaturated because water rapidly percolates through the sand to underlying rocks.

In most places, the valley-fill materials contain ground water under both confined and unconfined conditions. The unconfined aquifers extend downward from shallow depth to levels where clay or other impermeable materials are sufficiently thick to inhibit vertical ground-water movement. Below this level confined ground water is present, and, in some places, two or more confined aquifers may be stacked one on top of another with intervening poorly permeable clay layers. The hydraulic heads may vary considerably between the various confined aquifers and the water-table aquifer. In the Salinas Valley, a water-table aquifer is underlain by three distinct confined aquifers.

The primary source of recharge to the ground-water basins is from precipitation. In areas of abundant precipitation, recharge from direct percolation of rainfall into permeable materials is significant. In parts of the southern Coast Ranges where soil moisture is generally deficient, the most significant source of recharge is downward percolation through permeable materials along stream courses. Imported surface water for irrigation, in some cases, provides recharge to the ground-water reservoirs through leakage from canals and ditches as well as percolation from heavily irrigated fields. Artificial recharge is practiced in many areas. In the southern part of the Coast Ranges, stormwater held back by retention dams is utilized for recharge. Elsewhere, percolation of treated sewage contributes to recharge.

## HYDROLOGIC PROBLEMS RELATED TO GEOLOGY AND GROUND-WATER WITHDRAWAL

The interaction of ground water with local geologic conditions has important consequences for engineering and land use. Landslides and earthquakes are common to the Coast Ranges and are closely related to geologic structures, lithology, and fluid pore-pressure.

Some hydrologic problems are more directly related to a combination of hydrogeologic conditions and water-resources utilization. Land subsidence and seawater intrusion are two significant hydrologic problems in Region 4 that are controlled by the hydrogeologic setting and result from ground-water withdrawals.

### Land subsidence

Areas affected by land subsidence in the region include much of the southern part of the San Joaquin Valley and smaller areas in the Sacramento Valley, the San Joaquin–Sacramento Delta, and in the Santa Clara Valley (Fig. 2). The largest volume of land subsidence in the world due to man's activities has occurred in the Central Valley of California. More than one half of the San Joaquin Valley, or about 13,500 km$^2$, has undergone land subsidence of more than 0.3 m. In some areas, more than 6 m of subsidence has been recorded, with a maximum of 9 m recorded near Mendota. By far, the largest volume of land subsidence is caused by ground-water pumpage and the resulting compaction of clays in the San Joaquin Valley. However, other processes have contributed to land subsidence locally (Poland and Evenson, 1966; Poland, 1984). Four processes that cause subsidence are: (1) compaction of peat soils following drainage of marshland; (2) hydrocompaction resulting from compaction of moisture deficient sediments following application of water; (3) compaction of sediments in petroleum reservoir rocks caused by withdrawal of fluids from oil fields; (4) compaction of fine-grained sediments in the aquifer systems resulting from heavy withdrawals of ground water.

Compaction of peat soils and subsequent land subsidence has occurred in an area of about 1,170 km$^2$ in the San Joaquin–Sacramento Delta area (Poland and Evenson, 1966). Originally

Figure 2. Areas of land-subsidence caused by withdrawal of ground water.

the delta area contained islands at or slightly above sea level; as a result of subsidence, these islands are now 3 to 5 m below sea level. This type of subsidence results from oxidation and compaction of peat soils following drainage of marshlands for agriculture.

Hydrocompaction refers to the compaction of moisture-deficient deposits above the water table following the first application of water. This type of subsidence has resulted in 2 to 3 m of subsidence in dry areas along the western and southern margins of the San Joaquin Valley (Poland and Evenson, 1966).

Compaction of sediments due to the withdrawal of oil and gas has caused land subsidence locally; however, the magnitude is uncertain. Subsidence of less than 0.3 m has been attributed to this process in the oil fields near Bakersfield (Lofgren, 1975).

Land subsidence due to withdrawal of ground water is caused mainly by compaction of fine-grained (clay and silt) materials within an aquifer system. When pumping causes the hydraulic head to decline below the preconsolidation stress level, the clays compact and release water. Compaction is caused by an increase in effective stress (effective overburden pressure or grain-to-grain load).

As defined by Terzaghi (1925) and Terzaghi and Peck (1967), the equation for effective stress is

$$P' = P - u_w,$$

where $P'$ is effective stress, $P$ is stress (geostatic pressure), and $u_w$ is pressure (fluid pressure).

In a confined sand and gravel aquifer, the geostatic pressure is not significantly changed as the artesian head is reduced, because the decreased pore pressure is balanced by an increase in the grain-to-grain load. Compaction of the sand and gravel is small and immediate and is largely recoverable, referred to as elastic. In contrast, the adjustment of pore pressure in the clayey confining beds and the clayey interbeds proceeds slowly because of the lower permeability and higher specific storage. The adjustment may take months to many years and is also dependent on the thickness of the individual clayey beds. The clayey beds are more compressible compared to the sands and gravels, and when the grain-to-grain load exceeds the preconsolidation stress, the compaction of the clayey beds is largely nonrecoverable, referred to as inelastic. The amount of inelastic compaction of a function of the compressibility of the sediments within the range of artesian-head decline and the magnitude of that change (Poland and Davis, 1969). In the San Joaquin Valley, head declines in the confined aquifer of as much as 150 m due to ground-water withdrawals caused inelastic compaction of the clayey beds and resulted in as much as 9 m of land subsidence (Fig. 3).

Once pumping ceases, artesian heads recover, and inelastic compaction of the clayey beds eventually ceases, though it may continue for some time. The recovery of heads to predevelopment levels is not accompanied by a recovery of total storage due to the inelastic compaction of the clayey beds. If pumping resumes, inelastic compaction will not recur until the heads in the clayey beds decline below the previous low levels.

The percentage of pumped water in the subsiding areas that is derived from compaction ranges from a few percent to more than 60 percent. For any given pumping regime, the severity of compaction is related largely to the percentage of fine-grained materials and the mineralogy of the interval penetrated by wells. The Los Banos–Kettleman City area, where maximum land subsidence of 9 m has been recorded, is characterized by the highest percentage of fine-grained material within the upper 600 m of the aquifer system in the San Joaquin Valley. The type of clay mineral present also influences the amount of subsidence. Montmorillonite, which is highly susceptible to compaction, is the predominant clay mineral in major subsiding areas of the San Joaquin Valley.

The measured compaction in relation to head decline at a well in a subsiding area from 1960 to 1980 is shown in Figure 4. At this site, the 1960s were marked by steady head decline and a

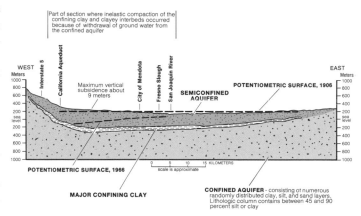

Figure 3. Generalized geohydrologic section of San Joaquin Valley aquifer system.

high rate of compaction. The decrease in ground-water pumping in the early and middle 1970s was accompanied by a steady recovery of water levels and a greatly reduced rate of compaction. The resumption of large ground-water withdrawals during the 1977–1978 drought was marked by a sharp decline in water levels and a short period of renewed compaction. Ireland and others (1984) reported that artesian heads generally declined 10 to 20 times as fast during the drought as during the period of long-term drawdown and compaction that ended in the late 1960s. Thus, much of the water pumped during the drought was probably supplied by elastic storage. Following the drought, recovery to predrought water levels was rapid and compaction virtually ceased. In fact, the negative compaction measured immediately after pumping returned to predrought levels indicates that part of the compaction during the drought was elastic.

### Seawater intrusion

Seawater intrusion, or the migration of seawater into freshwater aquifers, is another phenomenon that occurs in response to ground-water withdrawal and has been identified along coastal areas and around San Francisco Bay (California Department of Water Resources, 1975).

Under predevelopment conditions, in areas adjacent to the ocean or brackish estuaries, a wedge-shaped interface separates saline water from freshwater contained in aquifers where hydraulic communication exists between the two. The interface is actually a transitional zone that varies in dimension, depending on the hydrodynamic processes that mix fresh and saline waters where they meet. The location of the transition zone with respect to the coast or estuary is primarily dependent on the hydraulic head in the freshwater aquifer. The greater the freshwater head, the farther the saline water can be held back from the land area. As hydraulic head is reduced by pumping, saline water moves landward within the aquifer.

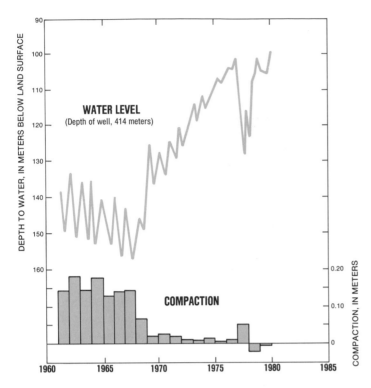

Figure 4. Hydrograph and compaction record for a 414-meter-deep well in the San Joaquin Valley, near Mendota, showing the relation between head decline and compaction. The rapid head decline during the period of increased pumpage in 1977–78 was due to the loss of storage in the aquifer system caused by compaction of fine-grained materials during the 1960s.

Lithology and geologic structure are important relative to the problem of seawater intrusion. Intrusion can occur only where exposures of permeable materials are in contact with saline water. Where estuaries or tidal flats are blanketed with thick layers of clay or silt, the intrusion of saline water is inhibited. Faults may act as barriers to the interchange of fresh and saline waters. Slumping of sediments in Monterey Bay associated with fault displacements is thought to expose fresh surfaces of permeable units to the landward intrusion of seawater (Greene, 1970).

Changes in base level have played a role in the problem of seawater intrusion. The eustatic rise in sea level that corresponds with the waning of Pleistocene glaciers has flooded the mouth of the Sacramento and San Joaquin Rivers. Prior to the latest rise, the lower base level allowed the rivers to cut their channels through sedimentary units that now are below sea level. The exposed edges of permeable sedimentary units now below sea level provide paths for seawater intrusion.

Many areas along the coast of this region have the potential for seawater intrusion; however, thus far, seawater intrusion has been identified only in a few areas around San Francisco Bay and the delta area, Monterey Bay, and Morro Bay (California Department of Water Resources, 1975). Minor areas of intrusion are recognized at the mouths of Petaluma, Sonoma, and Napa Valleys along the north side of San Francisco Bay. Here, seawater has intruded the developed aquifers by infiltrating downward from tidal channels rather than inflow to the aquifer in sub-sea level exposures. To the east of these areas, around Fairfield and Pittsburg, seawater intrusion has been exacerbated by the diminished flow of freshwater, due to diversions from the Sacramento and San Joaquin rivers, allowing daily influxes of seawater to migrate farther into the delta area. Heavy pumping during the 1930s in Alameda County, on the east side of the San Francisco Bay and in the Santa Clara Valley, along the southern side of the San Francisco Bay, created a wide area of seawater intrusion as well as land subsidence. The problem has been partly alleviated due to reduced pumping as the local economy moved from agriculture to industry and because much of the water supply is now imported. The agricultural areas near the mouths of the Pajaro and Salinas rivers have areas of seawater intrusion that are still expanding due to continued pumping for irrigation (Johnson, 1983; Showalter and others, 1984).

## SUMMARY

Region 4 includes about 160,000 km² of coastal and central California. The region can be divided into two distinctly different geologic and hydrologic subregions: the Coast Ranges and the Central Valley. Geologic evolution of the two subregions is closely tied to the large-scale processes of plate tectonics active in the region during post-Jurassic time. The Central Valley represents the forearc basin that received marine sedimentation from Jurassic through early Cenozoic time. During later Cenozoic time, sedimentation varied between marine and continental environments. The Coast Ranges are formed, in part, from the oceanic crust and trench sediments that were uplifted as thrust faulting accommodated the compressional forces associated with subduction of the Pacific plate along the western continental margin of North America, but mostly by Cenozoic large-scale strike-slip displacement of terrains along faults of the San Andreas system and by the associated deformations.

Deformation during Cenozoic time has been minimal in the Central Valley relative to that in the Coast Ranges. The thick (15,000 m) nearly flat-lying pile of sediments and sedimentary rocks occupy what is topographically one large valley. The Central Valley is virtually one large ground-water basin with hydraulic continuity throughout. It contains a vast quantity of fresh ground water in its upper 300 m. In contrast, extensive post-Jurassic faulting in the Coast Ranges has broken the subregion into numerous small alluvium-filled valleys hydrologically separate from one another. Ground water in the mountainous parts of the Coast Ranges is contained primarily in fractured rocks, and the availability of water is greatly restricted; areally extensive, productive aquifers are found within the sedimentary fill of the larger valleys.

Ground-water withdrawal from Central Valley aquifers has been the cause of extensive land subsidence affecting an area of about 13,000 km². Subsidence is greatest in areas underlain by the thickest sections of clay beds and has reached as much as 9 m in parts of the San Joaquin Valley. Ground-water withdrawals from some coastal valleys in the Coast Ranges has reduced hydraulic head sufficiently to allow intrusion of seawater into freshwater aquifers. These adverse impacts caused by ground-water withdrawals are controlled by lithology and geologic structures within the region.

## REFERENCES CITED

Bailey, E. H., Irwin, W. P., and Jones, D. L., 1964, Franciscan and related rocks, and their significance in the geology of western California: California Division of Mines and Geology Bulletin 183, 172 p.

Blake, M. C., Jr., and Jones, D. L., 1981, The Franciscan Assemblage and related rocks in northern California; A reinterpretation, *in* Ernst, W. G., ed., Geotectonic development of California: Englewood Cliffs, New Jersey, Prentice-Hall, p. 306–328.

California Department of Water Resources, 1975, Seawater intrusion in California; Inventory of coastal ground-water basins: Bulletin No. 63-5, 394 p.

Davis, G. H., Green, J. H., Olmsted, F. H., and Brown, D. W., 1959, Ground-water conditions and storage capacity in the San Joaquin Valley, California: U.S. Geological Survey Water-Supply Paper 1469, 287 p.

Dickinson, W. R., 1981, Plate tectonics and the continental margin of California, *in* Ernst, W. G., ed., Geotectonic development of California: Englewood Cliffs, New Jersey, Prentice-Hall, p. 1–28.

Durbin, T. J., Kapple, G. W., and Freckleton, J. R., 1978, Two-dimensional and three-dimensional digital flow models of the Salinas Valley ground-water basin, California: U.S. Geological Survey Water-Resources Investigations 78–113, 134 p.

Etter, S. D., Fritz, D. M., Gucwa, P. R., Jordan, M. A., Kleist, J. R., Lehman, D. H., Raney, J. A., Worrall, D. M., and Maxwell, J. C., 1981, Geologic cross sections, northern Coast Ranges to northern Sierra Nevada, and Lake Pillsbury area to southern Klamath Mountains: Geological Society of America Map MC-28N, scale 1:250,000.

Fox, K. F., Jr., 1983, Mèlanges and their bearing on Late Mesozoic and Tertiary subduction and interplate translation at the west edge of the North American plate: U.S. Geological Survey Professional Paper 1198, 40 p.

Greene, G. H., 1970, Geology of southern Monterey Bay and its relationship to the ground-water basin and saltwater intrusion: U.S. Geological Survey Open-File Report, 50 p.

Hackel, O., 1966, Summary of the geology of the Great Valley, *in* Bailey, E. H., ed., Geology of northern California: California Division of Mines and Geology Bulletin 190, p. 217–238.

Ireland, R. L., Poland, J. F., and Riley, F. S., 1984, Land subsidence in the San Joaquin Valley, California, as of 1980: U.S. Geological Survey Professional Paper 437-I, 93 p.

Johnson, M. J., 1983, Ground water in North Monterey County, California, 1980: U.S. Geological Survey Water-Resources Investigations Report 83–4023, 32 p.

Kahrl, W. L., ed., 1978, The California water atlas: Prepared by the Governor's office of Planning and Research in cooperation with the California Department of Water Resources, p. 113.

Lofgren, B. E., 1975, Land subsidence due to ground-water withdrawal, Arvin-Maricopa area, California: U.S. Geological Survey Professional Paper 437-D, 55 p.

McLaughlin, R. J., and Nilsen, T. H., 1981, Neogene non-marine sedimentation and tectonics in small pull-apart basins of the San Andreas fault system, Sonoma County, California: Proceedings of 2d International Fluvial Conference, Keele, England, September 1981, 18 p.

Miller, G. A., and Evenson, R. E., 1966, Utilization of ground water in the Santa Maria Valley area, California: U.S. Geological Survey Water-Supply Paper 1819-A, 24 p.

Norris, R. M., and Webb, R. W., 1976, Geology of California: New York, John Wiley and Sons, 365 p.

Olmsted, F. H., and Davis, G. H., 1961, Geologic features and ground-water storage capacity of the Sacramento Valley, California: U.S. Geological Survey Water-Supply Paper 1497, 241 p.

Page, B. M., 1981, The southern Coast Ranges, *in* Ernst, W. G., ed., Geotectonic development of California: Englewood Cliffs, New Jersey, Prentice-Hall, p. 329–417.

Page, R. W., 1986, Geology of the fresh ground-water basin of the Central Valley, California, with texture maps and sections: U.S. Geological Survey Professional Paper 1401-C, 54 p.

Poland, J. F., ed., 1984, Guidebooks to studies of land subsidence due to groundwater withdrawal: UNESCO Studies and Reports in Hydrology, no. 40, 305 p.

Poland, J. F., and Davis, G. H., 1969, Land subsidence due to withdrawals of fluids, *in* Varnes, D. J., and Kiersch, G., eds., Reviews in Engineering Geology, v. 2: Boulder, Colorado, Geological Society of America, p. 187–269.

Poland, J. F., and Evenson, R. E., 1966, Hydrogeology and land subsidence, Great Central Valley, California, *in* Bailey, E. H., ed., Geology of northern California: California Division of Mines and Geology Bulletin 190, p. 239–247.

Repenning, C. A., 1960, Geologic summary of the Central Valley of California with reference to disposal of liquid radioactive waste: U.S. Geological Survey Open-File Report, 69 p.

Showalter, P. A., Akers, J. P., and Swain, L. A., 1984, Design of a ground-water-quality monitoring network for the Salinas River basin, California: U.S. Geological Survey Water-Resources Investigations Report 83–4049, 74 p.

Suppe, J., 1979, Cross section of southern part of northern Coast Ranges and Sacramento Valley, California: Geological Society of America Map MC-25B, scale 1:250,000.

Terzaghi, K., 1925, Principles of soil mechanics; IV. Settlement and compaction of clay: Engineering News–Record, p. 874–878.

Terzaghi, K., and Peck, R. B., 1967, Soil mechanics in engineering practice [2d ed.]: New York, John Wiley and Sons, 729 p.

MANUSCRIPT ACCEPTED BY THE SOCIETY MARCH 6, 1987

The Geology of North America
Vol. O-2, Hydrogeology
The Geological Society of America, 1988

# Chapter 8

# *Region 5, Great Basin*

**M. D. Mifflin**

*Mifflin and Associates, Inc., 2700 Sunset, Suite C-25, Las Vegas, Nevada 89120*

## INTRODUCTION

The Great Basin is a 500,000 km$^2$ hydrographically defined region of the southwestern United States (Fig. 3, Table 2, Heath, this volume). Its hydrogeology is unique when compared to any other region of similar size in North America, owing to a combination of basin-and-range structure, arid climates, and interior drainage. Plate 3 (chiefly a U.S. Geological Survey compilation and interpretation of data from many sources) depicts in considerable detail ground-water relationships, such as patterns of flow, flow-system divides, estimates of ground-water budgets, and the distribution of ground-water discharge for a large part of the Great Basin.

A repetitious pattern of north-trending mountain ranges and intermontane structural basins establishes the hydrogeology of the region. Climates are generally a function of altitude and, to lesser extent, latitude and rain shadow effects. The climates range from the very dry and hot desert climate of Death Valley in the lowest basin, through cooler semiarid steppe-like climates in the higher basins, to the more humid alpine climates of the highest mountains. The entire region is characterized by interior drainage of surface water and ground water, with the exception of a small area of Nevada tributary to the Colorado River. The following are recognized as the dominant factors in creating the special hydrogeologic conditions common to the region:

I. Extensional faulting has resulted in marked topographic differences and lithologic contrasts between geologically diverse range blocks and intermontane sedimentary basins.

II. Regional patterns of atmospheric circulation transport winter moisture from the Pacific Ocean in eastward tracks across the Great Basin, producing humid alpine climates in the high mountains as a result of orographic influences, while the intermontane basins remain arid or semiarid.

III. Surface drainage and ground-water flow systems are confined to hydrologically closed regions from which external drainage does not occur. The boundaries of surface-water basins and ground-water flow systems do not always coincide due to combinations of aridity and transmissive character of the consolidated rocks.

IV. Quaternary climatic variations produced more effective moisture and associated paleohydrology that left abundant evidence of shorelines, spring and saline deposits, and basin deposits related to lacustrine and paludal environments. Locally, the paleohydrologic history determines the pattern of ground-water flow, distributions of discharge, and water quality in the hydrographically closed basins.

## GEOLOGIC CONTROLS

Geology imparts strong subregional and local influences on the hydrogeology. The Great Basin region displays the record of a long and active history of intermittent marine sedimentation and large-scale compressive deformation, island-arc plutonism and volcanism, bimodal basaltic and silicic volcanism, extensional tectonics, and terrestrial sedimentation. Rock types, ages, and deformational structures range through much of the known spectrum, and in many areas impressive diversities exist in juxtaposed rock types. Subsurface conditions are highly varied and generally complex, and bedrock geology beneath alluvial basins is only approximately known in much of the region.

The Great Basin has been subdivided hydrogeologically into two provinces related to either the abundance or lack of relatively permeable consolidated rock types. These provinces have been termed the "low permeability rock province" and the "carbonate rock province" (Fig. 1). In areas underlain by thick sequences of Paleozoic limestones and dolomites, ground-water flowpaths are not always concordant with the hydrographic and structural basins. Hydrologic evidence suggests extensive circulation of ground water between some structural basins, and such circulation may also control the distribution of known geothermal resource areas (Fig. 1) and anomalously low heat flow centered in east-central Nevada. In the low permeability rock province, the structural basins commonly constrain ground-water circulation to intrabasin flow configurations, and known geothermal resource areas are abundant.

The basin sediments are only approximately known in the subsurface. Maxey and Jameson (1948) were the first to demonstrate that complex histories of basin growth and sediment accumulation, with marked variations in structural and hydrologic histories, combine to vary the total thickness, age, composition,

Mifflin, M. D., 1988, Region 5, Great Basin, *in* Back, W., Rosenshein, J. S., and Seaber, P. R., eds., Hydrogeology: Boulder, Colorado, Geological Society of America, The Geology of North America, v. O-2.

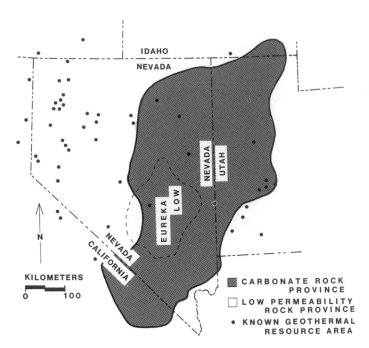

Figure 1. Carbonate rock province and distribution of known geothermal resource areas (in part after Mifflin, 1968; Mifflin and Hess, 1979).

and distribution of relatively transmissive sediments of the basin fills (Fig. 2). Semiconsolidated and unconsolidated basin deposits may range in age from Eocene to Holocene, with lithologic characteristics varying significantly and local deformation present (see Davis, Chap. 34.). The last major episode of the structural basin development and associated sedimentation probably began in middle or late Miocene time and continues to the present. In the southerly part of the region, mountainous relief and basin filling to greater than current levels had occurred by late Miocene or early Pliocene time. The northern two-thirds and the eastern and western margins of the region are marked by continuing basin-and-range fault movements; the range-margin faults of the southwesterly part appear to be less uniformly active, and many pediments have developed on both the bedrock and older basin sediments along the flanks of the elongated mountain ranges.

## CLIMATIC CONTROLS

Markedly varied climates and the associated distribution of moisture available for runoff and recharge greatly influence the hydrogeology. The availability of moisture (timing, amount, state, and distribution of precipitation) and high evaporation and transpiration rates combine to impart strong controls on the distribution and amount of surface runoff and ground-water recharge, the pattern and amount of ground-water flow, and ultimately, the distribution and character of surface-water and ground-water

sinks. Climates are generally arid to semiarid in the many basins of the Great Basin. Depending upon latitude and altitude, they range from cool steppe climates in the basins with altitudes of more than 1,200 m (much of the northern two-thirds of the province) to the warm desert climates in the lower basins of the southwestern one-third of the area. Strong seasonal and diurnal temperature variation is common throughout the region. High variability of the annual precipitation is also typical of the arid and semiarid climates of the region.

In the northern and highest basins and the higher mountain terrane, the minimum annual potential evaporation is about 0.6 m; in the low southwesterly basins, the annual potential evaporation may range to more than 3 m. When compared to average annual basin precipitation of 100 to 200 mm (the latter in some of the highest northern basins) the moisture deficit is substantial (ranging from about five to thirty times more potential evaporation than annual precipitation). However, the distribution of moisture, particularly in the more humid alpine climates in the high mountainous areas, permits moisture to persist at land surface, and even in more restricted conditions of time and space, permits moisture runoff and ground-water recharge to occur.

The range of climates associated with the mountainous terrane of the Great Basin is greater than that of the arid basins. The lower mountains in the southwesterly part are as arid as northerly basins with similar altitudes. Somewhat more precipitation falls on southern mountain ranges than on the adjacent basins owing to orographic effects. Also, the southern basins and ranges receive more precipitation from summer convectional storms. The few high ranges in the south, such as the Spring Mountains and the Sheep Range near Las Vegas, and the Panamint Range west of Death Valley, receive more than 500 mm of precipitation in the crestal areas. Similar amounts of precipitation occur on many of the high ranges in the northern and central part of the Great Basin. The resultant snowpacks store much of the winter moisture. This accumulated moisture provides a prolonged meltwater runoff. Much of this moisture infiltrates in the mountain source areas. Some reaches the basin margins and infiltrates before being lost to evapotranspiration. The quantitative importance of high-terrane winter moisture to total recharge remains uncertain but is most likely the principal source of recharge. Perhaps the only exception in the region occurs in the Mojave Desert, which consists of low mountains and extremely arid terrane. In this area, snowpacks do not form, and recharge is derived from infiltration along dry washes during the rare heavy precipitation events that produce runoff.

## HYDROLOGY

The region has more than 100 major hydrographic basins (Plate 3), which can be subdivided into more than 200 hydrologic basins (as viewed from a combination of surface water and ground-water basin perspectives). All ground water and surface water ultimately is evaporated or transpired within the region except for a minor amount that discharges to the Colorado River.

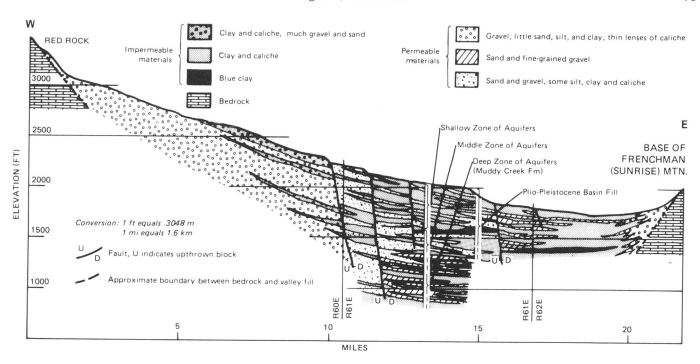

Figure 2. Lithologies, facies changes, and structure of Las Vegas Valley (as interpreted by Bell, 1981, after Maxey and Jameson, 1948).

Typically, perennial streams originate in the high mountains. Some of the larger streams traverse several structural basins until dissipated by infiltration, artificial diversions, natural evapotranspiration, and evaporation in terminal sinks (playas, marshes, playa lakes, and terminal basin lakes). All Great Basin streams become losing streams within the basin environments (decreasing total stream discharge downstream). For most streams, the position of balanced gains and losses in stream discharge will vary in geographic position with the season, the stage of runoff, the antecedent runoff in prior months and years, and the pattern and timing of artificial diversions. In general, the aridity of the intermontane basin(s) through which the stream flows permits little or no net ground-water contributions to the stream. However, the discharge of Great Basin streams typically varies greatly, both seasonally and annually. Therefore, important short-term interchanges between river and ground-water systems occur. The majority of ground-water recharge in irrigated basins is derived from streamflow during high runoff or infiltration of diverted irrigation water.

The high mountainous terrane bounding the Great Basin on the west, north, and east provides abundant water supplies for many of the basins. Along the western boundary, the high Sierra Nevada and the extreme southern portion of the Cascades supply perennial streamflow to the westernmost basins. To the north, near the Nevada-Oregon boundary, the Jarbridge Mountains, the Ruby Mountains, and the East Humboldt Range are major source areas for the Humboldt River, which flows to the west and traverses essentially the width of Nevada. Along the northeastern and eastern flank of the Great Basin, basins receive heavy runoff from the Wasatch and Uinta Mountains. To the southeast, in south-central Utah, the high plateaus and mountains (the Wasatch, Aquarius, Paunasaugunt, Sevier, and Kolob plateaus, and the Escalante, Tushar, and Pine Valley mountains) are major sources of moisture for the easternmost basins. All of these areas receive annual precipitation in excess of 500 mm, with the highest parts of the Wasatch and Uinta Mountains and plateaus, as well as the Sierra high country, receiving as much as 1,500 mm or more.

Most surface water within the region is diverted for irrigation within the basins. As a result, the usually limited streamflow is spread over extensive areas within the alluvial basins. The net effects of ubiquitous streamflow diversion have been increases in recharge and in evapotranspiration in the irrigated areas, and decreases in surface water that reaches the natural terminal sinks. These net effects of streamflow diversion result in profound impacts on the hydrology of irrigated basins, the lower reaches of streams, and terminal sinks. Below major diversion canals in irrigated basins, water tables are frequently shallow, and extensive water-logged areas, seeps, and ponds or marshes may be common. Extensive areas of phreatophytes are common. Surface-water quality in drainage and stream channels is adversely affected. The natural flows in small streams are dried up in the basinward reaches by upstream diversions. In the terminal basins, markedly increased upstream consumptive use has historically produced major water-level declines and/or loss of terminal basin lakes (e.g., Mono Lake, Pyramid Lake, Walker Lake, Great Salt Lake, and Winnemucca Lake).

The hydrogeologic characteristics of most of the basins are also influenced by the paleohydrologic histories of the closed basins. Figure 3 illustrates the latest Pleistocene plenipluvial distribution of drainage and pluvial lakes (about 14 to 18 ka), with the northeastern and northwestern bolsons totally inundated by Lakes Bonneville and Lahontan, respectively, and many other basins occupied by smaller lakes. Paleohydrologic influences on basin hydrogeology include the distribution of well-sorted river and delta deposits, forming favorable aquifers (Arnow and others, 1970); the distribution of fine-grained lacustrine sediments and associated broad depositional plains; and areas of extensive accumulations of evaporites and associated distribution of saline ground water (Mifflin, 1968). In some southerly bolsons, pluvial lakes failed to develop because of the warmer, more arid climate. In other bolsons, fine-grained sediments rich in secondary carbonates accumulated due to marked increases in ground-water discharge, supporting springs, marshes, and wet meadows (Mifflin and Wheat, 1979).

The terminal sinks of surface-water and ground-water systems vary in character. Variations in surface-water systems are dependent upon a combination of factors, including paleohydrologic history of the basin, rate of local tectonic movement, terminal-basin size and configuration, history of stream capture, and sedimentation rates. Deep-water terminal lakes, such as Pyramid and Walker Lakes in northwestern Nevada, apparently persist to the present due to a combination of three factors: (1) high rates of tectonic movement at the terminal basin; (2) Quaternary history of stream capture concentrating runoff to the terminal basins; and (3) the terminal-basin sizes and configurations. When the combination of basin configuration, sedimentation rate, and tectonic basining produce broad, low-relief bolsons with relatively large perennial surface-water supplies, broad but shallow terminal lakes form, such as Mono Lake and the Great Salt Lake. If the perennial surface-water supply to the basin is more limited, large but shallow playa lakes occur, such as Honey Lake in northeastern California. This lake is normally perennial, but historically has gone dry. There are all degrees of persistence and size displayed by playa lakes in the Great Basin. Playas—flat areas of clay and silt devoid of vegetation—probably are adjusted to ephemeral surface-water runoff of the Holocene or present climates. The playas occasionally fill to depths from a few centimeters to more than a meter of water. However, many of the largest playas in the northern Great Basin (e.g., the Black Rock and Smoke Creek deserts, and large playas peripheral to the Great Salt Lake) are in part relict depositional plains of former pluvial lakes and as such, with the present climatic conditions, may never be totally inundated.

Hydrogeologically, the playas range between two extreme types: (1) phreatic or wet playas (normally moist near or at the playa surface owing to the shallow regional zone of saturation), and (2) vadose playas (normally dry because the regional zone of saturation is too deep for the capillary fringe to reach the surface of the playa). Gradation from one condition to the other may occur within a given playa area. Phreatic playas commonly have areas of puffy or salty surfaces; vadose playas are normally composed entirely of smooth pans of clay and silt.

Quantitatively, the most common cause of discharge of ground water in the Great Basin is not direct evaporation from a phreatic playa surface, but rather transpiration by phreatophytes. Phreatophytes line the floodplains of perennial streams, surround phreatic playas, and may extend over large areas of the bolsons. Throughout the Great Basin, phreatophytes mark extensive areas where the water table is no more than about 15 m below land surface. Once their hydrogeologic significance was recognized by the early settlers of the arid West, the settlers' ability to develop shallow-well water supplies was greatly improved.

## HYDROGEOCHEMISTRY

The physical and chemical characteristics of ground water in the highly varied Great Basin environments have provided very useful supplementary evidence of ground-water circulation patterns. Many large regions have only sparse information on fluid-potential relationships. These data have often been supplemented by physical and chemical parameters, which may be used to indicate circulation paths.

Ground water of the region generally begins as a CaMg $HCO_3$ type in the recharge areas, and its chemistry evolves along the flowpaths as a function of the lithologies encountered. Typically, along the longer flowpaths in most types of terrane, including the basin fills, $SO_4$ and Cl ions tend to increase in relative concentrations. Base exchange is common in basin sediments, with Ca and Mg ion concentrations decreasing, and Na and K ions increasing along flowpaths.

Within both the bedrock and basin sediments, an important control on the concentrations of the highly soluble species is the distribution of evaporite deposits in Mesozoic and Cenozoic strata. In the young sediments of the bolsons, the presence or absence of hydrographic closure (both of surface-water and ground-water) determines the distribution of salt concentrations in surface and ground water. Indeed, Langbein (1961) observed that the absence of both evaporites and poor-quality water in a bolson was the strongest proof for the absence of hydrologic closure. Typically, vadose playas display only minor concentratoins of salts; phreatic playas, as congruent surface-water and ground-water sinks, display concentrations of evaporites.

Water chemistry of the many large springs (discharge rates equal or greater than 28 l/sec) in the carbonate-rock province has been used to aid in the interpretation of flow-system configurations. Figure 4 illustrates a spring classification based on increasing concentrations of Na, K, Cl, and $SO_4$ ions with apparent length of flowpath. Springs classified as regional are believed to represent points of discharge for regional flow (frequently long flow paths that are interbasin in configuration) whereas the local- and small-local-class springs characterize the discharge of flow

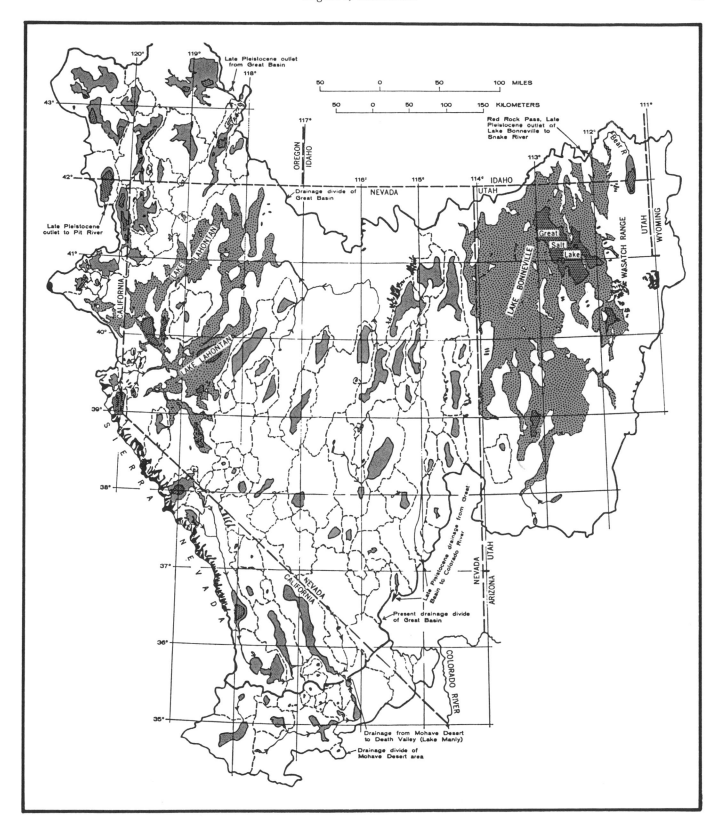

Figure 3. Plenipluvial Pleistocene lakes in the Great Basin (after Morrison, 1965; modifications from Mifflin and Wheat, 1979).

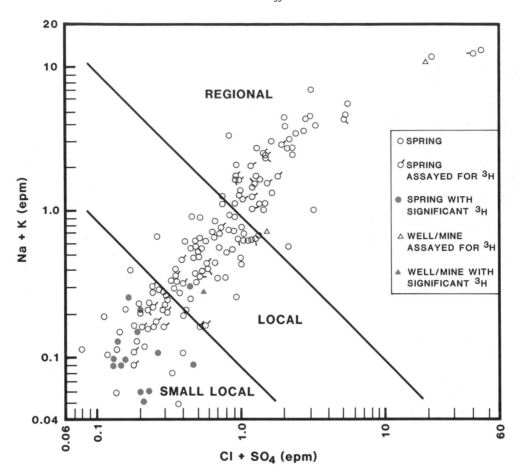

Figure 4. Flow system classification of large springs in the carbonate rock province of Nevada based on water chemistry and tritium (after Mifflin, 1968).

systems generally confined to the hydrographic basins. Only the small-local- and local-class springs are known to display major fluctuations in discharge, in seasonal or greater cycles, and small-local springs commonly demonstrate tritium concentrations indicating discharge of water recharged after 1954.

In the Great Basin, the majority of the large springs occur within the intermontane basins of the carbonate rock province and are often closely associated with nearby outcrops of carbonate rock. In south-central and southern Nevada, most of the major springs in the basins display water chemistries characteristic of the regional class. These spring areas are widely separated by extensive areas displaying little or no evidence of ground-water discharge (Plate 3). The springs have nearly constant discharge and probably represent points or zones of principal discharge from large regional flow systems. Accurate delineations of these regional systems are yet to be established in some areas; however, approximate delineations have been proposed (Eakin, 1966; Mifflin, 1968; Winograd and Thordarson, 1975; Harrill and others, 1988).

The water chemistry and temperature data at any given

position in the flow systems provide indications about the upgradient flowpaths. In the low-permeability bedrock province, there occurs a relatively small, although significant and concentrated, flow of deeply circulated ground water, as indicated by the widely distributed but highly localized discharge of thermal water. This deep ground-water circulation is believed to be enhanced by mountain-block recharge at high initial fluid potentials, which in turn, allows for deep percolation, initially within the mountain block along fractures or faults. Some of the thermal waters are moderately to highly mineralized.

Great Basin heat flow is high (Lachenbruch and Sass, 1978), and thermal ground water and geothermal resource areas are common in the low-permeability bedrock province. However, the paucity of these resource areas and the occurrence of the anomalous "Eureka Low" in the carbonate rock province of the Great Basin (Fig. 1) may be related to the hydrology of the carbonate terrane. If so, local, deeply circulated hydrothermal water may be masked by the transmissive carbonate aquifers that incorporate the hydrothermal water into regional flow. The missing heat in the "Eureka Low" may be advected to the large springs (those discharging in the basin environments are typically 5°C to 15°C above mean annual temperature) where heat flow has not been measured. The data base remains insufficient to fully test these hypotheses.

The ground-water temperature data is indicative of flow patterns and the location of recharge areas. Shallow circulating ground water has temperatures within a few degrees of mean annual air temperature. In the Great Basin, this relation gives some measure of the proximity of the recharge and the depth of circulation. Figure 5 illustrates ground-water temperature in water-supply wells finished in the alluvial aquifers of three selected basins. In the Truckee Meadows (Reno, Nevada, area), deeply circulated thermal waters issuing up along a range-front fault zone paralleling the Carson Range mix with the alluvial-fill waters recharged directly to the alluvium from extensive irrigation diversions. The histogram illustrates the two sources and the effective mixing (in some cases within the sampled wells). In the Quinn River Valley in north-central Nevada, the majority of ground water is recharged nearby from infiltrating streamflow on the bajada of the Santa Rosa Range. Most wells display temperatures compatible with shallow circulation. A lesser amount passes from the indurated rocks of the range to the flanking alluvium and displays higher temperatures of deeper flow paths.

The example of alluvial aquifers in Las Vegas Valley, where the majority of recharge is believed to occur high in the carbonate rock terrane of the Spring Mountains, indicates the majority of flow apparently has uniform depth of circulation from the adjacent mountains. The systematically decreasing number of higher temperature occurrences also suggests that some deeper flow lines are likely, and in the southerly extents of the valley (the most distant area from the highest mountain terrane), thermal water is present and may be associated with a low-permeability bedrock boundary.

Radiocarbon dates and tritium concentrations generally indicate relatively "old" ground water in most of the basin environments, with tritium concentrations commonly below postbomb levels and apparent radiocarbon ages in the thousands of years. These observations are compatible with the large volumes of water in storage and small annual flux rates in the ground-water systems. The stable isotopes $^{18}C$ and $^{2}H$ are currently being used to study the position of recharge (mountain or basin environments) and age relationships (modern climate or pluvial climate).

## CONFIGURATION OF FLOW SYSTEMS

Meinzer (1916) was the first to understand and accurately describe the typical hydrogeologic relationships of ground-water flow in the Great Basin. Figure 6 is his map of the ground-water discharge area of the Big Smoky Valley in Nevada, illustrating phreatophytic growth surrounding the playas and several typical spring-line areas at the toe of the bajada near the playa margins. He discussed the rapidly increasing depths to saturation under the steeply rising bajada and attempted a water budget by measuring the streamflow (a major portion of his recharge estimates) issuing

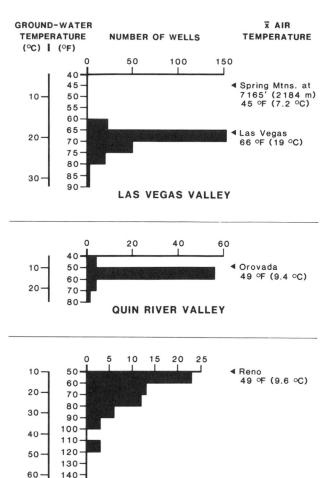

Figure 5. Histograms of ground-water temperature from wells in three ground-water basins in Nevada (after Mifflin, 1968).

from the mountain canyons onto the alluvial fans. Once on the alluvial fans, he observed that streamflow is generally lost before reaching the basin lowlands due to a combination of infiltration and evapotranspiration. Meinzer also estimated water use by phreatophytes (in the discharge area). He discussed the principal sources of recharge, the distribution and function of the discharge area, the associated increasing fluid potential of ground water with depth in the discharge area (giving rise to flowing wells), the former presence and distribution of large pluvial lakes and the associated influence on the lithology of the basin fill, and the concentration of salts in the discharge area due to ground-water discharge and lake evaporation. Even the varying hydrologic roles of phreatic and vadose playas were noted. He also discussed

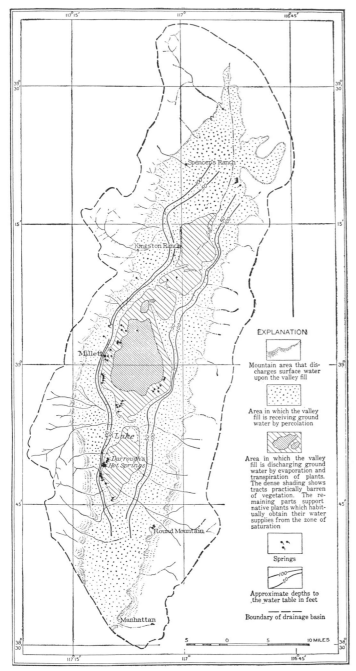

EXPLANATION:

Mountain area that discharges surface water upon the valley fill

Area in which the valley fill is receiving ground water by percolation

Area in which the valley fill is discharging ground water by evaporation and transpiration of plants. The dense shading shows tracts practically barren of vegetation. The remaining parts support native plants which habitually obtain their water supplies from the zone of saturation

Springs

Approximate depths to the water table in feet

Boundary of drainage basin

Figure 6. Meinzer's hydrogeologic map of Big Smoky Valley (after Meinzer, 1916).

the concept of safe yield for the ground-water basins, a topic of continuing interest in the water-short regions, and a prime stimulus for modern studies of flow-system configurations.

Figure 7 presents an idealized depiction of the typical flow-system relationships as indicated by fluid-potential relationships in boreholes. Mifflin (1968) recognized four important zones of

the flow systems with carefully collected water-level data: (1) recharge, with important downward flow; (2) lateral flow, usually quite extensive; (3) upward flow (usually within the basins adjacent to areas of active ground-water discharge); and (4) active discharge, with upward flow. There are notable differences between a "typical" Great Basin flow system and those of more humid terranes where ground-water discharge is widespread and the upper surface of the zone of saturation is typically a subdued replica of topography. Vast areas of lateral flow and great depths to the zone of saturation occur in most basins. These conditions commonly extend into foothills and low mountain areas. Perennial water is, therefore, limited over broad areas of the Great Basin, and many a truly "dry hole" has been dug by would-be water developers.

Through the efforts of many investigators in the past 30 years, a fascinating variety of flow-system configurations have been recognized within the Great Basin. Figure 8, developed by Eakin and others (1976), presents several. Hydrographically closed or undrained basins typical of the low-permeability rock province have ground-water discharge from the alluvial basin in the vicinity of the bolson playa. In partly drained basins, some discharge occurs within the hydrographically closed basins, but some ground water flows through permeable rock to another basin. In drained hydrographically closed basins, no ground-water discharge occurs within the basin due to subsurface flow out through permeable rock. In a basin that is a regional sink, there is discharge of a regional flow system with an important part of flow derived from adjacent basins. Not depicted in Figure 8 is the common configuration of ground-water discharge concentrated via large springs associated with carbonate rock terrane within the basins. In addition, the following discussion deals with postulated more complex flow patterns in the carbonate rock province.

In the central and northerly extent of the carbonate rock province, interbasin flow may be common but very complex, provided the previously discussed spring classification based on water chemistry is valid. Here the regional (interbasin) spring category cannot be tested by independent evidence of fluid potential relationships. Small-local-, local-, and regional-class springs frequently discharge within the same intermontane basin. When viewed from the water-chemistry and water-budget perspective, some of the flow-system configurations may follow a pattern of local, intermediate, and regional flow-system circulation cells as proposed by Toth (1962, 1963) and modeled by Freeze and Witherspoon (1966, 1967).

Most important to the multicell configurations of flow are relatively transmissive lithologies at depth. With such conditions, the models indicate that important flux may move under shallow circulation cells and, in effect, move under fluid potential "barriers" normally assumed to define flow-system boundaries. Most data on ground-water fluid potential are limited to the upper part of the phreatic zone throughout the region; hence, even a detailed knowledge of that configuration may be inadequate to ascertain where some waters enter the system in areas underlain at depth

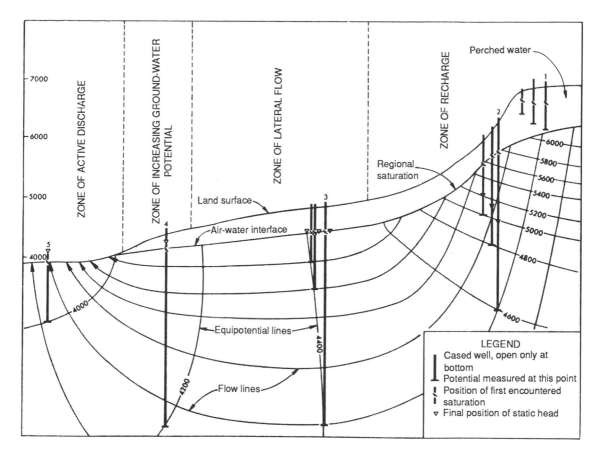

Figure 7. Idealized sketch of fluid potential relationships in Great Basin flow systems (after Mifflin, 1968).

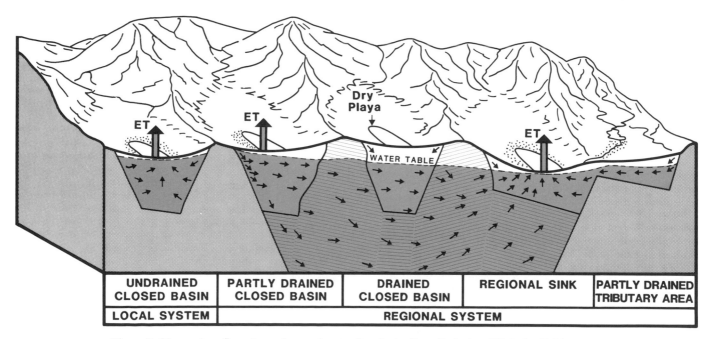

Figure 8. Observed configurations of ground-water flow in the Great Basin (modified after Eakin and others, 1976). ET, evapotranspiration.

*M. D. Mifflin*

by extensive transmissive zones. Two principal lines of evidence indicate that regional and local flow-system cells, as depicted by the conceptual models, may indeed exist. Imbalances in hydrologic budgets, where estimated recharge is established from precipitation in hydrographic basins and compared with estimates of ground-water discharge (Maxey and Eakin, 1949; Rush and others, 1971), and the hydrogeochemistries of large springs, indicating marked differences in total flowpath length, led Mifflin (1968) and Harrill and others (1988) to postulate complex circulation cells and interbasin flow controlled by important zones of transmissivity at depth in the extremely thick sequences of carbonate rock (perhaps as much as 20,000 m in some areas).

## CONCLUSIONS

Great Basin hydrogeology is complex and only sketchily documented. The data base that exists demonstrates that the geology and climate markedly impact the distribution of ground-water recharge, the patterns of flow, and the final discharge areas for both the surface-water and ground-water systems. Plate 3 summarizes the geographic distribution of these relationships, including the estimated water budgets. Marked variations can be seen in sizes of the ground-water flow systems, and the relative distances between recharge areas and the terminal discharge areas within multitudes of arid bolsons. Areas with interbasin-flow configurations are the most common in the carbonate rock province. When traversing the region, many of the unusual aspects of hydrogeology can be recognized, such as vast extents of shrubby phreatophytes, wet playas where much of the ground-water flow is dissipated, the playas and terminal-basin lakes of the closed basins, the large springs in the deserts hundreds of kilometers from the areas of recharge in the mountains, and vast arid areas where perennial water at land-surface is totally absent. These and other hydrogeologic phenomena are gradually being put into the context of the geologic and climatic influences as they occur in the Great Basin.

## REFERENCES

Arnow, T., Feth, J. H., and Mowers, R. W., 1970, Ground water in the deltas of the Bonneville Basin, U.S.A.: International Association Scientific Hydrology Symposium, v. 2, p. 396–407.

Bell, J. W., 1981, Subsidence in Las Vegas Valley: Nevada Bureau of Mines and Geology Bulletin 95, 84 p.

Eakin, T. E., 1966, A regional interbasin ground-water system in the White River area, southeastern Nevada: Water Resources Research, v. 2, no. 2, p. 251–271.

Eakin, T. E., Price, D., and Harrill, J. R., 1976, Summary appraisals of the nation's ground-water resources; Great Basin Region: U.S. Geological Survey Professional Paper 813–G, 37 p.

Freeze, A. R., and Witherspoon, P. A., 1966, Theoretical analysis of regional groundwater flow; 1, Analytical and numerical solutions to the mathematical model: Water Resources Research, v. 2, no. 4., p. 641–656.

—— , 1967, Theoretical analysis of regional groundwater flow; 2, Effect of water-table configurations and subsurface permeability variation: Water Resources Research, v. 3, no. 2., p. 623–634.

Harrill, J. R., Gates, J. S., and Thomas, J. M., 1988, Major ground water flow systems in the Great Basin area of Nevada, Utah, and adjacent states: U.S. Geological Survey Hydrologic Investigations Atlas 694–C, scale 1:500,000.

Lachenbruch, A. H., and Sass, J. H., 1978, Models of an extending lithosphere and heat flow in the Basin and Range Province, *in* Smith, R. B., and Eaton, G. P., eds., Cenozoic Tectonics and Regional Geophysics of the Western Cordillera: Geological Society of America Memoir 512, p. 209–250.

Langbein, W. B., 1961, Salinity and hydrology of closed lakes: U.S. Geological Survey Professional Paper 412, 20 p.

Maxey, G. B., and Eakin, T. E., 1949, Ground water in White River Valley, White Pine, Nye, and Lincoln Counties, Nevada: Nevada State Engineer's Office Water Resources Bulletin 8, 59 p.

Maxey, G. B., and Jameson, C. H., 1948, Geology and water resources of Las Vegas, Pahrump, and Indian Springs Valleys, Clark and Nye counties, Nevada: Nevada State Engineer's Office Water Resources Bulletin 5, 121 p.

Meinzer, O. E., 1916, Ground water in Big Smoky Valley, Nevada: U.S. Geological Survey Water Supply Paper 375, p. 85–116.

Mifflin, M. D., 1968, Delineation of ground-water flow systems in Nevada: Desert Research Institute Technical Report Series H–W, no. 4, 111 p.

Mifflin, M. D., and Hess, J. W., 1979, Regional carbonate flow systems in Nevada: Journal of Hydrology, v. 43, p. 217–237.

Mifflin, M. D., and Wheat, M. M., 1979, Pluvial lakes and estimated pluvial climates of Nevada: Nevada Bureau of Mines and Geology Bulletin 94, 57 p.

Morrison, R. B., 1965, Quaternary geology of the Great Basin, *in* Wright, H. E., Jr., and Frey, D. G., eds., The Quaternary of the United States: Princeton, New Jersey, Princeton University Press, p. 265–285.

Rush, F. E., Scott, B. R., Vanderburg, A. S., and Vasey, B. J., compilers, 1971, State of Nevada water resources and inter-basin flows: Nevada Division of Water Resources Map, scale 1:500,000.

Toth, J., 1962, A theory of ground-water motion in small drainage basins in central Alberta, Canada: Journal of Geophysical Research, v. 67, no. 11, p. 4375–4387.

—— , 1963, A theoretical analysis of ground-water flow in small drainage basins: Journal of Geophysical Research, v. 68, no. 16, p. 4795–4812.

Winograd, I. J., and Thordarson, W., 1975, Hydrogeologic and hydrogeochemical framework, south-central Great Basin, Nevada-California with special reference to the Nevada Test Site: U.S. Geological Survey Professional Paper 712–C, 126 p.

Manuscript Accepted by the Society March 15, 1988

## Chapter 9

# *Region 6, Coastal Alluvial Basins*

**W. F. Hardt**

*U.S. Geological Survey (retired), 801 E. Glendale Avenue, Orange, California 92665*

## INTRODUCTION

The coastal basins of southern California and adjacent parts of Mexico (Fig. 1) cover about 31,000 km$^2$ (Fig. 3, Table 2, Heath, this volume), and are characterized by irregularly shaped mountains adjacent to and surrounding sediment-filled basins. In the U.S. part of this region (Jahns, 1954) the Transverse and Peninsular Ranges drain to the Pacific Ocean, and the Colorado Desert drains to the Salton Sea. In Mexico this region is composed primarily of Baja California, the Sonoran Desert and the Sinaloa Coast.

The flat alluvial basins of the California coastal region range in elevation from sea level at the Pacific Ocean to 610 m about 40 to 80 km inland. East of the mountains, in the desert, the basin floors range in elevation from 150 m in the Coachella Valley to minus 30 m in the Imperial Valley. Most of the mountain peaks are from 1,500 to 2,200 m in elevation, with the maximum of 3,500 m in the San Jacinto Mountains. Less than 80 km southeast is the Salton Sea, the lowest point at minus 75 m.

Average annual temperatures range from about 16°C in the coastal area to 21°C in the desert. The coastal area has a Mediterranean-type climate, and the desert has temperatures higher than 38°C in the summer. Precipitation occurs chiefly in the winter and ranges from 150 mm/yr in the desert, to between 380 to 510 mm/yr along the coast, and 760 to 1270 m/yr in the mountains.

The Colorado desert is a basin of middle and late Cenozoic sedimentation, with a thick section of fine- to coarse-grained nonmarine strata, and some volcanic rocks exposed in its marginal parts. This section rests on igneous and metamorphic rocks of pre-Cenozoic age. Most of the sedimentary rocks were deposited as aluvial fans and lacustrine beds, but they include a sequence of marine beds that accumulated in a shallow, northward-extending arm of the Gulf of California (Golfo de Cortes) during Pliocene time.

The predominant northwest-southeast fault pattern in southern California is dominated by the San Andreas, San Jacinto, Elsinore, Newport-Inglewood, San Gabriel, and Imperial faults. The land-surface expression of these faults is clearly visible in the consolidated rocks of the mountains, but is generally not observable in the alluvial basins.

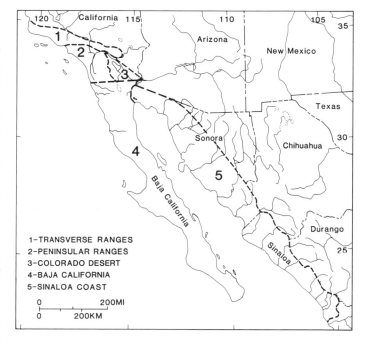

Figure 1. Map of the area of Region 6 showing major geographic features.

The mountain areas of southern California consist of metasediments, metavolcanics, and igneous rocks. These high-elevation areas are sources of recharge from precipitation and surface-water runoff to the lower elevation alluvial basins. The consolidated rocks in the mountains have aquifer transmissivities of less than 185 m$^2$/day, and hydraulic conductivities of less than 3 m$^2$/day. Larger aquifer characteristics are generally the result of fracture permeability or near-surface weathering of the rocks.

## HYDROGEOLOGIC FRAMEWORK

### *Transverse Ranges*

The Transverse Ranges are characterized by a series of mountains composed of igneous, metasedimentary, and metavolcanic rocks with permeable ground-water alluvial basins along the Pacific Ocean. The mountains are more important as major

Hardt, W. F., 1988, Region 6, Coastal alluvial basins, *in* Back, W., Rosenshein, J. S., and Seaber, P. R., eds., Hydrogeology: Boulder, Colorado, Geological Society of America, The Geology of North America, v. O-2.

sources of recharge than the alluvial basins because of the greater amounts of precipitation occurring in the high elevations. On the other hand, the mountains are less important for storage of large quantities of water. The base of the fresh water in the basin aquifers ranges from 60 to 1,220 m below the land surface and extends into the upper Pliocene and lower and upper Pleistocene sediments. All ground-water movement in the province is toward the Pacific Ocean. In the interiors of the basins, faults such as the San Jacinto, Newport-Inglewood, and others have extremely low permeabilities as compared to the adjacent permeable aquifers. Consequently, these faults act as barriers to ground-water movement. Prior to heavy ground-water pumping, the areas upgradient of the faults were natural discharge areas and commonly contained swamps, lakes, and flowing wells.

### Peninsular Ranges

In the southeastern part of the region, ground water is less available than in the Transverse Ranges because of the geologic conditions. The alluvial basins of Holocene age are small (sedimentary fill less than 100 m thick; width less than 5 km), irregular, fingerlike stream channels extending 8 to 15 km inland from the coast. Underlying the alluvial stream channels are sedimentary rocks of Pliocene and Pleistocene age. These rocks consist of interbedded sandstone, conglomerates, and sand and clay beds. The sediments yield less water than the alluvium, cover a much greater area, and have a maximum thickness of 600 m.

Most of the inland area consists of mountains underlain chiefly by igneous and metamorphic rocks of pre-Cenozoic age. The dominant rock type is granite, which yields water primarily from fracture systems less than 100 m below the land surface. Where isolated remnants of granite easily weather, the disintegrated granite yields water to wells less than 30 m deep. Regionally, the occurrence and movement of ground water in aquifers is largely controlled by topography.

The Peninsular Range extends into Baja California, which is a peninsula about 1,700 km long and about 50 to 100 km wide. Baja California is primarily a desert to semi-desert area. Its mountain ranges are large granitic batholiths flanked by Mesozoic metamorphic and sedimentary rocks and by Cenozoic volcanic and sedimentary rocks. The peninsula is a great fault block that became separated from mainland Mexico. Aquifers are almost nonexistent owing to lack of infiltration, the result of sparse rainfall on small catchment areas and low permeability of the rocks.

### Colorado Desert

The Coachella and Imperial Valleys are parts of the Colorado Desert. The valleys are on the north and south sides of the Salton Sea in a structural and topographic depression known as the Salton Trough. The trough, which is 210 km long, and as much as 100 km wide, is a landward extension of the depression of the Gulf of California. The trough is separated from the Gulf by the broad, fan-shaped subaerial delta of the Colorado River. The San Andreas fault system traverses the eastern part of the trough. Ground water moves southeast in the Coachella Valley, and northwest in the Imperial Valley and discharges into the Salton Sea.

The hydrogeology of the Coachella Valley is controlled by about 1,000 m of unconsolidated deposits of late Pleistocene and Holocene age. The flow systems of the alluvial basin are compartmentalized by several faults that influence ground-water flow. In the southern part of the valley, adjacent to the Salton Sea, fine-grained sediments at depth cause confined conditions that result in flowing wells near the sea.

In the Imperial Valley, the basement complex is as much as 6,100 m below the land surface. Overlying the basement rocks in the central part of the valley are mostly nonmarine sedimentary rocks that range in age from Eocene to Holocene. A marine unit, the Imperial Formation (Miocene or Pliocene age), is exposed in the western part of the valley. Overlying these deposits is a thick sequence of heterogeneous nonmarine silt, sand, and clay. Most of these fine-grained sediments were deposited by the Colorado River and contrast with the locally derived coarse sand and gravel along the margins of the valley. The sedimentary and volcanic rocks of Eocene and Miocene age, exposed in the bordering mountains, are moderately to strongly deformed and are semi-consolidated to consolidated.

The fine-grained material in the center of the valley is several thousand meters thick. This area coincides with the intersection of the East Pacific Rise and the North American continent. Heat flow through the basin sediments is greater than the world-wide average, and several areas have heat flow 4 to 10 times greater than the average. Consequently, fluid temperatures greater than 200°C occur at depths of 1,800 to 4,500 m.

The Sonoran desert is an extension of the Colorado desert into Mexico. Owing to lack of rainfall and minimal basin interflow, aquifers are of limited extent in this area. Farther to the south along the coast, the rivers that discharge into the Pacific Ocean and Gulf of California have deposited alluvium that locally contains aquifers.

### SUMMARY

The geology of this region is exceedingly complex. The complexities of the geology are compounded by the presence of many fault systems. Hydrologically, the area is less complex, with most of the ground water occurring in alluvial basins of Pliocene, Pleistocene, and Holocene age. The mountains consist of consolidated sediments and igneous, metasedimentary, and metavolcanic rocks. Comparatively little ground water occurs in these rocks. Topography is an important factor in controlling the presence and movement of ground water. Faults in the alluvial basins act as barriers to ground-water movement and can have major effects on pumping patterns in highly developed basins, whereas faults in the mountains have less hydrologic influence.

### REFERENCE CITED

Jahns, R. H., ed., 1954, Geology of Soutn California: California Division of Mines Bulletin 170, v. 1, ch. 1–10.

## Chapter 10

# *Region 7, Central Alluvial Basins*

**T. W. Anderson**
*U.S. Geological Survey, Federal Building, FB-44, 300 W. Congress Street, Tucson, Arizona 85701*
**G. E. Welder**
*U.S. Geological Survey, Western Bank Building, 505 Marquette, N.W., Room 720, Albuquerque, New Mexico 87102*
**Gustavo Lesser and A. Trujillo**
*Estudios Geotecnicos, S.A., Cuernavaca No. 89-A, Mexico, D. F. 06140*

## INTRODUCTION

The Central Alluvial Basins of North America encompass about 700,000 $km^2$ in southwestern United States and north-central Mexico (Fig. 3; Table 2, Heath, this volume). The land form of the area is characteristic of Basin and Range physiography with sharply rising mountains separated by broad alluvial plains (Fenneman, 1931). The structural basis of the present-day topography was created by regional extension and block faulting along steeply dipping normal faults during the middle to late Tertiary Period (Eberly and Stanley, 1978; Seager and others, 1984; Scarborough and Peirce, 1978; Shafiqullah and others, 1980). Mountains have a general north-to-northwest trend and divide the area into many basins. The term "basin" refers to the major sediment-filled graben that lies between the mountains, which may have either closed or through-flowing drainage. The extreme southern part of the region, the Central Mesa of Mexico, is separated from the remainder of the region by a major structural feature that trends east to west south of Torréon. The mountains that separate the region are an extension of the range of mountains that lies east of the cenetral alluvial basins within Mexico (Region 7, Fig. 1).

Altitude of the land surface in the basins ranges from less than 100 m above sea level in the southwestern part of Arizona to 2,000 m above sea level in the San Luis Basin of Colorado. Within Mexico, land surface in the basins ranges in altitude from 1,000 to 1,300 m above sea level. Land-surface slopes within the basins are moderate. Mountains range in altitude from about 500 m to more than 4,000 m above sea level; land-surface slopes are steep. The climate of the area is arid to semiarid. Precipitation is related to the altitude of the land surface, and the annual average precipitation ranges from 90 mm near Yuma, Arizona, to 1,200 mm in the high-altitude mountain areas near the headwaters of the Gila and Rio Grande Rivers. Within Mexico, average rainfall ranges from less than 400 mm to more than 800 mm annually.

Water resources in the region are derived from precipitation or from surface-water inflow. The Colorado River is the only river that enters from outside the region; it enters the region in northern Arizona. Major streams that drain the region flow to the Gulf of Mexico and the Gulf of California and include the Colorado, Rio Grande, Gila, and Pecos rivers in the United States and the Nazas and Conchos rivers in Mexico. The limited surface-water resources of the area are almost completely appropriated and, in recent years, ground water has been the major supplemental source used to meet increasing demands. Ground water generally has been overexploited, so that severe depletion of storage has occurred in many basins. The volume of water stored in the thick alluvial sediments remains many times the annual volume of water withdrawn, but the annual rate of natural recharge to or discharge from the basins is only a fraction of the annual volume withdrawn in moderately to heavily developed basins. Continued exploitation of ground-water resources at present rates, therefore, eventually will deplete the recoverable volume of stored water.

## HYDROGEOLOGIC FRAMEWORK

### *Rock Types and Ages*

Rocks in the Central Alluvial Basins region include igneous, metamorphic, and sedimentary types and can be divided into two broad age groups—pre-Cenozoic and Cenozoic. The rocks of the two age groups differ in structure, general rock type, and water-bearing character. In general, rocks of pre-Cenozoic age form the mountains that border the alluvial basins. The mountains are made up of one or more of the following rock types: granite of Precambrian age; schist, gneiss, phyllite, slate, and quartzite of Precambrian through early Tertiary age; consolidated sedimentary carbonate rock, sandstone, siltstone, and shale of Paleozoic and Mesozoic age; and several types of intrusive and extrusive

Anderson, T. W., Welder, G. E., Lesser, G., and Trujillo, A., 1988, Region 7, Central Alluvial Basins, *in* Back, W., Rosenshein, J. S., and Seaber, P. R., eds., Hydrogeology: Boulder, Colorado, Geological Society of America, The Geology of North America, v. O-2.

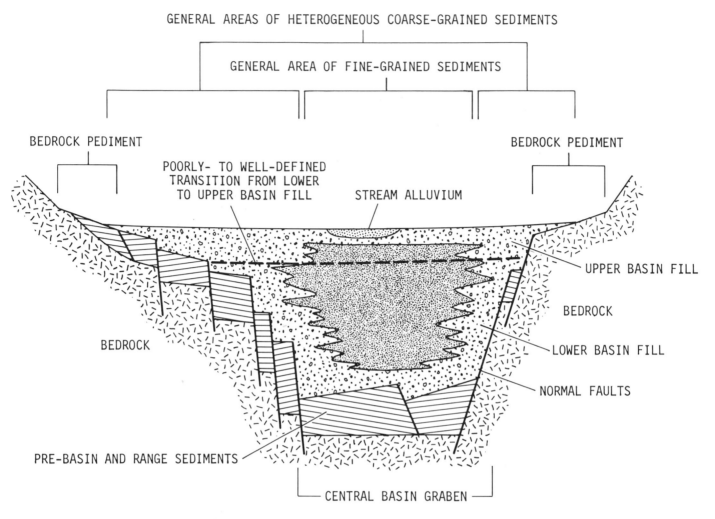

Figure 1. Typical basin structure and sequence of basin deposits.

igneous rocks of Late Cretaceous age (Wilson and others, 1969; Dane and Bachman, 1965). Although these rocks are not significant parts of the aquifer systems, they provide the source of detritus that fills the basins. Also, their lithologic and structural characters influence runoff and infiltration, and hence the water supply to the basins.

Rocks of Cenozoic age fill the basins and consist of unconsolidated to consolidated clastic sedimentary deposits and interbedded volcanic rocks. The sediments were deposited mainly during and following basin-and-range faulting.

The oldest basin deposits are of early to middle Tertiary age and may be several thousand meters thick in places. The areal extent of these deposits is poorly known. Subsequent deposition followed extensive volcanism during middle Tertiary time. The sediments were deposited in north- to northwest-trending closed basins created by the extensional tectonics. Alluvial fans at the base of the mountains coalesced to form bajadas that extended basinward. Playa-type deposition occurred near the center at the lowest altitude of the basin. This type of depositional environment produced two general facies of basin deposits—fine grained and well sorted near the basin center and coarse grained and porly sorted near the mountain fronts (Fig. 1). The fine-grained facies is generally basinward of the major basin-bounding faults. Evaporites in these closed basins are typically associated with the fine-grained facies; the most common evaporite is gypsum, which is disseminated within the deposits or in thin layers. Massive anhydrite and halite deposits are present in some of the deepest basins and may be thousands of meters thick (Peirce, 1976). The coarse-grained facies is highly heterogeneous, and the sediments range from silt and clay to cobbles and boulders.

Areal extent of basin deposition was altered as a result of a change in regional extensional stresses during middle to late Miocene time. Northerly oriented, deep-seated, normal faulting progressed from west to east during the period and was accompanied by basaltic volcanism (Seager and others, 1984). This period of faulting created the present-day distribution of basins and ranges.

Deposition in the basins continued into the Quaternary Period until through-flowing drainage was established in some basins. Formation of through-flowing or integrated drainages began in the western part of the region and progressed to the east as tectonic activity waned. The earliest integrated-drainage deposits may be about 10 m.y. old in southwestern Arizona (Eberly and Stanley, 1978).

Several basins in southern California, western Texas, southern New Mexico, Arizona, and northern and central Mexico have closed surface-water drainages. Most of these basins also have closed ground-water systems. Basin-center deposits in these basins are largely lacustrine and consist mainly of silt, clay, and evaporites.

Major aquifers of the area are composed of basin sediments of Cenozoic age. The aquifers are contained within the grabens between the mountain blocks, and the mountains act as hydrologic boundaries. The sediments in a basin generally are hydraulically interconnected and form a single aquifer. Locally, separation into one or several confined units and an upper unconfined unit occurs as a result of intervening fine-grained units and, in some places, interbedded volcanic rocks. Basin sediments are estimated to be 9,100 m thick in the San Luis Valley of Colorado (West and Broadhurst, 1975, p. 13). Thicknesses of 2,000 to 3,000 m are typical, and some basins may contain less than 300 m of sediments.

### Geologic Structure

The structural development of basins has been similar throughout the area. Typically, mountain blocks are upthrown horsts and basins are downthrown grabens. Most basins consist of a deep central graben flanked by a zone of en echelon normal faults that extends to the mountain blocks. According to Seager and others (1984, p. 87), the horst-and-graben development moved progressively eastward from southeastern California and culminated in the formation of the Rio Grande Rift and other fault-block structures in western Texas, New Mexico, and northern Mexico. The graben structures generally have many smaller high and low substructures within the main graben. Differential fault movement on the sides of the grabens resulted in many basins being structurally asymmetric (Fig. 1).

## GROUND-WATER HYDROLOGY

Vast quantities of ground water are in storage within the interstices of the sediments that fill the alluvial basins. Storage within individual basins ranges from $2.5 \times 10^{12}$ m$^3$ in the San Luis Basin (Kelly, 1974, Sheet 1) to about $1,200 \times 10^6$ m$^3$ in the upper 400 m of sediments in several basins in western Arizona (Freethey and Anderson, 1986, Sheet 1). Ground water has been withdrawn in recent years to meet the increasing demand for agricultural, municipal, industrial, commercial, mining, rural, domestic, and stock supplies. Near Buckeye, west of Phoenix in central Arizona, and in Wellton, in southwestern Arizona,

ground water is pumped to prevent waterlogging of agricultural land. Withdrawals in the early 1980s totaled more than $11,000 \times 10^6$ m$^3$/yr for the region; of this amount about $5,000 \times 10^6$ m$^3$/yr were withdrawn from basins in Mexico and about $6,000 \times 10^6$ m$^3$/yr were withdrawn from basins in the United States.

### Hydrologic Setting

The hydrologic setting of a typical basin generally is one of two types: closed or through flowing (Fig. 2). In closed basins, such as Willcox Basin in Arizona, Tularosa Basin in New Mexico, and those that make up the Central Mesa area of Mexico, surface-water or ground-water flow is not known to leave the basin. Before development, the discharge mechanism was evaporation from the playa area and transpiration by vegetation, which occurred near the center of the closed basins. Under present conditions of development, most discharge is through wells.

Where through flowing, several basins may be hydraulically interconnected, and the outflow from one basin may be part of the inflow to another basin in the downgradient direction. Many of the alluvial basins are the through-flowing type and are included in a regional system of basins that generally are aligned in the same flow direction as the surface-water drainage system.

Ground-water flow in each basin is controlled by the volume and areal distribution of recharge and discharge and by hydraulic properties of the materials in the aquifer. Recharge is related to the amount of precipitation in the surrounding mountains, which, in turn, increases with altitude. Recharge occurs mainly in the area of the hardrock–alluvium contact at the mountain front and along the main surface drainage near the central axis of the basin. Underflow into and out of basins occurs in through-flowing basins but generally is small. Underflow, however, may be a large part of total recharge in the more arid basins of western Arizona. Under predevelopment conditions, discharge from basins with through-flowing drainage systems occurs as underflow, evapotranspiration, and streamflow.

### Climatic Influences

The region is the driest area in the United States and Mexico, and large areas are classified as arid and semiarid. Annual precipitation generally is related to altitude of the land surface. More precipitation occurs in high-altitude areas, such as the mountains of southeastern Arizona and central New Mexico. In parts of western Arizona and southeastern California, however, precipitation in the mountains does not greatly exceed the precipitation within the basins.

Recharge to the ground-water systems is dependent on the amount of precipitation that occurs in the surrounding mountains. In the high-altitude areas, such as the upper San Pedro Basin in southern Arizona, recharge along the mountain fronts is the most significant source of inflow to the basin. In this basin, water-level contours are nearly parallel to the mountain fronts. In basins with much less recharge, the water-level contours are

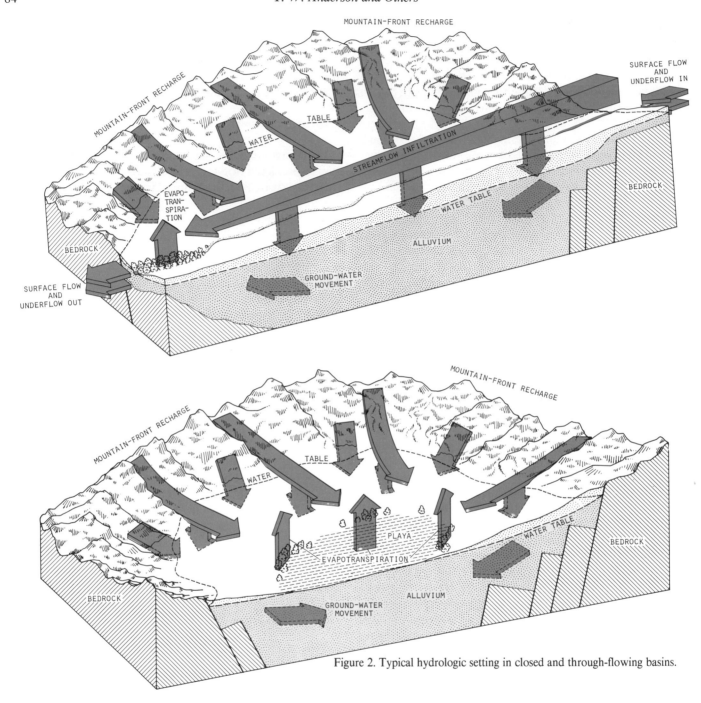

Figure 2. Typical hydrologic setting in closed and through-flowing basins.

nearly normal to the mountain fronts or have only a slight curvature. The magnitude of mountain-front recharge to the alluvial basins can be evaluated in qualitative terms in accordance with the shape of the water-level contours.

Because of the area's dependency on ground water, the impacts of droughts on water availability will generally be small. Extended droughts will result in decreased recharge to the ground-water systems, but this effect is not readily apparent, in part because of the immensity of the ground-water reservoir.

## Geologic Influences

Geologic influence on the hydrology of the alluvial basins is important in determining the occurrence and movement of ground water. The lateral and vertical boundaries of the aquifer system in a basin are controlled by lithologic and permeability changes, which are related to modes of alluvial deposition and to geologic structure.

Pre-Cenozoic rocks and Cenozoic volcanic rocks generally

are of low permeability and hinder movement of ground water. Some ground-water movement may occur in carbonate rocks that have not been significantly structurally disturbed, such as those near the Colorado Plateau in central Arizona, along the Rio Grande Rift, and in the Central Mesa of Mexico.

The alluvial material that forms the main aquifer in the basins was eroded from the adjacent mountains by steep-gradient streams and deposited as alluvial fans. Generally, the fan deposits become progressively finer toward the centers of the basins; however, some coarser-grained layers may extend into the basin centers. The basin deposits generally consist of pre-Basin and Range deposits, upper and lower basin fill, and stream alluvium.

Hydraulic conductivity of the alluvial material tends to decrease with increasing degrees of deformation and consolidation and with increasing proportions of fine-grained material. The pre-Basin and Range deposits are interbedded with volcanic rocks, are structurally deformed, generally are moderately to highly consolidated, and are poor aquifers. The textural character and areal distribution of lower basin-fill sediments are not well known in Arizona; however, on the basis of samples from wells that penetrate the unit, the sediments tend to be coarse grained in southwestern Arizona and fine grained in central Arizona. Throughout much of Arizona, the lower basin fill has been highly faulted and tilted but is less disturbed in southwestern Arizona. Lower basin fill near the Rio Grande Rift has been disturbed by normal faulting but has not been highly tilted and is a significant aquifer. Near the centers of basins along the Rio Grande Rift, basin fill is fine grained and consequently does not yield large amounts of water.

Upper basin fill that accumulated in the basins forms the major aquifers of the area. Fine-grained sediments generally are found in the basin centers. Overall, the upper basin fill is coarser grained, has less extensive fine-grained facies, and is a more productive aquifer than lower basin fill. Some upper basin fill near the Rio Grande Rift lacks a fine-grained facies. Fine-grained facies may exist in deltaic and lacustrine deposits along the Rio Grande River. Alluvial deposits along the present streams typically consist of coarse-grained sand and gravel, and they store and transmit large volumes of ground water where saturated.

The flow of ground water within the basins is influenced by geologic structure and hydraulic properties of the sediments. Structural features within the basins, such as faults and high bedrock areas, result in low transmissivity and locally steep hydraulic gradients. Shallow bedrock near the outflow area of structural basins generally results in areas of steep hydraulic gradient. Gradients are steep also at the inflow area of some basins and are associated with the movement of ground water in the direction of rapidly increasing cross-sectional area. The result is that the up-gradient basin is effectively hydraulically disconnected from the downgradient basin in the zone of steep hydraulic gradient. Steep hydraulic gradients are also associated with flow through older, less permeable sediments commonly found near the perimeters of basins. Flow is diverted around, over, or beneath the fine-grained facies in the basin centers. Water-table conditions prevail in the uppermost aquifers in most basins. Confined or locally confined conditions are found in deeper aquifers near the centers of some basins. The fine-grained facies also result in the occurrence of perched ground water in some areas.

### *Geochemical Influences*

Chemical quality of the ground water generally is controlled by composition of the basin sediments, evapotranspiration, and composition of recharge water. The quality of ground water generally is good near the mountain-front recharge areas and deteriorates downgradient as the water interacts with the basin sediments. In many basins, the shallow ground water contains less than than 600 mg/L of dissolved solids. However, areas of discharge by evapotranspiration generally contain water of poor quality because of the concentration of dissolved constituents locally in the ground water; in these areas, dissolved solids may exceed 3,000 mg/L.

The mineralogy of the bedrock on the basin perimeters affects the chemical character of ground water only to the extent that the basin sediments were derived from the basin-bounding bedrock and make up the matrix in which the ground water occurs. The surface water does not remain in contact with the surface bedrock sufficiently long for dissolution to occur. Pre-Cenozoic evaporites are a source of large concentrations of dissolved solids, sodium, and sulfate in areas of New Mexico. Areas of high fluoride concentration are associated with Tertiary volcanic rocks.

The water quality within the basins tends to deteriorite with the increased presence of fine-grained sediments and evaporites. Good-quality water is found at depth in many basins in western Arizona that have relatively thin intervals—300 to 500 m—of fine-grained sediments and evaporites. Temperature or chemical differences in ground water probably have little effect on the flow of water in the basins. Local temperature anomalies do exist near major normal faults. Extensive quantities of saline water—more than 1,000 mg/L of dissolved solids—exist in basins in New Mexico (Kelly, 1974). Encroachment of saline water into zones of fresh water could result from pumping such zones.

## SIGNIFICANT HYDROLOGIC PROCESSES, FEATURES, AND PROBLEMS

The Central Alluvial Basins, although diverse in physical character, are geologically and hydrologically similar. A typical basin contains a general pattern of four units that form the basin-fill material: (1) pre-Basin and Range rocks, (2) lower basin fill, (3) upper basin fill, and (4) stream alluvium. Differences in extent and character of the units are known to exist, and general areal trends in thickness and character exist in part of the area (Freethey and others, 1986). Similarities in basin hydrology are related to the overall water budget of a basin and to the mechanisms and magnitude of the individual inflow and outflow components. The magnitude of inflow components appears to be related to the altitude of the mountains around the basin and to a lesser extent to the altitude of the valley floor.

### Interrelations of Surface Water and Ground Water

Along the Colorado and Rio Grande Rivers, the ground-water system is dominated by the presence of the surface-water system. The rivers are underlain by sediments that readily store and transmit large quantities of water. The aquifer is hydraulically connected to the river, and ground-water levels in the flood plain material reflect the stage in the river. Depending on local gradients and volumes of aquifer material, the rivers may contain both gaining and losing reaches within a single basin. Withdrawal of ground water in the flood plains of these rivers represents an indirect withdrawal of water from the river because of the high degree of interaction between the ground-water and surface-water systems.

### Problems

Several significant hydrologic problems in the Central Alluvial Basins are related to man's development and utilization of the ground-water resources. In many basins, withdrawal commonly exceeds inflow, and the result is a depletion of ground water in storage. Water-level declines of nearly 150 m near Stanfield, Arizona (Konieczki and English, 1979), more than 120 m east of Phoenix, Arizona (Laney and others, 1978b), and 100 m in Reeves and Pecos Counties, Texas (West and Broadhurst,

1975, p. D19) have occurred since ground-water development began. In heavily pumped areas, the maximum annual declines were about 3 m per year in the mid-1960s; in many developed basins in central Arizona, declines of 0.5 to 2 m/yr were common. From about 1970 through 1980, the rate of water-level decline in many of these basins decreased significantly because of reductions in pumping and (or) increases in recharge.

Land subsidence of several centimeters to more than 4 m has occurred in some basins in Arizona (Laney and others, 1978a). The subsidence is a result of large-scale water-level declines and the compaction of fine-grained sediments. Surface cracks and earth fissures can develop in response to horizontal stresses induced by subsidence. Vertical offsets also occur across the fissures. Subsidence and earth fissures represent hazards to engineered structures such as highways, pipelines, canals, and buildings.

Pollution of ground water has been a potential problem since development began. The hazards have been much more in the forefront in recent years because of the increased use of ground water, greater awareness of and concern for the potential health effects of the pollutants, and more sophisticated or sensitive sampling and analysis techniques. The most widespread potential source of pollution is irrigation. Evapotranspiration concentrates dissolved salts in the fraction of applied water that infiltrates past the plant-root zone. At the same time, the water also dissolves and carries with it part of the fertilizer and pesticide applied to the field.

## REFERENCES CITED

Dane, C. H., and Bachman, G. O., 1965, Geologic map of New Mexico: U.S. Geological Survey map, scale 1:500,000.

Eberly, L. D., and Stanley, T. B., 1978, Cenozoic stratigraphy and geologic history of southwestern Arizona: Geological Society of America Bulletin, v. 89, no. 6, p. 921–940.

Fenneman, N. M., 1931, Physiography of western United States: New York, McGraw-Hill, 534 p.

Freethey, G. W., and Anderson, T. W., 1986, Predevelopment hydrologic conditions in the alluvial basins of Arizona and adjacent parts of California and New Mexico: U.S. Geological Survey Hydrologic Investigations Atlas HA–664, 3 sheets, scale 1:500,000.

Freethey, G. W., Pool, D. R., Anderson, T. W., and Tucci, P., 1986, Description and generalized distribution of aquifer materials in the alluvial basins of Arizona and adjacent parts of California and New Mexico: U.S. Geological Survey Hydrologic Investigations Atlas HA–663, 4 sheets, scale 1:500,000.

Kelly, T. E., 1974, Reconnaissance investigation of ground water in the Rio Grande drainage basin, with special emphasis on saline ground-water resources: U.S. Geological Survey Hydrologic Investigations Atlas HA–510, 4 sheets.

Konieczki, A. D., and English, C. S., 1979, Maps showing ground-water conditions in the lower Santa Cruz area, Pinal, Pima, and Maricopa Counties, Arizona—1977: U.S. Geological Survey Water-Resources Investigations 79–56, scale 1:125,000.

Laney, R. L., Raymond, R. H., and Winnika, C. C., 1978a, Maps showing water-level declines, land subsidence, and earth fissures in south-central

Arizona: U.S. Geological Survey Water-Resources Investigations, 78–83, 2 sheets.

Laney, R. L., Ross, P. P., and Littin, G. R., 1978b, Maps showing the ground-water conditions in the eastern part of the Salt River Valley area, Maricopa and Pinal counties, Arizona—1976: U.S. Geological Survey Water-Resources Investigations 78–61, maps.

Peirce, H. W., 1976, Tectonic significance of Basin and Range thick evaporite deposits: Arizona Geological Society Digest, v. 10, p. 325–339.

Scarborough, R. B., and Peirce, H. W., 1978, Late Cenozoic basins of Arizona, in Callender, J. F., Wilt, J. C., and Clemmons, R. E., eds., Land of Cochise: New Mexico Geological Society Guidebook, 29th Field Conference, p. 253–259.

Seager, W. R., Shafigullah, M., Hawley, J. W., and Marvin, R. F., 1984, New K-Ar dates from basalts and the evolution of the southern Rio Grande rift: Geological Society of America Bulletin, v. 95, p. 87–99.

Shafiqullah, M., Damon, P. E., Lynch, D. J., Reynolds, S. J., Rehrig, W. A., and Raymond, R. H., 1980, K-Ar geochronology and geologic history of southwestern Arizona and adjacent areas: Arizona Geological Society Digest, v. 12, p. 201–260.

West, S. W., and Broadhurst, W. L., 1975, Summary appraisals of the nation's ground-water resources; Rio Grande region: U.S. Geological Survey Professional Paper 813–D, 39 p.

Wilson, E. D., Moore, R. T., and Cooper, J. R., 1969, Geologic map of Arizona: Arizona Bureau of Mines map, scale 1:500,000

The Geology of North America
Vol. O-2, Hydrogeology
The Geological Society of America, 1988

## Chapter 11

# *Region 8, Sierra Madre Occidental*

**J. Joel Carrillo R.**
*C.F.E. Geohidrologia, Oklahoma 85, 8th Floor, Colonia Napoles, Mexico, D.F. 03810*

The Sierra Madre Occidental covers approximately 310,000 $km^2$. This mountainous chain is more than 1,500 km long and is essentially parallel to the west coast of Mexico (Fig. 3, Table 2, Heath, this volume). It extends from the international boundary near the village of Agua Prieta in a southeasterly direction to the volcanic belt region at 20° north latitude; its average width is about 220 km. The Sierra Madre Occidental is characterized by a high plateau of extrusive rocks with wide structural depressions striking north-northwest to south-southeast; among the gently sloping mountains are mesas and minor plateaus. The most prominent peaks are more than 3,700 m above sea level, and the low areas are less than 500 m. This region is dissected by canyons that are transverse or parallel to the regional strike and are formed by strongly eroding rivers discharging either to the Pacific or into the region of the Central Alluvial Basins. In some areas an abrupt change in slope is caused by relatively more resistant rocks.

Although climatological data are lacking, a qualitative understanding of the climate in the Sierra Madre Ocidental can be made. The climate ranges from humid temperate, with rainfall occurring all year in the higher parts of the Sierra, to a dry steppe climate to the east in the lower parts of the Sierra. The first type of climate produces forests of gimnosperms, especially conifers, while the second produces only grasses and shrubs. On the western side of the Sierra the climate is hot and humid with rain in the summer and principally forms prairies or meadowlands.

The precipitation of the Sierra Madre Occidental is principally of the orogenetic type. In the southern part and near the Pacific the rain is more than 1,600 mm per year; in the central part the average rainfall is slightly less. In the eastern part the precipitation decreases to about 600 mm. Maximum precipitation is during June to September and the minimum during March to May. The mean annual temperature ranges from 12°C in the high area to 22°C in the low areas of the Pacific watershed; near the region of the Central Alluvial Basins, the mean temperature is 16°C. Conclusive data about the actual evapotranspiration does not exist, but the potential evaporation exceeds 2,000 mm per year.

The rivers that discharge in the Pacific are virtually perennial and drain basins of great length; for example, the Rio Yaqui is 680 km long, and the Rio Culiacan is 340 km. The major rivers that flow toward the Central Alluvial Basins are intermittent; the Rio Conchos has a length of 480 km, and the Rio Nazas has a length of about 400 km.

The large amount of water that drains the Pacific watershed results from high rainfall on rocks with a wide range of permeability. The stream flow is augmented by base flow from the rocks of low permeability that crop out in these large drainage basins. The major rivers are the Rio Yaqui, with an average annual stream flow of $27 \times 10^8$ m³/year; the Rio Culiacan, with $35 \times 10^8$ m³/year; the Rio Nazas, with $11.8 \times 10^8$ m³/year; and the Conchos, with $51 \times 10^8$ m³/year.

All the important rivers have dams to control the flow and to achieve maximum water conservation and usage. The narrow canyons and the relatively impermeable rocks continue to provide suitable dam sites. Filling of the reservoirs caused few problems because the land was not primarily used for agriculture or settlements. The relatively high precipitation and the numerous reservoirs provide a direct contribution to the ground-water system. This recharge affects springs that are tens of kilometers from the reservoirs.

The topography of the region is formed principally by dissection of igneous rocks. Among the outcrops of major importance are the more than 2,000 m of rhyolite, ignimbrite, and tuff. Basalts and andesites and other intermediate rocks cover small areas. These units are interstratified with continental sediments and volcaniclastic materials ranging in age from Paleocene to Recent. These units overlie limestone, graywacke, and shale of the Upper Cretaceous that have been folded and intruded by Eocene batholiths (granite, grandodiorite, diorite) of the Laramide Revolution. In places, the intrusions are quartz-monzonite dikes of the Late Cretaceous. The batholiths also intruded Upper Triassic (?) lavas of basic and intermediate composition. Stratigraphically below these units are undifferentiated groups of crystalline metamorphic rocks such as marble, quartzite, shist, metavolcanic rocks with a composition of rhyolite or andesite, and porphyritic dacite; many of these are associated with the intrusive igneous rocks of the Eocene(?). These units are above the gneiss and shist that have been cut by pegmatite, granodiorite, and diorite of Precambrian (?) age.

In general, the outcrops of rhyolitic rocks, ignimbrites, and

Carrillo R., J. J., 1988, Region 8, Sierra Madre Occidental, *in* Back, W., Rosenshein, J. S., and Seaber, P. R., eds., Hydrogeology: Boulder, Colorado, Geological Society of America, The Geology of North America, v. O-2.

tuffs of the Miocene form the principal cover of the region. The younger units are in small patches in topographically high areas; the older units are at the base of the sierras, forming windows of small areal extent. The stratigraphy and geologic history are shown on Table 1 (adapted from Clark, 1981). The geology of this region is complex, and it is difficult to make a practical separation between the permeable lithologic units that compose the aquifers and the impermeable units that form aquitards. Because of the sparse habitation, there are few wells and hydrologic records. This lack of important information makes it difficult to assign to the different lithologic units values of hydraulic conductivity or storage coefficients from field tests or specific capacity data. Nevertheless, based on the experience obtained from areas with analogous lithology, a regional differentiation for the aquifers and nonaquifers can be made.

Low-permeability rocks that occur throughout the geologic section compose the aquitards. The wide extent of these rocks, with great thicknesses, and their topographic position relative to the regional ground-water level make them regional aquitards, which permit only small local supplies for small villages. The fractures and faults produced by the great forces of the Laramide orogeny are of special relevance to the regional setting, as they permit the occurrence and flow of ground water.

The numerous hot springs are associated with these low-permeability rocks. The bicarbonate nature of many of the thermal springs, their relatively low temperatures, and total solid content ($>2,000$ mg/l) indicate that the abundant rainfall has infiltrated and flowed through calcareous rocks and discharged along fractures. Only a small amount of water is of volcanic origin (White, 1957).

Rocks of higher permeability are common in the youngest units of the stratigraphic column. The most important materials of regional extent and thickness are the rhyolites, ignimbrites, and tuffs (Tomig) that compose a regional aquifer system.

In some places where the Tomig Unit has a great thickness, it is impermeable to the middle part, which has not undergone strong alterations that would have produced secondary permeability. This type of rock forms confining layers and local impermeable layers that form perched aquifers. This vertical variation of permeability forms a multisystem aquifer so that when erosion forms a topographic depression or gorge, there are seeps and some important springs. Also, there exist low-temperature thermal springs, with sulfate-type water, that are salty (total solids greater than 1,000 mg/l). The sulfate-type spring is related to the basaltic rocks of more recent age and with those thermal springs related to Late Miocene to Pleistocene magmatic centers.

The location of thermal springs associated with the rocks from the Precambrian(?) up through the Pleistocene show an alignment that extends from Aguascalientes to the northwest following a belt parallel to the Sierra Madre Occidental. This is the same direction as the fractures at the surface and in the crustal rocks of the region.

The potentiometric surface of the Sierra Madre Occidental is controlled primarily by the permeability distribution of the rocks. The hydraulic gradients are comparatively lower in the more permeable rocks, such as the youngest basaltic rocks, and in aquifers of alluvium and clastic sediments. The gradients are characteristically higher in the older rhyolites, ignimbrites, tuffs, and andesites, which are comparatively less permeable than the younger rocks in which the permeability is restricted to fractures. The higher gradients are associated with the higher topographic position of the rhyolites, ignimbrites, and tuffs where the principal recharge is by rainfall. The water infiltrates, flows through the joints and fractures, and either (1) is integrated with the regional flow to a lower gradient to recharge the aquifers of the valleys or to discharge as a base flow of some rivers, or (2) flows to greater depths, is heated, and acquires the chemical characteristics of thermal water and discharges in the form of springs. The presence of reservoirs whose infiltrations affect the flow of the ground water, the perched aquifers, and water of other origins are complicating factors on the position of the potentiometric surface of the Sierra Madre Occidental.

Materials resulting from erosion of the rocks composing the aquifer system are deposited at lower elevations in the small intermontane valleys of the Sierra Madre Occidental. Alluvium in the valleys constitutes the only aquifers that, because of their accessibility, have been exploited. These water-table aquifers are recharged by local rainfall that is added to the recharge from the subterranean aquifer system of the igneous rocks, which contain cold as well as thermal water. It is common for wells in these alluvial valleys to have capacities of 100 l/s.

## REFERENCES

Clark, K. F., 1981, Seccion Geologica-Estructural a traves de la parte sur de la Sierra Madre Occidental, entre Fresnillo y la costa de Nayarit: Asociacion de Ingenieros Mineros Mexicanos Geologos y Metalurgistas, Memorias Tecnicas, XIVV Congreso, p. 74–101.

White, D. E., 1957, Magmatic, connate, and metamorphic waters: Geological Society of America Bulletin, v. 68, p. 1659–1682.

Secretaria de Rucursos Hydraulicos (SRH), 1975, Atlas del Aqua de la Republica Mexicana: Cincuentenario de la Comision de Irrigacion Precursora de la S.R.H.

Printed in U.S.A.

# Chapter 12

# *Region 9, Sierra Madre Oriental*

**Juan M. Lesser**
*Cuernavaca 89-A, Colonia Condesa, Mexico 11, D.F.*
**Gustavo Lesser**
*Estudios Geotecnicos, S. A., Cuernavaca 89-A, Mexico, D.F. 06140*

## INTRODUCTION

The infiltration, circulation, and occurrence of ground water in the Sierra Madre Oriental is directly influenced by the lithology and structure of the rocks. This region (Fig. 3, Table 2; Heath, this volume) is composed primarily of a series of folded calcareous rocks of Mesozoic age that have a general northwest to southeast orientation. The region is characterized by arid and semiarid zones, except in its eastern portion; there it joins with the Gulf Coastal Plain, and the climate is transitional to humid tropical. The region is divided into three subregions (Fig. 1), the north, central, and south, respectively, which are designated: Sierra del Burro, Cuenca de Ojinaga–Monclova–La Paila Basin, and Sierra Torreon-Monterrey-Tamazunchale.

## SIERRA DEL BURRO

### *General setting*

This subregion is south of the Big Bend of the Rio Bravo (Rio Grande) and has elevations about 600 m above sea level. It is an extensive uplift of limestones of Cretaceous age that form an anticline striking northwest; flank dips are very gentle, normally 3° to 5°. Although the rocks of this region are chiefly carbonates, in places they are shaley. The rocks were deposited in a shallow marine environment known as the Laguna de Maverick (Maverick Basin) that was surrounded by the Stuart City Reef of Cretaceous age (Fig. 2). The lithology of the rocks and the gentle folding to which they have been subjected have given them different hydrogeologic characteristics than those of the rocks underlying the surrounding areas.

The rocks of the Lower and Middle Cretaceous are essentially highly soluble, medium-bedded limestones with a small amount of shale. These rocks grade laterally into the Stuart City Reef, which has a high primary porosity and many fractures that increase the permeability. The rocks of the Upper Cretaceous contain more shale and have lower permeability.

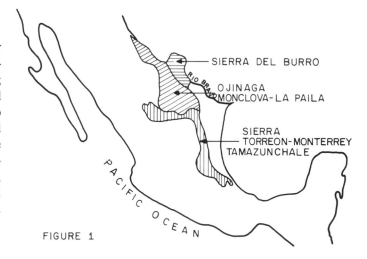

FIGURE 1

Figure 1. Map showing three subdivisions of the region.

The eastern flank of the Sierra del Burro is underlain by alluvial fill and conglomerates of Tertiary and Quaternary age—products of the weathering and erosion of the topographically higher regions. These deposits are only about 50 m thick. However, they are highly permeable and contain important volumes of ground water.

### *Geochemistry*

Two zones can be identified according to the chemical composition of the water. One occurs in the western two-thirds of the Sierra del Burro and is characterized by the presence of potable water. The second zone occurs in the eastern part of the Sierra and contains nonpotable water. The boundary between the two types of water is designated the "bad-water line" (Fig. 3). This "bad-water line" continues in an easterly direction into the

Lesser, J. M., and Lesser, G., 1988, Region 9, Sierra Madre Oriental, *in* Back, W., Rosenshein, J. S., and Seaber, P. R., eds., Hydrogeology: Boulder, Colorado, Geological Society of America, The Geology of North America, v. O-2.

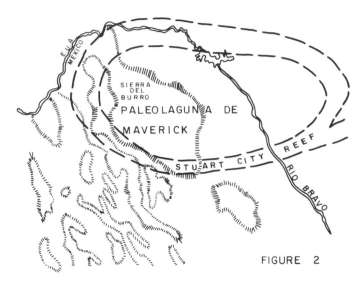

FIGURE 2

Figure 2. Map showing location of the Maverick Basin.

Five different hydrologic zones can be differentiated as follows:

*Zone 1.* At the foot of the Sierra, the water is a calcium carbonate type of good quality, and contains less than 1,000 mg/L dissolved solids. The aquifer is composed of Cretaceous-age limestone that was deposited in the Laguna de Maverick (Fig. 3).

*Zone 2.* To the southeast, where the water is of bad quality, it contains more than 1,000 mg/L dissolved solids and is chiefly a calcium sulfate type. This zone occurs within the Maverick paleolagoon where the rocks characteristically contain evaporites, disseminated pyrite, and carbonaceous material.

*Zone 3.* In the extreme southeast, where little ground water occurs at depth or where the water is of bad quality.

*Zone 4.* The zone in which ground water exists only in isolated locations.

*Zone 5.* Areas in which the water has more than 1,000 mg/L dissolved solids and a high sulfate content related to the amount of gypsum in the valley-fill deposits. Isolated outcrops of limestone also occur and, where present, the limestone aquifers have water of good quality.

*Hydrology*

The various rock units underlying the subregion contain confined aquifers. Some of the tight confining beds are breached by tractive systems that improve the degree of hydraulic interconnection between aquifers. The partial hydraulic connection among the near-horizontal aquifers causes significant local and areal variations in water levels, well depths, ground-water flow, and well yields. These aquifers are recharged by rainwater that infiltrates in the higher parts of the Sierra del Burro and flows generally to the north and east.

The southeastern prolongation of Sierra del Burro, known as the Peyotes Anticline, is an anticline with gently dipping flanks. Here the Austin Formation, a calcareous rock with a little shale, crops out. A shallow aquifer of small potential occurs in the fractured part of the formation. The plain east of the Sierra del Burro is underlain by permeable granular alluvium 30 to 50 m thick. The alluvium is recharged by the limestone aquifer in the Austin Formation. Springs with average flows of 600 L/s issue from the highly permeable alluvium.

## CUENCA (BASIN) OJINAGA–LA PAILA

*General Setting*

This subregion is underlain by calcareous rocks of Cretaceous age that have been folded into a series of anticlines that form sierras oriented N20°W. The sierras emerge as isolated forms within the extensive plains formed by alluvial material. These sediments of Cretaceous age were deposited in the open sea on a platform environment and generally are dense micrites with little or no primary permeability. These sediments do not form aquifers except in folded zones that have been fractured where

aquifer composed of the Edwards and associated limestones of south Texas (Jorgensen and others, this volume).

The nonpotable water is characterized by a high concentration of dissolved solids and is a sodium bicarbonate, calcium sulfate, and sodium chloride type that contains hydrogen sulfide, $H_2S$. The occurrence of this water is thought to be restricted to the paleolagoon of the Maverick Basin, and its origin is a consequence of the dissolution of evaporitic and calcareous material from the McKnight Formation, and dissolution of pyrite from the Eagle Ford and Buda Formations. The water temperature also is higher than normal ground water, and in this region the aquifer has low permeability. This delineation between poor-quality and good-quality water has been mapped to the vicinity of Monterrey, but the area of most accurate delineation is in the Sierra del Burro and south Texas (Back and others, 1977).

The aquifer with potable water is in the Aurora and Edwards Formations and their equivalents. At greater depths the water is highly saline and is a calcium-bicarbonate type.

The position of the "bad-water line" is influenced significantly by ground-water flow, which is controlled by the distribution of fractures associated with the uplift of the Sierra del Burro. The position of the Stuart City Reef marks the limit of the bad-water line. This line is continuous to the south, where its existence is associated with the presence of gypsum. Marked hydrogeologic differences occur within the zones of water of good and bad quality. The porosity and permeability of the aquifer are secondary; for example, in the fresh-water zone the fabric of the carbonate rock is altered, and the water is strongly oxygenated and has low dissolved solids. In the zone of nonpotable water, the aquifer has primary porosity and permeability; the fabric of the carbonate rock is only slightly altered, the water has high dissolved solids and is in a reducing environment.

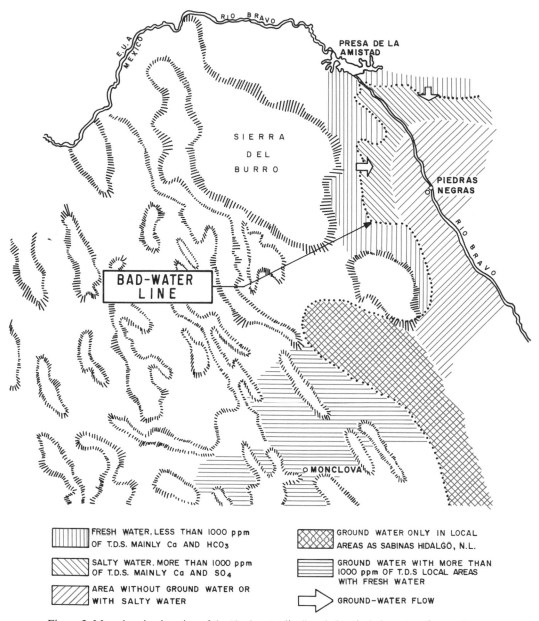

Figure 3. Map showing location of the "bad-water line" and chemical character of ground water.

dissolution along fracturing resulted in development of good permeability. Conditions for this process to take place are not favorable.

The subregion is underlain by two types of aquifers: (1) those that occur in the granular sedimentary fill, and (2) those that occur in the limestones. The aquifers in the granular sediments of the valleys and basins consist of low-permeability sand and shale. These aquifers contain saline water derived from dissolution of gypsum of Jurassic age that crops out in the high parts of the sierras. The aquifers occur in the Lower and Middle Cretaceous limestone.

The aquifers of the second type occur in the lower and middle parts of the limestones of Cretaceous age. Where these limestones outcrop in the sierras, the aquifers are recharged directly from rainfall. This recharge moves downdip toward the valleys. Wells drilled on the flanks of the sierras penetrate the aquifers between depths of 1,000 and 2,000 m and generally flow during the rainy season because of the high potentiometric surface. However, during the dry season, water levels decline as much as 100 m due to pumping, and wells cease to flow.

In some places near the valley floors, fractures that penetrate the aquifers give rise to large springs; among these are the springs at Muzquiz, Monclova, and Cuatro Cienegas, with flows of 1,000 L/sec.

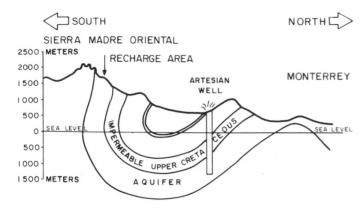

Figure 4. Cross section showing effects of folding on recharge and occurrence of ground water.

## Geochemistry

The potable water that occurs in the limestone aquifer has low dissolved solids and is a calcium-bicarbonate type. The nonpotable water has a high salinity content composed of calcium and sulfate, temperatures higher than ambient, and a high content of $H_2S$. Generally the nonpotable water is present in the deeper zones that are a considerable distance downdip from the recharge areas. Occasionally water from the deeper zones moves upward through fractures and faults and contaminates the shallow aquifers.

## SIERRA TORREON-MONTERREY-TAMAZUNCHALE

### General Setting

A series of continuous folds in carbonate rocks of Cretaceous age have formed the Sierra Madre Oriental chain. Its northern end is in the Mapimi area where the folding strikes N40°W, and continues southeast toward Torreon, where the folding trends east-west to the latitude of Monterrey and then changes to N20°W. The Sierra has altitudes greater than 4,000 m, a length of about 1,200 km, and a width of 200 km. In the Sierra are spectacular canyons with vertical walls exposing outcrops of rocks ranging in age from Precambrian to Quaternary; the principal exposures are limestone of Cretaceous age.

The climate is humid tropical in the eastern part of the Sierra where the mountains receive precipitation and humid winds from the Gulf of Mexico. A few kilometers to the west the Sierra has altitudes of 2,000 m with a variable climate. Rainfall diminishes westward where the Sierra becomes semiarid to arid.

### Torreon

Within the Sierra are intermontane valleys and basins composed of alluvial materials that form aquifers. The Laguna region of Torreon is an area of arid climate, and is one of the major agricultural regions of Mexico. About 3,000 wells for irrigation in

this region have overexploited the alluvial aquifers. To the west and south are carbonate aquifers greater than 1,000 m thick. These limestones contain little or no shale, are not dolomitized, and have been subjected to dissolution. This aquifer helps recharge the overexploited alluvial aquifers of the Laguna region, as confirmed by isotopic studies (Latorre and others, 1981).

### Monterrey

In the greater part of the Sierra there are reef complexes whose primary permeability, augmented by fracturing, forms highly permeable zones; at higher elevations these are recharge zones and at lower elevations they are aquifers. The anticlines that form the sierras of this region have steeply dipping flanks that are vertical to overturned in places. The steep bedding enhances the infiltration of rainfall and, combined with the primary porosity (reefal) and secondary porosity (fracturing and dissolution), permits the development of aquifers of great potential. On the flanks and plunging noses of these anticlinal structures, water is at depths of about 1,000 m and is under artesian conditions (Fig. 4). This type of aquifer is primarily exploited in the Monterrey-Saltillo area where the abundant yields satisfy the needs of Monterrey, the third largest metropolitan area of Mexico.

### Tamazunchale

In the southern part of the Sierra Madre Oriental, water infiltrates the higher elevations of the Sierra and is discharged through great springs at the base of the Sierras adjacent to the coastal plain. Patterns of carbonate sedimentation play a major role in aquifer development. The extensive El Abra-Doctor reef complex, which formed on the margin of the Valles-San Luis Potosi Platform in Albian-Santonian time, contains carbonates with high permeability. In the high Sierra are numerous dolines and sink holes where water infiltrates, flows laterally, and discharges as springs at the base of the Sierras (Fish, 1977). These springs, such as Coy and Frio, are among the largest in the world and have discharges of 25 m³/sec.

Within the Sierra are local and regional flow patterns. Water of local flows is recently infiltrated and of a calcium-bicarbonate type, whereas water of regional flows contains additional quantities of magnesium and sulfates.

## REFERENCES CITED

Back, W., Lesser, J. M., and Hanshaw, B. B., 1977, Structural and stratigraphic occurrence of "Bad Water" in Coahuila, Mexico: Geological Society of America Abstracts with Programs, v. 9, p. 885.

Fish, J. E., 1977, Karst hydrogeology and geomorphology of the Sierra de El Abra and the Valles-San Luis Potosi Region, Mexico [Ph.D. thesis]: Hamilton, Ontario, Canada, McMaster University, 469 p.

Latorre, C., Lesser, J., Quijano, M., and Payne, B., 1981, Isotopos ambientales aplicados al estudio de la interconexion de los acuiferos calizos y de rellanos en la colombia Region Lagunera de Coahuila-Durango, Mexico: Bogata, Columbia, Interamerican Symposium on Isotope Hydrology, p. 135–148.

MANUSCRIPT ACCEPTED BY THE SOCIETY JUNE 26, 1987

The Geology of North America
Vol. O-2, Hydrogeology
The Geological Society of America, 1988

# Chapter 13

# *Region 10, Faja Volcanica Transmexicano*

**Ruben Chavez**
*S.A.R.H. Subdireccion Geohidrologia, San Luis Potosi No. 199-3rd. piso Mexico, D.F.*

## REGIONAL FRAMEWORK

### *Location and extent*

The hydrogeologic region known as the Faja Volcanica Transmexicano (Transmexican Volcanic Belt) (Fig. 3, Table 2, Heath, this volume) is approximately coincident with the physiographic province of the same name. It is in central Mexico, covers an area of about 105,000 km², and extends partially or totally over several states and the Federal District of Mexico. This region is elongated and irregularly shaped and has a length of 950 km east-west and an average width of 110 km (Fig. 1).

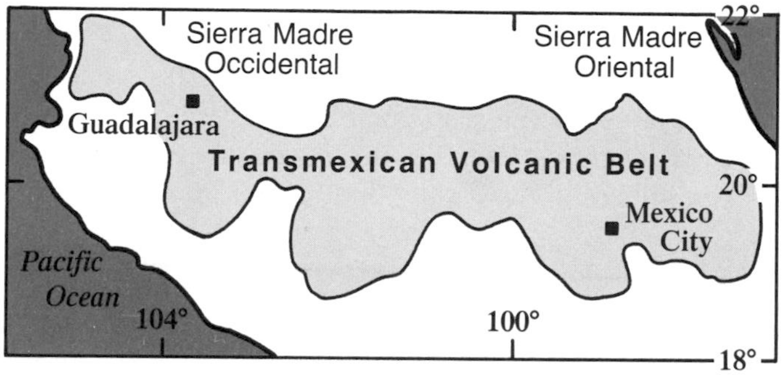

### *Physiography*

The region is predominantly mountainous and composed of some of the most extensive and spectacular manifestations of volcanism in North America. It is characterized by great ranges among which occur some of the highest peaks in the country, such as Citlaltepetl (5,720 m), Popocatepetl (5,540 m), Iztacclhuatl (5,200 m), and La Malinche (4,400 m). These mountains rise 1,500 to 3,500 m above the adjacent valley floors.

Other relevant physiographic features are the lacustrine valleys between the mountain chains; among the more notable are those of Mexico, Toluca, Puebla, and Chapala. In some of the valleys, there remain remnants of great lakes, such as Chapala, Cuitzeo, Patzcuaro, and Texcoco. Numerous rivers and arroyos drain this region. Major flows are in the Lerma and Panuco Rivers, whose basins are among the more extensive of Mexico.

### *Climate*

The Transmexican Volcanic Belt has a great variety of climatic conditions owing to its mountainous topography and wide range of elevations. The dominant climates are: (1) tropical and humid-temperate in the western portion, (2) subhumid temperate and arid temperate in the central mesa, and (3) humid-to-arid cold in mountainous areas and in the northeast portion.

Precipitation varies between 300 and 4,000 mm/yr and is controlled primarily by orography. The greatest part of the region receives more than 500 mm/yr of precipitation; only a few areas in the northern portion receive less. Precipitation in mountainous

Figure 1. Simplified geologic map of the Transmexican Volcanic Belt.

regions is over 1,200 mm/yr, and in the higher sierras it exceeds 2,000 mm/yr. The rainy season is the summer and early fall; winter is a time of isolated rains over the region and abundant snowfall on the higher peaks. Average annual temperatures vary with altitude between 13° and 23°C. Lower temperatures are recorded in mountainous areas; higher temperatures in the lower elevations of the basins. In the summer, the average daily temperature ranges from 15° to 26°C, with highs of 36°C. During winter, the average daily temperature is from 12° to 20°C, with lows of –5°C in the mountains.

These conditions result in potential evaporation values of 1,600 to 2,000 mm/yr. Although these average potential evaporation values are greater than precipitation, there is excess precipitation during the rainy season, which generates flows and recharge to the aquifers.

### *Geology*

During Middle Cretaceous to late Tertiary time (Oligocene to Miocene), tectonism and episodes of volcanism generated the basic framework upon which were formed the geologic structures of the Transmexican Volcanic Belt. Numerous extrusive structures, many of which reflect violent eruptions, were the product of Pliocene-Quaternary volcanic activity. Decreasing volcanic activity has continued into the Recent Epoch, as shown by localized manifestation in the latest centuries; in our time, the birth of

Chavez, R., 1988, Region 10, Faja Volcanica Transmexicano, *in* Back, W., Rosenshein, J. S. and Seaber, P. R., eds., Hydrogeology: Boulder, Colorado, Geological Society of America, The Geology of North America, v. O-2.

Paricutin Volcano is one example of the igneous activity that has formed this region.

Tectonism has determined the orientation of the more relevant structural features; volcanic activity is concentrated along faults and fractures. On a regional scale, there is an east-west orientation, but distinct zones defined by basement tectonism and aligned extrusive structures can be recognized. The dominant orientations are: (1) northwest-southeast in the western part, (2) east-west in the central part, and (3) northeast-southwest in the east.

Uplift of the mountains and subsidence of adjacent areas formed closed basins. These basins were subsequently filled by alluvium and volcanics from succeeding phases of erosion, sedimentation, and volcanism. Drainages in other basins were closed by subsequent volcanic extrusions. Lakes formed in the lower elevations of these closed basins, and great volumes of fine-grained sediments, interfingered with igneous rock and alluvial materials, were deposited. Some lakes disappeared owing to subsequent fluvial development and drainage, while others have persisted to the present.

The mountain masses are composed principally of extrusive igneous rocks (Fig. 1). In the higher sierras, flows and ashes of rhyolitic and andesitic composition (rhyolites, dacites, ignimbrites, andesites, etc.) of Tertiary age predominate. It is probable that acidic rocks crop out along gigantic fissures or fractures. Pliocene to Recent rocks of basaltic composition form the majority of the extrusive structures of the area.

The valley fill is composed of igneous and sedimentary rocks whose ages range from mid-Tertiary to Late Quaternary. Their thickness is highly variable, from a few dozen meters in fluvial valleys up to thousands of meters in deeper troughs where exploratory wells have drilled sequences containing alluvial materials, lacustrine deposits, and igenous rocks without reaching the base of the fill. In the basins, the lower parts of the fill are composed principally of andesitic lavas; at intermediate depths, the rocks are rhyolitic; and above this are interbedded basaltic lavas, lake deposits, and alluvium. Unconsolidated alluvium generally forms the upper part and crops out on valley floors.

## HYDROGEOLOGY

### Hydraulic characteristics of the rocks

The hydraulic characteristics and hydrogeologic behavior of the rocks depend on their lithology, stratigraphy, and structure. Lithologically, the oldest igneous rocks that form the mountains are essentially impermeable; nevertheless, those that have been tectonically fractured or altered by weathering may have significant permeability. Basalt flows, the most widespread rocks of the Transmexican Volcanic Belt, are characterized by high permeability; their permeability varies over a wide range and is controlled primarily by their degree of fracturing, and also by the presence of lava tubes and cooling joints.

In the subsurface of valleys, consolidated and unfractured rocks below the regional level of saturation are barriers to underground water flow, such as where andesitic flows form geohydrologic basements at depths of hundreds of meters in valleys of tectonic origin. Unfractured flows interbedded with alluvial fill at various depths form independent local aquifers in adjacent rocks. In contrast, most beds of rhyolitic composition, found in the northern part of this hydrogeologic region and which extend into the alluvial basins to the south, form aquifers of great thickness and moderate permeability because of their fracturing. Young basalt flows with vesicular texture and fractures that display weathering are aquifers of great permeability and porosity, although some dense and negligibly fractured basalts may function as aquitards.

In the lower parts of basins, aquifers composed of fractured volcanics are covered by lacustrine and alluvial deposits of lesser permeability. For this reason, the aquifers are confined or semi-confined and characterized by low storage coefficients owing to their virtual incompressibility.

The hydrogeologic characteristics of pyroclastics are controlled by their grain size, fracturing, and degree of cementation or compaction. Fine-grained materials (ash, lapilli, etc.) have a high porosity, although their permeabilities are low; in the zone of saturation, they function as aquitards, which on a regional scale may yield or transmit great quantities of water to adjacent aquifers. Abundant at the base of the volcanics are thick, uncemented pyroclastics, less porous and more permeable than the above, which are highly permeable aquifers beneath the regional water level.

Widely distributed in the subsurface of the extrusive Pliocene plains are shaley lacustrine deposits, very porous and of low permeability, which form aquitards some hundreds of meters thick and store great volumes of water. On a regional scale, these deposits contribute large volumes of water to adjacent aquifers, either naturally or from pumping.

Unconsolidated alluvial materials, interbedded gravels, sands, silts, and shales, are widely exposed in valleys and form young aquifers in the shallow subsurface. Their permeability and transmissivity vary according to their grain size and thickness. The coefficient of transmissivity of alluvial aquifers varies from $5,10^{-4}$ to $5,10^{-2}$ m$^2$/sec. Thick and highly permeable clastics are common in stream beds; they also form old stream channels that function as semiconfined aquifers. On the extensive flood plains, medium- to fine-grained clastics of moderate to low permeability are more abundant.

### Geologic Control of Hydrogeologic Processes

The described characteristics and geologic structure control to a great degree the hydrogeologic processes such as subsurface recharge, circulation, and aquifer discharge. The capacity for infiltration of surficial rocks is the geologic factor that controls access of water to the subsurface. In this region, rocks of high infiltration capacity are vesicular basalts, columnar basalts, and thick pyroclastics whose outcrops are excellent recharge areas. Also availa-

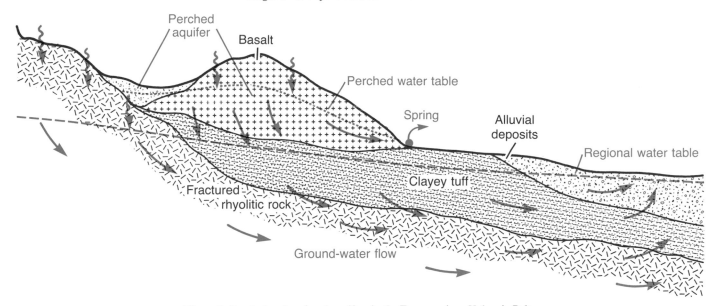

Figure 2. Perched and regional aquifers in the Transmexican Volcanic Belt.

ble for infiltration are the coarse-grained deposits on mountain flanks and in stream channels. Likewise, the permeability and geologic structures of the rocks control the descent of the water through the zone of aeration.

Subsurface flow is controlled by the intrinsic character of the rocks, their stratigraphy and geologic structure. In the Transmexican Volcanic Belt, the complex geology, typical of volcanic terranes, has given rise to a great variety of flow systems, some of which are shown in Figures 2, 3, and 4.

Frequently in elevated mountainous areas, fractured and altered rocks overlie less permeable rocks and form perched aquifers independent of the regional system. Young fractured basalts overlying pyroclastics, less permeable lavas, or lacustrine deposits are sequences forming perched aquifers in the hydrogeologic region. Depending on the amount of recharge, their dimensions, and hydraulic characteristics, these aquifers may be permanent or transitory. They discharge as springs on the mountain flanks or on the walls of deep ravines. Other perched systems form in intermontane valleys where fluvial deposits form small aquifers recharged by surficial flows. In some areas, the perched aquifers are recharged slowly but continuously by the regional systems.

Deep wells on mesas and elevated valleys intersect volcanic rocks, which possibly shows the existence of a regional phreatic level beneath the mountainous massifs. From this we may infer the presence of flow systems that interconnect the mountains and the adjacent valleys. In these systems, the mountains are the natural recharge areas and transmit the infiltrated water to the lower parts of the basins. The depth at which the phreatic level is encountered in mountainous areas depends on topographic, geologic, and hydrogeologic factors. The dominant factor is the transmissive capacity of the rocks, which, in turn, is dependent on their permeability and saturated thickness. The hydraulic gradient is inversely proportional to the transmissivity of the medium; as

such, water levels are deeper under mountain masses composed of fractured rocks of high permeability than in those where the rocks are of low permeability.

The recharge forms mounds on the potentiometric surface, inducing a descending flow in the regionally saturated zone. The flow paths in the subsurface are very tortuous and are determined greatly by geologic factors. Apart from the hydraulic character of the strata, the relative stratal position and structural features such as faults, large fractures, and lava tubes are dominant influences. Away from recharge and discharge areas, subsurface flow is parallel to the stratification of the more permeable aquifers, but the existence of aquitards and aquifers causes refraction of flow lines on passing from one to the other; these directional changes are directly proportional to the relation between the permeabilities of the strata in contact. In the subsurface of the valleys and plains of the Transmexican Volcanic Belt, it is common that fractured lava flows function as collectors of overlying fill of lesser permeability, such as alluvium, lacustrine deposits, and pyroclastics. Since this is a region characterized by active tectonism, faults and structural features play an important role in the hydrogeologic systems. In general, disruption of stratal continuity impedes subsurface flow. Their hydrogeologic behavior is quite varied; when faults are sealed by metamorphism or fine-grained mineral deposits, they form virtually impermeable barriers leading to independent flow systems or poorly cemented systems. Faults filled with coarse detritus or breccia form zones of preferential circulation, which facilitates recharge, intraformation flow of water, and springs. In some valleys of tectonic origin, such as Cuitzeo, numerous thermal springs occur along faults.

Under the control of the above factors, downward-infiltrating waters are incorporated into the local and regional systems. The water preferentially circulates through the more permeable strata, which present less flow resistance and tend to cross less permeable strata via the shortest route. When these are

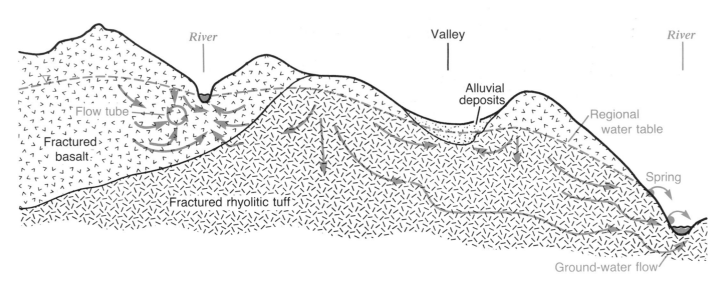

Figure 3. Hydrologic connection between mountains and valleys.

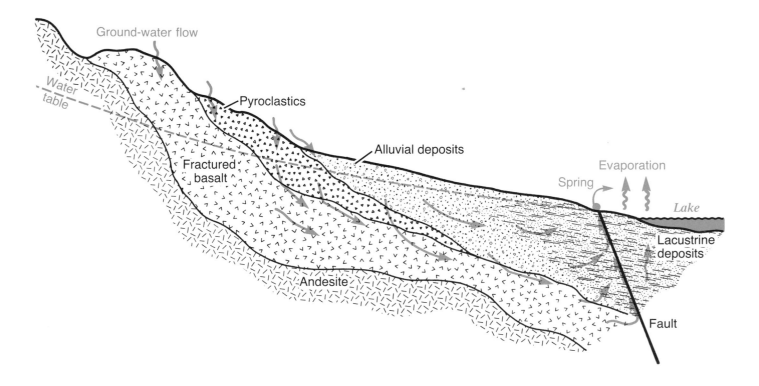

Figure 4. Typical flow system in the lacustrine basins.

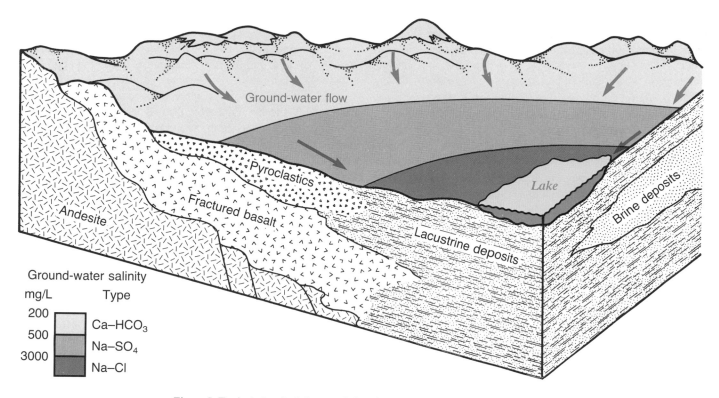

Figure 5. Typical chemical characteristics of ground water in the lacustrine basins.

found in the upper part of the stratigraphic sequence, the majority of the infiltrating waters are incorporated into local systems. This is frequently the case in the Transmexican Volcanic Belt. The basalt flows and thick pyroclastics that form the bulk of the young volcanics form excellent recharge areas and rapidly transmit the water to depth, nourishing the aquifers deep beneath the adjacent valleys. In the subsurface of the valleys and lacustrine plains, the water moves to discharge areas. In accord with observations and theory, it appears there is hydraulic continuity of the saturated zone; that is, the fill constitutes a single flow system and all the strata are hydraulically interconnected. Nevertheless, owing to the heterogeneity and anisotropy of the fill, there are significant differences in hydraulic charge, temperature, and water quality in the vertical dimension.

The natural discharge of the aquifer system is controlled partially by geologic factors. The volume of discharge is directly proportional to the transmissivity; its rate of decline also is directly proportional to this characteristic and inversely proportional to the potential volume of water that the system may discharge through its natural drains. In this hydrogeologic region, volcanic aquifers generate springs on the flanks and at the base of the mountains; this discharge ranges from a few liters to several cubic meters per second. The springs fed by basaltic mountain masses are generally free flowing and permanent; less extensive, low-transmissivity systems and perched aquifers give rise to in-

termittent springs of generally low volumes. In nature, the major surface streams receive a large part of the natural discharge of the aquifers. The base flow of rivers from extrusive aquifers of high transmissivity is a few cubic meters per second at low flow.

Geologic factors also are a dominant influence on the artificial discharge; to a great degree, the aquifer characteristics determine quantity of the flow and the efficiency of storage. In general, wells in Quaternary basalts have specific capacity greater than 20 lps/m, depending on their transmissivity. Wells in rhyolitic aquifers and alluvium are characterized by specific capacity in the range of 1 to 15 lps/m.

### Geochemical influence

The physiochemical characteristics of the ground water are a consequence of subsurface hydrogeochemical processes. In its underground travels, the water dissolves soluble rock components. The rock surface exposed to dissolution by water and the time they remain in contact are the dominant factors in this process. In turn, the contact time is directly proportional to length of flow and inversely proportional to flow velocity.

Figure 5 shows the variability of ionic concentrations and distributions of ground water in the more common systems of the Transmexican Volcanic Belt. On passing through rocks of basaltic composition, predominant in the recharge area, water gains cations of Ca and Mg from the dissolution of calcic feldspars and ferromagnesian minerals. Favoring this process is the large surface area of these rocks developed through fracturing, porosity, and alteration. Despite this, the water generally has dissolved solids less than 500 mg/l owing to its rapid circulation and short residence time in the subsurface. Rocks of andesitic and rhyolitic composition, also abundant in recharge areas, contribute Na and K cations derived from feldspar dissolution to the water. The water contained in these rocks also has low dissolved solids.

From the consolidated rocks that form the mountain masses, the ground water passes to the fill of the fluvial valleys and tectonic troughs in its flow path to discharge areas. The detritus contributes ions to the water in greater quantities because of the great surface area subject to dissolution and through its lesser flow velocity than in the fractured rocks. In addition to the solution of minerals, other processes that control the ionic concentration and distribution in the ground water are adsorption and ion exchange, which takes place in thick, clayey strata of lacustrine and alluvial origin. Ground water in fluvial valleys where the fill is thin and composed primarily of medium- to coarse-grained material may have dissolved solids less than 1,000 mg/l and is calcium-sodium-bicarbonate. In lacustrine valleys, the water salinity is relatively high owing to long residence time in the subsurface and its frequent circulation through or near evaporites; dissolved solids here are commonly greater than 1,000 mg/l. Deep saline aquifers and brines are found in some basins. Dominant ions of the more shallow water aquifers are sodium-calcium-bicarbonate and calcium-magnesium-bicarbonate. The ionic proportions of chlorine and sodium increase with depth; deep salty aquifers contain water of the sodium-chlorine-sulfate type.

## SIGNIFICANT SPECIAL PROCESSES

Within the Transmexican Volcanic Belt there are commonly thermal processes associated with various heat sources. One of these is the upward heat flow from depth, the geothermal gradient (2° to 4°C per 100-m depth). Heat sources for the volcanic features are the magma chambers still manifested by recent volcanism. Active faults are features of tectonic regions.

Infiltrating water that reaches great depths and circulates close to volcanic and tectonic heat sources transports and redistributes the heat, forming convective hydrothermal systems. The significant thermal gradients, which also induce subsurface flow, may modify the hydrodynamic flow system controlled by the hydraulic gradients. In these systems, water temperature is controlled by the temperature of the heat source, the depth and circulation velocity of the water, and the thermal conductivity of the rocks surrounding the system.

Geothermal manifestations may be observed in diverse areas of this hydrogeologic region; they range from moderate thermal effects on the water to vapor and hot-water geothermal systems. Thermal effects noted in wells and springs range from a few degrees above mean annual air temperatures to boiling. Thus, for example, springs along the southern border of Lake Cuitzeo have temperatures of 30° to 90°C. In this case, the heat is probably derived from geothermal and tectonic sources; the water is heated on circulating through fractured igneous rocks and ascends the fault zone that forms the southern margin of the lake.

The more important geothermal systems are in the western portion of the region. Notable, among others, are the fields known as "Los Azufres," "Los Negritos," and "Ixtlan de los Hervores" in the state of Michoacan. The "Los Azufres" field, found in rhyolitic and andesitic rocks, is a stream and hot-water system whose temperatures range from 230° to 260°C. Apparently, this is a system of "forced convection" fed by meteoric waters through fractures, although the origin and location of the heat source has not been located.

Manuscript Accepted by the Society February 23, 1988

# Chapter 14

# *Region 11, Sierra Madre del Sur*

**Ricardo Riva Palacio**
*Sumidero No. 3 La Lomita, Tuxtla Gutierrez, Chiapas, Mexico D.F.*

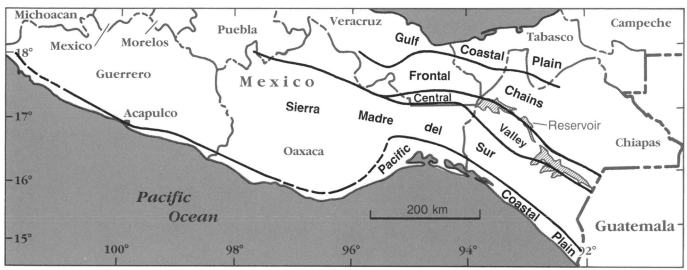

Figure 1. Map showing hydrogeologic subdivisions of the Sierra Madre del Sur region.

This hydrogeologic province (Fig. 3, Table 2, Heath, this volume) takes its name from the mountains that are the most prominent physiographic feature of the region. To the south of the Sierra Madre Mountains is the Pacific Coastal Plain, and to the north the small Central Valley, another mountain range—the Frontal Chains—and the Gulf Coastal Plain (Plate 1; Fig. 1). This region has the greatest rainfall in North America, with a mean annual rainfall ranging from 5,500 mm at the southern Mexican–Guatemalan border to 4,500 mm in central Chiapas. Ground-water resources are quite limited in this province, but the sites for surface-water impoundments and reservoirs are numerous.

The Sierra Madre del Sur is a mountain chain composed primarily of a basement of metamorphic rocks that were displaced during Jurassic time and a batholith that constitutes the Oaxaca Massif and the Sierra Madre de Chiapas. This area also is underlain by other impermeable formations composed of dolomite, recrystallized limestone, and red beds composed of shale, sand, and silt of Paleozoic, Triassic, and Jurassic age.

The Pacific Coastal Plain is a narrow strip, 10 to 25 m wide, parallel to both the mountains and the Pacific Ocean. It is composed of Quaternary alluvium, primarily silt derived from the erosion of the Sierra Madre del Sur and deposited by streams flowing to the ocean. This fine-grained alluvium constitutes aquifers of extremely low yield. Some of the major hydrologic features of the Coastal Plain are numerous lagoons that contain fine-grained material and abundant organic debris.

The Central Valley is a graben-like depression composed mainly of sedimentary rocks of Cenozoic age. The topography is gently rolling to flat. The low permeability of the sediments causes them to yield only small amounts of water.

The Gulf Coastal Plain is a swampy area composed of fine-grained sediments with low potential for ground-water occurrence.

The Frontal Chain contains the most permeable rocks of the province and, consequently, the greatest occurrence of ground water. It is composed of a series of mountain ridges and valleys.

Palacio, R. R., 1988, Region 11, Sierra Madre del Sur, *in* Back, W., Rosenshein, J. S. and Seaber, P. R., eds., Hydrogeology: Boulder, Colorado, Geological Society of America, The Geology of North America, v. O-2.

The mountains are formed primarily by Cretaceous limestone and dolomite, which are highly karstified. The plateaus and high valleys are composed of continental sedimentary rocks of Tertiary age, Holocene alluvium, and thick deposits of Quaternary pyroclastic sediments. The drainage area of this mountainous area is quite complex because of the wide range in permeability of the rocks and the numerous tectonic structures. Commonly the streams are lost into limestone cavities and reappear as springs that form magnificent waterfalls down the ridges. In other places the streams reappear to form lagoons that serve as holding ponds for the headwaters of rivers that continue on downslope. All of this water eventually drains into the major rivers, either as overland runoff or subsurface flow, as they meander to the Gulf Coastal Plain before discharging into the Gulf of Mexico. The limestones and dolomites of this area have a combined thickness of more than 2,800 m and consist of alternate sequences of lithographic limestone with chert, dolomitized limestone, dolomite, and shaley limestone. The porosity and permeability are mainly secondary and formed by fracturing and dissolution. In places the calcareous rocks are overlain by impermeable terrigeneous material of Tertiary age. This area has undergone great tectonic activity in which normal faults have produced horst-like anticlines of calcareous rocks and graben synclines of the soft terrigenous rocks.

The Frontal Chains and the Central Valley are separated by both lateral and normal faults. The tectonically dominated morphology reflects at least two stages of folding and fracturing that have generated long anticlines that are aligned with the northwest–southeast trends. In the calcareous highlands is a well-developed karstic terrain with numerous peaks, towers, and sinkholes. With the exception of the few major streams in the deeper valleys, the few other streams are short. As is typical of other karst terrains, many of the valleys are poorly formed and end abruptly in sinks or swallow holes. The ground-water flow is controlled primarily by the fracture pattern, most of which is discharged to the streams. Ground water is produced only in the recharge areas because the excessive subsurface drainage in other areas causes the water table to be extremely deep.

MANUSCRIPT ACCEPTED BY THE SOCIETY SEPTEMBER 10, 1987

## Chapter 15

# *Region 12, Precambrian Shield*

**R. N. Farvolden**
*Department of Earth Sciences, University of Waterloo, Waterloo, Ontario NL2 3G1, Canada*
**O. Pfannkuch**
*108 Pillsbury Hall, University of Minnesota, Minneapolis, Minnesota 55455*
**R. Pearson**
*Golder Associates Incorporated, 3772 Pleasantdale Road, Suite 165, Atlanta, Georgia 30340*
**P. Fritz**
*Department of Earth Sciences, University of Waterloo, Waterloo, Ontario NL2 3G1, Canada*

## INTRODUCTION

The Precambrian Shield represents the largest petrologically homogeneous and contiguous province of the North American continent. Crystalline metamorphic rocks underlie most of the Shield, and structural provinces are defined on the basis of age, lithology, and structural features. In places, relatively unmetamorphosed igneous rocks or indurated sedimentary rocks are predominant, but all have more or less the same hydrogeologic characteristics in that fractures control the occurrence and flow of ground water.

The extent of the Precambrian Shield Hydrogeologic Region is shown on Figure 1. The boundaries are defined on Fig. 3 and Table 2 of Heath (this volume). In general, the boundaries are rather easy to determine, and this portion of the Precambrian Shield is a well-defined ground water region.

## PHYSIOGRAPHY AND CLIMATE

The physiography of the region is a direct consequence of the bedrock and surficial deposits and their interaction with the climate through weathering and erosional processes, including glaciation, during long intervals of geologic time. For the most part the landscape is rolling and monotonous, with relief less than 50 m. Locally, however, differential erosion along structural and lithologic lineaments has resulted in steep cliffs and rugged topography, and greater relief. Surficial deposits have smoothed the surface by filling in many deep bedrock depressions (mostly lineaments), and some end moraines are prominent features. Ground surface elevations are about 200 to 300 m in the south and 50 to 100 m at the northern boundary of the region.

The drainage is mostly into Hudson Bay and to the Great Lakes–St. Lawrence River system. Drainage patterns are controlled by differential erosion along major structural features such as folds, faults, and intrusive contact zones. Glacial erosion and deposition disrupted preglacial drainage systems forming many lakes and swamps and causing the poor drainage of most of the region (Fig. 2). In localities such as this the preponderance of surface water tends to mask or obscure ground-water phenomena such as dry-weather flow, springs, seeps, and phreatophytes. In other areas, undrained flatlands underlain by glacial lake deposits support extensive wetlands. One of the largest contiguous peatlands in the world, the Red Lake patterned peatlands, is located in the Precambrian Shield region of northern Minnesota.

Climatic zonation is more or less along lines of latitude with the mean annual temperature decreasing from 11°C in the south to about 1°C in the north. The climate can be classified as humid continental, with mesoscale local effects caused by the Great Lakes and Hudson Bay. Precipitation ranges from 1,200 mm in the south to less than 500 mm in the extreme northwest, about 25 percent of which is snow. Evaporation increases from 200 mm in the north to about 750 mm in the southwest. Mean annual runoff increases from about 25 mm in the west to 500 mm in the east (Pfannkuch and others, 1983; Fisheries and Environment Canada, 1978). Quality of surface water is generally excellent—the water is low in dissolved solids and slightly acidic. Organic color is common in many poorly drained catchments.

Hydrographs of observation wells show that ground-water recharge is mostly caused by the annual spring snowmelt and infiltration from cyclonic and frontal rainfall. Sporadic convective summer storms are of lesser importance. On the other hand, isotope data show the ground water to be the same as a composite sample of annual precipitation.

## GENERAL GEOLOGY

The Precambrian Shield Hydrogeologic Region includes the Southern structural province and parts of Churchill, Superior,

Farvolden, R. N., Pfannkuch, O., Pearson, R., and Fritz, P., 1988, Region 12, Precambrian Shield, *in* Back, W., Rosenshein, J. S. and Seaber, P. R., eds., Hydrogeology: Boulder, Colorado, Geological Society of America, The Geology of North America, v. O-2.

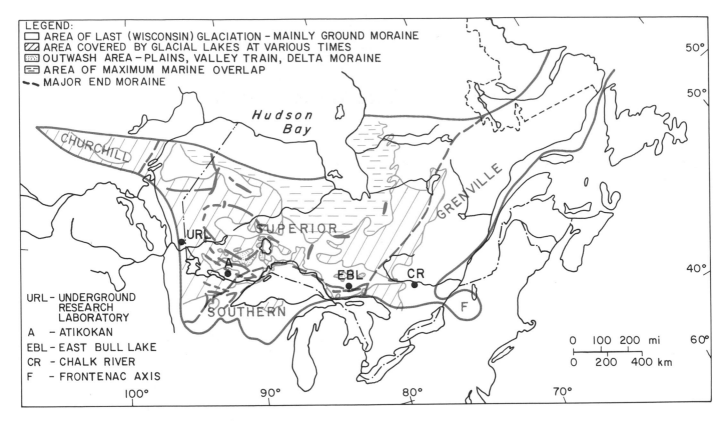

LEGEND:
☐ AREA OF LAST (WISCONSIN) GLACIATION - MAINLY GROUND MORAINE
▨ AREA COVERED BY GLACIAL LAKES AT VARIOUS TIMES
▦ OUTWASH AREA – PLAINS, VALLEY TRAIN, DELTA MORAINE
▤ AREA OF MAXIMUM MARINE OVERLAP
⌒ MAJOR END MORAINE

URL – UNDERGROUND
        RESEARCH
        LABORATORY
A   – ATIKOKAN
EBL – EAST BULL LAKE
CR  – CHALK RIVER
F   – FRONTENAC AXIS

Figure 1. The Precambrian Shield Hydrogeological Region with structural provinces, and locations cited in text.

and Grenville structural provinces (Fig. 1). Each province and subprovince of the Shield is identified by lithology, age, and characteristic structural features. Most of the information in the following paragraphs is from Stockwell and others (1970).

Churchill Province (Fig. 1) is underlain by Archean rocks comprising mainly metamorphosed volcanic flows and pyroclastics with lesser amounts of flysch-type sediments. Extensive areas of granitic terrain are probably metamorphosed equivalents of these ancient volcanics and sediments. These Archean rocks are intruded by granite and overlain by conglomerates, sandstones, siltstones, shales, and dolomites. The entire sequence has been folded and metamorphosed. Sedimentary facies changes suggest a geosynclinal environment. Granite intrusions of the Hudsonian orogeny are dated at 1,650 to 1,850 Ma. In the southern portion of Churchill Province, diabase rocks dated at 1,200 Ma intrude into these older rocks. Numerous faults, some major, cut all the rock types. Brecciated zones are commonly associated with these faults.

The history of Superior Province is similar to that of Churchill Province, but stratigraphic correlation between these provinces is not possible. Superior Province may be described as a granite gneiss terrain engulfing elongated pods of highly metamorphosed volcanic and sedimentary strata, commonly referred to as greenstone belts. The greenstones represent Archean volcanic rocks and associated flysch-type sediments. The granitic gneisses are from the Kenoran orogeny (about 2,500 Ma).

Younger intrusions in the form of alkaline plutons and diabase dikes have ages ranging from 2,200 to 1,000 Ma. As in Churchill Province, many faults with brecciated zones or shear zones appear to be associated with the intrusions and sedimentary basins.

The main distinctive feature of Grenville Province is the tight, complex folding, on a small scale, of a wide range of lithologic types. Many varieties of volcanic flows and pyroclastics are represented. Metamorphic equivalents of sandstones, graywackes, shales, and carbonate rocks are common in sequences perhaps 6,000 m thick or more. The soluble marble units are of special interest with respect to hydrogeology.

Major amounts of mafic rock (anorthosite) and minor amounts of other rock, including felsic, mafic, and intermediate types, were intruded prior to the major deformations caused by the Grenvillian orogeny (1,000 Ma). The Adirondack outliner in northern New York State is directly connected to the main part of Grenville Province, by the Frontenac Axis (Fig. 1).

Southern Province is characterized by sedimentary rocks of Aphebian age or younger, lying unconformably on older rocks of Archean age. The Aphebian rocks are relatively undeformed in Canada, and moderately folded in the U.S. The sediments range from basal conglomerates and sandstones to shales, limestones, and tillites. All are moderately metamorphosed and intruded. Gabbroic intrusions in the Sudbury region and elsewhere are 1,700 to 2,150 Ma. The sills that intrude the Aphebian sedimentary rocks in the Lake Superior region are about 1,000 Ma and may be related to the Keweenawan basalts that lie unconformably on Aphebian and Archean rocks. They are the youngest major rock type of the region.

## POST—PRECAMBRIAN WEATHERING

Large parts of the Precambrian Shield were previously covered by Paleozoic and Cretaceous strata. These strata were subsequently removed by erosion, and tectonic stability surely allowed the development of a deeply weathered regolith on the basement surface during the long interval preceding the onset of Pleistocene glaciation. Extensive weathering must have taken place over the entire Precambrian Shield, but the regolith was almost entirely removed by the continental glaciers, except for the southernmost extension in Minnesota. There, the regolith is as much as 20 m thick and preserved under glacial cover. Similar occurrences can be expected elsewhere but have not been described. Extensive surface sands over the extreme western portion of Churchill Province may have been derived from this regolith but have clearly been transported or disturbed by glacial action.

## PLEISTOCENE DEPOSITS

The glacial deposits overlying the Precambrian Shield region are all products of the Laurentide ice sheets and their different advances. In particular, the multiple glaciation in four major glacial episodes during the Pleistocene Epoch resulted in erosion of the regolith and differential erosion of the bedrock and then deposition of eroded material by the ice sheet and by associated meltwater streams and lakes (Fig. 1).

Over a wide area of central Minnesota, discontinuous patches of Cretaceous shales and sandstones overlie Precambrian rocks and are in turn overlain by thick Pleistocene deposits. These Cretaceous outliers do not warrant being considered as a distinct hydrogeologic unit because they exert little hydrologic influence. They react hydraulically with the regolith and are appropriately grouped with these materials. From a hydrogeochemical point of view, however, their influence is strong, in that they dominate ground-water quality. The western margin of the ground-water region is placed just east of the Red River of the North, where the Cretaceous and Mesozoic sedimentary rocks thicken to more than 100 m, and form a distinct hydrogeologic unit.

Of the four glacial episodes, only the deposits of the latest, the Wisconsinan, will be discussed because earlier drift was either eroded or buried. Wisconsinan deposits constitute virtually all of the surficial material of the region.

An interesting interplay exists between the bedrock and the drift in that the Shield materials—bedrock and regolith—provided the source material for the glacial deposits. Furthermore, the configuration of the bedrock topography, especially the location and trends of broad lowlands, controlled the pathways of lobate advances and the intricate pattern of lobe movement and interaction. This interaction is especially the case near the Great Lakes where bedrock topography controlled the direction of ice

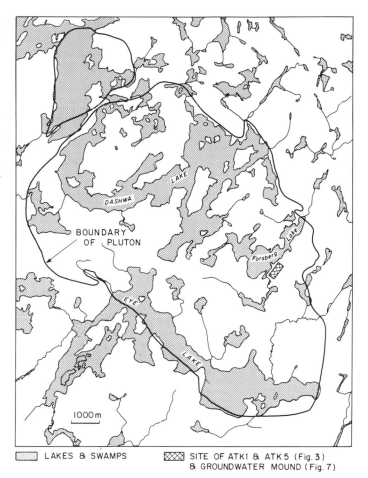

LAKES & SWAMPS    SITE OF ATK1 & ATK5 (Fig. 3) & GROUNDWATER MOUND (Fig. 7)

Figure 2. Drainage network in the vicinity of the Eye-Dashwa Lakes Pluton, near Atikokan, Ontario.

advances from the northwest, the north, and the northeast. Repeated advances and retreats from these different directions, plus the deposition of end moraines, created intricate drainage patterns by mutual blocking and diversion. One of the important results was the creation of large outwash regions and glacial lakes. The deposition of moraines by glaciers overriding each other from different directions resulted in some of the greatest thicknesses of glacial deposits—in central Minnesota they are as much as 200 m thick over a large area. Locally, drift thickness over buried bedrock valleys is even greater, but the areal extents of these features are limited (Wright and others, 1983).

Major terminal moraines in northern Ontario are locally important features of the landscape. Some end moraines northwest of Lake Superior are 30 m high and over 100 km long in places (Fig. 1). Most of the material is locally derived, so it is granular and coarse and rather permeable. These features appear to be local recharge zones, though little detailed hydrogeologic information is available. Some terminal moraines were deposited at ice fronts in proglacial lakes.

Extensive proglacial lakes such as Lake Ojibway were a feature of ice retreat on the Precambrian Shield (Fig. 1). Deposits of glacial Lake Agassiz cover an area of almost 500,000 km$^2$, albeit much of this is not on the Shield region. However, the extensive wetlands along the boundary of the Shield region in northwestern Minnesota and southeastern Manitoba are related to the occurrence and properties of Lake Agassiz clays. In some sectors of the Shield, notably the southwestern part of Superior Province, glaciofluvial sands are an important hydrogeologic unit. Outwash sand underlies many swamps, shallow lakes, and streams, in places to depths of more than 50 m.

This sand is an important factor in the hydrologic response of the drainage systems in the region because of its high storage capacity and high hydraulic conductivity relative to the rock. In some channels a considerable portion of runoff is by means of underflow.

Geophysical soundings indicate that the bedrock topography under some lakes has been significantly modified. Here, too, deeply eroded fracture zones and depressions in incompetent rock have been filled, and smoothed or partially obscured by outwash and lake sediments.

For the most part, uplands or interstream areas are bare rock, or have only a thin cover of till or outwash. The hydrologic role of this surface cover is not known, but Bredehoeft and Maini (1981) suggest that if a cover is more permeable than underlying rock, the flow pattern in the upper unit will govern that in the medium below.

## GROUND-WATER OCCURRENCE ON THE PRECAMBRIAN SHIELD

Ground-water occurrence on the Precambrian Shield has not been studied except in a very general way because the region is sparsely populated and bountifully endowed with surface water of good quality. Charron (1967) reports that of the 112 communities on the Canadian portion of the Shield with a population of more than over 1,000, 27 use ground water at the rate of 50,000,000 liters per day (L/day). Much of this is from aquifers in surficial deposits.

Charron (1967) provides abundant evidence for the sporadic occurrence of ground water in the rocks of the region. Some 53 mines on the Canadian part of the region pump an average of about 1,500,000 L/day from depths generally between 600 and 1,000 m. He notes remarkable evidence of the low permeability of some crystalline rocks; extensive underground workings only 30 m below a lake bottom produce only 1,000,000 L/day. This low rate of flow may be in part owing to sediments of low permeability on the lake bottom.

For the U.S. portion of the Shield, conditions of ground-water occurrence in the crystalline rocks are quite similar. No major ground-water withdrawals from the bedrock are made. The few wells in bedrock serve domestic needs, and extensive aquifer testing has not been done. The only data of any hydrogeologic significance available are well yields and specific capacity. In general, these are less than 0.5 1/sec and 0.2 1/sec/m, respectively (Pfannkuch and others, 1983).

Mineral deposits on the Precambrian Shield are commonly associated with geologic anomalies, in particular, contact zones, faults, or fracture zones and greenstone belts. Many of the communities on the Precambrian Shield are near mines and consequently near geologic anomalies, so the scant information that is available is highly biased and does not represent ground-water conditions generally. The iron mines of the Mississabi Range and quarries for building stone are perhaps exceptions in that the ore body is representative of the country rock.

## LAKES AND WETLANDS

Numerous lakes and extensive wetlands are the most striking hydrographic features of many parts of the Shield. Two basic types of lakes are recognized: bedrock scour lakes and lakes on glacial deposits. The first are found predominantly in the northern part of the Shield region. They have a somewhat radial orientation around Hudson Bay, aligned with the general drainage direction of the streams.

The role of so-called seepage lakes on glacial drift and their interconnection with the ground-water flow field has been the subject of many studies in Minnesota and elsewhere over the last decade (Winter, 1976; Pfannkuch and Winter, 1984; Lee, 1977; Frape and Patterson, 1981). Ground water is one of the most important components in sustaining these lakes, which in turn are important boundary conditions for the ground-water flow field and profoundly influence the boundaries and geometry of flow systems defined by Tóth (1963).

The most important finding for single lakes is that seepage into and out of the lake is concentrated in a fairly narrow near-shore zone, so that a large fraction of the total volumetric exchange with the ground-water reservoir takes place over a very small fraction of the lake bottom (Lee, 1977). The interactions depend on the geometry of the flow field, specifically the ratio of the lake width to the thickness of the aquifer with which the lake is interacting, and on the anisotropy of the aquifer.

In settings with multiple lakes at different elevations on a regional slope, local ground-water flow systems that discharge to the lakes may be superimposed on a regional flow system. In this case, stagnation points and other relations between hydraulic head and depth, as described by Tóth (1963), can be expected.

The conditions required for wetlands to form are (1) poor drainage and (2) a close balance between all inflow (direct precipitation, surface water, and ground water) and all abstractions (evapotranspiration and perhaps slow surface-water drainage), so that water depth remains low. Two basic types of wetlands are those that have formed by infilling or terrestrification of modern lakes or ponds and those on extensive ancient lake beds. The latter are usually very extensive, perhaps associated with open water, as for example the Upper and Lower Red Lakes in Minnesota and Lake of the Woods. These ancient lake beds are characterized by a flat surface and low surface gradients. They are underlain by lacustrine clays of low permeability, and are poorly drained. Near the ancient shorelines, where the gently sloping shield surface breaks to join the flat lake bed surface, conditions are created that cause the upward seepage of regional ground-water flow that sustains the wetland condition. Siegel (1981) has shown that under conditions of such low hydraulic gradient, intermediate and even regional flow systems are extremely sensitive to small topographic changes and concomitant changes in the configuration of the water table and in distribution of hydraulic head.

## FIELD RESEARCH AT ATIKOKAN

During the past several years (1976–1985), research related to the Canadian Nuclear Fuel Waste Management Program (CNFWMP) has provided important new insight into the hydrogeology and hydrochemistry of the Shield region, with emphasis on rocks of low hydraulic conductivity, particularly massive plutons and granitic terrain.

Extensive field studies have been made to find the best way of identifying individual fractures, fracture patterns, and their role in ground-water flow.

An example is provided by research on and around the Eye-Dashwa Lakes Granitic Pluton in the Superior Structural Province near Atikokan, Ontario (Fig. 1). The pluton is located toward the northern edge of an east-west, elongated, regional drainage basin. Several small catchment basins drain the pluton area (Fig. 2), locally to the south but eventually to Hudson Bay.

The pluton trends north–west; it is an elliptical body, 13 by 8 km, composed of medium- to coarse-grained, hornblende-biotite granite of Archean age, that intrudes tonalite-granodiorite gneisses (Brown and others, 1980). It is generally undeformed, except for joints, faults, dikes, and veins, collectively called fractures. Large gneissic inclusions occur within the pluton. Secondary foliations, lineations, folds, and evidence of penetrative ductile deformation are rare in the pluton but are prevalent within the surrounding gneissic country rock.

Ground-water flow in the area occurs through both the granitic and metamorphic bedrock and the overburden deposits. The near-surface bedrock has low primary hydraulic conductivity ($<10^{-10}$ ms$^{-1}$), but zones of localized, relatively high, secondary hydraulic conductivity ($10^{-6}$ to $10^{-8}$ ms$^{-1}$) and low interconnected porosity ($10^{-1}$ to $10^{-3}$), are associated with the rock-mass fracture system. In contrast, the surficial deposits, mostly of glacial origin, have low (clay) to high (sand) primary hydraulic conductivity ($10^{-11}$ to $10^{-3}$ ms$^{-1}$) but high primary porosity ($2 \times 10^{-1}$ to $5 \times 10^{-1}$).

A thin ($<1$ m) veneer of bouldery sand till covers a large portion of the area. The till appears fissured in many places and probably possesses enhanced secondary permeability in the order of $10^{-8}$ ms$^{-1}$ (Grisak and Cherry, 1975). Typically, the lowlands contain sandy outwash and other sediments (Ridgway and Pearson, 1985). The Eagle-Finlayson and Steep Rock moraines of gravelly sand and till form prominent east-west–trending ridges across the study area. Some lake beds are underlain by sediments exceeding 50 m in thickness, which rest on fractured eroded rock and mask the irregularities of the bedrock surface (Ridgway and Pearson, 1987).

## GROUND-WATER FLOW SYSTEMS

Considerable research has been done on the nature of recharge and discharge and on identification of ground-water flow systems at several sites on the Canadian Shield, selected as research areas in the CNFWMP. They include the Underground Research Laboratory (URL) and the Atikokan, East Bull Lake, and the Chalk River research areas (Fig. 1).

Near Atikokan on the Eye-Dashwa Lakes Pluton, hydrogeological testing of deep boreholes (Lee and others, 1983; Fig. 3), fracture analysis (Dugal and others, 1981; Fig. 4), and monitoring of specific fracture zones have revealed the presence of three regimes. A local, shallow, fresh-water regime to depths of 150 m and possibly deeper, is controlled by the hummocky surface topography, local lakes and swamps, and a locally intense network of fracturing. Discrete, high-permeability, flat-lying, and high-angle sheared fracture zones act as controls on the scale of the local flow systems. An intermediate, or transition, regime is related to the discrete shear zones mentioned above at depths from 150 m to 600 m. A zone below 600 m, with generally higher hydraulic head values than the intermediate regime (Dickin and tohers, 1984; Black and Chapman, 1981; Fig. 4), has been identified in several deep test holes.

The ground-water regimes identified in the studies on the Eye-Dashwa Lakes Pluton are very similar to those found at the three other research areas (Fig. 1). At all test sites, strong structural control of flow patterns exists, owing to flat-lying shear zones or dikes (Raven and others, 1985; Bottomley and others, 1984; Davison and Guvanasen, 1985; Fig. 5). Extensive and detailed testing of individual fractures and fracture systems has indicated strong anisotropy within the planes of fractures and fracture systems (Davison and Guvanasen, 1985; Fig. 6). The causes for this can be attributed to shear movement, to the intersection of fractures, and to chemical deposition within the fracture planes.

At Atikokan, shallow observation wells, completed to identify the water-table configuration on a small bedrock upland, show a ground-water mound typical of a recharge zone (Fig. 7). Shallow minipiezometers (Lee and Cherry, 1979) in surficial sand indicate that vertical ground-water gradients are present, and suggest recharge and discharge zones. Multilevel piezometers set at depths as much as 60 m into the bedrock prove that these vertical gradients are manifestations of a flow system in the bedrock rather than merely phenomena related only to ground-water flow within the surficial deposits. Of some 100 such piezometers, each isolated by packers to measure hydraulic head over several meters of the borehole, more than 90 function well enough to provide good data on hydraulic head and good samples for geochemical and isotope analyses (Wingrove and others, 1984).

The evidence is clear that shallow flow systems such as those described by Tóth (1963), Meyboom (1966), and Freeze and Witherspoon (1966, 1967, 1968) occur in these crystalline rocks

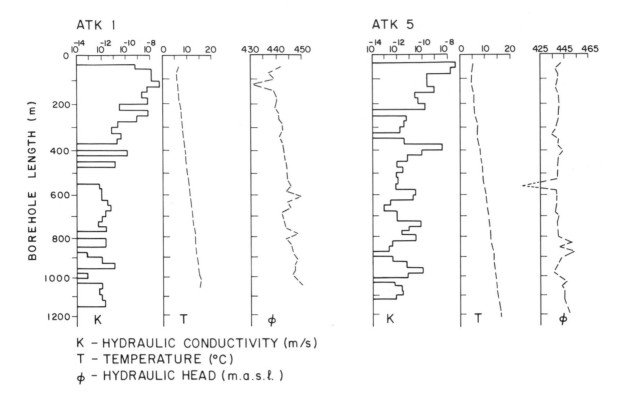

K – HYDRAULIC CONDUCTIVITY (m/s)
T – TEMPERATURE (°C)
φ – HYDRAULIC HEAD (m.a.s.ℓ.)

Figure 3. Hydrogeologic logs of two test-holes on the Eye-Dashwa Lakes Pluton near Atikokan, Ontario (from Lee and others, 1983).

and that, as for porous media environments, the water table is a subdued replica of the topography. Vertical fracturing in the near-surface bedrock is an important factor in recharge and ground-water flow.

Steep gradients suggest that hydraulic conductivity is low, even though fractures appear to be ubiquitous. Effective porosity must also be very low because in most places not enough ground water is discharged to be recognized by surface observations.

Further evidence of water-table configuration was obtained from ground-water investigations on the URL (Fig. 1) being constructed for research on nuclear-waste disposal. Here, too, the water table is a subdued replica of the surface topography. Recharge and discharge zones are evident; once again, this suggests an active flow system and hydraulic continuity. During controlled ground-water pumping to dewater the URL shaft for construction, measurements in observation wells reveal a drawdown cone that is similar in terms of size and geometry to that expected in a porous medium (Davison, 1984).

Before about 1977, when detailed testing began in the Canadian Shield log-linear decrease of hydraulic conductivity with depth was inferred (Davis and Turk, 1964; Snow, 1968). The inference was based on data from relatively shallow (<400 m) randomly drilled water wells and construction site boreholes and was attributed to a general tightening of fractures due to increasing vertical stress. Testing for the CNFWMP and studies of deep mines in the Canadian Shield together with information from

elsewhere, including Swedish, Swiss, and USSR sources, has modified this picture somewhat.

While a case can be made for a log-linear decrease in hydraulic conductivity with depth in the upper 100 to 400 m at a particular site, the flat-lying fracture zones mentioned above are the major controls on regional flow. These zones have been encountered at various depths between surface and at least 1,000 m, with a generalized vertical spacing of one to 300 m and a thickness of 10 to 50 m. Within these zones, permeability is highly anisotropic; distinct flow channels in the fracture plane are a controlling feature (Davison and Guvanasen, 1985). Some 10 to 30 percent of a given fracture zone can have permeabilities several orders of magnitude greater than the background fracture zone and the intact bedrock.

In the research areas (Fig. 1), the flat-lying fracture zones are associated with steep fault zones that penetrate to great depth and often appear as long lineaments at the surface (Raven and Gale, 1986). These zones also appear to be highly anisotropic, with permeability controlled by both fracture fillings and regional stresses. The spacing of these steep fault zones varies from less than 1 km to greater than 5 km. Regional stresses are important in the hydraulic properties of these faults. Where the maximum horizontal compressive stress is normal to the strike, these faults tend to be closed or tight and chemically filled. The opposite has been observed where regional stresses are parallel to the zone. A similar correlation of regional stress and permeability has been

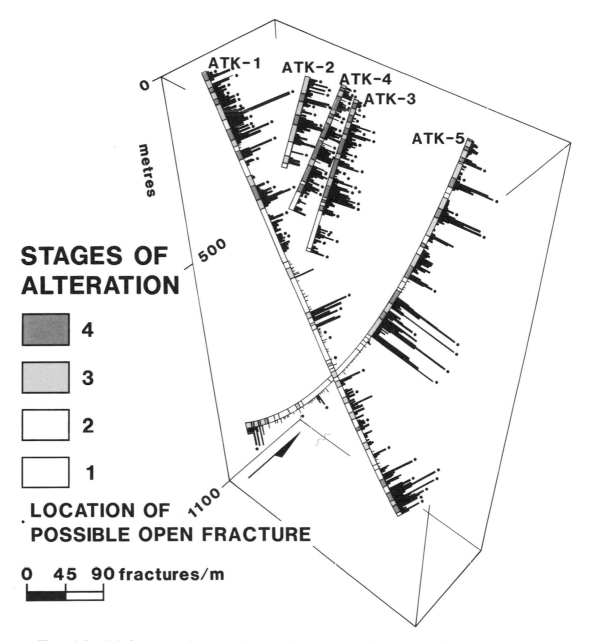

Figure 4. Borehole fracture logs from test holes near a fault on the Eye-Dashwa Lakes Pluton.

noted in work at the Stripa Mine in Sweden (Doe, personal communication, 1985).

On the other hand, in the relatively intact rock blocks between major fault zones, the rock is tight. All fractures are chemically filled and, except for the deep semihorizontal zones of high permeability  mentioned above, ground-water flow is minimal and no systematic flow pattern has been measured or observed.

The evidence to date from CNFWMP research indicates that, other than for horizontal exfoliation sheeting joints probably produced by glaciation, in the upper 100 m or so, all fractures were formed congruent with or shortly after the emplacement of the plutonic rock. Since that time these fractures have been rejuvenated many times (Brown and others, 1980). The implication is

that deep flow systems are controlled by these structures, with perhaps some influence by regional slope but little influence by local topography.

## GROUND-WATER QUALITY

Surface waters and shallow ground waters of the Canadian Shield region are generally very low in dissolved constituents. Surface water is typically soft, neutral to slightly acidic, colored but not turbid. In the western portion of the region, glacial deposits contain carbonate clasts from the area of provenance and these carbonate materials have a local influence on water quality. Similarly, where glacial-lake sediments occur, ground water tends to carry more $Na^+$, $Ca^{2+}$, and $HCO_3^-$ and is harder.

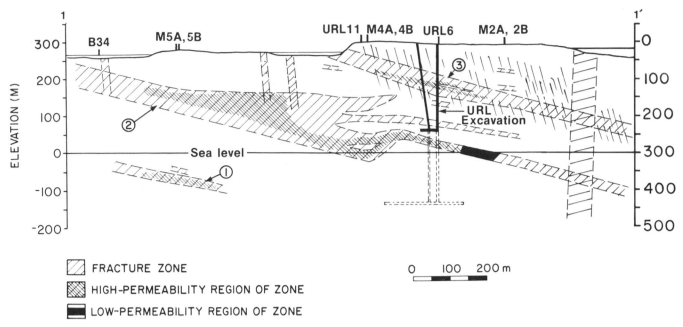

Figure 5. Schematic cross section through part of the Lac du Bonnet Batholith at the URL site illustrating the presence of flat-lying fracture zones (from Davison and Guvanasen, 1985).

Shallow ground water is similar to the surface water, but Frape and Fritz (1982) show that the average composition of several hundred samples increases in dissolved constituents within the first 150 m and increases by an order of magnitude within the first 300 m of depth (Fig. 8). Fewer samples are available from greater depths, but trends are rather clear. Ground waters below about 800 m are Ca-Na-Cl brines with dissolved solids exceeding 100,000 mg l$^{-1}$. The waters are depleted in $HCO_3^-$ and $SO_4^{2-}$ and, to a lesser degree, in $Mg^{2+}$.

Brackish water is generally encountered at intermediate depths. Kameneni and others (1982) believe that the minerals in fracture fillings and fracture walls react with ground water to produce the chemistry observed in boreholes in the Eye-Dashwa Lakes Pluton. They account for the trends in $Na^+$ and $Mg^{2+}$ content by exchange with clay minerals. Low $Ca^{2+}$ concentrations in shallow ground water are attributed to lesser abundance of calcite fillings in fractures in shallow zones than at depth.

In the Lake Nipigon–Lake Superior area, some small springs and seeps discharge ground water of high salinity—up to $10^4$ mg/L. They are known locally as "moose licks" and have been reported elsewhere on the Shield. The origin of the highly saline water is not clear, but may be caused by solution of Precambrian evaporites such as the Sibley Formation. However, some "moose licks" are found in regions where these rocks do not occur. Also, geochemical considerations preclude major contributions from evaporite solution (Blackmer and others, 1987).

Instead, isotope data suggest that these waters are depleted in $^{18}O$ and $^2H$ (deuterium) and are distinct from other local ground waters, indicating that these are "old" waters recharged under cooler climatic conditions. The salinity of these waters may be due to either mixture of surface water with deep saline ground waters encountered at depth across the Shield or derived through rock-water interaction, perhaps involving leakage of fluid inclusions, dissolution of grain boundary salts, and highly saline pore fluids—the residuals of former brines permeating these rocks. Strontium isotope data suggest that rock-water interaction dominates their geochemical history.

Ground waters from depths exceeding 650 m are fluids that are distinct chemically and isotopically from those in the shallower zones. These deep ground waters are exceedingly saline and are either enriched in deuterium, depleted in $^{18}O$, or both, if compared to normal meteoric waters. These characteristics and the fact that the subdued relief of this part of the Canadian Shield is thought to be insufficient to establish deep, active flow systems strongly suggest that these fluids are old and have undergone substantial rock-water interactions.

Chemical data for Canadian Shield ground waters have been summarized by Frape and others (1984). The most salient feature of these brines is not their high salinity but their chemical composition, which is Ca-Na-Cl dominated. Much higher salinities are known for sedimentary basin brines, but the strong Ca dominance is rarely encountered.

This difference is documented in Figure 9 where Na-Ca-Mg molar percentages are compared on a triangular diagram. Little or no overlap exists between the compositions of these Shield brines and those of evaporated ocean waters or other natural saline systems.

It is also noteworthy that for the most concentrated brines, this Ca dominance is independent of the type of host rocks, which range from very mafic to felsic. On the other hand, magnesium

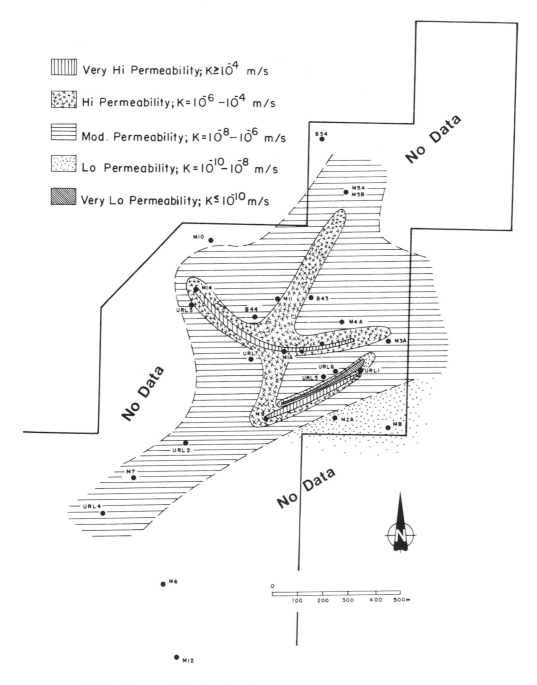

Figure 6. Variations in permeability in a flat-lying fracture zone at the URL site (from Davison and Guvanasen, 1985).

concentrations in these brines are more independent of salinity (and thus chlorinity) and rather closely linked to rock type. For example, the elevated magnesium concentrations in brines from a mine at Thompson, Manitoba, are most likely related to magnesium-rich ultramafic rocks with which those fluids had interacted (Frape and others, 1984).

A strong relationship between host rock and brine is also seen in the strontium data. Strontium isotopic composition suggests that "equilibration" between solid and aqueous phases has occurred in most of these very saline systems.

However, isotope data, and specifically $^{18}O$ and $^2H$ concentrations, indicate that brine samples collected to date may not represent the most saline fluids. Figure 10 shows increasing salinities with increasing $^2H$ contents and, to a lesser degree, $^{18}O$ contents. These relationships most certainly reflect mixing of brine fluids with local, nonsaline ground waters. In addition, in disturbed environments such as those created by operating mines and where deep cones of depression result from mine dewatering, local surface waters can penetrate to great depth. Water from the surface has been found in discharges in exploration boreholes

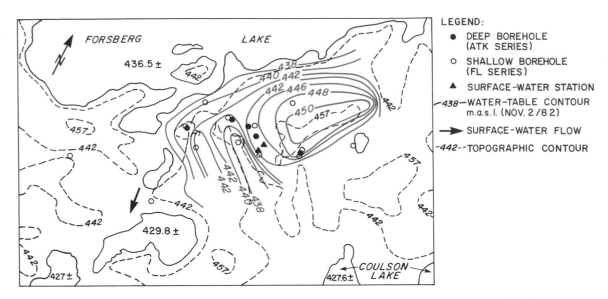

Figure 7. Ground-water mound under a rock knoll on the Eye-Dashwa Lakes Pluton near Atikokan, Ontario. Note the apparent influence of the faults (from Lee and others, 1983).

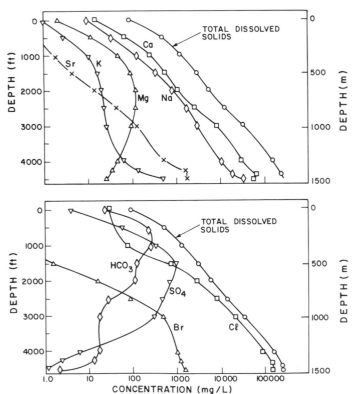

Figure 8. Average values for the concentrations of major ions in ground waters of the Canadian Shield.

drilled from mine levels, at depths exceeding 1,500 m. Thus hydraulic continuity does exist in many vertical "fracture" systems and cannot necessarily be attributed to man's activities (Frape and others, 1984; Gascoyne, personal communication, 1985).

The salinity-isotope regression lines can be extrapolated for each site and, if combined with geochemical considerations, permit the "definitions" of the isotopic composition of a "source" brine as it may exist across the Canadian Shield. The shaded area in Figure 10 shows the results; note that the extrapolated composition indicates somewhat higher $^{18}O$ and $^2H$ concentrations than measured values. Yet this extrapolation agrees with estimates based on brine chemistries and where it was assumed that the source brines would be at halite saturation (Pearson, 1986). Halite is a common fracture mineral and is frequently present in fluid inclusions of fracture minerals in crystalline rocks.

Isotope and geochemical data clearly show that the genesis of the deep brines in the Canadian Shield cannot be discussed in terms of the "origin" of the fluids. Rock-water interactions have almost certainly obliterated most primary characteristics. These brines should be viewed as fluids that are directly linked to the geochemical (and hydrologic) evolution of deep crystalline rocks. As such, the brine would have to be called "crustal fluids" whose primary origin cannot be uniquely defined.

## TECHNIQUES FOR HYDROGEOLOGIC STUDIES ON PRECAMBRIAN TERRAIN

Techniques for ground-water studies of Precambrian terrain are similar to those developed for other terrain in terms of the overall approach. They differ in some details since, for most applications, the prime targets in exploration are fractures and fracture zones and associated dikes and sills of various sizes

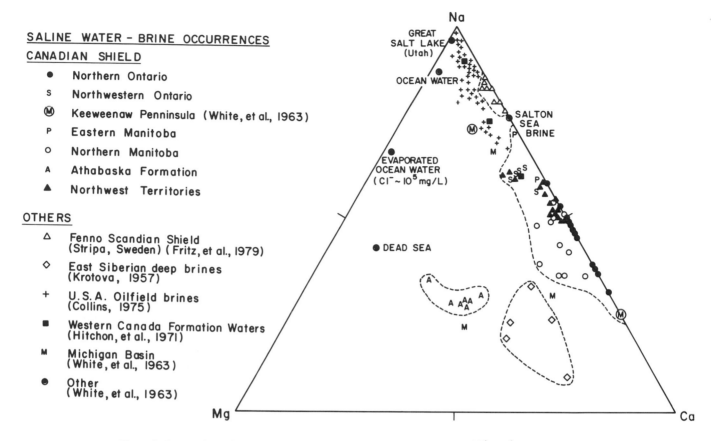

Figure 9. Comparison of chemical characteristcs (mole percent) of brines ($>10^4$ mgl$^{-1}$ TDS) with brines and saltwater from other environments (from Frape and Fritz, 1982).

within the rock mass. Some are very costly and unlikely to be used in water-resource projects. Topography, surface drainage, overburden, vegetation, and lake coverage are generally controlled by rock structure and/or surficial deposits and so may provide important hydrogeologic information.

Airborne techniques are invaluable as the first tools for investigating the regional scale. Satellite imagery and airphoto coverage, including infrared analyses (Lee and Tracey, 1984), can delineate not only vegetation cover and rock/overburden type but potential zones of discharging ground water and high velocity flow or potential discharging ground water in lakes and rivers.

Geophysical airborne surveys such as gravimetric, gamma spectrometry, and very-low-frequency electromagnetic (VLF-EM) surveys can be used to delineate rock type, major structural features, and areas of thick overburden (Soonawala, 1984). Gravity surveys provide an initial indication of the shape and possible depth of rock bodies. Gamma spectrometry can delineate plutonic rock from country rock due to the higher thorium, uranium, and potassium content of the former. Aeromagnetic techniques (Hood and others, 1976) not only aid in distinguishing plutonic from country rock and in identification of dikes and weathered zones, but also in identification of linear conductors that may be related to steep fractures and faults.

As is the case everywhere, a first-order surface description of

the lithological and structural framework, including surficial deposits and consequent vegetation and surface-water drainage patterns, together with adequate topographic control at the scale of interest, are required as basic information in a hydrogeological assessment.

Useful land-based techniques include all standard methods developed for geological, geophysical, and hydrological assessments with major emphasis on structure and regional stress assessments. Geophysical methods that have been used successfully are seismic reflection, VLF-EM, sonar, and radar tools. Seismic reflection holds great promise for detecting major flat-lying fracture zones (Green and Mair, 1983), while VLF-EM can help in interpretation of clay-based overburden materials. Sonar methods, when used from boats on lakes, are extremely valuable in determining depth of lakes and depth of sediments in lakes and thus in confirming the shape of water bodies and continuity of major lineaments (Holloway, 1983). Radar is useful in determining overburden thickness on land.

Seasonal analyses of surface-water body temperature and chemistry can be useful in identifying recharge and discharge zones and in detecting perennial and ephemeral surface discharge events (Ridgway and Pearson, 1987). Included in this work should be analysis for major ions and natural radioactivity (Larocque and Gascoyne, 1986) from potential discharge areas and

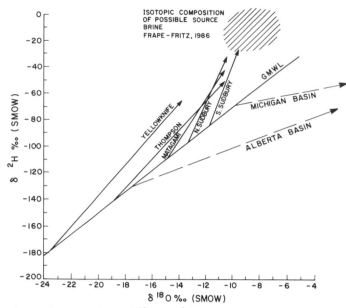

Figure 10. Composition of $^{18}O$ and $^2H$ contents in ground waters and brines of the Canadian Shield and representative adjacent sedimentary basins. Arrows are regression lines for data points and indicate direction of increasing salinity.

for tritium in both recharge and discharge areas. Other considerations include attempting to find effluent and influent conditions in rivers and streams focussing on the intersection of structural controls, since, although most base flow and stream flow is from overburden and lake storage, there have been strong indications of discrete deeper bedrock discharges. Related to this work is the need, in conjunction with lake sonar work, to examine the beds of lakes at structurally controlled points for submarine discharge. Here scuba diving and lake-bed drags fitted with temperature and conductivity measurement instruments are useful (Lee, 1985).

Techniques used in borehole assessment include identification of permeable fracture zones and chemical variations using many borehole techniques. They include through-the-bit hydraulic testing during drilling with single-packer systems, borehole television, acoustic television, core, gamma, neutron, fluid resistivity, temperature, sonic and radar logging, and straddle-packer testing and sampling of selected borehole zones (Lau, 1983; Lee and others, 1983; Davison and others, 1982; Davison, 1984; Paillet and others, 1985; Bottomley and others, 1984). In addition, several methods to establish continuity of permeable zones are useful. These include interborehole hydraulic (Black and Kipp, 1981) and tracer tests (Novakowski and others, 1985; Webster and others, 1970) over distances of 10s to 100s of meters in conjunction with geophysical techniques such as cross-hole seismic (Wong and others, 1982) and down-hole radar (Olsson and others, 1982; Holloway and others, 1985) and geological assessment of continuity (McEwan and Hillary, 1985). Two important considerations in this work are isolation of all test sections in boreholes used in hydrogeological work soon after drilling to stop intraborehole flows along the length of the borehole, and

proper completion of the borehole in terms of extracting all drilling fluids and ensuring that boreholes (and fractures) are "clean."

The first of these considerations is met by installation of borehole isolation systems such as multilevel piezometers or multipoint casings (Wingrove and others, 1984; Davison, 1984). The second is a matter of correct borehole test procedures and includes the requirement that zones of interest are pumped long enough to remove drilling fluids.

When these devices are installed, time and effort can be expended on establishing both ground-water chemistry and hydraulic-head values and fluctuations. In ground-water sampling it is important that the pumping or bailing be sufficient to ensure the obtaining of formation fluid. On the other hand, care must be taken that the radius of influence of the test does not tap other systems.

## CONCLUDING REMARKS

It should be clear from this paper and the recent dates on most references that knowledge of the hydrogeology of the Precambrian Shield region is in a state of very rapid growth. We have tried to present some of the important observations made during the past decade or so because these have provided information not available in earlier publications.

As is common in a review paper, we have presented mainly our interpretations and those of colleagues without much basic data. Supporting data are to be found in the references.

## REFERENCES CITED

Black, J. H., and Chapman, N., 1981, In search for nuclear burial grounds: New Scientist, v. 91, p. 402–404.

Black, J. H., and Kipp, K. L., 1981, Determination of hydrogeological parameters using sinusoidal pressure tests; A theoretical appraisal: Water Resources Research, v. 17, no. 3, p. 686–692.

Blackmer, A. J., Frape, S. K., Fritz, P., and McNutt, R. H., 1987, Geochemistry and hydrogeology of two salt-water springs in the Nipigon regon: Canadian Journal of Earth Science (in press).

Bottomley, D. J., Ross, J. D., and Graham, B. W., 1984, A borehold methodology for hydrogeochemical investigations in fractured rock: Water Resources Research, v. 20, no. 9, p. 1277–1300.

Bredehoeft, J. D., and Maini, T., 1981, Strategy for radioactive waste disposal in crystalline rocks: Science, v. 213, no. 4505, p. 293–296.

Brown, P. A., Kameneni, D. C., Stone, D., and Thivierge, R. H., 1980, General geology of the Eye–Dashwa Lakes Pluton, northwest Ontario; Current Research Activities, Part A: Geological Survey of Canada Paper 80-1A, p. 379–384.

Charron, J. E., 1967, The Canadian Shield hydrogeological region, in Brown, I. C., ed., Groundwater in Canada: Geological Survey of Canada Economic Geology Report no. 24, p. 120–130.

Collins, A. G., 1975, Geochemistry of oilfield waters: Amsterdam, Elsevier, 485 p.

Davis, N., and Turk, L. S., 1964, Optimum depth of wells in crystalline rocks: Groundwater, v. 2, no. 2, p. 6–11.

Davison, C. C., 1984, Hydrogeological characterization at the site of Canada's Underground Research Laboratory, *in* Proceedings of the International Groundwater Symposium on Groundwater Resources Utilization and Contaminant Hydrogeology: Atomic Energy Canada, v. 2, p. 310–335.

Davison, C. C., and Guvanasen, V., 1985, Hydrogeological characterization, modelling and monitoring of the site of Canada's underground research laboratory: Atomic Energy of Canada Limited Paper 8676, 26 p.

Davison, C. C., Keyes, W. S., and Paillet, F. L., 1982, Use of borehole-geophysical logs in hydrologic tests to characterize plutonic rock for nuclear waste storage: Atomic Energy of Canada Limited Report, WNRE-7810, 103 p.

Dickin, R., Frape, S. K., Fritz, P., Leach, R.E.J., and Pearson, R., 1984, Groundwater chemistry to depths of 1,000 m in low permeability granite rocks of the Canadian Shield, *in* Proceedings of the International Groundwater Symposium on Groundwater Resource Utilization and Contaminant Hydrogeology: International Association of Hydrogeologists, p. 357–371.

Dugal, J.J.B., Pearson, R., and Stone, D., 1981, Hydrogeological testing and fracture analysis of the Eye–Dashwa Lakes granitic pluton at Atikokan, Ontario: Atomic Energy of Canada Limited Paper TR-7363, p. 8.4.1–8.4.16.

Fisheries and Environment Canada, 1978, Hydrogeological atlas of Canada.

Frape, S. K., and Fritz, P., 1982, An initial summary of the major-minor element chemistry of the groundwaters from the Canadian Shield: Atomic Energy of Canada Limited Report TR-210, p. 43–46.

—— , 1986, Geochemical trends for ground waters from the Canadian Shield, *in* Fritz, P., and Frape, S. K., eds., Saline waters and gases in crystalline rocks: Geological Association of Canada, Special Paper 33, p. 19–38.

Frape, S. K., and Patterson, R. J., 1981, Chemistry of interstitial water and bottom sediments as indicators of seepage patterns in Pearch Lake, Chalk River, Ontario: Limnology and Oceanography, v. 26, p. 500–517.

Frape, S. K., Fritz, P., and McNutt, R. H., 1984, Water-rock interaction and chemistry of groundwaters from the Canadian Shield: Geochimica et Cosmochimica Acta, v. 48, p. 1617–1628.

—— , 1985, Water-rock interactions and the percipitation of gypsum fracture fillings in the Canadian Shield: Atomic Energy of Canada Limited Report TR-299, p. 315–333.

Freeze, R. A., and Whitherspoon, P. A., 1966, Theoretical analysis of regional groundwater flow; 1. Analytical and numerical solutions to the mathematical model: Water Resources Research, v. 2, p. 641–656.

—— , 1967, Theoretical analysis of regional groundwater flow; 2. Effect of water-table configuration and subsurface permeability variation: Water Resources Research, v. 3, p. 623–634.

—— , 1968, Theoretical analysis of regional groundwater flow; 3. Quantitative interpretations: Water Resources Research, v. 4, p. 581–590.

Fritz, P., and Frape, S. K., 1982, Saline groundwaters in the Canadian Shield—A first overview: Chemical Geology, v. 36, p. 179–190.

Fritz, P., Barker, J. F., and Gale, J. E., 1979, Geochemistry and isotope hydrology of groundwaters in the Stripa granite. Swedish-American Co-operation Program on Radioactive Waste Storage in Mine Caverns in Crystalline Rock: Lawrence-Berkeley Laboratory Publication 8285, 135 p.

Green, A. G., and Mair, J. A., 1983, Sub-horizontal fractures in a granitic pluton; Their detection and implications of radioactive waste disposal: Geophysics, v. 48, p. 1428–1449.

Grisak, G. E., and Cherry, J. A., 1975, Hydrogeologic characteristics and response of fractured till and clay confining a shallow aquifer: Canadian Geotechnical Journal, v. 12, p. 23–43.

Hitchon, B., Billings, G. K., and Klovan, J. E., 1971, Geochemistry and origin of formation waters in the western Canada sedimentary basin—II, Factors controlling chemical composition: Geochemica et Cosmochimica Acta, v. 35, p. 567–598.

Holloway, A. L., 1983, Sonar profiling in lakes near the Atikokan Research Area, Ontario, *in* Proceeding of the 17th Information Meeting of the Canadian Nuclear Fuel Waste Management Program: Atomic Energy of Canada Limited Report TR-299, v. 2, p. 622–633.

Holloway, A. L., Soonawala, N. M., and Collett, L. S., 1985, Three-dimensional fracture mapping in a granitic excavation using ground-penetrating radar, *in* Proceedings of the 87th Annual General Meeting of the Canadian Institute of Mining, Vancouver, p. 14.

Hood, P., Kornick, L. J., and McGrath, P. H., 1976, The aeromagnetic gradiometer; A new geophysical took for geophysical mapping programs [abs.]: International Geological Congress, v. 2, p. 391–392.

Kameneni, D. C., and Dugal, J.J.B., 1982, A study of rock alteration in the Eye–Dashwa Lakes Pluton, Atikokan, northwest Ontario, Canada: Chemical Geology, v. 36, p. 35–57.

Kameneni, D. C., Stone, D., Dugal, J.J.B., and Brown, P. A., 1982, Geochemistry of the Eye–Dashwa Lakes Pluton, Atikokan, northwestern Ontario: Atomic Energy of Canada Limited Technical Record 201, p. 21–41.

Krotova, V. A., 1957, Conditions of formation of calcium chloride waters in Siberia: Petroleum Geology, v. 2, p. 545–552.

Laroque, J.P.A., and Gascoyne, M., 1986, A survey of the radioactivity of surface water and groundwater in the Atikokan area, northwestern Ontario: Atomic Energy of Canada Limited Report, TR-379, 12 p.

Lau, J.S.O., 1983, The determination of true orientations of fractures in rock cores: Canadian Geotechnical Journal, v. 20, p. 221–227.

Lee, D. R., 1977, A device for measuring seepage flux in lakes and estuaries: Limnology and Oceanography, v. 22, p. 140–147.

Lee, D. R., and Cherry, J. A., 1979, A field exercise on groundwater flow using seepage metres and mini-piezometers: Journal of Geological Education, v. 27, no. 1, p. 6–10.

Lee, D. R., and Tracey, J. P., 1984, Identification of groundwater discharge locations using thermal infrared imagery: Proceeding of the 9th Canadian Symposium on Remote Sensing, p. 301–308.

Lee, P. K., Pearson, R., Leech, R.E.J., and Dickin, R., 1983, Hydraulic testing of deep fractures in the Canadian Shield: International Association of Engineering Geologists Bulletin, no. 26–27, p. 461–465.

McEwan, J., and Hillary, E., 1985, Early fracture evolution within the Eye–Dashwa Lakes Pluton, Atikokan, Ontario, Canada: Journal of Structural Geology, v. 7, no. 5, p. 591–603.

Meyboom, P., 1966, Unsteady groundwater flow near a willow ring in hummocky moraine: Journal of Hydrology, v. 4, p. 38–62.

Novakowski, K. S., Evans, G. V., and Raven, K. G., 1985, Field experiments investigating radionuclide migration in plutonic rock, *in* Hydrogeology of rocks of low permeability: International Association of Hydrogeologists, 17th International Congress Proceedings, p. 345–357.

Olsson, O., Duran, O., Jamtlid, A., and Stenberg, L., 1982, Geophysical investigations for the characterization of a site for radioactive waste disposal, *in* Proceedings of a Workshop on Geophysical Investigations in Connection with Geological Disposal of Radioactive Waste: Paris, France, Nuclear Energy Agency, Office of Economic Cooperation and Development, p. 45–56.

Paillet, F. L., Keyes, W. S., and Hess, A. E., 1985, Effects of lithology on televiewer-log quality and fracture interpretation: Transactions of the Society of Petroleum and Well Loggers 26th Annual Logging Symposium, 30 p.

Pearson, J. F., 1986, Models of mineral controls on the composition of saline groundwaters of the Canadian Shield, *in* Fritz, P., and Frape, S. K., eds., Saline waters and gases in crystalline rocks: Geological Association of Canada Special Volume 33, p. 39–51.

Pearson, R., 1984, Geoscience research activities in 1983, *in* Proceedings of the 16th Information Meeting of the Canadian Nuclear Fuel Waste Management Program: Atomic Energy of Canada Limited Report TR-218, p. 30–50.

Pfannkuch, H. O., and Winter, T. C., 1984, Effect of anisotropy and groundwater system geometry on seepage through lakebeds; 1. Analog and dimensional analysis: Journal of Hydrology, v. 75, p. 213–237.

Pfannkuch, H. O., Edgar, D., Van Luik, A., and Harrison, W., 1983, Hydrology, *in* Harrison, W., ed., Geology, hydrology, and mineral resources of crystalline rock areas of the Lake Superior region, United States: Argonne, Illinois, Argonne National Laboratory, ANL/ES-134, pt. 1, p. 184–302.

Raven, K. G., and Gale, J. E., 1986, A study of the surface and subsurface structural and groundwater conditions at selected underground mines and excavations: Atomic Energy of Canada Limited Report TR-177, 81 p.

Raven, K. G., Smedley, J. A., Sweezey, R. A., and Novakowski, K. S., 1985, Field investigations of a small groundwater flow system in fractured monzonitic gneiss, *in* Hydrogeology of rocks of low permeability: International Association of Hydrogeologists, 17th International Congress Proceedings, p. 72–86.

Ridgway, W. R., and Pearson, R., 1987, Regional hydrogeological characterization studies at Atikokan, northwest Ontario: Atomic Energy of Canada Limited Paper (in press).

Siegel, D. I., 1981, Hydrogeologic settings of the Glacial Lake Agassiz Peatlands, northern Minnesota: U.S. Geological Survey Water Resources Investigation 81-24, 33 p.

Snow, D. T., 1968, Hydraulic characterization of fractured crystalline rocks of the Front Range and implications to the Rocky Mountain Arsenal Well: Golden, Colorado School of Mines Quarterly, v. 63, no. 1, p. 167–199.

Soonawala, N. M., 1984, An overview of the geophysics activity within the Canadian nuclear fuel waste management program: Geoexploration, v. 22, p. 149–168.

Stockwell, C. H., and 7 others, 1970, Geology of the Canadian Shield, *in* Stockwell, C. H., ed., Geology and economic minerals of Canada: Geological Survey of Canada Economic Geology Report no. 1, p. 43–150.

Stone, D., 1984, Sub-surface fracture maps predicted from borehole data; An example from the Eye–Dashwa Lakes Pluton, Atikokan, Ontario, Canada: International Journal of Rock Mechanics, Mining Science, and Geomechanics Abstracts, v. 21, no. 4, p. 183–199.

Toth, J., 1963, A theoretical analysis of groundwater flow in small drainage basins: Journal of Geophysical Research, v. 68, p. 4795–4812.

Webster, D. S., Proctor, J. F., and Marine, I. W., 1970, Two-well tracer test in fractured crystalline rock: U.S. Geological Survey Water Supply Paper 1544-1, 26 p.

White, D. E., Hem, J. D., and Waring, G. A., 1963, Geochemical composition of subsurface waters: U.S. Geological Survey Professional paper 440-F, 67 p.

Wingrove, T. R., Rudolph, D. L., and Farvolden, R. N., 1984, Field evidence for groundwater flow in Precambrian terrain near Atikokan, Ontario, *in* Proceedings of the International Groundwater Symposium on Groundwater Resource Utilization and Contaminant Hydrogeology: International Association of Hydrogeologists (Canadian chapter), v. 2, p. 580–593.

Winter, T. C., 1976, Numerical simulation analysis of the interaction of lakes and groundwater: U.S. Geological Survey Professional Paper 1001, 45 p.

Wong, J., Hurley, P., and West, G. F., 1983, Crosshole seismology and seismic imaging in crystalline rocks: Geophysics Research Letters, v. 10, no. 8, p. 686–689.

Wright, H., Jr., Goldstein, B., Harrison, W., Pfannkuch, H. O., Edgar, D., and Van Luik, A., 1983, Physiograph and surficial deposits, *in* Harrison, W., ed., Geology, hydrology, and mineral resources of crystalline rock areas of the Lake Superior region, United States: Argonne, Illinois, Argonne National Laboratory, ANL/ES-134, pt. 1, p. 167–183.

MANUSCRIPT ACCEPTED BY THE SOCIETY MAY 15, 1987

The Geology of North America
Vol. O-2, Hydrogeology
The Geological Society of America, 1988

## Chapter 16

# *Region 13, Western Glaciated Plains*

**D. H. Lennox**
*Inland Waters/Lands Directorate, Environment Canada, Ottawa, Ontario K1A 0E7, Canada*
**H. Maathuis**
*Saskatchewan Research Council, Saskatoon, Saskatchewan, Canada*
**D. Pederson**
*Department of Geology, University of Nebraska, Lincoln, Nebraska*

## INTRODUCTION

The Western Glaciated Plains region (Fig. 3; Table 2, Heath, this volume) occupies an area of nearly two million km$^2$ in southern Canada and the northern United States. About three quarters of this area lies in Canada, extending across portions of the four Canadian provinces of British Columbia, Alberta, Saskatchewan, and Manitoba. There it forms the region referred to by Brown (1967) as the Interior Plains hydrogeological region. It is thus the hydrogeological equivalent of that part of the western Canada sedimentary basin that lies south of the permafrost zone and is identical to the Southern Interior Platform area identified by Douglas and others (1970). In the U.S. the region extends across parts of Montana, North Dakota, South Dakota, Minnesota, Nebraska, Iowa, and Kansas.

The climate throughout most of the region is subhumid continental except for that part that lies in Montana, southwestern Saskatchewan, and southeastern Alberta, where it is classified as arid to semiarid (Brown, 1967). Total precipitation varies from 250 to 750 mm annually in most parts of the region, exceeding this figure only along the western, eastern, and southeastern boundaries.

The region is flat to rolling with local areas of hummocky terrain with local relief of 30 to 60 m, except in flat glacial lake plains such as at Regina, Saskatchewan, and the Red River Valley of Manitoba, North Dakota, and Minnesota (Heath, 1984; Brown, 1967). Along the southwestern edge of the region, eroded bedrock terrain is only slightly modified by a thin, in places discontinuous, veneer of glacial sediment that generally lacks constructional topography. Over most of the region, however, the eroded bedrock surface is completely obscured by a blanket of glacial sediment that in places is more than 300 m thick.

Two pronounced west-to-east downward steps in the terrain are formed by the Missouri Coteau and the Manitoba Escarpment (Bostock, 1970). The former runs from South Dakota across North Dakota and into southern Saskatchewan, becoming indefinite and disappearing toward the northwest. The latter is a

significant feature in southwestern Manitoba and eastern North Dakota, rising as much as 300 m above the adjoining plain. In addition to these major features there are local topographic irregularities, such as deeply dissected river valleys with depths as great as 125 m and prominent steep-sided, flat-topped hills as much as 50 km across and rising 60 to 200 m above the surrounding plains. Across the region the elevation above sea level varies from about 1,200 m in the west to about 200 m in eastern Manitoba. Along the edge of the region in northeastern Saskatchewan, elevations in places exceed 400 m.

The Western Glaciated Plains region is underlain by nearly horizontal strata of Paleozoic, Mesozoic, and Tertiary age consisting primarily of sandstone, shale, limestone, and dolomite (Heath, 1984; Meyboom, 1967). These strata, which rest on the Precambrian crystalline basement, reach a thickness of more than 5,000 m in the deepest parts of the Alberta basin and 3,500 m in the Williston basin (Hitchon, 1969a).

The bedrock is almost everywhere concealed under a mantle of Pleistocene deposits including till, glacial lacustrine and glacial fluvial sediments, and loess (Heath, 1984; Meyboom, 1967). In Canada, according to Meyboom, about 60 percent of the plains area is covered by till and about 40 percent by lake deposits. Less than one percent is covered by glacial fluvial deposits. Although the entire region is blanketed by a thin veneer of windblown silt (Clayton and others, 1976), extensive areas of thick loess occur only in the southern part of the region in Kansas, Nebraska, and Iowa.

The region is thus both areally extensive and geologically diverse. Hydrogeologically, it provides the setting for a complex network of shallow, intermediate, and deep ground-water flow systems, which extend throughout the region from the land surface to the Precambrian basement. Model studies by Toth (1962, 1963) and by Freeze and Witherspoon (1966, 1967, 1968) have indicated that the spatial disposition of flow systems should depend on land surface topography and lithological variations

Lennox, D. H., Maathuis, H., and Pederson, D., 1988, Region 13, Western Glaciated Plains, *in* Back, W., Rosenshein, J. S., and Seaber, P. R., eds., Hydrogeology: Boulder, Colorado, Geological Society of America, The Geology of North America, v. O-2.

within the geological flow medium. Individual flow systems might be local in nature, might extend across the whole region, or might fall somewhere between these two extremes.

The existence in the Western Glaciated Plains region of both shallow, local ground-water flow systems and deeper, underlying, but still localized, intermediate flow systems has been confirmed in field studies (Freeze, 1969; Meyboom, 1966; Toth, 1966). The systems studied by these investigators had maximum lateral dimensions and depths of a few tens of kilometers and of a hundred or so meters, respectively. At the other end of the scale, van Everdingen's (1968) geochemical investigations and Hitchon's (1969a, 1969b) fluid-potential analyses for deep formation waters of the western Canada sedimentary basin provided convincing evidence for the existence of deep regional flow systems. All these studies also provide strong support for model predictions concerning the controlling effects of land topography and geology. In addition, Hitchon's fluid-potential analyses demonstrated that major rivers, such as the South Saskatchewan River, act as local drains, in some areas producing drawdown effects to depths of more than 1,500 m. This particular influence had not been considered in the model studies.

The earlier model studies, field investigations, and geochemical and fluid-potential analyses have been followed by investigations of broader, deeper intermediate flow regimes (Hitchon, 1984; Toth, 1978). The flow systems considered by these authors typically extend as much as 200 km laterally and a few thousand meters vertically. They generally overlie the deeper regional systems described by Hitchon (1969a, 1969b) and van Everdingen (1968) and modify the flow patterns within these deep systems accordingly. These deeper intermediate systems provide clear evidence for the effects at depth of topographic surface elevation differences ranging up to 600 m and more. Such elevation differences are characteristic of the topographic relief between many of the major upland areas and the adjoining lowlands within the Western Glaciated Plains region.

The more recent studies by Toth (1978) and Hitchon (1984) have also shown a convincing correlation between hydrocarbon accumulation and the rising limbs of some of the deeper ground-water flow systems. In addition, Hitchon has drawn attention to the influence of the ground-water flow regime on regional heat flow, a geophysical variable that may also be related to hydrocarbon accumulation. Majorowicz, Jones, and their colleagues have carried out an extensive series of investigations into the geothermal heat flow field in western Canada. Among the questions investigated was the correlation between heat flow, basin hydrodynamics, and hydrocarbon occurrences (Majorowicz and others, 1986). Their results have provided further support for the conclusions of van Everdingen (1968) and Hitchon (1969a, 1969b) on the regional flow system and those of Toth (1978) and Hitchon (1984) on hydrocarbon accumulation.

Details of the complex regional flow regime in the Western Glaciated Plains region are still being ascertained, in both Canada and the U.S. (Gosnold, 1985). However, the regime still remains to some extent ill-defined except in a broad regional sense. Certain aquifers and aquifer systems are, nevertheless, relatively well known. This chapter focuses on a number of examples from this category, selected on the basis of ready availability of published literature and on the personal experience of the authors. This approach has inevitably led to the omission from the discussion of a number of regionally important aquifers, such as the Milk River aquifer of northern Montana and southern Alberta (Meyboom, 1960).

Our discussion identifies three major generic groups of aquifers on the basis of their geological setting, with one to three aquifers included in each group. The first group is found where the sedimentary sequence thins against the Precambrian rocks of the Canadian Shield along the northeastern edge of the region and against the mountains in the southwest. In these areas, Paleozoic carbonate rocks are near enough to the surface to serve as aquifers. The Winnipeg carbonate aquifer is typical of the lower Paleozoic limestone and dolomite that serve as an important aquifer across northern Alberta, Saskatchewan, and Manitoba, and as far south as eastern Minnesota. Late Paleozoic carbonates, such as the Madison Formation, are equally important in places in the southwestern part of the region.

Vast wedges of clastic continental sediment that spread eastward from the rising Rocky Mountains during the Late Cretaceous and Early Tertiary, along with the marine sandstone deposited at the seaward edges mostly during the Early Cretaceous, constitute the second generic grouping of aquifers. The continental sequences that make up the wedges consist typically of sandstone and coal layers interbedded with siltstone and shale. The Dakota aquifer in the south, the aquifers in the Fox Hills to Fort Union formations in the central part of the region, and the Judith River aquifer in the north are examples of this type of aquifer.

The third generic grouping of aquifers includes those that are found in the glacial sediments that occur throughout the region and that give it its unity. Although individually named aquifers are recognized within the glacial sequence, they are generally much smaller and more localized than the more regionally extensive bedrock aquifers. The discussion of this group of aquifers is broadly generic with reference to numerous specific examples.

## AQUIFERS IN CARBONATE ROCK

### Winnipeg carbonate aquifer

The Winnipeg carbonate aquifer has been described by Render (1970). The aquifer consists primarily of the limestone and dolomite of the Ordovician Red River Formation immediately underlying the surficial deposits in the Winnipeg, Manitoba, area. It has two distinct zones (Fig. 1). The upper aquifer zone occurs in the top 15 to 30 m of the Silurian and Ordovician rocks, which range in thickness in the Winnipeg area from 76 to 230 m. Toward the east, this permeable zone occurs in dolomitic limestone of the Red River Formation of Late Ordovician age; toward the west it extends upward into dolomite, limestone, and

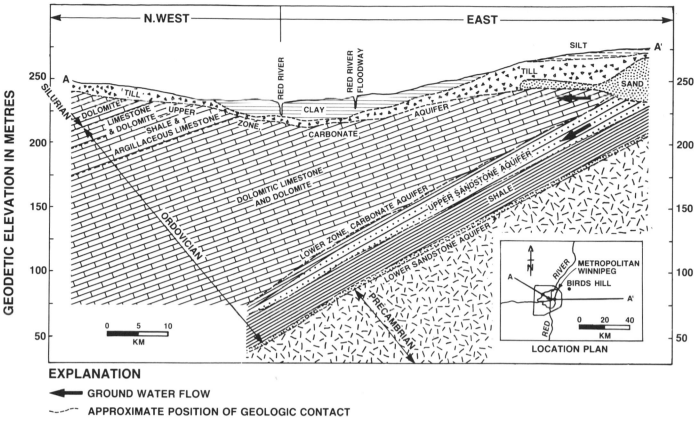

Figure 1. Hydrogeological cross section, Winnipeg carbonate aquifer (after Render, 1970).

argillaceous limestone of the overlying Silurian Stony Mountain Formation.

The lower aquifer zone occurs in the bottom 7.5 to 15 m of the Red River Formation, which is here underlain by the shale and sandstone of the Winnipeg Formation. Over most of the area it is separated from the upper zone by a sequence of slightly permeable carbonate rock. Toward the east, however, where the gently dipping beds of the Red River Formation rise toward the surface and subcrop beneath the surficial deposits, the two zones merge into one.

***Origin and character of porosity and permeability.*** The upper zone of the carbonate aquifer is characterized by a network of fractures, joints, and bedding planes, which are responsible for its high permeability (Render, 1970). These openings have been enlarged by solution. Large solution cavities, typical of karst topography, have been encountered during test drilling and have been utilized in well development. Fracture openings are largest at the bedrock surface and decrease in size with depth so that the upper 7.5 m of the carbonate rock constitutes the major permeability zone and is the section of most active ground-water flow. The transmissivity of the upper zone ranges approximately from 25 to 2,500 $m^2$/day and the storage coefficient from $1 \times 10^{-6}$ to $1 \times 10^{-3}$ (Render, 1970).

The permeability of the lower zone of the carbonate aquifer has not been affected by the development of secondary porosity

and is, therefore, much less than that of the upper zone. Its maximum transmissivity is probably less than 60 $m^2$/day (Render, 1970).

***Ground-water flow.*** Flow in the Winnipeg carbonate aquifer is complex (Render, 1970). The natural flow system that existed prior to European settlement responded to a number of factors, including (1) geology, (2) location at the end of a major flow system, and (3) the relationship to local recharge and drainage. These factors still play an important role, but the natural system has been transformed significantly as a result of human activities. These include extensive utilization of the aquifer for water supplies, construction of the Red River Floodway, and establishment of hydraulic connections with deeper saline aquifers.

Local recharge through overlying tills and fluvial deposits appears to be the predominant origin of ground water in the upper zone of the carbonate aquifer (Render, 1970). Some deterioration in quality toward the southwest might be due to intermixing with saline waters at the end of a continental flow system (van Everdingen, 1968). Birds Hill, an important topographic feature, with some 30 to 40 m of relief above the surrounding plain, is the major source of freshwater recharge (Fig. 1). The core of Birds Hill consists of outwash sands and gravels, which extend down to the carbonate aquifer. The surrounding surficial deposits are made up of glacial till and lacustrine clay, with considerably less infiltration capacity. Ground water originating in the Birds Hill recharge area spreads out laterally in all directions through the upper zone of the aquifer. To the east it moves updip toward discharge in the outcrop area of the Red River Formation. To the west, movement is toward discharge into the Red River, which runs roughly south to north through the Winnipeg area and appears to act as a local drain (Fig. 1). Groundwater flows in the upper zone of the aquifer in the area to the west of the Red River are also directed toward discharge in the river. Most likely the Birds Hill recharge area and discharge into the Red River have always been important characteristics of the carbonate aquifer flow system, both prior and subsequent to European settlement.

*Chemistry of water.* Water quality in the Winnipeg carbonate aquifer varies strikingly from east to west. Waters from the eastern and northwestern flow regions are fresh in the recharge areas and slightly brackish and very hard in the discharge area (Render, 1970). Total dissolved solids range from 300 to 1,500 mg/ℓ, and chloride ion concentrations range from 10 to 500 mg/ℓ. To the southwest the flow system contains brackish and saline water. Render postulates that this poor quality may be due to flow across formations containing evaporite deposits or to intermixing with saline water discharging into the system from a continental flow system (van Everdingen, 1968). Total dissolved solids in the southwest range from 2,000 to 10,000 mg/ℓ. Chloride ion values under north-central Winnipeg exceed 500 mg/ℓ, but Render states these values are due to contamination from wells drilled into sandstone aquifers of the Winnipeg Formation.

## Madison aquifer

The Madison aquifer is predominantly a carbonate rock of Mississippian age called the Madison Limestone or, where subdivided, the Madison Group consisting of (from bottom to top) the Lodgepole Limestone, the Mission Canyon Limestone, and the Charles Formation. The primary importance of the Madison aquifer is as a source of recharge to overlying aquifers in Cretaceous sandstones (Downey, 1984; Bredehoeft and others, 1983).

The three stratigraphic units were described by Miller (1976). The Lodgepole Limestone is mainly thin-bedded, argillaceous, dense to crystalline limestone, dolomitic limestone, or dolomite. In places, it is massive and contains large reef complexes. The Lodgepole Limestone ranges from dark-colored limestone interbedded with organic shale in the Williston basin to dolomite in extreme southeastern Montana. The Mission Canyon Limestone is mainly dense to crystalline, thick-bedded to massive limestone or dolomitic limestone. In the Williston basin, the basal Mission Canyon Limestone consists of dark-colored, partly argillaceous limestone. In central Montana, the basal Mission Canyon Limestone consists of dense to crystalline marine limestone that contains beds of coarsely crystalline, or oolitic and pisolitic, or bioclastic limestone. This unit grades upward to a massive evaporite section containing fine to coarsely crystalline porous dolomite and interbedded anhydrite and dolomitic limestone. In general

this upward gradation is common in the Mission Canyon Limestone. The Charles Formation consists mainly of restricted-marine sediments. Rock type ranges from thick anhydrite and halite beds with interbedded dense dolomite and limestone in the Williston basin to dense to coarsely crystalline normal-marine carbonates containing oolitic and bioclastic zones in central Montana. Anhydrite occurs throughout the three units.

Karst surfaces are widely developed on the top of the Madison Group. Relict karst surfaces have also been reported at the upper surfaces of the Mission Canyon and Lodgepole limestones.

*Origin and character of porosity and permeability.* Intergranular permeabilities of the individual rocks types within the Madison aquifer are insignificant as compared to the permeabilities due to fracturing, brecciation, or karst development (Thayer, 1983). Karstification and solution brecciation created significant porosity and permeability throughout the upper part of the Mission Canyon Limestone and, perhaps, the Charles Formation in northwestern North Dakota.

*Ground-water flow.* Recharge to the Madison aquifer is primarily in the outcrop areas of the Black Hills and the Rocky Mountains (Swenson, 1968; Miller, 1976; Downey, 1984). Rahn and Gries (1973), however, suggest that rates of recharge to deep flow systems in the Madison from Black Hills outcrop areas are considerably lower than previously estimated (Brown, 1944). They determined that recharge in these areas contributes instead almost exclusively to shallow flow systems that both begin and end in the Black Hills. In some cases, stream erosion has cut off a portion of the Madison from the main regional aquifer so that direct recharge to the deep flow systems is not possible. In others, where the Madison beds dip beneath the younger, sandy shale beds of the Spearfish Formation, limestone dissolution in the outcrop area of the surficial Madison–Spearfish contact line creates a zone of unusually high permeability in the Madison. The confined portion of the aquifer beneath the Spearfish is not affected. The end result is the creation of a ground-water dam, which forces most or all of the water recharging the aquifer in the outcrop area back to the surface to discharge as springs upstream of the surficial boundary between the two formations. The process has been described in some detail by Huntoon (1985) for a location farther to the west in Wyoming.

Water-table conditions exist only in the outcrop areas. In general, water moves outward (direction depends on location) from the recharge areas. The rate of movement is determined by the degree of leakage upward through aquitards, into overlying aquifers, rather than by the transmissivity of the Madison aquifer (Miller, 1976). Rates of movement are very slow through the brine areas of the Williston basin where little upward leakage occurs. Leakage to shallower aquifers is the major discharge of the Madison aquifer.

*Chemistry of water.* As would be expected, the best quality water is found in the recharge areas. In these areas, calcium, magnesium, and sulphate ions, in milliequivalents per liter, constitute more than 75 percent of the dissolved constituents. The chemistry evolves along the flow system into the Williston basin

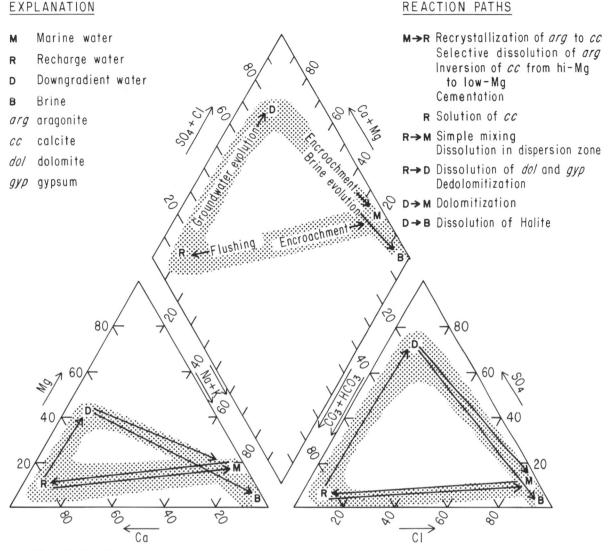

Figure 2. Reaction paths showing evolution of chemistry of ground waters in carbonate aquifers (after Hanshaw and Back, 1979).

where sodium, potassium, and chloride ions constitute more than 75 percent of the total. Brines in the central part of the basin exceed 200,000 mg/ℓ (Downey, 1984), and rates of water movement are very slow (Miller, 1976). Where greater leakage into overlying units occurs, the water is generally fresher than where leakage is minimal.

A conceptual model (Fig. 2) described by Hanshaw and Back (1979) depicts the changes in ground-water chemistry of carbonate aquifers from the time of their inception in the shallow marine environment, through their functioning as emergent aquifer systems. Some of the most profound changes occur when the carbonate sediments first emerge from the marine environment and undergo flushing by fresh water. Depending on the type of recharge water, the position within the flow system, and the possibilities for mixing, different evolutionary sequences occur in the ground-water chemistry and in the diagenesis of the rock

aquifer (Fig. 2). Dedolomitization has been demonstrated in the Madison aquifer (Pahasapa Formation) near the Black Hills, South Dakota. This appears to be the major geochemical process now operating in the Madison aquifer system.

## AQUIFERS IN MARINE MARGIN CLASTIC WEDGES

### Dakota aquifer

The Dakota aquifer has been generally considered to be the type example of a classic artesian system. Upham (1895) and, later, Darton (1905) proposed that recharge enters sandstone beds that are exposed in the elevated mountains in the west, with the water discharging in the east where the sandstone outcrops or subcrops beneath glacial drift (Fig. 3).

Very early this stratigraphically and hydrologically simplistic concept of a continuous system was challenged on a number of points: (1) the sandstone beds lack continuity, (2) anomalous lateral differences occur in the salinity of the water, (3) the salinity profile with distance does not match that of an aquifer system recharged with rainwater, (4) the volume of recharge does not equal the volume of discharge, and (5) the head gradient over the system is not uniform (Russell, 1928; Meinzer and Hard, 1925). Despite these difficulties with the original concept, the system does have hydraulic continuity, with the exception of part of the Denver basin.

The name Dakota Group was first applied by Meek and Hayden (1862) to the basal Cretaceous sandstone and interbedded shale cropping out in the bluffs along the Missouri River in Dakota County, Nebraska. Various units have been correlated with the type Dakota over the years by different authors. Figure 4 is a current hydrogeological interpretation of the complex regional stratigraphy by Jorgensen and others (1986). They propose that the Dakota aquifer be renamed the Great Plains aquifer system and that its upper and lower units be referred to as the Maha and Apishapa aquifers, respectively. Since this terminology change has not been generally adopted, this chapter will continue to refer to the Dakota aquifer or aquifer system.

The Dakota aquifer system includes rocks beyond the limits of the Dakota Group; in addition to the Dakota Group the system includes the Swan River Formation of Manitoba and eastern Saskatchewan, the Inyan Kara Formation and the Omadi Sandstone of the plains states, and the Mannville Group of south-central Alberta and Saskatchewan. Discussion of the Dakota aquifer in this chapter will be limited to the area overlain by the Western Glaciated Plains, with brief reference to other stratigraphic units as they impact water movement and water chemistry.

***Origin and character of porosity and permeability.*** The type Dakota Formation is mostly continental in origin and consists of coastal plain, fluvial, and flood plain sediments deposited as Cretaceous shorelines migrated to the west (Schoon, 1971). The Dakota aquifer is characterized by primary intergranular porosity that is reduced in places by slight compaction and cementation. Hydraulic conductivity ranges from about 0.3 to about 15 m/day in the U.S. portion of the glaciated plains area (Gries and others, 1976; Burkhart, 1984; Meinzer, 1929). A slightly wider range has been reported for the Mannville Group in Saskatchewan (Maathuis, 1986). In some areas, evidence exists for enhanced permeability as a result of the development of fracture porosity (V.H. Dreeszen, Nebraska Conservation and Survey Division, personal communication).

***Ground-water flow.*** Modern thinking about the source of ground water in and movement through the Dakota aquifer is largely in conflict with the original proposal of Darton (1905) that flow in the aquifer was analogous to pipe flow with an inlet in the mountains and an outlet in the discharge area to the east.

Swenson (1968) attributed both the source of water and the high pressures in the Dakota Sandstone in the "Great Artesian Basin" to leakage from the Madison Group. He found little loss of

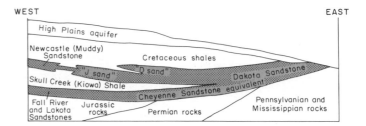

Figure 3. Dakota aquifer, Darton's early regional concept (Darton, 1905).

head as the water moves through the cavernous limestone of the Mississippian Pahasapa Limestone away from the recharge area on the flanks of the Black Hills. Where pre-Cretaceous erosion has removed most of the intervening shale units, the Dakota Sandstone is in hydraulic connection with units of the Madison Group. Schoon (1971) proposed other sources of leakage including sandstone of the Inyan Kara Formation. The concept that other hydraulically connected stratigraphic units are major sources or sinks for the Dakota aquifer is further supported by ground-water flow modeling studies that indicate that vertical flux to and from the Dakota aquifer exceeds horizontal flow (Milly, 1978; Neuzil, 1980). Most workers now agree that the Madison Group is a significant source of recharge to the Dakota Sandstone; only the amount is debated (Bredehoeft and others, 1983; Downey, 1984). Kolm and Peter (1984) mapped a lineation pattern that they related to patterns of anomalous ground-water chemistry in the Dakota aquifer. On the basis of this pattern, which they believe is related to discontinuities in the Precambrian basement, they argue for the existence of fracture-controlled vertical leakage. A number of investigators have pointed out that, even without fracturing, significant amounts of water can move vertically through confining layers with relatively low hydraulic conductivity because of the very large areas underlain by the Dakota aquifer (Swenson, 1968; Helgesen and others, 1984; Rutulis, 1984).

***Chemistry of water.*** Water quality of the Dakota aquifer cannot be adequately explained using Darton's simple model (1905) of ground-water flow. One would expect increasing dissolved solids and distinct ground-water facies changes from west to east. Instead, extreme differences exist in water quality throughout the aquifer system. These extreme differences are found even on a very local scale (Lawton and others, 1984; Rutulis, 1984).

Leonard and others (1984) state that water quality at any site is determined by: presence of connate water; location with regard to outcrop areas (recharge potential); leakage from overlying and underlying geologic units; lateral discontinuities; and mixing, natural softening, or ion filtration. They report that dissolved solids range from less than 500 mg/ℓ in the eastern outcrop areas to more than 100,000 mg/ℓ in the Denver basin and that water chemistry of the Dakota aquifer in the Western Glaciated Plains

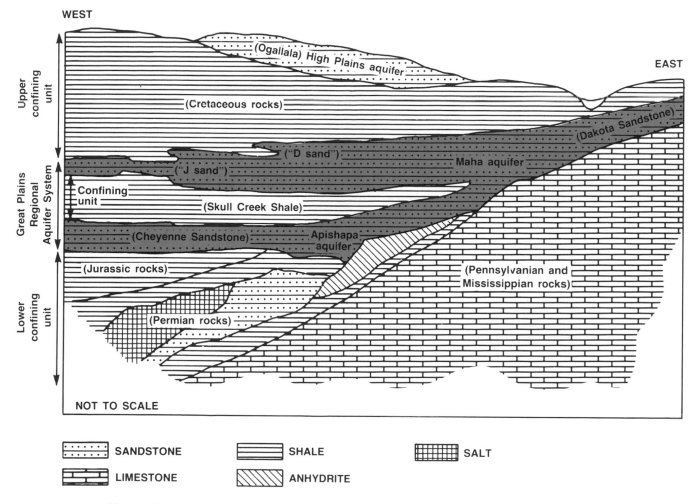

WEST

EAST

Upper confining unit

Great Plains Regional Aquifer System

Lower confining unit

(Ogallala) High Plains aquifer

(Cretaceous rocks)

(Dakota Sandstone)

("D sand")

Maha aquifer

("J sand")

Confining unit

(Skull Creek Shale)

(Cheyenne Sandstone)

Apishapa aquifer

(Jurassic rocks)

(Pennsylvanian and Mississippian rocks)

(Permian rocks)

NOT TO SCALE

SANDSTONE    SHALE    SALT

LIMESTONE    ANHYDRITE

Figure 4. Dakota aquifer, schematic cross section according to current concepts (after Jorgensen and others, 1986).

is characterized predominantly by sodium chloride, bicarbonate, and sulphate, and by calcium sulphate and bicarbonate.

### Judith River aquifer

The Judith River Formation (Belly River Formation) of Alberta, western Saskatchewan, and northern Montana, is typical of the regionally extensive clastic wedges of Late Cretaceous and Early Tertiary age that form important aquifers in the western part of the Western Glaciated Plains region (Kewen and Schneider, 1979; Whitaker, 1982; McLean, 1971; U.S. Geological Survey, 1985). This unit, which averages 275 m to 300 m in thickness, is overlain and underlain by marine shale of Late Cretaceous age (Fig. 5). The formation thins eastward from the Rocky Mountain foothills, where it is more than 600 m thick, to a feather edge in central Saskatchewan (Douglas and others, 1970). In limited areas of Alberta, and more extensively in Saskatchewan, the overlying Bearpaw Formation has been eroded and Pleistocene glacial sediment rests directly on the Judith River Formation.

Lorberg (1983) and Maathuis (1986) indicate that the Judith River Formation consists of gray to greenish gray, thick-bedded, feldspathic sand and sandstone interbedded with gray clayey siltstone and gray and green mudstone. The rocks are generally noncalcareous, although in places, zones cemented with calcium carbonate occur. The rocks are carbonaceous, with major coal-bearing zones present in Alberta and western Saskatchewan. Concretionary ironstone beds are commonly encountered. The sand and sandstone beds that distinguish the Judith River Formation from the overlying and underlying units make up no more than half the entire thickness of the formation.

***Origin and character of porosity and permeability.*** The Judith River Formation in Saskatchewan was deposited in a nonmarine, deltaic, near-shore environment during a period of marine regression. Consequently the basal portion is deltaic, relatively sandy, and moderately permeable. In contrast, the upper portion consists of alluvial plain sediments with a greater proportion of silt and clay beds and is less permeable. Marked lateral and vertical variations in lithology are characteristic of the formation. Within the main body of the formation, few units can be

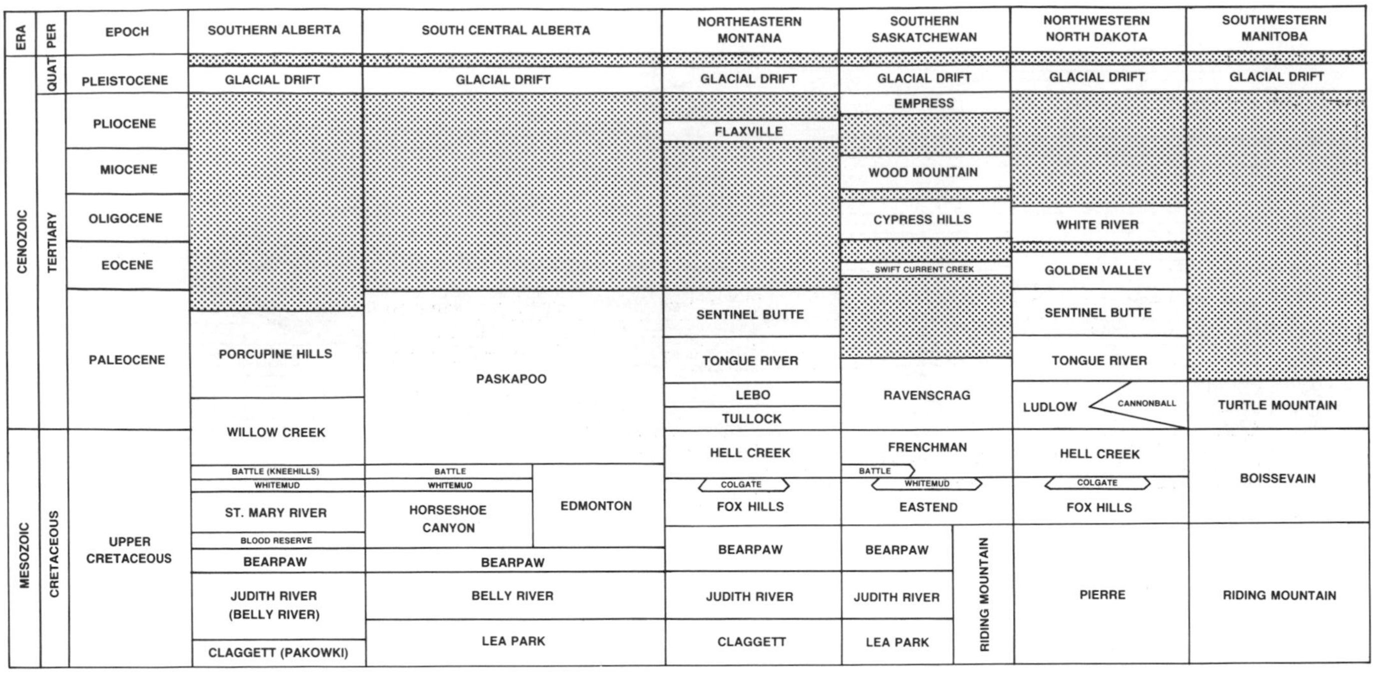

| ERA | PER | EPOCH | SOUTHERN ALBERTA | SOUTH CENTRAL ALBERTA | NORTHEASTERN MONTANA | SOUTHERN SASKATCHEWAN | NORTHWESTERN NORTH DAKOTA | SOUTHWESTERN MANITOBA |
|---|---|---|---|---|---|---|---|---|
| CENOZOIC | QUAT | PLEISTOCENE | GLACIAL DRIFT | GLACIAL DRIFT | GLACIAL DRIFT | GLACIAL DRIFT / EMPRESS | GLACIAL DRIFT | GLACIAL DRIFT |
| CENOZOIC | TERTIARY | PLIOCENE | | | FLAXVILLE | | | |
| CENOZOIC | TERTIARY | MIOCENE | | | | WOOD MOUNTAIN | | |
| CENOZOIC | TERTIARY | OLIGOCENE | | | | CYPRESS HILLS | WHITE RIVER | |
| CENOZOIC | TERTIARY | EOCENE | | | | SWIFT CURRENT CREEK | GOLDEN VALLEY | |
| CENOZOIC | TERTIARY | PALEOCENE | PORCUPINE HILLS | PASKAPOO | SENTINEL BUTTE | | SENTINEL BUTTE | |
| CENOZOIC | TERTIARY | PALEOCENE | PORCUPINE HILLS | PASKAPOO | TONGUE RIVER | RAVENSCRAG | TONGUE RIVER | |
| CENOZOIC | TERTIARY | PALEOCENE | PORCUPINE HILLS | PASKAPOO | LEBO | RAVENSCRAG | LUDLOW < CANNONBALL | TURTLE MOUNTAIN |
| CENOZOIC | TERTIARY | PALEOCENE | WILLOW CREEK | PASKAPOO | TULLOCK | RAVENSCRAG | LUDLOW < CANNONBALL | TURTLE MOUNTAIN |
| MESOZOIC | CRETACEOUS | UPPER CRETACEOUS | WILLOW CREEK | | HELL CREEK | FRENCHMAN | HELL CREEK | |
| MESOZOIC | CRETACEOUS | UPPER CRETACEOUS | BATTLE (KNEEHILLS) | BATTLE | | BATTLE | | BOISSEVAIN |
| MESOZOIC | CRETACEOUS | UPPER CRETACEOUS | WHITEMUD | WHITEMUD | COLGATE > | WHITEMUD > | COLGATE > | BOISSEVAIN |
| MESOZOIC | CRETACEOUS | UPPER CRETACEOUS | ST. MARY RIVER | HORSESHOE CANYON / EDMONTON | FOX HILLS | EASTEND | FOX HILLS | BOISSEVAIN |
| MESOZOIC | CRETACEOUS | UPPER CRETACEOUS | BLOOD RESERVE | | | | | |
| MESOZOIC | CRETACEOUS | UPPER CRETACEOUS | BEARPAW | BEARPAW | BEARPAW | BEARPAW | PIERRE | RIDING MOUNTAIN |
| MESOZOIC | CRETACEOUS | UPPER CRETACEOUS | JUDITH RIVER (BELLY RIVER) | BELLY RIVER | JUDITH RIVER | JUDITH RIVER (RIDING MOUNTAIN) | PIERRE | RIDING MOUNTAIN |
| MESOZOIC | CRETACEOUS | UPPER CRETACEOUS | CLAGGETT (PAKOWKI) | LEA PARK | CLAGGETT | LEA PARK | PIERRE | RIDING MOUNTAIN |

Figure 5. Stratigraphic correlation chart, Upper Cretaceous and Tertiary formations, Western Glaciated Plains region (after Garven, 1982; Whitaker and others, 1978).

followed laterally for more than a few kilometers, and individual beds rarely exceed a thickness of several meters.

Kewen and Schneider (1979) and Whitaker (1982) report that the horizontal hydraulic conductivity of Judith River aquifer sands in Saskatchewan varies by an order of magnitude from about 0.17 to about 1.7 m/day with an average of about 0.6 m/day. Lorberg (1983) reported that bulk horizontal and vertical hydraulic conductivities for the Judith River Formation in southeastern Alberta range from $1 \times 10^{-4}$ to $1 \times 10^{-2}$ m/day and from $1 \times 10^{-10}$ to $1 \times 10^{-7}$ m/day, respectively. Maathuis (1986) cites a probable bulk hydraulic conductivity range of about $1 \times 10^{-7}$ to $1 \times 10^{-5}$ m/day for the underlying Lea Park Formation, about 4 to 6 orders of magnitude lower than in the Judith River. The bulk hydraulic conductivity of the overlying Bearpaw Formation is believed to range over a similar set of values. Till deposits that directly overlie the Judith River Formation in parts of Saskatchewan are estimated by Maathuis to have a bulk vertical hydraulic conductivity in the $1 \times 10^{-5}$ to $1 \times 10^{-3}$ m/day range.

*Ground-water flow.* Because of the contrast in hydraulic conductivity in Saskatchewan between the aquifer and the underlying and overlying aquitards, flow in the Judith River aquifer is primarily subhorizontal, paralleling the regional dip (Maathuis, 1986). The underlying aquitard has such a high vertical hydraulic resistance that it can be considered an impermeable base. The aquifer receives recharge via vertical flow through the overlying aquitard. As demonstrated by Whitaker (1982), the Cypress Hills upland area is a major recharge area beneath which a vertical downward gradient of 0.5 exists. However, the actual amount of recharge to the aquifer by this route will be small because the

Bearpaw Formation is very thick in this area. In southwestern Saskatchewan, flow in the aquifer is away from the Cypress Hills upland area to the west, toward the South Saskatchewan River in Alberta, and south toward the Milk River in Montana. To the east, northeast, and north, flow in the Judith River aquifer is into various preglacial buried valley aquifers (Meneley, 1972). These aquifers ultimately discharge into surface-water bodies such as the North Saskatchewan River and Last Mountain Lake.

*Chemistry of water.* Water quality in the Judith River aquifer in Saskatchewan varies with location. In those parts of southwestern Saskatchewan where the overlying Bearpaw Formation is thick, water in the Judith River is generally of the sodium bicarbonate, sodium sulphate, or sodium chloride type (Whitaker, 1982). For these waters, chloride concentration generally increases with depth. Where the Bearpaw is thin or absent in southwestern Saskatchewan, water in the upper portion of the Judith River is of the calcium/magnesium sulphate type. Presumably this same phenomenon may characterize similar areas farther to the north, although Kewen and Schneider (1979) fail to mention it in their report on the Judith River aquifer in west central Saskatchewan.

In west-central Saskatchewan, Kewen and Schneider found that water from the Judith River aquifer is generally of the sodium bicarbonate or sodium sulphate type. The bicarbonate concentrations vary little, and consequently, the relative concentration of sulphate determines the type. Sulphate concentration tends to decrease with increasing depth within the aquifer as a result of reduction by bacteria. Moderate concentrations of organic carbon, giving the water a distinct yellowish brown color, are encountered throughout the aquifer. The intensity of the coloring appears to be unrelated to the organic carbon concentration, but rather is probably a function of the chemical composition of the organic compounds, which are likely derived from coal seams within the formation.

Dissolved-solids concentrations for the Judith River aquifer in southwestern Saskatchewan were determined to be in the 1,300 to 8,000 mg/ℓ range. In west-central Saskatchewan the corresponding range was 700 to 5,400 mg/ℓ, but observed values were generally less than 2,500 mg/ℓ. Gases have been noted in the Judith River waters in both southwestern and west-central Saskatchewan.

### Aquifers in Fox Hills to Fort Union Formations

The Late Cretaceous Fox Hills and Hell Creek Formations and the Early Tertiary Fort Union Formation (the Fort Union Group in western North Dakota) and their equivalents form a continuous geological sequence in eastern Montana, southeastern Alberta, southern Saskatchewan, western North Dakota, northwestern South Dakota and northeastern Wyoming (Fig. 5). Except where otherwise noted, information on this sequence for Saskatchewan is taken from Maathuis (1986). In southern Saskatchewan, the units equivalent to the Fox Hills, Hell Creek, and Fort Union Formations are the Eastend, Frenchman, and Ravenscrag Formations, respectively (Fig. 5). Two other units—the Whitemud and Battle Formations—are recognized in the Canadian sequence. Throughout all or most of the U.S. portion of the region, the upper Hell Creek Formation is considered to be a confining unit (Lobmeyer, 1985) and the sequence is considered to include two principal aquifers. The first of these is found in the Fox Hills–Lower Hell Creek section, and the second is primarily in the Fort Union. Where the Fort Union is overlain by the Wasatch Formation, the second aquifer can extend upward into the Wasatch. The Fox Hills–Hell Creek–Fort Union sequence is directly underlain by the Bearpaw Shale and the Pierre Formation, both of which are relatively impermeable and severely restrict the downward movement of water. The top of the sequence lies at depths ranging up to 300 m and more in Montana (U.S. Geological Survey, 1985), but it outcrops or subcrops beneath the glacial deposits throughout most of the area where the sequence is found in North Dakota and Saskatchewan. In southern Saskatchewan the glacial deposits, where present, are generally thin so that most of the aquifer is essentially unconfined.

The Fox Hills Formation and its Eastend equivalent in Canada consist generally of sandstone, with some siltstone and coal (U.S. Geological Survey, 1985). The Whitemud Formation of southern Saskatchewan is composed of kaolinitized white sand and clay separated by carbonaceous zones, and is overlain in parts of southwestern Saskatchewan by shale of the Battle Formation. The Hell Creek Formation and its Frenchman equivalent are composed of sandstone, siltstone, some claystone, and shale. The Fort Union Formation and the Ravenscrag equivalent contain sandstone, siltstone, clay, and lignite. Although the forma-

tions in the Canadian sequence can be distinguished individually in exposures, it is virtually impossible to discriminate between them in the subsurface. Christiansen (1983) has suggested that they be considered as one geological unit. In conformity with this suggestion, Maathuis considers the Eastend to Ravenscrag Formations in southwestern Saskatchewan to constitute a single aquifer. This conforms as well to the Canadian observation that there does not seem to be a continuous low-permeability layer to isolate water in the Ravenscrag Formation from water in the lower Frenchman and Eastend.

***Origin and character of porosity and permeability.*** According to Whitaker and others (1978), the sediments of the Eastend, Frenchman, and Ravenscrag Formations were deposited in an advancing delta and in the following alluvial deltaic plain. In southwestern Saskatchewan, the lower part of this sequence is characterized by a blanket sand that thins in an eastward direction. The upper part of the sequence is characterized by discontinuous sand units that were deposited in channels formed by streams that meandered across a swampy flood plain (Christiansen, 1983). Meneley (1983) states that the hydraulic conductivity of the sand beds in this sequence ranges typically from about 1.0 to about 10 m/day. No reliable information exists on porosity, specific yield or storage coefficient.

***Ground-water flow.*** The piezometric surface determined by Lobmeyer (1985) for the Fort Union aquifer in the Northern Great Plains area of the northern U.S. generally follows the regional topographic surface, indicating that flow in this aquifer is, for the most part, topographically controlled. Only in southeastern Montana and northeastern Wyoming is there some evidence for geological control of flow in this aquifer. In general, Lobmeyer finds the major recharge areas for the aquifer to be in the topographically high areas of east-central Montana and western North Dakota. Major discharge areas seem to be along the Missouri, Yellowstone, and Powder rivers.

The potentiometric surface determined for the Fox Hills–Lower Hell Creek aquifer shows the same general trends as that for the Fort Union aquifer (Lobmeyer, 1985). Because the horizontal permeability of this aquifer is much greater than its vertical permeability, Lobmeyer states that flow in the aquifer is probably horizontal, except in outcrop areas. This contrasts with the Fort Union aquifer in which significant vertical flow components occur in the vicinity of major stream valleys. Because of potentiometric head differences of as much as 60 m between the two aquifers in northeastern Wyoming, water probably moves downward from the Fort Union aquifer into the Fox Hills–Lower Hell Creek aquifer in that area. Potentiometric heads for the Fox Hills–Lower Hell Creek aquifer are generally highest in the south and west, indicating a regional flow toward the north, northeast, and east. Locally, however, deviations exist from this general flow pattern. Flow in the southeast portion of the Powder River basin is toward the southeast, and local recharge areas are associated with several topographic highs in north-central Montana. Discharge from the Fox Hills–Lower Hell Creek aquifer in the Northern Great Plains is inferred by Lobmeyer to be along the stream valleys in the eastern part of the Powder River structural basin and along the Yellowstone and Missouri rivers.

Recharge of the equivalent Eastend to Ravenscrag aquifer sequence in southern Saskatchewan is directly by infiltrating precipitation. Virtually the entire area of the aquifer in Saskatchewan is considered to be a recharge area as a consequence of its topographical and geological setting. Discharge is limited to the periphery of the outcrop area and to major incised preglacial and fluvial valleys. Where flow is into the deposits filling a preglacial valley, eventual discharge is commonly to the present-day river system. A notable example is discharge via the Estevan and Weyburn buried valley aquifers into the Souris River (Meneley and Whitaker, 1970).

*Chemistry of water.* The National Water Summary 1984 (U.S. Geological Survey, 1985) reports that dissolved solids in the Fort Union aquifer in Montana do not generally exceed 1,800 mg/$\ell$. Water in the same aquifer in North Dakota is reported to be generally soft and of the sodium sulphate bicarbonate type. The National Water Summary gives no figures for dissolved solids for the Fort Union in North Dakota but notes that sulphate concentrations in the aquifer range up to 9,600 mg/$\ell$. For the Fox Hills–Lower Hell Creek aquifer the National Water Summary indicates dissolved-solids concentrations of 1,200 mg/$\ell$ or less for Montana.

A hydrogeochemical model that was developed by Cherry (Moran and Cherry, 1977) and later refined by Koob and his associates (Groenewold and others, 1983) accounts for the observed chemical characteristics of subsurface water in arid to semiarid regions of North America such as North Dakota. The key to the model is alternate shallow wetting and drying within a few meters of the ground surface. Qualitatively the hydrogeochemical model (Fig. 6) assumes that oxygen, along with other atmospheric gases, is brought into the solum along with water during precipitation events. This process initiates the oxidation of carbon and sulphur with the formation of $SO_4^{2-}$, $CO_2$, and $H^+$ and, in turn, leads to the dissolution of calcite and dolomite, which yields $Ca^{2+}$, $Mg^{2+}$, and $HCO_3^-$. Evaporation and/or transpiration leaves the ions in the upper few meters as calcium and/or magnesium sulphates and bicarbonates. On the infrequent occasion that sufficient water is present to carry the ions below the root zone in a recharge event, some of the divalent ions exchange on the sodium-montmorillonitic clays that predominate throughout the area, releasing $Na^+$. This combination of processes is sufficient to account for the alkaline, sodium sulphate ground waters common to the Northern Great Plains.

In southern Saskatchewan, ground water in the Eastend to Ravenscrag aquifer is of the sodium bicarbonate type in the western part of the province where dissolved-solids concentrations are relatively low. Farther east, aquifer waters are characterized by both sodium bicarbonate and sodium chloride. Dissolved-solids concentrations range from about 500 to over 4,000 mg/$\ell$.

## AQUIFERS IN GLACIAL SEDIMENTS

### Introduction

The complex glacial and interglacial history of the Western Glaciated Plains region has resulted in a distribution of surficial sand and gravel deposits that seems to some degree unpredictable. At the land surface, the probable presence of permeable water-sorted sand and gravel is commonly made evident by the characteristic geomorphic forms of eskers, kames, outwash plains, beaches, and deltas. More deeply buried surficial sand and gravel does not, however, generally have any surface expression. For example, the courses of buried paleodrainage systems developed in the bedrock have only become relatively well known through systematic ground-water and geological investigations. However, information is still lacking to predict whether a buried valley contains sand and gravel along a given reach or whether it is completely filled with glacial till, lake clay, or silt. The presence or absence of isolated sand and gravel deposits interbedded with tills or lake sediments is equally unpredictable in the absence of detailed stratigraphic studies of the drift sequence.

The final retreat of the continental glacier left the Western Glaciated Plains region covered primarily with till or lacustrine sediment. Outwash deposits and loess also underlie the region but their areal extent is relatively minor. Drift cover is absent in a few unglaciated areas such as the Cypress Hills of southern Alberta and Saskatchewan. Where drift is present, the thickness tends to increase with distance from the mountain belt. Maximum thicknesses are typically associated with the courses of buried valleys.

The permeabilty of glacial till of the Western Glaciated Plains region is an important factor in the rates of recharge through these materials to both shallow and deep flow systems. In general, where till is both unfractured and unmodified by biological activity, such as root development and burrowing, flows are intergranular and hydraulic conductivities are low. However, fracturing is believed to be common although not always evident visually (Keller and others, 1986). High conductivity values, such as those reported by Schwartz (1975) for till in southeastern Alberta, suggest that flow through such till is actually strongly influenced by fractures.

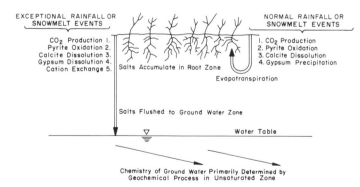

Figure 6. Schematic diagram of chemical processes and salt movement in the arid to semiarid plains region (after Moran and Cherry, 1977).

### Buried-valley aquifers

Prior to continental glaciation during the Pleistocene, drainage systems in the Western Glaciated Plains region were probably mature and well integrated (Farvolden, 1963b). Many valleys had been cut into the Cretaceous and Tertiary sediment that formed the bedrock surface. Subsequently, the bedrock surface underwent further fluvial dissection during interglacial periods, and additional valleys were formed that became intermingled with the preglacial set (Meyboom, 1967). At the end of the Pleistocene, the alluvial deposits at the base of the preglacial and interglacial valleys were overlain with glacial drift, which generally tended to fill in the valleys and to obscure or conceal the earlier drainage networks. Nevertheless, a number of major meltwater channels cut into bedrock during the final stages of deglaciation still remain open and from place to place are occupied by segments of the present-day drainage (Christiansen, 1963).

Aquifers occur both in the alluvial deposits at the base of the buried preglacial and interglacial valleys and within the overlying glacial drift. For Saskatchewan, our information on these aquifers is based chiefly on Maathuis (1986). There the Quaternary stratigraphic sequence has been subdivided into three groups. The oldest of these is the Empress Group, directly overlying the Cretaceous or Tertiary bedrock and comprising sand, gravel, silt, and clay of fluvial, lacustrine, and colluvial origin. The Empress is overlain by the Sutherland Group and the Sutherland by the Saskatoon Group. Both the Sutherland and the Saskatoon consist primarily of till but also contain layers of stratified drift made up of sand, gravel, silt, and clay. Contemporaneous with the Empress Group, and regarded by Meneley (1972) as part of that group, are the blanket sand aquifers. These are laterally extensive, range up to 30 m in thickness, and tend to be located along the flanks of major buried valleys, in particular the Hatfield Valley. The intertill stratified drift layers, the Empress blanket sands, and the basal Empress alluvial deposits all are aquifers. Individually, their importance depends on the continuity of sand and gravel strata within the buried valleys and along their flanks and on the spatial relationships between the buried valleys and the present-day drainage network.

In Alberta and Saskatchewan, sand and gravel are found in those preglacial valleys having their headwaters in the mountain areas or in areas of the plains where coarse materials were deposited during intervals of aggradation dating back to Oligocene time (Christiansen, 1963; Farvolden 1963b). Sand and gravel were carried down from these upland areas and deposited along the stream beds until gradients became too gentle for their continued transport. In Alberta (Farvolden, 1963b) the regional erosional base level is now as low as, or lower than, it has been at any time since the Oligocene. Consequently, buried valleys in that province may intersect present-day streams, pass beneath them, or outcrop or subcrop above stream level along the present-day valley walls. In Saskatchewan (Christiansen, 1963) the preglacial erosional base level is some 30 to 90 m below the present level.

Deposits of the Empress Group in Saskatchewan, lying at the base of the buried valleys, are saturated and the ground water in them is confined. These deposits, where sufficiently coarse and permeable, are good aquifers. The same is true of the similar deposits in those Alberta buried valleys whose courses run below the existing drainage level, for example, in the Sand River area of east-central Alberta (Andriashek and Fenton, 1988). Where buried-valley courses run above this level, the buried valleys commonly drain into the present-day valleys, and permeable buried-valley sand and gravel, if present, can be dry or contain little water. This condition is characteristic, for example, of buried valleys intersecting the Oldman River valley at locations near Lethbridge in southwestern Alberta (Meyboom, 1967).

More interesting hydrogeological situations can be found where present-day streams cut into permeable buried-valley deposits. At Medicine Hat in southeastern Alberta (Meyboom, 1963), the South Saskatchewan River intersects a layer of medium-sized to coarse preglacial gravel ranging in thickness from about 6 to about 13 m and extending to a maximum depth of almost 13 m below the river. For southern Alberta, this stream and aquifer interconnection presents an exceptional situation with unusual potential for ground-water development. In contrast, the thickness of present-day alluvial gravel in Alberta rarely exceeds a meter or so. Buried-valley sand and gravel deposits elsewhere in Alberta can range up to 13 m in thickness (Farvolden, 1963a). In Saskatchewan, buried-valley aquifer thickness can range up to 100 m. The typical thickness range for the Hatfield buried-valley aquifer is 30 to 50 m.

Within the buried valleys of Saskatchewan and Alberta, good hydraulic connections exist in some places between the permeable deposits of the Empress Group and layers of stratified drift in the overlying Sutherland and Saskatoon groups (Andriashek and Fenton, 1988). Good hydraulic connection may also exist in places between the buried-valley aquifers and bedrock aquifers, such as the Eastend to Ravenscrag aquifer, where these bedrock aquifers underlie the buried valley or abut its walls. Development of a buried-valley aquifer in these settings is equivalent to the development of a complex interlinked system made up of the aquifer being directly exploited and an assortment of nearby surficial and bedrock aquifer units. The major Hatfield Valley aquifer system, for example, which traverses the province of Saskatchewan in a northwest–southeast direction for some 550 km, includes a number of intertill stratified-drift aquifers and the Empress Group blanket sands, in addition to the main Hatfield Valley aquifer (i.e., the permeable alluvial deposits lying at the base of the buried valley). The Estevan Valley aquifer system in the extreme south of the province links the main buried-valley aquifer with a subsidiary buried-valley aquifer and with the Eastend to Ravenscrag aquifer.

Probably the largest and most productive buried-valley aquifer in the northern U.S. portion of the Western Glaciated Plains region is the Spiritwood aquifer, which extends completely across the eastern part of North Dakota in a southeasterly direction for a distance of about 400 km. In places, the buried valley is as much as 15 km wide. Generally the bottom of the aquifer is less than 30 m deep, but in places reaches a depth of nearly 170 m (Downey and Armstrong, 1977). The aquifer consists of lenticular deposits of sand and gravel interbedded with clay and silt deposited in preglacial bedrock channels. Facies changes common with this type of bedrock channel fill give rise to associated permeability changes. Although the overall aquifer itself can be readily mapped, estimation of water in storage and of transmissivity based on extrapolation of known values is prone to serious error. Downey and Armstrong cite a transmissivity of about 400 $m^2$/day and a storage coefficient of about 0.02 from an aquifer test in Griggs County, North Dakota.

*Ground-water flow.* Recharge to the buried-valley aquifers of Alberta and Saskatchewan is commonly derived from local precipitation, including snowmelt, which has infiltrated the overlying aquitards. Recharge mechanisms are similar in the northern U.S. Highest water levels in the Spiritwood aquifer occur after spring snowmelt. In addition, recharge may alsoo be by lateral flow from adjoining bedrock or surficial aquifers. Discharge is commonly into major streams of the present-day drainage network, which exerts a strong controlling influence on the buried-valley flow systems. Where, as in the case of the Hatfield Valley aquifer, there are several distinct discharge areas along the length of a buried valley, the valley is characterized by a number of discrete flow systems separated by piezometric surface divides. Where basal or intertill sand and gravel are sufficiently permeable and continuous and a good hydraulic connection exists with adjacent bedrock and surficial deposits, the buried valley also acts as a local drain (Le Breton, 1963; Meneley, 1972).

Grain size analyses and information from a limited number of in-situ tests indicate that horizontal hydraulic conductivities for Empress Group sands are in the 1.0 to 13.0 m/day range for the Hatfield Valley aquifer, and values for the Estevan Valley aquifer probably fall within a similar range. A much larger hydraulic conductivity value of 140 m/day was found by Meyboom (1963) for the coarse to medium-sized preglacial gravels of the Police Point aquifer at Medicine Hat, Alberta.

*Chemistry of water.* As has already been stated, the buried-valley aquifers of Saskatchewan and Alberta are commonly recharged by precipitation that has filtered down through overlying glacial till. The quality of the ground water in these aquifers is determined in the first instance by the geochemical properties of the original rainfall or snow pack, in particular its acidity, and by geochemical processes within the till, such as ion exchange and chemical dissolution. As the ground water moves along a buried-valley aquifer, its composition is further modified in conformity with the nature and chemical reactivity of the aquifer materials. For the basal Empress Group sediments of the Hatfield buried valley, consisting mainly of quartz sand and minor amounts of feldspar and mafic rock fragments, little such modification occurs with movement along the valley.

Water quality data for buried-valley aquifers in Alberta and Saskatchewan are not adequate for the identification and delineation of the shallow flow systems to which they contribute. For the Hatfield Valley aquifer, dissolved solids typically range from about 1,000 to about 3,000 mg/$\ell$. Values up to 5,500 mg/$\ell$ are observed locally. These waters range from the sodium bicarbonate through the calcium/magnesium bicarbonate to the calcium/magnesium sulphate or sodium sulphate type. Water in the Estevan Valley aquifer is typically of the sodium bicarbonate type, with dissolved solids in the 1,500 to 3,000 mg/$\ell$ range. In this case, Maathuis states that sulphate content is low because of bacteriological activity. Water from the Spiritwood aquifer in North Dakota is also of the sodium bicarbonate type, although locally the water can be a sodium sulphate or calcium sulphate type (Downey and Armstrong, 1977).

## REFERENCES CITED

Andriashek, L. D., and Fenton, M. M., 1988, Quaternary stratigraphy and surficial geology, Sand River map sheet (73L), Alberta: Alberta Research Council Bulletin (in preparation).

Bostock, H. S., 1970, Physiographic subdivisions of Canada, *in* Douglas, R.J.W., ed., Geology and economic minerals of Canada: Geological Survey of Canada Economic Geology Report 1, 5th ed., p. 10–30.

Bredehoeft, J. D., Neuzil, C. E., and Milly, D.C.D., 1983, Regional flow in the Dakota aquifer; A study of the role of confining layers: U.S. Geological Survey Water-Supply Paper 2237, 45 p.

Brown, C. B., 1944, Report on an investigation of water losses in streams flowing east out of Black Hills: U.S. Soil Conservation Service Special Report 8, 45 p.

Brown, I. C., ed., 1967, Groundwater in Canada: Geological Survey of Canada Economic Geology Report 24, 228 p.

Burkart, M. R., 1984, Availability and quality of water from the Dakota aquifer, northwest Iowa: U.S. Geological Survey Water-Supply Paper 2215, 65 p.

Christiansen, E. A., 1963, Hydrogeology of surficial and bedrock valley aquifers in southern Saskatchewan, *in* Proceedings, Hydrology Symposium No. 3, Groundwater: National Research Council, Canada, p. 49–66.

—— , 1983, Geology of the Eastend to Ravenscrag formations, Saskatchewan, *in* W. A. Meneley Consultants Ltd., Hydrogeology of the Eastend to Ravenscrag formations in southern Saskatchewan: Report prepared for Saskatchewan Environment, 17 p.

Clayton, L., Moran, S. R., and Bickley, W. B., Jr., 1976, Stratigraphy, origin, and climatic implications of late Quaternary upland silt in North Dakota: North Dakota Geological Survey Miscellaneous Series 54, 15 p.

Darton, N. H., 1905, Geology and underground water resources of the central Great Plains: U.S. Geological Survey Professional Paper 32, 433 p.

Douglas, R.J.W., Gabrielse, H., Wheeler, J. O., Stott, D. F., and Belyea, H. R., 1970, Geology of western Canada, *in* Douglas, R.J.W., ed., Geology and economic minerals of Canada: Geological Survey of Canada Economic Geology Report 1, 5th ed., p. 366–488.

Downey, J. S., 1984, Geohydrology of the Madison and associated aquifers in parts of Montana, North Dakota, South Dakota, and Wyoming: U.S. Geological Survey Professional Paper 1273G, 47 p.

Downey, J. S., and Armstrong, C. A., 1977, Ground-water resources of Griggs and Steele counties, North Dakota: North Dakota Geological Survey Bulletin 64, part III, 33 p.

Farvolden, R. N., 1963a, Bedrock topography, Edmonton–Red Deer map-area, Alberta, *in* Early contributions to the groundwater hydrology of Alberta: Research Council of Alberta Bulletin 12, p. 57–62.

—— , 1963b, Bedrock channels of southern Alberta, *in* Early contributions to the groundwater hydrology of Alberta: Research Council of Alberta Bulletin 12, p. 63–75.

Freeze, R. A., 1969, Regional groundwater flow; Old Wives Lake drainage basin, Saskatchewan: Environment Canada, Inland Waters Branch Scientific Series 5, 245 p.

Freeze, R. A., and Witherspoon, P. A., 1966, Theoretical analysis of regional groundwater flow; 1. Analytical and numerical solutions to the mathematical model: Water Resources Research, v. 2, p. 641–656.

—— , 1967, Theoretical analysis of regional groundwater flow; 2. Effect of water-table configuration and subsurface permeability variation: Water Resources Research, v. 3, p. 623–634.

—— , 1968, Theoretical analysis of regional groundwater flow; 3. Quantitative interpretations: Water Resources Research, v. 4, p. 581–590.

Garven, E., 1982, Groundwater hydrology of the Pine Lake research basin, Alberta; A preliminary analysis: Alberta Research Council Earth Science Report 82-2, 153 p.

Gosnold, W. D., 1985, Heat flow and groundwater flow in the Great Plains of the United States: Journal of Geodynamics, v. 4, p. 247–265.

Gries, J. P., Rahn, P. H., and Baker, R. K., 1976, A pump test in the Dakota Sandstone at Wall, South Dakota: South Dakota Geological Survey Circular 43, 9 p.

Groenewold, G. H., Koob, R. D., McCarthy, G. J., Rehm, B. W., and Peterson, W. M., 1983, Geological and geochemical controls on the chemical evolution of subsurface water in undisturbed surface-mined landscapes in western North Dakota: North Dakota Geological Survey Report of Investigations 79, 151 p.

Hanshaw, B. B., and Back, W., 1979, Major geochemical processes in the evolution of carbonate aquifer systems: Journal of Hydrology, v. 43, p. 287–312.

Heath, Ralph C., 1984, Ground-water regions of the United States: U.S. Geological Survey Water-Supply Paper 2242, 78 p.

Helgesen, J. G., Jorgensen, D. G., Leonard, R. B., and Signor, D. C., 1984, Regional study of the Dakota aquifer (Darton's Dakota revisited), *in* Jorgensen, D. G., and Signor, D. C., eds., Geohydrology of the Dakota aquifer: Dublin, Ohio, National Water Well Association, p. 38–40.

Hitchon, B., 1969a, Fluid flow in the western Canada sedimentary basin; 1. Effect of topography: Water Resources Research, v. 5, p. 186–195.

—— , 1969b, Fluid flow in the western Canada sedimentary basin; 2. Effect of geology; Water Resources Research, v. 5, p. 460–469.

—— , 1984, Geothermal gradients, hydrodynamics, and hydrocarbon occurrences, Alberta, Canada: American Association of Petroleum Geologists Bulletin, v. 68, p. 713–743.

Huntoon, P. W., 1985, Gradient controlled caves, Trapper–Medicine Lodge area, Bighorn basin, Wyoming; Ground Water, v. 23, p. 443–448.

Jorgensen, D. G., and Signor, D. C., eds., 1984, Geohydrology of the Dakota aquifer; Proceedings, First C. V. Theis Conference on Geohydrology: Dublin, Ohio, National Water Well Association, 247 p.

Jorgensen, D. G., Leonard, R. B., Signor, D. C., and Helgesen, J. O., 1986, Central Midwest regional aquifer-system study, *in* Sun, R. J., ed., Regional aquifer-system analysis program of the U.S. Geological Survey; Summary of projects, 1978–84: U.S. Geological Survey Circular 1002, p. 132–140.

Keller, C. K., van der Kamp, G., and Cherry, J. A., 1986, Fracture permeability and groundwater flow in clayey till near Saskatoon, Saskatchewan: Canadian Geotechnical Journal, v. 23, p. 229–240.

Kewen, R. J., and Schneider, A. T., 1979, Hydrologic evaluation of the Judith River Formation in west central Saskatchewan: Saskatchewan Research Council Geology Division Report G79-2, 78 p.

Klausing, R. L., 1983, Ground-water resources of Logan County, North Dakota: North Dakota Geological Survey Bulletin 77, part III, 42 p.

Kolm, K. E., and Peter, K. D., 1984, A possible relation between lineaments and leakage through confining layers in South Dakota, *in* Jorgensen, D. G., and Signor, D. C., eds., Geohydrology of the Dakota aquifer: Dublin, Ohio, National Water Well Association, p. 121–134.

Lawton, D. R., Goodenkauf, O. L., Hanson, B. V., and Smith, F. A., 1984, Dakota aquifers in eastern Nebraska, aspects of water quality and use, *in* Jorgensen, D. G., and Signor, D. C., eds., Geohydrology of the Dakota aquifer: Dublin, Ohio, National Water Well Association, p. 221–228.

Le Breton, E. G., 1963, Groundwater geology and hydrology of east-central Alberta: Research Council of Alberta Bulletin 13, 64 p.

Leonard, R. B., Signor, D. C., Jorgensen, D. G., and Helgesen, J. D., 1984, Geohydrology and hydrochemistry of the Dakota aquifer, central United States, *in* Jorgensen, D. G., and Signor, D. C., eds., Geohydrology of the Dakota aquifer: Dublin, Ohio, National Water Well Association, p. 229–237.

Lobmeyer, D. H., 1985, Freshwater heads and ground-water temperatures in aquifers of the Northern Great Plains in parts of Montana, North Dakota, South Dakota, and Wyoming: U.S. Geological Survey Professional Paper 1402-D, 11 p.

Lorberg, E., 1983, Groundwater component, South Saskatchewan River basin planning program: Alberta Environment Hydrogeology Branch report, 28 p.

Maathuis, H., 1986, Hydrogeology of southern Saskatchewan: Saskatchewan Research Council Geology Division unpublished internal report, 56 p.

Majorowicz, J. A., Jones, F. W., and Jessop, A. M., 1986, Geothermics of the Williston basin in Canada in relation to hydrodynamics and hydrocarbon occurrences: Geophysics, v. 51, p. 767–779.

McLean, J. R., 1971, Stratigraphy of the Upper Cretaceous Judith River Formation in the Canadian Great Plains: Saskatchewan Research Council Geology Division Report 11, 96 pp.

Meek, F. B., and Hayden, F. V., 1862, Descriptions of new Lower Silurian (Primordial), Jurassic, Cretaceous, and Tertiary fossils collected in Nebraska Territory. . . , with some remarks on the rocks from which they were obtained: Philosophical Academy of Science Proceedings, 1861, v. 13, p. 415–447.

Meinzer, O. E., 1929, Problems of the soft-water supply of the Dakota Sandstone, with special reference to the conditions at Canton, South Dakota: U.S. Geological Survey Water Supply Paper 597, p. 147–170.

Meinzer, O. E., and Hard, A. H., 1925, The artesian water supply of the Dakota Sandstone in North Dakota, with special reference to the Edgeley Quadrangle: U.S. Geological Survey Water Supply Paper 520, p. 73–95.

Meneley, W. A., 1972, Groundwater; Saskatchewan, *in* Water Supply for the Saskatchewan–Nelson Basin: Saskatchewan–Nelson Basin Board Report, App. 7, Sec. F, p. 673–723.

—— , 1983, Hydrogeology of the Eastend to Ravenscrag Formation in southern Saskatchewan: Report prepared for Saskatchewan Environment, 21 p.

Meneley, W. A., and Whitaker, S. H., 1970, Geohydrology of the Moose Mountain upland in southeastern Saskatchewan: Saskatchewan Research Council Geological Division Circular 3, 52 p.

Meyboom, P., 1960, Geology and groundwater resources of the Milk River sandstone in southern Alberta: Research Council of Alberta Memoir 2, 89 p.

—— , 1963, Induced infiltration, Medicine Hat, Alberta, *in* Early contributions to the groundwater hydrology of Alberta: Research Council of Alberta Bulletin 12, p. 88–97.

—— , 1966, Unsteady groundwater flow near a willow ring in hummocky moraine: Journal of Hydrology, v. 4, p. 38–62.

—— , 1967, Interior Plains hydrogeological region, *in* Brown, I. C., ed., Groundwater in Canada: Geological Survey of Canada Economic Geology Report 24, p. 131–158.

Miller, W. R., 1976, Water in carbonate rocks of the Madison Group in southeastern Montana; A preliminary evaluation: U.S. Geological Survey Water-Supply Paper 2043, 51 p.

Milly, C. D., 1978, Mathematical models of ground water flow and sulphate transport in the major aquifers of South Dakota [B.S. thesis]: Princeton, New Jersey, Princeton University, 88 p.

Moran, S. R., and Cherry, J. A., 1977, Subsurface-water chemistry in mined-land reclamation; Key to development of a productive post-mining landscape, *in* Proceedings, Annual General Meeting, 2nd, 1977: Edmonton, Alberta, Canadian Land Reclamation Association, p. 1–29.

Neuzil, C. E., 1980, Fracture leakage in the Cretaceous Pierre Shale and its significance for underground waste disposal [Ph.D. thesis]: Baltimore, Maryland, Johns Hopkins University, 167 p.

Rahn, P. H., and Gries, J. P., 1973, Large springs in the Black Hills, South Dakota and Wyoming: South Dakota Geological Survey Report of Investigations 107, 46 p.

Render, F. W., 1970, Geohydrology of the metropolitan Winnipeg area as related to groundwater supply and construction: Canadian Geotechnical Journal, v. 7, p. 243–274.

Russell, W. L., 1928, The origin of artesian pressure: Economic Geology, v. 23, p. 132–157.

Rutulis, M., 1984, Dakota aquifer system in the Province of Manitoba, *in* Jorgensen, D. G., and Signor, D. C., eds., Geohydrology of the Dakota aquifer: Dublin, Ohio, National Water Well Association, p. 14–21.

Schoon, R. A., 1971, Geology and hydrology of the Dakota Formation in South Dakota: South Dakota Geological Survey Report of Investigations 104, 55 p.

Schwartz, F. W., 1975, Hydrogeologic investigation of a radioactive waste management site in southern Alberta: Canadian Geotechnical Journal, v. 12, p. 349–361.

Swenson, F. A., 1968, New theory of recharge to the artesian basin of the Dakotas: Geological Society of America Bulletin, v. 79, p. 163–182.

Thayer, P. A., 1983, Relationship of porosity and permeability to petrology of the Madison Limestone in rock cores from three test wells in Montana and Wyoming: U.S. Geological Survey Professional Paper 1273-C, 29 pp.

Toth, J., 1962, A theory of groundwater motion in small drainage basins in central Alberta: Journal of Geophysical Research, v. 67, p. 4375–4387.

—— , 1963, A theoretical analysis of groundwater flow in small drainage basins: Journal of Geophysical Research, v. 68, p. 4795–4812.

—— , 1966, Mapping and interpretation of field phenomena for groundwater reconnaissance in a prairie environment, Alberta, Canada: Bulletin of the International Association of Scientific Hydrology, v. 11, no. 2, p. 20–68.

—— , 1978, Gravity-induced cross-formational flow of formation fluids, Red Earth region, Alberta, Canada; Analysis, patterns, evolution: Water Resources Research, v. 14, p. 805–843.

U.S. Geological Survey, 1985, State summaries of ground-water resources, *in* National Water Summary 1984: U.S. Geological Survey Water-Supply Paper 2275, p. 117–458.

Upham, W., 1895, The glacial Lake Agassiz: U.S. Geological Survey Monograph 25, 658 p.

van Everdingen, R. O., 1968, Studies of formation waters in western Canada; Geochemistry and hydrodynamics: Canadian Journal of Earth Sciences, v. 5, p. 523–543.

Whitaker, S. H., 1982, Groundwater resources of the Judith River Formation in southwestern Saskatchewan: Saskatoon, Silverspoon Research and Consulting Ltd., report prepared for Saskatchewan Environment, 61 p.

Whitaker, S. H., Irving, J. A., and Broughton, P. L., 1978, Coal resources of southern Saskatchewan; A model for evaluation methodology: Geological Survey of Canada Economic Geology Report 30, 151 p.

MANUSCRIPT ACCEPTED BY THE SOCIETY APRIL 17, 1987

The Geology of North America
Vol. O-2, Hydrogeology
The Geological Society of America, 1988

## Chapter 17

# *Region 14, Central Glaciated Plains*

**N. C. Krothe**
*Department of Geology, Indiana University, Bloomington, Indiana 47405*
**J. P. Kempton**
*Illinois State Geological Survey, 615 E. Peabody Drive, Champaign, Illinois 61820*

## INTRODUCTION

The glaciated terrain of the north-central United States constitutes one of the most significant ground-water regions of North America. Extensive sand and gravel aquifers occur within the glacial drift (Fig. 1), which averages as much as 60 m thick around the southern margins of the Great Lakes. The underlying bedrock (consolidated rock) also contains extensive aquifer systems.

The primary focus of early research within the region on glacial and glacially derived sediments was to establish the geologic sequence of events; e.g., Leverett and Taylor's (1915) classic monograph on the Pleistocene of Indiana and Michigan. Continued progress in understanding Quaternary stratigraphy during the past several decades provided part of the basic geologic framework required to develop regional hydrogeologic characteristics of glacial terrains. This basic framework and ground-water flow theory has provided significant insight into the rate and direction of ground-water flow and regional ground-water discharge and recharge patterns in glacial deposits.

The contribution of hydrogeologists to understanding glacial deposits of the region includes three-dimensional mapping of materials rather than features, and emphasizes distribution, thickness, and lithologic characteristics and thickness rather than age. The study of the stratigraphy and physical and mineralogic characteristics of glacial materials has continued during the past few decades. The information and subsurface methods developed for these studies have further contributed to defining the distribution and properties of sand and gravel aquifers as well as those of glacial tills and other glacially derived materials.

## REGIONAL GEOLOGIC CONTROLS AND RELATIONSHIPS

The hydrogeologic settings that occur throughout the Central Glaciated Plains Region (Heath, this volume, Fig. 3 and Table 2) include areas of both thin and thick drift over sedimentary bedrock. The indurated bedrock underlying the glaciated parts of the region ranges in age from Cambrian to Permian and consists primarily of sandstone, shale, coal, limestone, and dolomite. The bedrock of the region is important as the source of the local debris incorporated into the drift, which in part determines the drift's lithic characteristics. The distribution of the various bedrock lithologies influences the bedrock topography, which in turn controls the thickness of the overlying drift. The thickness, distribution, and permeability of the drift have a significant impact on ground-water occurrence and movement.

The bedrock lithologies reflect regional structure and lithology related to the Paleozoic history of the region modified by later Mesozoic and Tertiary erosion. From central Ohio eastward the rocks consist mainly of Devonian to Pennsylvanian clastics that dip and thicken toward the center of the Appalachian Basin. Older Paleozoic rocks occur around the Cincinnati, Kankakee, and other structural arches and at the southern end of the Great Lakes in New York State; these rocks are mainly carbonates, sandstones, and shales. The Michigan and Illinois Basins are capped by younger Paleozoic rocks—of Devonian through Permian age in Michigan, and Devonian through Pennsylvanian in Illinois—that are predominantly clastic sandstones and shales. Mississippian limestones crop out in southern Indiana and Illinois.

The topography of the bedrock surface and the bedrock structure dictate the subcrop pattern of the bedrock. A combination of preglacial stream erosion, glacial scour, ice marginal, and late- and post-glacial stream erosion have formed the present bedrock topography. It contains a series of valleys, well-defined uplands and lowlands, and some deeply entrenched valleys. This bedrock surface forms the basement for the sequence of glacial drift deposits. The drift fill of the bedrock valleys has been the focus of attention for locating ground-water resources.

The regional bedrock surface pattern has been described by Horberg and Anderson (1956), following Horberg's (1950) mapping of the bedrock topography of Illinois. Wayne (1956) described the bedrock topography and drift thickness of Indiana north of the Wisconsinan-glacial boundary, and more recently, a bedrock topography of Michigan was published (Western Michi-

Krothe, N. C., and Kempton, J. P., 1988, Central Glaciated Plains, *in* Back, W., Rosenshein, J. S., and Seaber, P. R., eds., Hydrogeology: Boulder, Colorado, Geological Society of America, The Geology of North America, v. O-2.

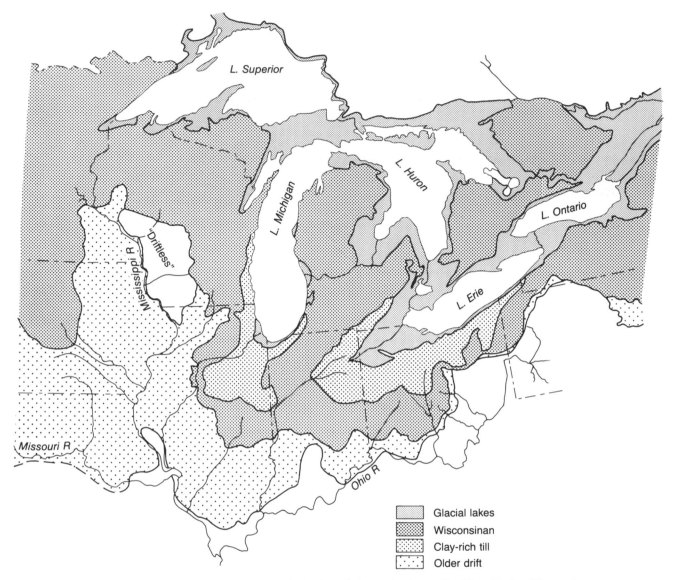

Figure 1. Distribution of glacial deposits in the Great Lakes region (generalized from National Research Council, 1959, and Richmond and Fullerton, 1983; sand and gravel areas not shown).

gan University, 1981). Gray (1982) published a detailed bedrock topography of Indiana and a companion drift thickness map (Gray, 1983). These have built on earlier studies of Fidler (1943) in Indiana and Stout and others (1943) in Ohio.

A variety of mechanisms has been employed to explain the various features displayed by these largely buried bedrock valleys, including ice marginal forebulge development (Frye, 1963; McGinnis, 1968) to explain the low gradients and potential reversal of gradient suggested in some valleys. It was into this rather complex bedrock valley system that some of the major glacial outwash aquifers were deposited.

Extensive local and regional studies of the glacial drift of the lower Great Lakes area, focusing on thickness and sedimentologic character, have provided the basis for geohydrologic characterization and predictive modeling of ground-water movement. Only

some general regional characteristics of the drift relative to hydrogeologic considerations are described here.

Studies in northeastern Ohio and northwestern Pennsylvanian (e.g., White, 1982) initiated systematic textural studies of glacial tills and related diamictons, although Krumbein (1933) may have been the first to suggest textural uniformity of individual till sheets. Information compiled from more recent soil surveys and geological field and subsurface studies have verified that the general distribution, thickness, and lithic character of tills can be mapped with some degree of accuracy (Fig. 1). The Quaternary Map of Illinois (Lineback, 1979) and more recent maps (e.g., Richmond and Fullerton, 1983; Berg and Kempton, 1988) have utilized such information for the tills as well as differentiating outwash lacustrine and related deposits for both surficial and three-dimensional maps.

A three-dimensional map of the Quaternary deposition east of the Rocky Mountains showing thickness and lithology has been compiled at a scale of 1:1,000,000 by Soller (1987, 1988). This map demonstrates the regional relation between thickness of the glacial sediments, which includes both Pleistocene and Holocene-age deposits, and patterns of ice flow during glaciation. The total thickness of the glacial sediments commonly is dependent upon bedrock lithology. In the Great Lakes region, ice flowed down the strike of bedrock, channeling into and eroding weakly resistant rocks in lowlands, and flowing around upland areas of more resistant rock. Where two or more ice lobes abutted in an interlobate area on uplands, thick sediment occurs. In one such area, in Michigan northwest of Saginaw Bay, thickness exceeds 80 m. In other interlobate areas and over buried valleys, maximum thicknesses of 30 to 65 m are more common. South of the Great Lakes, bedrock control of the ice flow apparently was diminished, and thick glacial sediment is present mostly as valley fill, commonly in buried valleys, and locally beneath moraines. On many uplands near the southern glacial margin, glacial sediment is thin and discontinuous, particularly beyond the Wisconsinan drift margin.

## HYDROGEOLOGIC CHARACTERISTICS AND RELATIONSHIPS

Although hydraulic properties of glacial deposits may vary drastically over short distances, extensive sand and gravel aquifers exist. Norris and Spieker (1966) cite an example of a buried valley aquifer near Dayton, Ohio, associated with the Great Miami River. This contains large amounts of highly permeable materials that are conducive to infiltration of rainfall. The Illinoian- and Wisconsinan-age glacial advances filled this area with interbedded till and outwash, and the upper sand and gravel is approximately 20 m thick.

This area of southwestern Ohio was covered by ice during at least two major glacial episodes. Before and between glaciations, a network of streams and valleys existed; the major river flowed past Dayton, Ohio, about where the Great Miami River flows today. This ancient valley and its tributaries contain a major aquifer system. The most recent (late Wisconsinan) glacier flowed over the area about 21,000 years ago, and most of the valleys were covered with till; however, in many valleys, sandy river sediments were not removed by the glacier and still exist as aquifers beneath the till. Because this major river valley was the course through which water was channeled from the melting glacier as it retreated, sandy sediments commonly occur from the land surface down to bedrock.

Many of the aquifer systems underlying this region are not associated with buried valleys and occur in the permeable glacial materials deposited by multiple advances and retreats of Pleistocene continental glaciers. For example, the glacial deposits of northwestern Indiana contain an extensive glaciofluvial aquifer system that underlies parts of Lake, Porter, LaPorte, and St. Joseph Counties. In Lake County, Indiana, this aquifer system (Rosenshein, 1962, 1967) consists of upper and lower confining beds (clay till with inter-till sand and gravel, the Valparaiso and Kankakee confining layers), and a glaciofluvial sand aquifer (the Kankakee aquifer), all of which are underlain by dolomitic limestone and dolomite of Silurian and Devonian ages. Rosenshein and Hunn (1968) have shown that cation exchange occurs as calcium-magnesium bicarbonate water moves downward from the Kankakee confining layer, the Kankakee aquifer, and through the Valparaiso confining layer to recharge the underlying bedrock aquifers. By the time the recharge reaches the bedrock aquifers, sodium is the dominant cation.

The Mahomet Bedrock Valley in east central Illinois (Heigold and others, 1983) contains one of the most extensive sand and gravel aquifers in Illinois. The aquifer, the pre-Illinoian Mahomet Sand (Stephenson, 1967; Kempton and Johnson, 1983) extends from just east of the Indiana-Illinois state line westward to near the center of Illinois, is up to 60 m (average 30 m) thick and averages about 20 km in width through a length of about 160 km. The top of the aquifer drops from an elevation of about 168 m on the east to 150 m at its junction with the Ancient Mississippi bedrock valley, and fines westward and upward from coarse gravel and sand to sand with some fine gravel. It is confined by 30 to 60 m of younger deposits, predominantly till.

Water levels in wells penetrating the Mahomet Sand range from elevations approaching 210 m in eastern Illinois to 180 m at its western junction, thereby indicating ground-water flow to the west. Recharge rates to the aquifer have been calculated at 107,000 gpd/mi$^2$ (Visocky and Schicht, 1969). Other hydrogeologic parameters have also been provided by Kempton and others (1982) and Cartwright and Visocky (1983), such as hydraulic conductivities ($2 \times 10^{-4}$ to $2 \times 10^{-3}$ m/s) transmissivities ($7 \times 10^{-4}$ to $5 \times 10^{-2}$ m$^2$s) and coefficients of storage ($2 \times 10^{-5}$ to $2 \times 10^{-3}$).

The Illinois Basin provides an example of hydrogeologic processes that occur in a structural basin. This basin lies within Illinois, Indiana, and Kentucky and attains a maximum thickness of 4,600 m. The basin is bordered by the Cincinnati Arch, the Mississippi River Arch, and the Ozark Dome. The basin's formation coincided with the formation of the Cincinnati Arch, as did the Michigan Basin. The Illinois Basin is an isolated basin that is hydrologically separated from adjacent areas (Bredehoeft and others, 1963). Bedrock crops out at the periphery of the basin and dips toward the basin's center in southern Illinois. Most of the Illinois Basin is covered by glacial drift. Recharge to the bedrock aquifer is impeded in many areas blanketed by glacial till and clayey silts (lacustrine deposits). The generally low intrinsic permeability of the Pennsylvanian strata also affects recharge potential.

Ground water upwelling from deeper strata is an additional possible source of recharge through fractures, faults, and oil and gas wells throughout the basin. Cartwright (1970) analyzed brine circulation patterns using geothermal gradient anomalies to show that the discharge zone for the Illinois Basin exists within a 160-km radius of the basin's center. Graf and others (1966) and Bredehoeft and others (1963) suggest that the major circulation pattern in the Illinois Basin is from basin margins, downdip toward the center of the basin, and then upward across shale membranes toward the surface at the basin center. The rate of movement was determined to be very low: $5 \times 10^{-10}$ ft/sec (Cartwright, 1970).

## CHEMISTRY OF GROUND WATER

Chemical evolution of the ground water in this area is complicated by numerous geochemical processes, ground-water withdrawal, and Pleistocene glaciation. The chemistry of ground water contained in glacial deposits is variable and depends on the

mineralogy of the materials derived from glacial erosion of preexisting bedrock and unconsolidated deposits.

In recharge areas of Wisconsin, southern Minnesota, northeastern Iowa, and north-central Illinois, the bedrock aquifers have a $Ca-Mg-HCO_3$-type water due to pyrite oxidation in the overlying Ordovician Maquaketa Formation. Supporting evidence for pyrite oxidation is undersaturation with respect to gypsum and negative $^{34}S$ $(SO_4)$ values for the water. From northwestern Iowa into the Illinois Basin, water changes from $Na-Ca-SO_4-HCO_3$ to NaCl along major flow paths. In the bedrock aquifers, highly mineralized water occurs in the southwestern, southern, and eastern parts of the area; these are regional discharge points or structurally low areas.

## REFERENCES CITED

Berg, R. C. and Kempton, J. P., 1988, Stack-unit mapping of geologic materials in Illinois to a depth of 15 meters: Illinois Geological Survey Circular 542, 23 p., 3 plates.

Bredehoeft, J. D., Blythe, C. R., White, W. A., and Maxey, G. B., 1963, Possible mechanism for concentration of brines in subsurface formations: Bulletin of the American Association of Petroleum Geologists, v. 47, p. 257–269.

Cartwright, K., 1970, Groundwater discharge in the Illinois Basin as suggested by temperature anomalies: Water Resources Research, v. 6, p. 912–918.

Cartwright, K., and Visocky, A., 1983, Hydrogeology of the Mahomet Valley Bedrock Valley deposits: Geological Society of America Abstracts with Programs, v. 15, no. 6, p. 540.

Fidler, M. M., 1943, The preglacial Teays Valley in Indiana: Journal of Geology, v. 51, p. 411–418.

Frye, J. C., 1963, Problems of interpreting the bedrock surface of Illinois: Illinois Academy of Science Transactions, v. 56, p. 3–11.

Graf, D. L., Friedman, I., and Meents, W. F., 1965, The origin of saline formation waters; 2, Isotopic fractionation by shale micropore system: Illinois Geological Survey Circular 393, 32 p.

Gray, H. H., 1982, Map of Indiana showing topography of the bedrock surface: Indiana Geological Survey Miscellaneous Map 35, scale 1:500,000.

—— , 1983, Map of Indiana showing thickness of unconsolidated deposits: Indiana Geological Survey Miscellaneous Map 37, scale 1:500,000.

Heigold, P. C., Kempton, J. P., and Cartwright, K., 1983, The Mahomet Bedrock Valley in Illinois, an update: Geological Society of America Abstracts with Programs, v. 15, no. 6, p. 594.

Horberg, Leland, 1950, Bedrock topography of Illinois: Illinois Geological Survey Bulletin 73, 111 p.

Horberg, Leland, and Anderson, R. C., 1956, Bedrock topography and Pleistocene glacial lobes in Central United States: Journal of Geology, v. 64, no. 2, p. 101–116.

Kempton, J. P., and Johnson, W. H., 1983, Stratigraphy of the sediments within the Mahomet Bedrock Valley in Illinois: Geological Society of America Abstracts with Programs, v. 15, no. 6, p. 610.

Kempton, J. P., Visocky, A. P., and Morse, W. J., 1982, Hydrogeologic evaluation of sand and gravel aquifers for municipal ground-water supplies in east-central Illinois: Illinois Geological Survey and Illinois Water Survey Cooperative Ground Water Report 8, 59 p.

Krumbeing, W. C., 1933, Textural and lithologic variations in glacial till: Journal of Geology, v. 41, no. 4, p. 382–408.

Leverett, F., and Taylor, F. B., 1915, The Pleistocene of Indiana and Michigan and the history of the Great Lakes: U.S. Geological Survey Monograph 53, 529 p.

Lineback, J. A., 1979, Quarternary deposits of Illinois: Illinois Geological Survey Map, scale 1:500,000.

McGinnis, L. D., 1968, Glacial crustal bending: Geological Society of America Bulletin, v. 79, p. 769–775.

National Research Council, Division of Earth Sciences, 1959, Glacial Map of the United States east of the Rocky Mountains, 1st edition: New York, Geological Society of America, scale 1:1,750,000.

Norris, B. E., and Spieker, A. M., 1966, Ground-water resources of the Dayton area, Ohio: U.S. Geological Survey Water-Supply Paper 1808, 167 p.

Richmond, G. M., and Fullerton, D. S., eds., 1983, Quaternary Geologic map of the Chicago $4° \times 6°$ Quadrangle, United States: U.S. Geological Survey Map I-1420 (NK-16), scale 1:1,000,000.

Rosenshein, J. S., 1962, Geology of Pleistocene deposits of Lake County, Indiana: U.S. Geological Survey Professional Paper 450–D, art. 157, p. D127–D129.

—— , 1967, Geohydrology of Pleistocene deposits and sustained yield of principal Pleistocene aquifer, Lake County, Indiana: U.S. Geological Survey Open-File Report, 99 p.

Rosenshein, J. S., and Hunn, J. D., 1968, Geohydrology and ground water potential of Lake County, Indiana: Indiana Division of Water Bulletin 31, 36 p.

Soller, D. R., 1987, A three-dimensional perspective of Quaternary sediments in the glaciated Great Lakes and Mississippi Valley areas, United States: Geological Society of America Abstracts with Programs, v. 19, p. 247.

—— , 1988, Map showing the thickness and character of Quaternary sediments in the glaciated United States and Rocky Mountains: U.S. Geological Survey Miscellaneous Investigation Series Map I–1970A, B, C, D, scale 1:1,000,000.

Stephenson, D. A., 1967, Hydrogeology of the glacial deposits of the Mahomet Bedrock Valley in east-central Illinois: Illinois Geological Survey Circular 409, 51 p.

Stout, W. E., Ver Stag, K., and Lamb, G. G., 1943, Geology of the water in Ohio: Ohio Geological Survey, 4th Series, Bulletin 44, 694 p.

Visocky, A. P. and Schicht, R. J., 1969, Ground-water resources of the buried Mahomet Bedrock Valley: Illinois Water Survey Report of Investigation 62, 52 p.

Wayne, W. J., 1956, Thickness of drift and bedrock physiography of Indiana, north of the Wisconsin glacial boundary: Indiana Geological Survey Report of Progress 7, 70 p.

Western Michigan University, Department of Geology, 1981, Hydrogeologic atlas of Michigan: Kalamazoo, Michigan, Western Michigan University, scale 1:500,000.

White, G. W., 1982, Glacial geology of northeastern Ohio: Ohio Geological Survey Bulletin 68, 75 p.

MANUSCRIPT ACCEPTED BY THE SOCIETY AUGUST 16, 1988

The Geology of North America
Vol. O-2, Hydrogeology
The Geological Society of America, 1988

# Chapter 18

# *Region 15, St. Lawrence Lowland*

**R. N. Farvolden and J. A. Cherry**

*Institute for Groundwater Research, University of Waterloo, Waterloo, Ontario N2L 3G1, Canada*

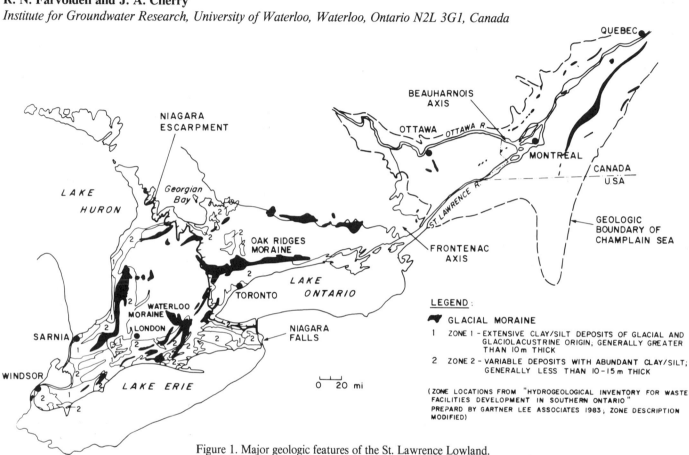

Figure 1. Major geologic features of the St. Lawrence Lowland.

## INTRODUCTION

The St. Lawrence Lowlands region in the southern part of Canada encompasses an area that extends from Windsor at the southern tip of the province of Ontario to the St. Lawrence River at Quebec City (Fig. 1). This region comprises only 100,000 km², or 2 percent of the land surface of Canada, but contains more than half the population of the country. Also it accounts for 75 percent of the industrial production and 40 percent of the agricultural income. Many of Canada's major cities, such as Montreal, Ottawa, Toronto, and Hamilton are in this region (Fig. 1).

The climate varies significantly across the 1,000 km from southwest to northeast but is a humid continental type with warm summers and cold winters. Annual precipitation varies from about 800 to 1,100 mm, and mean annual temperature varies from about 9°C in the southwest to 5°C in the northeast.

The St. Lawrence Lowlands region is bounded on the northwest by the uplands of the Canadian Shield and on the southeast by faults and structures of the Appalachian Folded Belt (Fig. 3; Table 2, Heath, this volume). Relatively undeformed Paleozoic sedimentary rocks from Cambrian to Mississippian age overlie the Precambrian basement, and the sedimentary rocks are more than 1,000 m thick at the southwest end of the region. The region is an extension of the Glaciated Central Region, and the local topography is dominated by the effects of Pleistocene Wisconsin glaciation. Ice flow was strongly controlled by the pregla-

Farvolden, R. N., and Cherry, J. A., 1988, Region 15, St. Lawrence Lowland, *in* Back, W., Rosenshein, J. S., and Seaber, P. R., eds., Hydrogeology: Boulder, Colorado, Geological Society of America, The Geology of North America, v. O-2.

cial drainage system now occupied in part by the Great Lakes; this control has had a strong influence on the glacial deposition, particularly the major moraines (Fig. 1) and outwash deposits, and consequently, the hydrogeology.

The region is naturally divided by the Frontenac Axis (Fig. 1), which brings Precambrian rocks to the surface in a lobe that extends across the region from the north and into New York State. Potsdam Sandstone (Nepean Formation) and Black River–Trenton shales and limestones (Ottawa Formation) of Cambrian and Ordovician age overlie the Precambrian and dip gently away from the Axis.

The northeast half of the region below the Frontenac Axis is further divided by the Beauharnois Axis at the confluence of the Ottawa and St. Lawrence rivers (Fig. 1). Precambrian rocks form the core of this axis but are not exposed.

Northeast of the Beauharnois Axis, local intrusions of Cretaceous age are exposed as the Monteregian Hills in the Montreal area, but their hydrogeologic significance is unknown. Water-supply wells tap the sandstone and siltstone at the base of the section and the fractured dolomites of the Beekmantown and Black River Formations.

In the broad basin structure between the Beauharnois and Frontenac axes, the rocks range form Cambrian to Upper Ordovician and are highly variable in permeability. In some localities southeast of Ottawa, even household supplies of potable ground water are difficult to locate in Upper Ordovician strata. Elsewhere, in Lower Ordovician and Cambrian strata, wells that produce 2 to 4 L/sec are common, and local yields of 20 to 40 L/sec are reported. The Ottawa Valley, which follows the Ottawa River (Fig. 1), is a graben, and faults exert a strong influence on ground-water flow and well productivity.

The Niagara Escarpment, capped by Middle Silurian dolomites (Lockport-Amabel-Guelph Formations) divides the southwestern half of the region into eastern and western sectors (Fig. 1). Between the Frontenac Axis and the Niagara Escarpment the near-surface rocks commonly have low well yields. Excessive chloride renders much of the deep ground water unsuitable for use.

West of the Niagara Escarpment, the dolomite of the Guelph-Amabel Formation (Lockport) is an excellent aquifer. The water is rather hard but usually acceptable chemically. To the southwest, and downdip, the Salina Formation overlies the Guelph-Amabel. While well yields may be appreciable, ground-water quality is subject to excessive sulphate concentrations. Throughout this sector the zones of high permeability seem to be restricted to the upper 5 to 10 m of the bedrock (Novakovic and Farvolden, 1974), and truly cavernous carbonates are uncommon. Farther west, good ground-water supplies are locally developed primarily in subcrop areas of the younger Devonian carbonates. Near Windsor these rocks are in turn overlain by shales and shaley limestones (Hamilton and Kettle Point formations) that are aquifers with poor yield and, in many places, poor quality.

The entire region is covered by glacial drift, and rock out-crops are restricted to steep slopes for the most part. Some good river-connected aquifers occur in the Quaternary alluvium along streams and rivers.

The preglacial topography had a profound effect on the direction of ice flow; this in turn accounts for the composition of surficial deposits at any site. Except for zones near the Precambrian Shield, the surficial deposits are rich in carbonate detritus. Glaciofluvial deposits are common, particularly west of the Niagara Escarpment, and granular deposits in kames and outwash are important aquifers. Sand plains, also associated with outwash and ancient Great Lakes shorelines, are important hydrogeologic features and generally provide good ground-water supplies.

The marine invasion of the St. Lawrence Valley, known as the Champlain Sea, was responsible for large areas of clay and for some sand in terraces along the lower Ottawa and St. Lawrence Valleys. The clay deposits form the rich soils of this part of the region. The clays are "sensitive," and slope failure by liquefaction is a serious hazard wherever they occur near steep slopes such as the edge of terraces, or cut banks, and also where tributary valleys are deeply incised at the confluences with main streams.

This chapter addresses four aspects of the hydrogeology of this region selected to demonstrate the manner in which geologic features control the origin, age, chemistry, yield, and discharge of ground water in the region. These topics are: (1) the origin and age of ground water in thick surficial clayey aquitards that overlie drift or bedrock aquifers in much of the region; (2) unusual features of two major drift aquifers; (3) the relic salinity of near-surface shale bedrock along the Niagara Escarpment; and (4) ground-water discharge and phenomena related to flow systems.

## THICK CLAYEY AQUITARDS

Surficial clay-rich Quaternary deposits exist in much of southwestern Ontario and in much of the area that was occupied by the Champlain Sea in eastern Ontario and Quebec (Fig. 1). These deposits are important because they severely restrict the vertical flow of recharge water into aquifers and because they usually have desirable characteristics for waste-disposal sites. In southwestern Ontario these deposits are primarily glacial till and glaciolacustrine, whereas in eastern Ontario and Quebec they are glaciomarine deposits of the Champlain Sea.

The hydrogeology of the thick clayey deposits in southwestern Ontario has been reported on by Desaulniers and others (1981, 1986) and Desaulniers (1986). The geological and geotechnical aspects of the deposits in the Montreal-Ottawa area have been described by numerous investigators, most notably Karrow (1972), Quigley (1980), and Quigley and others (1983). Hydrogeologic studies have been reported by Lafleur and Giroux (1983) and Desaulniers (1986). The discussion that follows is based on these sources unless otherwise noted.

The clay deposits exhibit a surficial-weathered zone, overconsolidated due to dessication effects, which is between 3 and 6 m thick. The water table occurs in this weathered zone, gener-

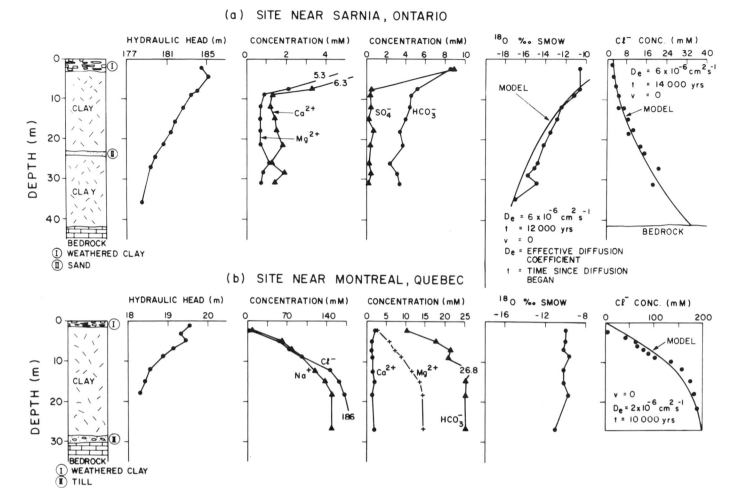

Figure 2. Hydraulic head, major ions, and oxygen-18 in thick clayey deposits at (a) a site near Sarnia, Ontario and (b) a site near Montreal (adapted from Desaulniers, 1986).

ally within a meter or two of ground surface. Continuous saturation exists beneath the water table. The weathered zone has many visible weathering features, such as secondary gypsum and calcite and brown to rust coloring due to oxidation of iron-bearing minerals. Fractures that are primarily vertical or near vertical are ubiquitous in this zone. Relic rootlets are common to depths of about 3 or 4 m. The weathering features and fractures probably formed during the relatively warm Altithermal climatic period that existed between about 9,000 and 3,000 years ago.

In many areas, the thickness of the clayey deposits in southwestern Ontario and in the area of the Champlain Sea exceeds 10 m. In the area between Sarnia, London, and Windsor (Fig. 1), the clayey deposits are commonly between 20 and 45 m thick. Beneath the weathered zone the clayey material is gray and relatively unfractured.

The hydraulic conductivity of the unweathered clayey deposits in both regions has been determined at many locations by means of piezometer-response monitoring and by consolidation and triaxial cell tests on undisturbed core samples. Values typically range from $4 \times 10^{-5}$ m/d to $9 \times 10^{-6}$ m/d. The surficial weathered zone commonly has a much higher bulk hydraulic conductivity caused by the fractures, root holes, and other secondary features. The hydraulic gradient in the unweathered zone is downward at nearly all locations where detailed head profiles have been measured. Examples of head profiles at a site in southwestern Ontario and at another site near Montreal are shown on Figure 2. The gradient is generally downward because the clayey deposits are underlain by fractured bedrock or, in some areas, extensive sandy deposits, which act as regional ground-water transmission zones. The average linear ground-water velocity downward through the clay is less than 1 cm per year. This velocity is obtained by dividing the calculated Darcy velocity by the total porosity, which for glacial till is about 0.4 and for glaciomarine clay is about 0.60.

The last Pleistocene glaciers receded from southwestern Ontario about 12,000 and 10,000 years ago. The Champlain Sea drained from the Montreal-Ottawa area approximately 9,800 years ago. From consideration of the geomorphic settings of the thick-clay areas, the magnitudes of the present-day hydraulic gradients in the various clay areas are not much different than the gradients that existed during the past 10,000 years. Thus, the calculated ground-water velocities are taken as evidence that ground-water circulation in the unweathered clay has been very slow during Holocene time.

Figure 2 shows vertical profiles of some major ions and environmental isotopes for a site in southwestern Ontario and for another site near Montreal. Similar profiles have been obtained for many other sites in southwestern Ontario. Figure 2a indicates that in southwestern Ontario, ground water in the weathered zone and in the several meters of the clay just below the weathered zone contains relatively high concentrations of $Mg^{2+}$, $Ca^{2+}$, $HCO_3^-$ and $SO_4^{2-}$. The $Na^+$ concentration, although not shown, is also high within and just below the weathered zone. $Cl^-$, which is nearly absent from the weathered zone, increases progressively with depth to the bottom of the clay. Below the shallow zone where weathering effects occur, $Na^+$ also increases with depth with the same trend as $Cl^-$.

The relatively high concentrations of major ions in and near the weathered zone are attributed to chemical weathering that took place primarily during Altithermal time when a warmer, drier climate probably caused the average water table to be 2 or 3 m deeper than the present-day water table. Dessication caused the fractures to form during this drier period. The invasion of air into this fractured unsaturated zone caused oxidation of pyrite ($FeS_2$) in shale fragments in the clayey till (Abbott, 1987). This acid-producing process induced dissolution of calcite ($CaCO_3$) and dolomite [$CaMg(CO_3)_2$] and precipitation of gypsum ($CaSO_4 \cdot 2H_2O$; Rodvang, 1987).

At the end of the Pleistocene Epoch, rain and snowmelt replaced glacial meltwater as the dominant contributor of water to the land surface. This change is reflected in the depth profiles for $^{18}O$ (Fig. 2a), which exhibit a gradual shift from relatively high values in the shallow zone that are typical of the present climate to lower values at depth believed to be representative of glacial meltwater. Deuterium profiles have a shape similar to the $^{18}O$ profiles. Carbon-14 dating of deep ground water in the clay in the Sarnia district indicates that the water is between 10,000 and 14,000 years old.

Desaulniers and colleagues developed the following explanation for the profiles of major ions and isotopes described above. At nearly all study sites, the vertical flow of ground water in the unweathered clay has a much smaller influence on the vertical migration of ions and isotopes than does molecular diffusion (Desaulniers, 1986; Desaulniers and others, 1986; Desaulniers and Cherry, 1987). Molecular diffusion causes each constituent to move in the direction of its concentration gradient. $Cl^-$, $Na^+$, and $CH_4$ are diffusing upward from the bedrock where the high concentrations of these constituents originate. Upward diffusion dominates over the downward flow (i.e., advection); therefore, the net movement of these constituents is upward. $Ca^{2+}$, $Mg^{2+}$, $HCO_3^-$, and $SO_4^{2-}$ are diffusing downward from the weathered zone where they originate. $^{18}O$ and $^2H$ (deuterium) are diffusing downward from the bottom of the weathered zone into the unweathered clayey till.

This interpretation has been used as a time scale for one-dimensional mathematical modeling of the development of the major-ion and isotope profiles. The modeling is based on Fick's Laws of diffusion. As an example, simulations of the development of the $^{18}O$ profile are shown in Figure 2a. In those simulations, the mean annual $^{18}O$ content of shallow recharge water is assumed to change quickly at the end of Pleistocene time from a value of about $-18^0/_{00}$ to the present-day average of $-10^0/_{00}$. Figure 2a indicates that the observed present-day profile would be achieved after about 12,000 years of downward $^{18}O$ migration governed by molecular diffusion. The coefficients of molecular diffusion for $^{18}O$ used in the simulations were obtained from laboratory diffusion tests on core samples (Desaulners, 1986). Other profiles from other sites in southwestern Ontario simulated in this manner provided similar results. The layer of clayey till lying directly on the bedrock surface originated more than about 16,000 years ago. Therefore, $Cl^-$ and $Na^+$ began diffusing upward into this deep till from the bedrock much earlier than the downward diffusion of $^{18}O$ and $^2H$ into the uppermost till. The simulations of upward $Cl^-$ diffusion closely match the field profile for diffusion times of 14,000 to 16,000 years. The studies by Desaulniers (1986) suggest that at most locations in southwestern Ontario where the thickness of surficial clayey deposits uninterrupted by significant sandy layers is greater than about 10 to 15 m, the clayey material beneath the weathered zone normally contains ground water that is many thousands of years old and that exhibits diffusion-controlled distributions of major ions and isotopes such as $^{18}O$ and $^2H$.

The magnitude of the downward ground water velocity below the weathered crust in the thick clay at the Montreal site is as low as at the sites in southwestern Ontario. The major ion and $^{18}O$ profiles (Fig. 2b), however, are much different than those in southwestern Ontario. These differences result from the pore water residing in the clayey sediment during the period of deposition. This original water was a mixture of about 35 percent sea water and 65 percent glacial meltwater and meteoric water, which provides $^{18}O$ values similar to modern-day precipitation at all depths in the clay. Except for $Ca^{2+}$, the major-ion profiles show a large increase in concentration with depth. The deep ground water is probably the original water from the Champlain Sea. Since drainage of the Champlain Sea, fresh shallow ground water derived from modern rain and snow has been mixing due to diffusion with this deep water. Desaulniers (1986) simulated the major-ion profiles using the one-dimensional approach described above. In these simulations, the diffusion is upward. The simulations support the conclusion that the mixing is governed by molecular diffusion and that the downward advection is unimportant. The major ion profiles will not achieve a steady-state shape until about 20,000 more years have passed. The thick deposits of Champlain Sea clay are slowly becoming desalinized as a result of molecular diffusion. The engineering properties of the clay depend significantly on the salinity. Diffusion models indicate that salinity depends on the clay thickness and on the salinity of the ground water in the bedrock or other deposits beneath the clay.

Areas in the St. Lawrence Lowlands region having thick

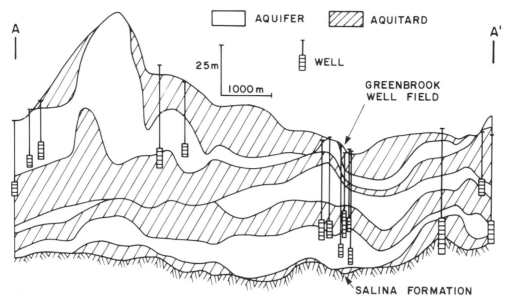

Figure 3. Cross section of multiaquifer systems of Greenbrook and adjacent well fields, Kitchener, Ontario (adapted from Rudolph, 1986).

deposits of clay are generally much better suited hydrogeologicaly than other areas for disposal of hazardous wastes. Ontario's only currently operating disposal site designated exclusively for hazardous waste is located on the Sarnia clay plain. At this site, the waste is buried below the weathered zone in a solidified form in deep pits (18 m; Johnson and others, 1987).

## UNUSUAL FEATURES OF TWO MAJOR AQUIFERS

Aquifers are associated with both the Waterloo and the Oak Ridges moraines (Fig. 1), and ground water is used for municipal supplies in these areas. The continuous search for additional supplies is difficult because both drillers and hydrogeologists report that in many test holes near productive wells, geologic deposits encountered do not have favorable aquifer characteristics, even in well fields that have produced 10 to 15 × 10⁶ liters per day for many years. The explanation given for this occurrence is that ground water flow is restricted to narrow zones of permeable gravel. However, a series of piezometer nests installed in one well field in the Waterloo region showed a drawdown cone with horizontal and vertical components of flow toward the pumping centers such as is normally expected in a continuous aquifer. Also, a quasi-three-dimensional steady-state computer model of flow in the aquifer system provides results entirely compatible with the existence of extensive hydrostratigraphic units (Fig. 3) and with the drawdowns measured near the well field (Fig. 4), which is in contrast to reports from test drilling (Rudolph, 1985).

The most plausible interpretation of the hydrogeology is that the aquifers have a rather low average hydraulic conductivity over a thick section (15 to 20 m), and that only a slight change in clay-silt content, which is very difficult to detect, causes large changes in transmissivity. Farvolden and others (1986) used a method of cross plots of borehole geophysical data to develop a

procedure for identifying hydrogeologic units in this sequence. The experiment demonstrated that good aquifers might not be recognized from drill cuttings and geophysical logs, even during carefully conducted test drilling, and that one of the best aquifers, when categorized using only drill cuttings and geophysical logs, is glacial till.

Responses to changes in pumping rates are transmitted rapidly throughout this aquifer system. No evidence of subsidence exists, which indicates that the clayey aquitards are consolidated, apparently due to overriding by Pleistocene glaciers. These hydrogeologic conditions meet the basic assumptions for the application of aquifer test analysis based on the Hantush and Jacob (1955) method for analyzing leaky layer artesian systems in which storage in the aquitard can be neglected.

The Oak Ridges moraine between Lake Ontario and Georgian Bay (Fig. 1) is also underlain and surrounded by regional sand and gravel aquifers. The granular deposits that make up the aquifer are mostly sand, as much as 65 m thick; but permeable gravel zones are present and important. The Oak Ridges moraine, like the Waterloo moraine, is capped by a clayey-silt till deposited when ice overrode the kame in a final stage of glacial activity. A perched water table is present in this till and as much as 20 m of unsaturated sand occurs between the bottom of the till and the "main" water table in the Oak Ridge aquifer. At least one stream (North Branch of the Holland River) and one permanent lake (Musselman Lake) are either perched or have strong downward gradients, and so are recharge zones rather than discharge zones, a rather unusual condition for permanent streams and lakes in a humid climate. Musselman Lake is about 1 km in length and has a fairly constant water level. Surface flow into the lake is minimal, and surface discharge from the lake occurs only during very high stages. Shallow piezometers show a surficial aquifer discharges to the lake along the northwestern shore, and this dis-

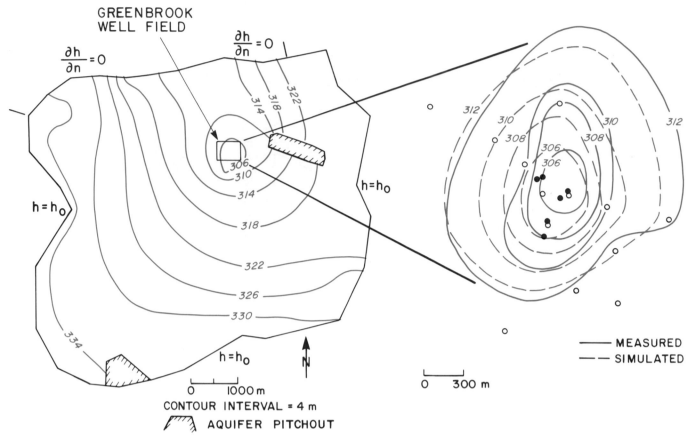

Figure 4. Flow in the Greenbrook aquifer, and detail of drawdown cone during heavy pumping (adapted from Rudolph, 1986).

charge provides inflow to the water balance for the lake. A few kilometers south on the lower slopes of the kame, the discharge zone for the main Oak Ridges aquifer is marked by a spring line, with rising streams and flowing wells.

A single continuous hydrograph for the main aquifer below the perched zone has a barometric response that would normally indicate a confined aquifer. The gaseous phase in the unsaturated zone must be responding to changes in barometric pressure that are transmitted to the saturated zone and the observation well.

## RELIC SALINITY OF NEAR-SURFACE SHALE BEDROCK ALONG THE NIAGARA ESCARPMENT

The Middle Silurian dolomite that caps the Niagara Escarpment is underlain by a thick sequence of shale containing thin-bedded limestone and sandstone. The shale, with a thin cover of drift, also underlies the terrain that extends from the steep middle part of the escarpment across a gently sloping plain to the western tip of Lake Ontario. The shale is horizontally bedded and has vertical and subvertical fractures as well as bedding-plane fractures. The fractures impart slight to moderate permeability to the shale. Ground water in the shale has high total dissolved solids, the dominant ion being $Cl^-$. At a depth of about 40 or 50 m, the $Cl^-$ content commonly approaches or exceeds that of sea water. At depths between about 10 and 50 m, the $Cl^-$

content of the ground water is generally greater than 5,000 mg/$\ell$. At some locations the $Cl^-$ content is even high at depths between 5 and 10 m (Hewetson, 1985).

Ground water circulates in the shale from the upper part of the escarpment to Lake Ontario, a distance of several kilometers. Circulation has probably been occurring in this manner since deglaciation. In spite of this circulation and even though leaching of $Cl^-$ from the shale has taken place continually during the past 10,000 years, $Cl^-$ still persists in the shale even at shallow depth. This 10,000-year persistence is attributed to the slow rate of release of $Cl^-$ from the shale to the ground water flowing in the fractures. The release occurs by slow movement from the interior of shale blocks to the fractures. The release of $Cl^-$ from the primary pore space to the fractures is probably governed by molecular diffusion. The porosity of the shale matrix is only about 3 to 6 percent. Therefore, the effective diffusion coefficient is low, and the diffusive release is slow.

It is not known whether the $Cl^-$ in the shale is from the original sea water in the marine basin in which the shale formed or whether it entered the shale from other stratigraphic zones at some later time. The persistence of $Cl^-$ through geologic time causes potable ground water to be a rare occurrence in the shale even in zones where the permeability is large enough to provide what would otherwise be useful well yields.

# GROUND-WATER DISCHARGE PHENOMENA AND FLOW SYSTEMS

Regional surveys of water resources in the St. Lawrence Lowlands have shown that water yield is about 50 percent of precipitation and that ground-water contributions to base flow make up about 40 percent of total stream flow (Sibul, 1969). Cherry and others (1976) used $^{18}O$ and tritium to show that even storm runoff in headwater catchments is mostly ground water— that is, water that was in the catchment prior to the rainfall event that produced the runoff. A mechanism for this was suggested (Sklash and Farvolden, 1979) and has now been confirmed in a series of laboratory and field studies, coupled with computer models of various types (Abdul, 1984; Abdul and Gillham, 1984; DeSilva, 1986). These later studies show that where the capillary fringe reaches the ground surface, a very slight amount of infiltration converts the pressure head from negative to positive almost instantly and provides a flow regime that drives ground water toward the nearest channel. The same phenomenon provides for rapidly expanding zones that contribute direct runoff to the channel, if the rate of precipitation exceeds the infiltration rate.

Along the south shores of Lakes Erie and Ontario and some river valleys, unconsolidated Quaternary deposits form cliffs as high as 85 m, and near-vertical slopes are maintained, for the most part, by toe erosion. Ground water causes high pore-water pressures and piping at the toe of the slope (Eyles and others, 1985), and discharge higher on the slope "softens" the clays (Quigley and others, 1977). Novakovic and Farvolden (1974) report headward erosion from piping in small valleys in the region.

Little evidence for regional ground-water flow has been found anywhere in the St. Lawrence Lowlands, perhaps because most potable ground water occurs at shallow depths, so data are not available for the depths at which evidence of regional flow might be found. Novakovic and Farvolden (1974) and Rudolph (1986) proved that in several typical watersheds, local flow predominates and no regional component can be identified. If this condition is general, then most ground water reaches the Great Lakes as stream flow originating from local ground-water flow zones. This hydrogeologic conditions requires careful consideration in any attempts to avoid contamination of the lakes through ground-water transport.

# REFERENCES

Abbott, D. E., 1987, The origin of sulphate, and the isotope geochemistry of sulphate rich shallow groundwater of the St. Clair till plain, southwestern Ontario [M.Sc. thesis]: Department of Earth Sciences, University of Waterloo (submitted).

Abdul, A. S., 1984, Experimental and numerical studies of the effect of the capillary fringe on streamflow generation [Ph.D. thesis]: Department of Earth Sciences, University of Waterloo, 224 p.

Abdul, A. S., and Gillham, R. W., 1984, Laboratory studies of the effects of the capillary fringe on streamflow generation: Water Resources Research, v. 20, p. 691–698.

Cherry, J. A., Fritz, P., Bottomley, D. J., Farvolden, R. N., Sklash, M. G., and Weyer, K. U., 1976, Storm runoff studies in upstream basins using $^{18}O$, *in* Proceedings of Canadian Hydrogeology Symposium 75, Winnipeg, August, 1975: National Research Council of Canada, p. 552–560.

Desaulniers, D. E., 1986, Groundwater origin, geochemistry, and solute transport in three major clay plains of East-central North America [Ph.D. thesis]: Department of Earth Sciences, Univeristy of Waterloo, 450 p.

Desaulniers, D. E., and Cherry, J. A., 1987, Origin and movement of groundwater and major ions in a thick deposit of Champlain Sea clay near Montreal, Quebec: Canadian Geotechnical Journal (in press).

Desaulniers, D. E., Cherry, J. A., and Fritz, P., 1981, Origin, age, and movement of pore water in clayey Pleistocene deposits on South-central Canada, *in* Perry, E. C., and Montgomery, C. W., eds., Isotope studies of hydrologic processes: DeKalb, Northern Illinois University Press, p. 45–55.

Desaulniers, D. E., Kaufmann, R. S., Cherry, J. A., and Bentley, H. W., 1986, $^{37}Cl$-$^{35}Cl$ variations in a diffusion-controlled groundwater system: Geochimica et Cosmochimica Acta, v. 50, p. 1757–1764.

DeSilva, C. J., 1986, Development and testing of a deterministic empirical model of the effect of the capillary fringe on near stream runoff [Ph.D. thesis]: Department of Earth Sciences, University of Waterloo, 156 p.

Eyles, N., Eyles, C. H., Lau, K., and Clark, B., 1985, Applied sedimentology in an urban environment; The case of Scarborough Bluffs, Ontario, Canada's most intractable erosion problem: Geosciences Canada, v. 12, no. 3, p. 91–104.

Farvolden, R. N., Greenhouse, J. P., Karrow, P. F., Pehme, P. E., and Ross, I. C., 1986, Subsurface Quaternary stratigraphy of the Kitchener–Waterloo Area using borehole geophysics: Final Report of Ontario Geological Research Fund Project 128, Ontario Geological Survey Open File Report no. OFR5623, 76 p.

Hewetson, J. P., 1985, An investigation of the groundwater zone in fractured shale at a landfill [M.Sc. thesis]: Department of Earth Sciences, University of Waterloo, 72 p.

Hantush, M. S., and Jacob, C. E., 1955, Nonsteady radial flow in an infinite leaky aquifer: EOS Transactions of the American Geophysical Union, v. 36, p. 95–100.

Johnson, R. L., Pankow, J. F., and Cherry, J. A., 1987, Diffusive contaminant transport in natural clay; A field example and implications for clay-lined waste disposal sites: Environmental Science and Technology (in press).

Karrow, P. F., 1972, The Champlain Sea and its sediments, *in* Legget, R. F., ed., Soils in Canada; Geological pedological and engineering studies: Royal Society of Canada Special Publication 3, p. 97–107.

Lafleur, J., and Girous, F., 1983, Permeabilite in-situ des Argiles Superficielles de la mer Champlain: Bulletin of the International Association of Engineering Geology, v. 26–27, p. 453–459.

Novakovic, B., and Farvolden, R. N., 1974, Investigations of groundwater flow systems in Big Creek and Big Otter Creek drainage basins, Ontario: Canadian Journal of Earth Sciences, v. 11, p. 964–975.

Quigley, R. M., 1980, Geology, mineralogy, and geochemistry of Canadian soft soils; A geotechnical perspective: Canadian Geotechnical Journal, v. 16, p. 261–285.

Quigley, R. M., Gelinas, P. J., Bon, W. T., and Packer, R. W., 1977, Cyclic erosion-instability relationships; Lake Erie North Shore Bluffs: Canadian Geotechnical Journal, v. 14, p. 310–323.

Quigley, R. M., Gwyn, Q.H.J., White, O. L., Rowe, R. K., Haynes, J. E., and Bohdanowicz, A., 1983, Leda clay from deep boreholes at Hawkesbury, Ontario; Pt. I, Geology and geotechnique: Canadian Geotechnical Journal, v. 20, p. 228–298.

Rodvang, J., 1987, Geochemistry of the weathered zone of a fractured clayey deposit in southwestern Ontario [M.Sc. thesis]: Department of Earth Sciences, University of Waterloo, 177 p.

Rudolph, D. L., 1986, A quasi three-dimensional finite element model for steady-state analysis of multiaquifer systems [M.Sc. thesis]: Department of Earth Sciences, University of Waterloo, 100 p.

Sibul, U., 1969, Water resources of the Big Otter Creek drainage basin: Ontario
	Water Resources Commission, Water Resources Report 1, 91 p.
Sklash, M. G., and Farvolden, R. N., 1979, The role of groundwater in storm and
	snowmelt generation: Journal of Hydrology, v. 43, p. 45–65.

MANUSCRIPT ACCEPTED BY THE SOCIETY JUNE 3, 1987

## Chapter 19

# *Region 16, Central Nonglaciated Plains*

**Donald G. Jorgensen**
*U.S. Geological Survey, 1950 Constant Avenue, Campus West, Lawrence, Kansas 66046*
**Joe Downey**
*U.S. Geological Survey, Mail Stop 416, Box 25046, Denver Federal Center, Denver, Colorado 80225*
**Alan R. Dutton**
*Bureau of Economic Geology, University of Texas at Austin, University Station, Box X, Austin, Texas 78713-7508*
**Robert W. Maclay**
*U.S. Geological Survey, North Plaza, Suite 234, 435 Ison Road, San Antonio, Texas 78216*

## INTRODUCTION

The Central Nonglaciated Plains of North America (Fig. 3; Table 2, Heath, this volume) extend from Montana to the Balcones Escarpment of central Texas (Fig. 1). Not all the regional aquifers that are found within the province are described in this chapter; specifically, the High Plains aquifer and the aquifers of alluvium along the major streams are described in Weeks and Gutentag (this volume) and Rosenshein (this volume).

Climate, especially precipitation and evapotranspiration, is widely recognized as the dominant factor controlling streamflow, which plays a major role in modifying landforms in the region. However, the role of climate in controlling the occurrence and movement of ground water is not as widely recognized. Nevertheless, climate is the primary control on the amount of recharge to aquifers in the region. Recharge to the water table from precipitation ranges from less than 25 mm/year near the Front Range of the Rocky Mountains to more than 300 mm/year in the Ozark Plateaus of Arkansas.

The dominant geologic characteristic of the Central Nonglaciated Plains is the extensive, nearly horizontal post-Precambrian sedimentary rocks that occur at nearly all locations (Fig. 2), although, the complete post-Precambrian stratigraphic sequence does not occur at any one location. Major geologic structural features are shown in Figure 3.

Geohydrologic units, as designated herein, are rock units that have similar characteristics. They are defined on permeability and their relationship to a reasonably distinct regional hydrologic system.

The flow system of most regional aquifers is controlled by altitude of recharge and discharge areas. Major recharge and discharge areas for regional aquifers typically correspond to large physiographic features. The features were used to delineate the geohydrologic units discussed in this chapter. The Central Non-glaciated Plains have been subdivided into hydrophysiographic regions because several regional ground water-water flow systems occur within them. The major hydrophysiographic regions include the Northern Great Plains, the Central Great Plains, the Ozark Plateaus, the Southern Great Plains, and the Edwards Plateau–Llano Hills region (Fig. 1).

The stratigraphic section of the Central Nonglaciated Plains can be divided into geohydrologic units based on hydrologic relationships and hydraulic characteristics. In turn, the hydraulic characteristics usually can be correlated with geologic formations. This correlation is the basis for describing geohydrologic units in a geologic framework. Figure 4 correlates geologic formations with geohydrologic units. Note that geohydrologic units cross geostratigraphic unit and geochronologic units.

## HYDROGEOLOGY OF THE NORTHERN GREAT PLAINS

Rocks of Paleozoic and Mesozoic age, which underlie the Northern Great Plains region (Fig. 1), contain at least five major water-yielding geohydrologic units (Fig. 4) and form one of the most extensive confined flow systems in the United States (Downey, 1982). The flow system extends more than 1,000 km from mountainous recharge areas in Montana, Wyoming, and South Dakota to discharge areas in the eastern Dakotas and the Canadian provinces of Manitoba and Saskatchewan. The total area of the flow system is more than 700,000 km$^2$.

Precambrian rocks form the basement confining unit of the hydrologic system. The deepest aquifers are in rocks of Cambrian and Ordovician age and are principally composed of limestone and dolostone of the Red River Formation (Upper Ordovician). Rocks of Silurian and Devonian age and the Bakken Formation

Jorgensen, D. G., Downey, J., Dutton, A. R., and Maclay, R. W., 1988, Region 16, Central Nonglaciated Plains, *in* Back, W., Rosenshein, J. S., and Seaber, P. R., eds., Hydrogeology: Boulder, Colorado, Geological Society of America, The Geology of North America, v. O-2.

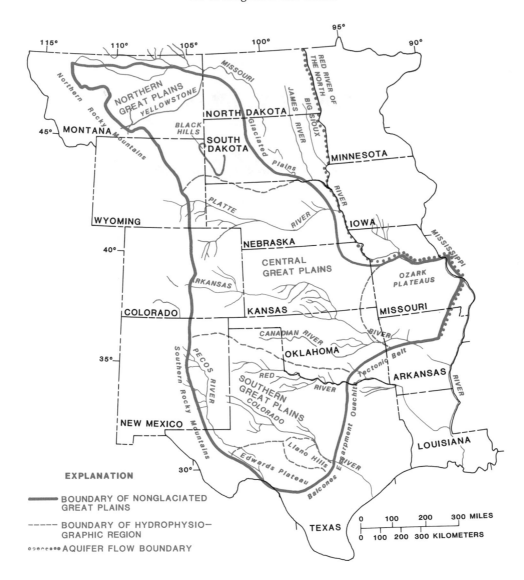

Figure 1. Hydrogeologic regions of the Central Nonglaciated Plains.

(includes Lower Mississippian) act as a confining system. The second major regional aquifer unit, the Madison aquifer, consists of parts of the Madison Limestone or Madison Group where it is divided into the Lodgepole and Mission Canyon limestones of Mississippian age. The Madison aquifer is composed principally of permeable limestone and dolostone but does contain minor amounts of bedded anhydrite and gypsum that restrict the flow of water (Peterson, 1981).

The Madison aquifer is overlain by a confining system that is composed of the Charles Formation and the Big Snowy Group and equivalents (Upper Mississippian). This confining system contains thick beds of halite in the Williston Basin that reduce vertical hydraulic conductivity to near zero and restrict leakage between the Madison aquifer and the overlying aquifer in Pennsylvanian rocks in the Williston Basin.

The third major regional water-yielding unit, the Minnelusa aquifer, is composed of permeable sandstone and limestone of the Minnelusa Formation of equivalents of Pennsylvanian age. This aquifer is present in western and central part of the Northern Great Plains. The Minnelusa aquifer is confined by low-permeability rocks of Permian, Triassic, and Jurassic age, which restrict vertical flow on a regional scale.

The fourth major water-yielding unit consists of a sequence of water-yielding sandstone and siltstone and is termed the Great Plains aquifer system in this chapter (Jorgensen, 1984). The aquifer system extends beyond the Northern Great Plains from northern Canada to New Mexico. Major aquifers of the Great Plains aquifer system are the sandstones found in the Lower Cretaceous rocks; these have been termed the Lower Cretaceous aquifers (Downey, 1984, p. 92) in the Northern Great Plains

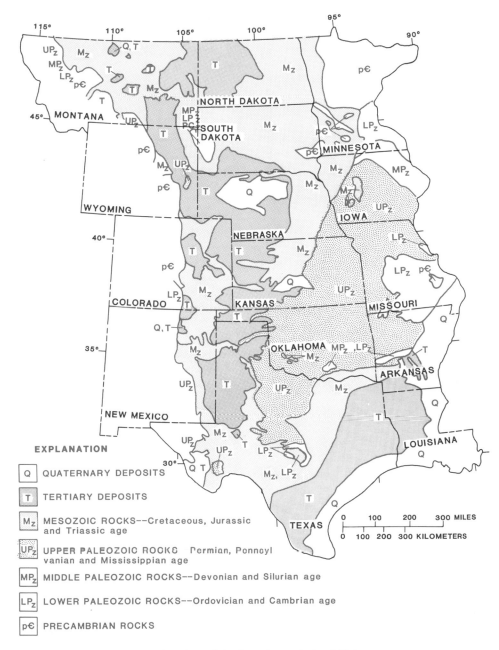

Figure 2. Regional geology of the Central Nonglaciated Plains.

**EXPLANATION**

Q    QUATERNARY DEPOSITS

T    TERTIARY DEPOSITS

Mz   MESOZOIC ROCKS--Cretaceous, Jurassic and Triassic age

UPz  UPPER PALEOZOIC ROCKS  Permian, Pennsylvanian and Mississippian age

MPz  MIDDLE PALEOZOIC ROCKS--Devonian and Silurian age

LPz  LOWER PALEOZOIC ROCKS--Ordovician and Cambrian age

pЄ   PRECAMBRIAN ROCKS

region. (Many other stratigraphic names also have been applied to the aquifers, causing much confusion.) The aquifer system has two separate regional aquifers: the lower unit is the Apishapa aquifer (also called the Inyan Kara aquifer by Butler, 1984; it has also been called the Dakota aquifer by a set of investigators), and the upper unit is the Maha aquifer (also called the Dakota by a different set of investigators). The two regional aquifers are separated at most locations by a confining unit (Skull Creek Shale in the Northern Great Plains and Kiowa Shale in the Central Great Plains).

The Apishapa aquifer is composed of permeable rocks of the Fall River Sandstone in the Northern Great Plains or of the Cheyenne Sandstone and equivalents in the Central Great Plains. The aquifer is not as extensive as the overlying Maha aquifer. The Apishapa aquifer crops out at many locations along the Rocky Mountains and the Black Hills but does not have an eastern outcrop area in the Northern Great Plains because it terminates in a subcrop to the overlying Maha aquifer.

The Maha aquifer extends from the eastern outcrops of Mesozoic rocks, shown in Figure 2, to the Rocky Mountains. The

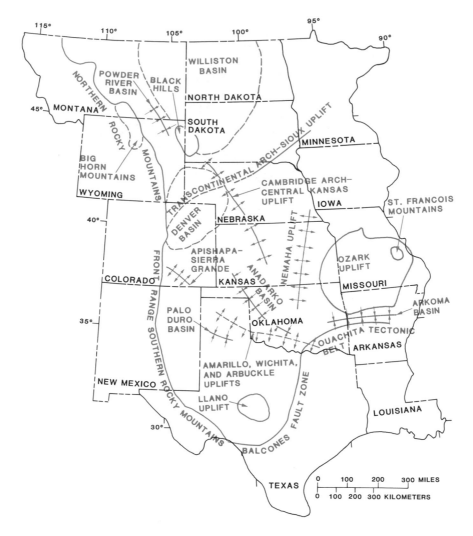

Figure 3. Regional geologic structures of the Central Nonglaciated Plains.

Maha aquifer comprises the sandstones of the Dakota Formation and equivalents.

A confining system overlies the Great Plains aquifer system. The system consists of several confining units, such as the shales of the Pierre, Carlile, and Graneros formations, and several aquifer units, such as those in the limestones of the Niobrara and Greenhorn Formations.

### Hydrogeologic history

The Northern Great Plains were covered intermittently with continental seas during Middle Cambrian through Mississippian time. Regionally continuous permeable sands were the first sediments to cover the Precambrian basement rocks. Overlying sediments were mostly calcareous muds. Uplifts and sea recessions interrupted the dominantly marine deposition. During uplifting, unconsolidated sediments were lithified, and rocks were frac-

tured. Erosion removed exposed sediments and rocks and the subsequent reduction of overburden load resulted in extensional fracturing. Weathering and subaerial dissolution, especially of carbonate rocks, greatly increased permeability of the near surface rocks. Ground-water flow systems were established and flushed out depositional and existing water from the exposed rocks. Dolomitization and additional dissolution of limestone ensued in a mixing zone near the coasts. In general, diageneses associated with uplift greatly increase permeability and are termed collectively herein "uplift diageneses."

Sedimentation processes also acted periodically during Middle Cambrian through Mississippian time. In general, sedimentation results in increased burial and a reduction of porosity and permeability by compaction. Increased depth of burial increases temperature. At temperatures greater than 50°C, usually present at depths of 1.2 km or more, thermal diagenesis of organic matter and clay minerals occurs at moderate rates. At temperatures greater than 130°C, usually present at depths greater than 3.5 km, thermal diagenesis of organic material occurs at a rapid rate. Thermal diagenesis of organic material produces $CO_2$ (carbon dioxide), water, and hydrocarbons. In general, thermal diagenesis of organic material will enhance secondary permeability locally. Thermal diagenesis of clay is also an important diagenetic process usually associated with burial. Smectite is also altered to mixed layered clay and illite while releasing water. Diagenetic processes associated with burial may locally increase permeability, but regionally the net effect is to reduce permeability. These processes are collectively termed herein "burial diageneses." Burial diageneses occurred intermittently during Middle Cambrian through Mississippian time. Because burial was not deep little thermal diagenesis occurred, the main effect being compaction.

At the end of Mississippian, nearly the entire area was exposed. Uplift diageneses, especially on the exposed carbonate rocks, increased permeability dramatically.

A trend of predominant clastic deposition started during the Pennsylvanian Period. This trend followed the trend of dominant carbonate deposition, which had occurred during the early Paleozoic. Permeable sand, calcareous mud, and clay were the dominant sediments of the Pennsylvanian. Sediments during the Permian, Triassic, and Jurassic Periods were deposited in environments ranging from marine to continental and consisted mostly of slightly permeable clay and calcareous mud and some permeable sand. These units (layers) collectively restricted flow in the vertical direction and have formed a confining system since deposition.

Continental seas returned during the Early Cretaceous. The trend of sea advances was interrupted by minor sea recessions. Permeable, subtidal, and supratidal sand was deposited over extensive areas near the shore, whereas marine clay was deposited offshore. Total sediment thickness ranges from near zero in eastern Minnesota to more than 300 m in western North and South Dakota. The additional sediments increased the depth of burial of the underlying rocks. The amount of rock material undergoing thermal diagenesis at a moderate rate increased. Burial diageneses ended with the retreat of the sea as the western part of the study area was uplifted.

A second period of intermittently advancing sea followed. As earlier, permeable nearshore sand was deposited contemporaneously with marine clay offshore.

Sediments, mostly clay, up to 3 km thick, were deposited during the Late Cretaceous. The ensuing burial diageneses reduced the regional permeability. In the deeper basins, thermal diagenesis of organic material was rapid and locally increased permeability.

The strong tectonic movement of the Laramide Orogeny started near the end of the Cretaceous and continued into Tertiary time. The orogeny was characterized by large-scale warping, magmatism, faulting, and uplifting. The Williston and Powder River basins were downwarped into much of their present position. The Black Hills, Big Horn Mountains, and Rocky

| System | Series | Northern Great Plains | Central Great Plains | Ozark Plateaus | Southern Great Plains |
|---|---|---|---|---|---|
| Quat. | | High Plains aquifer | High Plains aquifer | ///// | High Plains aquifer |
| Tertiary | | High Plains aquifer | High Plains aquifer | ///// | High Plains aquifer |
| Cretaceous | U | Confining system | Great Plains confining system | ///// | (High Plains aquifer system) |
| Cretaceous | L | Great Plains aquifer system | Great Plains aquifer system (Dakota) | ///// | Aquifer in Edwards–Trinity |
| Jurassic | U | Confining system | Western Interior Plains confining system | ///// | (High Plains aquifer system) |
| Jurassic | M | Confining system | Western Interior Plains confining system | ///// | |
| Jurassic | L | Confining system | ///// | ///// | |
| Triassic | U | Confining system | Confining system | ///// | Aquifer in Dockum group |
| Triassic | M | Confining system | Confining system | ///// | |
| Triassic | L | Confining system | ///// | ///// | |
| Permian | U | | Western Interior Plains confining system | ///// | Evaporite aquitard |
| Permian | L | | Western Interior Plains confining system | ///// | Evaporite aquitard |
| Pennsylvanian | U | Minnelusa aquifer | Western Interior Plains confining system | Confining system | Aquifer in upper Paleozoic rocks |
| Pennsylvanian | M | Confining system | Western Interior Plains confining system | Confining system | |
| Pennsylvanian | L | Confining system | Western Interior Plains confining system | Confining system | Granite wash aquifer |
| Mississippian | U | Madison aquifer | Western Interior Plains aquifer system | Springfield Plateau aquifer | Aquifer in lower Paleozoic carbonate rocks |
| Mississippian | L | Madison aquifer | Western Interior Plains aquifer system | Ozark c.u. | |
| Devonian | U | Confining system | Western Interior Plains aquifer system | Ozark c.u. ///// | Aquifer in lower Paleozoic carbonate rocks |
| Devonian | M | Confining system | Western Interior Plains aquifer system | (Ozark Plateaus aquifer system) | |
| Devonian | L | Confining system | Western Interior Plains aquifer system | (Ozark Plateaus aquifer system) | |
| Silurian | U | Confining system | Western Interior Plains aquifer system | (Ozark Plateaus aquifer system) | |
| Silurian | M | Confining system | Western Interior Plains aquifer system | (Ozark Plateaus aquifer system) | |
| Silurian | L | Confining system | Western Interior Plains aquifer system | (Ozark Plateaus aquifer system) | |
| Ordovician | U | Aquifers in Cambrian and Ordovician rocks | Western Interior Plains aquifer system | Ozark aquifer | Aquifer in lower Paleozoic carbonate rocks |
| Ordovician | M | Aquifers in Cambrian and Ordovician rocks | Western Interior Plains aquifer system | Ozark aquifer | |
| Ordovician | L | Aquifers in Cambrian and Ordovician rocks | Western Interior Plains aquifer system | Ozark aquifer | |
| Cambrian | U | Aquifers in Cambrian and Ordovician rocks | ///// | St. Fran. c.u. / St. Fran. c.u. | Aquifer in C Ss |
| Cambrian | M | ///// | ///// | ///// | ///// |
| Cambrian | L | ///// | ///// | ///// | ///// |
| PC | | Basement confining unit | Basement confining unit | Basement confining unit | Basement confining unit |

(Vertical labels: "High Plains aquifer system", "Ozark Plateaus aquifer system", "Deep-basin brine aquifer system")

Abbreviations used: aq = aquifer, Ss = sandstone, C = Cambrian, PC = Precambrian, Quat. = Quaternary, St. Fran. = Saint Francois, C.U. = confining unit, U = upper, M = middle, L = lower

Figure 4. Geologic and geohydrologic units of the Central Nonglaciated Plains.

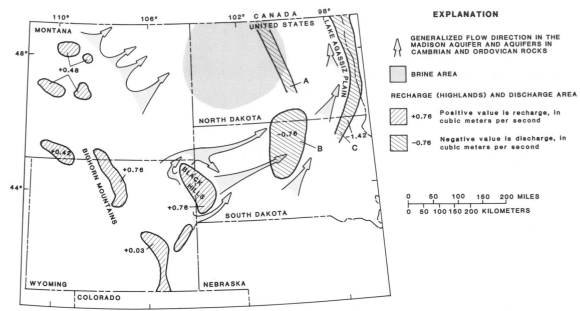

Figure 5. Calculated recharge and discharge for the Madison aquifer and aquifers in Cambrian and Ordovician rocks: A. area of flow from Madison aquifer to aquifer in Cambrian–Ordovician rocks; B. area of leakage from Minnelusa aquifer to Great Plains aquifer system; C. area of discharge from aquifers in Cambrian and Ordovician rocks.

Mountains were the products of the uplifts (Fig. 3). Sediments from eroded uplifts were deposited in the downwarped basins.

Post-Laramide time was characterized by erosion, fracturing, and extensive fluvial deposition. Regional ground-water flow systems, somewhat similar to the present ones, developed. The fresh ground water mixed with and flushed pre-existing water from the aquifers. The water in the Cambrian and Ordovician rocks was not flushed to the extent that the water in the Madison aquifer was because, in part, these rocks were less permeable and because they were deeper. The rocks of the Maha and Apishapa aquifers, although permeable, were not flushed to the extent of the Madison because the volume of recharge water was less. The present day ground-water flow continues diagenetic alteration, such as dedolomitization of the Pahasapa (Madison equivalent) near the Black Hills (Back and others, 1983).

### Present geohydrologic conditions

All five of the regional water-yielding geohydrologic units crop out and receive some recharge in the highland areas of the Rocky Mountains and around the Black Hills. Large volumes of recharge through the soil at outcrop areas are not typical because outcrop areas are of limited extent, and because annually only a few centimeters of recharge occur. However, recharge from streams is important. Virtually all eastward-flowing streams draining the highland areas lose part of their flow as they cross aquifer outcrops. (Swenson, 1968; Wyoming State Engineer's Office, 1974). Although available surface-water data indicate that

large quantities of water enter the aquifers along the outcrop areas in the western highlands, not all of this water recharges the deep, regional-flow system (Rahn and Gries, 1973). Most of the water entering the aquifer outcrops is discharged only a short distance away in springs and seeps along the flanks of the mountainous areas. Only a fraction of the total recharge that remains enters the regional-flow system.

Discharge from the Madison and also the Minnelusa aquifers is to adjacent and overlying aquifers along the eastern subcrop of both units. Discharge from the aquifers in the Cambrian and Ordovician rocks is to adjacent aquifers, to springs and seeps in the Lake Agassiz Plain of North Dakota, and to the land surface in Canada where aquifers in the Cambrian and Ordovician rocks are exposed (Fig. 5).

The Madison aquifer does not crop out in its discharge area. The formations comprising the aquifer terminate (pinch out) in the subsurface and are overlain by younger rocks, consisting mostly of shale. Thus, the discharge from the Madison aquifer is by vertical leakage to overlying units, such as to the Great Plains aquifer system, and by horizontal leakage to the aquifers in the Cambrian and Ordovician rocks. Ground-water discharge is concentrated along the eastern, rather than the northern, limits of the Madison aquifer. Vertical leakage through confining units appears to be one of the major sources of discharge for all five of the aquifer units in the Northern Great Plains.

Discharge from the Great Plains aquifer system is mainly upward to overlying units in eastern South Dakota and along the subcrop of the aquifer system to sediments in the Lake Agassiz

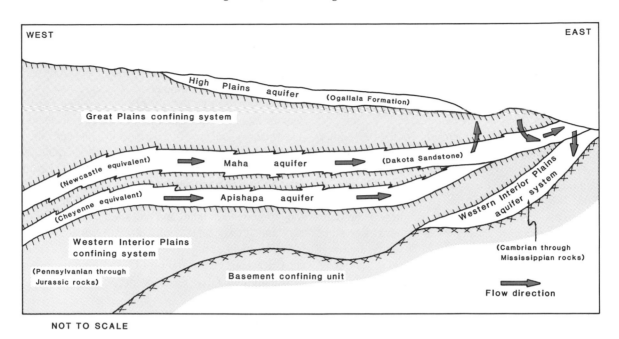

Figure 6. Regional geohydrologic units in the Central Great Plains, northern Nebraska.

Plain of North Dakota. At the present time, considerable water is discharged from the Great Plains aquifer system through wells along the Missouri and James River valleys in South Dakota (Fig. 1).

Geologic structure is an important control on the rate and direction of water movement in the major aquifers and through confining units. The major fault and lineament systems in the region channel ground-water flow into broad paths north and south around the Williston Basin to discharge areas in eastern South Dakota, the Lake Agassiz Plain of North Dakota, and along Lake Winnepeg and Lake Manitoba in the Canadian province of Manitoba. Fracturing of rock along lineament zones through confining layers also provides a path for vertical leakage of water between the aquifers.

The aquifers in Cambrian and Ordovician rocks and the Madison aquifer contain water with a dissolved-solid concentration greater than 300,000 mg/L in the deeper parts of the Williston Basin. The brine is very nearly static and may still be adjusting to past hydrogeologic conditions, such as "loading" resulting from past glaciers or paleotopography (Downey, 1984, p. 98).

## HYDROGEOLOGY OF THE CENTRAL GREAT PLAINS

Two important regional geohydrologic units are present in the Central Great Plains: the Western Interior Plains aquifer system and the Great Plains aquifer system. Flow in both systems is from recharge areas in the west, eastward to discharge areas in the

east. Underlying crystalline basement rocks in the region are fractured and water yielding locally, such as at the Rocky Mountain Arsenal in Denver, Colorado. However regionally, the rocks do not transmit water vertically through their thickness; thus, they form a confining unit.

The Western Interior Plains aquifer system overlies the basement confining unit. This regional aquifer system extends from the front ranges of the Rocky Mountains to the Missouri River or to the Ozark Plateaus (Fig. 1) and comprises mostly lower Paleozoic carbonate rocks. To the north, the aquifer system is separated from the aquifers in the Cambrian and Ordovician rocks of the Northern Great Plains by the Transcontinental arch in Nebraska. The permeability of the material in the deep basins, such as the Anadarko, is very low and typically is not considered aquifer material. Water in this part of the aquifer moves very slowly and could be said to be nearly stagnant.

A confining system overlies the Western Interior Plains aquifer system (Fig. 6). This thick hydrogeologic unit includes shale, salt, and anhydrite, all of which have very low permeability and are effective confining units. Aquifers are present in sandstone, arkose, and limestone strata of Pennsylvanian and Permian age, as well as in sandstone strata of Triassic and Jurassic age. At most locations, the overlying confining system effectively separates the underlying Western Interior Plains aquifer system from the overlying Great Plains aquifer system.

The Great Plains aquifer system, which has been described previously in this chapter, includes two regional aquifers, the Apishapa and Maha, and extends from the Rocky Mountains eastward to the outcrop area, which is approximated by the extent of the Mesozoic rocks (Fig. 2). The Great Plains aquifer system, in general, is confined by the overlying Great Plains confining system, which consists mostly of very low permeability Upper Cretaceous shales.

### Hydrogeologic history

The present water-yielding characteristics of the rocks differ greatly from the characteristics of the original sediments. Present characteristics are largely the result of postdepositional diagenesis and weathering, although some characteristics have been preserved since deposition in some units.

During the Late Cambrian–Mississippian interval, the Central Great Plains was covered many times by seas depositing marine sediments of mostly calcareous mud. Sedimentation was interrupted by uplifts and sea-level recession. During periods of sea-level recession, uplift diageneses resulted in fracturing, erosion, weathering, lithification, flushing, and coast line dolomitization and markedly reduced permeability.

Pennsylvanian tectonic activity was frequent and widespread, and large compressive and ancillary stresses, probably associated with plate tectonics, developed. The stresses were relieved by downwarping in an arcuate belt (Ouachita tectonic belt, Fig. 3) as well as by extensive thrust faulting in southern Oklahoma (Edwards and Downey, 1967, p. 3). Downwarping and basin filling were very rapid. At certain locations, such as in central Oklahoma, it is probable that very low permeability clays buried the more permeable sand layers at such a rapid rate that water was trapped with the sands. The trapped water expanded as the temperature increased with the increased depth of burial. Pressure increased dramatically in the trapped water, creating a geopressured zone. The hydrothermal pressure increased porosity of the entrapped sand layers and the adjacent clay layers, thus reducing the thermal conductivity of the mass of water, sand, and clay. The mass of low-thermal conductivity material acted as an insulator, reducing heat flow across the geopressured zone. Contemporaneously, uplifts, such as the Amarillo and Wichita, developed. Rapid uplift also occurred in Colorado and northeastern New Mexico (Apishapa–Sierra Grande) as well as along the front range of the Rockies. These uplifts were high above sea level and were the source area of extensive clastic sediments. Thick arkose sections were deposited near the uplifts and near the Front Range. Sand, silt, and clay were deposited away from these sediment sources. Arkose in Colorado had moderate permeability; however, the arkose or granite wash in Oklahoma and Texas panhandle had very high permeability and was an avenue for meteoric water from the uplifted area to recharge deeply buried aquifer material.

The downwarping and rapid sedimentation resulted in burial of pre-Pennsylvanian rocks (material of the Western Interior Plains aquifer system) to depths of nearly 6 km, and burial diageneses ensued. The weight of the newly deposited sediments compacted clay to shale and reduced porosity and permeability as it squeezed out relatively fresh water. Primary porosity of limestone layers probably was greatly reduced or virtually eliminated by plastic deformation. Thus, the deeply buried Mississippian limestone beds became confining units in the basin areas. Primary porosity of the less ductile dolostone and sandstone beds probably was reduced but not eliminated.

The Permian Period was characterized by restricted marine sedimentation of clay, sand, and evaporites, including dolomite. The extensive and nearly impermeable halite and anhydrite layers formed exceptionally tight confining units. Uplifts, such as the Arbuckle and Wichita, were active; however, in general, tectonic intensity was at a reduced level compared to the intensity during Pennsylvanian time. The Permian marked the beginning of a long-term trend of uplift and erosion that continued through Triassic and Jurassic times. Uplift diageneses increased permeability.

The hydrogeologic history of the Central Great Plains for the Cretaceous Period and post-Cretaceous time is similar to that of the Northern Great Plains, as previously discussed.

### Present geohydrologic conditions

Many of the highly permeable zones in the Western Interior Plains aquifer system are probably paleokarsts. For example, in eastern and central Kansas, the dominant water-yielding zone is apparently a karst zone developed on top of the uppermost dolostone bed. The aquifer system is recharged along the uplifts in southern Oklahoma, the Texas panhandle, southeastern Colorado, along the Front Range (where permeable zones such as granite wash and arkose overlie the aquifer), and along the front range of Wyoming and the Black Hills. The aquifer system discharges to the Missouri River in Nebraska and Missouri and possibly to the water table near the land surface around the Ozark Plateaus in Oklahoma and Kansas (Figs. 6 and 7).

Dissolved solids in the Western Interior Plains aquifer system range from less than 10,000 mg/L in north-central Kansas to more than 200,000 mg/L in central Oklahoma.

The Apishapa and the Maha aquifers of the Great Plains aquifer system are recharged in southeastern Colorado and to a minor degree along the front range of the Rocky Mountains (Helgesen and others, 1982; Leonard and others, 1983). Extensive recharge does not occur along most of the Front Range because faulting offsets the continuity of the aquifers in most areas, the outcrop band is narrow, and the quantity of water recharged from precipitation annually along the Front Range is small. Discharge is mostly to overlying aquifers near the eastern outcrop area (eastern extent of Mesozoic rocks, Fig. 2). Recharge and discharge through the underlying Western Interior Plains confining system and the overlying Great Plains confining system is minimal, except where the confining system is less than 200 m thick. Dissolved-solids concentrations range from less than 2,000 mg/L in the recharge area in northeastern New Mexico and southeastern Colorado to more than 100,000 mg/L in part of the Denver basin where water movement is slow. The water in the far majority of the aquifer is less than 10,000 mg/L.

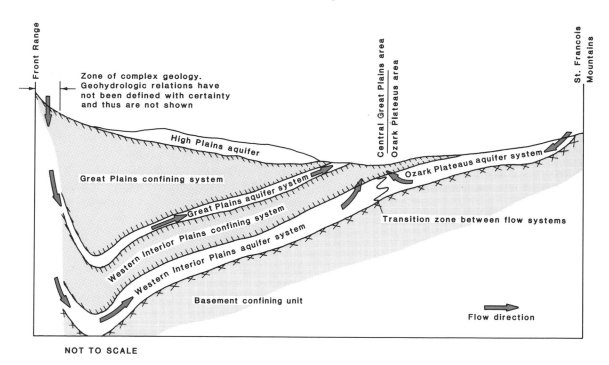

NOT TO SCALE

Figure 7. Regional geohydrologic relations for Central Great Plains and Ozark Plateaus.

## HYDROGEOLOGY OF THE OZARK PLATEAUS

The Ozark Plateaus area (Fig. 1) includes the St. Francois Mountains, the Salem Plateau, and the Springfield Plateau. Three regional aquifers, the St. Francois, the Ozark, and the Springfield Plateau, are present in the area. Collectively, with separating confining units, they form the Ozark Plateaus aquifer system (Figs. 4 and 7). Water in the aquifer system flows radially outward from the central area of the Ozark Plateaus toward the peripheral boundaries.

Crystalline rocks, largely of Precambrian age, underlie the St. Francois aquifer. These basement rocks typically are fractured and may yield water to wells. However, water does not move vertically through the basement rocks; therefore they are considered to be the lower confining unit to the overlying Ozark Plateaus aquifer system.

The St. Francois aquifer is the basal aquifer and is found in the permeable parts of the Lamotte Sandstone. The St. Francois confining unit overlies the St. Francois aquifer and consists of beds of low-permeability dolostone, limestone, and sandstone as well as low-permeability shale. The confining unit is composed of Upper Cambrian rocks, including the Derby and Doe Run Dolomites, the Davis Formation, and the Bonneterre Dolomite.

The Ozark aquifer, which overlies the St. Francois confining unit, is composed of fractured dolostone and some sandstone. The formations that make up the aquifer include the St. Peter Sandstone, the Everton Formation (sandstone), the Powell Dolomite, the Jefferson City Dolomite, the Roubidoux Formation (dolo-

stone and sandstone), and the Gasconade Dolomite including the Gunter Sandstone member, all of Ordovician age. The aquifer also includes the Eminence Dolomite and Potosi Dolomite of Cambrian age. The Ozark aquifer is the most permeable unit of the Ozarks Plateaus aquifer system and is the host to most of the lead and zinc deposits in central Missouri.

The Ozark aquifer is overlain by the Ozark confining unit. The confining unit consists of very low permeability shale, such as the Chattanooga and the younger Northview. The Ozark confining unit is relatively thin and is not present at all locations.

The Springfield Plateau aquifer consists of permeable, fractured, and porous limestones, such as those of the Boone Formation and its St. Joe Limestone member of Mississippian age. The aquifer material is also the host to the lead and zinc deposits of the tri-state area of Missouri, Kansas, and Oklahoma.

Shale, limestone, sandstone, and coal layers, if not eroded, cover the Springfield Plateau aquifer and form a confining system. Shale is the predominant lithology; shales form effective confining units as they restrict flow to the underlying Springfield Plateau aquifer. The limestone and sandstone layers commonly transmit water and, thus, are aquifers.

### Hydrogeologic history

The present water-yielding characteristics of the rocks in the Ozark Plateaus are almost entirely the result of diagenesis and weathering that has taken place throughout geologic time. For example, the Ozark aquifer in Missouri comprises intensely fractured dolostones, not the calcareous sediments that were deposited during Cambrian and Ordovician time.

The Ozark Plateaus region was the site of intermittent and repeated sea-level transgression and recession during Cambrian time. Advance and retreat of seas were even more characteristic of Ordovician time. Erosion, including karstification, occurred during periods when the region was above sea level. During periods of emergence, fresh water enriched with carbon dioxide selectively dissolved the carbonates of certain textural facies or fabric along fracture paths, thus increasing secondary permeability and porosity. Mixing zones of fresh and saline water developed near the coast lines. In the mixing zone, additional dissolution occurred because of the increased solubility of the mixed water. Of even more hydrologic importance was the widespread dolomitization that was probably associated with the mixing zone. Thus, it is probable that nearly all of the limestone was diagenetically changed to more permeable dolostone during the numerous sea advances and retreats.

Significant tectonic activity associated with the Ozark uplift started during Ordovician time. The uplifting, which over several geologic eras would ultimately reach 1,500 m (McCracken, 1967, p. 20), resulted in fracturing and folding. Faulting and fracturing occurred both parallel and orthogonal to the axis of the Ozark uplift, which runs approximately N70°E. The fractures, especially joints, became the dominant paths for ground-water flow. Solution along the fracture paths greatly enhanced the permeability, especially as compared to the permeability in the nonfractured part of the rock. The fracture permeability was highly anisotropic.

Very low-permeability marine clay of Late Devonian and Early Mississippian age was deposited unconformably over the dolostone of the Ozark aquifer at most locations. During the Mississippian, after the deposition of the marine clay (which restricted leakage of ground water), extensive carbonate materials were deposited; later, uplift, lithification, fracturing, and dissolution erosion (uplift diageneses) operated through the end of Mississippian time, increasing permeability.

During Pennsylvanian time, sediments associated with the fluctuating seas were deposited. The sediments included calcareous material, clay, sand, and supralittoral material. In addition, frequent and large-scale tectonic activity, which had started in a mild manner during Mississippian time, occurred. Sediment thickness exceeding 3 km in the basins around the Ozark Plateaus initiated burial diageneses. The weight of the thick layers of sediments (overburden) resulted in compaction and lithification of buried clay beds into shale bed, such as the Chattanooga Shale. Also, the beds of Mississippian carbonate sediments were altered to limestones by compaction, pressure dissolution, and recrystallization. Thermal diageneses of organic material and clay occurred in the deeper parts of the basins. During Middle Pennsylvanian time the Ozarks were submerged, and a series of normal faults was formed parallel to the axis of Ozarks (Huffman, 1958, p. 109).

A highland area, the Ouachita uplift, probably several km above sea level, was uplifted during the Middle Pennsylvanian. Regional ground-water flow from the mountains or highlands of the Ouachita uplift toward the lower Ozarks was probably initiated during Late Pennsylvanian and Early Permian.

After Pennsylvanian time, another long period of emergence and erosion developed as the result of regional tilting upward to the east, which included the Ozark area. Additionally, the Ozark Plateaus were independently uplifted. The uplift of the Ozarks and the lowering of the Ouachita uplift by rapid erosion terminated the ground-water flow from the Ouachitas to the Ozarks. The uplift, erosion, and unloading resulted in fracturing, which created permeability in the Lower Mississippian limestones and the underlying lower Paleozoic dolostones and sandstones of the Ozark and St. Francois aquifers.

The erosion first exposed the uppermost aquifer material (St. Francois aquifer) near the St. Francois Mountains as Pennsylvanian-age confining material was removed. Further erosion exposed additional underlying aquifer material. A ground-water flow system, similar to the present system but not as extensive, developed. Ground water entered the aquifers at higher elevation near the center of the uplift and discharged downdip at lower elevation along the edges of the uplift (Fig. 7). Secondary permeability was enhanced greatly in the areas of exposed carbonate rock. Both the lateral and vertical extent of the circulation of the fresh ground-water system were largely controlled by vertical permeability. At the edge of the uplift in particular, flow upward to streams and the water table was controlled by vertical permeability. The low-leakage Pennsylvanian-age shale restricted the upward discharge of water and thus limited the areal extent of the flow system.

### *Present geohydrologic conditions*

The Ozark Plateaus is an area of large amounts of precipitation. Average precipitation exceeds 1,000 mm. Annual recharge to the water table through the soil zone of about 300 mm is expected. Some of this recharge is discharged to streams a short distance from the point of recharge. Additionally, some precipitation recharges the aquifers directly in karst areas. At many locations the water table is well below the bottom of a stream; water is lost from the stream in these locations, recharging the underlying aquifer.

The Ozark area is drained by numerous streams that have extensively dissected the plateaus, thus forming steep hills. Dolostone and limestone are the dominant bedrock in the area. The interfluves (plateaus) often are covered with residuum. The residuum has high porosity and moderate permeability and is a large reservoir of water that slowly recharges the underlying bedrock aquifers. At locations with karst, recharge is rapid.

The Ozark area is characterized also by numerous springs that discharge the aquifers. More than 165 springs have been mapped that have minimum discharge rates of more than 28 L/sec. Most springs have recharge areas on the upland plateaus and discharge areas in or near streams in the valleys. The springs

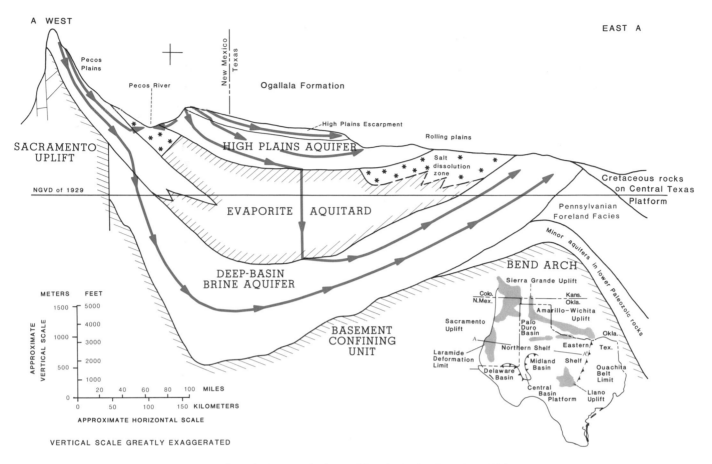

Figure 8. Hydrogeologic conditions in the Southern Great Plains.

typically emerge from large solutional openings that follow fractures downward to the water table at approximately stream level. However, many springs have developed solutional openings well below the water table. Some springs are believed to have circulation paths as deep as 240 m below land surface and, possibly at some locations, circulation is developed downward to the basement confining unit. Additionally the ground-water drainage area of a spring may or may not correspond to the drainage divide for surface water. The Missouri and Mississippi rivers are discharge boundaries to the Ozark flow system. Study of flow characteristics of these streams implies that several hundred cubic meters per second of water probably is discharged from the aquifers in this manner.

An interesting condition exists in western Missouri, southeast Kansas, and northeast Oklahoma where the slow-moving saline water of the Western Interior Plains aquifer system meets the rapidly moving fresh water of the Ozark Plateaus aquifer system. A transition zone of mixed water results (Fig. 7). The location of the zone is controlled by a broad topographic low and by the vertical leakage through the overlying confining system, since discharge from both aquifer systems is dominantly vertically upward. The transition zone leakage, which is inversely propor-

tional to the thickness of the confining material and proportional to the permeability of the confining material, is upward through the thin section of microfractured shale. Permeability of the shale decreases rapidly downward as the fractures are closed with depth. Thus, the thickness of the overlying confining system is a strong control on the location of the transition zone. The location of the transition zone is believed to have migrated radially outward through geologic time as the thickness of the overlying confining material was reduced by erosion from the Ozark uplift.

## HYDROGEOLOGY OF THE SOUTHERN GREAT PLAINS

The ground-water system of the Southern Great Plains (Fig. 1) comprises three heterogeneous geohydrologic units (Figs. 4 and 8): the "deep-basin brine" aquifer system, the "evaporite" aquitard, and the "High Plains aquifer system" (Bassett and Bentley, 1983).

The deep-basin brine aquifer system includes aquifers in Pennsylvanian and Lower Permian shelf and shelf-margin carbonates, aquifers in lower Paleozoic shallow-marine carbonates, and aquifers in upper Paleozoic arkosic or granite-wash sandstones deposited in fan-delta environments flanking the Amarillo-Wichita uplift (Dutton and others, 1982). Basin and shelf deposits of shale form confining units within the aquifer system.

The evaporite aquitard (Fig. 4) is a major confining unit that separates the shallow High Plain aquifer system from the deep-basin brine aquifer system. The evaporite aquitard consists of deposits of halite, anhydrite, limestone, dolostone, and red mudstone of Leonardian to Ochoan age, with an average permeability on the order of $10^{-22}$ m$^2$ ($10^{-8}$ darcies) (Kreitler and others, 1985).

In the Southern Great Plains, the surficial High Plains aquifer system includes the High Plains aquifer, which mostly comprises Tertiary sediments of the Ogallala Formation, as well as underlying aquifers in fluvial, deltaic, and fan-delta sandstones of the Dockum Group (Triassic) and in marine limestones and sandstones of the Edwards Limestone and underlying Trinity Group (Cretaceous). The underlying aquifers are hydraulically connected to the High Plains aquifer at some places (Knowles and others, 1984). In some areas of the Texas Panhandle, confining units separate the aquifers in Mesozoic rocks from the High Plains aquifer. Cretaceous rocks are generally absent in the Palo Duro basin area.

### Hydrogeologic history

In most cases, the original hydraulic properties of rocks have been altered by diagenesis. Porosity and permeability of pre-Leonardian, shelf-margin carbonate rocks were increased by dolomitization; however, hydraulic properties of back-shelf carbonates generally were not greatly changed. In Paleozoic fan-delta deposits, diagenetic history of distal marine-reworked sandstones differs from that of proximal, non-reworked sandstones. During Pennsylvanian and Early Permian time, ground water of meteoric origin flowed from subaerially exposed proximal facies down the hydraulic gradient toward distal facies of the fan system. The recharge area moved to the west as the Amarillo-Wichita uplift was buried; meteoric ground water then flowed along the depositional strike through the delta sandstone facies of the alluvial system (Dutton and Land, 1985). With longer flow paths and greater depth of burial, the salinity of water in the delta facies increased, and a variety of cements precipitated.

Following the burial of the deltas beneath an evaporite basin during the Late Permian, evaporite brine moved downward into fan-delta facies and caused albitization of plagioclase and precipitation of celestite and anhydrite cements.

Porosity and permeability of carbonate sediments deposited in the Permian evaporite basin during marine transgression also were reduced markedly by halite and anhydrite cements. The Upper Permian rocks formed large oil reservoirs where marine facies were unaffected by evaporite cementation across the Central Basin Platform and Northern Shelf area (Ramondetta, 1982).

The western part of the Permian Basin was raised during the Miocene Epoch by basin-and-range tectonism. After this event, the prevalent direction of ground-water flow must have been to the east, as it is now, and was controlled by topography. During early Quaternary time, dissolution of Permian salt beds and subsidence of sediment into solution basins influenced the geomorphic evolution of the Pecos River Basin, the Canadian River Basin, and the retreat of the Caprock Escarpment (Morgan, 1941; Gustavson and Finley, 1985). Enlargement of the Pecos River Basin cut off the High Plains aquifer system from recharge areas in the southern Rocky Mountains.

### Present geohydrologic conditions

The present direction of ground-water flow is generally eastward in each geohydrologic element of the deep-basin brine aquifer system (McNeal, 1965). The direction of flow is modified by variations in permeability between the different sedimentary facies. The dolomitized, shelf-margin carbonate is an aquifer, and the basinal shale is a confining unit that strikes north in central Palo Duro basin. This juxtaposition of facies of high and low transmissivity diverts ground-water flow paths northward or northeastward toward thick granite-wash sandstone that flanks the Amarillo-Wichita uplift. The uplift may be a barrier to flow between the Palo Duro and Anadarko basins. At the uplift, ground-water flow is diverted southeastward along the depositional strike through the granite-wash facies.

Discharge areas for the deep-basin brine aquifer system lie in the broad outcrop of Pennsylvanian and Lower Permian formations in north-central Texas and southwestern Oklahoma (Fig. 2). Permeable fluvial-deltaic sandstone and shelf-margin carbonate banks provide narrow paths across the discharge area for fluid movement out of the Midland basin. Vertical hydrologic continuity is restricted by thin, widespread beds of carbonate mudstone and interfluvial shale deposits. The fresh ground-water system of meteoric origin circulates only to a shallow depth in the regional discharge zone.

The San Andres Formation, in the Upper Permian, is a permeable zone with solutionally enlarged porosity below the Pecos Plains (Theis, 1965, p. 328–330) and is confined to the Roswell, New Mexico area. In the arid Southern Great Plains, ground-water recharge occurs during winter and early spring when storm runoff drains downward from river beds. Ground-water recharge that enters the San Andres Formation may leak downward and recharge the deep-basin brine aquifer system below the Pecos Plains. Ground-water flow in the San Andres Formation past the Pecos River is small because of discharge to the river and because of the low permeability of San Andres rocks in the Palo Duro basin east of the Pecos River (Summers, 1981). In the Palo Duro basin, the San Andres Formation contains thick salt beds and is a major part of the evaporite aquitard (Fig. 4).

The aquifer in Triassic rocks (Fig. 2) is recharged at its outcrop west of the High Plains of Caprock Escarpment (Fig. 8) by precipitation and stream loss, and below the High Plains possibly by downward leakage of water from the High Plains aquifer. The general southeastward direction of ground-water flow is modified adjacent to the Canadian River Valley and around the upper reaches of the Red River (Fig. 1). Regional flow in rocks of the Dockum Group probably also is influenced by the location of sandstone and conglomerate deposits. Water is discharged from rocks of the Dockum Group at springs and seeps in canyons incised in the eastern edge of the Caprock Escarpment. Water also is believed to move downward into the evaporite aquitard along the High Plains Escarpment and in the Canadian River Valley (Gustavson and others, 1980).

Downward flow of water, as shown in Figure 8, from the High Plains aquifer system in the Palo Duro basin area, might contribute as much as 33 to 50 percent of the flow in the deep-aquifer system (Kreitler and others, 1985). The potentiometric surface of the deep-basin brine aquifer system is below that of the High Plains aquifer because: (1) fluid pressure in the deep-aquifer system is controlled by land-surface elevation in recharge and discharge zones and not by depth below the High Plains land surface, and (2) the aquifer in the granite wash may act as a hydraulic sink by lowering fluid pressure in the rest of the deep-basin brine aquifer system.

The potentiometric surface of the aquifer in the Dockum Group is below the water table in the High Plains aquifer in the central part of the Texas panhandle (Fink, 1963), which indicates that there is potential for a downward-directed component of flow in the High Plains aquifer system. Downward flow from these aquifers into the upper part of the evaporite aquitard causes solution of halite, gypsum, and carbonate rocks. Fractured carbonate rocks and sandstone transmit water and salt from the dissolution zone to saline springs around the Southern High Plains.

Water in the deep-basin brine aquifer system is a brine (65,000 to 215,000 mg/L), generally is equilibrated with respect to calcite, and is slightly undersaturated with respect to halite. Such brine extends from the Palo Duro basin, Northern Shelf, and Midland basin areas eastward to discharge areas for the deep-basin brine aquifer system, where it is found at depths as shallow as 50 m.

## HYDROGEOLOGY OF THE EDWARDS PLATEAU, LLANO HILLS, AND BALCONES FAULT ZONE

The Edwards Plateau and the Llano Hills are included in the Central Nonglaciated Plains; however, the Balcones fault zone lies within the Coastal Plains province (Fig. 1). The Balcones Escarpment, which faces southeastward and separates the two provinces, is from 60 to 120 m high at most places (Fig. 9).

The Edwards Plateau is underlain by beds of limestone, dolomitic limestone, marl, shale, sandstone, and conglomerate, all of Cretaceous age (Holt, 1959). The rocks dip gently toward the south or southeast. Water occurs under water-table conditions in the permeable Edwards Limestone, termed herein the Edwards Plateau aquifer. Other aquifers, under artesian conditions, underlie the Edwards Plateau aquifer and are in the Hosston Formation, the Sligo Formation, the Travis Peak Formation, and the

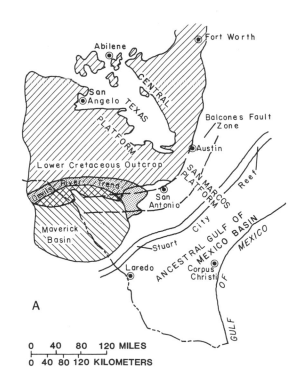

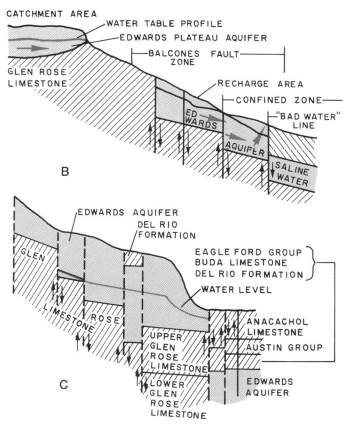

Figure 9. Hydrogeologic conditions of the Edwards aquifer: A. depositional regions; B. nonbarrier faulting; C. barrier faulting.

Glen Rose Limestone, all Lower Cretaceous. These aquifers have low permeability and poor hydraulic connections to the Edwards Plateau aquifer.

Water in the Edwards Plateau aquifer flows southeastward and is discharged by springs located in the reentrant valleys along the edge of the plateau. This discharge contributes significant flow to streams, which in turn recharge the Edwards aquifer in the Balcones fault zone (Fig. 9b). In general, there is no direct hydraulic connection between the Edwards Plateau aquifer and the Edwards aquifer, which is within the Gulf Coast Plains province.

In the Balcones fault zone, the Edwards Limestone and its stratigraphically equivalent rocks are complexly faulted and spatially isolated from the outcrops of Edwards rocks of the Edwards Plateau by other Cretaceous rocks. The Edwards Limestone and its stratigraphic equivalents in the Balcones fault zone form a hydraulically independent regional aquifer system that has very great transmissive capacity within the artesian zone: it is herein termed the Edwards aquifer. The aquifer is multilayered, consisting of highly solutioned and porous honeycombed strata separated by hard, dense, massive limestone. Openings along fractures have provided passageways for water to move vertically from one solutioned stratum to another. The aquifer is cut by many normal faults that generally parallel the Balcones Escarpment. The displacements along some segments of many faults are sufficient to vertically offset the entire thickness of the aquifer and thereby produce barriers to ground-water flow (Fig. 9c).

In the central part of the Llano Hills where Precambrian rocks crop out, fractured rocks form low-permeability aquifers. Although water of low mineralization usually is obtained, the amount available is small. Cambrian, Ordovician, and Pennsylvanian rocks dip away from the Llano uplift—a structural dome that forms the Llano Hills. Faults are common in the Llano Hills, and vertical fault displacements greater than 600 m are known. Faulting exists predominantly in the pre-Cretaceous strata, and at most locations low-permeability aquifers exist within sandstone and limestone layers in these formations. Water-table conditions generally occur in outcrop areas; however, artesian conditions exist in individual aquifers, down-dip, wherein water is confined by adjacent, less-permeable strata.

### Hydrogeologic history

The present-day hydrogeologic conditions of the Edwards Plateau, the Llano Hills, and the Balcones fault zone reflect the various depositional phases and subsequential diagenesis in different environments that have taken place through geologic time. Those depositional phases most directly related to ground water occurred during Cambrian, Ordovician, Pennsylvanian, Permian, Cretaceous, and Quaternary Periods.

During Late Cambrian time, a sequence of permeable nearshore marine sand, very low permeability clay, and carbonate mud was deposited on a surface of Precambrian metamorphic and igneous rocks. Toward the close of Cambrian time and during the Ordovician Period, thick sequences of carbonate sediments were deposited. The resulting overburden altered clay to shale, and carbonate mud was compacted and lithified by pressure dissolution and reprecipitation into limestone.

Prior to the deposition of Pennsylvanian sediments, an extended period of emergence, except for relatively brief periods of minor deposition during Devonian and Mississippian time, resulted in erosion that removed part of the earlier sediments. Extension fracturing and other uplift diageneses, which markedly increased permeability, resulted when erosion reduced the overburden load. Fresh ground water displaced saline water in the rocks near the coastline. Mixing zones of fresh and saline water near the coast were sites of dolomitization and additional dissolution of carbonates.

During Pennsylvanian time, deposition of sediments in widespread seas produced a sequence of marine sand, shale, and limestone over part of the Edwards Plateau, Llano Hills, and Balcones fault zone. This deposition was accompanied by regional subsidence. The area of subsidence and deposition is called the Permian Basin, which lies partly within the Edwards Plateau. Most of the Early and Middle Mesozoic was a period of emergence and erosion, resulting in increased porosity and permeability related to dissolution and fracturing.

During the Cretaceous Period, seas advancing north from the ancestral Gulf of Mexico first deposited permeable coarse sand and later carbonate material, clay, fine sand, and silt across a landmass that had been reduced to low relief by erosion. The ancient landmass was underlain by Paleozoic strata and was bordered on the south and at depths by metamorphic rocks within the present Balcones fault zone.

The sediments making up the Edwards Limestone or its stratigraphic equivalents were deposited on the margin of the Central Texas platform, a low-lying carbonate surface, by a shallow, transgressing and regressing Early Cretaceous sea (Fig. 9a). The now deeply buried Stuart City reef, a rudistid barrier reef, formed the offshore margin of this platform (Rose, 1972). The Devils River trend, another rudistid barrier reef, developed around the Maverick basin during a later period of deposition (Fig. 9a). The Maverick basin was the site of continuous deposition during most of Early Cretaceous time.

A broad lagoonal area behind the Stuart City reef in the vicinity of the San Marcos platform became the site of cyclic deposition. Cyclic deposition began with transgression of the sea followed by progradation of mostly permeable subtidal, intertidal, and supratidal sediments from the north and west. Nearly impermeable evaporites, the final stage of the sedimentation cycle, were deposited on the hot, supratidal flats and, subsequently, were wholly or partly removed by circulating ground water. Within these deposits, burrowed honeycombed rocks and collapse breccias subsequently were preferentially solutioned and produced high porosity and permeability. Near the end of Early Cretaceous time, subaerial erosion removed 30 m or more of the deposits from the San Marcos platform, resulting in extensive karstification of the limestone and dolostone. Shorter periods of subaerial exposure occurred between several cycles of carbonate deposition on the platform. During each emergence, meteoric ground water circulating through the rocks selectively leached or cemented the sediments. During these periods of erosion, much of the early texture-controlled secondary porosity of the Edwards Plateau aquifer was developed.

During Miocene time, tensional stresses resulting from the subsidence of the Gulf of Mexico were relieved by normal faulting and extensive fracturing. The faulting initiated the Balcones Escarpment and separated the aquifer material of the Edwards Limestone into the Edwards Plateau aquifer and the Edwards aquifer in the fault zone. The intense faulting created barrier faults to water flow at some locations within the Edwards aquifer. Since Miocene time to present, simultaneous dissolution and cementing has redistributed porosity in the freshwater zone of the Edwards aquifer.

### Present geohydrologic conditions

Recharge to the Edwards aquifer in the Balcones fault zone occurs at the outcrop (recharge area) of the Edwards Limestone (Fig. 9b). Streamflow draining the Edwards Plateau aquifer on the Edwards Plateau infiltrates the highly permeable limestone of the Edwards aquifer in the Balcones fault zone along the stream channels in the outcrop area and recharges the Edwards aquifer (Maclay and Small, 1976). Lesser amounts of recharge occur in the interstream areas, where part of the precipitation infiltrates directly to the water table through the fractured carbonates.

Recharge is calculated as the difference between the total inflow above the recharge area and the total outflow below the infiltration area. Most of the inflow and all of the outflow is measured at gaging stations. Runoff from ungaged areas and runoff within the recharge area are computed from estimates of unit runoff and rainfall distribution. The annual recharge to the Edwards aquifer in the San Antonio area of the Balcones fault zone ranged from about $54 \times 10^6 m^3$ (44,000 acre-feet) in 1956 to more than $2100 \times 10^6 m^3$ (1,700,000 acre-feet) in 1958. From 1934 through 1981, the average annual recharge was $755 \times 10^6 m^3$.

The regional direction of water movement within the Edwards aquifer is toward the east and northeast; however, locally the gradient may be other than east and northeast because the gradients are controlled by location of barrier faults. Higher hydraulic heads occur in the outcrop (recharge) areas in the Balcones fault zone, where water enters the aquifer, and lower hydraulic heads occur in the eastern part of the San Antonio area, where most of the natural discharge occurs.

All of the major springs of the Edwards aquifer occur along major faults. Some of the faults, particularly those that nearly or completely vertically displace the entire thickness of the aquifer, divert the direction of ground-water flow (Fig. 9c). Segments of faults act as internal barriers. At places they function to divert and restrict the direct flow of water from the recharge area to the artesian zone.

The Edwards aquifer is divided areally into freshwater and saline-water zones. The position of the southern boundary of the freshwater zone is determined by the occurrence of water containing 1,000 mg/L or more of dissolved solids. Rock characteristics and water chemistry in the freshwater zone differ markedly from those of the saline-water zone. The rocks in the saline-water zone are typically dark, dolomitic, have texturally controlled porosity, and contain unoxidized organic material. The rocks in the freshwater zone are usually light colored and have a dense recrystallized matrix associated with well-developed secondary porosity. Water from the freshwater zone is oxidizing, and it has an ionic ratio of magnesium to calcium of less than 0.5. Water from the saline-water zone is reducing, and it has a magnesium to calcium ratio of about 1.0

Within the freshwater zone, the rocks include nonporous and impermeable mudstone and wackestone as well as moderately or highly porous and permeable grainstone and leached dolomitized packestone that have been selectively dissolved by circulating ground water. The rocks in the freshwater zone show diagenetic changes related to carbonate facies that control selective dissolution and recrystallization. Significant dissolution has developed honeycombed rocks in dolomitized and burrowed intertidal deposits, in supratidal deposits containing beds of soluble gypsum and anhydrite, and in rudistid reefs. Many of the tectonic fractures within the freshwater zone are open and show evidence of ground-water circulation that resulted in both dissolution and recrystallization.

The high permeability of the aquifer within the freshwater zone results mostly from the selectively dissolved facies producing well-developed and coarsely vugular porosity along certain strata. However, the rock matrix of most of the rocks in the freshwater zone is less porous than the rock matrix in the saline-water zone because recrystallization and cementation has obliterated or filled the interparticle spaces. Tectonic fractures generally are closed or only slightly open within the saline-water zone, and the capacity of the aquifer to transmit water is relatively restricted because of the small openings.

## REFERENCES CITED

Back, W., Hanshaw, B. B., Plummer, L. N., Rahn, P. H., Rightmire, C. T., and Rubin, M., 1983, Process and rate of dedolomitization; Mass transfer and $^{14}$C dating in a regional carbonate aquifer: Geological Society of America Bulletin, v. 94, p. 1415–1429.

Bassett, R. L., and Bentley, M. E., 1983, Deep brine aquifers in the Palo Duro basin; Regional flow and geochemical constraints: The University of Texas at Austin, Bureau of Economic Geology Report of Investigations 130, 59 p.

Butler, R. D., 1984, Hydrogeology of the Dakota aquifer system, Williston basin, North Dakota, *in* Jorgensen, D. G., and Signor, D. C., eds., Geohydrology of the Dakota aquifer, Proceedings, First C. B. Theis Conference on Geohydrology: Worthington, Ohio, National Water Well Association, p. 99–110.

Downey, J. S., 1982, Geohydrology of the Madison and associated aquifers in parts of Montana, North Dakota, South Dakota, and Wyoming: U.S. Geological Survey Open-File Report 82-914, 120 p.

——, 1984, Hydrodynamics of the Williston basin in the northern Great Plains, *in* Jorgensen, D. G., and Signor, D. C., eds., Geohydrology of the Dakota aquifer, Proceedings, First C. V. Theis Conference on Geohydrology: Worthington, Ohio, National Water Well Association, p. 92–98.

Dutton, S. P., and Land, L. S., 1985, Meteoric burial diagenesis of Pennsylvanian arkosic sandstones, southwestern Anadarko basin, Texas: American Association of Petroleum Geologists Bulletin, v. 69, p. 22–38.

Dutton, S. P., Goldstein, A. G., and Ruppel, S. C., 1982, Petroleum potential of the Palo Duro basin: The University of Texas at Austin, Bureau of Economic Geology Report of Investigations 123, 87 p.

Edwards, A. R., and Downey, M. W., 1967, Stratigraphic cross secton of Paleozoic rocks; Oklahoma: American Association of Petroleum Geologists, 3 p.

Fink, B. E., 1963, Ground-water potential of the northern part of the southern High Plains of Texas: High Plains Underground Water Conservation District No. 1, Report 163, 76 p.

Gustavson, T. C., and Finley, R. J., 1985, Late Cenozoic geomorphic evolution of the Texas Panhandle and northeastern New Mexico; Case studies of structural controls on regional drainage: The University of Texas at Austin, Bureau of Economic Geology Report of Investigations 148, 42 p.

Gustavson, T. C., Findley, R. J., and McGillis, K. A., 1980, Regional dissolution of Permian salt in the Andarko, Dalhart, and Palo Duro basins of the Texas Panhandle: The University of Texas at Austin, Bureau of Economic Geology Report of Investigations 106, 40 p.

Helgesen, J. O., Jorgensen, D. G., Leonard, R. B., and Signor, D. C., 1982, Regional study of the Dakota aquifer (Darton's Dakota revisited): Ground Water, v. 20, no. 4, p. 410–414.

Holt, C.L.R., Jr., 1959, Geology and ground-water resources of Medina County, Texas: U.S. Geological Survey Water-Supply Paper 1422, 213 p.

Huffman, G. G., 1958, Geology of the flanks of the Ozark uplift, northeastern Oklahoma: Oklahoma Geological Survey Bulletin 77, 271 p.

Jorgensen, D. G., 1984, Editorial-aquifer names, *in* Jorgensen, D. G., and Signor, D. C., eds., Geohydrology of the Dakota aquifer, Proceedings, First C. V. Theis Conference on Geohydrology: Worthington, Ohio, National Water Well Association, p. 4–7.

Knowles, T., Nordstrom, P., and Klemt, W. B., 1984, Evaluating the ground-water resource of the High Plains of Texas: Austin, Texas, Texas Department of Water Resources Report 288, 113 p.

Kreitler, C. W., Fisher, R. S., Senger, R. K., Hovorka, S. D., and Dutton, A. R., 1985, Hydrology of an evaporite aquitard; Permian evaporite strata, Palo Duro basin, Texas, *in* Simpson, E. S., Chairman, Hydrology of rocks of low permability: Proceedings, 17th International Congress, International Associ-

ation of Hydrogeologists, January 7–12, 1985, Tucson, Arizona.

Leonard, R. B., Signor, D. C., Jorgensen, D. G., and Helgesen, J. O., 1983, Geohydrology and hydrochemistry of the Dakota aquifer, central United States: Water Resources Bulletin, v. 19, no. 6, p. 903–911.

Maclay, R. W., and Small, T. A., 1976, Progress report on geology of the Edwards aquifer, San Antonio area, Texas, and preliminary interpretation of borehole, geophysical, and laboratory data on carbonate rocks: U.S. Geological Survey Open-File Report 76-627, 94 p.

McCracken, M. H., 1967, Mineral and water resources of Missouri; Structure: 90th Congress 1st Session, Document 19, 399 p.

McNeal, R. P., 1965, Hydrodynamics of the Permian basin: American Association of Petroleum Geologists Memoir 4, p. 308–326.

Morgan, A. M., 1941, Solution phenomena in New Mexico, *in* Symposium on relations of geology to the groundwater problems of the southwest: American Geophysical Union Transactions, 23rd Annual Meeting, pt. 1, p. 27–35.

Peterson, J. A., 1981, Stratigraphy and sedimentary facies of the Madison Limestone and associated rocks of parts of Montana, North Dakota, South Dakota, Wyoming, and Nebraska: U.S. Geological Survey Open-File Report 81-642, 92 p.

Rahn, P. H. and Gries, J. P., 1973, Large springs in the Black Hills, South Dakota and Wyoming: South Dakota Geological Survey Report of Investigations 107, 46 p.

Ramondetta, P. J., 1982, Genesis and emplacement of oil in the San Andres Formation, northern shelf on the Midland Basin, Texas: The University of Texas at Austin, Bureau of Economic Geology Report of Investigations 116, 39 p.

Rose, P. R., 1972, Edwards Group, surface and subsurface, central Texas: The University of Texas at Austin, Bureau of Economic Geology Report of Investigations 74, 198 p.

Summers, W. K., 1981, Ground-water head distribution in the third dimension of the Pecos River basin, New Mexico: New Mexico Geology, v. 3, no. 1, p. 6–12.

Swenson, F. A., 1968, New theory of recharge in the artesian basin of the Dakotas: Geological Society of America Bulletin, v. 79, p. 163–182.

Theis, C. V., 1965, Ground water in the southwestern region: American Association of Petroleum Geologists Memoir 4, p. 327–341.

Wyoming State Engineer's Office, 1974, Underground water supply in the Madison Limestone: Cheyenne, Wyoming, 117 p.

Manuscript Accepted by the Society March 6, 1987

The Geology of North America
Vol. O-2, Hydrogeology
The Geological Society of America, 1988

# Chapter 20

# *Region 17, High Plains*

**John B. Weeks and Edwin D. Gutentag**
*U.S. Geological Survey, Mail Stop 412, Box 25046, Denver Federal Center, Denver, Colorado 80225*

## INTRODUCTION

The High Plains, Region 17 (Fig. 3; Table 2, Heath, this volume), includes about 450,000 km$^2$ of the central United States east of the Rocky Mountains in the southern part of the Great Plains. The High Plains aquifer, which underlies this region, is the principal source of water in one of the major agricultural areas in the United States. During 1980, about 170,000 wells pumped 22,000 hm$^3$ of water from the aquifer to irrigate 54,000 km$^2$ (Heimes and Luckey, 1983). The following discussion is limited to those geologic units that comprise, or are in contact with, the High Plains aquifer.

### Physiography

The High Plains extends from southern South Dakota to northwestern Texas and includes parts of eight States. The Platte, Arkansas, and Canadian Rivers are the only streams crossing the High Plains that originate in the Rocky Mountains. The area is characterized by flat to gently rolling terrain, which is a remnant of a vast plain that originally extended from the mountains eastward beyond the Missouri River (Trimble, 1980). Regional uplift caused streams to cut downward and erode the plains. Erosion isolated the plains from the mountains and formed escarpments that typically mark the boundary of the High Plains. Only in southeastern Wyoming does the plain still extend to the mountain front.

Wind-blown sand and silt, derived from the beds of rivers that eroded the plains, were deposited throughout large areas of the High Plains. The largest expanse of wind-blown sand deposits and dune topography in the western hemisphere (about 52,000 km$^2$) is in west-central Nebraska. Within these sand hills, lakes and meadows occur where the water table is at or near land surface.

Many lakes also occur on the High Plains outside of Nebraska. Most of these lakes are shallow depressions or playas that collect and store water during periods of runoff; some of the deeper playas hold water throughout the year. Playas are most prevalent south of the Arkansas River. Because playas in the High Plains generally are perched above the regional water table, the water surfaces in the playas do not reflect the altitude of the water table in the underlying aquifer.

### Climate

Most of the High Plains has a middle-latitude, dry continental climate. The climate is one of abundant sunshine, moderate precipitation, frequent winds, low humidity, and a high rate of evaporation. Mean annual precipitation increases eastward across the High Plains by about 25 mm every 40 km, from less than 400 mm in the western High Plains to about 700 mm in eastern Nebraska and central Kansas. Typically, about 75 percent of the precipitation falls as rain. However, much of the rain results from local thunderstorms, so large variations in rainfall are common.

Persistent winds and high summer temperatures cause high rates of evaporation in the High Plains. Mean annual evaporation from Class A pans ranges from about 1,500 mm in northern Nebraska and southern South Dakota to about 2,700 mm in western Texas and southeastern New Mexico. Because of large evapotranspiration demand, little precipitation is available to recharge the ground-water system. Except in sand dune areas where water can readily percolate down to the water table, most of the water that enters the soil is returned to the atmosphere by evapotranspiration.

## GEOLOGIC FRAMEWORK

The High Plains aquifer consists of near-surface deposits of Tertiary and Quaternary age. Bedrock units in contact with the High Plains aquifer range in age from Permian to Tertiary. The location of the bedrock units is shown in Figure 1. Although some sandstones of Triassic, Jurassic, and Cretaceous age yield water to wells, the bedrock units typically are much less permeable than the High Plains aquifer, which is the most productive aquifer in the region.

### Bedrock Units

**Permian rocks.** The oldest rocks in contact with the High Plains aquifer are of Permian age. Permian rocks directly underlie

Weeks, J. B., and Gutentag, E. D., 1988, Region 17, High Plains, *in* Back, W., Rosenshein, J. S., and Seaber, P. R., eds., Hydrogeology: Boulder, Colorado, Geological Society of America, The Geology of North America, v. O-2.

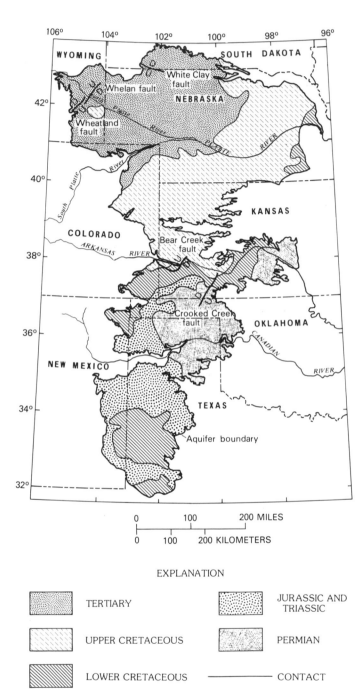

Figure 1. Geologic units underlying the High Plains aquifer (modified from Weeks and Gutentag, 1981).

EXPLANATION

| | | |
|---|---|---|
| TERTIARY | | JURASSIC AND TRIASSIC |
| UPPER CRETACEOUS | | PERMIAN |
| LOWER CRETACEOUS | ———— | CONTACT |

12 percent of the High Plains aquifer mainly in Kansas, Oklahoma, and Texas (Fig. 1). These rocks consist of red beds and evaporites that were deposited in extensive, shallow, brackish-to-saline seas that were subject to periodic influxes of marine water. Generally, Permian rocks contain mineralized water and have low permeability relative to the overlying High Plains aquifer.

*Triassic and Jurassic rocks.* Rocks of Triassic and Jurassic age directly underlie 16 percent of the High Plains aquifer in Colorado, Kansas, New Mexico, Oklahoma, and Texas (Fig. 1). These rocks are of continental origin and consist mainly of sandstone and shale. Locally, some of the sandstone beds are relatively permeable and contain water of suitable quality for irrigation, particularly between the Arkansas and Canadian Rivers in parts of Colorado, Kansas, Oklahoma, New Mexico, and Texas.

*Cretaceous rocks.* Lower Cretaceous rocks directly underlie 12 percent of the High Plains aquifer in parts of Colorado, Kansas, Nebraska, New Mexico, Oklahoma, and Texas (Fig. 1). The rocks were deposited in shallow seas that covered the area intermittently throughout Early Cretaceous time. North of the Canadian River, Lower Cretaceous rocks were deposited near sea level or in adjoining lowlands and consist mostly of sandstone and shale. The Dakota Sandstone is the most extensive bedrock aquifer in this area. South of the Canadian River, Lower Cretaceous rocks were deposited in deeper water and consist mainly of beds of shale that grade into limestone. Cretaceous sandstone and shale beds typically have low permeability and storage relative to the High Plains aquifer.

Upper Cretaceous rocks directly underlie 32 percent of the High Plains aquifer in large areas in Colorado, Kansas, and Nebraska and small areas in South Dakota and Wyoming (Fig. 1). During Late Cretaceous time, shale, chalk, limestone, and sandstone beds were deposited in the Rocky Mountain trough that formed a vast seaway nearly 1,600 km wide and about 4,800 km long. Within this sequence of rocks, only the Niobrara Chalk or Formation is known to contain permeable zones where fractures and solution openings have developed.

*Tertiary Rocks.* Rocks of Oligocene age directly underlie 28 percent of the High Plains aquifer in Colorado, Nebraska, South Dakota, and Wyoming (Fig. 1). Rocks in the White River Group are the oldest Tertiary deposits present under the High Plains. The Chadron and Brule Formations comprise the White River Group. The Chadron Formation is the lower unit consisting of consolidated beds of clay and silt, and stream-channel deposits of sandstone and conglomerate. The Chadron is overlain by the Brule Formation, which is predominantly a massive siltstone containing consolidated beds of volcanic ash, clay, and fine sand. Both the Chadron and Brule Formations generally have low primary permeability.

### Aquifer Units

The High Plains aquifer consists mainly of hydraulically connected geologic units of late Tertiary or Quaternary age. The principal geologic units in the High Plains aquifer are shown in

Figure 2. The Tertiary rocks consist of the Brule Formation (of the White River Group), Arikaree Group, and Ogallala Formation. The Quaternary deposits consist of alluvial, dune-sand, and valley-fill deposits. Except for dune sand in Figure 2, the Quaternary deposits are combined and shown only in areas where they do not overlie Tertiary aquifer units. In northern Texas, some collapse structures, filled with Triassic, Jurassic, and Lower Cretaceous rocks that have secondary permeability, are considered part of the High Plains aquifer; however, they are minor and are not shown in Figure 2.

*Brule Formation.* The Brule Formation crops out in or underlies much of western Nebraska, northeastern Colorado, southwestern South Dakota, and southeastern Wyoming. Maximum thickness of the Brule is about 180 m. In some places, permeability of the Brule Formation has been increased by secondary openings—joints, fractures, and solution cavities. The Brule Formation is considered part of the aquifer only in areas where it contains saturated zones resulting from interconnected secondary openings (Fig. 2). Where secondary openings have not developed, the top of the Brule Formation is considered the base of the High Plains aquifer.

*Arikaree Group.* The Arikaree Group (or Formation) includes all Upper Tertiary deposits between the underlying Brule Formation and overlying Ogallala Formation. The Arikaree is included in the High Plains aquifer in large areas in western Nebraska, southwestern South Dakota, and southeastern Wyoming (Fig. 2). The Arikaree is mainly a massive, very-fine- to fine-grained sandstone, but contains localized beds of volcanic ash, silty sand, and sandy clay. Maximum thickness of the Arikaree is about 300 m in western Nebraska and in adjacent parts of Wyoming. Secondary openings also occur in the Arikaree Group.

*Ogallala Formation.* The Ogallala Formation includes all Upper Tertiary rocks in the study area that are younger than the Arikaree Group. The Ogallala Formation is the principal geologic unit in the High Plains aquifer and underlies about 347,000 $km^2$ of the region (Fig. 2). Maximum thickness of the Ogallala is about 215 m. When the Ogallala was deposited, aggrading streams filled, and buried valleys eroded into pre-Ogallala rocks. Braided streams flowed eastward from the mountains and deposited a heterogeneous sequence of clay, silt, sand, and gravel beds.

Within the Ogallala, zones cemented with calcium carbonate are resistant to weathering and form ledges in outcrops. The most distinctive of these layers, the Ogallala cap rock (commonly called caliche or mortar bed), is near the top of the Ogallala Formation. The cap rock underlies large areas in Texas and New Mexico, and may be as thick as 18 m.

*Quaternary Rocks.* Unconsolidated alluvial deposits of Quaternary age that are in hydraulic connection with Tertiary deposits are part of the High Plains aquifer. Much of the gravel, sand, silt, and clay in the alluvial deposits was reworked from the Ogallala Formation. These Quaternary alluvial deposits have a maximum thickness of about 90 m. Alluvial deposits compose the High Plains aquifer in eastern Nebraska and central Kansas (Fig. 2). In many areas of the High Plains, Quaternary alluvial

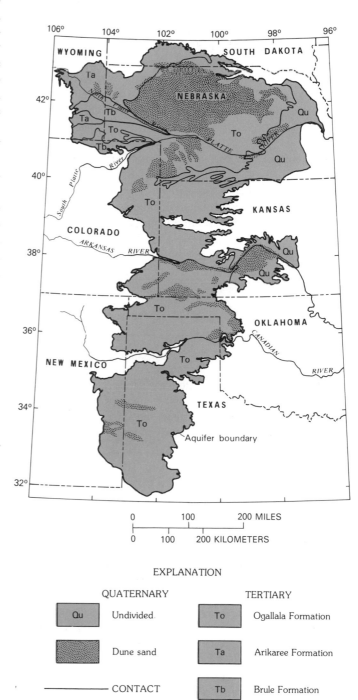

EXPLANATION

QUATERNARY

Qu    Undivided.

Dune sand

———— CONTACT

TERTIARY

To    Ogallala Formation

Ta    Arikaree Formation

Tb    Brule Formation

Figure 2. Principal geologic units in the High Plains aquifer (after Gutentag and other, 1984).

deposits directly overlie the Ogallala Formation to form one aquifer.

Dune-sand deposits of Quaternary age, consisting predominantly of very fine to medium wind-blown sand, are part of the High Plains aquifer. About 85,000 km² of the High Plains is covered by dune sand, as shown in Figure 2. The most extensive area of dune sand is in west-central Nebraska where the deposits cover an area of 52,000 km². Large areas also are covered by dune-sand deposits in Colorado and Kansas. Throughout the High Plains, dune sands are important recharge areas for the aquifer.

In Nebraska, the age of these wind-blown sand deposits ranges from 12 ka to 0.5 ka (Ahlbrandt, Swinehart, and Maroney, 1983). The source of the sand was local Tertiary rocks and the flood plains of streams that eroded the northern boundary of the High Plains. Based on dune orientation, the primary eolian transport direction was southeastward. The dunes are virtually stabilized by vegetative cover; however, the dunes may become unstable during severe drought. The thickness of the eolian sand ranges from 25 m or less in interdune areas to as much as 125 m beneath dunes.

Valley-fill deposits consist of unconsolidated gravel, sand, silt, and clay associated with the most recent cycle of erosion and deposition along present streams. These deposits are as much as 18 m thick. Valley-fill deposits that are hydraulically connected to underlying Tertiary or Quaternary deposits are part of the High Plains aquifer. The valley-fill deposits and associated streams form stream-aquifer systems that link the High Plains aquifer to surface streams, particularly along the Platte, Republican, and Arkansas Rivers.

## GROUND-WATER HYDROLOGY

### Recharge

Recharge to the aquifer, which generally is under water-table conditions, is entirely from precipitation and seepage from streams. Because evapotranspiration demand greatly exceeds precipitation, little recharge occurs except in sandy soils with large infiltration rates and low field capacities. Estimated rates of recharge from precipitation range from less than 1 mm per year in parts of Texas to 150 mm per year in sand-dune areas in Kansas and Nebraska.

Recharge to the aquifer also is derived from streamflow. During periods of normal precipitation, little runoff collects in channels of intermittent streams. However, during and after intense rains, the streams may transport large quantities of water. During these flow periods, water infiltrates into channel and flood-plain deposits and percolates downward to recharge the aquifer. In western Kansas, Jordan (1977) estimated the average streamflow loss during flood flows was 0.8 percent of the flow per kilometer of stream channel.

Recharge rates in the High Plains are quite variable. With the exception of areas of sandy soil and stream courses, recharge

to the High Plains aquifer is extremely small and probably results from infrequent periods of high precipitation that cause relatively large quantities of recharge during relatively short periods of time. Based on computer model simulations (Luckey and others, 1986), the annual volume of recharge to the aquifer is about 7,000 hm³ or 15.5 mm over the area of the aquifer. About 80 percent of the recharge occurs north of 39° latitude where evapotranspiration demand is least and dune sands are prevalent.

### Flow

Ground-water flow in the High Plains aquifer generally is from west to east in response to the slope of the water table. Based on average values of hydraulic gradient and aquifer characteristics, the velocity of water moving through the aquifer is about 0.3 m/day, which is typical of sand and gravel aquifers.

Hydraulic conductivity and specific yield are the two principal aquifer characteristics that control ground-water flow in a water-table aquifer. Both hydraulic conductivity and specific yield depend on the sediments that comprise the aquifer; therefore, they vary horizontally and vertically according to the variation in sediment types. Both fluvial and eolian sediments form the aquifer. The fluvial sediments that dominate most of the aquifer were deposited by braided streams, resulting in a virtually random sequence of coarse-grained and fine-grained sediments. The hydraulic conductivity of the aquifer averages 18 m/day, although the values range from less than 2 to 100 m/day for individual lithologic units. The specific yield of the aquifer averages 15 percent, although the values range from 3 to 35 percent for individual lithologic units within the aquifer.

### Discharge

Water from the High Plains aquifer discharges naturally to streams and springs, and seeps along the eastern escarpment. Streams that originate in the High Plains generally are ephemeral in their upstream reaches and perennial where their channels are incised below the water table. Although greatly exaggerated in vertical scale, the east–west sections in Figure 3 show the eastward slope of the water table and aquifer base. As shown on the north–south section (102°W longitude), the North Platte River at Lake McConaughy and the Arikaree, Arkansas, and Canadian Rivers have cut their channels through the aquifer into bedrock. These streams in addition to the South Platte, North Loup, Elkhorn, and North Fork Solomon Rivers are gaining streams at the locations shown. At other locations, these and most other rivers in the High Plains are losing streams, such as the Beaver and Cimarron Rivers in Figure 3, and may provide recharge to the aquifer during periods of runoff.

Discharge from the aquifer also occurs by evapotranspiration where the water table is near land surface. Ground-water discharge by evapotranspiration is greatest in areas where phreatophytes such as salt cedar, willows, cottonwoods, and sedges grow along stream valleys. In many places in the sand hills in Ne-

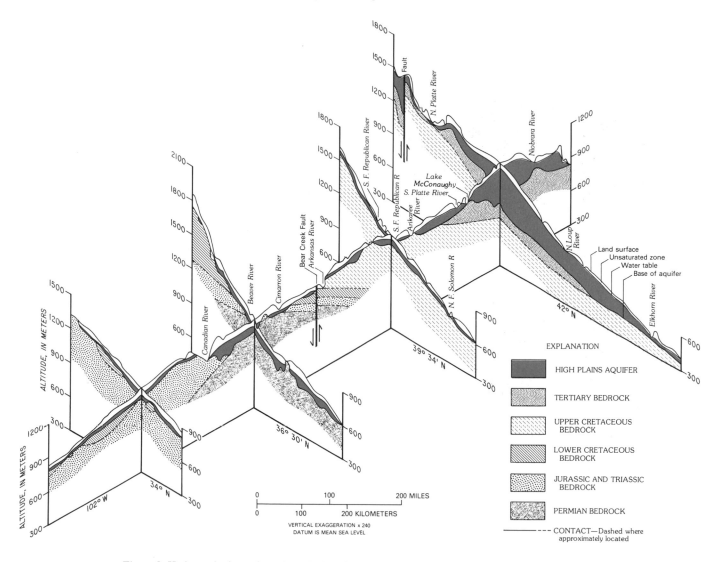

Figure 3. Hydrogeologic sections through the High Plains aquifer (after Weeks and Gutentag, 1981).

braska, the water table is so close to land surface that ground-water discharge by evapotranspiration may equal or exceed that of seepage into streams.

Ground-water flow in the sand hills is complex because of the relationships among streams, lakes, recharge, and discharge. Although the surface infiltration of precipitation in the sand hills is 50 to 125 mm per year, most of this water is discharged locally to streamflow and by evapotranspiration from lakes and meadows in interdune areas. As a result, many perennial streams drain the eastern part and many alkaline lakes occur in the western part of the sand hills. Using a computer model, Luckey and others (1986) estimated that, of the water that infiltrates, less than 40 mm per year becomes recharge to the regional ground-water flow system.

Discharge from the High Plains aquifer has been estimated by computer model simulations (Luckey and others, 1986). The average annual discharge calculated by the computer model was 2,800 hm³. About 85 percent of the discharge occurs north of 39° latitude in the Niobrara, Platte, and Republican River basins where recharge is greatest. The average annual discharge is considerably smaller than the average annual recharge because of pumping from wells.

### Storage

Saturated thickness of the aquifer varies in response to changes in recharge and discharge. Saturated thickness ranges from zero where the deposits comprising the High Plains aquifer are unsaturated to about 300 m in west-central Nebraska. The average saturated thickness of the High Plains aquifer is about 60 m. Areas where the saturated thickness of the aquifer exceeded to 60 m in 1980 are shown in Figure 4. About 46 percent of the

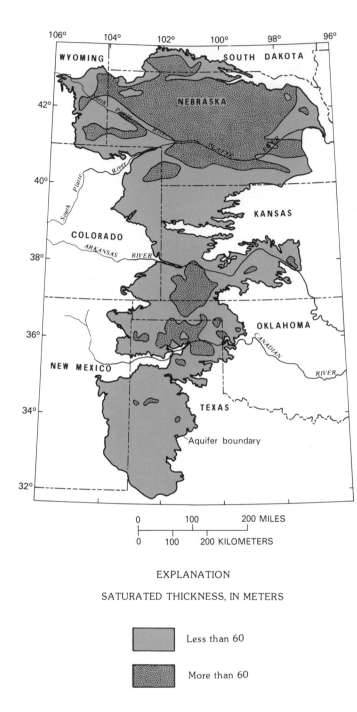

0        100        200 MILES
0    100        200 KILOMETERS

EXPLANATION

SATURATED THICKNESS, IN METERS

Less than 60

More than 60

Figure 4. Saturated thickness of the High Plains aquifer in 1980.

area of the High Plains aquifer has less than 30 m of saturated thickness, whereas only 5 percent has more than 180 m of saturated thickness. The saturated thickness of the aquifer is greater than 180 m only in Nebraska and Wyoming.

Water in storage in the High Plains aquifer depends on the saturated thickness and specific yield of the sediments that comprise the aquifer. The variability in the distribution of water in storage can be inferred from the saturated-thickness map shown in Figure 4 and the geohydrologic sections shown in Figure 3. The total volume of drainable water in storage (product of specific yield, area, and saturated thickness) in the High Plains aquifer in 1980 was estimated to be 4 million $hm^3$ (Gutentag and others, 1984). About 65 percent of the water in storage is in Nebraska, where recharge and aquifer thickness are greatest because of the presence of dune sands. Large volumes of water are also in storage where the aquifer is thick in southwestern Kansas, Oklahoma, and northern Texas (Fig. 4).

### Water quality

As a result of geochemical and biochemical interactions between water and the geologic materials through which it flows, ground water contains a variety of dissolved constituents. The concentration of dissolved solids in most of the water in the High Plains aquifer is less than 500 mg/L, as shown in Figure 5. About 27 percent of the volume of water in the aquifer contains less than 250 mg/L dissolved solids, and 15 percent of the volume of water in the aquifer contains more than 500 mg/L dissolved solids. Only 3 percent of the volume of water in the aquifer contains more than 1,000 mg/L dissolved solids, most of which is in Texas. Typically, the largest concentrations of dissolved solids in water from the High Plains aquifer are less than 3,000 mg/L (Krothe and others, 1982).

The concentration of dissolved solids in water in the High Plains aquifer is less than 250 mg/L dissolved solids in parts of Colorado, Kansas, Nebraska, and Wyoming (Fig. 5). A large part of this area is covered by sandy soils and dune sands (Fig. 2). The concentration of dissolved solids in the area covered by sand is less than 250 mg/L because recharge to the aquifer from precipitation is relatively large and most soluble material has been removed from the sand.

In most of the High Plains, water in the aquifer contains 250 to 500 mg/L dissolved solids, and the concentration generally increases from north to south and west to east (Fig. 5). Except south of the Canadian River, the age of the geologic units underlying the High Plains aquifer also increases from north to south and from west to east (Fig. 1). The age of the bedrock ranges from Tertiary to Permian and, in general, the older rocks contain more soluble minerals than the younger rocks. Generally, the chemistry of the water in the High Plains aquifer is affected by the mineral composition of the bedrock units.

In a small area of Colorado, the concentration of dissolved solids in water from the High Plains aquifer exceeds 1,000 mg/L (Fig. 5). In this area, the aquifer is underlain by the Smoky Hill

Marl Member of the Niobrara Formation of Late Cretaceous age, which contains gypsum (calcium sulfate). Water in the aquifer contains sulfate as the dominant anion, probably caused by solution of gypsum at the bedrock contact or from bedrock material contained in the aquifer.

In parts of Kansas, Oklahoma, and Texas, the High Plains aquifer is underlain by Permian age bedrock (Fig. 1) containing salt beds and saline water. In part of this area, the concentration of dissolved solids in water in the aquifer generally is 500 to 1,000 mg/L and exceeds 1,000 mg/L in one area (Fig. 5). The water contains large and nearly equal proportions of sodium and chloride indicating that water from the Permian bedrock is entering the aquifer.

In the southern High Plains of New Mexico and Texas, the aquifer is underlain by Triassic, Jurassic, and Lower Cretaceous bedrock (Fig. 1). The Lower Cretaceous rocks were deposited in a deep-water marine environment, and water in these rocks is very mineralized. Water in the overlying aquifer typically contains 500 to more than 1,000 mg/L dissolved solids (Fig. 5) and large proportions of magnesian, sodium, chloride, and sulfate caused by the movement of solutes from the marine bedrock into the aquifer.

### Geologic Controls

Faults are important because they affect aquifer thickness and productivity. The Wheatland and Whelan faults are of particular significance to the High Plains aquifer in Wyoming (Fig. 1). Along the faults, the bedrock has been displaced as much as 300 m, and the Tertiary fill material on the down-thrown side is part of the High Plains aquifer (Fig. 3). The White Clay fault is near the Nebraska–South Dakota state line (Fig. 1). This normal fault has caused about 150 m of displacement in the aquifer base.

Salt dissolution and collapse have a pronounced effect on the High Plains aquifer. From central Kansas to Texas, Permian salt deposits are being dissolved by circulating ground water. Saline water from Permian bedrock flows through the High Plains aquifer or the underlying bedrock to discharge along the Beaver, Cimarron, and Arkansas Rivers in Kansas and Oklahoma and all streams that drain the High Plains in Texas. Subsequent collapse and filling of dissolution zones has affected the quality of ground water, aquifer thickness, and ground-water flow.

Faulting and collapse structures associated with salt dissolution have been forming since Permian time. Collapse structures have been filled by younger material in large areas in southwestern Kansas, Oklahoma, and northern Texas (Figs. 3 and 4). Faulting caused by dissolution of evaporites in Permian deposits and collapse of the overlying rocks occurs along Bear Creek and Crooked Creek faults in Colorado, Kansas, and Oklahoma (Figs. 1, 3, and 4). Movement along these faults occurs as the edge of the dissolution zone is dissolved slowly by ground water. Collapse along these faults has caused about 60 m of displacement in the altitude of the aquifer base.

Salt dissolution has affected all of the rocks overlying the

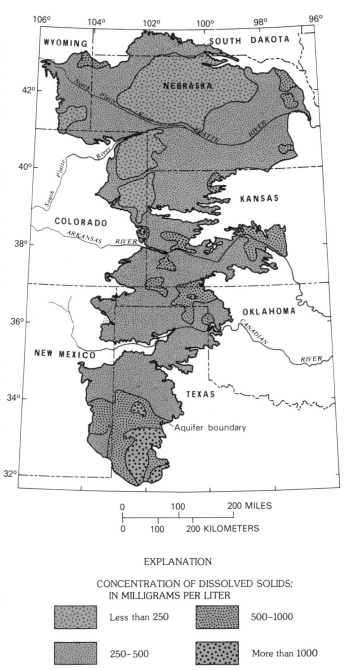

Figure 5. Concentration of dissolved solids in water from the High Plains aquifer (modified from Krothe and others, 1982).

Permian evaporite deposits. Where salt dissolution has occurred under competent rocks, collapse has caused chaotic structures in the now fractured rocks. The resulting surface depressions or sinkholes have been filled with fluvial sediments. Sinkhole formation and infilling has continued from Permian time to the present, creating complex and chaotic sedimentary structures. Consequently, the sinkhole deposits do not have the same ground-water flow characteristics as similar lithologic material deposited horizontally in areas without sinkholes. The chaotic bedding in the sinks greatly impedes regional ground-water flow.

## CONCLUSIONS

Because evaporative demand greatly exceeds precipitation, recharge to the High Plains aquifer is small except in areas of dune sand or sandy soil. Recharge rates range from less than 1 to 150 mm per year and average about 15 mm per year throughout the 450,000-km$^2$ area. Consequently, discharge from the aquifer to streams is small, and most streams draining the High Plains are intermittent.

The eolian and fluvial sediments that compose the aquifer have relatively large storage properties. The specific yield ranges from 3 to 35 percent and averages about 15 percent. The total volume of drainable water in the High Plains aquifer during 1980 was estimated to be 4 million hm$^3$. Because of the occurrence of dune sands, higher recharge, and lower evaporative demand, the High Plains aquifer generally has greater thickness and storage north of 39° latitude than to the south.

Dissolution of evaporites in Permian bedrock units has had a pronounced effect on the High Plains aquifer. Dissolution of salt by circulating ground water has affected the quality of water in the overlying aquifer and in the streams draining the area underlain by Permian bedrock. Collapse and infilling have caused chaotic sedimentary structures that affect aquifer thickness and ground-water flow.

## REFERENCES CITED

Ahlbrandt, T. S., Swinehart, J. B., and Maroney, D. G., 1983, The dynamic Holocene dune fields of the Great Plains and Rocky Mountain basins, *in* Brookfield, M. E., and Ahlbrandt, T. S., eds., Eolian sediments and processes: Amsterdam, Elsevier Science Publishers, p. 379–406.

Gutentag, E. D., Heimes, F. J., Krothe, N. C., Luckey, R. R., and Weeks, J. B., 1984, Geohydrology of the High Plains aquifer in parts of Colorado, Kansas, Nebraska, New Mexico, Oklahoma, South Dakota, Texas, and Wyoming: U.S. Geological Survey Professional Paper 1400-B, 63 p.

Heimes, F. J. and Luckey, R. R., 1983, Estimating 1980 ground-water pumpage for irrigation on the High Plains in parts of Colorado, Kansas, Nebraska, New Mexico, Oklahoma, South Dakota, Texas, and Wyoming: U.S. Geological Survey Water-Resources Investigations 83-4123, 36 p.

Jordan, P. R., 1977, Streamflow transmission losses in western Kansas: Proceedings of the American Society of Civil Engineers, Journal of the Hydraulics Division, v. 103, no. HY8, Paper 13156, p. 905–919.

Krothe, N. C., Oliver, J. W., and Weeks, J. B., 1982, Dissolved solids and sodium in water from the High Plains aquifer in parts of Colorado, Kansas, Nebraska, New Mexico, Oklahoma, South Dakota, Texas, and Wyoming: U.S. Geological Survey Hydrologic Investigations Atlas HA-658.

Luckey, R. R., Gutentag, E. D., Heimes, F. J., and Weeks, J. B., 1986, Digital simulation of ground-water flow in the High Plains aquifer in parts of Colorado, Kansas, New Mexico, Oklahoma, South Dakota, Texas, and Wyoming: U.S. Geological Survey Professional Paper 1400-D, 57 p.

Trimble, D. E., 1980, The geologic story of the Great Plains: U.S. Geological Survey Bulletin 1493, 55 p.

Weeks, J. B., and Gutentag, E. D., 1981, Bedrock geology, altitude of base, and 1980 saturated thickness of the High Plains aquifer in parts of Colorado, Kansas, Nebraska, New Mexico, Oklahoma, South Dakota, Texas, and Wyoming: U.S. Geological Survey Hydrologic Investigations Atlas HA-648.

MANUSCRIPT ACCEPTED BY THE SOCIETY MARCH 10, 1988

The Geology of North America
Vol. O-2, Hydrogeology
The Geological Society of America, 1988

# Chapter 21

# *Region 18, Alluvial valleys*

**J. S. Rosenshein**

*U.S. Geological Survey, 414 National Center, Reston, Virginia 22092*

## INTRODUCTION

Alluvial-valley aquifer-stream systems (Fig. 1) occur in parts of many of the regions discussed in Chapters 3 to 27 and Table 2, Heath this volume. This type of aquifer system is treated separately in this volume because of the unique aspects of its hydrogeology and its relative importance to the hydrogeology of large parts of North America—in particular, large areas of the conterminous United States and smaller areas in Alaska and Canada. The flow system typically associated with aquifer-stream systems is dynamic because of the relatively rapid interaction of surface water and ground water in these systems. In most of the aquifer-stream systems, changes affecting one component are reflected within short periods in changes in the other component. These changes can occur quickly because of the rapid and large change in stage and other changes that affect the surface-water component of these systems—the stream—and that can be transmitted to the associated sand and gravel deposits comprising the aquifer.

McGuinness (1963, pl. 1) delineated three types of alluvial-valley aquifers on his map of productive aquifers in the conterminous United States. These aquifers were classified as: (1) Water-course aquifers—that is, alluvial-valley aquifers that are traversed by perennial streams from which recharge could be induced; (2) surficial alluvial-valley aquifers no longer traversed by perennial streams; and (3) buried alluvial-valley aquifers. The alluvial-valley aquifer-stream systems discussed in this chapter generally are restricted to the first type of aquifer delineated by McGuinness (1963).

Heath (1984, p. 58–61) pointed out that alluvial-valley aquifer-stream systems generally occur in stream valleys in areas that were covered by ice sheets during the Pleistocene or in stream valleys that received meltwaters from ice sheets or mountain glaciers that existed during the Pleistocene. Heath (1984, p. 59) presented three hydrogeologic criteria for differentiating alluvial valleys from other valleys and, therefore, alluvial-valley aquifer-stream systems from other aquifer systems underlying stream valleys. Alluvial valleys and related aquifer-stream systems selected for inclusion in this regional synthesis (Fig. 1) reflect the criteria stated below, which generally were adopted from Heath (1984). (1) The alluvial valleys contain thick, extensive sand and gravel deposits that form an extensive aquifer. (2) The aquifer is hydraulically connected with a stream that generally has been considered perennial, although no flow may occur in the stream at times as a result of human effects on the hydrogeologic system. (3) The width of aquifer system is narrow (generally less than 4 km wide) in comparison to its length and generally is restricted to the width of the alluvial valley in which the system occurs.

The aquifer-stream systems in Region 18 are subject to a considerable range of climatic conditions from semiarid to humid. This range in climatic conditions has a marked effect on the discharge-recharge relations between the aquifers and the streams. Except for those aquifer-stream systems that are highlighted, the climatic aspects affecting a particular aquifer-stream system can be judged by referring to the discussion of the hydrogeology of the particular region through which the stream flows and in which the alluvial aquifer was deposited.

## HYDROGEOLOGIC ENVIRONMENT OF AQUIFER-STREAM SYSTEMS

### Depositional setting

Most of the aquifer-stream systems in Region 18 were formed as a direct or indirect result of either continental or mountain glaciation that took place during Pleistocene time. Most of the aquifer-stream systems that are north of the southern margin of continental glaciation (Fig. 1) have been subjected to the effects of advances and retreats of several ice sheets. Therefore, many of the systems have been formed in a complex depositional environment. In many, the types of glaciofluvial deposits within the aquifer consist of a combination of valley train, kame, kame terrace, and outwash alluvium. The glaciofluvial deposits can range from poorly sorted to well-sorted material; they can be nonstratified, poorly stratified, or well stratified; and locally, they can be interbedded with till or lacustrine silt and clay. Because the streams receiving meltwater from both advancing and retreating phases of the ice sheets transported large sediment loads, braided

Rosenshein, J. S., 1988, Region 18, Alluviual valleys, *in* Back, W., Rosenshein, J. S., and Seaber, P. R., eds., Hydrogeology: Boulder, Colorado, Geological Society of America, The Geology of North America, v. O-2.

Figure 1. Major alluvial-valley aquifer-stream systems (modified from Heath, 1984, Fig. 46).

stream patterns must have been a major factor in the characteristics of the stratification that developed in the fine-to-coarse–grained materials making up the aquifer systems.

Many stream systems beyond the margins of the continental ice sheets and the mountain glaciers also received meltwater discharge and served as major drainage ways for this discharge; as for example, the Mississippi and Ohio River drainage systems. These drainage systems conveyed large volumes of meltwater at rapid velocities with large loads of coarse sediment. The coarse-grained glaciofluvial material deposited in the stream valleys formed the alluvial-valley aquifers underlying Region 18. In many of these aquifers, it was common for the earliest material deposited to be coarse-grained sand and gravel (see Sharp, Fig. 2, this volume). As the margins of the ice sheets retreated, finer-grained glaciofluvial material was deposited on the coarse sand and gravel. Also, it was common for the material in major alluvial-aquifer systems deposited beyond the margins of the continental ice sheets and the major mountain glaciers to be finer-

grained in the downstream reaches of the aquifer-stream systems and to be coarser grained in the upstream reaches.

As pointed out by Sharp (this volume), most streams beyond the southern margin of continental glaciation have gone through a major cycle of sedimentation, controlled first by a braided-stream depositional pattern (Fig. 2) that gradually changed to a meandering-stream depositional pattern (Fig. 3). These depositional patterns have a controlling effect on the characteristics of the material making up the aquifer systems underlying the alluvial valleys as well as the hydraulic conductivity of these materials. Characteristically, braided-stream deposits: (1) contain channel bedding, festoon-type channel laminations, high-angle cross-stratification, and ripple cross-lamination; (2) have a considerable range in grain size (clay to cobbles); and (3) have a large areal distribution of coarse sand to cobbles, with clay and silt restricted to cut-off channels and, where present, thin flood-plain cover (Doeglas, 1962). The dominant bed forms are longitudinal gravel bars having poorly defined horizontal bedding

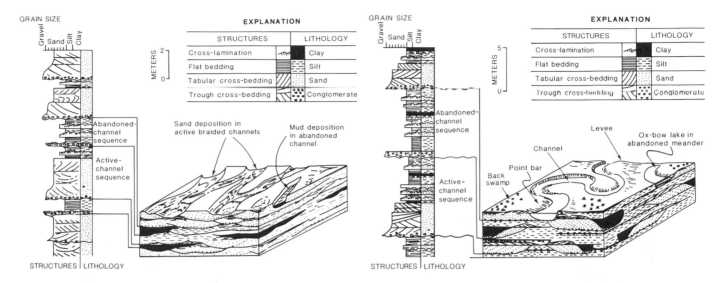

Figure 2. Depositional and geomorphic features of a braided-stream system that affect hydrogeologic characteristics of alluvial-valley aquifers (modified from Selley, 1982, Fig. 132).

Figure 3. Depositional and geomorphic features of a meandering-stream system that affect hydrogeologic characteristics of alluvial-valley aquifers (modified from Selley, 1982, Fig. 130; after Visher, 1965).

and wedge-shaped, cross-stratified, and ripple cross-laminated sand units (Rust, 1972). As a result, sedimentation occurred in a shifting complex of channels, with fine-grained deposits occurring in abandoned channels (Fig. 2). The fluvial processes involved in development of braided-stream channels are well defined in Leopold and others (1964, p. 281–295).

The latter phases of the depositional pattern are those typical of a meandering stream. As pointed out by Selley (1982), meandering streams—as compared with braided streams—are characterized by abandoned channels, overbank deposits, fine grain size, large ratio of silt and clay to sand and gravel, and a more homogeneous sedimentary deposit that becomes finer grained upward in the depositional sequence (Fig. 3). The fluvial processes involved in development of meandering channels also are well defined in Leopold and others (1964, p. 295–317).

Because of their complex depositional environment (Figs. 2 and 3), alluvial-aquifer systems have marked changes in grain size both laterally and vertically. Individual beds tend to be discontinuous and difficult to trace both laterally and vertically and can have marked variations in horizontal and vertical permeability and hydraulic conductivity. The effect of these variations on permeability and hydraulic conductivity can be modified in part by cross-stratification. However, based on gross depositional characteristics, the sediments form permeable hydrogeologic units that have an overall large hydraulic conductivity (Table 2, Heath, this volume).

### Relation to buried-valley systems

In the areas subjected to continental glaciation, pre-Pleistocene stream valleys were overridden by ice sheets, and their drainage ways disrupted or modified. Many of the drainage ways were partly or completely filled with glacial till. In others, glaciofluvial materials were deposited and then buried beneath glacial till or lacustrine deposits laid down by the next ice-sheet advance and retreat. During interglacial stages, the existing glacial deposits were eroded, and coarse-grained materials deposited in the interglacial drainage ways. The interglacial drainage ways were in turn overridden by the next ice-sheet advance and buried beneath the later glacial-till and lacustrine deposits. These systems of buried pre-Pleistocene and interglacial valleys containing coarse-grained deposits occur throughout southern Canada and part of the north-central United States, as for example in parts of Iowa, Illinois, Indiana, and Ohio. These buried-valley systems generally are not included as part of the alluvial-valley aquifer-stream systems of Region 18. However, in some places in the north-central United States (for example, parts of Indiana along the Wabash River), owing to late and post-Pleistocene erosion and redeposition, alluvial-valley aquifer-stream systems follow the same course as the buried valley for relatively short distances; they are in direct contact with buried-valley systems and form integrated hydrogeologic units (Fig. 4).

### Relation to other major aquifer systems

In many parts of Region 18, the alluvial-valley aquifer-stream systems are either underlain or bounded by geologic units that are relatively impermeable. These may consist of either clayey deposits or relatively dense or unfractured bedrock. This relationship occurs throughout much of the northeastern and north-central United States. In these areas, the alluvial-valley

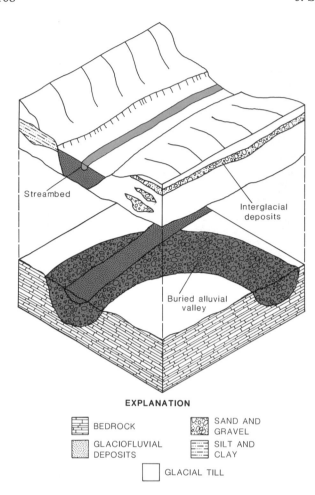

**EXPLANATION**

| | |
|---|---|
| ▦ BEDROCK | ▨ SAND AND GRAVEL |
| ▦ GLACIOFLUVIAL DEPOSITS | ▤ SILT AND CLAY |
| ▢ GLACIAL TILL | |

Figure 4. Hydrogeologic relations between alluvial-valley aquifer-stream systems, buried-valley systems, and other types of glacial and interglacial deposits.

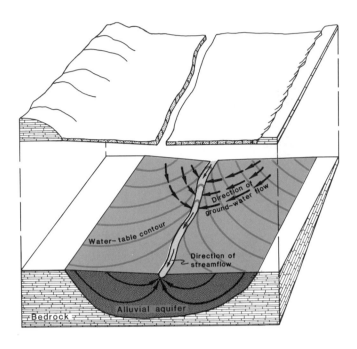

Figure 5. Gaining-stream conditions in alluvial-valley aquifer-stream systems.

aquifer-stream systems form a unique, separate, and clearly definable hydrogeologic system. In many other parts of Region 18, this relationship is not as well defined because the sand and gravel of the alluvial-valley aquifer-stream systems were deposited in valleys eroded into more permeable materials. This situation occurs, for example, where some of the alluvial-valley aquifer-stream systems cross the High Plains regional aquifer of Region 20. In the areas where this relationship occurs, the two aquifer systems form one integrated hydrogeologic unit, and the unique aspects of the alluvial-valley aquifer-stream system become less discernible.

## AQUIFER-STREAM INTERRELATIONS

In aquifer-stream systems, a distinctive hydrogeologic aspect is the exchange of water between the aquifer and the stream. Under natural conditions in a humid environment, ground water generally moves from aquifers to streams. Under these circumstances, the streams are gaining streams (Fig. 5). It is this ground water that maintains the flow of a gaining stream during dry-

weather conditions and provides a large fraction of the total flow in the stream. Typically, contours on the water table or potentiometric surface of the aquifers show a hydraulic gradient toward the streams, and where the contours cross the streams, their inflection will be upstream. Unless the stream has eroded deeply into the aquifer, part of the flow in the aquifer will move in a downstream direction.

Under natural conditions in a semiarid to arid environment, the water level in the aquifers adjacent to streams may be at or below the bed of the stream. Under these conditions, water will move from the stream into the aquifer, and the stream is a losing stream (Fig. 6). Under these conditions, contours on the water table of the alluvial-valley aquifer-stream system will indicate a hydraulic gradient from the stream toward the aquifer, and where the contours cross the stream, their inflection will be downstream. If the streambed is relatively permeable, the loss in transit per length of the streambed may be significant. For example, along the Arkansas River in Kansas, this loss may be 1 percent or more of the flood flow per kilometer length of the stream (Jordan,

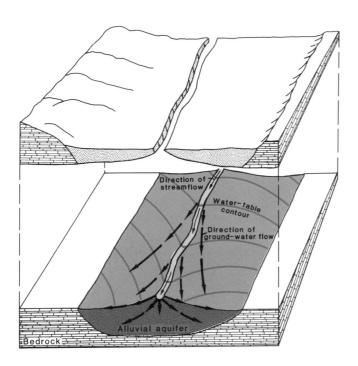

Figure 6. Losing-stream conditions in alluvial-valley aquifer-stream systems.

1977). During periods when runoff to a stream so affected is less than normal, the loss may be sufficient to temporarily dry up the stream. Commonly, streams in this type of environment receive much of their flow from runoff of precipitation (snowmelt and rainfall) in mountainous areas or flow of tributaries that drain chiefly semiarid areas.

Longitudinally extensive alluvial-valley aquifer-stream systems may extend through arid or semiarid environments to humid environments. Some of these extensive systems can be expected to change from gaining streams to losing streams (where affected by human activities, such as withdrawals from associated aquifers for irrigation or other uses) and back to gaining streams as they flow from semihumid, to semiarid, to humid areas.

## DYNAMICS OF AQUIFER-STREAM SYSTEMS

Aquifer-stream systems are hydrogeologically dynamic systems. Recharge to the systems varies, depending on whether the stream associated with the aquifer is a gaining or losing stream. In

aquifer-stream systems in semiarid areas, the natural recharge that takes place generally represents movement of water from stream to aquifer. The flow of the stream under semiarid conditions may be derived chiefly from water transported by the stream, commonly from a long distance. The flow of the stream generally is maintained by snowmelt and precipitation that has occurred in mountainous areas near the headwaters of the stream. This flow is supplemented by runoff of tributary streams that drain semiarid areas. Under normal conditions in semiarid areas, precipitation produces relatively small quantities of runoff to the streams. As a result, the rate of loss of flow in transit determines whether the stream maintains flow or becomes intermittent. In Region 18, this loss of flow is usually the result of human activities. Where the stream becomes intermittent, recharge to the associated aquifer becomes negligible, except during periods of major flooding. Annual variations in precipitation—snowfall and rainfall—are, therefore, important to the dynamics of the aquifer-stream system. Aquifer-stream systems in semiarid areas are in delicate balance with respect to recharge and discharge. Human activities that affect these systems under semiarid hydrologic conditions have changed parts of some streams associated with the aquifers into intermittent streams and have changed the natural recharge-discharge characteristics of the associated aquifers.

Most streams are dynamic water bodies. They are subject to rapid and sudden changes as reflected by marked natural variation in flow from season to season and year to year. Because of the hydrologic interrelation between the aquifer and the stream, these changes affect the related aquifer both directly and indirectly. Flooding, scouring, sediment transport, sediment deposition, and natural changes in stream channel and streambed all affect the hydrogeology of the combined system and contribute to the dynamic aspects of the aquifer-stream systems.

## UNIQUE HYDROLOGIC ASPECTS OF AQUIFER-STREAM SYSTEMS

Aquifer-stream systems have some hydrogeologic aspects that are not common to most other aquifer systems. These unique hydrogeologic aspects are related to the dynamics of these systems. Knowledge about these aspects has been derived chiefly from studies undertaken to evaluate the water resources of aquifer-stream systems and to evaluate the scientific factors controlling the development of aquifer-stream systems. Work by ground-water scientists, such as C. V. Theis, M. I. Rorabaugh, J. G. Ferris, and M. S. Hantush, as well as many others, has laid the foundation for understanding the dynamics of these systems.

### Permeability of streambed and banks

The movement of water from an aquifer to a stream and from a stream to an aquifer is dependent on the permeability of the streambed and banks. This permeability is controlled to some extent by changes in flow of the stream and the characteristics of the bedload. Those sections of the streambed that are undergoing continuous scour can be expected to be the most permeable. During periods of low flow, the chances are greatest for deposition of fine-grained material; that is, silt and clay and accumulations of organic matter. This material will, in turn, decrease the permeability of the streambed and, therefore, the rate of water movement from the aquifer to the stream or from the stream to the aquifer.

The geometry of the streambed and the depth to which the streambed is incised into the aquifer also affect the movement of water from the aquifer to the stream and from the stream to the aquifer. The greater the wetted perimeter of the stream, the more surface area for movement of water to and from the stream. The more steeply sloping the channel side, the less likely the permeability of that part of the channel will be decreased by fine-grained deposits. The deeper the stream channel is eroded into the aquifer, the more aquifer flow the stream will directly intersect and the less effect partial penetration of the channel will have on the quantity of intersected flow.

### Hydraulic-head difference

Movement of water between aquifers and streams also is affected by the vertical hydraulic gradient across the streambed. To move water from the aquifer to the stream, the water level in the aquifer must be higher than the water level in the stream (that is, the stage of the stream). The difference in water levels causes a hydraulic-head difference across the streambed and results in a vertical hydraulic gradient being established. In response to this hydraulic gradient, water moves from the aquifer through the streambed into the stream. The quantity of water flowing from the aquifer to the stream is a function of the vertical hydraulic conductivity of the streambed, the thickness of the streambed deposits, the hydraulic-head difference between the water level in the aquifer and that in the stream, and the area of the streambed across which the hydraulic-head difference occurs.

The hydraulic-head difference between the aquifer and the stream can change abruptly within a short time as the result of storm runoff to the stream. A rapid increase in stream stage reverses the hydraulic gradient between the aquifer and the stream (reverses the relation between the water levels in the aquifer and in the stream) and reverses the flow. In aquifer-stream systems in semihumid and humid areas, these changes between recharging and discharging conditions can occur frequently as a result of natural climatic conditions.

### Sensitivity of aquifer-stream systems to change in hydrogeologic relationships

Aquifer-stream systems are sensitive to a number of natural geologic processes; that is, processes that can change the hydrogeologic relationships between the aquifer and the stream and their recharge-discharge interrelations. Most aquifer-stream systems are relatively narrow in width. Much of the area underlain by the aquifer may be the flood plain of the associated stream.

Under moderately high-flow conditions, only the lower part of the flood plain may be flooded. This type of flooding may cause erosion of fine material and deposition of coarse-grained material in the active channel, thereby improving the recharge-discharge relationship. On rarer occasions, a relatively significant hydrologic event may cause flooding of most of the area underlain by the aquifer. This type of flooding may have significant effects on the aquifer-stream system. Fine-grained material may be deposited over a large area of the aquifer, and the channel may be altered; that is, the channel may be straightened or its course changed, and the hydrogeologic characteristics of the streambed material changed. Under these conditions, the recharge-discharge relations may be changed. These changes could be significant, particularly if the aquifer was mostly unconfined prior to the major flooding.

Sedimentation and scouring the active channel can change the recharge-discharge relations of the aquifer-stream systems. The streambed is an active component of the stream system. The exchange of water between the two components of the flow system is a direct function of the hydraulic conductivity of the streambed and its thickness. Both of these hydrologic characteristics can undergo rapid change as a result of changes in flow conditions that may result in scour in one part of the channel and deposition in another part. Insight into the fluvial processes that affect deposition and scouring in stream channels and flood plains, and, therefore, that affect recharge and discharge relations of aquifer-stream systems, can be found in Leopold and others (1964, p. 198–332).

Although the principal purpose of this volume is to emphasize natural hydrogeologic processes and principles, it is important to note that aquifer-stream systems are particularly sensitive to human activities. Streams historically have been a focal point for settlement. The stream valleys have been sites for cities and for agricultural and industrial development; the flood plains and the associated streams have been used for disposal of wastes. Because of the interrelation between the aquifer and the stream and the dynamics of the flow system, aquifer-stream systems are among the most sensitive of the hydrogeologic systems to human activities, particularly those related to changes in water quality. Although aquifer-stream systems are of limited areal extent (long but narrow) in comparison to most aquifer systems, they are among the most intensely used and, historically, ancient and modern, most important to people and their development, both culturally and economically.

### Sinusoidal hydraulic-head fluctuations

When an increase in flow takes place in a stream because of precipitation, this change in flow causes a rise in the stage of the stream. This rise continues until the inflow to the stream begins to decrease. The rise then peaks, and the stage gradually declines to a level near the previous stage. This rise and decline produces a sinusoidal-like hydraulic-head change, a change that may take place within less than a day to several weeks. The sinusoidal

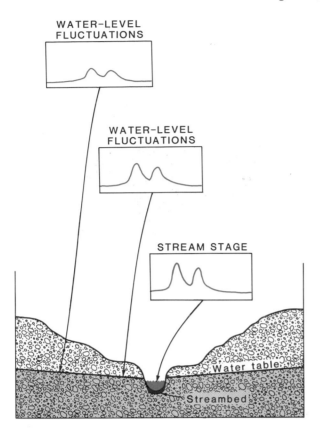

Figure 7. Water-level changes in an aquifer owing to changes in stage of an associated stream.

fluctuation is transmitted to the associated aquifer, and the water level in the aquifer responds in a similar manner to that of the stream, with the magnitude of increase in hydraulic head decreasing with distance from the stream. In addition, the fluctuations of water level in the aquifer are delayed with distance from the stream (Fig. 7). As has been previously indicated, the magnitude of the fluctuation in the aquifer is a function of the hydrogeologic characteristics of the aquifer, which are in turn related to the hydraulic conductivity, thickness, and storage characteristics of the materials making up the aquifer. A number of ground-water scientists have addressed the determination of hydrogeologic properties through analysis of water-level changes in an aquifer as the result of changes in stage of an associated stream. These changes are characteristic of aquifer-stream systems, in particular. Changes in stage in a stream and the relations of those changes to water levels also have been used to evaluate bank-storage effects, as well as to obtain regionalized values of transmissivity and storage, and estimates of evapotranspiration from aquifer-stream systems.

The analysis of fluctuations in water levels in an aquifer as the result of changes in stage of a stream can be a useful tool to help the geologist unravel depositional characteristics or litho-

logic composition of deposits, or both. The use of this type of analysis along with aquifer tests can be likened to use of more traditional geophysical techniques (seismic and resistivity methods) to aid in solution of geologic problems. These hydrogeologic tools can provide information on boundary conditions, areal extent of an aquifer, and aquifer properties; thereby providing insight into gross lithologic characteristics, streambed characteristics, as well as a host of other geologic information useful not only to ground-water geology but also to basic geology and economic and engineering geology.

As an example, Ferris and others (1962, p. 133–135) provided the basis for use of cyclic fluctuations in a stream and response of water levels in an associated aquifer to determine hydrogeologic properties of an aquifer. The equations they developed related fluctuation in aquifer water levels to stream-stage fluctuation, distance from the stream to point of observation in the associated aquifer, period of stage fluctuation, storage coefficient of the aquifer, and transmissivity of the aquifer. Ferris and others (1962) also developed an equation that related the time lag between the occurrence of the maximum or minimum water level in an aquifer after a similar occurrence in the associated stream and the distance to the stream contact with the aquifer—the subcrop of the aquifer offshore. In an earlier publication, Ferris (1951) gives details on the development of the equations, presents a graphic solution for application of the method, and provides an example of the application of the method to a relatively complex hydrogeologic aquifer-stream system. In spite of the limiting assumptions to the method, Ferris (1951) was able to obtain useful estimates of the range in transmissivity and the distance from the observation point in the aquifer to the effective offshore distance to the aquifer subcrop.

More sophisticated approaches also are available in the ground-water literature for determining aquifer characteristics using alluvial-aquifer response to fluctuations in river stage. The potentially useful application of this methodology to solving hydrogeologically related problems has been further indicated in papers by Grubb and Zehner (1973), Pinder and others (1969), and Stallman and Papadopulos (1966).

### Viscosity of water

In analyzing of properties of aquifers, the effects of changes in viscosity of water on the hydraulic conductivity generally are not taken into consideration. Part of the reason is the fact that the temperature of water in an aquifer only varies through a small range, and the effect on the hydraulic conductivity usually is small. However, in the case of aquifer-stream systems, this effect can be significant, particularly with respect to movement of water from the stream to the aquifer. On the other hand, the temperature of stream water will vary through a considerable range from season to season. Changes in temperature can result in a marked decrease in viscosity during warm weather and a marked increase in viscosity during cold weather. The changes in viscosity can, in turn, markedly affect the hydraulic conductivity of the streambed.

Therefore, temperature changes affect the flow both from the stream to the aquifer and from the aquifer to the stream. It is not uncommon for the temperature in a stream to range from near freezing to 30°C for an extended period. Based on the kinematic viscosity of water at these temperatures, the hydraulic conductivity of the streambed could vary by almost a factor of two. In other words, at higher temperatures, nearly twice as much water could move from the stream through the streambed to the aquifer, or vice versa, than at lower temperatures, assuming other factors affecting flow, including the hydraulic gradient, remain constant.

## HYDROGEOLOGY OF SELECTED AQUIFER-STREAM SYSTEMS

Space precludes a comprehensive discussion of the hydrogeology of aquifer-stream systems of Region 18. However, a few aquifer-stream systems have been selected for highlighting (Fig. 1). The aquifer-stream systems selected demonstrate the principles and concepts discussed in the preceding part of the chapter, particularly those hydrogeologic aspects unique to aquifer-stream systems. All of these aquifer-stream systems are used extensively for water supply, although use of the Ohio River valley alluvial-aquifer system in the Louisville, Kentucky, area has decreased markedly since the 1950s. The Potowomut-Wickford aquifer system of New England was chosen because its hydrogeology characterizes many of the glaciofluvial aquifer-stream systems in the glaciated part of North America. The Ohio River valley alluvial-aquifer system was chosen because it demonstrates many features typical of significant proglacial alluvial-valley aquifer-stream systems and in particular hydrogeologic aspects related to river infiltration. The Arkansas River alluvial-valley aquifer-stream system was chosen because its hydrogeology characterizes aquifer-stream systems that traverse the full range of climatic conditions from semiarid to humid, with particular emphasis on the hydrogeologic aspects of this aquifer-stream system in a semiarid climate.

### New England aquifer-stream system

Many of the alluvial valleys in New England contain aquifer-stream systems. These alluvial-valley aquifers are not as extensive as those in other parts of Region 18; that is, the valleys are not as long. As in much of the remainder of New England, many of the alluvial valleys in Rhode Island are underlain by aquifers hydraulically connected with the associated streams that traverse the alluvial valleys. The alluvial-valley aquifers generally are bounded at their edges by almost impermeable igneous and metamorphic bedrock or bedrock covered by glacial till.

The Potowomut-Wickford aquifer system in the southeastern part of Rhode Island occurs in the valley of the Potowomut River (Fig. 1, area 1). The aquifer-stream system consists of the Potowomut-Wickford aquifer and the associated Potowomut River and its tributaries, which drain the area. The hydrogeology of the aquifer-stream system (Fig. 8) has been studied by Rosen-

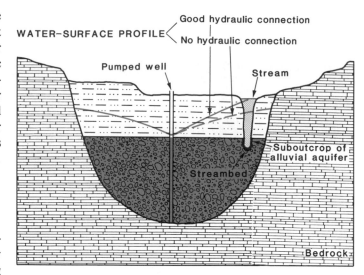

Figure 8. Idealized section showing hydrogeology of the Potowomut-Wickford aquifer-stream system.

shein and others (1968). The geologic materials underlying the area consist chiefly of: (1) stratified sand or gravel interbedded with very fine sand and silt; (2) glacial till, a poorly sorted mixture of silt, sand, and gravel with some clay; (3) stratified sand or gravel interbedded with various thicknesses of glacial till; and (4) bedrock consisting of crystalline and metamorphosed sedimentary rocks. The alluvial valley ranges in width from 1.6 to 3.2 km and is underlain by saturated material more than 30 m thick. The thickest material occurs along the western edge of the valley.

The hydraulic conductivity of the glaciofluvial material making up the Potowomut-Wickford aquifer ranges from less than 15 to more than 140 m per day. As a result, the transmissivity of the aquifer ranges from less than 120 $m^2$ per day to more than 3,700 $m^2$ per day. As generally is typical of aquifer-stream systems, the maximum transmissivity of the aquifer occurs in relatively narrow, elongated areas that reflect the effects of deposition and bedload transport in the meltwater streams that formed the deposits.

The associated stream—the Potowomut River—may discharge as little as 43 cm of runoff annually in exceptionally dry years and more than 96 cm in exceptionally wet years. The river derives as much as 50 percent of its flow from ground-water discharge. During exceptionally dry years, ground-water discharge may be as little as 40 percent of the total discharge of the stream.

Analysis of aquifer tests using wells adjacent to the Potowomut River system indicated a good hydraulic connection between the aquifer and the streams in part of the area. The sediments composing the streambeds form a veneer over the aquifer; the thickness of the veneer averages about 0.6 m. Measured streambed hydraulic conductivity ranges from 0.03 to 4.6 m per day at 15.6°C and averages about 0.7 m per day; the smallest values are associated with dense sediment containing large proportions of organic matter. The vertical hydraulic conductivity of that part of the aquifer in contact with the streambed ranges from about 0.9 to about 6 m per day. Based on hydrogeologic relations, streambed infiltration rates were estimated by Rosenshein and others (1968, Table 6) to range from about 0.02 to about 0.03 $m^3$ per second (1,340 to 2,620 $m^3$ per day) per 1,000 $m^2$ of streambed for each 1-m difference in hydraulic head between stream stage and aquifer water levels. The range in streambed-infiltration rates is least in January and greatest in July owing to changes in viscosity of the water, related to temperature variation.

### Ohio River valley aquifer-stream system

The Ohio River valley aquifer-stream system (Fig. 1, area 2) was formed as a result of advances and retreats of continental glaciers during Pleistocene time. This aquifer-stream system is typical of systems that occur along major streams. Meltwaters from the early advances and retreats of the ice sheets changed the pre-Pleistocene drainage system and eroded a broad, deep valley in the bedrock of the present Ohio River valley (Walker, 1957). This Pleistocene valley ranges in width from 1.6 to 16 km. During the later stages of continental glaciation, glaciofluvial sand and gravel were deposited in the valley. Alluvial clay, silt, and fine sand, as much as 21 m thick, were deposited during the late Pleistocene by meltwater from the last phases of the retreating ice sheet. The channel of the Ohio River has been eroded in most places below the bottom of the fine-grained alluvium of late Pleistocene age. The streambed now is 25 to 35 m below the original level of the Pleistocene fill, and the streambed rests on the underlying glaciofluvial sand and gravel. The saturated part of this sand and gravel thickens from about 8 m at the eastern edge of area 2 to about 33 m at the western edge.

Near Louisville, Kentucky, the valley was eroded into bedrock consisting of dolomite, limestone, and shale of Ordovician and Silurian age. About 25 m of glaciofluvial sediments were deposited in the valley by meltwaters. These outwash deposits underlie about 5 to 9 m of silt, clay, and fine soil of Holocene age. The outwash deposits form the Ohio River valley aquifer and the overlying deposits form a confining layer.

The Ohio River flows through the central part of the alluvial valley and ranges in width from 900 to 2,600 m. The alluvial valley near Louisville, Kentucky, is about 2,000 m wide and is surrounded by uplands formed in limestone of Silurian and Devonian age. The river and the aquifer are in hydraulic connection, and water levels in the aquifer respond to changes in river stage as well as to other factors, such as changes in barometric pressure and precipitation. Flow from the aquifer to the river under natural conditions is about 475 $m^3$ per day$^{-1}$ per km of river length. Silting of the river bed can occur during periods of low flow and can affect the movement of water from the aquifer to the stream.

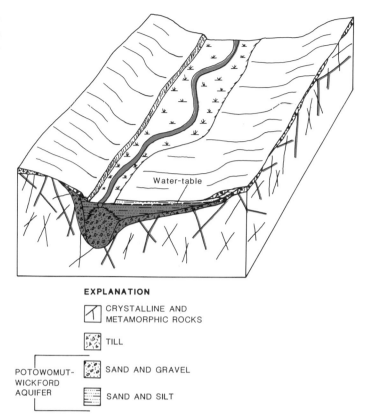

EXPLANATION

⬒ CRYSTALLINE AND METAMORPHIC ROCKS

⬚ TILL

POTOWOMUT-WICKFORD AQUIFER ⬚ SAND AND GRAVEL

⬚ SAND AND SILT

Figure 9. Idealized section showing hydrogeologic relations established by Rorabaugh (1956) confirming that the Ohio River was in good hydraulic connection with the alluvial aquifer.

The hydrogeologic properties of the aquifer-stream system, including river infiltration, were defined by Rorabaugh (1956) through use of a specially designed aquifer test. The purpose of the aquifer test was to determine if the river and the aquifer were hydraulically connected and to determine the aquifer's transmissivity and the effective distance from the river edge to where the river water entered the aquifer. Interpretation of the aquifer test by Rorabaugh established that a steeper hydraulic gradient developed on the river side toward the pumping well than on the landward side, as was to be expected according to theory (Fig. 9). This difference in gradient confirmed movement of water from the stream to the aquifer and, therefore, that a good hydraulic connection existed between the aquifer and the stream. The steep slope of the hydraulic gradient on the river side indicated that the river was a recharge boundary limiting the expansion of the water-level decline in the aquifer. The effective distance from the river bank to where the water entered the aquifer was about 120 m (distance to line source). The quantity of water moving to the aquifer varies according to the water temperature because of the effect of temperature on viscosity. A decrease in water temperature of 1°C decreases the flow by about 2.5 percent. This decrease in flow is significant because of the considerable range in river temperature.

### Arkansas River aquifer-stream system

The Arkansas River aquifer-stream system extends from the Rocky Mountains of Colorado to the confluence of the Arkansas River with the Mississippi River in Arkansas. The aquifer-stream system traverses semiarid, semihumid, and humid areas. In eastern Colorado and western Kansas, the aquifer-stream system is subjected to semiarid conditions, and in eastern Kansas, Oklahoma, and western Arkansas to semihumid and humid conditions. As such, the aquifer-stream system is affected by a considerable range of climatic and hydrologic environments. The Arkansas River flows through areas in which the average annual runoff ranges from about 0.25 to more than 40 cm (Bedinger and Sniegocki, 1976, Fig. 3, p. H4).

In eastern Colorado, much of the natural flow of the stream is derived from snowmelt from the mountainous areas in the headwaters of the stream. As a result, the aquifer-stream system is in a relatively delicate balance. Part of the time, under normal conditions, the aquifer discharges water to the stream. During extended dry weather, when aquifer discharge ceases, the stream loses water to the aquifer.

In eastern Colorado and western Kansas (Fig. 1, area 3), the valley of the Arkansas River is underlain by alluvium that ranges in thickness from less than 8 to more than 60 m. The alluvium consists chiefly of permeable sand and gravel with locally interbedded layers of silt and clay. In western Kansas, where the alluvial valley crosses the High Plains aquifer system, the aquifer-stream system locally is in hydraulic connection with the High Plains aquifer, and they form an integrated aquifer system. In the 32-km reach of the Arkansas River east of the Colorado-Kansas state line, the valley is eroded into relatively impermeable bedrock (limestone, shale, and sandstone) of Cretaceous age. The valley ranges from 4.8 to 8.0 km in width, and the alluvial fill (as much as 38 m thick) consists chiefly of well-sorted loose sand and gravel of Pleistocene and Holocene age. The hydraulic conductivity of this sand and gravel ranges from 150 to 370 m per day. The valley contains as much as 30 m of saturated sand and gravel, with the thickest saturated zone underlying the Arkansas River. The thickness of the saturated zone decreases markedly toward the valley edge. This saturated alluvial fill forms the aquifer of the aquifer-stream system. The associated stream, the Arkansas River, is in good hydraulic connection with the aquifer, as is indicated by a streambed hydraulic conductivity that ranges from about 0.2 to 0.5 m per day.

In this 32-km reach, the water level in the aquifer under natural conditions sloped toward the stream, and the aquifer discharged to the Arkansas River. During dry weather, this discharge formed most of the base flow of the stream. During extremely dry weather, discharge from the aquifer ceased, and for short periods the river had little or no flow. However, the delicate balance between the alluvial aquifer and the Arkansas River has been disturbed by human activity, and for about 24 km east of the Colorado-Kansas state line, the river has no flow in its channel for extended periods (multiple years).

Under existing hydrogeologic conditions, the aquifer no longer discharges water to the stream for significant periods. Instead, the aquifer receives much of its recharge from loss of flow in the stream. The hydrogeologic significance of this change is indicated by a computer-model analysis (Barker and others, 1983) of the aquifer-stream system in the 32-km stream reach east of the Colorado-Kansas state line. This analysis for 1975 to 1979 describes the delicate balance in which the aquifer may discharge $0.04 \text{ m}^3$ per second ($1.23 \text{ m}^3$ per year) to the stream and may receive $0.59 \text{ m}^3$ per second (18.5 million $\text{m}^3$ per year) of recharge from the river. The analysis also indicates that the aquifer-stream system is more affected by decreases in streamflow (which is regulated) than changes in average annual precipitation, indicating how sensitive this aquifer-stream system is to changes in the hydrologic environment.

## SUMMARY

Alluvial-valley aquifer-stream systems form long but narrow systems that occur in many parts of the conterminous United States and in smaller areas in Alaska and Canada. The alluvial-valley aquifers are in hydraulic connection with associated streams, and the systems are typified by the relatively rapid interaction of surface water and ground water. Although the aquifer-stream systems are of limited areal extent in comparison to most aquifer systems, they are among the most intensively used and most historically important to people and their development, both culturally and economically. The hydrogeology of aquifer-stream systems reflects a complex depositional environment. However, based on gross depositional characteristics, the sediments making up the aquifer form permeable hydrogeologic units that have, overall, a large hydraulic conductivity.

# REFERENCES CITED

Barker, R. A., Dunlap, L. E., and Sauer, C. G., 1983, Analysis and computer simulation of stream-aquifer hydrology, Arkansas River valley, southwestern Kansas: U.S. Geological Survey Water-Supply Paper 2200, 59 p.

Bedinger, M. S., and Sniegocki, R. T., 1976, Summary appraisals of the Nation's ground-water resources—Arkansas-White-Red Region: U.S. Geological Survey Professional Paper 813-H, 31 p.

Doeglas, D. J., 1962, Structure of sedimentary deposits of braided rivers: Sedimentology, v. 1, p. 167–190.

Ferris, J. G., 1951, Cyclic fluctuations of water level as a basis for determining aquifer transmissivity: International Union of Geodesy and Geophysics, International Association of Scientific Hydrology Assembly, Brussels, 1951, v. 2, p. 148–155.

Ferris, J. G., Knowles, D. B., Brown, R. H., and Stallman, R. W., 1962, Theory of aquifer tests: U.S. Geological Survey Water-Supply Paper 1536-E, 174 p.

Grubb, H. F., and Zehner, H. H., 1973, Aquifer diffusivity of the Ohio River alluvial aquifer by flood-wave response method: U.S. Geological Survey Journal of Research, p. 597–601.

Heath, R. C., 1984, Ground-water regions of the United States: U.S. Geological Survey Water-Supply Paper 2242, 78 p.

Jordan, P. R., 1977, Streamflow transmission losses in western Kansas: Proceedings, American Society of Civil Engineers, Hydraulics Division Journal, v. 103, no. HY8, p. 905–919.

Leopold, L. B., Wolman, G. M., and Miller, J. P., 1964, Fluvial processes in geomorphology: San Francisco and London, W. H. Freeman and Company, 522 p.

McGuinness, C. L., 1963, The role of ground water in the national water situation: U.S. Geological Survey Water-Supply Paper 1800, 1121 p.

Pinder, G. F., Bredehoeft, J. D., and Cooper, H. H., Jr., 1969, Determination of aquifer diffusivity from aquifer response to fluctuations in river stage: Water Resources Research, v. 5, no. 4, p. 850–855.

Rorabaugh, M. I., 1956, Ground water in northeastern Louisville, Kentucky, with reference to induced infiltration: U.S. Geological Survey Water-Supply Paper 1360-B, 169 p.

Rosenshein, J. S., Gonthier, J. B., and Allen, W. B., 1968, Hydrologic characteristics and sustained yield of principal ground-water units, Potowomut-Wickford area, Rhode Island: U.S. Geological Survey Water-Supply Paper 1775, 38 p.

Rust, B. R., 1972, Structure and process in a braided river: Sedimentology, v. 18, p. 221–245.

Selley, R. C., 1982, An introduction to sedimentology: New York, Academic Press, 408 p.

Stallman, R. W., and Papadopulos, I. S., 1966, Measurement of hydraulic diffusivity of wedge-shaped aquifers drained by streams: U.S. Geological Survey Professional Paper 514, 50 p.

Visher, G. S., 1965, Use of vertical profile in environmental reconstruction: American Association of Petroleum Geologists Bulletin 49, p. 41–61.

Walker, E. H., 1957, The deep channel and alluvial deposits of the Ohio valley in Kentucky: U.S. Geological Survey Water-Supply Paper 1411, 25 p.

MANUSCRIPT ACCEPTED BY THE SOCIETY MARCH 6, 1987

The Geology of North America
Vol. O-2, Hydrogeology
The Geological Society of America, 1988

Chapter 22

# Region 19, Northeastern Appalachians

**Allan D. Randall**
*U.S. Geological Survey, P.O. Box 1669, Albany, New York 12201*
**Rory M. Francis**
*Department of Community and Cultural Affairs, P.O. Box 2000, Charlottetown, Prince Edward Island C1A 7N8, Canada*
**Michael H. Frimpter**
*U.S. Geological Survey, Suite 1001, 150 Causeway Street, Boston, Massachusetts 02114*
**James M. Emery**
*BCI Geonetics, P.O. Box 529, Airport Road, Laconia, New Hampshire 03247*

## HYDROGEOLOGIC FRAMEWORK

The Northeastern Appalachian region (Fig. 3; Table 2, Heath, this volume) consists largely of hills and highlands, which commonly reach altitudes of 200 to 600 m in the mainland interior and western Newfoundland. In most coastal areas the topography is gently rolling and much lower in altitude. Dense igneous and metamorphic bedrock predominates, but porous unmetamorphosed sedimentary rocks and soluble carbonates and evaporites underlie some valleys and lowlands. Glacial erosion and deposition produced minor changes in drainage and topography and left a nearly continuous layer of till over bedrock. Bands of stratified drift, chiefly sand and gravel, follow the larger valleys. Blankets of stratified drift, chiefly fine sand, silt, and clay, mantle some lowlands.

Figure 1 depicts typical hydrogeologic units and patterns of recharge and discharge in an idealized locality that incorporates many features of the region. Table 1 provides information on hydraulic properties of the hydrogeologic units. The most productive aquifers consist of coarse-grained stratified drift, but bedrock is more widely used and in some places provides large yields.

## GEOLOGIC CONTROLS ON HYDRAULIC PROPERTIES

### Bedrock

Fractured nonporous bedrock is the most abundant lithologic unit underlying the region (Fig. 2). It includes many types of igneous and metamorphic rock and well-indurated Paleozoic sedimentary rock that have similar hydraulic properties. Fractures control both hydraulic conductivity and storage capacity of the rock mass. Subvertical joints and faults commonly predominate (Cross, 1974; Lin, 1975) and may impart horizontal anisotropy. Fractures tend to be more extensive and permeable in homogeneous aluminum-deficient rocks than in micaceous rocks with well-developed mineral fabric, because the latter are less brittle and their weathering products have a high clay content that reduces fracture permeability. Similarly, fracture permeability tends to be greater in alkalic than in calcic igneous rocks because feldspars rich in potassium and sodium weather more slowly than those rich in calcium and produce half as much clay per volume of feldspar weathered (R. Hoag, BCI Geonetics, written communication, 1985). Above-average yields from wells that intersect the contacts between bedrock types suggests preferential fracturing along contacts (Porter, 1982).

Porous bedrock includes the Mesozoic red beds of southern New England and extensive upper Paleozoic and Mesozoic sandstone and claystone in Atlantic Canada (Fig. 2). Fractures control hydraulic conductivity, but storage in fractures is significantly augmented by primary porosity, which averages 5 percent in arkoses of New England (Heald, 1956) and 16 percent or more in sandstones of Canada (Francis, 1981; Trescott, 1969). Horizontal bedding-plane separations are commonly the major hydraulic conduits (Francis, 1981; Van de Poll, 1983; Parsons, 1972) and, in combination with lithologic layering, can create high vertical/horizontal anisotropy that strongly influences ground-water flow. Horizontal anisotropy controlled by fracture geometry also has been demonstrated (Francis and others, 1984; Peters, 1977; Ryder and others, 1981). Well yields from porous bedrock are greater than from nonporous bedrock, and some deep wells yield as much as wells in stratified drift.

Soluble bedrock includes Cambrian and Ordovician carbonate rocks in western New England and eastern New York, principally the Stockbridge Limestone, and the Mississippian Windsor Group (limestones, gypsum, salt, red beds) in Nova Scotia and vicinity. Large well yields are reported more frequently from soluble bedrock than from other bedrock types,

Randall, A. D., Francis, R. M., Frimpter, M. H., and Emery, J. M., 1988, Region 19, Northeastern Appalachians, *in* Back, W., Rosenshein, J. S., and Seaber, P. R., eds., Hydrogeology: Boulder, Colorado, Geological Society of America, The Geology of North America, v. O-2.

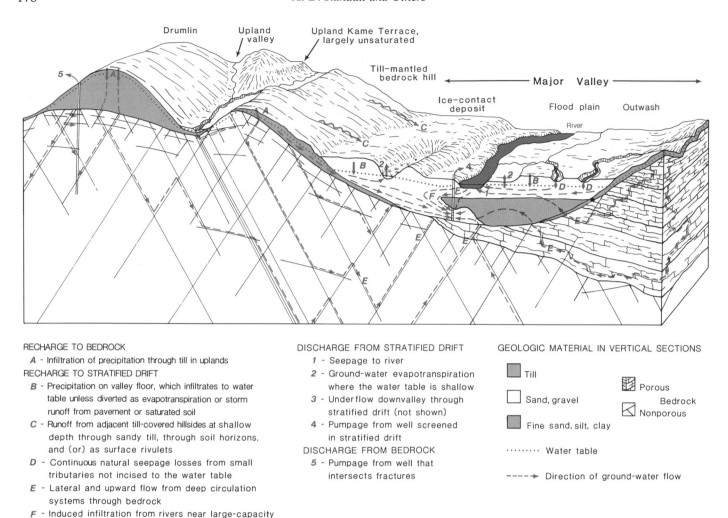

Figure 1. Idealized distribution of geologic units and ground-water flow. Not to scale; major valleys actually occupy only 5 to 30 percent of large basins.

RECHARGE TO BEDROCK

*A* - Infiltration of precipitation through till in uplands

RECHARGE TO STRATIFIED DRIFT

*B* - Precipitation on valley floor, which infiltrates to water table unless diverted as evapotranspiration or storm runoff from pavement or saturated soil

*C* - Runoff from adjacent till-covered hillsides at shallow depth through sandy till, through soil horizons, and (or) as surface rivulets

*D* - Continuous natural seepage losses from small tributaries not incised to the water table

*E* - Lateral and upward flow from deep circulation systems through bedrock

*F* - Induced infiltration from rivers near large-capacity wells, where the water table is lowered by pumping

DISCHARGE FROM STRATIFIED DRIFT

1 - Seepage to river

2 - Ground-water evapotranspiration where the water table is shallow

3 - Underflow downvalley through stratified drift (not shown)

4 - Pumpage from well screened in stratified drift

DISCHARGE FROM BEDROCK

5 - Pumpage from well that intersects fractures

GEOLOGIC MATERIAL IN VERTICAL SECTIONS

▪ Till

□ Sand, gravel

▪ Fine sand, silt, clay

▨ Porous

◩ Bedrock Nonporous

·········· Water table

-----→ Direction of ground-water flow

perhaps because of solution enlargement of fractures, or perhaps because the prevalence of fractures and permeable overburden in valleys often coincides with the distribution of soluble bedrock. Solution enlargement is probably not extensive, because sinkholes, caverns, and large springs are rare, and many wells yield no more than those in other types of bedrock.

The vertical and areal distribution of fractures are functions of stress as well as lithology; fracture distribution with depth has received the most study. Ellis (1909) measured fractures in many quarries and compiled depths and yields of wells. He concluded that fractures decrease in number and size with depth and are much less abundant below 75 m depth. Subsequent studies have drawn similar conclusions from different evidence. Gale (1975) concluded from theoretical and experimental studies that fracture frequency and aperture should decrease with depth because of increased stress. Francis (1981) demonstrated a progressive decrease in fracture aperture and frequency with increased depth to 60 m in porous bedrock on the basis of hydraulic-conductivity

profiles and logs of fractures in cores (Fig. 3). Sylvestre (1981) tested permeability at several depths in deep wells and reported a rapid decrease below the uppermost 5 m. Simard (1977) concluded from isotopic and chemical analyses and digital-model simulation of ground-water flow in southeastern Quebec that hydraulic conductivity is greatest and ground-water flow most rapid in the uppermost 15 to 35 m of the bedrock. Frimpter and Maevsky (1979) tested 12 deep drill holes that penetrated an average of 235 m of indurated sedimentary rock; their caliper and temperature logs detected 15 fractured intervals, all but two of which were within the top 50 m of bedrock. They also computed transmissivity from specific capacity at small drawdown, and if their median transmissivity (0.70 cm$^2$/s) were assumed to result solely from fractures in the top 70 m, median hydraulic conductivity would be $1 \times 10^{-4}$ cm/s, comparable to values derived by others for typical wells in fractured nonporous bedrock (Table 1).

Some evidence, however, does not support the concept of minimal fracture permeability at depth. Plots that show well yield

**TABLE 1. CONDUCTIVITY AND STORAGE PROPERTIES OF HYDROGEOLOGIC UNITS**

| Unit | | | Storativity[*] | Hydraulic Conductivity (cm/s)[†] | Method or Basis for Computing Median Hydraulic Conductivity and Storativity Values | References[§] |
|---|---|---|---|---|---|---|
| Fractured Bedrock | Nonporous: | top 30 - 170 m | 0.0002 | $3.2 \times 10^{-4}$ | Aquifer tests of 32 industrial/municipal wells; most intersect linear fracture zones by design or chance | 3, 8, 16, 17, 20, 21 |
| | | top 10 - 100 m | —— | $2 \times 10^{-4}$ | Specific capacity of 250 wells, adjusted to small drawdown | 11 |
| | Porous: | top 30 - 150 m | 0.0003 | $16.5 \times 10^{-4}$ | Aquifer tests of 52 industrial/municipal wells | 2,4,5,9,10,16,18,22,23 |
| | | top 10 - 200 m | —— | $1.1 \times 10^{-4}$ | Specific capacity of 401 wells, as reported by drillers | 14 |
| | Soluble | | - - - - - - - - - - - - - - - Insufficient Data - - - - - - - - - - - - - - - | | | |
| Stratified Drift | Coarse-grained (sand and gravel, ice-contact deposit or outwash) | | —— | $600 \times 10^{-4}$ | Aquifer tests or specific capacities adjusted for partial penetration for 157 screened wells, median aquifer thickness 17 m. | 7, 12, 13, 19 |
| | | | 0.36 | —— | 25 undisturbed samples | 1, 11 |
| | Fine-grained (fine sand, silt, and clay, marine or lake-bottom) | | 0.29 | $0.5 \times 10^{-4}$ | 6 undisturbed samples, oriented vertically, lake silts | 11 |
| | | | | $0.002 \times 10^{-4}$ | 6 undisturbed samples, marine silty clays | 14 |
| Till | Loose upper till | | 0.28 | $9.4 \times 10^{-4}$ | 10 undisturbed samples | 1, 6, 11 |
| | Compact lower till | | 0.04 | $0.2 \times 10^{-4}$ | 4 undisturbed samples (vertical) | 11 |
| | Till, not subdivided | | —— | $3.3 \times 10^{-4}$ | Slug tests of 28 wells | 6 |

[*]Specific yield for unconsolidated deposits.
[†]For well tests, transmissivity divided by distance penetrated into hydrostratigraphic unit; the resulting values describe an isotropic homogeneous medium hydraulically equivalent to the real unit, which is heterogeneous and which in bedrock may be strongly anisotropic, with flow limited to fractures or fracture zones that intersect a small fraction of borehole length.

[§]**References:**

1. Baker and others, 1964
2. Callan, 1978
3. Carr, 1967
4. Francis, 1981
5. Hennigar, 1972
6. Huntley and Black, 1979
7. Lang and others, 1960
8. Lin, 1975
9. Peters, 1977
10. Randall, 1964
11. Randall and others, 1966
12. Rosenshein and others, 1968
13. Ryder and others, 1970
14. Ryder and others, 1981
15. Sammel and others, 1966
16. Trescott, 1968
17. Trescott, 1969
18. Vaughn and Somers, 1980
19. Wilson and others, 1974
20. BCI Genetics, unpub.
21. Dunn Geoscience Corp., unpub.
22. PEI Water Res. sect., unpub.
23. U.S. Geological Survey, unpub.

or specific capacity to be inversely related to well depth appear in many reports, but such correlations probably reflect two biases: (a) deep wells are deep because little water was obtained at shallow depth, in which case low yields demonstrate merely that at some sites fractures are sparse at any depth, not that fractures become less frequent with depth; and (b) deep wells enable water levels to be drawn far below shallow fractures, in which case the calculated specific capacity is much less than the maximum value that truly represents the site. Ryder and others (1981) observed that specific capacity increased in proportion to well depth in the porous bedrock of central Connecticut, which suggests a uniform frequency of fractures to depths of at least 200 m. Some wells throughout the region have penetrated productive fracture zones (up to 2,000 L/min) at depths of 100 to 650 m in nonporous bedrock. However, no conceptual model incorporating these occurrences has yet been formulated.

The intensity of fracturing varies areally as much as it does vertically, but areal variations are poorly understood. In parts of Prince Edward Island, bedrock is so extensively and uniformly fractured that water levels respond to pumping as they do in porous media (Francis and others, 1984; Callan, 1978). In general, however, large differences in yield and response to pumping among nearby wells are commonplace. Many reports treat well yield from bedrock as a matter of random probability, and present graphs that show the percentage chance of obtaining any particular yield from a new well (e.g., Cervione and others,

1972). Others (Caswell, 1979; Hoag, 1985; Emery and Cook, 1984; Randall and others, 1966) present evidence of scattered intensely fractured zones in which many wells yield at least 100 L/min. Large quarries consistently reveal multiple zones of close fracturing separated by intervals with much greater distances between fractures (Ellis, 1909). Zones of tensional fracture and zones perpendicular to the bedrock fabric are more likely to be open and bear water than zones of shear fracture or zones parallel to the bedrock fabric.

Ground-water storage in fractures is so small that large continuous withdrawals from nonporous bedrock can be sustained only if equally large recharge is continuously available, usually from storage in the overburden. Hence, overburden type influences water availability from bedrock. Sustained yields tend to be smaller where till or silt overlie bedrock than where saturated sand and gravel overlie bedrock (Wilson and others, 1974; Emery and Cook, 1984) and should be still less where overburden is absent or unsaturated.

## Till

Till mantles the bedrock throughout the Northeastern Appalachian region, although it is discontinuous and interrupted by bedrock outcrops in many localities. Till is typically 3 to 9 m thick, but beneath stratified drift in lowlands it is generally thinner (Mazzaferro and others, 1979). Till more than 30 m thick

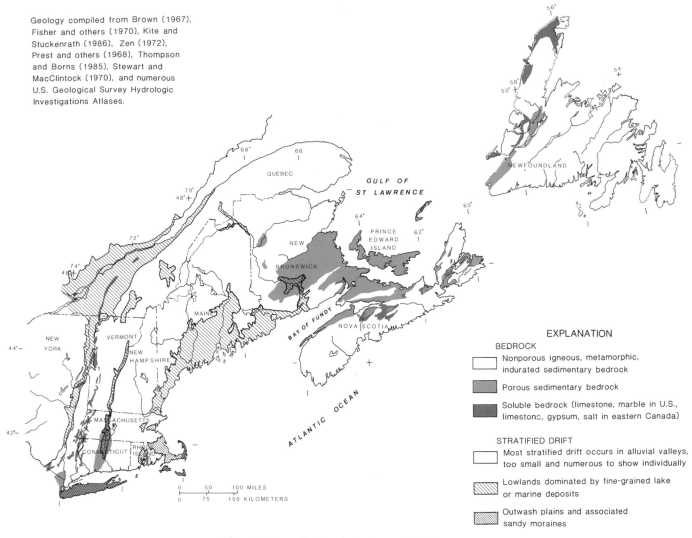

Geology compiled from Brown (1967),
Fisher and others (1970), Kite and
Stuckenrath (1986), Zen (1972),
Prest and others (1968), Thompson
and Borns (1985), Stewart and
MacClintock (1970), and numerous
U.S. Geological Survey Hydrologic
Investigations Atlases.

Figure 2. Generalized major hydrogeological terranes.

is found on scattered drumlins and the flanks of some bedrock hills in areas of moderate relief (Mazzaferro and others, 1979; Randall and others, 1966; Bradley, 1964) and in some valleys in mountainous areas (Stewart and MacClintock, 1969). Very few valleys in this region are buried or mantled by till to the point that they are unrecognizable.

Geologic studies report three widespread, discontinuous till facies. The uppermost is weakly consolidated, has a matrix chiefly of sand, and commonly contains deformed sand lenses and crude stratification; it is inferred to be a resedimented ablation deposit. Some lower till is unoxidized, well consolidated, compact, rarely stratified, and has a matrix of sand with some silt; it may be a lodgment or meltout till. Other lower till is similar in texture and lack of stratification but is deeply oxidized, generally even more compact, and abundantly jointed; these properties and other evidence point to deposition from an earlier ice advance (Pessl and Schafer, 1968; Koteff and Pessl, 1985).

Data on hydraulic properties of till in this region are scant

(Table 1). The upper till should be more permeable than the lower, compact tills, and a few permeameter tests indicate vertical intergranular hydraulic conductivity of upper till to be about 50 times that of lower till (Randall and others, 1966). However, the prominent jointing in the older lower till doubtless increases its permeability (Baker and others, 1964; Grisak and others, 1976).

In the uplands, some water supplies are obtained from till, but bedrock is the principal aquifer; the till functions as a filter and storage reservoir for the bedrock aquifer. Specific yield of till (Table 1) far exceeds specific yield of fractured nonporous bedrock, which is $< 0.005$ and probably $< 0.0005$ (Trainer and Watkins, 1975; Snow, 1968). Therefore, specific capacity of bedrock wells in uplands should decrease sharply when the water table declines below the till. Water that follows intergranular pores in till travels more slowly and contacts more mineral surfaces than water in fractured bedrock, which may explain why wells show more evidence of pollution where the drift is thin (Randall and others, 1966).

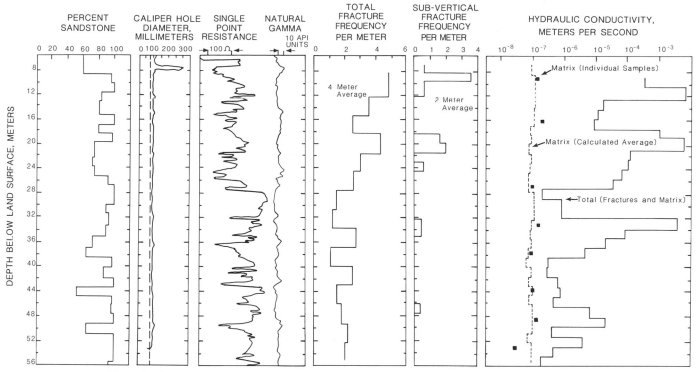

Figure 3. Lithological and geophysical logs, fracture frequency, and hydraulic conductivity profiles from corehole in porous sandstone, Prince Edward Island (Francis, 1981).

### Stratified drift

Stratified drift was transported by meltwater and deposited in bodies of standing water or, less commonly, along stream channels. Therefore, stratified drift is found chiefly in valleys or lowlands. Most of these areas contain similar types of deposits that formed successively as the ice melted, such as;

1. Ice-contact deposits, predominantly gravel and sand, that were laid down when the valley or lowland was still choked with ice. The distribution is biased toward localities of shallow bedrock (typically the valley sides).

2. Fine-grained deposits, predominantly clay, silt, and very fine sand, which were laid down somewhat later in extensive lakes or bodies of marine water. At the same time, coarse sands and gravel were deposited in deltas or subaqueous fans where streams entered the water body.

3. Outwash and alluvium, predominantly coarse sand and gravel, that were laid down late in deglaciation atop the fine-grained deposits. They are commonly thin and above present stream grade.

All these deposits occur in most lowlands, but their relative proportions differ according to physiographic setting, as described below.

*Alluvial valleys.* The ribbons of stratified drift that follow the larger valleys are ordinarily less than 30 m thick and rarely more than 120 m thick. Stratified drift commonly exhibits a succession of overlapping (shingled) depositional profiles termed

morphosequences (Koteff, 1974), each of which heads further north than its predecessor. In valleys that drained away from the ice, water was ponded behind earlier stratified drift downstream, and the downstream ends of successive morphosequences merge near lake level. In valleys that drained toward the ice, water was ponded behind saddles on the divide, and successively lower morphosequences formed as the retreating ice uncovered lower saddles. Ideally, each morphosequence grades from coarse-grained heterogenous ice-contact deposits at the head to a delta or valley train capped by pebbly coarse sand that progrades over lake-bottom silts (Stone and others, 1982). Idealized transverse and longitudinal sections are shown in Figure 4A. In narrow, shallow valleys, where ponds were small and meltwater velocities correspondingly high, coarse-grained deposits predominate; most fines were carried elsewhere. Greater depth to bedrock generally resulted in correspondingly greater thickness of very fine sand, silt, and clay; coarse-grained stratified drift rarely exceeds 30 m in thickness, regardless of depth to bedrock.

*Lake-dominated lowlands.* Several broad lowlands contained extensive, long-lived water bodies during deglaciation—either freshwater lakes created when preglacial valleys were blocked by drift, or ocean water that invaded lowlands temporarily depressed from the weight of the ice (Fig. 2). Fine-grained sediments predominated in these large water bodies, and aquifer distribution is distinctive in at least three respects:

1. The early coarse-grained ice-contact deposits tend to form widely separated multiple knolls or ridges, deposited along

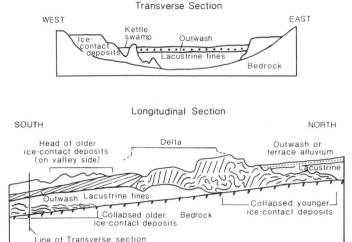

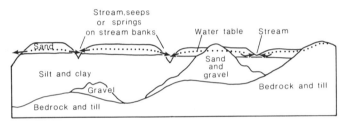

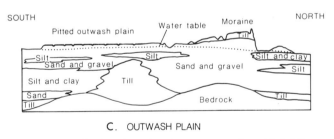

Figure 4. Idealized sections showing geometry of stratified drift.

meltwater drainage lines within the ice and along ice-frontal positions; they are overlapped and commonly buried by fine-grained sediment (Fig. 4B).

2. The fine-grained sediment ordinarily is capped by sand or fine gravel of various origins (outwash, regressive shorelines, postglacial streams). Many streams are now incised into the fine-grained sediment, leaving mesalike terraces capped by a thin aquifer that is not in contact with streams and is recharged solely by precipitation (Fig. 4B).

3. Deltas formed where large streams entered the water bodies. The best aquifers occur near the former shore, where delta foresets are thickest and may overlie older ice-contact deposits, and where recharge from stream seepage is readily available. In Maine, most deltas lie near the inland limit of marine incursion; hills near the coast protruded above the sea as islands, but streams

on these islands were too small to build deltas. Therefore, coastal Maine generally lacks "windows" where coarse sediment extends continuously from stream channel to the base of the stratified drift and provides an avenue for recharge to ice-contact deposits beneath marine sediments.

*Outwash plains.* A few areas of low relief received such large volumes of drift, deposited in part as successive end moraines but mostly as outwash beyond the moraines, that the preglacial topography was largely buried (Fig. 4C). The drift is predominantly sand-size quartz and feldspar. These areas include Long Island in New York and several areas in southeastern New England (Fig. 2). Depth to the water table is commonly substantial, reaching 70 m beneath one hill in eastern Massachusetts. Nearly all runoff consists of ground-water discharge, either to the ocean or to small streams originating within the outwash plains, and at relatively constant rates. Where outwash plains abut the ocean, the lower part of the outwash generally contains salty water, although fresh water extends 100 m below sea level beneath a clay layer on the northern shore of Cape Cod in eastern Massachusetts.

## GEOLOGIC AND CLIMATIC CONTROLS ON GROUND-WATER FLOW PATTERNS

### Position of the water table

Precipitation on the uplands can readily infiltrate the soil and sandy surficial till, but because compact till and bedrock of lesser permeability commonly lie at shallow depth, water levels in most wells open to till rise to within 2 m of land surface each spring (Frimpter, 1980), and periodic saturation virtually to land surface could probably be demonstrated in many localities (Patric and Lyford, 1980). Where the bedrock is relatively permeable, as in parts of Atlantic Canada, the till may remain unsaturated all year. Elsewhere, ephemeral streams fed by ground-water seepage appear each spring in even the smallest upland swales, then disappear in summer because of water-table declines of 2 to 6 m and consumption by evapotranspiration of all water that can flow through the till to points of discharge. Despite these variations, the water-table configuration in uplands nearly replicates the topography throughout the region. Although local transfer of ground water beneath minor topographic divides can result from sand lenses (Lin, 1975), permeable bedrock layers (Trescott, 1970), or fracture zones (Hoag, 1985), interbasin flow systems involving significant flux have not been shown to exist.

In stratified drift, the water table slopes gently toward streams regardless of local topographic irregularities; it steepens where permeability or saturated thickness is small. Water levels beneath the valley floor fluctuate with stream stage; the range of fluctuation is less than 1.5 m along all but the largest streams, and is repeated each year. Near ground-water divides and valley walls remote from streams, water levels fluctuate 1.5 to 3 m each year (Frimpter, 1980), but net gains in wet years result in a long-term range that exceeds the typical annual fluctuation.

Water levels normally reach their highest point each year during the spring freshet (March in southern New England, April–May in northern New England and Atlantic Canada, June or later at high altitudes in Newfoundland) then decline during the growing season. Slow percolation through thick unsaturated sand can delay the peak by a few weeks or even months. A secondary peak commonly occurs near year's end, followed by a midwinter decline when most precipitation is frozen. Annual minimum water levels and stream flow are reached in midwinter in a few northern highlands (Shawnigan and MacLaren, 1968). Intermittent recharge may take place in winter in coastal areas and southern New England, resulting in an irregular rise.

### Recharge and discharge

Rates of ground-water recharge and discharge have been calculated for several watersheds in the Northeastern Appalachian region, chiefly by hydrograph-separation techniques. However, most watersheds contain stratified drift, till, and one or more types of bedrock, each of which has its own pattern of recharge and discharge, so rates for individual aquifers are difficult to derive from the composite discharge hydrograph.

At least three approaches have been used to estimate recharge to stratified drift under natural conditions. (1) Some investigators defined ground-water runoff per unit area from a watershed by hydrograph separation, accounted for changes in storage, and treated the result as recharge per unit area of stratified drift (Rosenshein and others, 1968). (2) Studies in Connecticut showed that ground-water runoff as defined by hydrograph separation, increases in proportion to the percentage of the basin covered by stratified drift (Mazzaferro and others, 1979) and that recharge to stratified drift could be calculated by comparing ground-water runoff from several basins differing in relative area of stratified drift. (3) Other investigators reasoned that because nearly all precipitation on stratified drift either evaporates or infiltrates to the water table, annual ground-water runoff from stratified drift should equal annual runoff from the basin times the percentage of basin area covered by stratified drift (Sammel, 1967). Some studies added an estimate of lateral recharge from adjacent areas of till (Morrissey, 1983). This last approach is most nearly correct in concept because some runoff from uplands is recycled as recharge to stratified drift (items 2 to 4, Fig. 1), but estimates of recycled recharge generally have been conservative. Furthermore, some studies ignored discharge by underflow and (or) by evapotranspiration of ground water. Thus, all these approaches may have significantly underestimated natural recharge to stratified drift. The potential for induced recharge is discussed below in the section "Interaction of aquifers and streams."

Recharge in till-covered uplands has proved difficult to define conceptually, and more difficult to measure. Analysis of stream-flow hydrographs from southern New England suggests that average ground-water discharge in uplands is about 35 percent of runoff (Mazzaferro and others, 1979). However, studies of timing and chemistry of storm runoff in eastern Canada (Sklash

and Farvolden, 1982) and studies of the role of the capillary fringe (Abdul and Gillham, 1984) suggest that ground-water discharge is a much larger fraction of runoff. Ground water was estimated to provide 40 percent of runoff from a small upland basin in northeastern New York, where till thickness averages 24.5 m, but nearly zero percent from a nearby basin of similar size (Newton, 1983; Newton and April, 1982). The slight ground-water runoff from the second basin was ascribed to extensive bare bedrock and eolian silt and to minimal storage in the till, which averages only 2.3 m thick and thus is quickly saturated during storms. This interpretation implies that water must infiltrate into till more easily than into the underlying fractured nonporous bedrock.

From a practical viewpoint, the important question is not how much water reaches the water table but how much reaches the bedrock, which is the only aquifer in uplands from which large withdrawals are possible (by clusters of wells or wells that tap fracture zones). Recharge rates estimated by hydrograph separation commonly reflect ground-water circulation through till and thus overestimate recharge to bedrock. In areas of porous sedimentary bedrock, such as parts of Atlantic Canada (Fig. 2), the water table is often below the overlying till, and ground-water levels fluctuate in response to recharge to and discharge from bedrock. The close correlation between fluctuations in stream flow and water levels (Fig. 5) suggests that ground-water discharge from the highly fractured bedrock accounts for 70 percent of annual stream flow. Pumping ground water from this type of bedrock aquifer can reverse gradients and significantly reduce base flow. Preliminary analysis of dissolved gases and environmental isotopes in Prince Edward Island ground water suggests that local postglacial recharge has penetrated to depths of at least 150 m (R. Francis, Prince Edward Island Department of Community and Cultural Affairs, communication, 1986).

## GEOLOGIC AND HYDROLOGIC CONTROLS ON GEOCHEMISTRY

Most of the bedrock in this region, and the glacial drift derived therefrom, is nearly insoluble. Consequently, concentrations of dissolved chemical species are low, even in deep wells. Median concentrations of dissolved solids in areas of fractured nonporous bedrock in New England and eastern New York are 115 mg/L both for wells that tap bedrock and for wells that tap stratified drift (J. Dysart, U.S. Geological Survey, oral communication, 1984). Concentrations rarely exceed 1,000 mg/L, and are less than 30 mg/L in some wells distributed throughout the region.

The porous sedimentary bedrock of southern New England and Atlantic Canada contains calcium carbonate and, in some formations, traces of gypsum. This results in generally higher dissolved solids. Median dissolved solids concentrations are 220 mg/L for wells tapping the lower Mesozoic Newark Supergroup of southern New England. Water from a few deep wells in central Connecticut exceeds 2,000 mg/L dissolved solids, with sulfate as

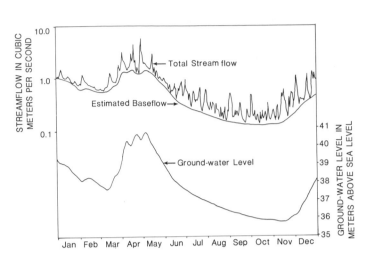

Figure 5. Flow of Winter River, Prince Edward Island, 1982, in relation to water level in bedrock.

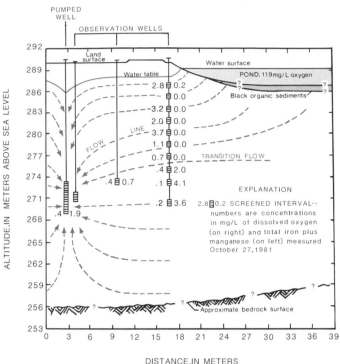

DISTANCE, IN METERS

Figure 6. Vertical distribution of dissolved iron resulting from infiltration through organic sediment in a pond in western Massachusetts. (Modified from Gay and Frimpter, 1984.)

high as 1,500 mg/L (Ryder and others, 1981; Weiss and others, 1982; Randall, 1964).

Carbonate rock of early Paleozoic age in eastern New York and western New England yields hard water containing about three times as much calcium and bicarbonate as water from noncalcareous bedrock; the median dissolved-solids concentration is 245 mg/L. Highly mineralized water with hardness of about 1,500 mg/L as $CaCO_3$ is typical of the evaporite-bearing Windsor Group in Atlantic Canada (Trescott, 1969, p. 36; Golder and Associates, 1983).

The most frequent water-quality problem in this region is excessive iron or manganese, which gives water a metallic taste and causes staining. Several reports ascribed localized high concentrations of dissolved iron to oxygen depletion resulting from ground-water flow through organic-rich sediment (Fig. 6) in alluvium, swamps, or ponds (Gay and Frimpter, 1984; Randall, 1964; Silvey and Johnston, 1977), or from oxidation of pyrite in bedrock (Randall and others, 1966).

The distribution of radon-222 and of arsenic in ground water have been studied because of possible health threats. Radon is more soluble than uranium or other radionuclides produced by decay of uranium-238, and does not combine chemically with other elements; therefore, it readily enters and remains in ground water. Natural radioactivity in ground water is about 1,100 pCi/L (picoCuries per liter) in the chlorite-grade metasediments that cover about half of Maine, but averages 22,000 pCi/L in granite and 13,000 pCi/L in sillimanite-grade metasediments of New Hampshire and southern Maine. Concentrations may differ widely among closely spaced wells but tend to be highest near pegmatites and in regional ground-water discharge areas (Lanctot and others, 1985).

High concentrations of arsenic occur locally in ground water, particularly in the Silurian Paxton Formation of New England, the Harvey porphyry of New Brunswick, and the Meguma Group in gold-mining areas of Nova Scotia. Water samples containing 110 and 560 $\mu$g/L arsenic were obtained from wells cased through thick till into the Paxton Formation beneath two drumlins in Leicester, Massachusetts. However, Boudette and others (1985) cite a lack of correlation of arsenic with sulfate, zinc, or copper, and also a positive correlation with other trace elements and with population density, as evidence that arsenic may be derived from human activities rather than sulfide minerals.

## SIGNIFICANT HYDROGEOLOGIC FEATURES

### Susceptibility to contamination

Soluble contaminants released at land surface commonly have a large and immediate influence on ground-water quality, for two reasons. (1) Most ground water in fractured nonporous bedrock or stratified drift in this region is a nonbuffered solution of low ionic strength, the product of weathering of silicate minerals. Therefore, its chemistry is easily altered by small admixtures of water of differing chemistry. (2) Most aquifers are close to land surface; very few wells have more than 60 m of casing, many less than 10 m. Although confining layers in the stratified drift are numerous, most are of small areal extent. Where overburden is thin and patchy, infiltration into and through fractured bedrock is rapid. Therefore, contaminants reach most aquifers quickly. Only beneath drumlins and other thick accumulations of till, and beneath extensive marine or lacustrine clays in a few lowlands, are aquifers well protected from surface contamination.

### Interaction between aquifers and streams

The interchange of water between stratified-drift aquifers and streams is the outstanding feature of the hydrogeology of this region.

The recharge potentially available to aquifers from streams by induced infiltration is a critical factor in the design of well fields. Stratified drift crossed by large streams can receive more recharge from this source than from all natural sources combined. Studies in Prince Edward Island have shown that recharge also can be induced by withdrawals from permeable, porous sedimentary bedrock. However, the precise measurement of stream-flow losses and substream heads required to quantify the properties that control infiltration have seldom been made in this region.

Discharge from stratified-drift aquifers to streams is equally significant. Upland till and surface storage may contribute appreciable water at near-median stream flow, especially in the relatively level uplands of Nova Scotia and Newfoundland, where extensive lakes and bogs yield water slowly to streams and sustain base flow for extended periods (Golder and Associates, 1983). In prolonged dry weather, however, when these sources dwindle and virtually cease, stream flow consists largely of discharge from stratified drift. Therefore, the extent of surficial coarse-grained stratified drift is the most significant basin characteristic that determines extreme low flow (Thomas, 1966; Mazzaferro and others, 1979). The low rate of flow from a basin mantled half by till and half by sand or gravel is exceeded 95 percent of the time and is more than 30 times that from a basin of equal size mantled entirely by till (Fig. 7). The increase in precipitation with altitude (Knox and Nordenson, 1955; Dingman, 1981) also strongly influences low flow. Permeable bedrock and thick drift in uplands favor increased low flow (Golder and Associates, 1983; Newton and April, 1982). Wetland area is negatively correlated with low flow because lakes and swamps permit evapotranspiration of water that would otherwise become stream flow. However, the effects of stratified drift and of altitude predominate.

## REFERENCES

Abdul, A. S., and Gillham, R. W., 1984, Laboratory studies of the effects of the capillary fringe on streamflow generation: Water Resources Research, v. 20, no. 6, p. 691–698.

Baker, J. A., Healy, H. G., and Hackett, O. M., 1964, Geology and ground-water conditions in the Wilmington-Reading area, Massachusetts: U.S. Geological Survey Water-Supply Paper 1694, 80 p.

Boudette, E. L., and 5 others, 1985, High levels of arsenic in the ground waters of southeastern New Hampshire; A geochemical reconnaissance: U.S. Geological Survey Open-File Report 85-202, 23 p.

Bradley, E., 1964, Geology and ground-water resources of southeastern New Hampshire: U.S. Geological Survey Water-Supply Paper 1695, 80 p.

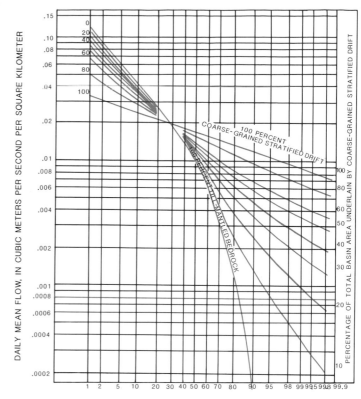

Figure 7. Duration curves showing effect of surficial geology on stream flow (modified from Thomas, 1966). These curves represent unregulated streams in Connecticut having a mean annual flow of $7.6 \times 10^{-3}$ $m^3/sec/m^2$ and are based on 1930 to 60 records.

Brown, I. C., ed., 1967, Groundwater in Canada: Geological Survey of Canada Economic Geology Report no. 24, 228 p.

Callan, D. M., 1978, Aquifer analysis, Winter River Basin, Prince Edward Island: Report prepared for Prince Edward Island Department of the Environment and for Environment Canada, 138 p.

Carr, P. A., 1967, Appalachian hydrogeologic region, *in* Brown, I. C., ed., Groundwater in Canada: Geological Survey of Canada, Economic Geology Report 24, 228 p.

Caswell, W. B., 1979, Ground-water handbook for the State of Maine: Maine Geological Survey, 145 p.

Cervione, M. A., Jr., Mazzaferro, D. L., and Melvin, R. L., 1972, Water resources inventory of Connecticut; Part 6, Upper Housatonic River basin: Connecticut Water Resources Bulletin no. 21, 84 p.

Cross, H. J., 1974, Natural and manmade variations in groundwater flow and chemistry in the Birch Cove and Sackville areas of Halifax County, Nova Scotia [M.Sc. thesis]: Halifax, Nova Scotia, Dalhousie University, 231 p.

Dingman, S. L., 1981, Elevation, a major influence on the hydrology of New Hampshire and Vermont, U.S.A.: Hydrological Sciences Bulletin, v. 26, no. 4, p. 399–413.

Ellis, E. E., 1909, Ground water in crystalline rocks of Connecticut, *in* Gregory, H. E., ed., Underground water resources of Connecticut: U.S. Geological Survey Water-Supply Paper 232, p. 54–103.

Emery, J. M., and Cook, G. W., IV, 1984, A determination of the nature of recharge to a bedrock fracture system: National Water Well Association, Eastern Regional Ground-Water Conference, 16 p.

Fisher, D. W., Isachsen, Y. W., and Rickard, L. V., 1970, Geologic map of New York: New York State Museum and Science Service, Map and chart series 15, scale 1:250,000, 5 sheets.

Francis, R. M., 1981, Hydrogeological properties of a fractured porous aquifer, Winter River basin, Prince Edward Island [M.Sc. thesis]: Waterloo, Ontario, University of Waterloo, 153 p.

Francis, R. M., Gale, J. E., and Atkinson, L. C., 1984, Characterization of aquifer zones in a fractured porous media: Proceedings, International Symposium on Groundwater Resources Utilization and Contaminant Hydrogeology, Montreal, Canada, v. 1, p. 33–43.

Frimpter, M. H., 1980, Probable high ground-water levels in Massachusetts: U.S. Geological Survey Water Resources Investigations Open-File Report 80-1205, 19 p.

Frimpter, M. H., and Maevsky, A., 1979, Geohydrologic impacts of coal development in the Narragansett basin, Massachusetts and Rhode Island: U.S. Geological Survey Water-Supply Paper 2062, 35 p.

Gay, F. B., and Frimpter, M. H., 1984, Distribution of polychlorinated biphenyls in the Housatonic River and adjacent aquifer, Massachusetts: U.S. Geological Survey Open-File Report 84-588, 34 p.

Gale, J. E., 1975, A numerical, field and laboratory study of flow in rocks with deformable fractures: Inland Waters Directorate, Water Resources Branch, Scientific Series no. 72, 145 p.

Golder and Associates, 1983, Hydrogeology of the Humber Valley area: Newfoundland and Labrador Department of Environment, Water Resources Report 2-5, 55 p.

Grisak, G. E., Cherry, J. A., Vonhoff, J. A., and Blumele, J. P., 1976, Hydrogeologic and hydrochemical properties of fractured till in the interior plains region, in Legget, R. F., ed., Glacial till: Royal Society of Canada Special Publication no. 12, p. 304–355.

Heald, M. T., 1956, Cementation of Triassic arkoses in Connecticut and Massachusetts: Geological Society of America Bulletin, v. 67, p. 1133–1154.

Hennigar, T. W., 1972, Hydrogeology of the Truro area, Nova Scotia: Nova Scotia Department of Mines, Groundwater Section Report 72-1, 127 p.

Hoag, R. B., Jr., 1985, An innovative technique for determining contaminant pathways in fractured bedrock, in Proceedings, 6th National Conference on Management of Uncontrolled Hazardous Waste Sites, Washington, D.C.: Silver Springs, Maryland, Hazardous Materials Control Research Institute, pub. no. 010085, p. 202–208.

Huntley, D. H. and Black, R. B., 1979, Determination of hydrologic parameters for glacial tills in Connecticut: University of Connecticut, Institute of Water Resources W79-07715, 18 p.

Kite, J. S., and Stuckenrath, R., 1986, Postglacial history of the upper St. John drainage basin, in Kite, J. S. and others, ed., Contributions to the Quaternary geology of northern Maine and adjacent Canada: Maine Department of Conservation Bulletin 37, 141 p.

Knox, C. E., and Nordenson, T. J., 1955, Average annual runoff and precipitation in the New England–New York area: U.S. Geological Survey Hydrologic Atlas HA-7, 6 p.

Koteff, C., 1974, The morphologic sequence concept and deglaciation of southern New England, in Coates, D. R., ed., Glacial geomorphology: Binghamton, State University of New York, Publications in Geomorphology, p. 121–144.

Koteff, C., and Pessl, F., Jr., 1985, Till stratigraphy in New Hampshire; Correlations with adjacent New England and Quebec, in Borns, H. W., Jr., and others, eds., Late Pleistocene history of northeastern New England and adjacent Quebec: Geological Society of America Special Paper 197, p. 1–12.

Lanctot, E. M., Tolman, A. L., and Loiselle, M., 1985, Hydrogeochemistry of radon in ground water [abs.]: National Water Well Association, Second Annual Eastern Regional Ground Water Conference.

Lang, S. M., Bierschenk, W. H., and Allen, W. B., 1960, Hydraulic characteristics of glacial outwash in Rhode Island: Rhode Island Hydrologic Bulletin no. 3, 38 p.

Lin, C. L., 1975, Hydrogeology and groundwater flow systems of the Smith's Cove area, Nova Scotia: Nova Scotia Department of Environment, Water Planning and Management Division Report 75-1, 44 p.

Mazzaferro, D. L., Handman, E. H., and Thomas, M. P., 1979, Water resources inventory of Connecticut; Part 8, Quinnipiac River basin: Connecticut Water Resources Bulletin no. 27, 88 p.

Morrissey, D. J., 1983, Hydrology of the Little Androscoggin River valley aquifer, Oxford County, Maine: U.S. Geological Survey Water-Resources Investigations 83-4018, 79 p.

Newton, R. M., 1983, Distribution and characteristics of surficial geologic materials in the ILWAS watersheds, in Tetra Tech; The Integrated Lake–Watershed Acidification Study, Proceedings of the ILWAS Annual Review Conference: Electric Power Research Institute, EPRI EA-2827 Research Project 1109-5.

Newton, R. M., and April, R. H., 1982, Surficial geologic controls on the sensitivity of two Adirondack lakes to acidification: Northeast Environmental Science, v. 1, no. 3–4, p. 143.

Parsons, M. L., 1972, Determination of hydrogeological properties of fissured rocks, in Proceedings, 24th International Geological Congress, Montreal, Canada, Section II, Hydrogeology, p. 89–99.

Patric, J. H., and Lyford, W. H., 1980, Soil-water relations at the headwaters of a forest stream in central New England: Harvard Forest, Petersham, Massachusetts, Harvard Forest Paper no. 22, 24 p.

Pessl, F., Jr., and Schafer, J. P., 1968, Two-till problem in Naugatuck-Torrington area, western Connecticut in Orville, P. M., ed., Guidebook for field trips in Connecticut: New England Intercollegiate Geological Conference, 60th Annual Meeting, trip B-1, p. 1–25.

Peters, L., 1977, Groundwater resources in the Chatham area, New Brunswick: New Brunswick Department of Environment, Water Resources Branch, 53 p.

Porter, R. J., 1982, Regional water resources of southwestern Nova Scotia: Halifax, Report to Nova Scotia Department of Environment, 87 p.

Prest, V. K., Grant, D. R., and Rampton, V. N., 1968, Glacial map of Canada: Geological Survey of Canada, scale 1:5,000,000.

Randall, A. D., 1964, Geology and ground water in the Farmington-Granby area, Connecticut: U.S. Geological Survey Water-Supply Paper 1661, 129 p.

Randall, A. D., Thomas, M. P., Thomas, C. E., Jr., Baker, J. A., 1966, Water resources inventory of Connecticut; Part 1, Quinebaug River basin: Connecticut Water Resources Bulletin no. 8, 102 p.

Rosenshein, J. S., Gonthier, J. B., and Allen, W. B., 1968, Hydrologic characteristics and sustained yield of principal ground-water units Potowomut-Wickford area, Rhode Island: U.S. Geological Survey Water-Supply Paper 1775, 38 p.

Ryder, R. B., Cervione, M. A., Jr., and Thomas, C. E., Jr., 1970, Water resources inventory of Connecticut; Part 4, Southwestern coastal river basin, Connecticut: Connecticut Water Resources Bulletin no. 17, 54 p.

Ryder, R. B., Thomas, M. P., and Weiss, L. A., 1981, Water-resources inventory of Connecticut; Part 7, Upper Connecticut River basin, Connecticut: Connecticut Water Resources Bulletin no. 24, 84 p.

Sammel, E. A., 1967, Water resources of the Parker and Rowley River basins, Massachusetts: U.S. Geological Survey Hydrologic Investigations Atlas, HA-247, scale 1:24,000, text 9 p.

Sammel, E. A., Baker, J. A., Brackley, R. A., 1966, Water resources of the Ipswich River basin, Massachusetts: U.S. Geological Survey Water-Supply Paper 1826, 83 p.

Shawnigan Engineering Company, Limited, and James F. MacLaren, Limited, 1968, Water resources study of the province of Newfoundland and Labrador: Atlantic Development Board, report 3591-1-68, 8 volumes.

Silvey, W. D., and Johnston, H. E., 1977, Preliminary study of sources and processes of enrichment of manganese in water from University of Rhode Island supply wells: U.S. Geological Survey Open-File Report 77-561, 33 p.

Simard, G., 1977, Isotopes naturels et systèmes d'écoulement souterrain bassin de la Rivere Eaton: Ministère des Richesses naturelles du Québec, Rapport H.G.-8, 87 p.

Sklash, M. G., and Farvolden, R. N., 1982, The use of isotopes in the study of high-runoff episodes in streams, in Perry, E. C., Jr., and Montgomery, C. W., eds., Isotope studies in hydrologic processes: DeKalb, Northern Illinois University Press, p. 65–73.

Snow, D. T., 1968, Rock fracture spacings, openings, and porosities: Journal Soil Mechanics and Foundation Division of American Society of Civil Engineering no. Sm-1, p. 5736.

Stewart, D. P., and MacClintock, P., 1969, The surficial geology and Pleistocene history of Vermont: Vermont Geological Survey Bulletin no. 31, 251 p.

—— , 1970, Surficial geologic map of Vermont: Vermont Geological Survey, scale 1:250,000, 1 sheet.

Stone, J. R., Schafer, J. P., and London, E. H., 1982, The surficial geologic maps of Connecticut illustrated by a field trip in central Connecticut, *in* Joesten, R., and Quarrier, S. S., eds., Guidebook for fieldtrips in Connecticut and south-central Massachusetts: New England Intercollegiate Geological Conference, 74th Annual Meeting, Connecticut Geologic and Natural History Survey Guidebook 5, p. 5–29.

Sylvestre, M., 1981, Perméabilité dans les milieure fracturés: Ministèr de l'Environnement du Québec, Rapport H.G.-14.

Thomas, M. P., 1966, Effect of glacial geology upon the time distribution of streamflow in eastern and southern Connecticut: U.S. Geological Survey Professional Paper 550-B, p. B209–B212.

Thompson, W. B., and Borns, H. W., ed., 1985, Surficial geologic map of Maine: Maine Geological Survey, 1 sheet, scale 1:500,000.

Trainer, F. W., and Watkins, F. A., 1975, Geohydrologic reconnaissance of the upper Potomac River basin: U.S. Geological Survey Water-Supply Paper 2035, 68 p.

Trescott, P. C., 1968, Groundwater resources and hydrogeology of the Annapolis-Cornwallis valley, Nova Scotia: Nova Scotia Department of Mines Memoir 6, 159 p.

—— , 1969, Groundwater resources and hydrogeology of the Winsor-Hantsport-Walton area, Nova Scotia: Nova Scotia Department of Mines, Groundwater Section Report 69-2, 58 p.

—— , 1970, Piezometer nests and groundwater flow system near Berwick, Nova Scotia: Nova Scotia Department of Mines, Groundwater Section Report 70-1, 29 p.

Van de Poll, H. W., 1983, Geology of Prince Edward Island: Prince Edward Island Department of Energy and Forestry, Energy and Minerals Branch, Report 83-1, 114 p.

Vaughn, J. G., and Somers, G. H., 1980, Regional water resources, Cumberland County, Nova Scotia: Water Planning and Management Division, Nova Scotia Department of Environment, 81 p.

Weiss, L. A., Bingham, J. W., and Thomas, M. P., 1982, Water-resources inventory of Connecticut; Part 10, Lower Connecticut River basin, Connecticut: Connecticut Water Resources Bulletin no. 31, 85 p.

Wilson, W. E., Burke, E. L., and Thomas, C. E., Jr., 1974, Water resources inventory of Connecticut; Part 5, Lower Housatonic River basin: Connecticut Water Resources Bulletin no. 19, 79 p.

Zen, E-An, 1972, A lithologic map of the New England states: U.S. Geological Survey Open-File Report 72-458, 17 sheets.

MANUSCRIPT ACCEPTED BY THE SOCIETY JULY 14, 1987

The Geology of North America
Vol. O-2, Hydrogeology
The Geological Society of America, 1988

## Chapter 23

# *Region 20, Appalachian Plateaus and Valley and Ridge*

**Paul R. Seaber***
*U.S. Geological Survey (retired)*
**J. V. Brahana and E. F. Hollyday**
*U.S. Geological Survey, Federal Courthouse, Nashville, Tennessee 37203*

## INTRODUCTION

The Appalachian Plateaus and Valley and Ridge Region is part of the Appalachian Sector, which lies in a long, narrow, curving band extending from Newfoundland, Canada, to central Alabama, United States (Heath, Fig. 3, Table 2, this volume). The region is subdivided into three subregions on the basis of hydrogeologic characteristics: the Valley and Ridge, the Appalachian Plateaus, and the Interior Low Plateaus (Fig. 1). Boundaries of the subregions used in this chapter coincide closely with the boundaries of three physiographic provinces of the same name designated by Fenneman (1938), Fenneman and Johnson (1946), and the U.S. Geological Survey (1970), except that herein the northern boundary of the region is the southern limit of significant Pleistocene glaciation.

## REGIONAL HYDROGEOLOGY

Geologically, the Appalachian Plateaus and Valley and Ridge Region encompasses part of two major tectonic domains: (a) the southern extent of the Appalachian Basin (Schneider and others, 1965, sheets 1–11; and Wyrick, 1968, sheet 2), and (b) the southeastern part of the Eastern Interior Basin (Craig and others, 1979, Fig. 10). The Appalachian Basin occurs in the eastern two-thirds of the region, and includes the Valley and Ridge Subregion and the Appalachian Plateaus Subregion. The Eastern Interior Basin occurs in the western part, and includes the Interior Low Plateaus Subregion (Fig. 1). The areal distribution of rocks in the region is shown on the geologic map of the United States by King and Beikman (1974), and the structural and tectonic configuration is shown on the tectonic map of North America (King, 1969).

The hydrogeologic framework is based on the generalized stratigraphic succession described in Table 1, with indurated sedimentary rocks of Paleozoic age forming the predominant units.

These rocks are thickest (10,000 to 12,000 m) in the eastern part of the region (Schneider and others, 1965, sheet 3) and thinnest (1,800 to 2,400 m) in the western part (Schwalb, 1982, Fig. 7). Paleozoic rocks outcrop at many locations. Where these rocks are not exposed at land surface, they are commonly overlain by a relatively thin layer of regolith (generally less than 30 m) or alluvium, and in the northern part of the region, by areally restricted deposits of glacial origin.

Compared to the more prolific regional aquifers of North America, aquifers in this region typically yield less water to wells. Although some aquifers within this region are highly permeable and porous on a local or intermediate scale, few are permeable on a regional scale of hundreds to thousands of square kilometers.

Ground-water flow paths are typically short, commonly extending no more than several tens of kilometers in their longest dimension. Because permeability in most of the indurated Paleozoic rocks is secondary, it decreases with depth. Thus, most of the active dynamic flow systems are formed within 100 m of land surface where permeability is greatest. Conceptual models of ground-water framework and occurrence are shown for each subregion (Fig. 1).

### *Valley and Ridge Subregion*

The Valley and Ridge Subregion is underlain mainly by Cambrian to Pennsylvanian sedimentary rocks (Table 1) that crop out in long, narrow belts of northeast-trending ridges and valleys (Schneider and others, 1965, sheet 3). Lithologically, the rocks are predominantly clastics and carbonates, and include sandstone, conglomerate, shale, siltstone, dolomite, and limestone.

Structurally, the subregion has been intensely deformed. Sedimentary rocks were tilted and faulted into a series of disharmonic sheets thrust from several kilometers to several tens of kilometers west or northwestward. At the surface, the northern part of the subregion is dominated by folds, whereas the southern part is dominated by thrust faults. In places in the southern part, faults are so closely spaced that no intervening folds are preserved

*Present address: Illinois State Geological Survey, Natural Resources Building, 615 Peabody Drive, Champaign, Illinois 61820.

Seaber, P. R., Brahana, J. V., and Hollyday, E. F., 1988, Region 20, Appalachian Plateaus and Valley and Ridge, *in* Back, W., Rosenshein, J. S., and Seaber, P. R., eds., Hydrogeology: Boulder, Colorado, Geological Society of America, The Geology of North America, v. O-2.

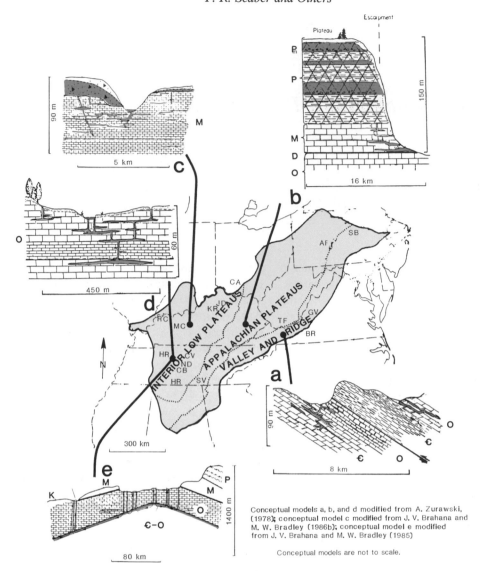

Conceptual models a, b, and d modified from A. Zurawski,
(1978); conceptual model c modified from J. V. Brahana and
M. W. Bradley (1986b); conceptual model e modified
from J. V. Brahana and M. W. Bradley (1985)

Conceptual models are not to scale.

# EXPLANATION

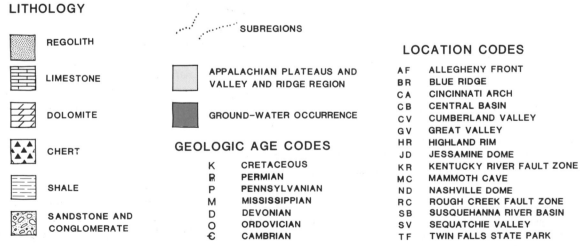

**LITHOLOGY**

- REGOLITH
- LIMESTONE
- DOLOMITE
- CHERT
- SHALE
- SANDSTONE AND CONGLOMERATE

............... SUBREGIONS

APPALACHIAN PLATEAUS AND
VALLEY AND RIDGE REGION

GROUND-WATER OCCURRENCE

**GEOLOGIC AGE CODES**

| K | CRETACEOUS |
| R | PERMIAN |
| P | PENNSYLVANIAN |
| M | MISSISSIPPIAN |
| D | DEVONIAN |
| O | ORDOVICIAN |
| Є | CAMBRIAN |

**LOCATION CODES**

| AF | ALLEGHENY FRONT |
| BR | BLUE RIDGE |
| CA | CINCINNATI ARCH |
| CB | CENTRAL BASIN |
| CV | CUMBERLAND VALLEY |
| GV | GREAT VALLEY |
| HR | HIGHLAND RIM |
| JD | JESSAMINE DOME |
| KR | KENTUCKY RIVER FAULT ZONE |
| MC | MAMMOTH CAVE |
| ND | NASHVILLE DOME |
| RC | ROUGH CREEK FAULT ZONE |
| SB | SUSQUEHANNA RIVER BASIN |
| SV | SEQUATCHIE VALLEY |
| TF | TWIN FALLS STATE PARK |

Figure 1. Location of the Appalachian Plateaus and Valley and Ridge Region, including generalized
conceptual models of ground-water occurrence in each of the three major subregions.

TABLE 1. GENERALIZED HYDROGEOLOGIC FRAMEWORK

| Age | Dominant Lithologies | Hydrogeologic Classification* | Occurrence Within Subregion[†] | | | Comments[§] |
|---|---|---|---|---|---|---|
| | | | VRS | APS | ILPS | |
| Pleistocene-Recent | Nonindurated sand, gravel, clay of glacial and alluvial deposits. | Aquifers | O[‡] | O | O[‡] | Terrace and alluvial deposits along streams; in Pennsylvania, Indiana, and Ohio occurs as outwash and drifts of glacial origin. |
| Pennsylvanian-Permian | Clastic rocks, shaley in the upper part, and coarse-grained sandstones and conglomerates in lower part with shales interbedded. | Aquitards and aquifuges (upper) Aquifers and aquitards (lower) | O | O• | O | Outcrop area of permeable aquifers is limited by overlying fine-grained rocks. |
| Mississippian | Siliceous clastic rocks in northern part of region; carbonate rocks, commonly limestone with chert in the south. | Aquifers, aquitards | O | O | O• | Where carbonate rocks have been exposed at land surface for long geologic time, they have developed into permeable aquifers. Karst features are common. |
| Devonian | Clastic rocks; shale, siltstone, sandstone. | Aquitards, aquifers | O• | O[‡] | O | Coarse-grained rocks are aquifers—fine-grained rocks are aquitards. |
| Silurian-Devonian | Carbonate rocks; limestone, siltstone. | Aquitards, aquifers | O• | O[‡] | O | Aquifers where dissolution has been active; aquitards elsewhere. |
| Silurian | Clastic rocks; shale, siltstone, sandstone. | Aquitards, aquifers | O• | O[‡] | O | Coarse-grained rocks are aquifers; fine-grained rocks are aquitards. |
| Cambrian-Ordovician | Carbonate rocks; dolomite, limestone and shale. | Aquifers, aquitards | O• | O[‡] | O• | Form some of most transmissive aquifers of VRS. May be aquifers in ILPS, but commonly aquitards. Karst features are common. |
| Cambrian | Clastic rocks; siliceous, detrital siltstone, sandstone. | Aquitards, aquifers | O• | S | S | Permeability is greater where fracturing has occurred. Little is known about these aquifers in APS and ILPS. |
| Precambrian | Crystalline metamorphic and igneous rocks. | Aquifuge | S | S | S | Form lower boundary of ground-water flow through region. |

*See Seaber, this volume.

[†]O = outcrop; S = subsurface; • = dominant outcrop within subregion; [‡] = minor outcrop within subregion.

[§]VRS = Valley and ridge subregion; APS = Appalachian Plateaus subregion; ILPS = Interior Low Plateaus subregion.

(King, 1969, p. 60). At depth, structure in the northern part of the subregion is similar to structure in the southern part, with surfaces of décollement occurring in various incompetent shale units that underlie the disharmonic structure above.

The topography of this subregion is characterized by a sequence of ridges and valleys that are controlled by the structure and the weathering characteristics of the different lithologies. The ridges generally are underlain by folded and faulted resistant rocks such as sandstone, cherty limestone and dolomite, and conglomerate; the valleys generally are underlain by nonresistant rocks, such as limestone, shale, and dolomite; the flanks generally are underlain by siltstone, shale, or other rocks of intermediate resistance. This subregion is topographically lower than the Blue Ridge Subregion, which lies to the east (see LeGrand, this volume), and the Appalachian Plateaus Subregion, which lies to the west. Altitudes are commonly in the range of 200 to 500 m above sea level, and local topographic relief seldom exceeds 200 m.

The conceptual model of ground-water occurrence in this subregion is shown in Figure 1a. The repeating lithology in combination with dip-oriented streams effectively compartmentalize most of the ground-water flow into adjacent, but isolated, shallow flow systems. Exceptions are the deep-circulating ground water of warm springs defined later in this chapter. Most ground-water flow is from ridge to valley, usually across strike, until the water either discharges directly to local streams or is intercepted and routed along strike by a highly permeable layer or zone. These highly permeable layers or zones typically are coarse-grained carbonate rocks with well-developed secondary permeability or are permeable fracture zones that act as collectors or conduits. The flow components that occur along strike commonly discharge to a spring or a master drain in the valley (Parizek and others, 1971; Hollyday and Goddard, 1979, p. 20). Some of the large springs flow at an average rate of as much as 245,000 m$^3$ per day (Sun and others, 1963, p. 11). Because flow is limited to dissolution openings in relatively impermeable carbonate rock, discharge commonly is concentrated at a few large springs rather than at many small springs or seeps.

The largest ground-water supplies (indicative of large ground-water storage) are produced from soluble carbonate rocks, especially where they are associated with thick regolith. In fact, thick regolith is important as a storage reservoir throughout the entire region, although areal variability of regolith thickness is common (DeBuchananne and Richardson, 1956). Regolith stores recharge that would otherwise be rapidly diverted to overland runoff, and slowly releases this water to underlying Paleozoic aquifers. Because of the (1) widespread distribution of carbonate rocks and associated regolith, (2) abundant precipitation in a humid climate, and (3) relatively steep hydraulic gradients, the Valley and Ridge Subregion is one of the major karstlands in the eastern United States.

Paleozoic rocks of this subregion are characterized by low primary porosity and permeability. Localized zones of secondary porosity and permeability, concentrated in the upper 200 m from land surface, are important in storing and transmitting water

throughout not only this subregion, but the entire region; fracturing and dissolution are the major causes of secondary permeability. In general, these zones of highest porosity and permeability are located along fracture traces or in areas of ground-water discharge such as stream valleys. Swingle (1959, p. 90) hypothesized that surface faults indicate areas with deep and numerous fractures that allow deep circulation, but few data exist to test this quantitatively. Clastic shale and siltstone formations generally have the lowest porosity and permeability, the carbonate rocks have the highest, and the sandstones are intermediate. Locally, however, shales can be aquifers where they are hard and brittle enough to maintain open fractures.

Permeable regional aquifers of this subregion include the Conasauga, Knox-Beekmantown, St. Paul, Chickamauga, and Helderberg, which are primarily carbonate aquifers, and the Clinton–Red Mountain, Ridgeley, and Oriskany, which are primarily sandstone aquifers. The Knox-Beekmantown carbonate sequence is the most transmissive aquifer in the subregion, occurring in the Great Valley in Virginia, Maryland, and Pennsylvania and in the Valley and Ridge of Tennessee, Georgia, and Alabama.

As commonly recognized, the chemical character of water is related to the hydrogeology of the aquifers. In the shallow flow systems, dissolved-solids concentration in water from springs and wells commonly is in the range from 50 to 500 mg/L (Brahana and others, 1986, Table 2), suggesting these are active, dynamic flow systems that are well flushed. The most frequently occurring water type in the shallow flow system is calcium magnesium bicarbonate, and the second most frequent is calcium magnesium sulfate chloride. Magnesium concentrations are higher in water issuing from dolomite than in water issuing from limestone. The deeper, slower ground-water flow systems typically produce water types similar to those in the shallow flow systems, but have about twice the dissolved solids. These facts suggest that mixing and dissolution are important processes in ground-water flow systems of shallow and intermediate depth. Tests for oil and gas show the occurrence of brine in paleoaquifers at depths greater than 1,500 m (M. W. Bradley, U.S. Geological Survey, Nashville, Tennessee, written communication, 1987). Brines indicate that fresh-water circulation and flushing are not dominant processes at these depths.

### Appalachian Plateaus Subregion

The Appalachian Plateaus Subregion is underlain by a thick sequence of Cambrian to Permian sedimentary rocks of variable lithology (Table 1). Pennsylvanian clastics are the predominant outcropping units in the southern part of the subregion, and Pennsylvanian and Permian clastics make up most of the outcrops in the north (Schneider and others, 1965, sheet 3). Ordovician to Mississippian rocks outcrop in relatively narrow bands around the margins of the subregion, and in some of the major valleys. The Pennsylvanian and Permian rocks of this subregion consist chiefly of sandstone, conglomerate, siltstone, and shale, although coal and limestone also occur within this stratigraphic interval. Glacial outwash deposits of Pleistocene age occur at the surface in the valleys of northern Pennsylvania, and along the Allegheny and Ohio Rivers.

Structurally, the rocks of this subregion are gently folded to nearly flat lying, as contrasted to the tilted and faulted rocks of the Valley and Ridge Subregion. Fracturing and jointing are common, and although faults and broad regional folds occur in this subregion, particularly in the northern half, structure does not dominate the hydrogeology as it does in the Valley and Ridge Subregion.

The topography of this subregion can be generalized as a series of sloping, uplifted, dissected plateaus that are capped by resistant layers, commonly sandstone (Fenneman, 1938, p. 281). For the most part, this subregion occurs at higher altitudes than the surrounding subregions. Major tributaries have eroded deep, steep-sided valleys, and local relief along valleys and escarpments may be greater than 300 m. Altitudes commonly are in the range of 500 to 600 m above sea level, although maximum values exceed 1,000 m.

The conceptual model of ground-water occurrence in the Appalachian Plateaus Subregion is shown in Figure 1b (Zurawski, 1978, p. L18). Regolith is commonly thin, providing little ground-water storage, and the outcropping Pennsylvanian and Permian rocks generally have low permeability (DeBuchananne and Richardson, 1956, p. 51). Secondary permeability due to jointing and stress-release fracturing (in major valleys) accounts for most of the porosity and permeability in these rocks, inasmuch as original intergranular porosity has been destroyed. Sandstones and conglomerates form shallow fresh-water aquifers at locations where these lithologies are recharged and where they have suitable fracture permeability. Flow systems that are active and dynamic commonly occur within 100 m of land surface, and seldom below 200 m. The older Paleozoic rocks (Table 1), which are deeply buried throughout the subregion, form a sequence of aquifers, aquitards, and aquifuges about which little is known. Springs emerge along the sides of the plateaus and from the Mississippian and older limestones exposed in major valleys (Zurawski, 1978, p. L15).

The geohydrologic characteristics of the Pennsylvanian and Mississippian rocks are controlled by the predominantly fine-grained, well-indurated lithologies of this subregion (Table 1), and few permeable regional aquifers have been documented (Wyrick, 1968, sheet 2). Along the eastern edge of the subregion in Pennsylvania and West Virginia, the Pocono Sandstone of Mississippian age is a regional aquifer, especially where it has been fractured and folded. Sandstone and conglomerate aquifers of Pennsylvanian age occur in a belt in the eastern part of the subregion, but their capacity as regional aquifers is untested in much of the subregion (Scheider and others, 1965, sheet 10). More common than regional aquifers are regional aquitards, such as the Dunkard Group of Permian age, which occurs in southwestern Pennsylvania and adjacent parts of Ohio and West Virginia.

The ground-water chemistry in this subregion is variable. Water in the regolith is slightly acidic and low in dissolved solids (Zurawski, 1978, p. L20), typical of shallow, well-flushed flow systems throughout North America. Ground water in this subregion often reflects the chemical character of the aquifer matrix; for example, ground water that has come in contact with sandstone and shale containing pyrite remains soft, but is more acidic and higher in concentrations of iron and hydrogen sulfide than in nonpyritic rocks. Water in limestone or calcareous aquifers usually is a calcium magnesium bicarbonate type. Sodium chloride and other brines occur below a depth of several hundred meters in all hydrostratigraphic units, and waters above this saline water are usually hard. The depth to saline water decreases west of the Allegheny Front (Feth and others, 1965, sheet 1), suggesting that very little fresh-water flushing is occurring at depth.

### Interior Low Plateaus Subregion

The Interior Low Plateaus Subregion is characterized by outcropping rocks from Ordovician to Pennsylvanian age that are underlain by older Paleozoic sedimentary rocks (Table 1). In the northern part of the subregion, deposits of glacial and alluvial origin locally may overlie the Ordovician to Pennsylvanian rocks. Lithologically, Ordovician through Mississippian age rocks are predominantly limestones, dolomites, and calcareous shales. Evaporites are not uncommon within restricted intervals of Mississippian age rocks (Craig and others, 1979, p. 99). Rocks of Pennsylvanian age in this subregion typically are shales, coals, sandstones, and siltstones.

Structurally, the rocks of this subregion are gently dipping, and broad regional structures control the outcrop distribution. Ordovician rocks are exposed along the eroded crests of anticlines and domes, including the Cincinnati Arch, the Jessamine Dome, and the Nashville Dome (Fig. 1), and Pennsylvanian rocks are preserved in the southern part of the Illinois Basin where the basin occurs in western Kentucky and southern Indiana (Craig and others, 1979, p. 97). Areas in the subregion that occur on the flanks or away from the axes of these regional structures are characterized by nearly flat-lying rocks of Silurian through Mississippian age. Orthogonal joint sets are common, and several major fault zones occur in the subregion (Fig. 1).

The topography of this subregion is defined by the outcrop distribution, which is controlled by structure and lithology. In general, a mature, well-rounded topography with several dominant areas of karst characterizes the subregion. The Interior Low Plateaus Subregion is lower than the Appalachian Plateaus Subregion that lies to the east, but it is higher than both the Central Glaciated Plains Region to the north (Krothe and Kempton, this volume), and the Western Gulf Coastal Plain Region to the west (Grubb and Carillo, this volume). Within the Interior Low Plateaus Subregion, land surface altitudes typically are in the range of 100 to 300 m above sea level, with the rocks of Mississippian age occupying the higher altitudes.

Three conceptual models of ground-water occurrence are

required to show the diversity of conditions that have been documented in the subregion. Figure 1c (Brahana and Bradley, 1986b, p. 19) and 1d (Zurawski, 1978, p. L20) show shallow, dynamic ground-water occurrence in rocks of Mississippian and Ordovician age, respectively, and Figure 1e shows the dynamic deep flow system of the Knox Group of Cambrian and Ordovician age in west-central Tennessee (Brahana and Bradley, 1985, p. 18).

The conceptual model of ground-water occurrence from three distinct facies common in the silicenous carbonate rocks of Mississippian age is shown in Figure 1c. Mississippian rocks show highly variable development of regolith thickness (from less than 10 to more than 35 m thick), variable nature of ground-water occurrence (from diffuse to conduit), and variable development of secondary permeability and storage (from very poorly developed to very well developed). The diverse effects are thought to be caused by differences in lithology, including percent purity of the carbonate rocks and original permeability (Burchett and others, 1980, p. 395). For the most part, dynamic ground-water flow systems in rocks of Mississippian age are limited to zones of secondary permeability that are concentrated within 100 m of land surface.

The conceptual model of shallow ground-water occurrence in the Ordovician rocks of the Interior Low Plateaus Subregion is shown in Figure 1d. Ordovician carbonate rocks are generally gently dipping to flat lying. Ground water moves through shallow, solution-enlarged vertical joints and horizontal bedding plane openings that are concentrated in valleys and within 100 m of land surface (Zurawski, 1978, p. L16). Ground-water distribution in rocks of Ordovician age in this subregion is highly variable. Except for the anistropic and nonhomogeneous development of secondary openings, much of the limestone has extremely low porosity and permeability and functions as an aquitard. Large solution openings are relatively uncommon, although karst is very well developed in the Kentucky Bluegrass Region near the Jessamine Dome and moderately well developed in the southern part of the Central Basin of Tennessee, near the crest of the Nashville Dome.

The conceptual model of occurrence and movement of water in the Knox aquifer (upper 70 m of Knox Group) is shown in Figure 1e. This aquifer occurs in a buried paleokarst at depths ranging from 100 to 460 m below land surface. It is part of an active, dynamic, regional fresh-water flow system, although its consistent productivity (0.3 liter per second) to wells suggests very low but homogenous transmissivity. Little is known about the hydrogeology of the rocks below the Knox aquifer, but evidence suggests that permeability in this deep part of the section is extremely low.

The siliceous carbonate aquifer within a facies of the Fort Payne Formation of Mississippian age is the most transmissive aquifer in this subregion, occurring in the Highland Rim of Tennessee and Alabama. Other formations that have locally developed zones of high transmissivity are the St. Louis Limestone, the Ste. Genevieve Limestone, and the Girkin Limestone, all in Ken-

tucky. Many aquifers of this subregion have developed secondary permeability on a local or intermediate scale, but for the most part, they are not considered regional aquifers.

Water chemistry of the shallow carbonate rock aquifers in the subregion is characterized by a calcium magnesium bicarbonate type water that is fresh (<1,000 mg/L) to depths ranging from about 70 to 100 m (Brahana and Bradley, 1986a, 1986b), reflecting water-aquifer interaction in a dynamic, fresh-water flow system. As it does throughout the rest of the region, salinity increases with depth, suggesting greatly restricted flow in the deeper aquifers. Feth and others (1965, sheet 1) show highly saline water at relatively shallow depths in the vicinity of the Nashville Dome. This shallow saline water is thought to be part of hydrologically isolated, almost stagnant, flow cells that may be in contact with evaporites (Brahana and Bradley, 1986b, p. 5).

The Knox aquifer in the Central Basin and western Highland Rim of Tennessee is an exception to saline water occurring at depth, and dissolved solids in the Knox aquifer have concentrations typically less than 1,000 mg/L (Brahana and Bradley, 1985, p. 3). The occurrence of fresh water below saline water suggests the Knox aquifer is part of a deep regional flow system.

## GEOLOGIC INFLUENCES ON HYDROGEOLOGY

The occurrence, movement, storage, and chemical character of ground water in the region is affected by numerous geologic controls, and three of these—lithology, topography, and structure—have been selected for closer scrutiny. In the four brief sections that follow, these three controls (singly and in total combination) are evaluated with respect to the hydrogeologic parameters of permeability and porosity. Although a single dominant control has been selected in each case, the hydrogeology of each of the four studies is the sum of many interrelated geologic factors. The nondominant factors are nonetheless important in influencing the hydrogeology, even though they may not control it. The studies cited are "sample windows"; they illustrate a single concept and are greatly restricted in scope from that of the original studies.

### Lithology. Regional aquifer coefficients

Regional aquifer coefficients of rocks in the Appalachian Plateaus and Valley and Ridge Subregions provide insight into the role of lithology as it affects permeability and porosity. Two regional aquifer coefficients that quantified lithologic types in terms of their capability and variability for transmiting and yielding water were derived by Seaber and Hollyday (1965, 1966a, 1966b). These are called coefficients of aquifer capability and aquifer variability. The generalized results presented in Figure 2b are based on an analysis of specific capacity data for 56 formations (lithostratigraphic units) that were classified into five major lithologic types in the Susquehanna River basin.

The specific capacity values (L/m) for each formation were adjusted to a common time period (180 days), arranged in order

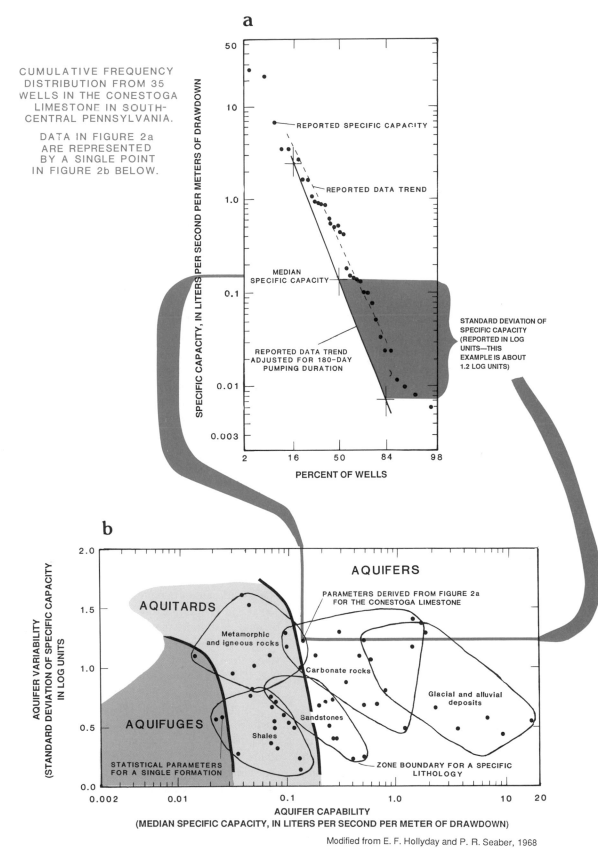

Figure 2. Relation between lithology and statistical measures of specific capacity for 56 formations in the Susquehannah River basin.

of magnitude, plotted on log-normal probability paper, and a straight line fitted to the data (Fig. 2a). The median and standard deviation of the specific capacities were determined from the log-normal distribution. The median value defines aquifer capability, and the standard deviation defines aquifer variability. A larger median value indicates higher transmissivities, and a larger standard deviation indicates a greater range in expected local transmissivities. Aquifer capability—expressed as the median specific capacity in real values or the mean specific capacity in log units—is directly related to the regional permeability and porosity. It is a measure of the volume of ground water that a rock unit will transmit in a specified time interval. Aquifer variability, expressed as the standard deviation of specific capacity in log units, is a measure of the magnitude of variation about the median value. The coefficient of aquifer variability is derived graphically, and is, in part, a measure of the heterogeneity and anisotrophy of the formation.

The five major lithologic types form distinct groupings in Figure 2b. The range of coefficients for relatively pure lithologic types essentially do not overlap. Carbonate rock coefficients do overlap with those of other rock types, but reflect the inclusion of metamorphic and sandy carbonates. In general, carbonate and metamorphic rocks have standard deviations >0.75, and sandstones and shales have standard deviations <0.75. Carbonates and sandstones generally have medians >0.20, and shales and metamorphic rocks have small medians <0.20. Glacial deposits have larger medians >0.70, but a wide range in standard deviations. Interbedded lithologies have intermediate coefficients depending on the heterogeneity of the mapped unit.

The regional aquifer coefficients are readily determinable for any rock unit for which specific capacity data are available and allow regional quantitative descriptions of the aquifer units in terms of their capability and availability for transmitting and yielding water. Lane and Lei (1950) constructed log-normal probability plots of the flow records of eastern U.S. streams and derived similar coefficients, which suggests a possible correlation between ground water, lithology, and streamflow.

Later studies in the Susquehanna River basin (Taylor and others, 1982, 1983; Taylor, 1984; and Taylor and Werkheiser, 1984) concluded, "Rock type is the most important factor in determining well yield because the occurrence of both primary and secondary porosity and permeability is ultimately controlled by lithology. Lithologic factors that control development of secondary openings consist of rock susceptibility to solution, rock susceptibility to fracturing, and the size and spacing of bedding plane openings." Topography, of course, is a significant hydrogeologic control, but in this area of fairly uniform climate, the topography is largely controlled by lithology. The lithologic distribution is largely controlled by the structure of the region.

### Topography. Stress-relief fracturing

Stress-relief fracturing in the Appalachian Plateaus Subregion provides insight into the role of topography as it affects permeability and porosity. In a study at Twin Falls State Park, West Virginia, Wyrick and Borchers (1981, p. 10–13) hypothesized that stress-relief fracturing was responsible for the development of enhanced permeability and porosity in typical valleys of the subregion. Stress-relief fracturing is an unloading phenomenon controlled not only by topography, but also by erosion and rock mechanics.

Stress relief, the removal of compressional stress on underlying rocks by erosion of overlying rocks, results in secondary fracturing and, thus, enhanced permeability and porosity in valleys. The fractures are mainly horizontal bedding-plane fractures under valley floors, and vertical and horizontal slump fractures along valley walls, where the rocks are nearly flat lying. Fractures pinch out away from the valleys. Wyrick and Borchers (1981, p. 12) explain this as follows: the buried rocks are subjected to compressional stress from the weight of the overlying rock material. After uplift, the compressional stress on rocks of a valley floor will be relieved where the overlying rocks are removed by erosion. Other rocks of the same geologic unit that form the valley floor remain under compressional stress beneath adjacent hills. One result of this imbalance of forces is the upward arching of rocks in the valley floor.

The arrows on Figure 3 indicate compressional stress and resultant stress, and the blue color highlights the orientation and occurrence of secondary permeability zones with respect to the topography. Unequal arching of beds, with shallower beds arching more than deeper beds, results in slipping and fracturing along horizontal planes of weakness. These planes of weakness are generally bedding planes, but may be any physical discontinuity in the rock. Arching also causes minor vertical fracturing near the axis of the arch (Fig. 3).

Another result of unequal stress distribution is the vertical and horizontal fracturing along the valley walls, which increases permeability and porosity (Fig. 3). Where material is eroded from a valley, its walls are subjected to unequal horizontal stresses. These stresses result in vertical tensile fractures along the valley walls. The vertical fractures allow the valley walls to slump downward, causing compressional fractures at the base The horizontal and vertical fracture systems are interconnected and thus become conduits for the movement of ground water from hills to valleys. The aquifer is under confined conditions beneath the valley floor, and under unconfined conditions along the valley wall. The valley walls form impermeable boundaries where the fractures pinch out laterally (Wyrick and Borchers, 1981, p. 30).

Wyrick and Borchers (1981, p. 6) conclude that stress-relief fractures constitute the most transmissive part of the aquifer and affect the surface-water hydrology of the valley. The unconfined aquifer is recharged along the valley walls where fractures intercept surface runoff. The confined aquifer is discharged where the fracture system occurs in the stream channel at a change in gradient at a waterfall. Runoff per square kilometer below the falls is eleven times greater during low stream flow, and six times greater during high stream flow, than it is above the falls. Wyrick and Borchers (1981) conclude the increased runoff below the falls was the result of ground-water discharge to the stream at the falls.

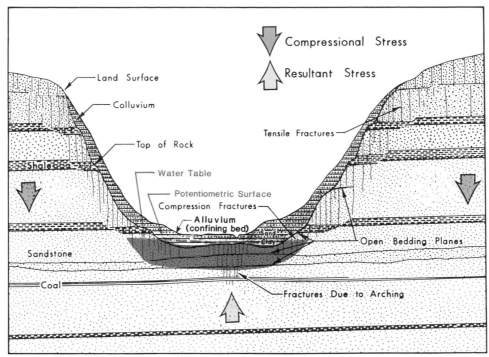

Modified from H. F. Ferguson (1974) and G. G. Wyrick and J. W. Borchers (1981, fig. 5.1.5-2)

Figure 3. Generalized geologic section showing features of stress-relief fracturing.

### Structure. Thermal springs

Thermal springs in the Valley and Ridge Subregion provide some understanding of the role of geologic structure as it affects permeability and porosity. In a study that encompassed the subregion from New York to Georgia, Hobba and others (1979) investigated the hydrology and geochemistry of thermal springs in nine areas.

In general, all warm springs have a deep component of circulation that is developed preferentially to the shallow component; favorable structure is essential for establishing a deep component of the circulation. The normal geothermal gradient provides the source of the heat, topographically high recharge areas provide the energy to drive the water through the system, and structure orients or provides enhanced vertical permeability and porosity. Meteoric water is recharged and moves deeply (from 250 to more than 1,600 m) beneath land surface, gaining heat slowly from the aquifer under "normal" geothermal gradient conditions. Fractures or folded, highly permeable zones allow the heated water to escape rapidly without thermal reequilibration as it moves from the source of the heat to the warm spring.

Warm-spring flow systems are controlled by the structure. Specifically, folding and tilting are necessary to orient existing zones of high permeability and porosity with dips that allow hydraulic continuity from land surface to depths great enough to acquire the heat. Additional structure, commonly faulting or fracturing, or suitably oriented lithology, is necessary to allow the

heated water to escape rapidly. Warm springs typically occur in valleys near the crest or flanks of an anticline and commonly discharge from sandstone and limestone at the land surface.

Two of many examples of structural conditions that favor warm springs (Hobba and others, 1979, Figs. 9 and 10) are shown in Figure 4. The first, dominated by folding, is typical of warm springs in Virginia. Examples are found in valleys at or near the crests of anticlines, where open bedding planes permit flowlines of upward-moving water to converge (Fig. 4a). Increased fracturing near the crests of the folds allows the water to move vertically across less permeable beds that, under other conditions, function as confining units. In addition to the folding and fracturing shown in Figure 4a, low-angle thrust faults at depth, vertical faults or fractures at the spring, and faults or fractures in the plane of the cross secton may also be present. Any such factors would contribute to the development of a system of conduits, permitting ground-water circulation to great depths.

The second selected example of structure associated with warm springs illustrates one role of faulting (Fig. 4b). Permeable faults can concentrate flow and transmit it along the fault plane in deep flow paths. Where fault planes intersect (Fig. 4b), rapid upward flow may result in thermal springs. Figures 4a and 4b also show that the hydrogeologic behavior of faults is quite variable, from an impermeable barrier (Fig. 4a) to a main conduit of flow (Fig. 4b).

Thermal springs are unknown along faults in the Appalachian Plateaus Subregion where the rocks are relatively flat lying

**a**

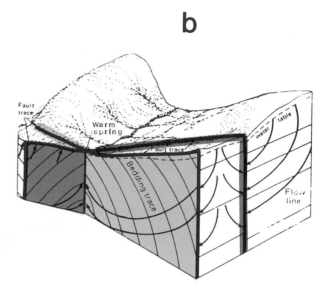

Possible movement of ground-water through multilayered folded, faulted, and fractured aquifers such as those in Warm Springs Valley, Va. Much of the flow (not indicated by arrows) is into the figure and upward beneath the hot spring.

**b**

Possible circulation of water to a warm spring located near the crest of an anticline at the intersection of two faults. Much of the flow is into the plane of the figure along open tension fractures, bedding, or fault planes and upward beneath the spring.

Modified from W. A. Hobba, Jr. and others (1979, figs. 9-10)

Figure 4. Generalized sections showing the relation of geologic structure to ground-water flow paths for selected thermal springs of the Appalachians.

and the relief is high. Similarly, many localities in the Valley and Ridge Subregion are sites of cold springs, even though these localities have similar lithology and relief as localities with warm springs. The warm springs have developed only where deeper circulation paths have developed preferentially to shallower circulation paths (Hobba and others, 1979, p. E4–E5). Deep circulation paths require enhanced vertical permeability. The fact that most springs in the area are cold indicates that truly deep flow paths are not common.

### Lithology, structure, topography. Mammoth Cave

Mammoth Cave, featuring development of conduit-type flow in a well-developed karst area of the Interior Low Plateaus Subregion, provides insight into the combined role of lithology, structure, and topography as controls on the development of secondary permeability and porosity. Based on the work of Quinlan and others (Quinlan and Rowe, 1977; Quinlan, 1981; Quinlan and Ray, 1981; Quinlan and others, 1983; Quinlan and Ewers, 1985), it has been established that Mammoth Cave and adjacent related cave systems form the largest single system of

caverns known in the world. Rocks, structure, and topography similar to Mammoth Cave exist throughout the subregion, but nowhere except Mammoth Cave are the geologic and hydrologic conditions so uniquely combined and the secondary permeability and porosity so well developed.

Lithologically, highly soluble limestones of Mississippian age (St. Louis, Ste. Genevieve, Girken), with evaporite facies in the lower part (St. Louis) occur at shallow depth. Structurally, these dip variably toward the northwest at approximately the same gradient as the water level (Fig. 5). Topographically, escarpments delineate the retreat of sandstone cap rock (Big Clifty) to the northwest. Although the cap rock has been breached in numerous locations, it has served to protect parts of the underlying subterranean drainage network from rapid destruction by surface erosion processes. The potentiometric surface lies within the soluble limestones (Fig. 5). It is speculated that zones of higher permeability were concentrated within facies and microfacies of the limestone units, and preferential flow paths caused greater dissolution to occur where flow was concentrated. Hydrologic circuits remained open for long periods; recharge, flow through, and discharge successfully transported the water and sediment load. Base level of the master surface streams that controlled

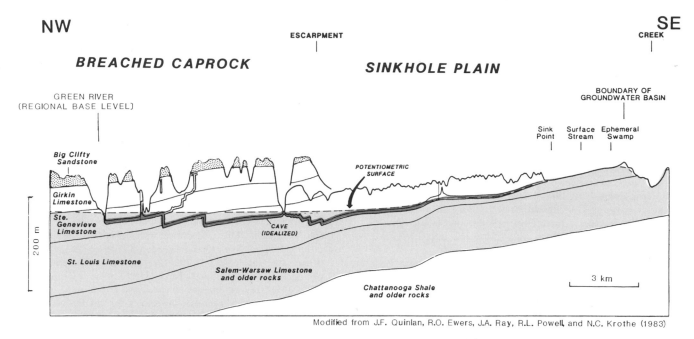

NW

**BREACHED CAPROCK**

ESCARPMENT

**SINKHOLE PLAIN**

SE
CREEK

Modified from J.F. Quinlan, R.O. Ewers, J.A. Ray, R.L. Powell, and N.C. Krothe (1983)

Factors controlling the occurrence and preservation of Mammoth Cave: (1) lithology--highly soluble, relatively pure limestones with high permeability zones occur at shallow depth; (2) structure--high permeability zones lie within region of active ground-water flow, and are oriented with similar gradients to the regional potentiometric surface; (3) topography--presence of resistant caprock retards surface erosional processes which would destroy cave. Topographic relief provides adequate head differential to drive ground water through the aquifer; (4) hydrology--stable, long-term duration of base level (Green River) allowed fully integrated subsurface drainage development, with recharge, flow through, and discharge.

Figure 5. Generalized section showing selected geologic and hydrologic aspects of the Mammoth Cave region.

discharge was relatively stable, allowing time for secondary permeability caused by aquifer dissolution to become well developed and integrated into a series of dendritic tributary conduits (Quinlan and Ray, 1981). Additional recharge was captured by the underground tributary network as it grew, including flow from swallets, sinkholes, sinkhole plains, sinking creeks, karst valleys, areal flow from the drains of vertical shafts, vertical spring discharge from perched aquifers, and direct infiltration through the soil.

Although hydrologic factors are important in the development and enhancement of secondary permeability and porosity and accompanying karst features in this area, it is felt that no single geologic factor was dominant in comparison to the other two evaluated.

## REFERENCES

Brahana, J. V., and Bradley, M. W., 1985, Delineation and description of the regional aquifers of Tennessee; The Knox aquifer in central and west Tennessee: U.S. Geological Survey Water-Resources Investigations Report 83–4012, 32 p.

—— , 1986a, Delineation and description of the regional aquifers of Tennessee; The Central Basin aquifer system: U.S. Geological Survey Water-Resources Investigations Report 82–4001, 35 p.

—— , 1986b, Delineation and description of the regional aquifers of Tennessee; Highland Rim aquifer system: U.S. Geological Survey Water-Resources Investigations Report 82–4054, 38 p.

Brahana, J. V., and Bradley, M. W., and Macy, J. A., 1986, Delineation and description of the regional aquifers of Tennessee; The East Tennessee aquifer system: U.S. Geological Survey Water-Resources Investigations Report 82–4091, 30 p.

Burchett, C. R., Brahana, J. V., and Hollyday, E. F., 1980, The occurrence of ground water in relation to facies within the Fort Payne Formation in Tennessee: Geological Society of America Abstracts with Programs, v. 12, p. 395.

Craig, L. C., and 24 others, 1979, Paleotectonic investigations of the Mississippian System in the United States: U.S. Geological Survey Professional Paper 1010, 559 p.

DeBuchananne, G. D., and Richardson, R. M., 1956, Ground-water resources of east Tennessee: Tennessee Division of Geology Bulletin 58, pt. 1, 393 p.

Fenneman, N. M., 1938, Physiography of Eastern United States: New York, McGraw-Hill, 714 p.

Fenneman, N. M., and Johnson, D. W., 1946, Physical divisions of the United States: U.S. Geological Survey Physiography Committee Special Map, scale 1:7,000,000.

Ferguson, H. F., 1974, Geologic observations and geotechnical effects of valley stress relief in the Allegheny Plateaus: American Society of Civil Engineers Annual Meeting, Los Angeles, California, 31 p.

Feth, J. H., and others, 1965, Preliminary map of the conterminous United States showing depth to and quality of shallowest ground water containing more than 1,000 parts per million dissolved solids: U.S. Geological Survey Hydrologic Investigations Atlas HA–199, scale 1:3,168,000.

Hobba, W. A., Jr., Fisher, D. A., Pearson, F. J., Jr., and Chermys, J. C., 1979, Hydrology and geochemistry of thermal springs of the Appalachians: U.S. Geological Survey Profesional Paper 1044–E, 36 p.

Hollyday, E. F., and Goddard, P. G., 1979, Ground-water availability in carbonate rocks of the Dandridge area, Jefferson County, Tennessee: U.S. Geological Survey Water-Resources Investigations Report 79–1263, 50 p.

Hollyday, E. F., and Seaber, P. R., 1968, Estimating cost of ground-water withdrawal for river basin planning: Ground Water, v. 6, no. 4, p. 15–23.

King, P. B., 1969, The tectonics of North America; A discussion to accompany the tectonic map of North America, scale 1:5,000,000: U.S. Geological Survey Professional Paper 628, 94 p.

King, P. B., and Beikman, H. M., 1974, Explanatory text to accompany the geologic map of the United States: U.S. Geological Survey Professional Paper 901, 40 p.

Lane, E. W., and Lei, K., 1950, Stream flow variability: American Society of Civil Engineers Transactions, v. 115, p. 1084–1134.

Parizek, R. R., White, W. B., and Langmuir, D., 1971, Hydrogeology and geochemistry of folded and faulted carbonate rocks of the Central Appalachian type and related land-use problems: Geological Society of America Guidebook, 1971, Annual Meeting, 193 p.

Quinlan, J. F., 1981, Hydrologic research techniques and instrumentation used in the Mammoth Cave region, Kentucky, *in* Roberts, T. G., ed., GSA Cincinnati '81 Field Trip Guidebooks: Washington, D.C., American Geological Institute, v. 3, p. 502–504.

Quinlan, J. F., and Ewers, R. O., 1985, Groundwater flow in limestone terranes; Strategy rationale and procedure for reliable, efficient monitoring of ground water quality in karst area: National Symposium and Exposition on Aquifer Restoration and Ground Water Monitoring, 5th, Columbus, Proceedings, p. 197–234.

Quinlan, J. F., and Ray, J. A., 1981, Groundwater basins in the Mammoth Cave region, Kentucky: Friends of the Karst Occasional Publication no. 1, map.

Quinlan, J. F., and Rowe, D. R., 1977, Hydrology and water quality in the Central Kentucky Karst; Phase I: University of Kentucky, Water Resources Research Institute Research Report no. 101, 93 p.

Quinlan, J. F., Ewers, R. O., Ray, J. A., Powell, R. L., and Krothe, N. C., 1983, Ground-water hydrology and geomorphology of the Mammoth Cave region, Kentucky, and of the Mitchell Plain, Indiana, *in* Shaver, R. H., and Sunderman, J. A., eds., Field Trips in Midwestern Geology: Bloomington, Indiana, Geological Society of America and Indiana Geological Survey, v. 2, p. 1–85.

Schneider, W. J., and 10 others, 1965, Water resources of the Appalachian region, Pennsylvania to Alabama: U.S. Geological Survey HA–198, 11 sheets: Sheet 1, Schneider, W. J., and Barksdale, H. C., Description, scale 1:7,000,000; Sheet 2, Schneider, W. J., Precipitation; Sheet 3, Meyer, G., Geology and mineral resources; Sheet 4, Schneider, W. J., Patterns of runoff; Sheet 5, Schneider, W. J., and Friel, E. A., Low flows; Sheet 6, Schneider, W. J., Floods; Sheet 7, Collier, C. R., and Whetstone, G. W., Quality of water; Sheet 8, Wark, J. W., Sediment loads; Sheet 9, Musser, J., Acid mine drainage; Sheet 10, Meyer, G., Wilmouth, B. M., and LeGrand, H. E., Availability of ground water; Sheet 11, Schneider, J. F., Surface water development. Scale, sheets 2–11, 1:2,500,000.

Schwalb, H. R., 1982, Paleozoic geology of the New Madrid area: U.S. Nuclear Regulatory Agency Commission, NUREG/CR–2909, 61 p.

Seaber, P. R., and Hollyday, E. F., 1965, An appraisal of the ground-water resources of the lower Susquehanna River basin: U.S. Geological Survey Open-file report, 75 p.

—— , 1966a, Statistical analysis of the regional aquifers [abs.]: Geological Society of America Abstracts with Programs, p. 196–197.

—— , 1966b, An appraisal of the ground-water resources of the Juniata River basins: U.S. Geological Survey Open-File Report, 125 p.

Sun, P-C., Criner, J. H., and Poole, J. L., 1963, Large springs of East Tennessee: U.S. Geological Survey Water-Supply Paper 1755, 52 p.

Swingle, G. D., 1959, Geology, mineral resources, and ground water of the Cleveland area, Tennessee: Tennessee Division of Geology Bulletin 671, 125 p.

Taylor, L. E., 1984, Groundwater resources of the upper Susquehanna River basin, Pennsylvania: Pennsylvania Geological Survey, 4th ser., Water Resources Report 58, 136 p.

Taylor, L. E., and Werkheiser, W. H., 1984, Ground water resources of the lower Susquehanna River basin, Pennsylvania: Pennsylvania Geological Survey, 4th ser., Water Resources Report 57, 130 p.

Taylor, L. E., and Werkheiser, W. H., DuPont, N. S., and Kriz, M. L., 1982, Groundwater resources of the Juniata River basin, Pennsylvania: Pennsylvania Geological Survey, 4th ser., Water Resources Report 54, 131 p.

Taylor, L. E., Werkheiser, W. H., and Kriz, M. L., 1983, Groundwater resources of the West Branch Susquehanna River basin, Pennsylvania: Pennsylvania Geological Survey, 4th ser., Water Resources Report 56, 143 p.

Wyrick, G. G., 1968, Ground-water resources of the Appalachian Region: U.S. Geological Hydrologic Atlas 295, 4 sheets.

Wyrick, G. G., and Borchers, J. W., 1981, Hydrologic effects of stress-relief fracturing in an Appalachian Valley: U.S. Geological Survey Water-Supply Paper 2177, 51 p.

U.S. Geological Survey, 1970, The national atlas of the United States of America: Washington, D.C., U.S. Geological Survey, 417 p.

Zurawski, A., 1978, Summary appraisals of the Nation's ground-water resources; Tennessee Region: U.S. Geological Survey Professional Paper 813–L, 35 p.

MANUSCRIPT ACCEPTED BY THE SOCIETY FEBRUARY 23, 1988

The Geology of North America
Vol. O-2, Hydrogeology
The Geological Society of America, 1988

# Chapter 24

# *Region 21, Piedmont and Blue Ridge*

**Harry E. LeGrand**
*331 Yadkin Drive, Raleigh, North Carolina 27609*

## INTRODUCTION AND GEOGRAPHIC SETTING

The Piedmont and Blue Ridge Region (Fig. 3; Table 2, Heath, this volume) is considered as a collective unit for hydrogeologic purposes, although some distinctions in topography and character of the rocks lead to separate descriptions in some cases. The region is bordered on the east and south by the Atlantic and Gulf Coastal Plain and on the west and north by the Appalachian Highlands region (Ridge and Valley and Appalachian Plateaus provinces). The region extends from New Jersey and southeastern Pennsylvania to eastern Alabama and includes large parts of Maryland, Virginia, North Carolina, South Carolina, and Georgia, as well as a small part of Tennessee. Igneous and metamorphic rocks, commonly having a mantle of residual soil, characterize the bedrock of the region. Several grabens filled with Triassic sedimentary rocks occur in the Piedmont (Fig. 1).

The many cities and towns of the region are closely interspersed; most of the rural areas are moderately populated. The distribution of population gives rise to the need for widespread ground-water supplies. Various engineering works and human actions change to some extent the hydrogeologic regime.

Precipitation reaches almost 200 cm a year in a small area of the southern Blue Ridge, but the average over the region is slightly more than 115 cm. The precipitation, chiefly in the form of rain, is fairly evenly distributed throughout the year.

The topography of the Piedmont Province is characterized by low, rounded hills and gentle slopes, but to the west the slopes of the Blue Ridge are steeper. On some mountain slopes bare rock is exposed, on others the soil zone is thin, and on others there is moderately thick soil and a thick vegetation cover. In the Piedmont, a heavy layer of residual soil and its veneer of vegetation cap the bedrock in most places. Drainage is facilitated by a close network of perennial streams (Fig. 2).

Several key factors prevent the Piedmont and Blue Ridge region from being considered in the same way as other ground-water regions. The more conventional type of sedimentary aquifer (which is composed of a nearly flat-lying blanket formation, has a distinct base, and is composed of only one medium of material) is not present in the Piedmont and Blue Ridge. Semblances of stratigraphic units are not widespread, and the term "hydrostratigraphic unit" could be only awkwardly applied. Nevertheless, ground water does occur and circulate. Thus, the study of ground-water occurrence in the region should not be patterned after that of other regions but should be treated singularly.

In this discussion, optimal use will be made of many generalizations, leading to a high level of knowledge about the region. These generalizations can then be melded and interwoven with key observations or data at specific sites to provide even greater knowledge about the subsurface hydrology at a particular setting or site. The generalizations are based on specific studies of local regions over many years by the author and colleagues.

The processes operating, and principles derived, are not fully documented in this text due to space limitations. Fortunately, Trainer (this volume) treats clearly some of the relations of factors discussed in this chapter. Other key references drawn upon for this chapter are Meyer and Beall (1958), Stewart (1962), Davis and Turk (1964), Hack (1966), LeGrand (1967, 1979), Poth (1968), Nutter and Otton (1969), Cressler and others (1983), Daniel and Sharpless (1983), and Heath (1984).

## GEOLOGIC SETTING

The igneous and metamorphic rocks of the Piedmont and Blue Ridge are more amply and specifically discussed in Hatcher and others (1988). The rocks have been studied and described locally throughout much of the region, but the local ranges in types of rock prevent description at a useful scale for the purpose of this discussion.

Gneisses and schists compose much of the Proterozoic core of the region. Granites of various ages, extending into late Paleozoic, are widespread, and mafic intrusives are also common. Altered volcanic rocks are interspersed with other rocks in certain belts. Two northeast-trending belts of metamorphosed sedimentary rocks are noteworthy. One of these, the Ocoee Series in the Blue Ridge, contains nonvolcanic clastic sedimentary rocks as much as 800 m thick (Fairbridge, 1975). The other, known as the Carolina Slate Belt, occurs in the Piedmont and contains a mixture of clastic sediments and altered volcanic rocks. Tectonic

LeGrand, H. E., 1988, Region 21, Piedmont and Blue Ridge, *in* Back, W., Rosenshein, J. S., and Seaber, P. R., eds., Hydrogeology: Boulder, Colorado, Geological Society of America, The Geology of North America, v. O-2.

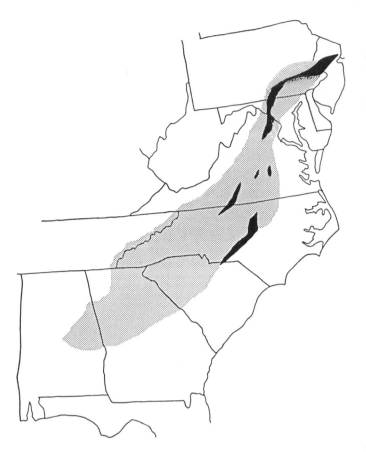

Figure 1. Map of the Piedmont and Blue Ridge region showing igneous and metamorphic rocks (stippled) and Triassic sedimentary rocks (darkened).

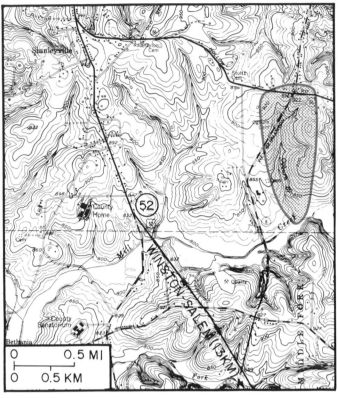

Figure 2. A part of the Rural Hall topographic map in North Carolina. The small Ogburn Branch drainage basin represents a typical ground-water unit, to which other small drainage basins or units and their integral parts can be compared (LeGrand, 1979).

actions, including deformation and igneous intrusions, have placed rocks of different types and age in close juxtaposition. The arrangement of rocks is that of thin bed-by-bed alternation in some places, of crosscutting rocks in other places, and of large distinctive rock masses in other places. Steeply dipping beds and foliation are more the rule than the exception.

Strips and patches of nonmarine sedimentary rocks of Triassic age extend through the Piedmont, as shown in Figure 1. They lie unconformably on the metamorphic and plutonic rocks. They were deposited in fault depressions after the main deformation in the Piedmont and Blue Ridge region. They are composed chiefly of arkose, conglomerate, and shale; these rocks are intruded by diabase dikes and sills, and these intrusives tend to extend into the neighboring igneous and metamorphic rocks.

A characteristic feature of the region is the prevailing mantle of residual soil and saprolite that covers the bedrock in most places (see Fig. 3). This mantle of weathered bedrock is composed chiefly of sandy clay, although fragments of solid rock are common near the bedrock surface. The thickness of this mantle frequently ranges from about 2 m to 20 m, though it is absent in

many places and thicker than 20 m in others. The mantle represents, in a sense, a special geologic unit that crosses various types of rock. The mantle is a true hydrogeologic unit that has an important impact on the ground-water conditions.

## THE HYDROGEOLOGIC SYSTEM

Two key factors prevent the Piedmont and Blue Ridge region from being considered similarly to other ground-water regions. First, there are no flat-lying or blanket-type formations that have characteristics of conventional sedimentary aquifers. In much of the region the metamorphic rocks are tilted and intercalated with granite or other igneous intrusions. Rocks of different types tend to have erratic outcrop distributions. Second, fracture-type permeability characterizes the rocks, and there is only a moderate range in degree of fracturing, according to the type of rock. The conventional approach of placing emphasis on the relation of ground water to a particular formation or type of rock generally has only moderate to low merit in regard to the Piedmont and Blue Ridge.

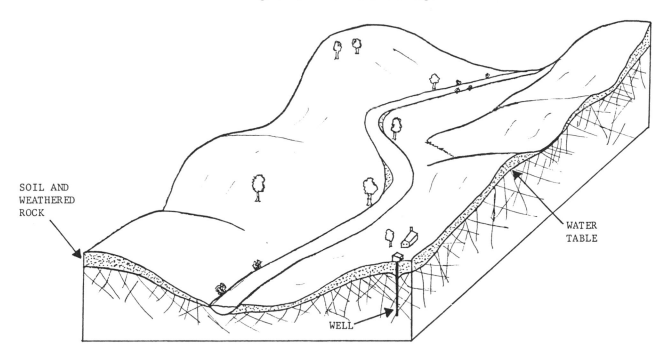

Figure 3. Subsurface cross section of a typical Piedmont and Blue Ridge terrain.

The mantle of residual soil and soft weathered rock that covers the underlying fractured bedrock in most places provides an intergranular medium through which recharge and discharge of water from the fractured rocks commonly occur. Thus, a composite two-media system characterizes the ground-water flow. Igneous and metamorphic rocks and consolidated sedimentary deposits of the region tend to be fractured to some extent where they occur no deeper than several hundred meters below the land surface (see Fig. 3). These fractures are the chief avenues for water movement.

The distribution and character of fractures can be viewed in several ways, according to (a) texture of permeability, (b) changes with depth, and (c) degree of enlargement by solution. As regards texture of permeability, the fractures may be evenly and closely spaced, unevenly and sparsely distributed, or unevenly spaced and elongated. As regards fracture distribution with depth, some fractures extend to considerable depth, but most show a decrease in size and number with increasing depth. Regarding enlargement of fractures by solution, rocks of low solubility—such as shales, slates, and granites—have fractures that are not enlarged by solution, whereas some mafic rocks near land surface and within the rapid ground-water circulation part of the system may have enlarged openings that resulted from solution along fractures (Le Grand, 1958).

In the region the fractured rocks are (1) exposed at the land surface, (2) covered by a layer of residual soil and decomposed rock, or (3) covered by alluvium in some valleys. Where moderately thick, the covering layer of unconsolidated materials may constitute an aquifer. Even where the unconsolidated materials are thin, they represent the upper part of the ground-water hydrologic system. Such a system is complex because water that enters as recharge or leaves as discharge passes through porous granular materials either after or before circulating through particular fractures in the rock.

A surface drainage basin that contains a perennial stream represents an entity that is useful for descriptions of ground-water conditions. Each basin in a general way is a unit that is isolated from, yet similar to, the adjacent basins. More specifically, some ground-water conditions are essentially repetitive from basin to basin. The small drainage basin occupied by Ogburn Branch in Figure 2 represents a ground-water entity. The generalizations in this paper apply almost equally to the Ogburn Branch drainage basin and to all other small drainage basins underlain by igneous and metamorphic rocks of the region. There are, of course, ranges of conditions noted within the generalizations that prevent ground-water conditions from being identical in all such basins.

## CHEMICAL CHARACTER OF GROUND WATER

The igneous and metamorphic rocks vary considerably in chemical composition and, consequently, in solubility. Some major rock groups of the region yield ground water having distinctive chemical character; enough contrast generally exists for chemical analyses of ground water to be an aid in geologic mapping where outcrops are scarce.

In their chemical character and that of the water derived

from them, the rocks can be divided into two groups (LeGrand, 1958, p. 178). The first includes granite, granite gneiss, mica schist, slate, and rhyolite flows and tuffs; these rocks resemble granite in composition. The second group includes diorite, gabbro, hornblende gneiss, and andesite flows and tuffs; these rocks resemble diorite in composition. The granite group yields a soft, slightly acidic water that is low in dissolved mineral constituents; the diorite group yields a hard, slightly alkaline water that is relatively high in dissolved material.

In contrast to the "granite group" of rocks, which commonly yield water having less than 100 mg/1 of total solids, the "diorite group" commonly yields water having more than 200 mg/1 of total solids. The water in the diorite group of rocks resembles limestone water; it is characterized as a calcium bicarbonate water. Anomalies in dissolved mineral constituents that are not due to differences in rock type may indicate either abnormal structural conditions, resulting in abnormally slow rates of circulation of water, or the presence of concentrated mineral deposits.

Most of the water in the Triassic sedimentary rocks is moderately low in total dissolved solids, similar to that in the igneous and metamorphic rocks. However, in some places, especially at depths greater than 100 meters, the water may be slightly mineralized or brackish, suggesting that it is old residual water that is not in the present circulation system.

Radioactivity of ground water in the region has not been fully explored. Some rocks of granite composition contain more uranium than other crystalline rocks; well water from granites tends to contain more radon and radium than water from other rocks. The radon and radium in well water from granites occur in concentrations greater than would be expected from the moderately low uranium content of the host rock. Therefore, there are indications that radium and radon can be accreted and concentrated around pumped wells, as ground water and subsurface air are agitated. At any rate, the radium and radon concentrations in some well waters greatly exceed the recommended levels. Studies thus far have not fully explored this potential problem.

## DEVELOPMENT OF THE HYDROGEOLOGIC SYSTEM

In contrast to the intrinsic and syngenetic permeability of unconsolidated sedimentary materials, the permeability of both the (1) soil and saprolite zone and (2) the fractured bedrock zone below are epigenetic. These zones make up the composite ground-water system. In relatively recent geologic time, changes in degrees of balance between the overall processes of weathering and erosion have resulted in the development of the soil and saprolite zone as we know it today. In the relatively recent geologic past, sets of systematic tension joints have developed or enlarged to provide the fractured-rock ground-water system.

The general processes of weathering, involving disintegration and decomposition of silicate minerals, are well known. It is sufficient here to point out that a certain combination of air and water penetrating the rock is needed to develop a thick soil and saprolite zone. Relatively high permeability, in the form of pervasive fractures, can lead to full penetration of air and water. The soil and saprolite zone does not develop readily where bare rock crops out or where the water table is shallow. Pronounced weathering does not commonly extend below the low seasonal stage of the water table.

Prior to the latest erosional history of an area, most of the present fractures were only hairline cracks at considerable depth. As the erosional history becomes more recent, the incipient fractures tend to be enlarged by solution or by other forces as erosion at the land surface removes overlying material. As overlying material is progressively removed by erosion, the rocks that were once deeply buried become close to the land surface and, in the process, develop tension fractures, and become part of the ground-water circulation system. Near the land surface the fractures tend to be larger or more pronounced than at deeper levels.

The continual erosion through geologic time causes a concomitant lowering of stream valleys, of interstream areas, and of the water table. Thus, fractures that at slightly earlier geologic time were below the water table ultimately are above the water table; as erosion progresses, rocks containing these fractures are destroyed by erosion, and these destroyed fractured rocks are replaced in relative space by fractured rocks that appear to have "risen with respect to the land surface" and are now above or slightly below the water table. The lowerings (1) of a stream valley, (2) of its nearby interstream area, and (3) of the water table are not necessarily in unison and are not at a steady and uniform rate. These imbalances from place to place lead to a range of topographic features, to a range in thickness of soil cover, and to a range in water-table gradients. Thus, the imbalances in erosion lead to a variety of geologic and hydrologic settings that are continually and dynamically changing.

## INFERENCES FROM HYDROGEOLOGIC FEATURES

### Introduction

The prevailing mantle of soil and vegetation prevents ready observations of the host rocks and their characteristics at specific places. The uncertainties of subsurface conditions at specific places in the region have led to a preponderance of test drilling and data-collecting programs.

Before massive data-collecting programs get under way, it is wise to review all pertinent principles. An overview of existing information should be made and the geologic and hydrologic history should be conceptually reconstructed. The stage of development in each segment of the study area should be interpreted so that approximations of the distribution of permeability can be made.

The locally sporadic changes in rock fractures and in soil and saprolite thickness do not allow for simple arithmetic interpolations. The uneven distribution of permeability and some unpatterned fracture openings can best be considered in the context of the principle of indeterminacy. The averages of certain features are determinate, but each specific case may be indeterminate. An individual case may be considered only in a statistical sense (Leopold and Langbein, 1963). There are too many unknowns in the range of combination of values of the interdependent variables to predict precisely certain features of hydrology in a setting. For example, the existence of a fracture at a certain place and depth is indeterminate prior to drilling.

The best way out of the dilemma of indeterminacy is to learn more about the processes operating in a variety of settings so that we can reduce the range of uncertainties. Some of the uncertainties of characterizing fracture patterns are being removed by the studies of LeGrand (1967), Daniel and Sharpless (1983), and Trainer (this volume). Fortunately, we can reach a high level of knowledge if we use existing data and make full use of inferences. Thus, it may be foolhardy and unduly expensive to demand precision in cases where indeterminacy is involved, especially if best inferences are likely to be successful. Reducing the range of uncertainties for the needed answers should be a major objective.

Following are certain features of the hydrology of the region that allow useful inferences.

*Soil and saprolite thickness.* If general observations in a local area show no signs of rock outcrops, it can be assumed that the soil and saprolite zone should be moderately thick. A thick soil and saprolite zone suggests that the underlying rock is fractured; a thin soil and saprolite zone suggests less than normal fracturing and, therefore, lower than normal permeability. A thick soil and saprolite zone also suggests that the water table is relatively deep.

*Type of rocks.* In most places the schists and gneisses contain more fractures and have thicker soils and saprolite than do the massive igneous rocks, such as granite or gabbro. Light-colored soils are characteristic of underlying granite and rocks that are chiefly silicates of potassium and aluminum; red and brown soils characterize areas underlain by gabbro, diorite, and other rocks high in calcium and in ferromagnesian minerals.

The gabbro, diorite, and associated rocks commonly yield water higher in total dissolved solids (chiefly calcium bicarbonate) than do granite and associated acidic rocks. Dissolution of the gabbro-type rocks, chiefly along the top of bedrock, commonly gives these rocks a lower topographic setting than adjacent granite-type rocks. Although no typical karst conditions occur, some basic rocks, chiefly hornblende gneiss, have fractures that have been enlarged by dissolution.

*Topographic expressions.* In addition to being surface-water divides, *ridges and hilltops* tend to be water-table divides. The majority of ridges and hilltops are composed of rocks that were poorly fractured at the immediate past stage of weathering and erosion; some fracturing may have since developed. Thin soil zones are not uncommon on ridges and hilltops. On the other hand, many *broad upland areas* are composed of fractured rocks and, thus, are likely to have a thick soil and saprolite zone. *Upland slopes,* like the ridges and hilltops, are recharge areas and

also are the central transmission parts of the water-table system. *Draws,* or *upland topographic sags,* are normally underlain by permeable fracture zones. Alignment of some draws and surface drainage systems, as depicted on topographic maps and air photos, suggest an alignment of some fractures. However, much of the drainage is dendritic and not related to rock structures.

Rivers and creeks are bordered commonly by *flood plains,* but in areal distribution they swell and pinch out sporadically along extensive stretches of each stream. A common vertical section of a flood plain is represented by a meter or more of clay, below which may be several meters of unsorted sandy, clay, and gravel. The base of these alluvial deposits rarely extends more than a meter or two below the base of the stream. The deposits are not normally productive aquifers.

The flood plain and its associated stream represent the discharge area for ground water that is recharged on the adjacent upland slopes. The discharge is through evapotranspiration on the flood plain and by diffused seepage from deposits into the stream.

Almost all perennial streams are in contact with bedrock in parts of their courses. Where not in contact with bedrock, the streams are underlain commonly by no more than 2 m of sand and gravel.

*The water table.* The water table is a somewhat subdued replica of the land surface topography. It lies deeper beneath uplands (commonly 6 to 15 m) than beneath valleys (commonly 1 to 4 m). The water table commonly lies close to the top of bedrock, more often than not lying in the saprolite. Below upland areas the water table is commonly 2 to 6 m higher in the wet spring season than in the dry fall season; the seasonal fluctuation in the lowland areas commonly ranges between 1 and 2 m.

*The ground-water system.* A complex two-media ground-water system prevails; the system includes the soil and saprolite medium and the fractured bedrock. The fractured bedrock generally grades downward into unfractured rock below a depth of about 100 m. Thus, the base of the ground-water system is indistinct (Fig. 4). The region contains countless separate small ground-water systems; these small systems are in a sense ground-water basins, almost congruent with small surface water drainage systems in each of which a perennial stream occurs. The small Ogburn Branch surface water basin, shown in Figure 3, is typical. Each ground-water basin is somewhat disjoined from an adjacent one at ridges or hilltops because the major surface drainage divides also tend to be water-table divides. The prevailing flow of ground water from beneath the upland area to the perennial stream allows one to make useful synthetic water-table maps prior to collection of specific data.

## EVALUATION APPROACH

Recognizing that there are many aspects of terrain evaluation, the approach is to consider the evaluation of a setting in the region in two stages. In the first stage, one applies to a particular setting an approximate generalization that may be considered as normal; such a statement of normality may be considered as

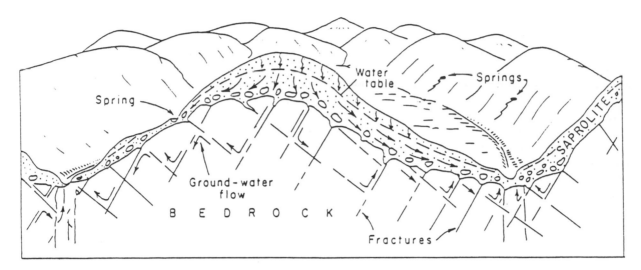

Figure 4. Typical subsurface cross section in the Piedmont and Blue Ridge region. Most of the fractures are confined to the upper portion of the bedrock zone. The distance from the land surface to the top of the bedrock commonly ranges from 6 to 23 m in this region (from Heath, 1980).

always true except where certain special conditions prevail (Kitts, 1963). In the second stage, a refinement is made in a particular statement of normality at a certain site by applying "key tendency" statements that are based on inferences of interrelated factors derived at the site from observation, discernment, and readily available data. The evaluation using the two stages gives a high level of knowledge that could otherwise be obtained only through massive data collection and interpretation. Some background generalizations that can lead to a specific terrain evaluation using the two-stage approach follow.

Several important inferences can be drawn from the relation of the water table and the top of rock. On an upland area, a thicker than normal weathered zone indicates that oxygen and carbon dioxide from the atmosphere and soil zone have moved downward below the land surface more easily than is normal; there is a further suggestion that the permeability is greater than normal (more fractured bedrock) and that a deeper and more arcuate flow of ground water exists between the upland recharge area and the lowland discharge area.

As an example of this terrain evaluation approach, a person wishes to know the depth of the water table and the likelihood of a productive well at a site on a particular hilltop prior to subsurface data collection. It is noted that the normal range of water-table depth on hilltops is between 6 and 15 m. The person observes evidence of some wet weather seeps and moderately hard weathered rock exposures in two places on intermediate slopes of the hill. One then uses the stated tendency that "upland seeps and rock outcrops suggest a shallow water table and poorly fractured rocks." One then infers that the depth to the water table is relatively shallow and close to the low end of the normal range. He also infers that the rocks are poorly fractured at the site and

that the chance of a large-production well is not great. The person rightfully ignores data on the two nearest wells—one well 300 m away on a lower slope and the other on a hilltop 700 m away. A warning should be made that a simple interpolation between data points more than a few meters or tens of meters apart is not a reliable approach.

## SUMMARY AND CONCLUSIONS

The gross ground-water system in the region is not represented by a semblance of an extensive continuum, as is the case in most regions. Instead, the region is composed of countless relatively small ground-water units, each unit almost confined to each small surface drainage basin in which a perennial stream occurs. Thus, in a sense, any small ground-water unit may represent a model to which other units in the region may be compared.

Following are concise statements (LeGrand, 1967) that can form a basis for a conceptual model of the settings of the Piedmont and Blue Ridge Region (Fig. 1) of the southeastern U.S., underlain by igneous and metamorphic rocks.

1. The climate is temperate and humid; average annual rainfall is slightly more than 100 cm and is fairly evenly distributed throughout the year.

2. The region is underlain by igneous and metamorphic rocks; the rocks range in chemical composition between that of granite (mainly silica and silicates of aluminum and potassium) and that of gabbro (chiefly silicates of aluminum, iron, magnesium, and calcium).

3. A layer of saprolite lies on the fresh rock in most places; the thickness of the saprolite ranges from a feather edge to slightly more than 40 m.

4. Water occurs in two types of media: (a) clayey granular weathered material; and (b) underlying fractures and other linear openings in the bedrock.

5. A close network of streams prevails, and in a few places on an interstream area is a perennial stream more than 1 km away. A hill and dale topography occurs, commonly with gentle slopes.

6. Toward each stream is a continuous flow of ground water. Some of the outflowing ground water is consumed as evapotranspiration in valleys; the remainder discharges as small springs and as bank and channel seepage into the stream.

7. Since all the perennial streams receive ground water from adjacent interstream areas, streams are the linear sinks in the water table. This part of the water table is directly observable. The topography of the water table is similar to that of the land surface, but its relief is less. Thus, it is easy to construct synthetic water-table maps and to predetermine the general direction of the natural movement of ground water.

8. The path of natural movement of ground water is relatively short and is almost invariably restricted to the zone underlying the gross topographic slope extending from the surface divide to the stream.

9. From a point source of infiltration, water, or waste that might be with it, extends as a fan or expansive trail down-gradient toward the stream; its dispersal depends on the kind and degree of permeability, on the hydraulic gradient, and on the distance to the stream.

10. Almost all recharge and discharge are through porous granular material (clayey soil or flood-plain deposits), but much of the intermediate flow between the recharge and discharge areas is through bedrock openings.

11. The saturated zone is not simple to define. Its top boundary is the water table, which lies in the clayey weathered material more often than not, but which becomes discontinuous where it lies in fractured bedrock. The lower boundary is irregular and indistinct; it is represented by the base of the zone in which interconnecting fractures exist. The saturated zone is absent where unfractured rocks crop out, but it is commonly 10 to 30 m thick. Water-yielding capacity within the zone ranges through several orders of magnitude; commonly it is less near the base of the saturated zone than near the top.

12. The water table is near land surface in valleys and as much as 8 to 20 meters below land surface beneath hills. The range of seasonal fluctuation of the water table is as little as 1 m in valleys and as much as 4 m beneath hills.

13. Bedrock fractures tend to decrease in size and number with depth, and in most places there is an insignificant storage and circulation of ground water below a depth of 125 m.

14. Many fractures are enlarged by the action of solution, especially in gneisses and schists containing silicates of calcium. Many of these enlarged fractures underlie "draws" or linear sags in the surface topography. These draws, representing zones of relatively high permeability in the bedrock, are sites for the best-producing wells.

15. The yields of wells range from less than 0.07 L/s to as much as 15.0 L/s; the sustained yield of most wells is less than 6.7 but more than 0.3 L/s. The cone of pumping depression of a domestic well does not generally extend to the cone of another pumping well a few hundred meters away.

16. The yield of a well cannot be predetermined, but experience developed by using past records of yield from various topographic and rock conditions allows useful predictions based on degrees of probability.

17. The distinctive chemical types of ground water are present. The first includes soft, slightly acidic water low in dissolved mineral constituents; water of this type comes from light-colored rocks resembling granite in composition, and includes granite, granite gneiss, mica schist, slate, and rhyolite flows and tuffs. The second includes a hard, slightly alkaline water relatively high in dissolved solids; water of this type comes from dark rocks, such as diorite, gabbro, hornblende gneiss, and andesite flows and tuffs.

An understanding of the hydrologic system of fractured rocks of the region leads to conceptual models that are fundamental to evaluation. The major approach taken here emphasizes inferential hydrogeology and relies on conceptual models based on best use of principles and past knowledge with continual upgrading in inferences as new information becomes available (LeGrand, 1979).

## REFERENCES

Cressler, C. W., Thurmond, C. J., and Hester, W. G., 1983, Ground water in the Greater Atlanta Region, Georgia: Georgia Geological Survey Information Circular 63, 144 p.

Daniel, C. C., III, and Sharpless, N. B., 1983, Ground-water supply potential and procedures for well-site selection, upper Cape Fear basin: North Carolina Department of Natural Resources and Community Development Report, 73 p.

Davis, S. N., and Turk, L. J., 1964, Optimum depth of wells in crystalline rocks: Ground Water, v. 2, no. 2, p. 6–11.

Fairbridge, R. W., ed., 1975, The encyclopedia of world regional geology; Part 1, Western Hemisphere (United States, Appalachian Region): Stroudsburg, Pennsylvania, Dowden, Hutchinson, and Ross, Incorporated, p. 522–529.

Hack, J. T., 1966, Circular patterns and exfoliation in crystalline terrane, Grandfather Mountain area, North Carolina: Geological Society of America Bulletin, v. 77, p. 975–986.

Hatcher, R. D., Jr., Viele, G. W., and Thomas, W. A., 1988, The Appalachian-Ouachita orogen in the United States: Boulder, Colorado, Geological Society of America, The Geology of North America, v. F-2 (in press).

Heath, R. C., 1980, Basic elements of ground-water hydrology with reference to conditions in North Carolina: U.S. Geological Survey Water Resources Investigations Open-File Report 80-44, 86 p.

—— , 1984, Ground-water regions of the United States: U.S. Geological Survey Water-Supply Paper 2242, 78 p.

Kitts, D. B., 1963, The theory of geology, in Fabric of geology: Reading, Massachusetts, Addision-Wesley, p. 49–68.

LeGrand, H. E., 1958, Chemical character of water in the igneous and metamorphic rocks of North Carolina: Economic Geology, v. 53, p. 178–189.

—— , 1967, Ground water of the Piedmont and Blue Ridge provinces in the southeastern States: U.S. Geological Survey Circular 538, 11 p.

—— , 1979, Evaluation techniques of fractured-rock hydrology, in Back, W., and

Stephenson, D. A., eds., Contemporary hydrology, The George Burke Maxey Memorial Volume: Journal of Hydrology, v. 43, p. 333–346.

Leopold, L. B., and Langbein, W. B., 1963, Association and indeterminacy in geomorphology, *in* Fabric of geology: Reading, Massachusetts, Addison-Wesley, p. 184–192.

Meyer, G., and Beall, R. M., 1958, The water resources of Carroll and Frederick counties: Maryland Department of Geology, Mines, and Water Resources Bulletin 22, 355 p.

Nutter, L. J., and Otton, E. G., 1969, Ground-water occurrence in the Maryland Piedmont: Maryland Geological Survey Report of Investigations 10, 56 p.

Poth, C. W., 1968, Hydrology of the metamorphic and igneous rocks of central Chester County, Pennsylvania: Pennsylvania Geological Survey Bulletin W25, 4th series, 83 p.

Stewart, J. W., 1962, Water-yielding potential of weathered crystalline rocks at the Georgia Nuclear Laboratory: U.S. Geological Survey Professional Paper 450-B, p. B106–B107.

MANUSCRIPT ACCEPTED BY THE SOCIETY MARCH 6, 1987

The Geology of North America
Vol. O-2, Hydrogeology
The Geological Society of America, 1988

# Chapter 25

# *Region 22, Atlantic and eastern Gulf Coastal Plain*

**Harold Meisler**
*(U.S. Geological Survey, retired) 32 Winding Way West, Morrisville, Pennsylvania 19067*
**James A. Miller**
*U.S. Geological Survey, 75 Spring St. SW, Suite 772, Atlanta, Georgia 30303*
**LeRoy L. Knobel**
*U.S. Geological Survey, P.O. Box 2230, INEL, CF 690, Room 164, Idaho Falls, Idaho 83401*
**Robert L. Wait**
*(U.S. Geological Survey, retired) 3084 Cleethorpes Drive, Lithonia, Georgia 30058*

## HYDROGEOLOGIC SETTING

The Atlantic and eastern Gulf Coastal Plain (Fig. 3; Table 2, Heath, this volume) is a gently rolling to flat region of about 400,000 km$^2$ extending from Long Island, New York, to eastern Mississippi. It is underlain by a wedge of unconsolidated to semi-consolidated, predominantly clastic sedimentary rocks that range in age from Jurassic to Holocene. The rocks consist mostly of sand, silt, and clay, but also contain lesser amounts of gravel and limestone. They thicken seaward from a feather edge at their updip limit, where they overlie older metamorphic, igneous, and consolidated sedimentary rocks of the Piedmont and other physiographic provinces, to 3,000 m in eastern North Carolina and 6,400 m in southern Alabama, thickening further beneath the Atlantic Ocean and Gulf of Mexico.

The Coastal Plain sediments generally dip gently seaward; however, broad folds, arches, and embayments are superimposed on the Coastal Plain wedge (Fig. 1), affecting the thickness and distribution of sediments. Structural basins such as the Southeast Georgia embayment have been the site of marine sedimentation since the Late Cretaceous. By contrast, the Cape Fear arch in North Carolina, in existence throughout most of Tertiary time, has served more as a source of sediments than as a depositional area.

The sediments are derived from adjacent upland areas, and were transported by streams and deposited in fluvial, deltaic, and marine environments. Coarsest-grained materials in a given stratigraphic unit are generally more prevalent near its updip limit; the amounts of clay and other fine-grained materials increase seaward. Seaward progression of facies, from fluvio-deltaic-strandline deposits (largely sand, silt, and gravel) to lagoonal–shallow marine deposits (argillaceous rocks with lesser amounts of interbedded sand) to midshelf deposits (predominantly calcareous clay and chalk) produces a general seaward decrease in permeability. Major marine transgressions and regressions and the resulting shifts of facies in either the landward or seaward direction are the chief determinants of the location of regionally extensive high- and low-permeability sediments. In addition, the arches and embayments exerted a significant influence on the distributions of regional facies and permeability; coarse, permeable materials such as sand and gravel generally are present on or near the arches, whereas fine-grained sediments are prevalent in basins and embayments. Accordingly, facies changes are rapid, and the patterns of interbedding of different rock types are complex.

Low-permeability sediments are generally thicker and more extensive in the eastern Gulf Coast than farther to the north and east—particularly those deposited during the Late Cretaceous and early Tertiary. The Gulf Coast Basin subsided at a more rapid rate than the Atlantic Coast Basin, allowing more extensive marine transgression. Also, the source of sediments in the eastern Gulf Coast was of low relief and a considerable distance from the basin.

Permeability distribution in the clastic sediments is closely related to their environment of deposition, which in turn reflects geologic age. Thus, based on age and depositional environment, the Coastal Plain sediments are subdivided into four major hydrogeologic units. These units—depicted in the sections in Figure 2 and described below—are, from oldest to youngest: (1) predominantly nonmarine Cretaceous rocks (mostly of Early and early Late Cretaceous age); (2) predominantly marine and marginal marine Cretaceous rocks (Late Cretaceous age); (3) lower Tertiary rocks (mostly shallow marine); and (4) Quaternary and upper Tertiary rocks (shallow marine to fluvial).

### *Predominantly nonmarine Cretaceous rocks*

Through most of Early Cretaceous and early Late Cretaceous time, the Atlantic and eastern Gulf Coastal Plain was a low,

Meisler, H., Miller, J. A., Knobel, L. L., and Waite, R. L, 1988, Region 22, Atlantic and eastern Gulf Coastal Plain, *in* Back, W., Rosenshein, J. S., and Seaber, P. R., eds., Hydrogeology: Boulder, Colorado, Geological Society of America, The Geology of North America, v. O-2.

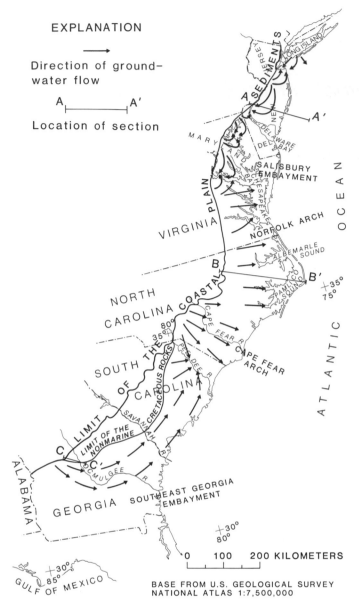

Figure 1. Location of arches and embayments and the direction of ground-water flow in the predominantly nonmarine Cretaceous rocks.

broad, flat plain cut by numerous shallow, wide, meandering streams. The streams built extensive coalescing deltas and drained into shallow seas. Structural basins, such as the Salisbury and Southwest Georgia embayments, received the greatest thicknesses of sediments. The sediments deposited in these fluvial and deltaic environments consist largely of interbedded sand, silt, and clay. The sands are fine to coarse in texture and are commonly feldspathic. The clays and silts commonly contain lignite. Because of the fluvio-deltaic environment in which they were deposited, the sand beds, while numerous, are not well connected. They are mostly lens shaped in cross section and long and sinuous in plan

view; individual sand bodies cannot be traced for any great distance.

### Predominantly marine and marginal marine Cretaceous rocks

During Late Cretaceous time, a widespread marine transgression that occurred over a long period of geologic time caused most of the Atlantic and eastern Gulf Coastal Plain to be covered by a shallow sea. Marine sediments were deposited on the nonmarine sediments. However, most of Virginia in the vicinity of the Norfolk Arch and part of Maryland remained above sea level, and no strata were deposited until Tertiary time (Brown and others, 1972). As subsidence in a basin generally kept pace with sediment input, thick sequences of sediments of similar lithology continued to be deposited in the same parts of the basin. For example, a thick sequence of highly permeable sands (Blufftown, Ripley, and Providence formations) in eastern Alabama and western Georgia was deposited in a recurring barrier-island complex (Reinhardt and Gibson, 1980).

Marine and marginal marine deposits generally do not contain as much sand as do deposits of nonmarine environments. However, marine and marginal marine sand bodies are more extensive, and the sediments are better sorted. Barrier-bar deposits and beach sands mark the landward limit of marine transgression. These deposits grade seaward into marginal marine and shallow-marine shelf deposits that consist largely of dark-colored, calcareous, fossiliferous clays interbedded with glauconitic sands. Farther seaward, deeper-water sediments consist of massive marine clays or a thick chalk sequence in central and western Alabama. These low-permeability sediments form a downdip barrier to ground-water flow.

### Lower Tertiary rocks

The widespread marine transgression that started in the Late Cretaceous continued into Tertiary time. Hence, lower Tertiary sediments are mostly of marine and marginal marine origin. They generally consist of glauconitic quartz sand that is commonly fossiliferous, and contain interbedded carbonaceous, calcareous clay. Sandy, fossiliferous layers of limestone occurring in an otherwise clastic sequence constitute important aquifers in North Carolina, Georgia, and Alabama. In Georgia and Alabama, lower Tertiary sandy strata grade seaward into massive marine clays or into carbonate rocks that are part of the Floridan aquifer system (Miller, 1986).

### Quaternary and upper Tertiary rocks

The widespread marine transgression that had continued into the Tertiary was interrupted in many parts of the Coastal Plain during the late Tertiary. Upper Tertiary sediments are a generally regressive sequence that includes deposition during local stillstands and marine transgressions. Hence, marginal ma-

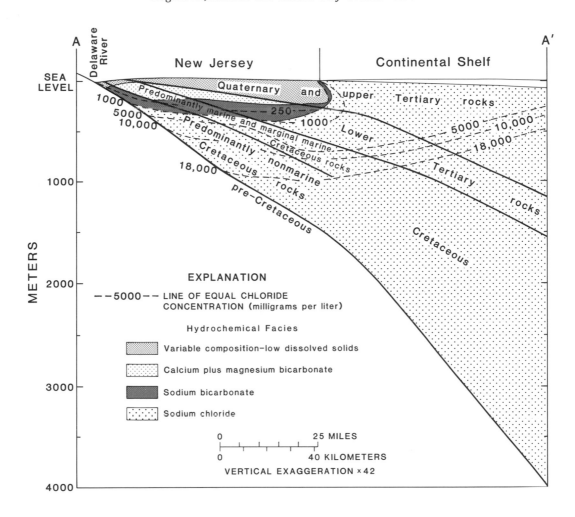

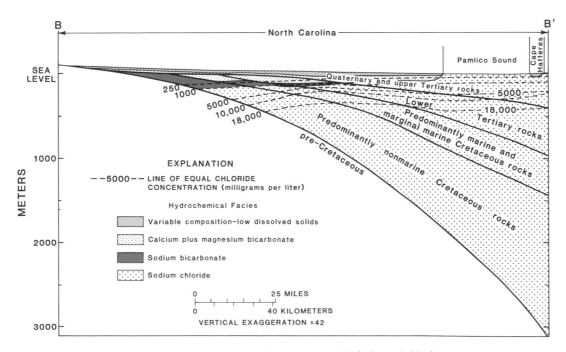

Figure 2. Hydrogeologic sections showing principal hydrochemical facies and chloride concentrations.

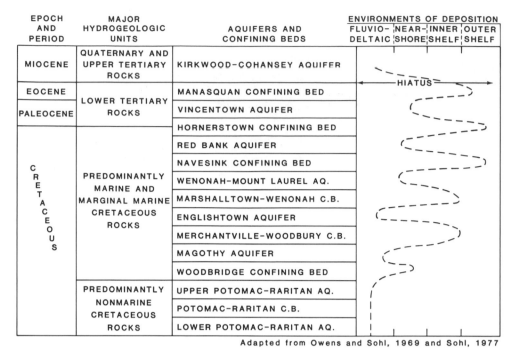

Figure 3. Relation of hydrogeologic units to environments of deposition.

rine beds consisting of calcareous, fossiliferous sand and minor amounts of clay are commonly overlain by predominantly fluvial deposits. The fluvial deposits consist largely of coarse-grained gravelly sand and are overlain in many places by marine terrace sands that are relics of fluctuating Pleistocene sea levels. Glacial outwash of Pleistocene age forms a highly permeable aquifer on Long Island, New York.

## RELATION OF HYDROGEOLOGIC UNITS TO DEPOSITIONAL ENVIRONMENT

The relation of Coastal Plain hydrogeologic units to changing depositional environments and to a series of marine transgressions and regressions may be illustrated by the aquifers and confining beds of New Jersey. Eight aquifers and seven intervening confining beds are depicted in Figure 3; most of these have been defined by Zapecza (1987).

The upper and lower Potomac–Raritan aquifers and the intervening confining bed are of fluvio-deltaic origin. Accordingly, the sediments are highly variable and discontinuous; individual sand and clay bodies cannot be traced over long distances. The sands are generally stream deposits, whereas the clays are deposited on swampy interfluves. Although the aquifers contain both sand and clay beds, the sand beds are more abundant. Clay beds dominate in the confining bed which was deposited in a low-energy environment consisting of shallow, discontinuous backswamp basins.

The Woodbridge confining bed, overlying the upper Potomac–Raritan aquifer, was partly deposited in a marginal ma-

rine swampy environment and includes the first marine deposits in the New Jersey Cretaceous (Olsson, 1980, p. 119). Deposition of the Woodbridge was followed by that of the Magothy aquifer which consists of beach deposits as well as fluvial deposits similar to those of the Potomac–Raritan aquifers. Following deposition of the Magothy, a series of major marine transgressions and regressions began. The Merchantville–Woodbury confining bed is the oldest massive shelf deposit that crops out in the Coastal Plain of New Jersey. It is also the oldest outcropping deposit to contain glauconite, a mineral that is indicative of a marine origin. Because of its deeper-water marine origin, it is a more widespread and continuous confining bed than either of the two lower confining beds.

The sequence of five confining beds and four aquifers—from the Merchantville–Woodbury confining bed to the Manasquan confining bed—represents a series of transgressions and regressions in which the confining beds generally consist of clayey to silty glauconitic sands accumulated on the Continental Shelf and the aquifers consist generally of sands that accumulated in a nearshore or coastal environment. Some of the sands of the Englishtown aquifer, however, are probably alluvial deposits (Owens and Sohl, 1969, p. 245).

A major lowstand of sea level during early Oligocene time resulted in a beveled erosional surface on Eocene rocks (Olsson, 1980, p. 125). This was followed during Miocene time, by accumulation of the sediments of the Kirkwood–Cohansey aquifer in a variety of predominantly nearshore and fluvio-deltaic environments. The deeper-water shelf deposition characteristic of the Late Cretaceous and early Tertiary never returned to the Coastal Plain of New Jersey.

The depositional environments of the New Jersey Coastal Plain sediments may be compared to those of the sediments penetrated in the COST B-2 well located about 120 km off the New Jersey coast (Rhodehamel, 1977, p. 16, Table 3). Interpretation of samples of the offshore sediments suggests five periods of predominantly nonmarine and nearshore deposition. Four of these periods correspond to periods of similar environmental conditions in New Jersey in which the four most extensive Cretaceous aquifers were deposited. These are the two Potomac–Raritan aquifers, the Magothy aquifer, and the Wenonah–Mount Laurel aquifer. Similarly, the confining beds of the New Jersey Coastal Plain, which reflect a more marine depositional environment, correspond to predominantly marine sediments from the COST B-2 well. However, the Englishtown, Red Bank, and Vincentown aquifers, which are local or pinch out downdip, reflect coastal or nearshore depositional conditions and also correspond to predominantly marine sediments from the COST B-2 well.

The continuity of Coastal Plain aquifers with permeable sediments offshore affects the occurrence and distribution of salt water in the emerged Coastal Plain. As discussed in the section on saltwater–freshwater relations, the source of much of this salt water is in the sediments offshore.

## RECHARGE, DISCHARGE, AND GROUND-WATER FLOW

Precipitation, the source of freshwater recharge to the Coastal Plain sediments, ranges from 1,020 to 1,420 mm per year in the Atlantic Coastal Plain and from 1,220 to 1,630 mm in the eastern Gulf Coastal Plain. Ground-water recharge varies considerably over the region, partly because of differences in precipitation but, more importantly, because of variability in the intake capacity of the sediments and a southward increase in evapotranspiration. Estimates of annual recharge to the more permeable sediments from North Carolina to Long Island, New York, range from about 300 to 510 mm, the higher recharge rates being most common in New Jersey and Long Island. Analysis of the baseflow component of streamflow in the Cretaceous and Tertiary clastic outcrop area from South Carolina to eastern Mississippi (Stricker, 1983) suggests annual recharge rates of up to 510 mm and an average of about 180 mm. Recharge on the more permeable sediments averages about 250 mm per year (Barker, 1986).

Layering of the Coastal Plain sediments, as well as the development of a complex stream system on the surface of the Coastal Plain, have had significant effects on the ground-water flow system and on the disposition of the recharge. Most of the recharge to an aquifer occurs in its outcrop area. Most water enters the shallow unconfined flow system and discharges into nearby streams. A small amount, however, enters the deeper confined part of the flow system and, moving greater distances, discharges into major rivers, estuaries, or the ocean. Recharge to the "regional flow system" within the Tertiary and Cretaceous clastic sediments from South Carolina to eastern Mississippi is estimated to average about 25 mm per year (Barker, 1986). This estimate is based on simulation of the flow system using a multilayer finite-difference computer model.

A detailed analysis of ground-water flow in two major drainage basins in South Carolina and Georgia indicates comparable recharge rates (R. E. Faye, U.S. Geological Survey, written communication, 1984). Results of a cross-sectional finite-difference model representing flow to the Ocmulgee River in east-central Georgia suggest that of an average annual recharge of 155 mm, about 7 percent (or 11 mm) enters the confined flow system and discharges into the Ocmulgee River. Both the unconfined and confined flow systems along this section are depicted in Figure 4. Computer simulation of a section in southern South Carolina representing flow to the Savannah River suggests that of an average annual recharge of 165 mm, about 18 percent (or 30 mm) enters the confined flow system and discharges into the Savannah River. These simulated values of recharge, however, may differ from average amounts of recharge in these basins, because the sections modeled may not fully represent basin conditions.

The direction of ground-water flow within the deepest hydrogeologic unit in the Atlantic Coastal Plain—the predominantly nonmarine Cretaceous rocks—is depicted in Figure 1. Ground water in these rocks in Georgia and South Carolina originates principally as recharge on outcrops in western Georgia; the rocks do not crop out in eastern Georgia or in South Carolina. Flow is eastward across Georgia and northeastward (parallel to the coast) toward the Pee Dee River in northern South Carolina. The flow system underlies and apparently has poor hydraulic connection with the confined flow system that contributes to the Ocmulgee River (Fig. 4). Flow paths in this hydrogeologic unit are several hundred kilometers in length. However, because of low hydraulic conductivities and hydraulic gradients, flow through this part of the system is probably very sluggish.

The predominantly nonmarine Cretaceous rocks crop out, or subcrop beneath a veneer of upper Tertiary sediments, along the inner margin of the Coastal Plain from North Carolina to Long Island. Recharge occurs principally in this outcrop or subcrop area. In North Carolina and Virginia, flow in the confined system is generally to the coast and to the major sounds and bays where discharge from the aquifer is upward through the overlying sediments. Some of the flow is upward to major rivers such as the Cape Fear and Pee Dee. Flow paths are as much as a few hundred kilometers in length.

From Maryland to Long Island, ground-water flow in the predominantly nonmarine Cretaceous hydrogeologic unit is strongly influenced by the large bays and estuaries that indent the coastline. Flow in the confined part of the unit is from upland areas, where the unit is at or near land surface, to lowland areas where it underlies the bays and estuaries. Most of the flow takes arcuate paths that do not extend very far downdip. The downdip limit of flow is marked by the occurrence of salt water (discussed later in this chapter). Typical flowpaths are generally less than 50 km in length.

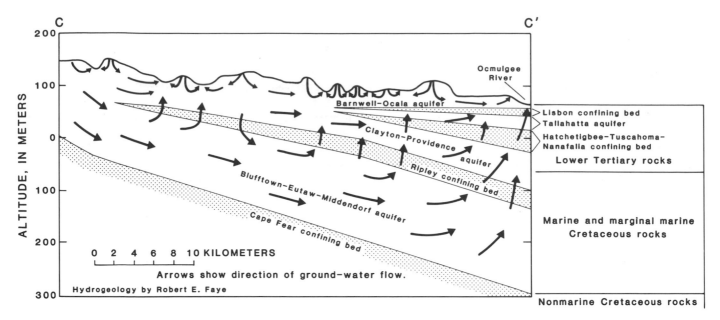

Figure 4. Generalized section showing ground-water flow in the Ocmulgee River Basin, Georgia.

## FRESHWATER GEOCHEMISTRY

The chemical variability of fresh ground water in the Coastal Plain stems from both mineralogical variations within the sediments and chemical changes that occur as the water moves through the sediments. The most striking differences in the mineralogy of the sediments is caused by the different depositional environments. The freshwater geochemistry of sediments deposited in both nonmarine and marine environments is discussed below.

### Predominantly nonmarine Cretaceous rocks

The fluvio-deltaic sediments of the predominantly nonmarine Cretaceous hydrogeologic unit generally consist of quartz sand with minor amounts of silicate minerals and interbedded silt and clay lenses. The clay lenses commonly contain pyrite and lignite. Atmospheric precipitation falling on the outcrops of these sediments is saturated with dissolved oxygen and has a low pH. As this water recharges the sediments, the dissolved oxygen reacts with lignite and pyrite to produce ferric hydroxide, dissolved carbon dioxide, and hydrogen and sulfate ions (Chapell, 1984). The dissolved carbon dioxide, in turn, dissociates into bicarbonate ions and additional hydrogen ions. The hydrogen ion is consumed by hydrolysis of silicate minerals, such as feldspar, to produce dissolved silica ($SiO_2$) and the following dissolved ions: calcium ($Ca^{2+}$), magnesium ($Mg^{2+}$), sodium ($Na^+$), potassium ($K^+$) and bicarbonate ($HCO_3^-$).

As the water flows through the sediments, $Ca^{2+}$ and $Mg^{2+}$ released by the weathering of the silicate minerals are removed from solution by cation exchange and replaced by $Na^+$. The sodium on cation exchange minerals (primarily clay minerals) is probably derived from loading of the exchange sites during episodes of saltwater intrusion or from formation of sodium-rich clay by weathering prior to its deposition.

The sulfate ion ($SO_4^{2-}$) produced by the dissolution of pyrite is reduced to sulfide or remains in the ground water as $SO_4^{2-}$, depending on local conditions. The precipitated $Fe(OH)_3$ is commonly observed as secondary mineralization (as a cement) in these sediments.

### Glauconitic marine sediments

Glauconite-rich deposits constitute a significant part of the marine sediments of Late Cretaceous and early Tertiary age. The Aquia aquifer of Paleocene age, in southern Maryland, illustrates the chemical processes that occur. The Aquia aquifer is composed primarily of quartz sand, glauconite, and shell debris. The shell material is composed of magnesium calcite and aragonite and is commonly a source material for secondary calcite cementation.

Glauconite has a high capacity to exchange ions adsorbed onto its surface for ions in ground water. Because glauconite is formed in a marine environment, the exchange sites are initially occupied by sodium. Analyses of paired samples of ground water and glauconite from the Aquia aquifer indicate that exchangeable cation composition of glauconite changes systematically in the direction of ground-water flow in a manner similar to changes in the concentrations of cations in the ground water (Chapelle and Knobel, 1983). These data suggest that cation exchange reactions simultaneously alter water and glauconite surficial composition along the flow path.

In fresh ground-water systems the order of selectivity for ions on exchange sites is $Ca^{2+} > Mg^{2+} > K^+ > Na^+$. Because the exchange reaction is rapid, $Ca^{2+}$ in solution displaces $Na^+$ on the exchange sites. The process continues until sodium is no longer available and $Ca^{2+}$ is the dominant ion both in the solution and on the exchange sites. The exchange front migrates through an aquifer as the $Ca^{2+}$ is transported by the flow system. The exchange front in the Aquia aquifer in southern Maryland has migrated more than 50 km along the flowpath (Chapelle and Knobel, 1983).

Ground water in the Aquia aquifer is calcium-magnesium bicarbonate in character in the outcrop area and changes to sodium bicarbonate in character downgradient. The chemical processes responsible for this change can be understood by inspection of Figure 5, in which concentrations of $Ca^{2+} + Mg^{2+}$, $Na^+$, and $HCO_3^-$ in the water are plotted as a function of distance along the flowpath. Concentrations of $Ca^{2+} + Mg^{2+}$ and $HCO_3^-$ increase downgradient to a peak at about 15 km and begin to decrease farther downgradient. The sharp increases are caused by the rapid dissolution of magnesium calcite shell material in the presence of $CO_2$. High-magnesium calcite has a higher solubility than low-magnesium calcite. The water becomes saturated with respect to the latter at about 15 km; farther downgradient, precipitation of calcite occurs simultaneously with dissolution of the high-magnesium calcite shell material. This process also accounts for the distribution of secondary calcite cementation beyond 15 km. $Na^+$ does not increase significantly within the first 50 km along the flowpath because, as already discussed, the exchange capacity of the glauconite has been exhausted. However, beyond 50 km, rising $Na^+$ concentrations and declining $Ca^{2+} + Mg^{2+}$ concentrations are primarily due to $Ca^{2+}$ and $Mg^{2+}$ being replaced by $Na^+$ in the ground water.

At about 80 km the ground water is characterized by very low $Ca^{2+} + Mg^{2+}$ and sharply increasing $Na^+$, caused by ion exchange, and by increasing $HCO_3^-$. The low concentrations of $Ca^{2+}$ and $Mg^{2+}$ cause the water to become undersaturated with respect to both high and low magnesium calcite. Dissolution of these minerals occurs, causing $HCO_3^-$ concentrations to increase. $Ca^{2+} + Mg^{2+}$ concentrations in the ground water, however, remain low as a result of continuing ion exchange.

### *Origin of carbon*

Ground water in parts of the Coastal Plain has $HCO_3^-$ concentrations that are several times higher than could be generated by reactions involving $CO_2$ gas dissolved in the water that recharges aquifer outcrop areas. Foster (1950) concluded that $CO_2$ was being generated within the aquifer system.

Stable carbon isotopes ($^{12}C$ and $^{13}C$) used in conjunction with major-ion water chemistry data are useful in helping to identify the sources of $HCO_3^-$ in ground water. Stable carbon isotope data for water in the Aquia aquifer were examined by Chapelle and Knobel (1985) who found that the $HCO_3^-$ becomes enriched with the heavier isotope ($^{13}C$) as distance along the flowpath increases. This increase is more pronounced at distances greater than 80 km and coincides with the rapid rise of $HCO_3^-$ concentration in the Aquia aquifer shown in Figure 5.

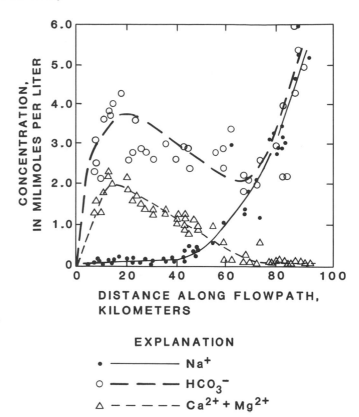

Figure 5. Relation of dissolved sodium, calcium plus magnesium, and bicarbonate to distance along the flowpath in the Aquia aquifer, Maryland.

The increases in both the heavier carbon isotope and the $HCO_3^-$ concentration require an aquifer-generated source of $CO_2$ that is enriched in the heavy isotope. A possible source of aquifer-generated $CO_2$ that would be enriched in the heavy isotope involves the bacterially mediated fermentation of lignitic material associated with bacterially mediated methanogenesis. These reactions tend to fractionate the stable carbon isotopes; the lighter isotope accumulates in the methane. The anoxic conditions required by these reactions occur in the Aquia aquifer at distances along the flowpath greater than 70 km (Chapelle and Knobel, 1985).

### *Hydrochemical facies*

The chemical processes controlling the freshwater and saltwater chemistry of the Coastal Plain have resulted in four identifiable hydrochemical facies (Seaber, 1965; Back, 1966) listed below. Their distribution in New Jersey and North Carolina is shown in the sections in Figure 2. The locations of the sections are shown in Figure 1.

*1. Variable composition.* This water contains low dissolved solids and has no dominant anion or cation. It commonly has high iron concentrations. The water is typical of recharge areas and unconfined aquifers.

*2. Calcium plus magnesium bicarbonate.* Calcium plus magnesium constitutes more than 50 percent of the cations and bicarbonate constitutes more than 50 percent of the anions. This water commonly occurs in the upgradient parts of confined aquifers.

*3. Sodium bicarbonate.* Sodium constitutes more than 50 percent of the cations, and bicarbonate constitutes more than 50 percent of the anions. This water generally occurs downgradient from zones containing calcium plus magnesium bicarbonate water.

*4. Sodium chloride.* Sodium continues to dominate the cation composition; however chloride constitutes more than 50 percent of the anions. This water is typical of the deeper parts of the aquifer system, particularly in coastal areas and near major estuaries.

## SALTWATER–FRESHWATER RELATIONS

Salty ground water underlies fresh water in much of the Atlantic and eastern Gulf Coastal Plain, particularly in the seaward parts of the area. Chloride concentrations generally increase with depth within a transition zone (Fig. 2) between the deepest fresh water and the underlying salt water. Depths at which chloride concentrations of 5,000 mg/ℓ or greater occur in the Atlantic Coastal Plain are shown in Figure 6 (Meisler, 1987; Lee and others, 1985). Locally in Georgia, however, ground water containing less than 5,000 mg/ℓ chloride is present below these depths.

The depth to salty ground water is related to the natural flow pattern of fresh ground water. "Theoretically, equilibrium between fresh water and sea water in a coastal region requires that the hydraulic head of the fresh water be at least high enough to balance the head of salt water. . ." (Upson, 1966, p. C242). Hence, salt water should be found at shallowest depths near ground-water discharge areas, where freshwater heads are lowest, and at greatest depths (or not at all) near recharge areas, where heads are highest. Areas where salt water is relatively shallow, do indeed, coincide with areas of natural ground-water discharge, such as Delaware Bay, lower Chesapeake Bay, Albemarle Sound, the Cape Fear River, and the Atlantic Ocean. Also, the greatest depth to salt water in Georgia occurs in the area of highest head; both heads and depths to salt water generally decrease northward along the coast from Georgia to North Carolina.

In New Jersey, however, salt water is deeper at the coast than it is farther inland, and the predevelopment heads along the coast do not appear to be high enough to account for the relatively great depths to the salt water or for the wedge of relatively fresh ground water (compared to sea water) that extends 90 km off the coast (Figs. 2 and 6).

In order to test the hypothesis that this distribution of salt water is related to long-term lower sea levels, the saltwater–freshwater transition zone in New Jersey and beneath the adjacent Continental Shelf, was analyzed by means of a cross-sectional finite-difference computer model (Meisler and others, 1984). Comparison of simulated steady-state saltwater–freshwater interfaces with observed chloride concentrations sug-

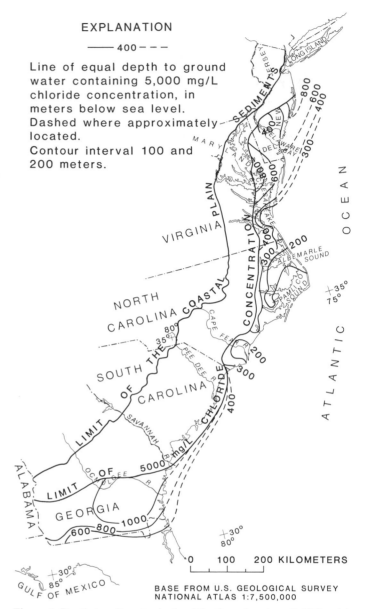

EXPLANATION

——— 400 – – –

Line of equal depth to ground water containing 5,000 mg/L chloride concentration, in meters below sea level. Dashed where approximately located. Contour interval 100 and 200 meters.

BASE FROM U.S. GEOLOGICAL SURVEY
NATIONAL ATLAS 1:7,500,000

Figure 6. Depth to salt water in the Atlantic and eastern Gulf Coastal Plain.

gests that the configuration of the saltwater freshwater transition zone in southern New Jersey and the occurrence offshore of relatively fresh ground water (compared to sea water) reflect average sea levels that were 15 to 30 m below present sea level from the Miocene to the Holocene.

Farther south, in southeastern Virginia and North Carolina, where the transition zone is generally much shallower and does not extend very far offshore (Figs. 2 and 6), the position of the transition zone probably reflects higher sea levels, relative to land surface, than in New Jersey. Indeed, the depositional environments of sediments of middle Miocene and younger age indicate that submergence and marine conditions were generally more prevalent here than in New Jersey. Hence, higher average sea levels during that period resulted in the transition zone being developed at shallower depths as well as closer to the present coastline.

The shallow occurrence of salt water in the vicinity of the Cape Fear River, in southeastern North Carolina, is of particular interest. Because of high heads in this area (18 to 30 m above sea level), little or no salt water theoretically should be present under equilibrium conditions. The salt water is probably residual water that infiltrated the aquifer during two marine transgressions that covered much of this area in the early Pliocene and early(?) Pleistocene (Zullo and Harris, 1979, p. 38). Apparently, insufficient time has elapsed for the salt water to have been flushed. Although these marine transgressions also covered northeastern South Carolina, similarly high concentrations of chloride do not occur there; higher heads (30 to 40 m) may have accelerated the flushing, pushing the transition zone offshore.

In southwestern Long Island, New York, salt water in which chloride concentrations range almost up to those of sea water occurs in the Magothy aquifer. The salt water, however, is underlain by fresh water in the Lloyd aquifer (Perlmutter and others, 1959). The wedge shape of the saltwater body in the Magothy aquifer and the distribution of chloride concentrations indicate that the salt water is hydraulically connected to Jamaica Bay and is unrelated to the deeper salt water in the aquifers of New Jersey.

The broad saltwater–freshwater transition zone depicted in Figure 2 probably developed as a result of saltwater circulation caused by the many large-amplitude sea-level changes that recurred through the late Tertiary and Quaternary (Vail and others, 1977; Zellmer, 1979). Repeated advance and retreat of the salty ground water caused the salt water and fresh water to mix, which resulted in saltier water predominating in the deeper and seaward parts, and fresher water predominating in the shallower and landward parts.

The saltwater–freshwater transition zone in the northern Atlantic Coastal Plain is characterized by increases in concentrations of calcium, magnesium, sodium, potassium, and sulfate concomitant with the increases in chloride concentration and depth. Bicarbonate, on the other hand, tends to decrease as chloride concentration and depth increase (Meisler and others, 1984). The generally linear relation of concentrations of these ions to chloride concentration suggests that the transition zone in North Carolina is primarily a mixture of sea water and fresh sodium bicarbonate water. From Virginia to New Jersey, the transition zone appears to be largely a mixture of fresh sodium bicarbonate water and sodium calcium chloride brine that has a chlorinity several times that of sea water. A similar brine occurs beneath the seawater–freshwater mixing zone in the vicinity of Pamlico and Albemarle Sounds in North Carolina (Manheim and Horn, 1968, Table 2).

The relation of chloride concentration to other ion concentrations in the ground water from Virginia to New Jersey suggests that geochemical processes other than fresh water–brine mixing are occurring. These include (1) mixing with a third end member—sea water—particularly in the upper part of the transition zone and (2) ion exchange, by which the solution becomes enriched with calcium in the concentrated water and with sodium and potassium in the dilute water. In southwestern Long Island, on the other hand, the relation of chloride concentration to other ion concentrations in the salty ground water from the Magothy aquifer indicates that the water is a simple mixture of sea water and fresh water, thus supporting the hypothesis that the salt water is unrelated to the regional salt water in New Jersey.

The origin of brines in the Atlantic Coastal Plain is not well understood. Two possible sources are (1) leaching of evaporitic strata and (2) concentration of dissolved solids through membrane filtration. Graf (1982) discusses three of the most important mechanisms in the development of the overpressures that are needed to drive membrane filtration. One of these—tectonic compression—has not been of sufficient magnitude to cause overpressuring in the Atlantic Coastal Plain. A second mechanism—abnormally high geothermal gradients—appears to be ruled out by the geothermal data available (Robbins, 1977, p. 44–45; Trapp and others, 1984, p. 17). The third mechanism—rapid accumulation of fine-grained sediments—has been evaluated by Bredehoeft and Hanshaw (1968) who concluded that "a sedimentation rate of 500 m/10$^6$ yr (reasonable for the Gulf Coast) will create fluid pressures approaching lithostatic in a sedimentary column that has a hydraulic conductivity of 10$^{-8}$ cm sec$^{-1}$, or lower." In the northern Atlantic Coastal Plain, however, sedimentation rates have generally been 25 to 50 m/10$^6$ yr in the outer Continental Shelf and considerably lower on the emerged Coastal Plain. Head measurements in Coastal Plain wells and reported pressures in several offshort oil tests do not indicate excess fluid pressures. Any excess pressure generated has probably been dissipated laterally updip or toward the Continental Slope and Rise. Hence, membrane filtration is probably not important in the Atlantic Coastal Plain.

Drilling and geophysical data (Mattic and Bayer, 1980, p. 8) indicate the presence of evaporitic strata of probable Early and Middle Jurassic age beneath the Atlantic Continental Shelf and Slope. Leaching of these strata is the most likely source of brine in the Atlantic Coastal Plain. Because of the predominantly marine environment of the Continental Shelf and Slope during most of the Early and Middle Jurassic, water of marine origin has probably been the principal leaching agent. Burial of the evaporites under several thousand feet of carbonate sediments prevented fresh water from reaching the evaporites during periods of emergence during and after the Late Jurassic. The brine, formed by leaching, advanced landward, initially possibly due to basinal compaction, and subsequently because of added head provided by major sea-level rises. Diffusion of the salt vertically through the sediments above the evaporites, as postulated by Manheim and Hall (1976, p. 699), probably provided a larger volume of brine for the lateral migration.

# REFERENCES CITED

Back, W., 1966, Hydrochemical facies and ground-water flow patterns in northern part of Atlantic Coastal Plain: U.S. Geological Survey Professional Paper 498–A, p. A1–A42.

Barker, R. A., 1986, Preliminary results of a steady-state ground water flow model of the Southeastern Coastal Plain regional aquifer system, *in* Proceedings of the Southern Regional Ground Water Conference, Sept. 18–19, 1985, San Antonio, Texas: Association of Ground Water Scientists and Engineers, Division of National Water Well Association, p. 315–338.

Bredehoeft, J. D., and Hanshaw, B. B., 1968, On the maintenance of anamolous pressures; I, Thick sedimentary sequences: Geological Society of America Bulletin, v. 79, p. 1097–1106.

Brown, P. M., Miller, J. A., and Swain, F. M., 1972, Structural and stratigraphic framework, and spatial distribution of permeability of the Atlantic Coastal Plain, North Carolina to New York: U.S. Geological Survey Professional Paper 796, 79 p.

Chapelle, F. H., 1984, The occurrence of dissolved oxygen and the origin of ferric-hydroxide cemented hardbeds in the Patuxent aquifer, Maryland, *in* Canadian-American Conference of Practical Applications of Ground-Water Geochemistry: Alberta Research Council and National Water Well Association Meeting, Banff, Canada, June 22–26, Proceedings, p. 41–46.

Chapelle, F. H., and Knobel, L. L., 1983, Aqueous geochemistry and the exchangeable cation composition of glauconite in the Aquia aquifer, Maryland: Ground Water, v. 21, no. 3, p. 343–352.

——, 1985, Stable carbon isotopes of $HCO_3^-$ in the Aquia aquifer, Maryland; Evidence for an aquifer-generated source of $CO_2$: Ground Water, v. 23, no. 5, p. 592–599.

Foster, M. D., 1950, The origin of high sodium bicarbonate waters in the Atlantic and Gulf Coastal Plains: Geochimica et Cosmochimica Acta, v. 1, p. 33–48.

Graf, D. L., 1982, Chemical osmosis, reverse chemical osmosis, and the origin of subsurface brines: Geochimica et Cosmochimica Acta, v. 46, no. 8, p. 1431–1448.

Lee, R. W., DeJarnette, S. S., and Barker, R. A., 1985, Distribution and altitude of the top of saline ground water in the Southeastern Coastal plain: U.S. Geological Survey Water-Resources Investigations Report 84-4109, 1 sheet.

Manheim, F. T., and Hall, R. E., 1976, Deep evaporitic strata off New York and New Jersey; Evidence from interstitial water chemistry of drill cores: U.S. Geological Survey Journal of Research, v. 4, no. 6, p. 697–702.

Manheim, F. T. and Horn, M. K., 1968, Composition of deeper subsurface waters along the Atlantic Continental Margin: Southeastern Geology, v. 9, no. 4, p. 215–236.

Mattic, R. E., and Bayer, K. E., 1980, Geologic setting and hydrocarbon exploration activity, *in* Scholle, P. A., ed., Geologic studies of the COST No. B-3 well, United States Mid-Atlantic Continental Slope area: U.S. Geological Survey Circular 833, p. 4–12.

Meisler, H., 1987, The occurrence and geochemistry of salty ground water in the northern Atlantic Coastal Plain: U.S. Geological Survey Professional Paper 1404-D (in press).

Meisler, H., Leahy, P. P., and Knobel, L. L., 1984, The effect of eustatic sea-level changes on saltwater–freshwater relations in the northern Atlantic Coastal Plain: U.S. Geological Survey Water-Supply Paper 2255, 28 p.

Miller, J. A., 1986, Hydrogeologic framework of the Floridan aquifer system in Florida and in parts of Georgia, South Carolina, and Alabama: U.S. Geological Survey Professional Paper 1403-B, 91 p.

Olsson, R. K., 1980, The New Jersey Coastal Plain and its relationship with the Baltimore Canyon trough, *in* Manspeizer, W., ed., Field studies of New Jersey geology and guide to field trips; 52nd Annual Meeting of the New York State Geological Association: Newark, New Jersey, Geology Depart-ment, Rutgers University, p. 116–129.

Owens, J. P., and Sohl, N. F., 1969, Shelf and deltaic paleoenvironments in the Cretaceous–Tertiary formations of the New Jersey Coastal Plain, *in* Subitzky, S., ed., Geology of selected areas in New Jersey and eastern Pennsylvania and guidebook of excursions; Geological Society of America and associated societies, November 1969, Annual Meeting, Atlantic City: New Brunswick, New Jersey, Rutgers University Press, p. 235–278.

Perlmutter, N. M., Geraghty, J. J., and Upson, J. E., 1959, The relation between fresh and salty ground water in southern Nassau and southeastern Queens Counties, Long island, New York: Economic Geology, v. 54, p. 416–435.

Reinhardt, J., and Gibson, T. G., 1980, Upper Cretaceous and lower Tertiary geology of the Chattahoochee River valley, western Georgia and eastern Alabama, *in* Frey, R. W., ed., Excursions in southeastern geology, v. 2: Geological Society of America, Annual Meeting, Atlanta Field Trip Guidebooks, p. 385–463.

Rhodehamel, E. C., 1977, Lithological descriptions, *in* Scholle, P. A., ed., Geological studies on the COST No. B-2 Well, U.S. Mid-Atlantic Outer Continental Shelf area: U.S. Geological Survey Circular 750, p. 15–22.

Robbins, E. I., 1977, Geothermal gradients, *in* Scholle, P. A., ed., Geological studies on the Cost No. B-2 well, U.S. Mid-Atlantic outer Continental Shelf: U.S. Geological Survey Circular 750, p. 44–45.

Seaber, P. R., Variations in chemical character of water in the Englishtown Formation, New Jersey: U.S. Geological Survey Professional Paper 498–B, p. B1–B35.

Sohl, N. F., 1977, Benthic marine molluscan associations from the Upper Cretaceous of New Jersey and Delaware, *in* Owens, J. P., Sohl, N. F., and Minard, J. P., eds., A field guide to Cretaceous and lower Tertiary beds of the Raritan and Salisbury embayments, New Jersey, Delaware, and Maryland: Annual American Association of Petroleum Geologists/Society of Economic Paleontologists and Mineralogists Convention, Washington, D.C., p. 70–91.

Stricker, V. A., 1983, Base flow of streams in the outcrop area of southeastern sand aquifer; South Carolina, Georgia, Alabama, and Mississippi: U.S. Geological Survey Water Resources Investigation Report 83-4106, 17 p.

Trapp, H., Jr., Knobel, L. L., Meisler, H., and Leahy, P. P., 1984, Test well DO-CE 88 at Cambridge, Dorchester County, Maryland: U.S. Geological Survey Water Supply Paper 2229, 48 p.

Upson, J. E., 1966, Relations of fresh and salty ground water in the northern Atlantic Coastal Plain of the United States: U.S. Geological Survey Professional Paper 550-C, p. C235–C243.

Vail, P. R., Mitchum, R. M., Jr., and Thompson, S., III, 1977, Seismic stratigraphy and global changes of sea level; Part 4, Global cycles of relative changes of sea level, *in* Payton, C. E., ed., Seismic stratigraphy; Applications to hydrocarbon exploration: American Association of Petroleum Geologists Memoir 26, p. 83–97.

Zapecza, O. S., 1987, Hydrogeologic framework of the New Jersey Coastal Plain: U.S. Geological Survey Professional Paper 1404-B (in press).

Zellmer, L. R., 1979, Development and application of a Pleistocene sea level curve to the Coastal Plain of southeastern Virginia [M.S. thesis]: Williamsburg, Virginia, College of William and Mary, 85 p.

Zullo, V. A., and Harris, W. B., 1979, Plio-Pleistocene crustal warping in the outer Coastal Plain of North Carolina, *in* Baum, G. R., Harris, W. B., and Zullo, V. A., eds., Structural and stratigraphic framework for the Coastal Plain of North Carolina: Carolina Geological Society and Atlantic Coastal Plain Geological Association, Field Trip Guidebook, Oct. 19–21, 1979, p. 31–40.

MANUSCRIPT ACCEPTED BY THE SOCIETY APRIL 17, 1987

## Chapter 26

# *Region 23, Gulf of Mexico Coastal Plain*

**Hayes F. Grubb**
*U.S. Geological Survey, 211 East 7th Street, Austin, Texas 78701*
**J. Joel Carillo R.**
*C.F.E. Geohidrologia, Oklahoma 85, 8th Floor, Colonia Napoles, Mexico, D.F. 03810*

## HYDROGEOLOGIC SETTING

The Coastal Plain Physiographic Province of the south-central United States and eastern Mexico (Fig. 3, Table 2, Heath, this volume) is a gently rolling to flat region of about 670,000 km$^2$ extending from the southern tip of Illinois to the Laguna de Términos near the southwestern terminus of the Yucatán peninsula (Fig. 1). It is underlain by a gulfward-thickening wedge of unconsolidated to semiconsolidated sedimentary rocks of Cenozoic age. These sediments overlie rocks of Mesozoic age and range in thickness from a few meters near their landward limit to more than 10,000 m in southern Louisiana. The Cenozoic rocks discussed herein consist mostly of sand, silt, and clay derived by erosion from nearby continental upland areas. Many regional aquifers have developed in these sediments. The older Coastal Plain sediments of Cretaceous age in Mexico predominately are rocks of marine origin with minimal permeability and do not contain regionally significant aquifers, due to: (1) the presence of areally extensive bentonite deposits; (2) the occurrence of coarse material in a matrix of fine-grained sediments; and (3) the predominance of sediments like shale and compact sandstone. Where significant permeability occurs in these rocks, the contained fluids are typically saline water or oil. Older Coastal Plain sediments of Cretaceous age in the United States are discussed in Jorgensen and others (this volume) and Meisler and others (this volume).

The Coastal Plain rocks dip gently toward the Gulf of Mexico except where regional structural features such as arches, growth faults, embayments, horsts, grabens, and uplifts affect the distribution and thickness of the sediments. The dip and, in some instances, the thickness of the sediments are affected locally by salt diapirs rising from deep underlying Jurassic salt beds, which have pierced the Cenozoic sediments. The most abrupt increase in thickness of individual units within short distances is associated with growth faults (faults forming contemporaneously with deposition; Bruce, 1972). Associated with the regional growth faulting, a zone of abnormally high fluid pressure (also called geopressure; Dickinson, 1953; Jones, 1969) has developed onshore in sediments of Eocene age in parts of coastal Louisiana, Texas, Nuevo León, and Tamaulipas. The top of geopressure or the transition from normal hydrostatic pressure to abnormally high pressure is considered the base of the groundwater flow system for purposes of this discussion.

The geopressured zone typically has been a horizon for the production of hydrocarbons and saline water associated with the withdrawal of the hydrocarbons. Water from the geopressured zone typically has a temperature greater than 100° Celsius and contains varying quantities of dissolved methane. Energy concerns of the past decade have led to increased attention to the geopressured zone and the waters that occur there (Papadopulos and others, 1975; Wallace and others, 1979, 1981). The depth to the geopressured zone varies from less than 2,000 m in parts of coastal Texas to more than 5,000 m in small areas of coastal Louisiana (Wallace and others, 1981). The depth to water with a temperature of 150° Celsius is less than 3,000 m in a large area of south Texas and is more than 4,500 m in most of coastal Louisiana (Wallace and others, 1979). The movement of water in the geopressured zone is slow, and pressure data indicate that movement generally is upward toward the meteoric zone. However, because of the negligible permeability of the sediments, the volume of water that flows out of the geopressured zone is very small relative to that volume of water that circulates in the sediments containing meteoric water (Hanshaw and Bredehoeft, 1968, p. 1113).

The rapid deposition of sediments in the slowly subsiding Gulf of Mexico basin has resulted in a large volume of undercompacted sediments near the present-day coast. In isolated areas, pumping of large quantities of fluids (both water and petroleum) from these sediments has accelerated the rate of compaction and has resulted in as much as 1 to 3 m of land-surface subsidence in parts of coastal Louisiana and Texas. Subsidence of more than a fraction of a meter due to pumping has not been detected in sediments older than Pliocene in the western Gulf Coastal Plain.

Grubb, H. F., and Carrillo R., J. J., 1988, Region 23, Gulf of Mexico Coastal Plain, *in* Back, W., Rosenshein, J. S., and Seaber, P. R., eds., Hydrogeology: Boulder, Colorado, Geological Society of America, The Geology of North America, v. O-2.

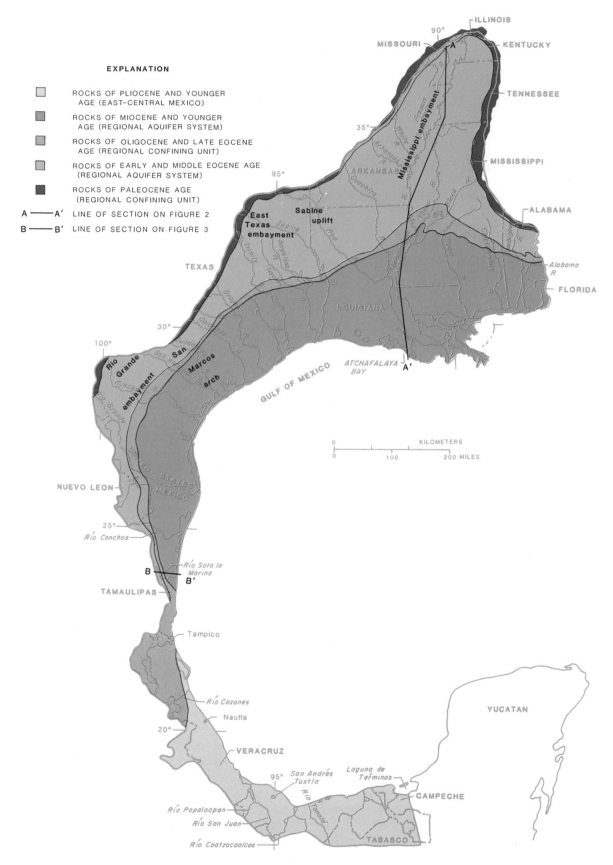

Figure 1. Outcrop area of hydrogeologic units and regional structural features.

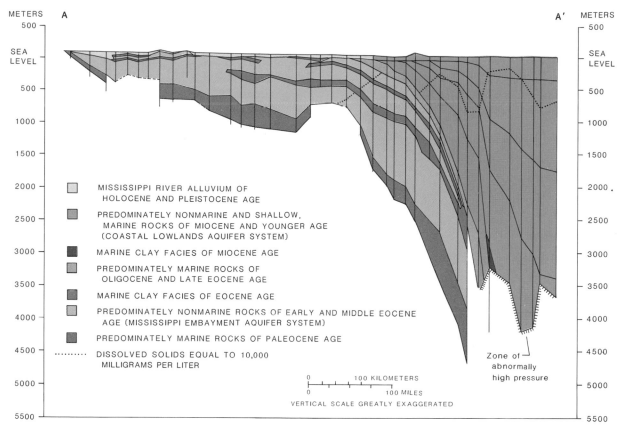

Figure 2. Coastal Plain rocks of Cenozoic age from southern Missouri to Atchafalaya Bay, Louisiana.

Sediment has been derived from adjacent upland areas, transported by streams, and deposited in fluvial, deltaic, and shallow marine environments, resulting in a complexly interbedded sequence of sand, silt, and clay with minor beds of lignite, gravel, and limestone. Regionally extensive sediments with either substantial or minimal permeability are the result of major marine transgressions and regressions. The shifting of facies both vertically and horizontally in response to marine transgression and regression results in complex interbedding of rock types both regionally and locally.

The water-bearing characteristics of the sediments were affected by the source rocks, weathering, tectonic phenomena, and the energy of the streams transporting the sediments to the western Gulf Coastal Plain. The contrast in source rocks is illustrated by the larger proportion of coarse-grained sediments in the Mississippi Embayment as compared to the fine-grained, upper(?) Tertiary sediments between the Coatzacoalcos and Tonala Rivers (Fig. 1). The sediments of the Mississippi Embayment were derived from intrusive and metamorphic rocks of the Appalachian Mountains in the southeastern United States, whereas the sediments at the southern end of the Coastal Plain were derived from the fine-grained sedimentary rocks of the Sierra Madre Oriental in east-central Mexico.

The sediments from near Río Soto La Marina, Tamaulipas, to the Laguna de Términos contain volcanic tuff and ash, which are abundant in some formations. Most of the surficial rocks of the western Gulf Coastal Plain between 20° and 19° north and in San Andrés Tuxtla, Veracruz, consist of impermeable andesite and diorite. These igneous rocks of Oligocene and Miocene age are partially covered by thin permeable basalt flows of Pliocene and Pleistocene age. The permeable basalt flows and alluvial sediments along the major streams constitute the major aquifers in this section of the Coastal Plain from near the Río Cazones to the Laguna de Téminos.

## HYDROGEOLOGIC UNITS

The complex sequence of Tertiary rocks is divided into four major hydrogeologic units in the United States and northeastern Mexico on the basis of sediment age and the predominate depositional environment. These units depicted in Figures 2 and 3 and Table 1 are: (1) rocks of Paleocene age, predominately of marine origin; (2) rocks of early and middle Eocene age, predominately of nonmarine origin; (3) rocks of late Eocene and Oligocene age, predominately of marine origin; and (4) rocks of Miocene and younger age deposited predominately in nonmarine and shallow marine environments. A fifth major hydrogeologic unit comprises volcanic rocks of Pliocene and Pleistocene age, and Quaternary alluvium associated with the major streams in east-central Mex-

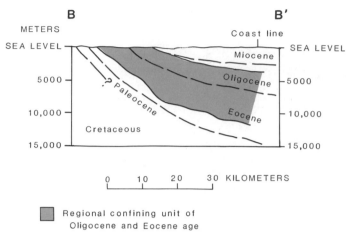

Figure 3. Coastal Plain rocks of northeastern Mexico.

ico. It is not present along the lines of section shown in Figures 2 and 3.

## Rocks of Paleocene age

Rocks of Paleocene age constitute the Midway Group that consists of regionally extensive massive clay in the western Gulf Coastal Plain. This unit represents a stage of marine transgression and is a regional confining unit that restricts the vertical movement of water between underlying rocks of Cretaceous age and overlying rocks of Eocene age. In the United States the unit has a maximum thickness of about 1,100 m in east-central Louisiana and averages about 240 m thick. In northeastern Mexico, available information (López-Ramos, 1982) indicates that the unit has a maximum thickness of more than 1,000 m.

## Rocks of early and middle Eocene age

Three major regressive stages of the sea, separated by at least two major marine transgressions, are represented by sediments in this unit, which is composed of the Wilcox Group (containing Paleocene rocks in the lower part) and the Claiborne Group in the United States. The regionally extensive sand strata of the Claiborne Group and the underlying upper part of the Wilcox Group represent the three regressive stages of deposition, which were dominated by fluvial systems and are shown as three major aquifers of the Mississippi Embayment aquifer system on Figure 2. The two stages of marine transgression within the Claiborne Group are represented by the regionally persistent clay beds (Fig. 2) that restrict the vertical movement of water between the sand facies. Numerous minor marine transgressions and regressions are represented by the major part of the Wilcox Group, typically consisting of thinly interbedded sand and clay. The unit composed of rocks of Eocene age (includes Paleocene rocks in the basal part) is approximately equivalent to two aquifer systems (Table 1), the Mississippi Embayment aquifer system and the

Texas Coastal Uplands aquifer system, defined by Grubb (1984). In the United States the unit reaches a maximum thickness of about 2,300 m, in east-central Texas, and generally is more than 300 m thick except for a large area around the margin of the Mississippi Embayment and a large area in the vicinity of the Sabine uplift (Fig. 1).

The unit continues into Nuevo León, Tamaulipas, and Veracruz, Mexico. Guzmán (1952, p. 1213) mapped what is thought to be a combined thickness of this unit and rocks of Paleocene age north of Tampico, and showed a maximum thickness of more than 6,000 m in the Rio Grande Embayment. The combined thickness of the Paleocene, and early and middle Eocene units in south Texas, as used in this report, is not directly comparable to thicknesses mapped by Guzmán. This is because the thickness mapped in this report in Texas and Louisiana does not include sediments in the geopressured zone; however, Guzmán's map does not distinguish between normal and geopressured zones. The depth to the top of the geopressured zone in Mexico has not been mapped. However, maps of the top of the geopressured zone in south Texas (Wallace and others, 1981) suggest that several thousand meters of sediments of Eocene and Paleocene age in northeastern Mexico lie below the top of geopressure. The early and middle Eocene unit typically contains moderately saline to very saline water in much of northeastern Mexico. Outcrops of igneous rocks of Oligocene and Miocene age near 20° north latitude mark the southern limit of the unit.

## Rocks of late Eocene and Oligocene age

This unit represents a regional transgression of the sea and its rocks predominately are massive marine clay. These rocks comprise the undifferentiated Jackson and Vicksburg Groups. They are of negligible permeability and constitute a regional confining unit in the western Gulf Coastal Plain that restricts the vertical movement of water between the underlying predominately nonmarine rocks of early and middle Eocene age and the overlying rocks of Miocene age. In the United States the late Eocene and Oligocene unit has a maximum thickness of about 2,000 m in southeastern Louisiana, and averages about 210 m thick. The unit's effectiveness as a regional confining unit extends southward into Veracruz. However, some locally permeable zones containing oil and gas are found in these rocks from Tampico southward to Nautla. These permeable zones typically are thin and consist of sandstone or coral reef materials.

## Rocks of Miocene and younger age

This unit is a thick sequence of rocks deposited in fluvial, deltaic, and shallow-marine environments. The resulting individual, interbedded sandstone and shale beds are not regionally extensive. Therefore, at a regional scale, aquifers in this unit are composed of about equal proportions of sediments with relatively substantial and minimal permeability. Two regionally extensive clay facies that represent major marine transgressions occur in the deep subsurface of coastal and offshore Louisiana and Texas, but they do not extend updip to the outcrop area.

TABLE 1. GEOLOGIC AND HYDROGEOLOGIC UNITS, WESTERN GULF COASTAL PLAIN

| Geologic Unit | | | | United States | Northeastern Mexico | East-central Mexico |
|---|---|---|---|---|---|---|
| Erathem | System | Series | Group | Hydrogeologic Unit | Hydrogeologic Unit | Hydrogeologic Unit |
| Ceno-zoic | Quat-ernary | Pleis-tocene and Holo-cene | | Predominantly nonmarine and shallow marine rocks. Coastal lowlands aquifer system | Predominantly nonmarine and shallow marine rocks. Multiaquifer system. Miocene part has minimal permeability | Predominantly igneous and volcanic rocks.. Aquifers are in basalt flows and alluvial valley sediments. |
| | | Plio-cene | | | | |
| | | Mio-cene | | | | |
| | Tertiary | Eocene and Oligo-cene | Jackson and Vicks-burg | Predominantly marine rocks. No regional aquifers; a regional confining unit. | Predominantly marine rocks. No regional aquifers; a regional confining unit | |
| | | Eocene | Clai-borne | Predominantly nonmarine rocks. Mississippi embayment aquifer system. | Predominantly nonmarine rocks. Multi-aquifer system containing mostly brakish to salty water | |
| | | ? | Wil-cox | Texas Coastal Uplands aquifer system. | | |
| | | Paleo-cene | Mid-way | Predominantly marine rocks. No regional aquifers; a regional confining unit. | Predominantly marine rocks. No regional aquifers; a regional confining unit. | |

Miocene and younger rocks have a maximum thickness of more than 3,000 m in a large area of southern Louisiana and a smaller area of south Texas, and average about 1,500 m thick. The unit extends to the vicinity of the Soto la Marina River in northeastern Tamaulipas where Guzmán (1952) mapped thicknesses of 600 to 3,000 m in an area of about 16,000 $km^2$. Sediments of Miocene age in the Veracruz area are typically calcareous sandstone, shale, and sandy limestone with minimal permeability.

The unit is approximately equivalent to the Coastal Lowlands aquifer system (Table 1) in the United States as defined by Grubb (1984), and has been subdivided locally in several ways by different investigators (Wood and Gabrysch, 1965; Jones and others, 1956; Jorgensen, 1975). The most recent subdivision suggested by Weiss and Williamson (1985) is based on the vertical hydraulic-head gradient in areas of ground-water pumping with continuity of subdivided units extended arbitrarily by maintaining more-or-less uniform thickness along strike between pumped areas, and extrapolating reasonable gulfward thickening downdip from pumped areas. This method of subdivision results in a regionally consistent geohydrologic framework with a maximum vertical hydraulic-head gradient between subdivisions and a minimum vertical hydraulic-head gradient within subdivisions.

### Rocks of Pliocene and younger age

This unit, which is restricted to Mexico, is a combination of a thin sequence of basalt flows with the alluvium of the major streams that flow from the Sierra Madre Oriental across the Coastal Plain to the Gulf of Mexico. The basalt deposits generally are several tens of meters thick and may cover several thousand square kilometers. The alluvial deposits generally are several tens of meters thick and occur in narrow bands parallel to the major streams. This is the principal hydrogeologic unit in coastal Veracruz and Tabasco, Mexico (Fig. 1).

## GROUND-WATER RECHARGE, FLOW PATTERNS, AND DISCHARGE

Precipitation, the ultimate source of all recharge to the sediments of the western Gulf Coastal Plain, varies on an annual basis, from a maximum of more than 1,600 mm in coastal areas of the extreme southern parts of Louisiana and Mississippi, to a minimum of less than 600 mm in a large area of south Texas and northeastern Mexico. Annual precipitation increases gradually southward from about 700 mm in central Tamaulipas to more than 2,500 mm in the vicinity of Río Coatzacoalcos in Mexico. However, only a small percentage of the annual precipitation percolates to the ground-water system after the demands of evaporation, transpiration, and surface runoff are satisfied.

Regional ground-water recharge, flow patterns, and discharge result from the interaction of several factors in the western Gulf Coastal Plain. In areas with an abundant supply of precipitation, predevelopment ground-water flow patterns primarily are controlled by topography, permeability, and the geometry of the aquifers. The three major aquifer systems in the western Gulf Coastal Plain of the United States (Table 1) are characterized respectively by: (1) regional discharge to a large area of low topography (Mississippi Embayment aquifer system); (2) regional

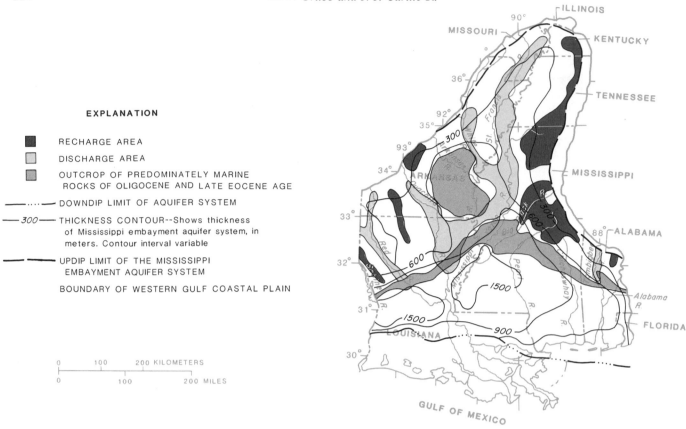

Figure 4. Mississippi Embayment aquifer system, regional recharge and discharge areas, and thickness.

discharge primarily controlled by large streams (Texas Coastal Uplands aquifer system); and (3) regional discharge to areas of low topography, to large streams, and to the Gulf of Mexico (Coastal Lowlands aquifer system).

### Discharge controlled by low topography

The rocks of early and middle Eocene age, which comprise the aquifers and confining units of the Mississippi Embayment aquifer system (Fig. 4), are exposed at land surface on the east side of the Mississippi Embayment. In the western one-half of the embayment, these rocks typically are covered with alluvium deposited by the Mississippi River and its major tributaries, except for an area on the eastern flank of the Sabine uplift in southwestern Arkansas and northwestern Louisiana. The dip of these rocks is toward the axis of the embayment; they are exposed in narrow bands parallel to the margin of the embayment. Topography is a principal control on regional recharge and discharge in this aquifer system. The major recharge areas are along the eastern margin of the system in parts of Mississippi, Tennessee, and Kentucky, where land surface altitudes generally are 60 to 90 m higher than in the Mississippi Alluvial Plain to the west. Some of the water that percolates to the water table in the recharge area moves toward small streams and rivers where it is discharged as

part of the local flow system, and some moves toward the axis of the embayment and becomes part of the regional flow system. Most of the water that becomes part of the regional flow system is discharged to the alluvial aquifer of the Mississippi Alluvial Plain and finally to the water table and to streams that are hydraulically connected to the alluvial aquifer. Gulfward movement of water beneath sediments of late Eocene and Oligocene age is restricted by these predominately marine rocks of minimal permeability. Only a small part of the regional flow in the underlying rocks of early and middle Eocene age discharges upward through the rocks of late Eocene and Oligocene age to the overlying rocks of Miocene and younger age.

### Discharge controlled by major streams

The rocks of early and middle Eocene age, which comprise the Texas Coastal Uplands aquifer system (Table 1 and Fig. 5), are exposed at land surface in narrow bands approximately parallel to the present-day coast except where they extend further inland in the Rio Grande Embayment of south Texas and in the East Texas Embayment. These rocks dip gently toward the Gulf of Mexico beneath sediments of late Eocene and Oligocene age. Regional recharge areas are in the topographically high areas along the inner margin of the Rio Grande Embayment and be-

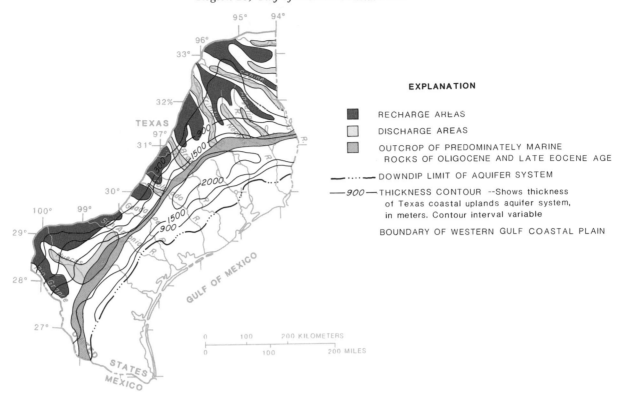

Figure 5. Texas Coastal Uplands aquifer system, regional recharge and discharge areas, and thickness.

tween the major streams. Most of the regional discharge is to the major stream valleys that generally are oriented perpendicular to the present-day Gulf coast. Flow paths generally are relatively short because major streams usually are less than 80 km apart. Between the Rio Grande and East Texas embayments, gulfward movement of water beneath sediments of late Eocene and Oligocene age is restricted by the predominately marine rocks of minimal permeability. Only a small part of the regional flow in the underlying rocks of early and middle Eocene age discharges upward through the rocks of late Eocene and Oligocene age to the overlying rocks of Miocene and younger age. The discharge from the thin basalt aquifers of Veracruz is also primarily controlled by major streams.

### *Discharge to streams, topographic lows, and the Gulf of Mexico*

The rocks of Miocene and younger age that comprise the Coastal Lowlands aquifer system (Fig. 6) crop out in bands parallel to the present-day Gulf coast. These rocks dip gently toward the axis of the Gulf of Mexico geosyncline. Regional recharge is in a band of topographically high areas between the major streams. The recharge areas extend gulfward about 30 to 120 km from the inland extent of Miocene sediments except east of the Mississippi River where the gulfward extent of recharge areas is much greater. Regional discharge is: (1) to the major streams that flow gulfward between the recharge areas: (2) to the water table and streams in an area of lower topography about midway be-

tween the recharge areas and the Gulf of Mexico; and (3) to the Gulf of Mexico. The oldest aquifers in sediments of Miocene age are disrupted downdip near the coast in Louisiana and Texas by the geopressured zone (Fig. 6). In younger aquifers the top of geopressure is located offshore; however, only a small part of the regional discharge is to the Gulf of Mexico, due to the relatively low topography in the outcrop areas and the large distance to the shoreline. In central Veracruz and Tabasco, movement of ground water is restricted in Miocene and younger rocks because the permeability of the rocks is minimal.

## GROUNDWATER GEOCHEMISTRY

The sediments of the western Gulf Coastal Plain typically are saturated with dilute meteoric waters in and near the areas of recharge. The water becomes increasingly saline downdip toward the Gulf of Mexico (Pettijohn, 1987). Because of the flat topography, there are large areas near the coast where the near-surface sediments contain slightly to moderately saline waters. During periods of low streamflow, saline water may move tens of kilometers upstream from the coast. Locally, pumping of ground water from aquifers near the coast can induce movement of saline waters landward as has occurred near Galveston, Texas.

The volume of fresh water in rocks of early and middle Eocene age seems to be related to the area of outcrop updip of the major regional confining unit in rocks of late Eocene and Oligocene age (undifferentiated Jackson and Vicksburg Groups) and to regional ground-water flow patterns. The Mississippi Embayment

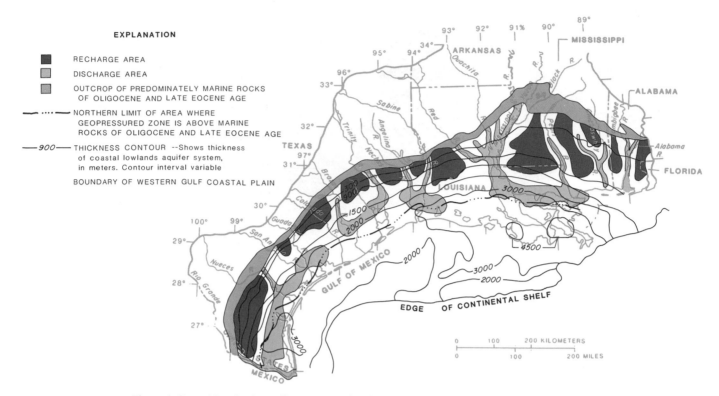

Figure 6. Coastal Lowlands aquifer system, regional recharge and discharge areas, and thickness.

aquifer system is characterized by a large extensive area (more than 10,000 km²) of fresh water and many thousand square kilometers of outcrop area updip from the regional confining unit of late Eocene and Oligocene age (Fig. 4). The Texas Coastal Uplands aquifer system is characterized by a relatively small area of fresh water in an elongate area about 50 km wide between the East Texas and Rio Grande embayments and updip from the regional confining unit of late Eocene and Oligocene rocks. Fresh water occurs throughout several thousand square kilometers in the east Texas and Rio Grande embayments where there is a more extensive outcrop area updip from the regional confining unit. The coastal uplands of northeastern Mexico are characterized by a relatively small area of fresh water and an outcrop area that typically is less than 50 km wide updip from the regional confining unit of late Eocene and Oligocene rocks.

The typical pattern of dilute meteoric waters in and near outcrop areas and increasing salinity toward the Gulf of Mexico is illustrated by the distribution of dissolved solids in water from rocks of the Claiborne Group in the Mississippi Embayment aquifer system (Fig. 7). Fresh to slightly saline water occurs in rocks of Claiborne age throughout most of the area (Pettijohn, 1987). However, the thickest part of the Mississippi Embayment aquifer system occurs downdip from the outcrop of predominately marine rocks of late Eocene and Oligocene age and is almost entirely within the area of moderately saline to very saline water (Grubb, 1986, p. 11). The distance downdip from slightly saline to very saline water is about 15 to 65 km. The updip

extension of saline water that is coincident with the axis of the Mississippi Embayment is related to the flow pattern from the recharge areas along the east margin of the embayment toward the discharge area centered in northeastern Louisiana (Fig. 4). Water moving from the recharge area in east-central Mississippi beneath the predominately marine rocks of late Eocene and Oligocene age has flushed out the more saline water from the rocks of the Claiborne Group (Fig. 7) throughout a large area of south-central Mississippi.

The extent of fresh water in the rocks of Miocene and younger age is related to the altitude of the outcrop area, distance from the Gulf of Mexico, climate, and regional ground-water flow patterns. Fresh water at depth extends to the coast in Alabama and Florida, and to the barrier islands of southeastern Mississippi. In this part of the coastal lowlands, the land surface rises to an altitude greater than 30 m within 15 km of the coast. Across southern Louisiana, the distance inland from the Gulf of Mexico to areas where land-surface altitudes are greater than 30 m typically is 50 to 90 km. Slightly saline water commonly is present in Miocene and younger rocks within the top meter of depth within a distance of 25 to 80 km of the coast.

Slightly saline water is present in the shallowest sediments within 15 km of the coast in south Texas. Fresh water occurs only locally in northeastern Mexico. The distance from the coast to areas of land-surface altitude greater than 30 m across the coastal lowlands of Texas is typically 50 to 90 km. The distance generally decreases from east to west across Texas and from north to

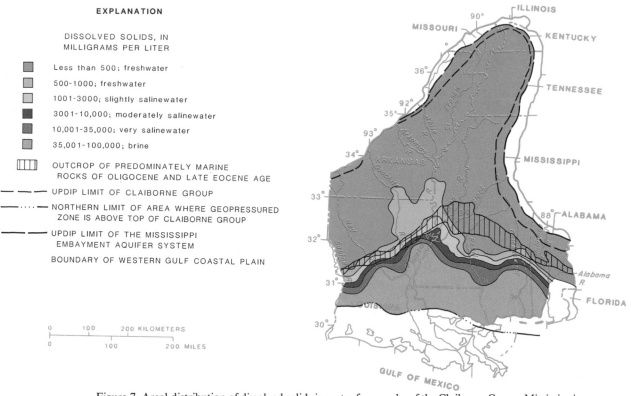

Figure 7. Areal distribution of dissolved solids in water from rocks of the Claiborne Group, Mississippi Embayment aquifer system. (Modified from Pettijohn, 1987.)

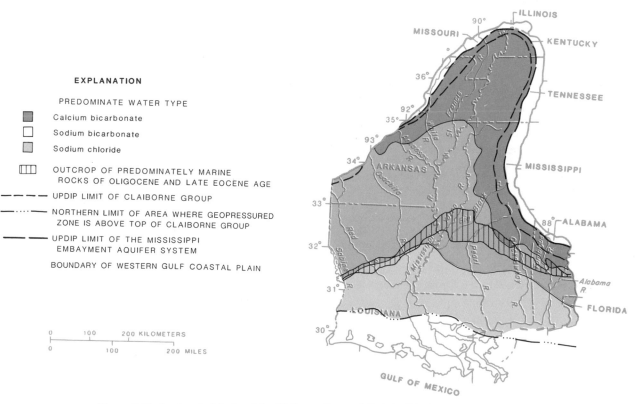

Figure 8. Hydrochemical facies of the Claiborne Group, Mississippi Embayment aquifer system. (Modified from Pettijohn, 1987.)

south in northeastern Mexico until the entire outcrop of Miocene and younger sediments is only about 60 km wide at the Río Conchos. These rocks are present only offshore at a latitude of about 24.5° north. The increasing salinity of water in the shallow sediments of south Texas and northeastern Mexico is not totally explained by topography, but probably is related to less precipitation and to less permeable sediments in the subsurface. Both of the latter factors would inhibit the flushing of saline water from the rocks.

The distribution of geochemical facies in waters of the Claiborne Group (Fig. 8) is typical of that occurring in other sediments in the humid northern parts of the western Gulf Coastal Plain (Pettijohn, 1987). Calcium bicarbonate is the predominate water type along the eastern margin of the Claiborne outcrop in Mississippi and in parts of western Kentucky and Tennessee in and near regional recharge areas. Sodium bicarbonate is the predominate water type in the middle third of the embayment. Sodium chloride is the predominate water type in the southern part of the embayment. The transition zone from slightly saline to very saline water is coincident with the change in water type from sodium bicarbonate to sodium chloride.

In this part of the western Gulf Coastal Plain, the principal geochemical processes in the meteoric part of the flow system are: (1) dissolution of calcium bicarbonate in the recharge areass, (2) cation exchange of sodium for calcium generally downdip from the recharge areas, and (3) mixing with saline waters in the deeper parts of the subsurface.

The mechanisms responsible for the brine (dissolved solids concentrations greater than 35,000 mg/l) found at depth in the southern part of the Mississippi Embayment aquifer system (Fig. 7) are not completely understood. Dissolution of evaporite minerals from the sediments, and ultrafiltration are two mechanisms that have been suggested for the origin of these brines (DeSitter, 1947; Hanor, 1979, 1986), but experimental data to test these or other theories are difficult to obtain because of the enormous expense of drilling to the required depths.

## SELECTED REFERENCES

Bruce, C. H., 1972, Pressured shale and related sediment deformation; A mechanism for development of regional contemporaneous faults: Gulf Coast Association of Geological Societies Transactions, v. 22, p. 23–31.

DeSitter, L. U., 1947, Diagenesis of oil-field brines: American Association of Petroleum Geologists Bulletin v. 31, p. 2030–2040.

Dickinson, G., 1953, Reservoir pressures in Gulf Coast Louisiana: American Association of Petroleum Geologists Bulletin, v. 37, p. 410–432.

Grubb, H. F., 1984, Planning report for the Gulf Coast Regional Aquifer-System Analysis in the Gulf of Mexico Coastal Plain, United States: U.S. Geological Survey Water-Resources Investigations 84-4219, 30 p.

—— , 1986, Gulf Coast Regional Aquifer-System Analysis; A Mississippi perspective: U.S. Geological Survey Water-Resources Investigations 86-4126, 22 p.

Guzmán, E. J., 1952, Sedimentary volumes in Gulf Coastal Plain of the United States and Mexico; Part V, Volumes of Mesozoic and Cenozoic sediments in Mexican Gulf Coastal Plain: Geological Society of America Bulletin, v. 63, p. 1201–1220.

Hanshaw, B. B., and Bredehoeft, J. D., 1968, On the maintenance of anomalous fluid pressures: II. Source layer at depth: Geological Society of America Bulletin, v. 79, p. 1107–1122.

Hanor, J. S., 1979, The sedimentary genesis of hydrothermal fluids, in Barnes, H. L., ed., Geochemistry of hydrothermal ore deposits (second edition): New York, John Wiley & Sons, p. 137–168.

—— , 1986, Evidence for large-scale overturn of pore fluids in the Louisiana Gulf coast: Geological Society of America Abstracts with Programs, v. 18, p. 627.

Jones, P. H., 1969, Hydrodynamics of geopressure in the northern Gulf of Mexico basin: Journal of Petroleum Technology, v. 21, p. 803–810.

Jones, P. H., Hendricks, E. L., Ireland, B., and others, 1956, Water resources of southwestern Louisiana: U.S. Geological Survey Water-Supply Paper 1364, 460 p.

Jorgensen, D. G., 1975, Analog-model studies of ground-water hydrology in the Houston District, Texas: Texas Water Development Board Report 190, 84 p.

López-Ramas, E., 1982 Geologia de Mexico v. II, (3rd edition): Mexico, D. F., Tesis Resendiz, S. A., 454 p.

Papadopulos, S. S., Wallace, R. H., Jr., Wesselman, J. B., and Taylor, R. E., 1975, Assessment of onshore geopressured-geothermal resources in the northern Gulf of Mexico basin, in White, D. E., and Williams D. L., eds., Assessment of geothermal resources in the United States, 1975: U.S. Geological Survey Circular 726, p. 125–146.

Pettijohn, R. A., 1987, Dissolved-solids concentrations and primary water types, Gulf coast aquifer systems, south-central United States: U.S. Geological Survey Hydrologic Investigations Atlas 706, scale 1:5,000,000, 2 sheets (in press).

Wallace, R. H., Jr., Kraemer, T. F., Taylor, R. E., and Wesselman, J. B., 1979, Assessment of geopressured-geothermal resources in the northern Gulf of Mexico basin, in Muffler, L.J.P., ed., Assessment of geothermal resources of the United States, 1978: U.S. Geological Survey Circular 790, p. 132–155.

Wallace, R. H., Jr., Wesselman, J. B., and Kraemer, T. F., 1981, Occurrence of geopressure in the northern Gulf of Mexico basin: Geopressured-Geothermal Energy Conference, 5th, 1979, New Orleans, Louisiana, p. 200–220.

Weiss, J. S., and Williamson, A. K., 1985, Subdivision of thick sedimentary units into model layers for simulation of ground-water flow: Ground Water, v. 23, no. 6, p. 767–774.

Wood, L. A., and Gabrysch, R. K., 1965, Analog model study of ground water in the Houston District, Texas: Texas Water Commission Bulletin 6508, 103 p.

MANUSCRIPT ACCEPTED BY THE SOCIETY MARCH 6, 1987

The Geology of North America
Vol. O-2, Hydrogeology
The Geological Society of America, 1988

## Chapter 27

# Region 24, Southeastern United States

**Richard H. Johnston**
*U.S. Geological Survey, 4311 9th St., East Beach, St. Simons Island, Georgia 31522*
**James A. Miller**
*U.S. Geological Survey, 75 Spring St. SW, Suite 772, Atlanta, Georgia 30303*

## REGIONAL SETTING

Unconsolidated to semiconsolidated sedimentary rocks of Jurassic to Holocene age underlie the Coastal Plain province of the southeastern United States (Fig. 3, Table 2, Heath, this volume). These strata, which include highly prolific carbonate aquifers, thicken seaward from a featheredge where they crop out against older crystalline rocks and consolidated sediments of the Appalachian and Piedmont provinces to a maximum thickness of more than 6,400 m in southern Alabama and more than 5,500 m in south Florida. Locally, the generally low-angle seaward dip of Coastal Plain rocks is interrupted by faults or gentle folds. The topography developed on these easily eroded sediments ranges from low rolling hills near their inner margin to flat, wide wetlands near the coast.

Southeastern Coastal Plain sediments can be grouped into two general categories: (1) clastic rocks that contain minor amounts of limestone and extend eastward and southward from the Fall Line (which marks the inland limit of the Coastal Plain) to the Atlantic Ocean and the Gulf of Mexico, and (2) a thick, continuous sequence of shallow-water platform carbonate rocks underlying the Florida peninsula, southeastern Georgia, and adjacent areas. This chapter describes the carbonate rocks that compose prolific aquifers practically everywhere they occur. Figure 1 shows the extent of the carbonate aquifers of the southeastern Coastal Plain and part of the adjacent sand aquifers described by Meisler and others elsewhere in this volume. The sand aquifers are generally of Cretaceous and early Tertiary age, whereas the carbonate aquifers are mostly middle Tertiary age and younger. In north-central Florida and southeastern Georgia, clastic and carbonate rocks interfinger with one another; accordingly, facies changes in this area are abrupt and complex.

The two major carbonate aquifers in the southeastern United States are the Biscayne aquifer and the Floridan aquifer system (Miller, 1986). The Biscayne aquifer underlies a small area in south Florida and includes several limestone and sand units of Pliocene and Pleistocene age. The underlying regionally extensive Floridan aquifer system consists mostly of carbonate strata of

Figure 1. Map showing boundary of the Coastal Plain and extent of the Floridan aquifer system, Biscayne aquifer, and southeastern Coastal Plain sand aquifers.

Eocene age but in places includes beds as old as Late Cretaceous or as young as early Miocene. The Floridan and Biscayne, whose extents are shown in Figure 1, are separated from each other by a clayey confining unit of mostly Miocene but locally also of Pliocene age. The vertical relation between these hydrogeologic units is shown on a generalized hydrogeologic section (Fig. 2).

Johnston, R. H., and Miller, J. A., 1988, Region 24, Southeastern United States, *in* Back, W., Rosenshein, J. S., and Seaber, P. R., eds., Hydrogeology: Boulder, Colorado, Geological Society of America, The Geology of North America, v. O-2.

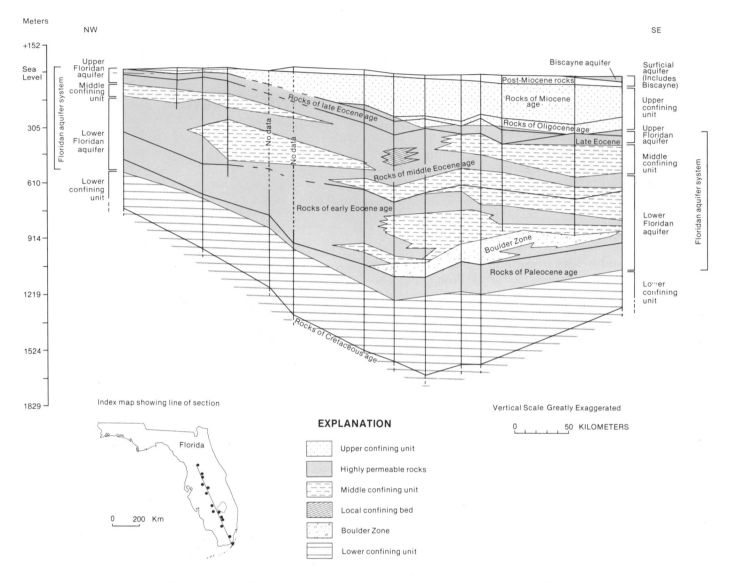

Figure 2. Generalized hydrogeologic section from north-central to southern Florida.

## FLORIDAN AQUIFER SYSTEM

### Hydrogeologic framework

The Floridan aquifer system is a vertically continuous sequence of carbonate rocks of generally high permeability that are mostly of middle and late Tertiary age. The aquifer system is hydraulically connected in varying degrees and its permeability is, in general, an order to several orders of magnitude greater than the rocks that bound the system above and below (Miller, 1986). The aquifer system generally consists of several highly permeable zones, commonly either solution riddled or fractured, separated by less-permeable rocks. Because the aquifer system is defined on the basis of permeability, the top and base of the system locally cross both formation and age boundaries (Fig. 2). Accordingly, the stratigraphic units that make up the aquifer system vary widely from place to place.

In most places the top of the Floridan aquifer system is at or near the top of the Suwannee Limestone of Oligocene age. However, where Oligocene rocks have been removed by erosion, the Floridan's top generally coincides with the top of the Ocala Limestone of late Eocene age. Locally, the Tampa Limestone of early Miocene age is highly permeable and is included in the Floridan. At a few places in central peninsular Florida, upper Eocene through lower Miocene rocks are either missing or are of low permeability; rocks of middle Eocene age, the Avon Park Formation, compose the top of the Floridan. Because the high permeability that distinguishes the Floridan from its overlying confining unit does not follow stratigraphic horizons everywhere, the top of the aquifer system does not everywhere coincide with the top of either rock- or time-stratigraphic units.

The time-stratigraphic units that mark the top and base of the Floridan aquifer system at various places as well as the units that make up the major permeability variations within the Floridan are shown on Table 1. In most places, the Floridan consists of rocks of middle Eocene and late Eocene age. Beds of early Eocene age and Oligocene age are part of the system in places, and locally the system includes beds as old as Late Cretaceous and as young as early Miocene. Regardless of age, the strata that compose the Floridan were originally limestones of various textures deposited on a carbonate platform. Much of this limestone has subsequently been dolomitized in varying degrees. Anhydrite and gypsum occur locally as beds and as pore-filling minerals in the Floridan, mostly in central and south Florida. Where such evaporites are bedded, they represent local tidal-flat or sabkha conditions.

The Floridan generally thickens to the south and southeast because carbonate platform conditions existed for a much longer time in peninsular Florida and southeastern Georgia than elsewhere, allowing thick sections of limestone and dolomite to accumulate. To the north and west, the Floridan grades into clastic beds that are time equivalents of the carbonate rocks that compose the system. Progressively younger carbonate beds of the Floridan generally extend farther to the north and west in an onlap relation.

The base of the Floridan is defined as the level below which there is no high-permeability carbonate rock. The system's base is generally either (1) calcareous clastic rock ranging in age from late Paleocene to late Eocene, or (2) massively bedded anhydrite in the lower part of the Paleocene Cedar Keys Formation in Florida. Like the top of the Floridan, the base of the system represents a permeability contrast that may not coincide everywhere with the top of a rock- or time-stratigraphic unit.

The Floridan aquifer system can be divided in most places into an Upper and a Lower Floridan aquifer that is separated by less-permeable rock (Miller, 1986). These less-permeable carbonate strata may vary in lithologic character and in age, but wherever they are present, they create hydraulically distinct zones within the aquifer system. In places, thick sequences of low-permeability rock may occur at several successive levels within the system (Table 1 and Fig. 2). In other places, the Floridan consists entirely of rocks of sufficiently high permeability to allow good interconnection, and the entire aquifer system is referred to as the Upper Floridan aquifer. Where the system is divided by low-permeability rocks, the Upper Floridan aquifer generally consists of rocks of Oligocene, late Eocene, and late middle Eocene age (Table 1); the Lower Floridan aquifer generally comprises rocks of early middle Eocene to late Paleocene age. The degree of hydraulic separation between the Upper and Lower Floridan aquifers is dependent on the thickness and character of confining units separating them.

The Upper Floridan aquifer is generally more permeable than the Lower Floridan aquifer, except where the latter contains extensive zones of paleokarst such as southern Florida's Boulder Zone. Because it lies at greater depths, the Lower Floridan is not as well known as the Upper Floridan. Deep-well data show, however, the Lower Floridan contains highly mineralized water in many places, suggesting that it moves sluggishly.

### *Development and distribution of permeability*

The platform carbonate rocks that make up the Floridan aquifer system were laid down in warm, shallow water in an environment similar to the modern Bahama Banks. The original texture of the limestone varied considerably, depending upon the local depositional environment; accordingly, the primary porosity and permeability of the limestone, which are largely determined by the primary rock textures, are highly variable. For example, the upper Eocene Ocala Limestone, a part of the Upper Floridan aquifer, consists in many places of a coquina of loosely cemented large foraminiferan and bryozoan fragments, deposited in warm, clear, shallow water. This coquina is highly permeable practically everywhere. By contrast, one of the major confining units within the Floridan aquifer system consists of gypsiferous dolomite of middle Eocene age that was deposited in a tidal flat (sabkha) environment. Gypsum was deposited penecontemporaneously within the granular dolomite framework, resulting in a hydrogeologic unit of very low permeability.

Postdepositional diagenesis can either increase or decrease limestone permeability. Dolomitization has been a major diagenetic process affecting Floridan permeability (Thayer and Miller, 1984). Some Floridan micrites have been recrystallized by the dolomitization process into a highly permeable mosaic of loosely interlocking dolomite crystals. Conversely, in places, finely crystalline dolomite has filled the openings of the loosely packed coarsely pelletal limestones, thereby creating an effective confining unit out of what was previously a highly porous and permeable rock. Dolomitization is not always complete, so some of the original limestone texture and porosity may be preserved.

Some of the extensive dolomitization that has affected the Floridan aquifer system may be the result of ground-water circulation. Hanshaw and others (1971) note that brackish or saline water with Mg/Ca ratios (milliequivalents) greater than 1 can dolomitize limestone. Such waters commonly occur within the zone of diffusion separating fresh ground water and saline water in deeper parts of the Floridan and in coastal areas. Sea-level fluctuations and long-term climatic changes will change the position of the freshwater-saltwater interface and thereby provide an opportunity for large volumes of rock to be dolomitized (Hanshaw and others, 1971). Because flow in the deeper parts of the Floridan aquifer system is sluggish and locally not responsive to head changes in the Upper Floridan aquifer (Bush and Johnston, 1987), it is more likely that dolomitization due to the mixing of saltwater and freshwater occurs in the Upper Floridan in coastal areas where the aquifer is thinly confined, or in other places where a vigorous flow system exists.

**Table 1.—**_Relation of chronostratigraphic units to carbonate aquifers and their confining units in Florida and southeastern Georgia. Column numbers key to locations shown on figure 1._

Circulating fresh ground water increases the permeability of limestone by dissolution. Secondary porosity, developed as the limestone was partially dissolved, ranges from pinpoint vugs to caverns whose diameter may be several meters. The larger solution conduits are generally well connected and greatly increase the local transmissivity of the Floridan. Solution features are expressed at land surface as karst topography and are best developed where the Floridan crops out or is thinly covered. However, the Floridan also contains significant zones of buried karst such as south Florida's Boulder Zone. The extremely high transmissivity of the Boulder Zone, which lies deep within the saline-water part of the Floridan, makes this zone important as the receiving zone for the storage of treated municipal wastewaters along Florida's southeast coast.

The distribution of permeability within the Floridan aquifer system is complex and depends partly on the original texture of the carbonate strata and partly on the postdepositional history of the rocks. Some generalizations can be made, however, about the relation between the geology of the aquifer system and its hydraulic properties.

High average rainfall (about 1,350 mm per year) and generally flat topography combine to provide abundant recharge to the Floridan, especially where it is unconfined or thinly confined. The removal of the Floridan's upper confining unit (composed principally of the Miocene Hawthorn Formation) during the Pleistocene is largely responsible for the distribution of present-day karst, as noted by Stringfield (1966). It is not surprising, therefore, that the distribution of transmissivity in the Upper Floridan is directly related to the thickness and lithology of the upper confining unit.

The areas where the Floridan is confined, thinly confined, and unconfined are shown on Figure 3. Transmissivities are highest (generally greater than 93,000 m$^2$/day) in the karst areas of central and northern Florida where the Floridan is generally unconfined or thinly confined. In these areas the upper part of the aquifer system contains numerous caves, sinkholes, pipes, and other types of solution openings, which account for the high transmissivity. Where the Floridan is thickly confined, its transmissivity is generally less than 23,000 m$^2$/d, and transmissivity variations are due to variations in the lithology and thickness of the aquifer system.

Low values of transmissivity (less than 5,600 m$^2$/day) occur in panhandle Florida and southernmost Florida where the aquifer is confined by thick beds of clay and contains thick beds of low-permeability limestone, and in the updip areas of Alabama, Georgia, and South Carolina where the aquifer is thinnest.

Table 2 shows the ranges of transmissivity in various localities of the Upper Floridan; the values are based on aquifer tests and simulation (Bush and Johnston, 1987). Note that high transmissivity is closely associated with thin confinement and solution-cavity development. Conversely, lower transmissivity generally is associated with thick confinement and lack of solution-cavity development. The distribution of transmissivity in the Upper Floridan ranges from less than 93 m$^2$/day in micritic limestone

Figure 3. Map showing confined and unconfined conditions in the Floridan aquifer system.

that is confined by several hundred meters of clayey sediments in western panhandle Florida to more than 93,000 m$^2$/day near major springs in the unconfined karstic areas of central Florida.

Aquifer tests in the Floridan aquifer system are often complicated by highly variable permeability within the limestone sequence and by complex boundary conditions. Other complications such as anistrophy and partial penetration may further decrease the chances of obtaining an interpretable response in a field test. Single well tests often provide questionable results. However, multiwell tests that utilize distant observation wells and test much larger volumes of aquifer have provided transmissivity values agreeing with regional values derived from computer simulation (Bush and Johnston, 1987). Presumably the scale of such multiwell tests is sufficient to test a representative volume of aquifer and thus provide "average" values of transmissivity. In any event, it would seem likely where cavities exist that are a few meters in length, aquifer tests should involve well spacings of hundreds of meters or more. As an alternative to traditional pumping tests, flow-net analyses and computer simulation have been used successfully to obtain estimates of regional transmissivity (Bush and Johnston, 1987).

### Regional ground-water flow

The existence of a regional flow system in the Floridan has been known since the early 1930s, when V. T. Stringfield (1936)

Table 2.—*Relation of transmissivity of the Upper Floridan aquifer to hydrogeologic conditions.*

| LOCALITY | | | Western Panhandle Florida | Southwest Georgia (Dougherty Plain) | Florida south of Lake Okeechobee | Savannah, Ga. to Jacksonville, Fla. coastal area | Central Florida, northern Florida, and adjacent Georgia | |
|---|---|---|---|---|---|---|---|---|
| | | | | | | | Outside springs area | Major springs area |
| UPPER CONFINING UNIT | Thin (<30m) | Sandy or breached | | ⊗ | | | | ⊗ |
| | | Clayey | | ⊗ | | | ⊗ | |
| | Thick with some clayey beds | | ⊗ | | ⊗ | ⊗ | ⊗ | |
| UPPER FLORIDAN AQUIFER | Thin (<61m) Solution cavities: | Major | | ⊗ | | | | |
| | | Minor | | ⊗ | | | | |
| | Thick Solution cavities: | Major | | | | | ⊗ | ⊗ |
| | | Minor | ⊗ | | ⊗ | ⊗ | ⊗ | |
| RANGE OF TRANSMISSIVITY (m²/d) | | | 93-2300 | 930-19,000 | 930-5600 | 2300-23,000 | Mostly 1900-23,000 / Locally 23,000-93,000 | >93,000 |

of the U.S. Geological Survey published his classic Water-Supply Paper, "Artesian Water in the Floridan Peninsula." In this paper, he identified for the first time a regional flow system in the carbonate rocks of Florida. He presented a potentiometric surface map showing the natural recharge and discharge areas in peninsular Florida and the general direction of ground-water movement. A major potential recharge area was shown in central Florida.

Following Stringfield's pioneer work, potentiometric surface mapping and related hydrologic studies in the 1940s through the 1960s (summarized by Parker and others, 1955; Stringfield, 1966) disclosed that this hydraulically interconnected carbonate system extends throughout peninsular Florida, northward through southeast Georgia and adjacent South Carolina, and westward into panhandle Florida and adjacent Alabama. Recently (1980), an aquifer-wide potentiometric surface map was prepared by the U.S. Geological Survey based on 2,700 nearly simultaneous measurements of water level or artesian pressure throughout the Floridan's extent (Fig. 4; adapted from Bush and Johnston, 1987).

The configuration of the potentiometric surface indicates that in South Carolina and Georgia, the direction of flow is generally eastward and southeastward from the topographically high outcrop areas toward the Atlantic Coast and Florida. In Alabama and west Florida, flow is generally from the outcrop areas southward toward the Gulf Coast. In peninsular Florida the general flow direction is from the central inland areas toward the Gulf and Atlantic coasts.

The characteristic of the system that most strongly influences the distribution of natural recharge, flow, and discharge is the degree of confinement of the Upper Floridan. The aquifer's flow system is most dynamic in unconfined and thinly confined areas (Bush and Johnston, 1987). Potentiometric contours that bend as they cross streams indicate discharge and typify unconfined and thinly confined aquifer conditions. Smoother contours and flat gradients are associated with thickly confined parts of the system.

Local structural features may interrupt the regional ground-water flow pattern. An example is a graben system in south-central Georgia within which low-permeability clastic rocks have been downfaulted opposite permeable carbonate beds of the Upper Floridan aquifer, creating a damming effect on the general southeastward movement of ground water. This retardation of flow is reflected by a marked steepening in the slope of the potentiometric surface of the Upper Floridan aquifer (band of closely spaced contours in south-central Georgian on Fig. 4). Large amounts of intergranular gypsum are present in the Upper Floridan aquifer just southeast of this area because the restricted ground-water flow has been insufficient to dissolve the gypsum and flush it from the limestone.

With the exception of a few cones of depression centered in areas of heavy withdrawal, the potentiometric surface is little changed from that mapped by Stringfield (1936), and the major features are believed similar to those that existed prior to development. The dominant feature of the Floridan flow system, both before and after ground-water development, is discharge from springs.

Nearly all the springs occur in unconfined and thinly confined parts of the aquifer in Florida. Today (early 1980s) the combined average discharge from about 300 known Upper Floridan springs ranges from about 350 to 370 m³ per second. This constitutes more than one-half the present-day discharge from the Floridan aquifer system, estimated to be about 680 m³ per second (Bush and Johnston, 1987).

The Floridan springs have long been the subject of hydrogeologic investigation. The concentration of 27 first-magnitude springs (discharge exceeds 2.8 m³/sec) within the northern half of Florida is unique in the United States and probably exceeds in number and discharge the springs of any country in the world (Rosenau and others, 1977). Most of the major Floridan springs, which have been termed "artesian springs," occur where the confining beds overlying the Floridan have been locally breached and the artesian head is above land surface.

The impact of ground-water pumping (about 120 m³/sec in 1980) is shown by the cones of depression in Figure 4. The steeper cones at Fort Walton Beach and Savannah are due primarily to lower transmissivity and, to a lesser extent, to confinement rather than to larger withdrawals as compared to other pumping centers. In contrast, larger withdrawals near Orlando, near Tampa, and seasonally in southwest Georgia—all located in unconfined or thinly confined areas of moderate to high transmissivity—have produced shallow localized cones of depression that cannot be shown at the scale of the potentiometric surface presented here. Pumping in 1980 accounted for about 17 percent of the total Floridan discharge; however, there remain large areas in northern Florida and adjacent southern Georgia where pumping is an insignificant part of ground-water discharge and nearly "virgin conditions" prevail. Adverse effects of development are limited primarily to the degradation of water quality in a few coastal areas. Pumping at Savannah, Georgia, and east of Orlando, Florida, has caused inland migration of saline water. At Brunswick, Georgia, there has been localized intrusion of saltwater that is possibly moving upward along fault zones. A minor effect of development in the karst areas of central Florida may be the hastening of sinkhole development due to water-level declines caused by pumping. Land subsidence resulting from water-level declines in the Floridan has reportedly occurred in the Savannah area. However, the maximum subsidence since the early 1930s has been only a few centimeters.

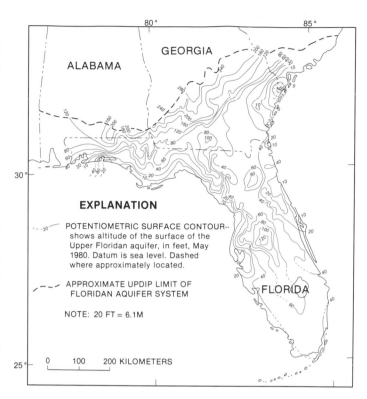

Figure 4. Map showing the configuration of the potentiometric surface of the Upper Floridan aquifer, May 1980.

Pumpage is supplied primarily by the diversion of natural outflow from the aquifer system and by induced recharge rather than from storage (Bush and Johnston, 1987). The aquifer system's response to changes in pumping rates dissipates within days or weeks in most areas. Thus on the average (that is, excluding the effects of seasonal changes in recharge and discharge), the aquifer system is approximately at equilibrium, except during periods following sustained increases in pumping.

Flow activity and the water chemistry of the Floridan aquifer system are closely related. Sprinkle (1987) noted that dissolved-solids concentrations are low in the unconfined or thinly confined areaa where flow is vigorous. Where the system is thickly confined, flow is more sluggish and concentrations are higher. In Florida south of Lake Okeechobee and in parts of the St. Johns River valley, saltwater thought to be residual from a marine transgression remains unflushed from the system, and dissolved-solids concentrations are greater than 1,000 mg/L. Concentrations also increase in coastal areas toward the present-day saltwater-freshwater interface. Inland throughout most of the Upper Floridan, dissolved-solids concentrations are less than 500 mg/L. This is the result of (1) the relatively simple mineralogy of the aquifer (calcite and dolomite, (2) the relatively low saturation concentrations of these minerals, and (3) the limited occurrence of more soluble minerals like gypsum.

## BISCAYNE AQUIFER

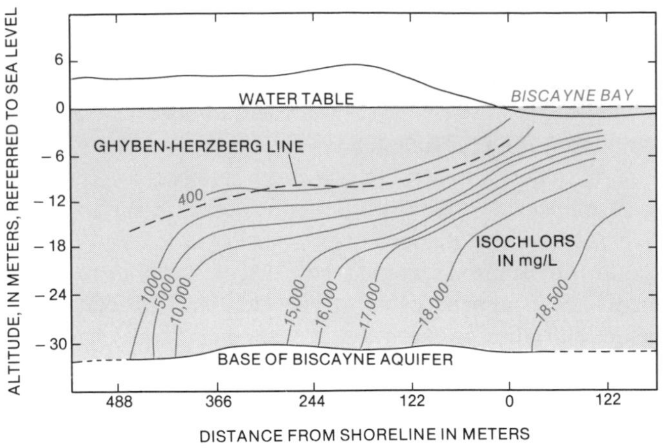

The major source of ground water in much of southeastern Florida is the Biscayne aquifer (Fig. 1), a sequence of Pliocene and Pleistocene limestones and minor sands that is generally highly permeable and contains water under unconfined conditions. The Biscayne is composed primarily of, in descending order, the Miami Oolite (oolitic limestone), the Fort Thompson Formation (fossiliferous limestone and sand), the Anastasia Formation (sandy limestone and coquina), and sandy limestone of the upper part of the Tamiami Formation. Locally, the Pamlico Sand, the Key Largo Limestone (coralline reef rock), and the Caloosahatchee Marl (sand, shell, silt, and clay) are included in the aquifer.

The Biscayne aquifer generally thickens and contains more limestone to the south; it becomes more sandy as it thins to the north and west. The limestone parts of the aquifer are highly permeable because of the development of solution channels and cavities (some of which are sand-filled). Permeability is lower in the sandy parts of the aquifer. As with the Floridan, the transmissivity is closely related to the occurrence of cavernous limestone rather than thickness. Southwest of Miami, the transmissivity is highest (93 to $186 \times 10^3$ m$^2$/day) because the aquifer is principally cavernous limestone (Klein and Hull, 1978). Along the coastal strip between Miami and Fort Lauderdale, the Biscayne is thickest but sandier, and the transmissivity is lower (19 to $46 \times 10^3$ m$^2$/day).

Saltwater encroachment is an ever-present threat to the Biscayne because the aquifer is unconfined, hydraulically connected to the sea, heavily pumped, and cut by a network of canals. Present pumpage of about 22 m$^3$/sec has caused water levels to drop below sea level near some well fields.

Detailed mapping of heads and salinities in the 1940s and 1950s showed that a zone of diffusion separates freshwater and

Figure 5. Zone of diffusion in the Biscayne aquifer (modified from Kohout, 1960).

seawater in the Biscayne aquifer (Fig. 5). Kohout (1960) noted that the position of the saltwater-freshwater interface, shown as the 10,000 mg/L line in Figure 5, is dynamically stabilized seaward of the theoretical position given by the Ghyben-Herzberg principle (hydrostatic equilibrium). Kohout proposed a theory of cyclic flow of seawater to account for this discrepancy. During periods of heavy recharge (and high water table), the freshwater and the interface moves seaward. However, during periods of little or no recharge (and low water table), saltwater in the lower part of the aquifer flows inland but some diluted seawater (in the zone of diffusion) continues to flow seaward. According to Kohout (1960), this cyclic flow pattern reduces saltwater encroachment because some saltwater always returns to the sea.

## REFERENCES

Bush, P. W., and Johnston, R. H., 1987, Ground-water hydraulics, regional flow, and ground-water development of the Floridan aquifer system in Florida and in parts of Georgia, South Carolina, and Alabama: U.S. Geological Survey Professional Paper 1403-C (in press).

Hanshaw, B.B., Back, W., and Deike, R. G., 1971, A geochemical hypothesis for dolomitization by ground water: Economic Geology, v. 66, p. 710–724.

Klein, H., and Hull, J. E., 1978, Biscayne aquifer, southeast Florida: U.S. Geological Survey Water-Resources Investigations 78-107, 52 p.

Kohout, F. A., 1960, Cyclic flow of saltwater in the Biscayne aquifer of southeastern Florida: Journal of Geophysical Research, v. 65, p. 2133–2141.

Miller, J. A., 1986, Hydrogeologic framework of the Floridan aquifer system in Florida and in parts of Georgia, South Carolina, and Alabama: U.S. Geological Survey Professional Paper 1403-B, 91 p.

Parker, G. G., Ferguson, G. E., Love, S. K., and others, 1955, Water resources of southeastern Florida, with special reference to the geology and ground water of the Miami area: U.S. Geological Survey Water-Supply Paper 1255, 965 p.

Rosenau, J. C., Faulkner, G. L., Hendry, C. W., and Hull, R. W., 1977, Springs of Florida: Florida Bureau of Geology Bulletin 31 (revised), 461 p.

Sprinkle, C. L., 1987, Geochemistry of the Floridan aquifer system in Florida and in parts of Georgia, South Carolina, and Alabama: U.S. Geological Survey Professional Paper 1403-I (in press).

Stringfield, V. T., 1936, Water in the Florida Peninsula: U.S. Geological Survey Water-Supply Paper 773-C, p. 116–195.

—— , 1966, Artesian water in Tertiary limestone in the southeastern states: U.S. Geological Survey Professional Paper 517, 226 p.

Thayer, P. A., and Miller, J. A., 1984, Petrology of lower and middle Eocene carbonate rocks, Floridan aquifer, central Florida: Gulf Coast Association of Geological Societies Transactions, v. 34, p. 421–434.

MANUSCRIPT ACCEPTED BY THE SOCIETY MAY 15, 1987

## ACKNOWLEDGMENTS

The authors thank P. D. Ryder, R. E. Krause, H. Meisler, and J. V. Brahana, all of the U.S. Geological Survey, for critically reviewing the manuscript.

Printed in U.S.A.

The Geology of North America
Vol. O-2, Hydrogeology
The Geological Society of America, 1988

## Chapter 28

# *Region 25, Yucatan Peninsula*

**Juan M. Lesser**
*Cuernavaca 89-A, Mexico 06140 D.F.*
**A. E. Weidie**
*University of New Orleans, New Orleans, Louisiana 70148*

### INTRODUCTION

The Yucatan Peninsula, in the eastern portion of Mexico, is bounded on the west and north by the Gulf of Mexico; on the east by the Caribbean Sea; on the southwest it merges with the Gulf Coastal Plain; and on the south it is bounded by the Sierra Madre del Sur (Fig. 3, Table 2, Heath, this volume). A humid tropical climate prevails over the peninsula with rainfall varying from 800 mm/yr in the northwest to 1,300 mm/yr on the east coast and 1,700 mm/yr on Cozumel Island off the east coast. Approximately 90 percent of the rainfall occurs in the May to October period. Mean annual temperature is 25°C, with highs in July and August and lows in December and January. High temperature and abundant vegetation cause about 85 percent of the rainfall to be lost through evapotranspiration; the remaining 15 percent infiltrates the subsurface; virtually no streams or surficial water bodies exist on the peninsula.

### GEOMORPHOLOGY

The Yucatan Peninsula, part of the Mexican Gulf Coastal Plain, has dimensions of 250 km by 300 km. The greater portion of the peninsula is from 0 to 50 m above sea level; elevations as high as 300 m occur in the southwest. Four distinct physiographic regions have been recognized: (1) Northern Pitted Karst Plain, (2) Sierrita de Ticul, (3) Southern Hilly Karst Plain, and (4) Eastern Block-Fault District (Weidie, 1985).

### GEOLOGY

The peninsula is covered by Tertiary carbonates whose maximum thickness is about 1,000 m. They are horizontal to subhorizontal and overlie Cretaceous carbonates and evaporites that have been penetrated by various wells (Weidie, 1985).

Surface studies are hampered by few roads, thick vegetative cover, near horizontal bedding, few outcrops, and extensive calichification. Hence, a detailed geologic map of the entire peninsula has yet to be drawn.

Bonet and Butterlin (1962) described seven (7) Cenozoic units whose ages range from Paleocene to Quaternary. Their "formations" (actually biostratigraphic units) were modified by Lopez Ramos in 1981. The thickness of the Cenozoic section ranges from 200 to 1,000 m; they are subhorizontal in the north and east, but dips up to 30° have been reported in the southwest. In general, younger rocks crop out on the periphery of the peninsula, and older rocks occur in the southern and central area.

The most areally extensive rocks and the major aquifers are in carbonates of Eocene and Mio-Pliocene age. The Eocene is composed mainly of dense, recrystallized, fine- to medium-grained limestones. The lower part of the section contains marls and calcareous shales, which grade laterally into dolomitic limestone, marls, gypsum, and anhydrite. Most lithofacies have good permeability. Permeability is better developed in the Mio-Pliocene carbonates, which are coquinas, fossiliferous packstones, and grainstones; there are occasional thin interbeds of marl and calcareous shale.

In the southern and central portions of the peninsula the Paleocene and lower Eocene carbonates are dolomitic and, in part, slightly silicified. These rocks are pervasively fractured, permitting rapid infiltration and flow of ground water; water levels in this region are often many tens of meters to 100 m below the surface, making extraction of fresh water difficult.

Surficial carbonates on the greater part of the Yucatan Peninsula are covered by a thin (0.1 to 0.5 m) zone of massive caliche or by 0.5 to 10 m thick zone of "saskab," a chalky and friable weathering product of the limestones in this humid, tropical environment. These white to tan strata with occasional red horizons ("terra rosas") have been studied and described by various authors (see Isphording, 1974).

### HYDROGEOLOGY

#### *Northern pitted karst plain*

This region (Fig. 1) occupies the northern portion of the peninsula. From the coastline, elevations gently increase inland to the south to about 35 to 40 m near the base of the Sierrita de

Lesser, J. A., and Weidie, A. E., Region 25, Yucatan Peninsula, 1988, *in* Back, W., Rosenshein, J. S., and Seaber, P. R., eds., Hydrogeology: Boulder, Colorado, Geological Society of America, The Geology of North America, v. O-2.

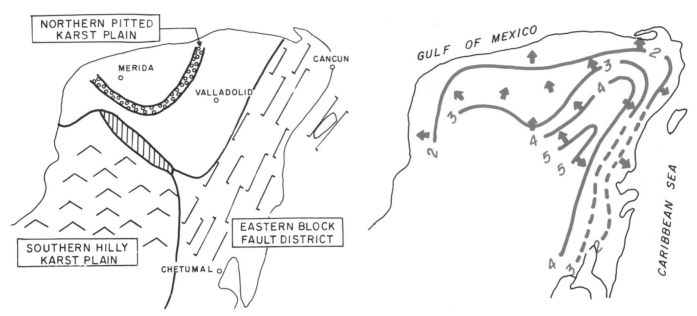

Figure 1. Map showing physiographic regions (Weidie, 1985).                    Figure 2. Map showing altitude of water table.

Ticul. Local relief rarely exceeds 10 m, and relief diminishes as one moves to lower elevations near the coast. This area is formed by marine Tertiary carbonates that have been subjected to extensive dissolution. Large solution holes and cavities have been formed by infiltration of rainfall and have formed a highly permeable aquifer. The high degree of karstification permits rapid infiltration, and there are no surface streams. In the absence of stream erosion, there is strong subsurface erosion resulting in the development of typical karst topography. Both mechanical and chemical erosion occur in the subsurface.

Mechanical erosion occurs near the surface where clays are washed into dissolution openings and deposited beneath the calichified surface. This absence of soil cover is one of the characteristics of the plain. Clays are found at depths of 0.5 to 1.5 m beneath the surface where plant roots tend to trap them.

Rapid infiltration of rainfall into the aquifer results in unsaturated waters retaining a high potential for dissolution. Dissolution enlarges the fissures and cavities, producing large openings and caverns. Collapse of rocks above the openings produces the dolines and sinkholes known locally as "cenotes," a word of Mayan origin. These karst features are of various types and have formed in response to fluctuations of the water table. The water table occurs at depths of 3 to 15 m in this region of 4 to 20-m surface elevations (Fig. 2). Most "cenotes" are circular with vertical walls and diameters of about 100 m where the depth to water is about 15 m (Fig. 3). Work by divers shows the "cenotes" narrow with depth and have a conical form.

Frequently the "cenote" roof has collapsed completely; in other cases it is hemispherical with small openings 1 to 3 m in diameter. These openings typically enlarge with depth, and their diameters may reach several tens of meters (Fig. 4). The formation of these "cenotes," as well as the other karst features, is a function of the depth of the water table. At this depth the water is

charged with carbon dioxide, promoting the formation of carbonic acid which dissolves the subsurface carbonates. Various levels of dissolution show that the water table has fluctuated, apparently in response to the gradual emergence of the peninsula.

In the northern part of the peninsula the Tertiary carbonates are horizontal and show significant lateral facies changes, which control the occurrence of zones of different degrees of karstification. The "cenotes" or sinkholes occur along the entire length of the peninsula, but they are more notable and numerous in certain zones (Fig. 4) where an elongated strip is characterized by abundant karst features, especially those of great size.

### Southern hilly karst plain

In this zone the topography is varied; there are isolated, low-relief, conical karst hills about 40 m above the surrounding land surface. Maximum altitude in this region is 300 m above sea level.

The plain is formed of Eocene carbonates including limestone, dolomitic limestone, and dolomite. Some of the carbonates show a slight degree of silicification. Attitude of the beds ranges from horizontal to small folds with 15° to 20° dips; and they are highly fractured.

The high fracture permeability of these carbonates is reflected in low gradients of the water table. In places the water table is 100 m beneath the surface, making ground-water exploitation difficult and costly. For this reason there are few large villages or agricultural development in this region of higher topography and deeper aquifers.

### Sierrita de Ticul

Between the extensive plains of the northern peninsula and

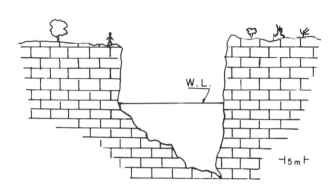

Figure 3. Sketch showing typical vertical walls of the cenotes.

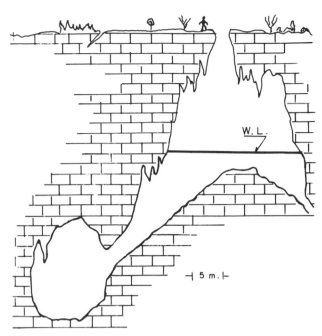

Figure 4. Sketch showing cenote formed by collapse of a cave roof.

the hills of the south there occurs an elongated topographic high known as the Sierrita de Ticul. It is composed of Eocene limestones, trends northwest to southeast, and is 160 km long and 15 km wide. The sierrita is a fault scarp formed by a normal fault whose down-dropped block is on the north. Altitudes in the region are 150 m above sea level and 100 to 110 m above the plain to the north. The elevations along the scarp are practically the only geomorphic feature interrupting the otherwise uniform countryside.

### Eastern block-fault district

This area, about 80 km wide, extends from Cabo Catoche, the northeastern tip of the peninsula, south to Belize. In this region the carbonate rocks have been faulted by a series of north-northeast-trending normal faults (the Rio Hondo fault zone), forming horsts and grabens. The faults are of varying lengths and displacements, and many of them are expressed at the surface.

Notable surficial expressions of these faults are along the Caribbean coast, including Cozumel Island, a horst block bounded on the east and west by large faults. On and near the coast there is a marked orientation and alignment of bays, such as Chetumal and Ascencion. Slightly inland, elongated depressions 10 to 15 m deep form lakes or marsh and swamp regions. Good examples of these long depressions are the lakes of Bacalar, which are oriented northwest to southeast and northeast to southwest and are formed by down-dropped blocks 8 to 10 m lower than the surrounding land surface. The lakes vary in width from as little as 2 to 10 m and may attain lengths up to 50 km. Further to the northwest near the coast, similar but smaller features occur only a few meters above sea level, giving rise to marshes and small lakes. In the coastal zone, ground-water discharge into the Caribbean dissolves the carbonates and forms coastal inlets ("caletas") and lagoons (Back and others, 1979).

## HYDROGEOCHEMISTRY

In the Yucatan Peninsula there are two sources of saltwater: (1) dissolution of evaporite deposits interbedded in the carbonate sediments, and (2) the sea water surrounding the peninsula.

In the northern and southeastern plains the freshwater aquifer is 30 to 70 m thick and overlies saline water, which is found at depths of 40 to 80 m. The principal source of saltwater is from the dissolution of subsurface beds of gypsum, anhydrite, and halite. Rainfall over the peninsula infiltrates, and the fresh ground water moves seaward toward the coastline. There is constant replenishment of fresh water in the upper part of the aquifer, and salinities increase at greater depths. Some "cenotes" and a few wells are deep enough to permit the detection and measurement of saltwater; Figure 5 shows the approximate position of the fresh and saltwater bodies.

Along the coastline there is saltwater intrusion into the fresh-water aquifers. Owing to the high permeability of the carbonates, the water table is only a few centimeters above sea level, and its altitude gently increases inland. Because of this, the saltwater interface is close to the surface, and the fresh water forms a thin wedge. Villages along the coast are forced to develop their water supplies from 15 km or more inland.

Sea-water intrusion occurs in annual cycles. No recharge during dry seasons, combined with exploitation of the aquifers, permits the advance of the saltwater front to as much as 12 km inland. During the rainy season, with its major recharge and lower pumping rates, there is a seaward retreat of the interface; the high permeability permits rapid movement of the interface.

A typical example of sea-water intrusion is on Cozumel Island off the eastern coast of the peninsula. Holocene and Pleistocene carbonates crop out and are underlain by Mio-Pliocene carbonates, similar to much of the northern peninsula. Sea water has intruded along the island margins, and only in the central part of the island is recharge sufficient to maintain a thin lens of fresh water about 20 m thick. Shallow wells about 10 m deep in the central region yield flows of 1.0 L/sec; greater rates of pumping lower water levels and permit greater sea-water intrusion (Fig. 6). High pumping rates can result in withdrawal of saltwater, and reduction in pumping results in near-instantaneous recovery of water levels and quality.

## STRUCTURE

The near-horizontal Cenozoic carbonates of the peninsula are fractured extensively. Two sets of vertical, perpendicular fractures facilitate the rapid infiltration of water into the subsurface.

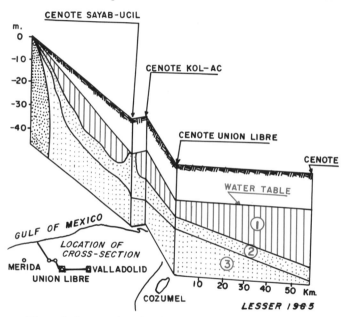

Figure 5. Cross section showing distribution of fresh saltwater.

Virtually all the carbonate rock types are affected by these fractures. Direct observation of fractures inland is difficult because of the lush tropical vegetation, but numerous aligned sinkholes, cenotes, and other karst features attest to their presence. Removal of vegetative cover through "slash and burn" agricultural techniques shows the fracturing to be pervasive. Fractures are easily observed in the less vegetated coastal areas and contribute significantly to coastal erosion by the formation of bays and caletas. These features vary in size from a few meters to tens of kilometers. The fractures are major conduits of seawater ground-water discharge and solution.

The ground water is normally saturated with respect to calcite, but near the coastline it mixes with saline water; the resultant mixture is undersaturated, causing carbonate dissolution and fracture enlargement leading to the formation of coastal bays and caletas (Back and others, 1979; Back, this volume).

The more dolomitic Eocene rocks are intensively fractured, sometimes presenting a "slaty" appearance. In these rocks, fracturing is the most important factor controlling the development of aquifers. In the more soluble limestones of Mio-Pliocene age that dominate the peninsula, aquifer formation is primarily a result of dissolution processes.

The horizontal limestones of the eastern portion of the peninsula are faulted extensively; the normal faults form a series of horsts and grabens recognizable on the surface and verified by drilling and geophysical methods. Topographic and structural lows in this region frequently are filled with more clayey material. These depressions are enlongated parallel to fault trends and frequently form lagoons or lakes. The depressions are from 2 to 10 km wide and as much as 50 km in length despite vertical displacements as small to 5 to 10 m along the faults.

The most notable lineaments associated with the faulting may be observed in: (1) the course of the Rio Hondo in the southeastern part of the peninsula and extending into Belize, where the river follows the northwest to southwest faulting for a considerable distance; (2) the Laguna de Bacalar, which is 5 km

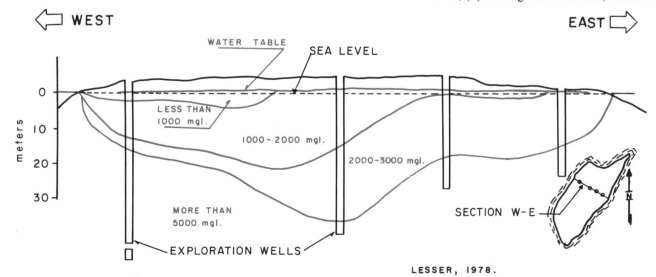

Figure 6. Cross section showing distribution of salinity of water on the island of Cozumel.

wide and 50 km long and occupies a graben depression; the lagoon and adjacent lows are maintained by rainy season precipitation. Some depressions are sites of ephemeral lakes, and most low-lying areas are oriented either northeast to southwest or northwest to southeast; and (3) northeast to southwest submarine fault scarps in the Caribbean adjacent to the eastern margin where water depths may reach 1,000 m a few kilometers offshore. The offshore island of Cozumel is a horst block of this fault system.

## REFERENCES CITED

Back, W., Hanshaw, B. B., Pyle, T. E., Plummer, L. N., and Weidie, A. E., 1979, Geochemical significance of groundwater discharge and carbonate solution to the formation of Caleta Xel Ha, Quintana Roo, Mexico: Water Resources Research, v. 15, p. 1521–1535.

Bonet, F., and Butterlin, J., 1962, Stratigraphy of the northern part of the Yucatan Peninsula, *in* Field trip to Peninsula of Yucatan: New Orleans Geological Society Guidebook, p. 52–57.

Isphording, W. C., 1974, Weathering of Yucatan limestones; The genesis of terra rosas, *in* Weidie, A. E., ed., Yucatan Guidebook: New Orleans, Louisiana, New Orleans Geological Society, p. 78–93.

Lopez Ramos, E., 1981, Geologia de Mexico, 2a: Mexico, D. F., Edicion, Tomo III, 446 p.

Weidie, A. E., 1985, Geology of Yucatan platform, *in* Ward, W. C., Weidie, A. E., and Back, W., eds., Geology and hydrogeology of the Yucatan and Quaternary geology of northeastern Yucatan peninsula: New Orleans, Louisiana, New Orleans Geological Society, 160 p.

The Geology of North America
Vol. O-2, Hydrogeology
The Geological Society of America, 1988

# Chapter 29

# *Region 26, West Indies*

**William Back**

*U.S. Geological Survey, 431 National Center, Reston, Virginia 22092*

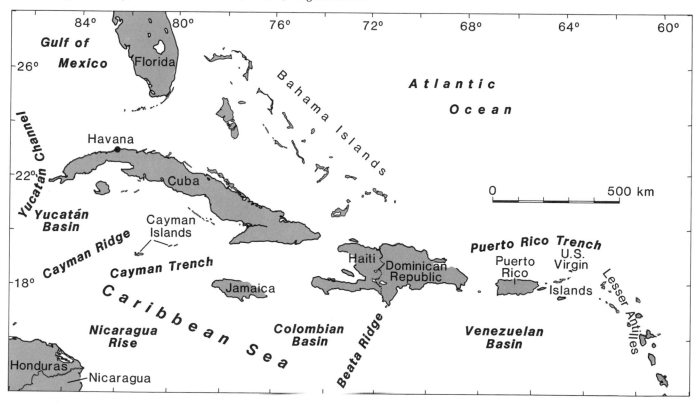

Figure 1. General index map showing position of islands and selected marine structural features of the West Indies.

## INTRODUCTION

In this chapter, the West Indies are considered to include the few thousand islands that compose the Greater Antilles and the Bahamas (Fig. 1; Fig. 3, Table 2, Heath, this volume). The Lesser Antilles were not included because of page limitation and scarcity of hydrogeologic information. The Greater and Lesser Antilles surround the deep (4,000 m) waters of the Caribbean Sea and separate it from the deep (8,000 m) waters of the Atlantic Ocean. That means the north shore of Puerto Rico, Hispaniola, and most of Cuba borders the Atlantic; the north shore of western Cuba is in the Gulf of Mexico. The Bahamas are entirely in the Atlantic Ocean; Jamaica and the Cayman Islands are in the Caribbean.

The Virgin Islands and Lesser Antilles generally are considered to be in the Caribbean.

The hydrogeologic setting of the Greater Antilles is discussed somewhat in order from northwest to the southeast and in order of decreasing size; the Bahamas are discussed last because their geologic origin is distinct from the other islands. The major political subdivisions of the Greater Antilles are Cuba, including more than 1,500 smaller islands with a land mass of 111,000 km[2]; Hispaniola, comprising the Dominican Republic (48,700 km[2]) and Haiti, including its several smaller islands (27,700 km[2]); Jamaica (11,000 km[2]), Puerto Rico (8,900 km[2]), U.S.

Back, W., 1988, Region 26, West Indies, *in* Back, W., Rosenshein, J. S., and Seaber, P. R., eds., Hydrogeology: Boulder, Colorado, Geological Society of America, The Geology of North America, v. O-2.

Virgin Islands (340 km$^2$), and Cayman Islands (260 km$^2$); and the Bahamas land area of (13,900 km$^2$).

Many geologic processes have combined to control the origin, distribution, and functioning of the aquifers of this region. The major aquifer systems and those of more restricted extent and production are composed of limestone and dolomite sequences, alluvial fill in coastal valleys, volcanic and intrusive rocks, and various types of coastal deposits such as coastal plain sands and gravels, beach ridges, and sand dunes. The geologic processes that have formed these aquifers include aspects of plate tectonics, which have contributed to folding and overthrust faulting, volcanic eruptions that deposited more ash than lava flows, fluvial sedimentation, deposition of carbonate material in all its various environments (deep sea, reef, fore-reef, and back-reef), in addition to onshore deposition of carbonates. These sediments have undergone many processes of diagenesis, including vadose and phreatic dissolution and cementation, mixing-zone dissolution and alteration of minerals, and processes of marine lithification.

The occurrence, movement, and chemical character of ground water are controlled by the lithology and permeability distribution of the aquifers and aquitards that make up the ground-water system, along with the climate and topography of the region. To understand the present characteristics and distribution of the aquifers, it is necessary to understand the origin and geologic evolution of the islands on which these aquifers are present. This region is a geologically complex and dynamic area that is presently undergoing both geologic evolution and an exceedingly rapid change in its interpretation. One of the definitive and noncontroversial statements that can be made about this region is that both these activities will continue.

## HYDROGEOLOGIC SETTING

Much of the information in this section is excerpted from United Nations (1976). Detailed hydrogeologic data are available in UNESCO (1986), including data for many islands that are not mentioned in this chapter.

The Caribbean Island arc can be viewed as a partially submerged cordillera. The nucleus of the Greater Antilles contains igneous rocks that form the basal complex of Mesozoic age, which is overlain in places by Cretaceous sediments of marine and volcanic origin. Cenozoic sediments, mainly limestones, some folded, cover large areas, and younger Tertiary and Quaternary formations of alluvium, sandstone, and reef limestone are abundant in coastal areas. Some islands, such as the Bahamas and the Cayman Islands, are essentially all limestone.

In the West Indies, the aquifers with highest transmissivity and, consequently, the largest well yields are composed of alluvium and karstified limestone; secondary aquifers are composed of limestone with a lower-degree dissolution and fractured igneous and metamorphic rocks. All the islands, with the exception of the Bahamas and Caymans, consist of volcanic and intrusive igneous rocks, which are overlain in places by limestone and alluvium. Major carbonate aquifers are present along the north

coast of Puerto Rico, in western Cuba, in the major part of central Jamaica, and in some of the structural valleys of Hispaniola.

Major alluvial aquifers occur along the south coast of Puerto Rico, including some that are alluvial fans; along present-day rivers in coastal areas, as in Cuba and the Dominican Republic; and on many of the islands as local coastal-plain sediments. In many places, the alluvium acts as a reservoir that recharges the underlying limestone or fracture-rock aquifers.

On the larger islands, ground water is relatively abundant in the alluvium of the river valleys and in the permeable limestones and volcanic rocks. On the smaller islands where neither limestone nor alluvium are present, such as the Virgin Islands, low-yielding aquifers are developed in the volcanic rocks. The water-yielding properties of Mesozoic volcanic rocks range from mediocre to poor. In those areas, most of the rainfall is lost by evaporation and by runoff to the sea. Quaternary volcanics are good aquifers in some areas, but if they are highly permeable, the rain water infiltrates to the water table and discharges rapidly to the sea. In the carbonate islands, a large amount of rainfall infiltrates to the water table; most of it also flows laterally and discharges to the sea. Fresh water is present in the form of a relatively thin lens, and water levels in wells are practically at sea level even as far as several kilometers inland. Sea water can intrude rapidly as a result of pumping.

The West Indies have a tropical climate with slight annual range in temperature and an abundant, but areally variable, rainfall in nearly all the islands. The islands lie within the zone of tradewinds that blow from the northeast and east. These winds, along with topography, control the wide variation in the distribution of rainfall. Rainfall on the windward slope is much greater than that on the leeward slopes, and higher elevations receive greater rainfall. Because of the warmth and constant movement of the tradewinds, evapotranspiration is high and a given amount of precipitation is less effective in producing runoff and groundwater recharge than it would be in the conterminous United States.

This variation in rainfall also is controlled by the wide range in elevation; much of the land surface of the West Indies is near sea level but in places attains heights of more than 3,000 m. For example, the terrain of Puerto Rico consists of rugged hills and mountains that rise to an average elevation of about 800 m above sea level in the interior; highest peaks are nearly 1,100 m above sea level. About 40 percent of Cuba is hilly to mountainous; the highest elevation is about 2,500 m. Most of Jamaica has an elevation of less than 600 m; the highest area, in the east, is about 2,000 m above sea level.

### Cuba

Cuba has a very rugged topography composed primarily of four mountain types: dome shaped, eroded horst blocks, eroded monoclines, and karst. Plains cover about two-thirds of the country, mainly in the coastal areas and in a broad intermontane area.

The morphology of much of Cuba is characterized by the vast extent and diversity of the karst features.

The oldest rocks (Lower and Middle Jurassic age) in the Greater Antilles are in western Cuba and include 5,000 to 6,000 m of metamorphic rocks, sandstone, and limestone. The Upper Jurassic has a thickness of 1,500 to 1,700 m of limestone and dolomite. Cretaceous limestone is present in central Cuba, and Eocene limestone and younger Tertiary limestone and marl are widely distributed in western Cuba.

Karstified Miocene limestones, which are 1,300 m thick and are the main aquifers of Cuba, contain good quality water. The values of hydraulic conductivity range from 10 to 250 m per day; most measured values are from 30 to 150 m per day. The high storativity of these karst aquifers maintains a constant discharge for the rivers that drain this area.

The rainy season extends from May to October; greatest rainfall occurs at the higher elevations and in the western part of the country. Havana has an annual rainfall of 1,800 mm.

## Hispaniola

Hispaniola has some of the highest elevation and most rugged topography of any island in the West Indies. Four physiographic-structural trends converge and meet in Hispaniola: These are the main axis of both the Greater Antilles and the Cayman Trench, the Nicaraguan Rise, the Beata Ridge, and the Bahamas-Cuba intersection (Fig. 1). The northwest-trending structural grain along with the physiography are the basis on which Lewis (1980) divided the island into 10 physiographic and geologic morphotectonic zones or terranes. For additional information on these units, the reader is referred to that paper.

***Dominican Republic.*** The basal complex of Cretaceous age in the Dominican Republic is composed of metamorphic rocks, volcanic tuffs, and intrusive rocks. This complex is overlain by several thousand meters of Paleocene and Eocene limestone conglomerate and shale. The major aquifers are the recent-alluvial-filled valleys composed of alluvium, alluvial fans, and terrace deposits; reef limestones; Upper Tertiary limestone, and Eocene limestone and sandstone.

The topography of the Dominican Republic is extremely diverse. For example, the elevation at the shore of an inland lake is 40 m below sea level; only 330 km away, the elevation is 3,175 m—the highest point in the Antilles.

The climate, which can be termed in a broad sense subtropical, is modified by the irregular influence of high-pressure systems over the tradewinds, which travel in a belt of tropical atmospheric depressions, and by the extreme topographical variations of the island; as a result the climate ranges from semiarid to very humid. The average annual precipitation ranges from 2,750 mm along the northwestern coast to 450 mm in the southwestern area. There are usually two rainy seasons per year; the first season extends from December through April, and the second season of heavier summer rains occurs from June to August.

***Haiti.*** Haiti occupies the western third of the island of Hispaniola. An extensive embayment lies between a northwest and southern mountainous peninsula. Much of northern Haiti is occupied by the Massif du Nord—a mountainous complex that extends from its Dominican border westward into the northwestern peninsula and whose peaks attain elevations of 1,200 to 1,500 m above sea level. An extensive coastal plain slopes northward to the sea from the base of the Massif; to its south is another lowland—the central plain. Extending to the south and east are other mountain ranges separated by broad plains.

Because of the orientation of the mountain chains and the intervening lowlands with respect to the rain-bearing northeast tradewinds, the quantity and areal distribution of rainfall in Haiti is extremely variable. Annual rainfall in the mountainous areas commonly exceeds 1,200 mm and can be as much as 2,700 mm. In the lowlands, annual rainfall commonly is less than 1,200 mm and can be as little as 550 mm.

Approximately 80 percent of Haiti is underlain by sedimentary rocks of early Cretaceous to Recent age. The mountain chains are essentially in compound anticlinal structures, whereas the larger plains, valleys, and other lowlands lie in downwarped synclinal structures. The most extensive aquifer system in Haiti is in the Eocene–Oligocene carbonate rocks. These are predominantly massive, but parts are chalky or well bedded; the carbonate rocks are present throughout much of the country. Underground drainage is well developed in several areas because of the presence of carbonate rocks, but the surface evidence of such drainage is not extensive. Numerous small ponds and lakes—some permanent and others ephemeral—occupy sinkholes in the limestone terrane. Many perennial streams on the well-watered windward mountain slopes discharge into the lowland plains and infiltrate to the aquifer.

In addition, the semiconsolidated sandstone conglomerate and limestone of the Miocene series are important aquifers because their permeability has not been decreased by folding or faulting. Also, productive wells have been completed in the Miocene conglomerate, sandstone, and limestone. Productive aquifers occur in the alluvial deposits of Recent and Pleistocene age in many of the plains.

## Jamaica

Jamaica has a mountainous area characterized by rugged terrain that extends in an east-west direction through the island. Much of the island is surrounded by coastal plains. The topography of Jamaica is composed of eastern mountain ranges, central limestone mountain ranges and dissected plateaus, and coastal plains and interior valleys. The central limestone ranges and plateaus generally are below 1,000 m and develop a characteristic rugged karst landscape with conical hills enclosing circular depressions. The flat coastal plains chiefly border the limestone and are best developed on the southern part of the island. Flat-bottomed interior valleys are developed within the limestone terrane by the coalescing of individual depressions. Many sinkholes exist in the limestone terrane and extend to the water table where they form open bodies of water.

The oldest rocks exposed at the surface are pyroclastics, intrusives, and porphyric rocks of Cretaceous to Lower Eocene age. These crop out in the eastern mountain range where later Tertiary sediments have been removed by erosion. Over the rest of the island, Lower Eocene mudstones, siltstones, and pyroclastics overlie the Cretaceous basement. These deposits are overlain by a thick series of Eocene to Pliocene calcareous sediments that are divided into the Yellow Limestone Formation, the White Limestone Group, and undifferentiated coastal formations. In general, the Yellow Limestone Formation is an aquitard, but, in some places, yields small amounts of water. The productive aquifers are the White Limestone Group and alluvium. The areas underlain by the White Limestone are characterized by an absence of surface drainage. The water readily infiltrates into limestone because of high secondary permeability developed by cracks, fissures, and solution openings. Alluvial deposits of Pliocene to Recent age—consisting of interbedded clay, silt, sand, and gravel—overlie limestone in the coastal plains.

### Puerto Rico

Much of Puerto Rico is underlain by volcanic rocks that have been folded and metamorphosed. Along the northern and southern coasts, these older rocks are mantled by water-bearing limestone and marl of Middle Tertiary age. Alluvium of Quaternary age partially fills valleys along the western and eastern coasts and forms a substantial coastal plain to the south. The principal aquifers of Puerto Rico are the alluvium, limestone of Middle Tertiary age on the north and south coasts, and bedrock of Late Cretaceous and Early Tertiary age in the interior.

### U.S. Virgin Islands

The U.S. Virgin Islands comprise about 40 islands, cays, and rocks; the three largest are St. Thomas and St. John, which consist almost entirely of volcanic rocks, and St. Croix, which consists of volcanic rocks and limestone. The volcanic rocks are poor aquifers, and the ground water is developed largely in the alluvium and in the weathered saprolite where present. The central part of St. Croix is underlain by more than 2,000 m of Miocene clay that serve as the hydrologic basement. Overlying this clay is about 100 m of Miocene limestone consisting of deep-water pelagic layers interbedded with transported shallow-water derived debris. The uppermost carbonates of Pliocene age are about 75 m thick and are composed of benthic foraminiferal packstones and reefal deposits (Gill and Hubbard, 1986). These hydrologically connected limestones form an aquifer of moderate productivity in central and southwestern St. Croix.

### Cayman Islands

The Cayman group comprises three islands: Grand Cayman, Little Cayman, and Cayman Brac. The islands are composed of Tertiary limestones that are massive, thick, and almost horizontal. They were slightly tilted to the west during the Quaternary period, and the main tectonic feature is the west-east orientation of fractures. The Quaternary limestone of various facies overlies the Tertiary limestone in some places.

Karst topography is well developed on the limestone surfaces, as characterized by numerous caves and sinkholes. The low elevation of the islands does not permit development of high freshwater head; consequently, the freshwater lens is quite thin, and many of the ponds, wells, and waterholes of the islands are brackish. Although the annual rainfall exceeds 1,250 mm, there is practically no runoff because of the high permeability of the limestone, and most of the rain water infiltrates to the water table or is lost by evapotranspiration.

### Bahamas

The Bahamas are a group of low islands that extend over an area of about 300,000 km$^2$ of ocean. The Bahamas consist of islands that occupy less than five percent of the area, shallow-water carbonate banks, and deep-water channels that indent and separate the various banks. The word Bahama comes from the Spanish word, Bajamars (shallow seas), referring to the water that generally is less than 10 m deep over the Bahama platform. Not only are the Bahamas the largest carbonate platform in the world, but with a maximum possible thickness of 10 km, they are also one of the thickest ever developed through geologic time. The Bahama platform consists of a series of shallow-water carbonate banks that developed along the subsiding continental margin of North America and extends more than 1,400 km from Florida to the Island of Hispaniola.

The islands are formed on a low platform of Pleistocene limestone partially mantled by lithified dunes and beach ridges. In coastal areas these deposits are commonly overlain by unconsolidated or partially consolidated calcareous sands of Recent age. The marine sediments consist of calcareous sand, silt, and mud that have been cemented, to a variable degree, after emergence above sea level.

A series of karst land forms have developed including shallow dolines, sinkholes, and collapsed caverns. Surface water bodies such as lagoons and lakes occur in topographical lows where the water table is locally above ground surface. The lagoons commonly are shallow, and the water in them is hypersaline. One of the most spectacular solution features are the deep, steepsided submarine sinkholes called "Blue holes," which are as much as 100 m in diameter and occur throughout the Bahamas.

The marine limestone is the main aquifer, with the Holocene carbonate sands locally yielding significant quantities of ground water. The aquifer material consists largely of carbonate bioclastic material deposited in the form of beach ridges and other shoreline features and eolianites deposited as sand dunes. The availability of ground water is controlled largely by the thickness of the freshwater lens. All systems in the Bahamas, as in most of the West Indies, have chemical boundaries—that is, the extent of fresh water is controlled by the position of the saltwater-freshwater interface both near the shoreline and at depth beneath the islands (Chidley and Lloyd, 1977).

## PLATE MIGRATION

Although there is substantive debate about significant problems relating to the mechanisms and timing of critical events, there is a consensus that the Caribbean Plate, of which all of the aforementioned islands (except the Bahamas) are a part, originated in the Pacific Ocean during Cretaceous time and migrated during late Cretaceous and Cenozoic time into its present position (Lewis and Draper 1988; Pindell, 1985). In pre-Mesozoic time, the continents of North and South America and Africa were joined; neither the Gulf of Mexico, Caribbean Sea, nor the Atlantic Ocean existed. With the rotation and eastward migration of Africa, the Bahama Platform remained part of the North American continent. By Middle Jurassic time, the Gulf of Mexico was opening, and the Yucatan Peninsula had rotated to its present position. Jurassic rocks of a continental facies deposited at that time are the sandstones of western Cuba, which are the oldest sedimentary rocks in the West Indies. Carbonate deposition occurred over Cuba and the eastern Bahama Platform in the later Jurassic and early Cretaceous.

As the Caribbean Plate moved from the Pacific Ocean, low-density basalts and gabbros were extruded onto the plate, giving it a thick, buoyant nature. Because of the difference in density between the crust of the Caribbean Plate and the surrounding dense oceanic crust, subduction was occurring; this gave rise to the Antillean arc structure which moved northeast from a probable original position near present-day Central America. The major tectonic and magmatic events that occurred in the Late Cretaceous and Early Tertiary culminated with the collision of the northwardly moving Greater Antillean arc with the Florida-Bahama Platform in the middle Eocene (Lewis and Draper, 1987). However, the relative position of island blocks was modified by east-west strike-slip movement along the Greater Antillean arc. The absence of subduction-related volcanism and the relative tectonic quiescence in the later Tertiary allowed the accumulation of thick limestones over the Cretaceous and Early Tertiary rocks.

Since early Eocene time, the Caribbean would have appeared much as it does today. Aquifers have developed in these rocks that were formed during Cretaceous and Eocene time, by the continual processes of fracturing, diagenesis, uplift, and weathering, all of which have influenced the present-day occurrence and flow pattern of ground water. In addition, in many areas, these formations are covered by later Tertiary and Holocene deposits in which additional aquifers have developed.

## VOLCANISM AND MAGMATIC INTRUSION

Although the processes of volcanism, intrusion of igneous rocks, and metamorphism are extremely important processes in the development of the West Indies Islands, they have had only indirect consequences on the development of major aquifers of this region. Volcanic and metamorphic rocks generally are dense or fine grained, with low permeability, and therefore, water occurs only in fractures and weathered zones. However, these rocks form the core of many of the islands around and on which the sediments that form the major aquifers were deposited. In addition, these rocks are the source of clastic sediments that comprise the Tertiary and Holocene aquifers. For example, the interior hydrogeologic province of Puerto Rico is composed of tuffs, volcanic sediments, and minor lava flows ranging from Early Cretaceous to Late Eocene that are intruded by Upper Cretaceous and Lower Tertiary igneous rock (Briggs and Akers, 1965).

Some volcanic rocks, such as the basalt of the Snake River Plain of Idaho (Chapter 5) and the Hawaiian Islands (Chapter 30), are highly productive aquifers. Those volcanic rocks are composed primarily of lavas that have a high primary porosity and permeability and have not been subjected to intensive alteration or folding. However, the volcanic rocks of the West Indies are not significantly water bearing because they are composed largely of volcanic ash and fragmental material, which although they may have had a high initial permeability, have had their permeability drastically reduced by the extensive tectonic activity such as folding and magmatic intrusion. These intrusions have caused metamorphism to some extent and thereby converted the volcanic debris to hard dense rocks of low permeability. Hydrologic characteristics of the volcanic and other igneous rocks that underlie many areas, and are exposed in other areas in the West Indies, resemble more closely the crystalline rocks of the Piedmont and Blue Ridge regions of the eastern United States than the extensive aquifers of Hawaii or the Snake River Plain (Heath, 1984).

The islands of St. John and St. Thomas in the Virgin Islands provide examples of the occurrence of ground water in West Indies volcanic rocks. These islands are composed of waterlaid tuffs and breccias, flow breccias, and minor lava flows. The volcanic angular rubble that became the lithified breccia originally had much pore space, which has since been almost entirely destroyed by the processes of tectonism and metamorphism, with the result that water now occurs in them primarily along fractures. Weathering is an important process because solution and alteration of the rocks opens the joints and allows water to pass along them more easily. The unweathered rock contains water in joints and fractures, which tend to decrease in number and size with depth. Therefore, relatively shallow wells are more productive than deeper wells in the fractured rock.

Three geohydrologic zones are generally recognizable in fractured crystalline rock aquifers: an uppermost zone composed of soil and saprolite (highly weathered rock); a zone of weathered but still recognizable volcanic rock; and a zone of hard, unweathered, fractured rock (Fig. 2) (Cosner, 1972). If the soil is relatively permeable, rainwater infiltrates into the saprolite and to the underlying rock. Saprolite generally has significant porosity and permeability, and its water-yielding properties are deter-

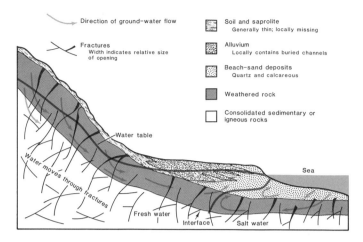

Figure 2. Schematic cross-section of a coastal area composed largely of noncarbonate aquifers showing bedrock physical boundary and salt-water chemical boundary of ground-water flow system. (Modified from Cosner, 1972).

mined almost entirely by its thickness. Therefore, in those areas where the saprolite has adequate thickness, it can be an aquifer suitable for small domestic supplies. Because of the processes of erosion and sediment transport, the soil and saprolite are generally much thinner on the higher slopes of the mountains than on lower slopes and can be relatively thick in the low-lying areas. However, on the islands of St. Thomas and St. John, saprolite generally is thin and not well developed. The weathered rock zone ranges in thickness from 1 or 2 meters to about 15 m but, in a few places, extends to depths of almost 60 m.

## FAULTING, FOLDING, AND FRACTURING

The role of faulting as a geologic process in the development of aquifers is demonstrated on the island of Hispaniola (Fig. 3). Hispaniola has evolved from Early Cretaceous time, and the oldest rocks are metamorphic rocks in the mountains. Volcanic rocks were deposited in the Late Cretaceous time and were intruded by later igneous rocks. The mountains are flanked by Tertiary limestone, and structural depressions in the mountainous areas are filled by thick sequences of Tertiary limestone and younger sediments.

The island consists of four mountain ranges separated by three valleys. In the central northwestern part of the island, the ranges trend northwest—parallel to the structural elements of central and eastern Cuba. These elongated morphologic units reflect the structural features of horsts and deep grabens characteristic of block movement in the structural evolution of the island. Fault zones in the igneous rocks serve as water conduits, and shattered zones contain ground water to depths of 100 to 150 m (Gilboa, 1980). Water that infiltrates into the fractured zones of the igneous and metamorphic rocks in the mountains provides the base flow for many streams.

Interpretation of head data along the northern coast of Puerto Rico demonstrates the control of ground-water flow patterns by faulting. This interpretation (Guisti and Bennett, 1976) shows the (Fig. 4) flow pattern if a postulated fault exists and the different patterns if it does not. Such a fault would move rock of low hydraulic conductivity against the artesian zones, which would block or impede the seaward flow and force the water to discharge upward. On the other hand, it is possible that fracturing associated with the theorized faulting could have produced a highly permeable vertical zone, which would create conditions favorable to vertical outflow. Interpretation of hydrogeologic data from a recent (1986) drilling program should help determine the existence and nature of this fault.

Another example of the role of tectonism in controlling the distribution and characteristics of aquifers is from western Cuba. The Sierra de los Organos are composed of Jurassic limestones that have been overthrust onto Cretaceous and Eocene sediments. The Jurassic limestones are highly karstified, form the highest parts of the mountains, and are productive aquifers in the valleys. If it were not for the folding and overthrusting, the limestone would be deeply buried beneath thick sequences of Cretaceous and Eocene sediments (Fig. 5). The deep burial would not have been conducive to development of secondary permeability by dissolution, and the limestone might not have evolved into aquifers.

The major drainage systems in the volcanic areas of the West Indies generally formed parallel to major fault zones, and minor valleys formed along the minor fault and major joint zones. In general, fault zones can act as conduits for ground-water movement from the head of the valleys to their mouths. Although the rock along the fault itself may be so pulverized that the nearly impermeable gouge acts as a barrier to ground-water flow, the rock generally is fractured a short distance away from the fault, thereby facilitating the movement of water. Most recharge in the areas where the volcanic rocks are near the surface generally is through the soil zone rather than through the stream channels.

## FLUVIAL SEDIMENTATION

Another geologic process that has a significant effect on the development of aquifers in the West Indies is the transport and deposition of fluvial sediments derived from the crystalline core of the islands. These sediments, which consist of sand and clay, have been deposited in marine and continental environments during the late Tertiary and Holocene time. Depending on grain size and sorting, these sediments can serve either as aquifers or as confining beds for the more permeable aquifers, particularly the limestones.

Holocene clastic sediments deposited either in stream channels or in alluvial fans locally serve as major sources of ground water. For example, the northeastern part of the interior hydrogeologic province of Puerto Rico contains a sequence of stratified sedimentary and volcanic rocks that range in age from Early Cretaceous to Early Tertiary. Tectonic activity that has fractured these rocks facilitates the weathering and transport of sediments,

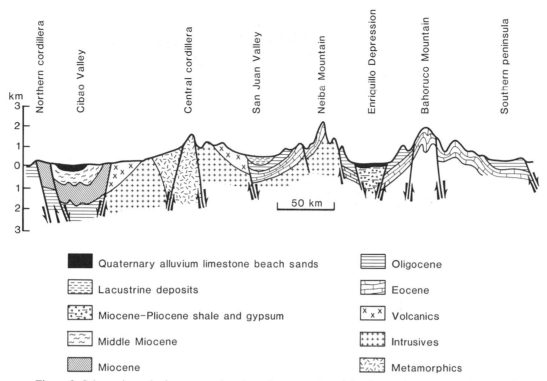

Figure 3. Schematic geologic cross section through western Dominian Republic showing influence of faulting on position of aquifers (from Gilboa, 1980).

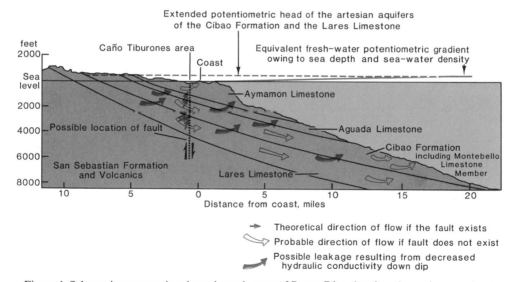

Figure 4. Schematic cross section through north coast of Puerto Rico showing change in ground water by the possible existence of a fault (after Guisti and Bennett, 1976).

which form alluvium in many of the stream channels; these channels serve as aquifers in areas where the alluvium has adequate thickness. The latest Cretaceous unconformity marks a distinct break in sedimentation and is related to the first emergence of a large part of Puerto Rico. The emergent area was a source of the lower Tertiary sediments, whose coarser facies comprise aquifers. Much of the finer-grain material is carried farther downstream

and deposited as delta or lagoon deposits that are essentially non-water bearing (Gomez-Gomez and Heisel, 1980).

The igneous and metamorphic rocks of the Dominican Republic are the source material for the alluvium present in small, separate areas in river and coastal valleys and in intermountain depressions. This alluvium composes some of the country's major aquifers. Coarse clastics containing some clay compose the aqui-

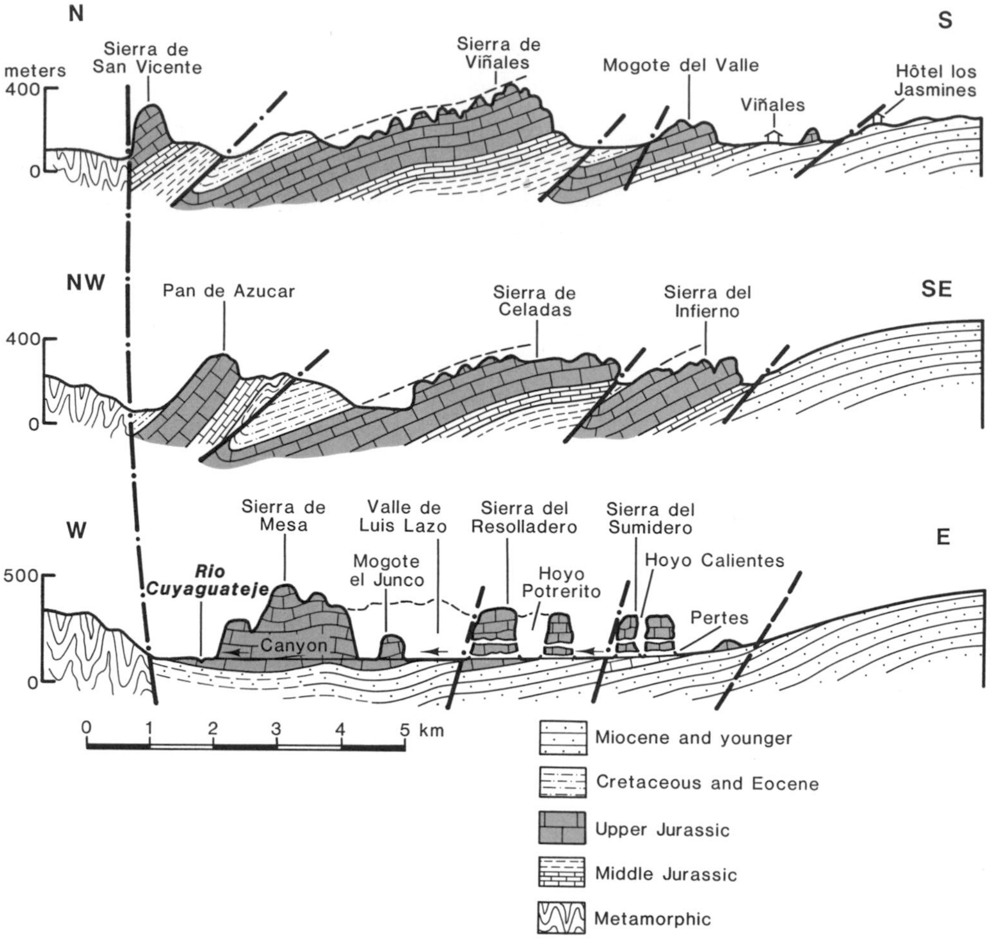

Figure 5. Schematic cross sections through the Sierra de los Organos in Western Cuba show effect of folding and faulting in development of carbonate aquifers (modified from Geze and Mangin, 1980).

fers and are present as (1) fluvial deposits developed in active river valleys that commonly contain several terraces, (2) alluvial deposits that cover the coastal plains where no river presently exists, and (3) poorly sorted alluvium of alluvial fans that accumulate along the foothills. Heterogeneity of these three types of alluvial aquifers is caused primarily by the clay content and by the degree of sorting of the clastic material (Gilboa, 1980).

## CARBONATE SEDIMENTATION AND DIAGENESIS

Many of the major aquifers of the West Indies are in sequences of carbonate rocks. Processes of carbonate sedimentation, diagenesis, tectonism, weathering, and subsequent deposition of sediments control the functioning of these carbonate aquifers. The environments of deposition and diagenesis largely control how these rocks will respond to tectonic forces. For example, the purity of the limestone, degree of recrystallization, and thickness of the beds influence the plasticity of the rocks and determine their response to tectonic forces. The karst of the Caribbean is discussed in the centennial volume on geomorphology by Gardner and others (1987).

Unfortunately, carbonate petrologic studies have not been done extensively enough on the aquifers of the West Indies to determine their environments of deposition. Generally these aquifers are known to have formed from processes that govern recent carbonate sedimentation, even if specific examples of aquifers that originated in particular environments of deposition cannot be given. After studies of carbonate petrology are completed, we will have a better understanding of the distribution of permeability and porosity, and the consequent functioning of the carbonate aquifer systems of the West Indies. Much of the next several paragraphs is excerpted from Hanshaw and Back (1979), in which they discuss the geochemical evolution of carbonate aquifer systems (Fig. 6). Carbonate rocks are deposited in relatively shallow marine environments, and most carbonate aquifers originate either in (1) rocks associated with reefs, or (2) environments to which carbonate sand was transported and accumulated, or (3) environments in which carbonate mud was generated, transported, and accumulated.

Carbonate minerals tend to form and accumulate primarily in warm, shallow lower-latitude parts of the world's oceans, such as in the West Indies. Although surface ocean water is supersaturated with respect to dolomite, calcite, and aragonite, most of the biogenic or inorganic calcium carbonate found near the surface dissolves in the deep ocean. Despite saturation with respect to dolomite, this mineral essentially is unknown as a primary mineral in the marine environment. In addition, based on thermodynamics, calcite ordinarily would be expected to precipitate directly from sea water, but it has been shown by experiments and by field observations that only aragonite precipitates, and this is not in significant quantities. The general understanding is that remains of organisms are overwhelmingly the major source of calcium carbonate in the present-day oceans.

Although reefs are the most spectacular form of carbonate mineral accumulation, they are volumetrically insignificant in modern, and apparently also in ancient, areas of carbonate accumulation. The best-known reef type is the tropical coral-reef environment, which includes the back-reef lagoons and the adjacent bank areas upon which carbonate mud accumulates. Usually these reefs drop off steeply at the continental shelf boundary and have good water circulation, which aids in the growth of algae that encrusts the framework coral and cements the reef into a solid, massive structure. The sand-pile reefs typically contain individual living corals and exhibit some aspects of coral-algal framework reefs when seen in place; however, drilling reveals coral skeletal fragments in an extensive pile of uncemented carbonate sand. Virtually all reefs in ancient rocks are good aquifers, because of their initial high porosity and permeability.

Normal reef growth can have a significant effect on the chemical character of subsequent ground water because sometimes the reefs generate salt ponds. Reefs growing laterally from adjacent headlands can eventually unite to form a continuous barrier across the bay mouth. Sediments accumulate landward from the reef barrier both from wave action that deposits coral debris and from streams that deposit terrigenous sediments. Because these salt ponds trap stream sediments, they are conducive to reef growth by maintaining clarity of the ocean water—a major requisite for healthy reefs. Some of the ocean water and stream water discharges seaward through the sediments, but much of it evaporates and generates a highly saline residual brine from which gypsum and sometimes halite precipitate. Continuous deposition can bury these evaporites; if they subsequently become part of an aquifer system, they are likely to dissolve, whereby porosity and permeability will be increased. Mixing of the saline water in these ponds with freshwater or normal ocean water can create another environment for dissolution and diagenesis of carbonate minerals. Such salt ponds are common hydrologic features throughout the West Indies; for example, the small island of St. John has about 25 (Robinson, 1974). In the Bahamas, salt ponds also are formed by dune accumulation such as in the central part of San Salvador.

Another type of sand-sized carbonate particle is the skeletal remains of marine invertebrates. Most calcite shells are coarse grained and break down readily into sand-sized material, whereas aragonitic shells typically disintegrate into mud-size particles. Carbonate-sand accumulations are volumetrically less important than accumulations of carbonate muds, whether in areas of modern marine carbonate deposition or in the ancient rock record; yet these cemented sands commonly are prolific aquifers. Sands tend to accumulate in areas with higher energy than the carbonate muds that evolve in, or are transported into, lower-energy environments. The sands typically accumulate as linear features such as bars, beaches, and, upon emergence, beach ridges.

Upon emergence, uncemented sand bodies are especially subject to extensive eolian action. During times of sea-level lowering, many marine oolite deposits are exposed and are swept into large dune deposits. Because oolites originate in a high-energy environment, finer particles are winnowed out; wind erosion and dune accumulation further enhances the segregation of sizes. Well-sorted oolitic sand forms highly permeable aquifers of limited extent.

As a general rule, cementation that occurs in the vadose zone is minor, whereas alteration below the water table leads to more extensive cementation. Slightly cemented carbonate sands are excellent aquifers, because they originate and accumulate in high-energy wind and wave environments, are well sorted, and contain essentially no evaporite minerals that would adversely affect the water quality.

Interpretation of carbonate depositional and diagenetic features is helpful in understanding paleoenvironmental conditions and provides clues to both the paleo and modern distribution of permeability within aquifers. Many diagenetic processes and reactions that occur during the early development of carbonate rocks have significant influence on the distribution of permeability and on the chemical character of ground water after the rocks have developed into a ground-water system. For example, most modern marine carbonate muds are composed of about equal amounts of aragonite and calcite; the calcite contains variable amounts of Mg in the lattice. During the ground-water-induced processes of recrystallization and cementation, aragonite neomorphoses to the stable form of low-Mg calcite. As marine water is flushed from the sediments and freshwater circulation is initially established, profound mineralogic and hydrologic changes occur. Most of the carbonate sediment is calcilutite, which initially can have high porosity composed of small pores and has low permeability. During the flushing process, fine particles recrystallize to larger grains, increasing both the size of the pores and permeability. These diagenetic processes can also affect the chemical character of ground water by releasing Mg from the high-Mg calcite and strontium from the aragonite because of the instability of both minerals in fresh water.

Even though the processes of lithification can continue in a ground-water regime, the emphasis is different from those processes of early diagenesis that occur in supratidal and nearshore emergent zones. For example, recrystallization is extremely important in both environments; crystal growth and repreciptation of cement are critical during early diagenesis, whereas dissolution and removal of material along with continued growth of porosity and increase of permeability are more important processes in the ground-water regime.

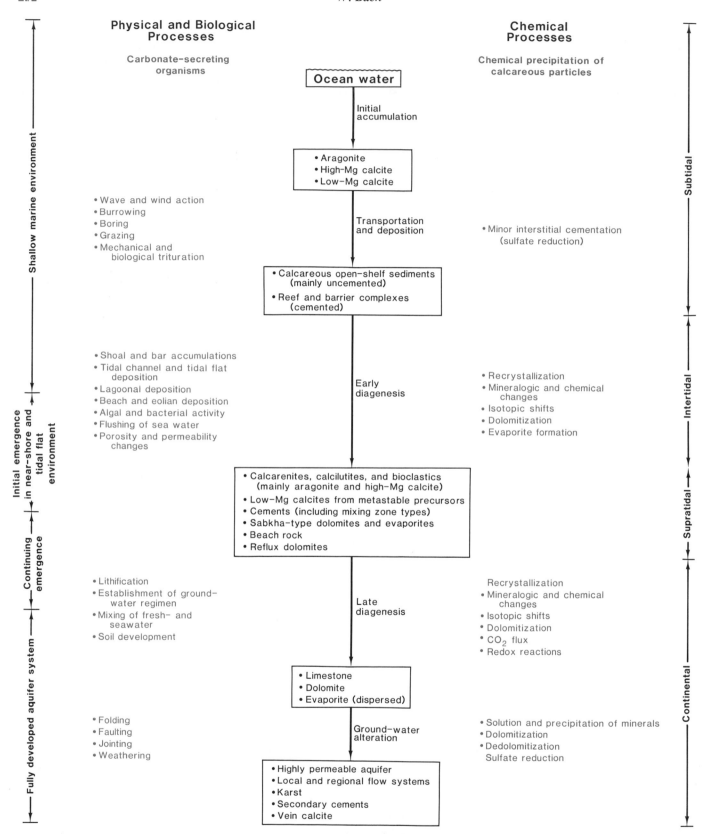

Figure 6. Processes and environments affecting the development of carbonate aquifers in the West Indies (from Hanshaw and Back, 1979).

All of the carbonate aquifers of the West Indies have accumulated in one or more of the environments discussed above and have undergone diagenetic changes. Extensive diagenesis is influenced by subsequent geologic processes, such as uplift, folding and faulting, mineral alteration, and the extensive dissolution caused by ground-water flow. Later erosion of nearby silicate rocks and sedimentation determines the amount of confinement of the carbonate aquifers and whether they function under artesian or water-table conditions.

Extensive and productive carbonate aquifers exist throughout much of the West Indies. For example, in Jamaica the transmissivity of the shallow-water facies of the White Limestone is about 6,048 m$^2$/d (White, 1985). The northern coastal plain of Puerto Rico, which is about 20 km wide and more than 110 km long, is underlain by more than 1,600 m of limestone and alluvium. The upper part of the limestone has been extensively modified by solution that has resulted in a high hydraulic conductivity (Bogart and others, 1964). Guisti (1978) estimated that the hydraulic conductivity of this upper part of the Aymamon Limestone is 163 m/d; the hydraulic conductivity decreases in the underlying limestone and is less than 0.3 m/d in the stratigraphically lowest formation—the Lares Limestone. Even with this extremely low permeability, the Lares can be an important aquifer because of high head generated by high elevation of the recharge area.

Much of Cuba is underlain by Oligocene and Miocene limestones and younger carbonate sediments that compose major aquifers. Other carbonate sequences of Jurassic and Cretaceous age exist in Cuba where much of the spectacular karst is developed. One area of karst is the coastal plain region around the Gulf of Batabano, which is characterized by solution features such as cenotes, lagoons, and drowned sinkholes and caves. The other area of karst along the northern coast, is characterized by mogotes—the erosional remnants of the mountain chains, particularly of the Sierra de los Organos and Sierra de Rosario (Geze and Mangin, 1980). Limestone of Eocene age covers much of Haiti and also forms the mountains south of the Cordillera Central in the Dominican Republic. Elsewhere, as in the Las Haitises area, Miocene limestones form tower karst and steep limestone pinnacles that rise above the surrounding alluvial plains. These limestones of Hispaniola are, in part, equivalent in age to limestones that form the major, gently dipping aquifer systems along the northern coast of Puerto Rico. Distribution in Hispaniola of the different facies that compose the Quaternary (Plio-Pleistocene) reef carbonate aquifer reflects cyclic sedimentation that shows (1) reef facies with white to cream, hard limestone composed of cemented coral and algae; (2) back-reef facies, clastic deposits derived from the reef, and sand and gravel from the land that have become cemented; and (3) lagoonal facies of clay and marl deposited in the inner lagoons. The reef and back-reef facies have the highest permeability, but this can be a disadvantage in coastal areas because the higher permeability permits more extensive saltwater encroachment (Gilboa, 1980).

The processes of carbonate sedimentation have played an especially important role in the geologic development in the Bahamas. The geologic history that most interests hydrologists began in Jurassic time with the deposition of evaporites (halite or anhydrite) and shallow-water carbonate facies on either a continental or oceanic crust. In part because of lack of migration of the Bahama platform, this deposition has been essentially uninterrupted up to the present time, and its rate has kept pace with the subsidence of this area. These aquifers are biogenetic and consist of coral reefs, fragments of algae, and animals associated with coral reefs. The steep slopes on the edge of the platform in the deep-water channels apparently have been maintained by submarine cementation, which stabilized the sediments. At any time during geologic history, the freshwater aquifers of the Bahamas would have been essentially as they are today—a thin lens of fresh water floating on an extensive body of salt water that underlies the low-lying islands.

## SEA-LEVEL FLUCTUATIONS

Another hydrogeologic process that has had a significant effect on the functioning of present-day aquifers of the West Indies is sea-level fluctuations. Sea-level elevation is a major control on both the hydraulic gradient and the nature and position of the downgradient boundary of aquifers. The position of the fresh water–salt water interface and thickness of the freshwater lens are controlled by the relation of freshwater head to sea level. The magnitude and distribution of freshwater head is controlled by size and elevation of the island, transmissivity of the carbonate aquifer, permeability of overlying material, and amount of precipitation. These factors, in turn, control the recharge rate, altitude of the potentiometric surface, and degree of confinement—all of which determine the amount and location of ground-water discharge, hydraulic gradient, and physical characteristics of the ground-water mixing zone.

As ground water flows toward discharge areas along the coast, it mixes with sea water that has encroached into the aquifer. This mixing generates brackish water, less dense and therefore more buoyant than sea water, that moves upward to discharge near the shore. The numerous submarine and nearshore brackish-water springs represent discharge from this mixing zone. This discharge requires a replacement of sea water, thereby setting up a cyclic flow system for water from the sea, through the coastal part of aquifers, upward along the interface, through the mixing zone, and back to the sea.

The mixing zone associated with this interface is a geochemically reactive area that has been a major influence in the development and distribution of porosity and permeability, particularly in carbonate rocks (Back and others, 1986). In most coastal areas, the fresh ground water has had long enough residence time to become saturated with respect to calcite, and normal surface ocean water is supersaturated with respect to the carbonate minerals. Therefore, these waters separately are incapable of dissolving limestone. However, because of the nonlinearity of certain geochemical parameters (e.g., activity coefficients, pH, partial pressure of carbon dioxide) as a function of salinity, the resulting mixture can become an aggressive water capable of limestone dissolution. All marine carbonates that now contain fresh water have been subjected to the geochemical reactivity of this mixing zone at least once, and, because of the repeated sea-level fluctuations of the Caribbean, these rocks have been subjected repeatedly to these reactions and processes. Every change of sea level, whether eustatic or tectonic, has caused the mixing zone to oscillate through the aquifer material during its geologic history with each landward or seaward migration of the interface, additional dissolution, and mineral alteration that occurred.

These processes have contributed to the neomorphism of the original carbonate minerals—aragonite and high-Mg calcite—to stoichiometric calcite and dolomite. Dissolution in this mixing zone not only has contributed to the development of porosity and permeability but also has been a significant geomorphic process in the formation of coastal caves and in shoreline retreat. Shore-line retreat has occurred by subterranean dissolution of the carbonate material, which leaves the overlying limestone unsupported; blocks then collapse into the sea. For example, the shoreline has been modified by the mixing-zone phenomena, as observed in the Yucatan where the roof blocks of nearshore sea-level caves have collapsed to form lagoons (Back and others, 1986). In Cuba the same process is hypothesized to have developed the large lagoon of Cienega de Zapata.

Sea-level fluctuations also have been a contributing factor in the formation of many alluvial-fill valleys in coastal areas of the West Indies. At times of lower sea level, rivers incised their valleys to the base level of the sea. After a subsequent rise of sea level, these valleys were drowned, and the rivers progressively filled the valley with alluvial deposits of sand and gravel. Although these aquifers are small, they are particularly important for the volcanic islands where the permeability of the volcanic ash layers is too low to yield water of usable quantity.

## REFERENCES CITED

Back, W., Hanshaw, B. B., Herman, J. S., and Van Driel, J. N., 1986, Differential dissolution of a Pleistocene reef in the ground-water mixing zone of coastal Yucatan, Mexico: Geology, v. 14, p. 137–140.

Bogart, D. B., Arnow, T., and Crooks, J. W., 1964, Water resources of Puerto Rico: Water Resources Bulletin, no. 4, 102 p.

Briggs, R. P., and Akers, J. P., 1965, Hydrogeologic map of Puerto Rico and adjacent islands: Hydrological Investigations Atlas, U.S. Geological Survey, HA-197.

Chidley, T.R.E., and Lloyd, J. W., 1977, The hydrogeological assessment of freshwater lenses in oceanic islands, in Morel-Seytoux, H. J., ed., Surface and subsurface hydrology: Fort Collins, Colorado, Water Resources Publication, p. 232–245.

Cosner, O. J., 1972, Water in St. John, U.S. Virgin Islands: U.S. Geological Survey Open-File Report 72–0078, 126 p.

Gardner, T. W., and others, 1987, Central America and the Caribbean, in Graff, W. L., ed., Geomorphic systems of North America: Boulder, Colorado, Geological Society of America, Centennial Special Volume 2, p. 343–402.

Geze, B., and Mangin, A., 1980, Le karst de Cuba: Paris, Revue de Geologie Dynamique et de Geographie Physique, v. 22, fasc. 2, p. 157–166.

Gilboa, Y., 1980, The aquifer systems of the Dominican Republic: International Association of Hydrological Sciences Bulletin, v. 25, no. 4, p. 379–393.

Gill, I. P., and Hubbard, D. K., 1986, Subsurface geology of the St. Croix carbonate rock system: St. Thomas, College of the Virgin Islands, Water Resources Research Institute, Technical Report no. 26, 71 p.

Gomez-Gomez, F., and Heisel, J. E., 1980, Summary appraisals of the nation's groundwater resources; Caribbean region: U.S. Geological Survey Professional Paper 813–U, 32 p.

Guisti, E. V., 1978, Hydrogeology of the karst of Puerto Rico: U.S. Geological Survey Professional Paper 1012, 68 p.

Guisti, E. V., and Bennett, G. D., 1976, Water resources of the North Coast Limestone area, Puerto Rico: U.S. Geological Survey, Water Resources Investigations 42–75, 42 p.

Hanshaw, B. B., and Back, W., 1979, Major geochemical processes in the evaluation of carbonate-aquifer systems, in Back, W., and Stephenson, D. A., eds., Contemporary hydrogeology, the George Burke Maxey memorial volume: Journal of Hydrology, v. 43, p. 287–312.

Heath, R. C., 1984, Ground water regions of the United States: U.S. Geological Survey Water Supply Paper 2242, 78 p.

Lewis, J. F., 1980, Resume of the geology of Hispaniola: Caribbean Geological Conference 9, Santo Domingo, Dominican Republic, Field Guide, p. 5–31.

Lewis, J. F., and Draper, G., 1988, Geologic and tectonic evolution of northern Caribbean margin, in Dengo, G., and Case, J. E., The Caribbean region: Boulder, Colorado, Geological Society of America, The Geology of North America, v. H (in press).

Pindell, J. L., 1985, Alleghenian reconstruction and subsequent evolution of the Gulf of Mexico, Bahamas, and proto-Caribbean: Tectonics, v. 4, no. 1, p. 1–39.

Robinson, A. H., 1974, Virgin Islands National Park; The story behind the scenery: Las Vegas, Nevada, KC Publications, 27 p.

UNESCO, 1986, First workshop on the hydrogeological Atlas of the Caribbean Islands, Santo Domingo, 7–10 October 1986: Montevideo, Uruguay, UNESCO, 229 p.

United Nations, 1976, Ground water in the Western Hemisphere: New York, United Nations, Natural Resources/Water Series no. 4, 337 p.

White, M. N., 1985, Groundwater movement and storage in karstic limestone aquifers in Jamaica: Journal of the Geological Society of Jamaica, v. 23, p. 1–16.

MANUSCRIPT ACCEPTED BY THE SOCIETY SEPTEMBER 9, 1987

## ACKNOWLEDGMENTS

The suggestions and advice given by Ralph Heath, U.S. Geological Survey, Raleigh, North Carolina; John Lewis, The George Washington University, Washington, D.C.; and David Aronson, U.S. Geological Survey, Reston, Virginia, significantly improved this chapter and their help is greatly appreciated.

The Geology of North America
Vol. O-2, Hydrogeology
The Geological Society of America, 1988

# Chapter 30

# *Region 27, Hawaiian Islands*

**Charles D. Hunt, Jr., and Charles J. Ewart**
*U.S. Geological Survey, P.O. Box 50166, Honolulu, Hawaii 96850*
**Clifford I. Voss**
*U.S. Geological Survey, National Center, MS 431, 12201 Sunrise Valley Drive, Reston, Virginia 22092*

## INTRODUCTION

Although the Hawaiian Islands (Fig. 3, Table 2, Heath, this volume) are surrounded by seawater, favorable circumstances cause the islands to be underlain by large quantities of fresh ground water. Foremost among these circumstances is the role of the island masses in causing orographic rainfall. Despite the presence of moist oceanic air, mean annual rainfall over the open ocean near Hawaii is only about 750 mm. The Hawaiian Islands obstruct oceanic winds, causing air to rise and moisture to precipitate. This orographic effect provides the mountainous uplands of the larger islands with as much as 7,000 to 10,000 mm of mean annual rainfall.

Favorable geologic conditions allow much of the abundant rainfall to accumulate as fresh ground water. Permeable soils and rocks permit easy infiltration and subsurface movement of water, and low-permeability geologic features impound large amounts of water in thick ground-water reservoirs. If geologic conditions were less favorable, more of the rainfall would run off to the sea and less water would be stored as ground water.

The geology of the Hawaiian Islands is varied and is the end result of many processes including volcanism, erosion, subsidence, sedimentary deposition, and even glaciation. Volcanic rocks predominate in volume, though sedimentary rocks are common and some metamorphic rocks are found in zones of geothermal activity. Thick sequences of lava flows form the largest and most important ground-water reservoirs; sedimentary and pyroclastic rocks form aquifers of secondary importance.

Geologic features at various scales control the occurrence and movement of ground water in Hawaii. Small-scale textural features of the various rocks determine their porosity and hydraulic conductivity. The behavior of ground water at local and regional scales is strongly influenced by features such as dikes, valley fills, ash beds, weathered zones, and coastal plain sediments. The low hydraulic conductivity of these features causes them to act as barriers or restraints to the flow of water in the lava aquifers. They control the direction of flow and impound water to greater heights than would be reached in their absence.

## REGIONAL SETTING AND GEOLOGIC PROCESSES

The Hawaiian Islands are the exposed parts of a large volcanic mountain range, the Hawaiian Ridge, most of which lies beneath the sea. The Hawaiian Ridge probably formed as the Pacific lithospheric plate drifted northwestward over a convective plume, or hotspot, in the mantle (Wilson, 1963). The plate motion and continuous volcanism at the hotspot thus produced an apparent southeasterly progression in mountain building, with the younger islands and present volcanic activity at the southeast end of the island chain. Each mountain mass is composed mainly of intrusive and extrusive volcanic rocks and contains thousands of individual lava flows. Thorough summaries of Hawaiian geology have been published by Macdonald and others (1983) and Stearns (1985), and much of the following discussion has been adapted from these sources.

### Shield-building volcanism

Hawaiian volcanoes are known as shield volcanoes because of their low, gently sloping profiles. Most of the activity that built the Hawaiian Islands was similar to the eruptions that take place on the active volcanoes of Mauna Loa and Kilauea, on the island of Hawaii. These eruptions of tholeiitic basalt typically begin with high fountaining of gas-charged lava from an elongate fissure. The firefountains soon diminish along most of the fissure, and activity becomes restricted to a single major vent. Continued effusion deposits large amount of tephra near the vent and feeds lava flows that spread rapidly down the flanks of the volcano. The high fluidity of tholeiitic lavas results in thin flows that range in thickness from several centimeters to a few tens of meters and average about 3 to 7 m.

Near the end of the shield-building stage of many Hawaiian volcanoes, lavas of alkalic or andesitic basalt were erupted and formed a cap over the earlier flows. These late-stage lavas were more viscous than the tholeiitic lavas, and eruptions were more explosive, producing more ash. The flows tend to be thicker and more massive; they are commonly 15 to 30 m thick. In some

Hunt, C. D., Jr., Ewart, C. J., and Voss, C. I., 1988, Region 27, Hawaiian Islands, *in* Back, W., Rosenshein, J. S., and Seaber, P. R., eds., Hydrogeology: Boulder, Colorado, Geological Society of America, The Geology of North America, v. O-2.

places, a gradual transition from tholeiitic lavas to alkalic lavas is seen, but in other places, minor erosional unconformities separate the two rock types.

Hawaiian shield-building eruptions emanate mainly from one or more elongate rift zones along the topographic crests of the volcanoes. Deep dissection of rift zones by erosion has revealed numerous feeder dikes within them. These sheets of intrusive rock have near-vertical attitudes and are arranged in subparallel patterns aligned roughly with the trends of the rift zones. In the central parts of these dike complexes, dikes may number a few hundred per kilometer and constitute 10 percent or more of the total rock. In marginal parts of the rift zones and on the outer flanks of the volcanoes, dikes are sparsely distributed or absent.

The summit of a Hawaiian volcano is typically occupied by a caldera. A variety of materials accumulates in the caldera, including tephra, talus breccia, thin-bedded lavas, and ponded lavas that are thicker and more massive those on the flanks of the volcano.

### Subsidence, erosion, and sedimentary deposition

Other geologic processes besides volcanism are active in Hawaii. All of the islands have been modified by subsidence, erosion, and sedimentary deposition, though such modifications are more pronounced on the older islands. Each island has subsided under its own weight, as much as a few thousand meters in some cases. Stream valleys have cut deeply into the shields and many of the valleys are partly filled with sediments. Coral reefs and sediments have accumulated on the flanks of the islands, forming coastal plains and submarine shelves.

### Post-erosional volcanism

After a long period of erosion, volcanic eruptions resumed on some islands, perhaps in response to isostatic adjustments along the Hawaiian Ridge (Jackson and Wright, 1970, p. 427). This rejuvenated activity occurred at scattered vents and produced lavas and pyroclastic rocks that typically lie on or are interbedded with soils and sediments. Rocks of this stage of activity are typically alkalic or nephelinic basalts.

## THE ROCKS AND THEIR WATER-BEARING PROPERTIES

### Lava flows

Although most Hawaiian volcanic rocks have similar chemical and mineralogic compositions, their modes of emplacement result in a variety of rock textures that determine their water-bearing properties. The rocks that predominate in volume and form the most productive aquifers are the thin-bedded basaltic lavas, which have a diverse assortment of porosity and permeability features. In a typical sequence of such lavas (Fig. 1A), there are elements of fracture permeability (joints and cracks), inter-

granular permeability (beds of clinker, or rubble), and cavernous permeability (lava tubes and interflow voids). The net effect is a high overall hydraulic conductivity of a few hundred to a few thousand meters per day. Thick and massive lavas, such as those of the late alkalic/andesitic stage, have lower hydraulic conductivity and tend to inhibit infiltration and increase surface runoff.

The two basic textural forms of basaltic lava in Hawaii are pahoehoe and aa. Pahoehoe lavas are extruded in a fluid state and spread out to form thin flows with smoothly undulating surfaces that are cracked and collapsed in places. Voids of various sizes are common in pahoehoe lavas and may impart a permeability that is similar to the secondary solution permeability of carbonate rocks. Interflow voids are gaps and spaces where the irregular upper surface of a flow is not completely filled in by the molten lava of subsequent flows. Lava tubes are voids formed when the top surface of the lava cools and hardens into a crust and the molten lava beneath drains out. Most tubes are less than a meter in diameter and of small lateral extent, but some are a few meters in diameter and extend several kilometers. Although thin-bedded pahoehoe typically contains many open cracks and voids, thick beds of massive pahoehoe form where lava is ponded in depressions or on gentle slopes. The hydraulic conductivity of massive pahoehoe is likely much lower than that of thin-bedded pahoehoe.

Pahoehoe lava commonly grades into aa lava with increasing distance from the eruptive vent. As flowing lava loses gas and becomes more viscous, the cooling crust on top of the flow breaks up into a mass of angular fragments known as aa clinker. Beneath the clinker is a still-molten but viscous lava mass that continues to advance, carrying the sheath of clinker along like a tractor tread. The interior part of the flow, known as aa core, has a massive texture when cooled. An aa flow unit is typified by a dense aa core a meter or more in thickness, with fragmental clinker beds both above and below the dense part.

The alternation of clinker with dense rock in aa sequences exerts a strong layering control on the flow of ground water. As determined from drilling and excavating, water-producing zones in aa lava typically correspond to clinker layers, with relatively little water being contributed by the dense aa cores. The hydraulic conductivity of clinker appears similar to that of a coarse, well-sorted gravel and likely ranges from several tens to several thousands of m/day. In contrast, the appearance of aa cores suggests that they may have far lower hydraulic conductivity, perhaps by one or two orders of magnitude, than either aa clinker or thin-bedded pahoehoe. The principal hydraulic conductivity of aa cores is provided by cooling joint fractures. In aa flows thicker than a meter or so, the joints are vertical cracks spaced about a meter apart. The intervening, crudely polygonal blocks of massive rock are virtually impermeable. Vertical bridges of clinker at the lateral margins of aa flows may provide important avenues for vertical movement of water; however, such bridges may be horizontally far apart, especially in thicker flows of broad lateral extent.

Gas vesicles are conspicuous in both aa and pahoehoe and

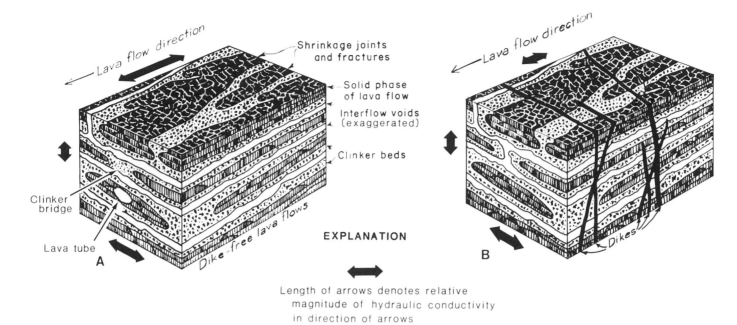

Figure 1. Structural features associated with lava flows (A) in which hydraulic conductivity is highest in the lava-flow direction, and (B) in which hydraulic conductivity is significantly lowered by dikes intruded at right angles to the lava-flow direction (modified from Takasaki and Valenciano, 1969).

may impart a high bulk porosity to the lavas. The vesicles generally are not interconnected and do not contribute appreciably to effective porosity or to hydraulic conductivity.

A sequence of thin-bedded lavas contains many individual pahoehoe and aa flows; their areal extent is much greater than their thickness. Flow widths of tens to hundreds of meters and lengths of several kilometers are common. The relative dimensions of the lavas and their conspicuous, near-horizontal bedding suggest that they are anisotropic, with greater horizontal than vertical hydraulic conductivity (Fig. 1A). Anisotropy in hydraulic conductivity has yet to be strongly affirmed or quantified by field testing. Sequences of thick flows might be expected to be more highly anisotropic than sequences of thin flows; various collapse features and disruptions in bedding that enhance vertical hydraulic conductivity appear to be less common in thicker flows.

Layered lava flows that make up an island mass constitute aquifers of great thickness. Permeable lavas of subaerial origin extend as much as a few thousand meters below sea level as a result of island subsidence. The nature and depth of the effective bottoms of volcanic aquifers are open to conjecture. Hydraulic conductivity probably decreases gradually with depth due to lithostatic compaction, and there may be a decrease in hydraulic conductivity associated with submarine-emplaced lavas and related sediments. The depth of submarine lavas varies with the subsidence history of each volcano; they are present at shallower depths within younger volcanoes and at greater depths within older ones. A geothermal well on Kilauea, Hawaii's youngest emergent volcano, entered submarine pillow lavas at about 250 m below sea level (Macdonald, 1976, p. 23).

## *Dikes*

Dikes are thin sheets of dense, intrusive rock with only fracture permeability. Most dikes are no more than a meter thick but can be areally extensive. Where dikes intrude lava flows, they inhibit the movement of ground water, especially if flow is perpendicular to the plane of the dike (Fig. 1B). Widely scattered dikes, like those in the marginal dike zones, generally impound water in large compartments of more permeable rock. Where dikes are numerous and intersect at various angles, as in dike complexes, the net effect is an overall reduction in rock porosity and hydraulic conductivity (Takasaki and Mink, 1985, p. 7).

## *Pyroclastics*

Pyroclastic rocks include ash, cinder, spatter, and larger blocks and bombs. They are essentially granular and have porosity and hydraulic conductivity similar to granular sediments. Fine-grained pyroclastics such as ash have lower hydraulic conductivity than coarse pyroclastics; hydraulic conductivity may be reduced further by weathering or by compaction to tuff.

## *Soil and saprolite*

The basaltic composition of volcanic materials, the warm climate, and generally abundant rainfall result in rapid chemical weathering of the rocks to soil and saprolite. Weathering creates clay minerals and destroys the primary permeability of the parent

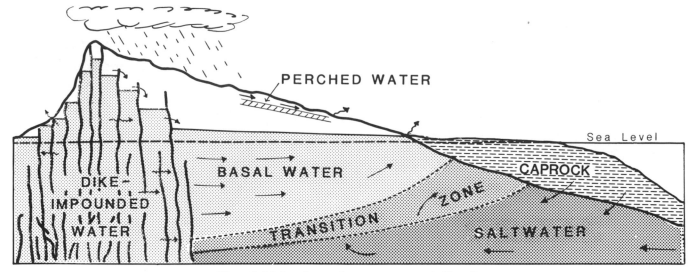

Figure 2. Modes of ground-water occurrence in Hawaii.

rocks, causing an overall reduction in hydraulic conductivity. The effects of weathering commonly extend several tens of meters below land surface. Basaltic saprolite has a complex heterogeneous structure that may include preferred zones of water movement and retention.

### Sedimentary rocks

Sedimentary rocks in Hawaii have wide ranges of composition, grain size, degree of consolidation, and hydraulic properties. Terrestrial sediments include beach and dune sands, colluvium, and alluvium that ranges from boulders to estuarine muds. Marine sediments are mostly calcareous and include coral-algal reefs, coralline rubble, sands, and marls.

Sedimentary rocks can be divided into calcareous and non-calcareous rocks with regard to their controls on the movement of ground water. Most calcareous rocks have moderate to high hydraulic conductivity due to a combination of primary depositional textures and secondary solution permeability. Cavernous limestone and coralline rubble are among the most permeable rocks in Hawaii, and in many instances are more permeable than basaltic lavas. Hydraulic conductivities of a few hundred to a few thousand m/day are typical, and many field estimates exceed 3,000 m/day. Low-permeability calcareous rocks include marls and indurated sandstones.

Noncalcareous sedimentary rocks are derived from basalt and commonly undergo rapid and thorough weathering that results in low hydraulic conductivity. The moderate to high hydraulic conductivity of coarse-grained aluvium or talus persists only in dry areas or in young deposits. Where valleys have cut deeply into volcanic aquifers, valley fills of fine-grained and weathered alluvium may act as barriers to ground-water flow.

Thick sequences of coastal plain sediments have accumulated at the margins of older islands. The successive layers of calcareous and noncalcareous sediments are collectively called "caprock" because they form an overlying confining bed for the volcanic aquifer beneath. Caprock has a bulk permeability much lower than that of the basaltic lavas, mainly because of the contained muds and other fine-grained sediments. However, permeable layers of limestone and coralline rubble are interbedded with the finer sediments as well.

## OCCURRENCE OF GROUND WATER

The occurrence of ground water in Hawaii is determined by the nature of the rocks and by the presence of geologic features such as dikes, valley fills, and caprock. Principal modes of ground-water occurrence are shown in Figure 2, and maps of ground-water occurrence for the major islands are shown in Figure 3.

### Basal water

The largest bodies of freshwater in Hawaii occur as basal ground water. Basal water refers to a body of freshwater that floats on saltwater within an aquifer. Synonymous terms for basal water include "freshwater lens," "basal lens," and "Ghyben-Herzberg lens."

The water table of a Hawaiian basal-water body typically is flat and lies no more than several meters above sea leve. Beneath the water table, freshwater extends to depths below sea level of about 40 times the water-table elevation. This situation results from the buoyant displacement of saltwater by freshwater and is similar to the way an iceberg floats in the sea with most of its mass below sea level. At the base of the basal lens, freshwater grades into saltwater in a transition zone of mixed water that may range from a few meters to a few hundred meters thick.

Basal ground water occurs in both volcanic and sedimentary aquifers in Hawaii, and can occur under confined or unconfined conditions. Most basal water in volcanic aquifers is unconfined

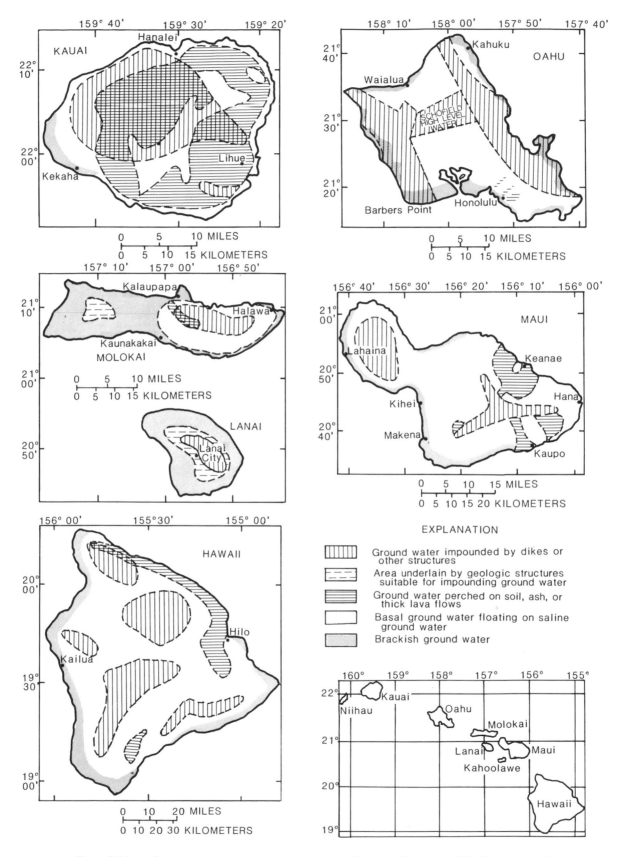

Figure 3. Maps of ground-water occurrence for the principal Hawaiian Islands (modified from Takasaki, 1977).

except near the coast, where the aquifers may be confined by caprock. Recharge to basal-water bodies occurs by direct infiltration of rainwater and by underflow from upgradient water bodies; direct infiltration probably has more temporal variability than underflow.

The size of a basal-water body depends on the amount of recharge received, the hydraulic conductivity of the aquifer, and the presence or absence of confinement. Basal-water bodies in lava aquifers are thin where there is no caprock to impede fresh-water discharge to the sea. In this situation the water table is less than 1 m above sea level at the shore, rising inland at about 0.2 to 0.4 m/km to a maximum height of perhaps 3 or 4 m above sea level. A freshwater lens of this type is generally less than 150 m thick and discharges to the sea by mostly unmeasurable seepage from submarine and shoreface springs.

Thick basal-water bodies are formed where caprock confines the lava aquifer at the coast. Confinement causes artesian heads of a few meters above sea level beneath the caprock and also has an impounding effect on the unconfined basal water inland of the caprock, causing higher water levels and a greater thickness of freshwater. Caprock-impounded lenses have fresh-water heads of about 3 to 12 m above sea leve, and have attained thicknesses of nearly 550 m. Discharge from an impounded lens is by measurable flow from prominent springs near the inland edge of the caprock and by diffuse, unmeasurable seepage through the semiconfining caprock. The greater amounts of water stored in impounded lenses causes them to be less severely affected by climatic variations such as droughts.

Basal water bodies in sedimentary aquifers generally are no more than 40 m thick, with water tables no more than 1 m above sea level. Most basal water in sedimentary rocks occurs in aquifers of limestone or sand in the uppermost parts of caprock sequences. Deeper pars of caprock are occupied by saltwater.

### Dike-impounded water

Dikes impound water to great heights within the rift zones of Hawaiian volcanoes. The occurrence and behavior of dike-impounded water is best known on Oahu, where erosion has exposed the interior parts of rift zones and where many water-development tunnels have been dug into dike-intruded aquifers (Stearns and Vaksvik, 1935; Takasaki and Mink, 1985). Water in dike reservoirs reaches heights of nearly 500 m above sea level on Oahu and nearly 1,000 m on Maui and Hawaii. Freshwater may extend far below sea level in dike reservoirs, but the principle of buoyant balance that governs basal water probably does not apply to freshwater impounded to such great heights. Instead, the rocks beneath dike reservoirs probably become progressively less porous and less permeable with depth as the number of intersecting dikes increases (Takasaki and Mink, 1985, p. 7).

Where valleys cut deeply into dike reservoirs, discharge of dike-impounded water is evident at springs and as gaining streams. Underflow from dike compartments also recharges downgradient water bodies. Water moves from dike reservoirs

through joints and gaps in the dikes and where impounded water overtops dikes. Dike reservoirs store large amounts of water and are thought to discharge water at a more or less constant rate. Thus, underflow from dike reservoirs may help smooth the effects of climatic variations on downgradient water bodies by providing them with a near-constant recharge component.

### Perched water and water in the unsaturated zone

The unsaturated zone in Hawaii is commonly tens to thousands of meters thick and includes layers of soil, saprolite, and weathered and unweathered rock. Although hydrologic processes in the unsaturated zone are poorly known, this zone is important because it is a source of solutes and is the pathway by which recharge water and surface contaminants reach deeper water bodies.

Perched water occurs in valley-fill deposits, in saprolite, and on ash beds and soils intercalated within lava sequences. Where older lavas are separated from younger lavas by an erosional unconformity, complex multiple-aquifer systems occur in lavas and alluvium that have filled valleys in the older surface. Within such aquifer systems, water is perched on massive valley-filling lavas or on weathered material (Stearns and Macdonald, 1942, p. 232). Perching also develops where widespread layers of ash have been buried by younger lavas. The lower permeability of ash causes water to be perched above it (Stearns and Macdonald, 1946, p. 264). Perched water occurs in smaller volumes than basal or dike-impounded water, and responds more rapidly to climatic stresses such as droughts and recharge events.

### Saltwater

Enormous bodies of saltwater fill the lava aquifers beneath the freshwater bodies. While the thickest known basal-water body extends about 550 m below sea level, saltwater in permeable lavas may extend as much as 2,000 to 3,000 m deeper. Few wells have been drilled to such depths and hydrologic processes in saltwater aquifers are poorly known. Information on saltwater beneath dike-impounded water is even more scarce. Freshwater/saltwater relations may be very complex within the rift zones of active volcanoes where hydrothermal processes may dominate.

Beneath a caprock-impounded basal lens on Oahu, saline ground water 400 m below sea level is 15° C warmer than ocean water at the same depth. The higher temperature has been attributed to slight geothermal heating of the relatively immobile salt-water within the aquifer (Visher and Mink, 1964, p. 126). Where caprock is absent, the temperature of saline ground water more closely resembles that of the open ocean, suggesting easier transfer of saltwater between the aquifer and the sea (Lau, 1967, p. 264).

Saltwater is present throughout most of the coastal plain sediments, except where thin freshwater bodies occur in the uppermost caprock aquifers. (For additional discussion concerning saltwater encroachment in coastal island areas, see Back, this volume.)

# INFLUENCE OF GEOHYDROLOGIC BARRIERS ON THE OCCURRENCE AND MOVEMENT OF GROUND WATER: OAHU EXAMPLE

Hydrologic studies have been more extensive on Oahu than on other Hawaiian Islands, and its geohydrologic framework is best known. Oahu is the eroded remnant of two volcanoes, the Waianae and the Koolau.

Figure 4 shows geohydrologic barriers that govern the occurrence and movement of water in the principal volcanic aquifers. Figure 5 shows the water-level relationships that guided the delineation of this geohydrologic framework. Generalized patterns of ground-water flow are inferred from the water levels and are shown by arrows. A map of ground-water occurrence on Oahu appears in Figure 3.

Hydraulic continuity is generally high within the broad areas between major geohydrologic barriers. Within the rift zones, however, dikes divide the aquifer into many small compartments. The degree of hydraulic continuity between adjacent areas depends on the nature of the geohydrologic barrier that separates them.

The rift zones and dike structures are deep-rooted structural elements of the volcanoes. They are effective hydrologic barriers, and hydraulic stresses cannot be transmitted across them from one area to another. The erosional unconformity and the valley fills are more surficial in nature and are less effective hydrologic barriers.

The erosional unconformity defines an eroded and weathered surface of the Waianae Volcano that was buried by younger Koolau lavas. This surface is near-horizontal, sloping perhaps 5° to 15°, and is likely mantled with low-permeability saprolite. This weathered zone forms a leaky or semiconfining bed between the Waianae and Koolau aquifers. It causes an abrupt discontinuity in water levels but is thought to transmit large amounts of water from one area to the next. Although hydraulic stresses in one aquifer do not strongly affect water levels in the other, changes in water-level relationships could conceivably change the amount or direction of water that passes through the semiconfining bed.

Valley fills also behave as partial barriers to flow in the volcanic aquifers. The weathered sediments in the fills have very low hydraulic conductivity, but the valleys do not penetrate the volcanic aquifers fully. The valleys act chiefly as barriers to freshwater flow, causing water-level discontinuities of a meter or two between adjacent areas. Where the valleys become shallower inland, appreciable amounts of freshwater may flow beneath the fills from area to area. The valleys are not deep enough to greatly affect the underlying saltwater, which may also allow hydraulic stresses to be transmitted between adjacent areas.

## GEOCHEMISTRY

Geochemical processes in Hawaii include the weathering and diagenesis of various volcanic and sedimentary rocks, and the deposition and alteration of minerals by geothermal gases and fluids.

Figure 4. Geohydrologic barriers in the principal volcanic aquifers of Oahu.

Rainfall on the Hawaiian Islands contains as little as 10 to 20 mg/l (milligrams per liter) dissolved solids. Under natural conditions, dissolved solids concentration increases to about 200 mg/l as rainwater moves through volcanic rocks to the water table. The unaltered appearance of rock samples and drill cuttings from basaltic aquifers suggest that little rock-water interaction occurs in the saturated zone and that most inorganic solutes are supplied by weathering near the land surface.

Natural sources of solutes in ground water are the sea, the atmosphere, and the rocks and overburden through which water moves. Agricultural application of irrigation water and fertilizer contributes additional amounts of dissolved constituents and increases concentrations to as much as several thousand mg/l. Still higher concentrations can be caused by seawater intrusion.

Seawater that intrudes into volcanic aquifers can differ chemically from water in the open ocean. As water passes through caprock sediments, changes in water chemistry may result from solution of carbonate rocks, reducing conditions, and ion exchange between saltwater and clays. Chloride concentration of the saltwater appears to be little affected during intrusion, but there is significant enrichment of magnesium and calcium, and significant depletion of sodium, potassium, and sulfate (Visher and Mink, 1964, p. 2–3).

## SUMMARY

The geohydrologic framework of the Hawaiian Islands exerts characteristic controls on the occurrence and movement of ground water. These controls result from small-scale hydraulic properties of the various rocks and from larger geohydrologic barriers of volcanic, erosional, and depositional origin. The major occurrences of ground water in Hawaii are: (1) mostly uncon-

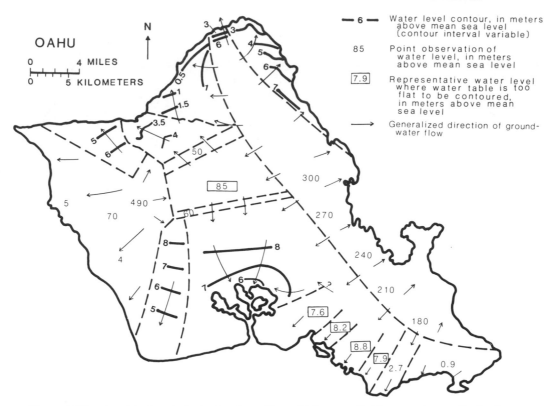

Figure 5. Water levels in the principal volcanic aquifers of Oahu, compiled from previous studies by the U.S. Geological Survey, and generalized directions of ground-water flow.

fined freshwater lenses floating on saltwater in highly permeable, layered lavas; (2) freshwater impounded to high elevations in permeable lavas that are intruded by impermeable, vertical dikes; (3) freshwater perched on poorly permeable ash beds or soils within lavas or sediments; and (4) saltwater in volcanic or sedimentary rocks.

## REFERENCES CITED

Jackson, E. D., and Wright, T. L., 1970, Xenoliths in the Honolulu Volcanic Series, Hawaii: Journal of Petrology, v. 11, pt. 2, p. 405–430.

Lau, L. S., 1967, Seawater encroachment in Hawaiian Ghyben-Herzberg systems: Proceedings of the National Symposium on Ground-Water Hydrology, American Water Resources Association, p. 259–271.

Macdonald, G. A., 1976, *in* Hawaii geothermal project well completion report, HGP-A: Honolulu, University of Hawaii Research Corporation, 23 p.

Macdonald, G. A., Abbott, A. T., and Peterson, F. L., 1983, Volcanoes in the sea; The geology of Hawaii (2nd ed.): Honolulu, University of Hawaii Press, 517 p.

Stearns, H. T., 1985, Geology of the State of Hawaii (2nd ed.): Palo Alto, California, Pacific Books, 335 p.

Stearns, H. T., and Macdonald, G. A., 1942, Geology and ground-water resources of the island of Maui, Hawaii: Hawaii Division of Hydrography Bulletin 7, 344 p.

———, 1946, Geology and ground-water resources of the island of Hawaii: Hawaii Division of Hydrography Bulletin 9, 363 p.

Stearns, H. T., and Vaksvik, K. N., 1935, Geology and ground-water resources of the island of Oahu, Hawaii: Hawaii Division of Hydrography Bulletin 1, 479 p.

Takasaki, K. J., 1977, Elements needed in design of a ground-water-quality monitoring network in the Hawaiian Islands: U.S. Geological Survey Water-Supply Paper 2041, 23 p.

Takasaki, K. J., and Mink, J. F., 1985, Evaluation of major dike-impounded ground-water reservoirs, island of Oahu: U.S. Geological Survey Water-Supply Paper 2217, 77 p.

Takasaki, K. J., and Valenciano, S., 1969, Water in the Kahuku area, Oahu, Hawaii: U.S. Geological Survey Water-Supply Paper 1874, 59 p.

Visher, F. N., and Mink, J. F., 1964, Ground-water resources in southern Oahu, Hawaii: U.S. Geological Survey Water-Supply Paper 1778, 133 p.

Wilson, J. T., 1963, A possible origin of the Hawaiian Islands: Canadian Journal of Physics, v. 41, p. 863–870.

MANUSCRIPT ACCEPTED BY THE SOCIETY MAY 15, 1987

## Chapter 31

# *Region 28, Permafrost region*

**Charles E. Sloan** (retired)
*Water Resources Division, U.S. Geological Survey, 4230 University Drive, Suite 201, Anchorage, Alaska 99508-4664*
**Robert O. van Everdingen**
*Arctic Institute of North America, University of Calgary, Calgary, Alberta, Canada*

## INTRODUCTION

### *Occurrence of permafrost*

Permafrost is defined as ground (rock or soil) with temperatures that have remained below 0°C continuously for two or more years. It is widespread in the arctic and subarctic regions of Alaska and Canada, and extends southward in the alpine zone of the western Cordillera of North America (Fig. 1). (Fig. 3 and Table 2, Heath, this volume). Permafrost influences the occurrence of ground water in about 20 percent of the land area of the world, including about 85 percent of Alaska and about 50 percent of Canada.

The occurrence of permafrost is controlled by the surface heat balance, which in turn is influenced by terrain factors such as relief, slope aspect, vegetation, snow cover, moisture content, soil and rock type, and the presence of surface-water bodies (Brown, 1970). Near the southern fringes of the permafrost region, the permafrost consists of thin, isolated masses. To the north, the areas in which permafrost occurs gradually become larger, and the depth to the base of the permafrost increases to the point where, in a broad band across northern Alaska and Canada, the permafrost is essentially continuous, and its thickness may exceed 500 m.

### *Properties of frozen ground*

Freezing of a water-bearing soil or rock affects a number of its physical properties. Electrical and hydraulic conductivities decrease by several orders of magnitude; apparent specific heat capacity decreases, and thermal conductivity increases (Anderson and Morgenstern, 1973). Any water remaining in the liquid phase (e.g., due to depression of its freezing point by dissolved-mineral content) will have increased viscosity.

*Ice and water content.* Permafrost, which is defined solely on the basis of temperature, is *not necessarily frozen.* It may be "dry," containing neither ice nor water. It may contain unfrozen water exclusively, if the freezing-point depression is large enough to prevent freezing at the prevailing negative temperature; it may contain a mixture of unfrozen water and ice; or it may contain ice exclusively. Segregated ice, in the form of lenses, layers, or ice wedges, may also be present.

Unfrozen water exists in many soil-water-ice systems as thin films adsorbed on the surfaces of soil (mineral) particles, even if the temperature is well below the initial freezing point of the soil water (see Fig. 2). The thickness of the water films is a function of temperature (50 Angstroms or more at 0°C; 9 Angstroms at –5°C; 3 Angstroms at –193°C); it is nearly independent of the total water content (liquid plus solid), except in very dry soils.

Adsorbed water is mobile under both electrical and thermal gradients. Because unfrozen-water content is primarily a function of temperatures below 0°C, a thermal gradient in a frozen porous medium is analogous to a water-content gradient in an unfrozen, unsaturated porous medium.

The term *cryopeg* describes zones of negative temperature in which *all* water is in the liquid phase (Fig. 3). Cryopegs are a common feature in permafrost; in many places, a water-bearing basal cryopeg, rather than ice-bearing permafrost, is found immediately above the permafrost base.

*Hydraulic conductivity.* Rates of water movement in frozen porous materials depend on the overall temperature of the system, on the thermal gradient, and on the available cross-sectional area of interconnected films of unfrozen water. The latter is likely to be largest in fine-grained materials (silt and clay sizes). Frozen ground, whether seasonal or perennial, exerts a retarding influence on water movement. However, frozen ground should not be regarded as an impermeable material, but rather as a confining material of very low hydraulic conductivity (a leaky confining bed).

Hydraulic conductivities of frozen and unfrozen parts of the same unsaturated soil may be similar when soil temperatures are close to 0°C. With decreasing temperature below 0°C, the hy-

Sloan, C. E., and van Everdingen, R. O., 1988, Region 28, Permafrost region, *in* Back, W., Rosenshein, J. S., and Seaber, P. R., eds., Hydrogeology: Boulder, Colorado, Geological Society of America, The Geology of North America, v. O-2.

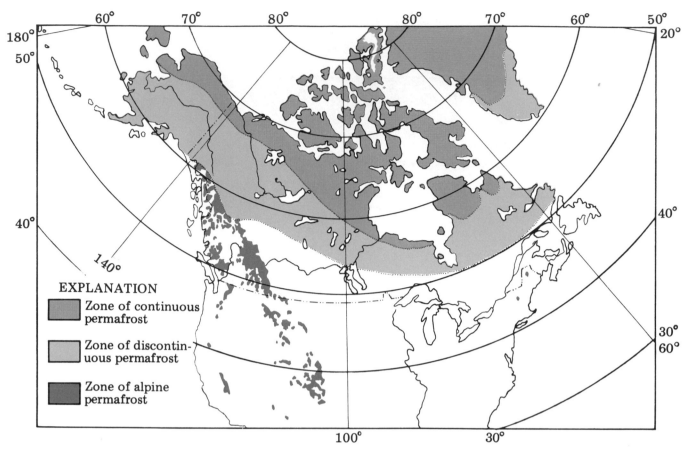

Figure 1. Distribution of permafrost in North America (adapted from Péwé, T. L., 1983). Isolated areas of alpine permafrost not shown on the map exist in high mountains and outside the map area in Mexico and Hawaii.

draulic conductivity of the soil decreases as the cross-sectional area of interconnected films of unfrozen water becomes progressively smaller.

Fracture systems can provide significant additional hydraulic conductivity to a frozen soil or rock mass. The presence of segregated ice, as layers or lenses, may decrease hydraulic conductivity to the point where the ground becomes effectively impermeable.

### Climate of the permafrost region

*Mean annual air temperature.* Mean annual air temperatures reaching as low as –25°C impose restrictions on recharge that do not exist in temperate regions. Air temperatures below 0°C keep the "active layer" frozen as much as ten months per year. (The active layer is the surficial layer of ground in permafrost areas that is frozen each winter and thawed each summer. Its thickness varies from several centimeters to more than a meter.) Thawing and subsequent recharge through the active layer may only be possible for as little as two months each year. Although recharge can take place through frozen ground under the appro-

priate conditions, the resulting recharge rates are usually one or more orders of magnitude lower than those for the same ground in unfrozen condition.

*Mean annual precipitation.* Mean annual precipitation is relatively low over much of the area of the permafrost region. It ranges from 500 to 600 mm per year in the southern, discontinuous part of the region in North America to less than 100 mm per year in parts of the Canadian Arctic Islands. The low precipitation severely restricts the amount of water available for recharge. In addition, potential evapotranspiration exceeds precipitation over a large portion of the permafrost region.

## HYDROGEOLOGY

The hydrogeology of permafrost areas has, until recently, received relatively little attention in North America. For large parts of Alaska and northern Canada the knowledge of hydrogeology is limited to observations of ground-water discharge phenomena, observations of karst recharge, and data on a small number of widely scattered and commonly shallow wells (Brandon, 1965; Heath, 1984; van Everdingen, 1974; Williams, 1965,

1970; Williams and van Everdingen, 1973; Williams and Waller, 1966; Zenone and Anderson, 1978).

Directions and rates of ground-water movement in the permafrost region are, in general, dependent on the same physical parameters as in areas without permafrost. In addition, however, ground-water regimes in permafrost are affected by climatic effects and by the presence of perennially frozen ground.

### Position of aquifers relative to permafrost

In discussing the hydrogeology of the permafrost region it is convenient to group aquifers under three major headings: (A) suprapermafrost aquifers, situated above the permafrost, with permafrost generally serving as a relatively impermeable basal boundary; (B) intrapermafrost aquifers found in unfrozen zones (taliks) within the permafrost; and (C) subpermafrost aquifers, situated below the permafrost, with permafrost serving as a relatively impermeable upper boundary. Various interconnections exist among these aquifer types as well as between them and the surface waters in lakes and streams (Fig. 3). Each of the three aquifer types may be found in either unconsolidated deposits or bedrock.

*Suprapermafrost aquifers.* Suprapermafrost aquifers are normally found in low-lying areas, in closed taliks beneath lakes and rivers, and in elevated areas with little relief; they are less common in sloping areas, where they are usually restricted to gentle, south-facing slopes.

Suprapermafrost aquifers are unreliable as a source of water supply. Most of these aquifers freeze during the winter, or their storage may be rapidly depleted by winter discharge. Commonly, the water has high organic content, and its mineralization may increase significantly during the winter. Suprapermafrost aquifers are primarily useful as a seasonal (summer) water supply. Water discharged through and into riverbed taliks via an open talik from subpermafrost aquifers (Fig. 3) can be used in winter as well, if the degree of mineralization is acceptable. Suprapermafrost aquifers are a source of fresh-water supply in the lowlands bordering the Beaufort Sea in Alaska and Canada (Williams and van Everdingen, 1973). Suprapermafrost water also plays a significant role in geotechnical problems such as frost heave, formation of icings and frost mounds, and instability of slopes.

Seasonally increased pressures, developing in saturated material between seasonal frost advancing from the surface and the top of the permafrost, may be responsible for some icings. Such increased pressures can also affect the stability of cut slopes, especially during the spring thaw when the pressures may be at their maximum while the thickness of the seasonally frozen layer decreases rapidly.

Seasonally increased pressures, in combination with gradual reduction in the areal extent of discharge areas (by encroaching frost) can cause development of "quick" conditions, and formation of seasonal frost mounds of the frost-blister type. Soil is in a "quick" condition when the upward seepage force exceeds a critical value at which there is no grain-to-grain pressure; the soil

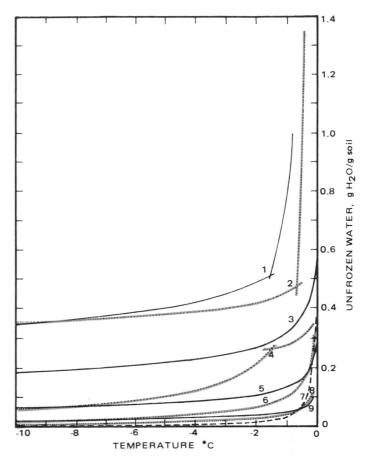

Figure 2. Unfrozen water content as a function of temperature for several typical soils and soil constituents (modified after Anderson and Morgenstern, 1973, Figs. 6 and 7). 1, Umiat bentonite; 2, Wyoming bentonite; 3, Hawaiian clay; 4, kaolinite; 5, Suffield silty clay; 6, Down Field silty clay; 7, basalt; 8, West Lebanon gravel $<100$ $\mu$; 9, Fairbanks silt.

will exhibit no shearing strength and will have the properties of a fluid.

Frost blisters develop through lifting of sections of frozen ground by water under increased hydraulic pressures. Part or all of the water in the blisters subsequently freezes. Frost blisters may rupture (sometimes explosively), drain, reseal, and resume growth several times during the winter. Their horizontal dimensions range from less than 10 m to more than 100 m; their height usually ranges between 0.5 and 5 m; vertical growth rates of more than 0.5 m/day have been recorded (van Everdingen, 1982).

*Intrapermafrost aquifers.* Intrapermafrost aquifers are not subject to seasonal freezing, and their extent is relatively constant, being affected primarily by long-term climatic (temperature) changes. During cooling trends, the extent of intrapermafrost taliks will be reduced by encroaching permafrost, whereas during warmer trends the reverse will take place.

Intrapermafrost aquifers can be found in open or "through" taliks, unfrozen zones that completely penetrate the permafrost; in lateral taliks, unfrozen layers within the permafrost; and in iso-

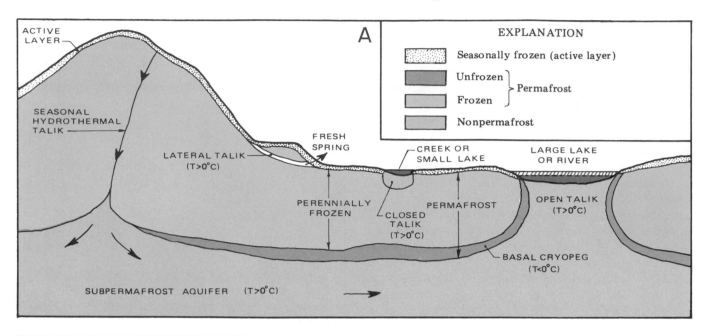

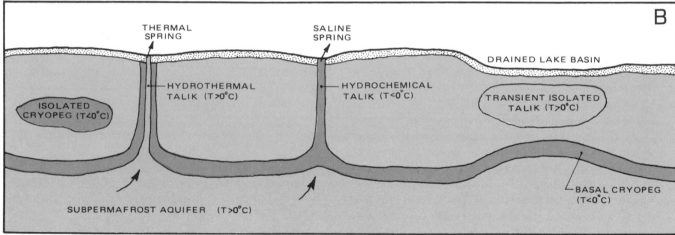

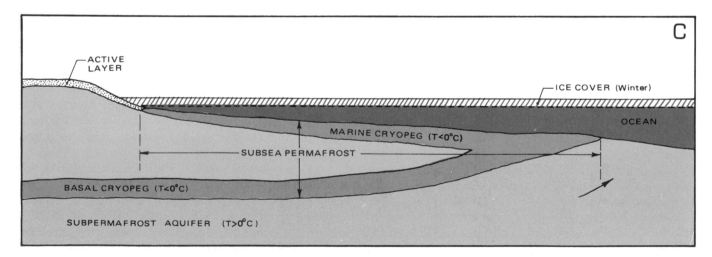

Figure 3. Aquifers and water conditions in permafrost areas. A. Suprapermafrost: active layer; closed taliks. B. Intrapermafrost: open taliks (lake, river, hydrothermal, hydrochemical); lateral taliks; transient isolated taliks; isolated, marine and basal cryopegs. C. Subpermafrost.

lated taliks, unfrozen zones completely surrounded by permafrost (Fig. 3).

Open taliks can occupy extensive areas below large lakes and major rivers. Their extent and incidence decrease with decreasing size of the surface-water bodies, and with increasing latitude; eventually open taliks give way to closed taliks that gradually decrease further in size. Many open taliks are maintained through convective heat transport by upward flow of slightly warmer and/or mineralized subpermafrost water (hydrothermal taliks with T >0°C and hydrochemical taliks with T <0°C). Some open taliks are maintained by downward flow of surface or suprapermafrost water (e.g., in karst areas).

Lateral taliks are sometimes found where coarse- and fine-grained deposits alternate. The temperature in the more permeable layers is kept above freezing by the heat content of moving water. The basal portion of thick permafrost is often unfrozen, due to freezing-point depression in mineralized pore water. Such negative-temperature taliks (Fig. 3) are called basal cryopegs (Tolstikhin and Tolstikhin, 1974).

Isolated taliks develop below the beds of recently drained lakes and abandoned river channels, where permafrost is actively developing, from the surface downward, into what were originally closed taliks. Such isolated taliks are transient phenomena that decrease in size with time. They contain what was originally suprapermafrost water. High pressures can develop in such shrinking taliks, sometimes giving rise to the formation of large pingos (perennial ice-cored frost mounds).

The chemical composition of intrapermafrost water depends on the nature of the taliks and on the composition of the source water. Waters in open and lateral taliks are either similar to suprapermafrost water, with slightly increased mineralization and somewhat attenuated organic content, or similar to subpermafrost water, which can be highly mineralized.

Water in isolated taliks below recently drained depressions is affected gradually by cryogenic metamorphism, which causes increasing mineralization, followed by changes in ionic composition as a result of precipitation of less soluble salts, ion exchange, adsorption, and possibly, biochemical processes (Tolstikhin and Tolstikhin, 1974).

Water temperatures in open and lateral taliks are usually just above 0°C; they can be much higher in taliks that carry thermal subpermafrost water. Temperatures below 0°C prevail in basal cryopegs and are found locally in open taliks that carry highly mineralized water. Initial above-zero water temperatures in isolated taliks below recently drained depressions may gradually decrease to slightly below 0°C before freezing, due to increasing mineralization of the water.

*Subpermafrost aquifers.* Subpermafrost aquifers (Fig. 3) are situated below the basal cryopeg part of the permafrost, and therefore have water temperatures higher than 0°C. In the discontinuous zone, where permafrost may be very thin, subpermafrost aquifers commonly occur in unconsolidated deposits; farther north, they are found predominantly in bedrock. Karst development has created major subpermafrost aquifers in some areas of soluble rock (carbonates, sulfates, halite). Examples of such aquifers are found in the Brooks Range and British Mountains in Alaska and Yukon (Williams and van Everdingen, 1973), and in the Franklin Mountains and the region west and northwest of Great Bear Lake in the Northwest Territories, Canada (van Everdingen, 1981). Karst aquifers are recharged through zones of fractured rock and through solution and collapse sinkholes. Several small rivers in the Great Bear Lake region flow into such sinkholes. Karst recharge can account for 15 to 30 percent of the annual precipitation, especially in areas where much of the precipitation falls as snow, recharging the karst during snowmelt and thus escaping summer evaporation.

Subpermafrost karst aquifers often show extreme differences between low water levels in late winter, and high water levels after snowmelt and early-summer rains. This change in water level is especially notable in mountainous areas, where groundwater flow velocities may be high (exceeding 50 m/day) and annual fluctuations in water level great (in excess of 150 m; Tolstikhin and Tolstikhin, 1974). Consequently, discharge rates also show wide fluctuations, similar to karst discharge rates in temperate regions.

The chemical composition of subpermafrost karst water depends on both the source of the recharge and the character of the karst. In carbonate terrain (especially in mountainous areas), calcium and bicarbonate are the major constituents. Significant concentrations of dissolved sulfate are found in areas where the karst is developed in gypsum (or anhydrite) beds. In areas of saline karst, developed in salt (halite) beds, high sodium and chloride concentrations are encountered. In some areas with sulfate or halite karst, however, ground-water flow is concentrated in zones of collapsed strata overlying the original solution cavities and channels. In such cases the chemical composition of the water does not reflect the nature of the original soluble strata (van Everdingen, 1981).

The regime of subpermafrost aquifers in igneous and metamorphic rocks is extremely varied. In the discontinuous permafrost zone, such aquifers may receive recharge from rain or snowmelt, or from surface water and suprapermafrost water in river valleys. In the continuous permafrost zone they receive recharge almost exclusively from rivers and from intrapermafrost aquifers, through open taliks.

The chemical composition of subpermafrost water in igneous and metamorphic rocks is usually similar to that of overlying surface water and suprapermafrost water. Highly mineralized subpermafrost water may reflect cryogenic metamorphism; high sulfate concentrations are found in the vicinity of sulfide mineralization.

Subpermafrost aquifers in alluvial deposits below river valleys are widely used as sources of water supply in the basins of the Yukon and Tanana rivers in Canada and Alaska (Williams and van Everdingen, 1973). Exploration there is often hampered by a lack of knowledge of the local distribution of permafrost, which is affected by the history of river-channel migration and the transient nature of surface-water bodies.

## Ground-water movement

*Infiltration and recharge.* The presence of extensive muskegs (peatlands) and numerous lakes and ponds in the generally water-deficient permafrost region shows that infiltration and recharge are restricted by permafrost. However, many of the larger lakes and rivers are underlain by unfrozen zones (taliks) of variable horizontal and vertical extent. If these taliks completely penetrate the permafrost, pathways are available for either recharge or discharge of subpermafrost water.

Recharge is least affected by permafrost in areas with bedrock exposures that present relatively large openings for infiltration of water, such as open fractures and joints in sedimentary rocks and solution conduits in soluble sedimentary rocks (carbonates, sulfates, salt), and in areas with thick sequences of coarse-grained unconsolidated deposits. The continuity of fracture systems with depth is normally more restricted in crystalline rocks than in sedimentary strata; consequently, recharge is more restricted by permafrost in areas with crystalline bedrock. Recharge is very limited in areas covered with thick sequences of medium- to fine-grained unconsolidated deposits.

*Lateral movement.* Lateral movement of ground-water in the permafrost region is confined to: (1) seasonal flow in the active layer during the frost-free season; (2) flow in taliks in permafrost; and (3) flow in unfrozen materials below the permafrost.

*Discharge.* The existence of active groundwater flow systems in the permafrost region is indicated by natural ground-water discharge phenomena similar to those observed in more temperate regions. In addition, the presence of open water in ice-covered rivers, the occurrence of icings (Carey, 1973), and the presence of certain types of frost mounds (Academia Sinica, 1975; van Everdingen, 1982) indicate ground-water discharge. Some of these phenomena can be utilized to derive quantitative information on ground-water flow.

Most of the discharge from aquifers within and below the permafrost takes place through taliks below lakes and stream beds, and through a relatively small number of large springs. The existence of open taliks below a river is often indicated by the presence of willows and poplars along the river channels and on the flood plain, and by numerous small springs. During the winter, discharge from open taliks is revealed by pools or reaches of open water, and by large icings for some distance downstream.

**Springs.** The discharge rates of springs and seeps provide reliable information on available ground water. The temperature and chemical composition of a spring's water provide clues regarding the source of the water, that is, supra-, intra-, or subpermafrost.

Perennial springs commonly discharge water from subpermafrost aquifers or, less commonly, from intrapermafrost aquifers. All thermal springs (T >10°C), most springs with discharge rates exceeding about 5 l/s, and most mineral springs (dissolved solids 1 g/l or more) derive their water from subpermafrost aquifers, via open taliks. Seasonal springs and seeps discharge water from suprapermafrost aquifers or, less commonly, from intrapermafrost aquifers.

**Baseflow.** The winter baseflow in rivers without lakes or reservoirs in their drainage areas reflects discharge of ground water. As in more temperate regions, a first approximation of the magnitude of the ground-water component in the surface runoff can be derived from the baseflow characteristics and from the seasonal changes in dissolved-solids concentrations in the river water (Brandon, 1965).

The ground-water contribution to baseflow in drainage basins in the discontinuous permafrost zone generally ranges between 1.0 and 5.0 l/s/km$^2$. In many basins in the zone of continuous permafrost, winter baseflow approaches zero, and most or all of the limited ground-water discharge is stored in icings.

**Icings.** During the winter, much of the ground-water discharge in the continuous permafrost region, which would normally maintain winter baseflow in streams, freezes instead and forms icings (Carey, 1973). The distance between the point of discharge and the point where icing formation starts is a function of the discharge rate, temperature, and dissolved-solids content of the ground water; the geometry and gradient of the discharge channel(s); and the variable meteorological conditions of air temperature, humidity, and wind direction and speed.

Virtually all icings are related to discharge of ground water. They represent temporary above-ground storage of the ground water discharged during the winter; the stored water is released by melting of the icings during the following summer. Icings thus cause interseasonal redistribution of water resources. For instance, Williams and van Everdingen (1973) reported that 123 million m$^3$ of icings formed in the Sagavanirktok River basin in Alaska during an 8-month freezing period. That volume of ice would represent an average ground-water discharge rate of about 6 m$^3$/s; melting of the icings during the following 2-month summer season would add about 24 m$^3$/s to streamflow from the basin. The rate of runoff from melting icings appears to be 1.5 to 4 times as high as the rate of ground-water discharge that forms the icings; the actual ratio depends on the shape and exposure of individual icings and on weather conditions.

The occurrence and size of icings can provide quantitative information on ground-water discharge rates as an indication of available ground-water resources.

The concentration of recharge and discharge in small areas and at a few points, because of the presence of permafrost, leads in many instances to high local recharge and discharge rates. For instance, during snowmelt the inflow into individual karst sinkholes in permafrost west of Great Bear Lake, Northwest Territories, can exceed 1.0 m$^3$/s; the discharge from springs in the South Fishing Branch of the Porcupine River, Yukon Territory, exceeds 11 m$^3$/s (van Everdingen, 1974).

## GEOCHEMISTRY

Permafrost can have a significant effect on the mineralization or chemical composition of ground water. Reaction and dissolution rates are reduced under seasonal and perennial low-temperature conditions prevailing in permafrost areas, but reduced rates of ground-water movement provide longer residence times for reactions between the ground water and the aquifer materials. Solubilities of calcium and magnesium bicarbonates are somewhat increased because of increased solubility of carbon dioxide at lower temperatures. Solubility of calcium sulfate also increases slightly with decreasing temperature.

### Effects of aquifer position relative to permafrost

The chemical composition of suprapermafrost water generally is influenced by rainfall, snowmelt, and surface runoff. It is commonly high in organic (humic acid) content derived from muskeg (peat) areas. Where suprapermafrost water is partly derived from intra- and subpermafrost aquifers, its dissolved-solids concentration may be extremely high.

Intrapermafrost water may be similar in chemical composition to either suprapermafrost water or subpermafrost water.

Subpermafrost water can range from fresh water of the Ca (Mg)-$HCO_3$ type, through brackish, sulfurous, and saline water, to sodium chloride or calcium/sodium chloride brines. As stated earlier, the chemical composition of subpermafrost water depends principally on the residence time of the water in the ground, and on the mineral composition of the aquifers. Waters having low dissolved-solids contents usually reflect rapid flow in carbonate karst or flow through fractured, insoluble bedrock.

In areas where permafrost is degrading, the water immediately below the permafrost, derived from melting of pore ice or ice lenses, should be unusually low in dissolved minerals, especially carbonates and sulfates that were precipitated during the initial freezing. Low solution rates at low temperature, and low solubility of the carbonates in the absence of free $CO_2$, tend to limit resolution of these precipitated minerals.

The chemical composition of spring water is a reliable indicator of the quality of available ground water. The quality of ground water may also be deduced from the composition of the baseflow component of streamflow, especially during winter when the surface-water component approaches zero. However, on small streams the quality of ground water contributed to streamflow may vary as the source changes and the contributions vary among suprapermafrost, intrapermafrost, and subpermafrost sources. These changes in chemical composition in the water of small streams introduce uncertainty into linking the composition to that of the ground water.

### Icings

The chemical composition of ice samples from icings is not a reliable indicator for the quality of the ground-water discharge that formed the icings, because the mineral precipitates, and mineralized liquid inclusions are unevenly distributed throughout the icing. During the gradual freezing of the water flowing over and around an icing, concentration of dissolved solids in the remaining water gradually increases until the saturation point is reached for the least soluble minerals. Precipitation of minerals starts at that point and continues, eventually involving more soluble minerals. The last dissolved material is precipitated, or incorporated as brine inclusions, at the point where the last water freezes. The colder the weather, the closer to the discharge source this sequence of events takes place. In addition, the continuously changing position of runoff channels on an icing tends to distribute both the water and the precipitates unevenly across the icing at any time.

In addition, the composition of meltwater from icings does not necessarily reflect that of the ground water from which the icing formed. During melting, the more soluble constituents in the ice (e.g., NaCl) go back into solution immediately, whereas calcite, dolomite, and gypsum redissolve much more slowly, forming a mineral slush on the surface and around the edges of a melting icing. Icing meltwater, compared to ground water that formed the icing, is thus generally lower in dissolved minerals and higher in its ratio of chloride to bicarbonate and sulfate.

## REFERENCES CITED

Academia Sinica, 1975, Permafrost: Lanchou, People's Republic of China, Research Institute of Glaciology, Cryopedology and Desert Research (Translation published by Institute for Scientific and Technical Translation, National Research Council of Canada, Ottawa, NRC Technical Translation No. 2006, 1981, 224 p.).

Anderson, D. M., and Morgenstern, N. R., 1973, Physics, chemistry, and mechanics of frozen ground; A review, *in* Permafrost, the North American Contribution to the Second International Conference, Yakutsk, U.S.S.R., July 16–28, 1973, Proceedings: Washington, D.C., National Academy of Sciences, p. 257–288.

Brandon, L. V., 1965, Ground-water hydrology and water supply in the District of Mackenzie, Yukon Territory, and adjoining parts of British Columbia: Geological Survey of Canada Paper 64-39, 102 p.

Brown, R.J.E., 1970, Permafrost in Canada; Its influence on northern development: Toronto, University of Toronto Press, 234 p.

Carey, K. L., 1973, Icings developed from subsurface water and ground water: U.S. Army Cold Regions Research and Engineering Laboratory, Cold Regions Science and Engineering Monograph III-D3, 67 p.

Heath, R. C., 1984, Ground-water regions of the United States: U.S. Geological Survey Water-Supply Paper 2242, 78 p.

Péwé, T. L., 1983, Alpine permafrost in the contiguous United States; A review: Arctic and Alpine Research, v. 15, no. 2, p. 145–156.

Tolstikhin, N. E., and Tolstikhin, O. N., 1974, Ground water and surface water in the permafrost region, *in* Melnikov, P. I., and Tolstikhin, O. N., eds., General permafrost studies, Chapter IX: Novosibirsk, U.S.S.R. Academy of Sciences (Translation published by Inland Waters Directorate, Ottawa, Technical Bulletin 97, 1976, 25 p.).

van Everdingen, R. O., 1974, Ground water in permafrost regions of Canada: Ottawa, Proceedings, Workshop Seminars on Permafrost Hydrology, Canadian National Committee, International Hydrological Decade, Environment Canada, p. 83–93.

——— , 1981, Morphology, hydrology, and hydrochemistry of karst in permafrost terrain near Great Bear Lake, Northwest Territories: Ottawa, National Hydrology Research Institute, Inland Waters Directorate, NHRI Paper 11,

53 p.

—— , 1982, Frost blisters of the Bear Rock spring area near Fort Norman, Northwest Territories: Arctic, v. 35, no. 2, p. 243–265.

Williams, J. R., 1965, Ground water in permafrost regions; An annotated bibliography: U.S. Geological Survey Water-Supply Paper 1792, 294 p.

—— , 1970, Ground water in permafrost regions of Alaska: U.S. Geological Survey Professional Paper 696, 83 p.

Williams, J. R., and van Everdingen, R. O., 1973, Ground water investigations in permafrost regions of North America; A review, *in* Permafrost, the North American Contribution to the Second International Conference, Yakutsk, U.S.S.R., July 16–28, 1973, Proceedings: Washington, D.C., National Academy of Sciences, p. 435–446.

Zenone, C., and Anderson, G. S., 1978, Summary appraisals of the nation's ground-water resources; Alaska: U.S. Geological Survey Professional Paper 813-P, 28 p.

Manuscript Accepted by the Society March 6, 1987

## Chapter 32

# *Nature of comparative hydrogeology*

**Stanley N. Davis**
*Department of Hydrology and Water Resources, University of Arizona, Tucson, Arizona 85721*

The term "comparative hydrogeology" was probably first used by LeGrand (1970). He included topography, precipitation, rock type, availability of recharge rich in carbon dioxide, and other factors in his discussion of "comparative hydrogeology." In this chapter, we will confine our comparison to primarily that of rock type. This does not imply, however, that we are redefining the useful concepts introduced by LeGrand. To the contrary, the application of comparative hydrogeology should consider all the physical and chemical variables that can be measured in a meaningful way. Our narrower coverage of the topic is dictated only by considerations of space in this volume.

The use of comparative hydrogeology has been implied in the writings of many authors, as exemplified by ground-water textbooks during at least the past 100 years (Daubree, 1887; Tolman, 1937; Davis and DeWiest, 1966). In contrast, most modern texts have either emphasized broader, large-scale geologic provinces (Meinzer, 1923; Thomas, 1952; McGuinness, 1963; Heath, 1984) or have concentrated on the basic mathematical, geophysical, and geochemical principles and tools needed for hydrogeologic studies (Freeze and Cherry, 1979; Fetter, 1980; Todd, 1980). Within the past ten years, however, the need to model hypothetical or "generic" repositories for radioactive waste and the desire to construct hydrogeologic maps for land-use planning have stimulated new interest in comparative hydrogeology (Aller and others, 1987).

Comparative hydrogeology helps to fill at least three general needs for the professional hydrologist. First, and most important, it allows predictions to be made of expected hydrogeologic conditions based on geologic and topographic information (LeGrand, 1970). Second, and roughly the inverse of the first, is the development of the ability to recognize hydrogeologic anomalies. This recognition can be used to check possible errors in reported data and to pinpoint areas needing further study, such as suspected contaminant plumes. Third, comparative studies are essential for many purposes of regional or local land-use planning. Money simply is not available for the hundreds of thousands of test holes needed to obtain basic data for land-use maps. Extrapolations of hydrogeologic characteristics must always be made between widely spaced measurement points. This extrapolation should be based on a knowledge of comparative hydrogeology.

Too much should not be expected from comparative hydrogeology, however. The statistical scatter of hydrologic characteristics of even a single rock type such as granite is too large to provide useful information for many purposes. For example, as Frank Trainer points out in the next chapter, the permeability of granite will decrease rapidly with depth for at least the first 100 m or so of penetration of the average drill hole. However, the chances of a reversal of this trend in any given hole are great, so the information is misleading unless the statistical variations are considered. The safety of a tunnel being drilled into granite cannot be assured on the basis of the compilation of hydrogeologic data from various parts of the world. The risk of a large inflow of water estimated from comparative hydrogeology alone is unacceptable even if the probability of being wrong is less than 5 percent. No substitute exists for detailed, site-specific investigations for most engineering projects.

## REFERENCES CITED

Aller, L., Bennett, T., Lehr, J. H., Petty, R. J., and Hackett, G., 1987, Drastic; A standardized system for evaluating ground-water pollution potential using hydrogeologic settings: Ada, Oklahoma, U.S. Environmental Protection Agency, 615 p.

Daubree, A., 1887, Les eaux souterraines, aux epoques anciennes et a l'epoque actuelle: Paris, Dunod, 3 volumes.

Davis, S. N., and DeWiest, R.J.M., 1966, Hydrogeology: New York, John Wiley and Sons, Incorporated, 463 p.

Fetter, C. W., Jr., 1980, Applied hydrogeology: Columbus, Ohio, C. E. Merrill Publishing Company, 488 p.

Freeze, R. A., and Cherry, J. A., 1979, Groundwater: Englewood Cliffs, New Jersey, Prentice-Hall, 604 p.

Heath, R. C., 1984, Ground-water regions of the United States: U.S. Geological Survey Water-Supply Paper 2242, 78 p.

LeGrand, H. E., 1970, Comparative hydrogeology; An example of its use: Geological Society of America Bulletin, v. 18, p. 1243–1248.

Davis, S. N., 1988, Nature of comparative hydrology, *in* Back, W., Rosenshein, J. S., and Seaber, P. R., eds., Hydrogeology: Boulder, Colorado, Geological Society of America, The Geology of North America, v. O-2.

McGuinness, C. L., 1963, The role of ground water in the national water situation: U.S. Geological Survey Water-Supply Paper 1800, 1121 p.

Meinzer, O. E., 1923, Occurrence of ground water in the United States with a discussion of principles: U.S. Geological Survey Water-Supply Paper 489, 321 p.

Thomas, H. E., 1952, Ground-water regions of the United States; Their storage facilities: U.S. 83rd Congress, House Interior and Insular Affairs Committee, The Physical and Economic Foundation of Natural Resources, v. 3, 78 p.

Todd, D. K., 1980, Groundwater hydrology (2nd edition): New York, John Wiley and Sons, Incorporated, 535 p.

Tolman, C. F., 1937, Ground water: New York, McGraw-Hill Book Company, 593 p.

MANUSCRIPT ACCEPTED BY THE SOCIETY MARCH 15, 1988

The Geology of North America
Vol. O-2, Hydrogeology
The Geological Society of America, 1988

# Chapter 33

# *Alluvial aquifers along major rivers*

**John M. Sharp, Jr.**
*Department of Geological Sciences, The University of Texas at Austin, Austin, Texas 78713-7909*

## INTRODUCTION

Alluvium associated with modern major river systems is an immense, yet relatively untapped source of ground water. Because large rivers provide an alternate source of abundant water, North American alluvial systems, in general, have been only slightly developed. In Europe, on the other hand, such aquifers are used more extensively because of river pollution and European preference for well water. Alluvial ground water is less turbid and more uniform in physical and chemical characteristics. As a source of water it is not greatly influenced by problems of low river stage, flooding, or freezing. It is protected from much of the pollution now prevalent in surface waters. Although river water is also commonly good in quality and does not require the expense of well-field development, in many areas of North America surface waters are totally allocated for designated uses. This problem is most acute in the arid areas of the southwestern and western United States, but restrictions on the use of surface water are spreading to the nonarid areas. Requirements to maintain minimum instream flow for navigation, recreation, fish and wildlife, agriculture, and power generation are now limiting surface water resources in nonarid areas (Davis and others, 1980).

The most important major river flood-plain alluvial systems are shown in Figure 1. They have common characteristics (Heath, 1984, p. 59; Rosenshein, this volume): (1) The valleys contain thick, productive sand and gravel deposits. Permeability of the alluvium is typically an order of magnitude or more greater than that of the adjacent strata. (2) The alluvial deposits are in contact with a major perennial stream. (3) The alluvial deposits occur in a clearly defined band, which does not normally extend beyond the flood plain or terrace deposits.

## FORMATION OF LARGE FLOOD-PLAIN ALLUVIAL SYSTEMS

This description of major alluvial valley aquifers, as defined here, is restricted mostly to streams affected by glacial meltwater, and primarily those tributary to the Mississippi River system. Exceptions include the Brazos and, locally, some other Texas rivers, the upper parts of some rivers draining the Rocky Moun-

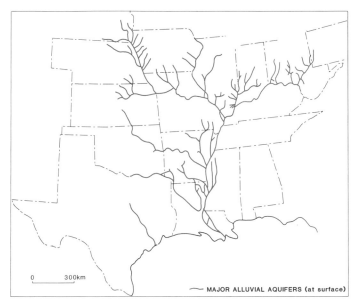

Figure 1. Major flood-plain alluvial aquifers of the United States (after Heath, 1984; Sharp, 1977).

tains, and certain rivers in Nova Scotia (i.e., the Antigonish; Trescott and others, 1970). These streams are underfit so that river-valley meander geometry is greater in scale than the geometry of the river meanders. The great river systems were profoundly influenced by Pleistocene events such as the 130-m drop in sea level during the peak of the Wisconsin glaciation, the subsequent rise of sea level, and transport of great amounts of sediment and meltwater during the advances and retreats of the ice margins.

The rivers were able to completely or partially erode valley sediment during drops in sea level. Abundant sediment and fluctuating discharge of meltwater led to the formation of braided streams in the valleys. Sediment bed load was quite coarse. Clasts the size of basketballs or larger were transported during periods of high discharge. When meltwater discharges declined, suspended loads were deposited. During the fall and winter, when the annual floods from meltwater receded, winds swept across the bare flood plains to transport the fine-grained material and deposit it on the surrounding uplands as loess.

Sharp, J. M., Jr., 1988, Alluvial aquifers along major rivers, *in* Back, W., Rosenshein, J. S., and Seaber, P. R., eds., Hydrogeology: Boulder, Colorado, Geological Society of America, The Geology of North America, v. O-2.

It was during these times of great discharge and lowered base level that the big rivers entrenched their present alluvial valleys. Some, like the pre-Ohio and pre-Missouri rivers, were deflected by glaciers and formed new valleys. The abandoned alluvium-filled channels were later covered with glacial drift. The outstanding example of a buried major river valley is the Teays Valley in Illinois, Indiana, and Ohio.

The effects of sea-level lowering and stream incision were felt far upstream: the Mississippi River was affected into southern Wisconsin and Minnesota; the Ohio River into Pennsylvania; and the Missouri River into South Dakota. These effects also were propagated into the tributaries for tens of miles or more. As streams incised, their knickpoints migrated upstream at rates controlled by the drop in major stream base level, discharge, and erosional resistance. The downcutting created straths or cut terraces in some systems, but in other valleys, terraces are much less evident or absent.

The rise in sea level concomitant with glacial retreat also created striking changes in these alluvial systems. Base level increased, and there was a reduction in both glacial meltwater discharge and sediment load, somewhat out of synchronization with the base-level changes. Maximum meltwater discharge and sediment transport probably occurred after sea level began to rise. The net result was a filling of the incised valleys. Presumably, valley filling began in downstream areas by vertical accretion and migrated upstream, although this process was strongly affected by local fluvial conditions. The gravel-cobble-sand facies of the braided streams were left in the bottoms of the trenches. As stream competency and mean size of sediment bed load both decreased, vertical accretion continued, depositing a sandy facies. Stream regimen changed from braided to meandering conditions because of decreasing channel slope, smaller and less flashy stream flow, and decreased sediment load.

This pattern is typical of incised major river systems in North America, but it is particularly representative of those fed by glacial meltwaters. The typical upward reduction of median grain size in valley alluvium is shown in Figure 2.

### Geometry of the bedrock valleys

The configuration of the bedrock valley or trench is dependent upon former stream competence, the history of flow and glacial conditions, and bedrock lithology. Generally, alluvium thickness and lateral extent increase downstream and with the size of the Pleistocene rivers. During incisement, valleys widened and deepened. Existing valley fill was scoured out and then replaced, perhaps several times, as the glaciers waxed and waned. The width of the alluvial valleys is also controlled by lithology. This control is clearly seen in the Missouri River valley. Downstream from Glasgow, Missouri (275 km above the confluence of the Missouri with the Mississippi), the river incises resistant Mississippian and Ordovician carbonates. The flood-plain width ranges between 1.5 to 2.5 km. Above Glasgow, where less resistant Pennsylvanian rocks are present, the flood-plain width ex-

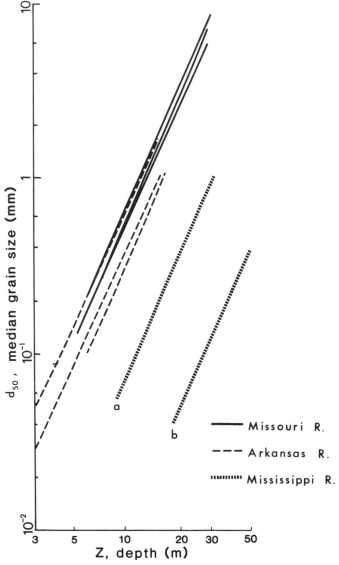

Figure 2. Regression lines of median grain sizes for the Missouri River (Moraes, 1969), the Arkansas River (Bedinger, 1961), and the Mississippi River (a, Mansur and Kaufman, 1956; and b, Johnson and others, 1966) showing the fining-upward trends. Mississippi River data are given as effective grain size, $d_{10}$. Figure is redrawn from Sharp (1977).

ceeds 15 km, and valley walls are gentler. Total alluvial thickness and substratum properties bear little relationship to the present-day stream regime.

### Controls of modern river regime on the hydrogeology

The flood regime and modern depositional processes control the nature of topstratum (Fisk, 1944) sediments. The lower 275 km of Missouri River is narrowly confined by Paleozoic carbonate bluffs, overbank deposition occurs under relatively rapid flow. The entire flood plain here is analogous to the natural levee environment (Schmudde, 1963). Only about 5 percent of the flood plain (at the downstream portions of meander loops, often called bottoms, or in areas of poor surface drainage) is covered with low-permeability backswamp deposits. The Lower Mississippi River flood plain, on the other hand, is much wider and has a much higher proportion of fine-grained, overbank deposits. Therefore, the alluvial aquifers in this area are much more likely to be confined conditions (see also Fig. 6).

### Effect of man on flood-plain aquifers

In the last 100 years, and the last 50 years in particular, man has significantly altered the flood-plain environment. These changes relate to three trends: navigational improvements, intensive cultivation of flood plains, and, in places, urbanization. The main navigational alterations have been the straightening and shortening of channels, man-made cutoffs, bank stabilization, dredging and spoil disposal, construction of locks and dams, and construction of wing dams or dikes. Winkley (1970) documented a shortening (by cutoffs) of the Mississippi River from Cairo, Illinois, to the Gulf of Mexico, from 1740 km to approximately 1,500 km. Consequently, channel slope steepened and a new set of sediment transport conditions was created. Pools now tend to be deeper and the crossovers shallower, and the lower Mississippi now tends toward wider, braided-type river conditions.

The bank stabilization prevents the natural lateral migration of the river channel. In places, dams have created still water conditions; elsewhere wing dams and dikes promote a fast, uniform stream flow in order to maintain minimal navigational channels. The results on the groundwater system of these navigational changes are: (1) altering river bank/bed permeability which, in turn, affects stream/ground-water interaction; (2) changing channel elevation and slope, which induces equivalent changes in the alluvial water table and associated changes in fluvial sediment transport; (3) more constant stream flow, stabilizing river/ground-water interactions, and, combined with levee systems, limiting overbank flows, reducing ground-water recharge; and (4) changing channel width and geometry.

Because flood plains are very fertile, they are intensely cultivated. This has altered the flood plain by eliminating wetlands and attaching islands to the shore for cultivation. The combined effects can be profound. For example, Funk and Robinson (1974) showed that, because of dike construction, bank stabilization, and meander cutoffs, the lower Missouri River (802 km above the confluence with the Mississippi) has lost 50 percent of its surface area between 1879 to 1972. The number of individual islands was reduced from 161 to 18, and island area was reduced from 98.8 km$^2$ to less than 1 km$^2$ during that same period. The river distance from Rulo, Nebraska, to the confluence with the Mississippi was reduced from 875 to 802 km. Recreational sites and important fish-and-wildlife habitats were largely eliminated. Of course, the use of agricultural chemicals has also affected ground-water and surface-water quality.

Finally, flood plains are being urbanized in some areas. Pumping and pollution by sewage and industrial wastes have accompanied urbanization. In areas where ground-water pumping has decreased or other factors have elevated the water table, serious damage to building foundations has been reported (e.g., Louisville, Kentucky, and the American bottoms, east of St. Louis).

## HYDROGEOLOGIC CHARACTERISTICS

### Physical nature of alluvium

These major alluvial aquifers, as shown in Figure 3, are generally subdivided into the fining-upward substratum and the overlying topstratum (Fisk, 1944; Dahl, 1961; Granneman and Sharp, 1979). The substratum sediment typically exhibits increasing mean grain size with depth. Permeameter tests on well cuttings have been analyzed by, among others, Moraes (1969), Mansur and Kaufman (1956), Bedinger (1961), and Johnson and others (1966). Because permeability generally increases with grain size, and sediment grain size increases with depth (Fig. 2), the greatest permeability is generally deep in the aquifer (Fig. 4).

The major rivers are for the most part only partially penetrating. Lateral erosion and deposition in the topstratum creates a great variability in sedimentary facies. A general description of the topstratum was provided by Fisk (1944)—the channel belt, the meander belt, and the flood basin. The channel belt comprises the active river channel. The meander belt is defined by the amplitude of the river meanders. It consists of a wide variety of levee, point-bar, crevasse splay, lacustrine, and overbank deposits. The rest of the flood plain is the flood basin. The flood basin has not been recently reworked by lateral river migration; it is inundated only in the large floods; and it is commonly capped with fine-grained silts and clays. Confined groundwater conditions are most prevalent in the flood basin. Minor geomorphic zones include terrace deposits, alluvial fans, and deposits of slope wash, which are restricted to the margins of the flood plain.

### Hydraulic properties of the alluvial aquifers

Compiled aquifer data for the alluvium along the Mississippi, Ohio, Missouri, Red, and Arkansas rivers are displayed in Figures 5 (hydraulic conductivity) and 6 (storativity). The data are somewhat biased by the large number of analyses for the lower Mississippi River alluvial aquifer. Hydraulic conductivities are averaged over the thickness of the alluvium, although the conductivity typically increases with depth. Measured hydraulic conductivities range from 1 to $10^3$ m/day; the median values approach $10^2$ m/day, close to that of a clean sand (Freeze and Cherry, 1979, p. 29; (Heath, Table 2, this volume).

Figure 6 shows the range in storativity data—$10^{-4}$ to 0.3. The cluster of storativity values between 0.05 and 0.3 clearly represents unconfined ground-water conditions. These data correlate well with Johnson's (1967) analysis of specific yield and Cohen's (1963) evaluation of specific yield for Humboldt River alluvium. Cohen's mean for 209 samples was 0.207. Sediments with higher porosity (clays and silts) have lower specific yields.

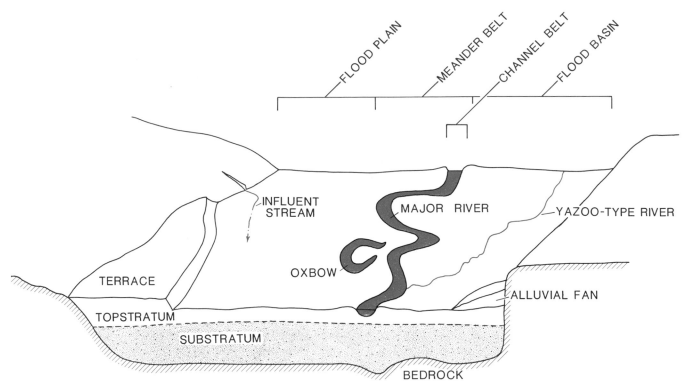

Figure 3. Flood-plain geomorphic zones as originally described by Fisk (1944).

With increasing lateral extent and thickness of fine-grained top-stratum sediment, storativity decreases until confined ground-water conditions exist. Foreman and Sharp (1981), in a detailed analysis of a pump test in Missouri River alluvium, calculated a specific yield of 0.21, but this test would have yielded a storativity of $4.6 \times 10^{-4}$ under confined conditions. The degree of confine-ment varies across the flood plain, but confined conditions are most common in wide flood plains and within the flood basin. Storativity also can change with time. Continual deposition of fine-grained materials by slow overbank flows would tend to create greater confinement, but occasional large floods are capa-ble of efficient scouring and refilling with more permeable river wash sediment. Storativity may be roughly delineated by soil types—sandy soils indicate unconfined ground-water conditions; clay-rich soils indicate confined ground-water conditions.

### Hydraulic properties of the topstratum

Topstratum deposition by lateral river migration and by overbank flows creates zones of high and low permeability. Two studies have analyzed topstratum permeabilities. Hydraulic con-ductivity values for Missouri River point-bar sediment (Stewart, 1982) ranged between 10 and $10^2$ m/day with a mean of 10.7 m/day. Pryor (1973) demonstrated a significant anisotropy in topstratum cross-bedded sands. Maximum permeability is paral-lel to the inclined stratifications. Erosional boundaries between cross-bedded units were shown to be zones of low vertical per-meability. Pryor's study of Wabash River sediments indicates a 1-

to 2-order of magnitude difference between maximum (hori-zontal) and minimum (vertical) permeabilities is possible. Finally, the horizontal permeability tends to be maximum in the direction parallel to the paleocurrent direction. The Wabash River trends were confirmed by field tests along the Missouri River (Stewart, 1982). These findings are contrary to the observed responses in the substratum. Foreman and Sharp (1981) found that the sub-stratum responds as a heterogeneous but isotropic system.

The concept of the "skin effect," where the river's bed and banks are lined with materials lower in permeability than the alluvial aquifer, has received a good deal of attention. Cun-ningham and Sinclair (1979) showed that the hydraulic conduc-tivity of the channel bottom is a crucial parameter in coupled ground-water/surface-water models. Where the skin effect exists, optimum well siting may be at some distance away from the river channel as has been shown by Kernodle (1977) and Weeks and Appel (1984). Scheurmann's (1970) study of the Isar River (West Germany) showed that the skin effect can be transient, and it also varies spatially. While some studies (Grubb, 1974; Norris and Fidler, 1969) have found streambed sediments to be several orders lower in permeability than the alluvial aquifer, Stewart (1982) found that cut banks in point-bar sediments, protected by riprap, maintained high hydraulic conductivities (on the order of $10^2$ m/day or greater). Channel-bed sediments, which have grain-size distributions similar to point-bar sediments, have similar hy-draulic conductivities. Lenses of silt and clay deposited by overbank flows or in oxbow lakes represent the opposite extreme—low permeability.

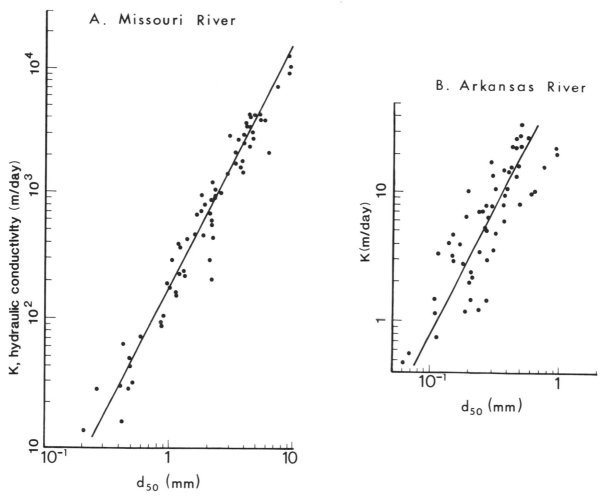

Figure 4. Laboratory determined hydraulic conductivities as a function of alluvium depth for the Missouri (Moraes, 1969) and Arkansas rivers (Bedinger, 1961). Figure is redrawn from Sharp (1977).

## Recharge/discharge relationships

The river is a principal control on the hydrogeology of the alluvial aquifer system. The river serves as a regional discharge zone; major rivers are gaining on an annual basis. The alluvium's potentiometric surface is generally close to but above average river level, although small-scale fluctuations are considerable. Ground-water recharge is from infiltration of river water, precipitation, runoff from terraces, irrigation return flow, infiltration from tributary streams and overland flow, and flow from the bedrock. The ground-water system discharges by flow into the major river and tributaries, and by evapotranspiration, pumping, and flow into adjoining bedrock aquifers.

Recharge to the alluvial aquifers occurs during times of rising river stages. The river water thus enters bank storage. After the stage declines, water in bank storage drains back into the stream. On narrow flood plains this process predominates. Recharge by flood waters can also be important locally and occasionally on a regional scale.

In wider flood plains, however, precipitation becomes most important. In the lower Mississippi River flood plain, it accounts for most of the recharge, as indicated by water quality data (Table 1). Where terrace deposits occur, recharge by precipitation also dominates. The potentiometric levels in the terrace areas are much less influenced by river-induced fluctuations. In some areas, piezometer heads are observed to rise significantly because of rapid influx of precipitation, which causes pressurization of air entrained between the infiltrating water and the water table. Irrigation return flow is, locally, a significant source of recharge.

Recharge from tributary streams is generally small because their beds and banks tend to be draped with low-permeability, fine-grained sediments deposited after rising main-stem waters and suspended loads backed up the tributaries. Interestingly, small streams are commonly a more important recharge source (Grannemann and Sharp, 1979) because they may be less affected by main-stem backwaters. Where small streams enter the flood plain, they flow above the water table and create local recharge mounds. Similarly, overland flow is a locally effective recharge mechanism where the flood-plain soils on which it descends are also permeable.

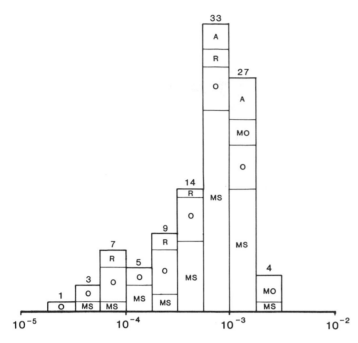

HYDRAULIC CONDUCTIVITY (m/sec)

Figure 5. Aquifer-test hydraulic conductivities for the following rivers: O, Ohio (Gallaher, 1963, 1964; Price, 1964a–d; Gallaher and Price, 1966; Grubb, 1974); MS, Mississippi (Krinitzsky and Wire, 1964; Boswell and others, 1968; Newcome, 1971); R, Red (Newcome and Page, 1962); A, Arkansas (Bedinger and others, 1963; and Boswell and others, 1968); and MO, Missouri (Fishel, 1948; Emmett and Jeffery, 1968, 1969b; Nuzman, 1969; Foreman and Sharp, 1981).

Finally, recharge from bedrock aquifers also occurs, but because alluvium is usually much more permeable, the effects are less pronounced. The major evidence for this type of recharge shows in the water chemistry. Some zones of anomalous water chemistry are attributed to this process in the alluvium of the Missouri River (Foreman and Sharp, 1981), the Ohio River (Gallaher and Price, 1966), and the Arkansas River (Whitfield, 1980). In smaller alluvial systems, recharge from the bedrock becomes proportionately more important, especially in carbonate bedrock terrains (Davey, 1982).

Discharge from alluvial aquifers is to the major river, and to a lesser extent, to its tributaries, through evapotranspiration and through pumping. Discharge into surrounding bedrock aquifers is also possible, especially where these underlying aquifers have been heavily pumped, such as in the Tertiary aquifers of Mississippi.

### Ground-water response zones

Certain recharge and discharge relationships have been observed in large flood-plain aquifers in North America and Europe, as would be expected from their similar geological histories and hydraulic properties. In separate study areas on the Missouri and Mississippi Rivers, fine patterns of ground-water response are identified (Fig. 7). These are: (1) zones of rapidly fluctuating ground-water levels; (2) zones of long-term stability/slow ground-water response; (3) zones of predominantly down-valley flow; (4) zones of persistent ground-water highs; and (5) zones of persistent lows in the potentiometric surface. These zones are readily distinguishable and should be identified in flood-plain analyses. The zones can be readily inferred from long-term observation well data in a given flood plain, but the exact boundaries are subjective.

The zone of rapidly fluctuating ground-water levels represents transient bank storage effects. This zone includes the entire channel belt and parts of the meander belt. Proximity to the river, permeability of river bed and banks, and river stage fluctuations are the controlling factors.

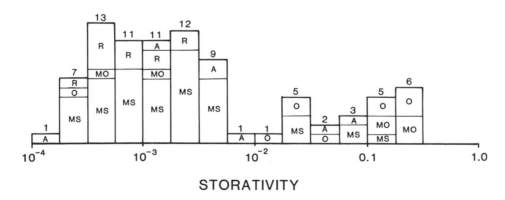

STORATIVITY

Figure 6. Storativities (including specific yields) for the Ohio, Mississippi, Red, Arkansas, and Missouri river alluvial aquifers. Data sources and abbreviations are the same as for Figure 5.

Zones of long-term stability or slow ground-water response are generally distant from the river. They are controlled by long-term changes in river stage, precipitation, and, to a lesser extent, recharge from upland sources. These zones may comprise parts of the meander belt and most of the flood basin. Terrace deposits, high bottoms (rarely flooded, elevated parts of the flood plain), and wide flood plains are in this zone. Precipitation recharge and evapotranspiration are the dominant processes.

Zones of predominantly down-valley flow are commonly observed in many areas of the lower Missouri River flood plain where the river parallels the valley walls for some distance.

Down-valley flow is enhanced by the presence of parallel tributary streams (Sharp, 1977), which are often artificially channelized adjacent to the valley wall. This parallelism of tributary streams within the flood plain is common; such streams are called Yazoo-type rivers. Their influence on the ground-water system is often greatly lessened by their low-permeability beds and banks.

Zones of persistent ground-water highs tend to be localized near the valley walls. They possess the highest water tables in the flood plain and serve as secondary recharge areas for alluvial systems. These generally represent infiltration of overland flow from adjoining uplands or small creeks that enter the flood plain

### TABLE 1. ALLUVIUM WATER QUALITY DATA

| River | Silica | Iron | Manganese | Calcium | Magnesium | Sodium | Potassium | Bicarbonate | Sulfate | Chloride | Fluoride | Nitrate | TDS* | pH | Reference |
|---|---|---|---|---|---|---|---|---|---|---|---|---|---|---|---|
| **Missouri** | | | | | | | | | | | | | | | |
| Median (16) | 28 | 2.5 | .56 | 114 | 26 | 22 | 5.4 | 451 | 36 | 3.2 | 0.3 | .2 | 497 | 7.7 | Emmett and |
| Minimum | 16 | .12 | .24 | 54 | 16 | 4.3 | 2.4 | 219 | 1.4 | 0.2 | .1 | .0 | 201 | 7.5 | Jeffrey, |
| Maximum | 36 | 4.2 | 5.6 | 190 | 56 | 60 | 9.1 | 836 | 442 | 70 | .5 | 4.1 | 978 | 8.1 | 1970 |
| Median (15) | 29 | 3.4 | .80 | 129 | 34 | 23 | 5.4 | 529 | 32 | 3.0 | .4 | .1 | 532 | 7.8 | Emmett and |
| Minimum | 20 | .14 | .04 | 67 | 18 | 4.9 | 2.0 | 288 | .2 | 0.8 | .3 | .0 | 270 | 6.2 | Jeffrey, |
| Maximum | 34 | 6.4 | 3.3 | 224 | 47 | 191 | 6.5 | 892 | 152 | 538 | .6 | 3.9 | 1450 | 8.1 | 1969a |
| Median (11) | 26 | 2.0 | .57 | 127 | 25 | 17 | 4.2 | 513 | 32 | 4.8 | .2 | .4 | 485 | 7.6 | Emmett and |
| Minimum | 21 | .12 | .16 | 53 | 14 | 5.5 | 3.0 | 236 | .2 | 0.7 | .0 | .0 | 246 | 7.4 | Jeffrey, |
| Maximum | 38 | 6.2 | 2.8 | 196 | 51 | 71 | 7.1 | 900 | 122 | 113 | .2 | 18 | 687 | 8.0 | 1969b |
| Median (12) | 26 | 3.4 | .59 | 131 | 36 | 9.4 | 5.6 | 542 | 45 | 3.4 | .2 | .0 | 530 | 7.2 | Emmett and |
| Minimum | 16 | .26 | .0 | 61 | 2 | 1.1 | 3.3 | 317 | .4 | 1 | .0 | .0 | 343 | 7.0 | Jeffrey, |
| Maximum | 32 | 5.2 | 4.4 | 172 | 46 | 38 | 8.2 | 856 | 117 | 46 | .5 | 13 | 794 | 7.9 | 1968 |
| **Mississippi** | | | | | | | | | | | | | | | |
| Median (82) | 30 | 5.2 | --- | 73 | 21 | 15 | 1.4 | 360 | 9.8 | 7.5 | --- | --- | 353 | 7.6 | Dalsin, 1978 |
| Minimum | 5.2 | .0 | --- | 10 | 6.1 | 5.8 | .0 | 34 | 0 | 1 | --- | --- | 153 | 6.5 | |
| Maximum | 50 | 26 | --- | 189 | 48 | 78 | 5.5 | 879 | 132 | 132 | --- | --- | 883 | 8.5 | |
| Median | 26 | 5.2 | --- | 66 | 18 | 21 | 2.0 | 291 | 11 | 19 | 0.2 | .7 | 331 | 7.5 | Boswell and |
| Minimum | 1.6 | .0 | --- | .1 | .3 | 2.2 | .0 | 0 | 0 | 0.3 | .0 | .0 | 18 | 4.3 | others, 1968 |
| Maximum | 7 | 62 | --- | 432 | 174 | 943 | 35.0 | 716 | 1000 | 1870 | .8 | 132 | 4190 | 8.9 | |
| **Red** | | | | | | | | | | | | | | | |
| Median | 21 | 1.3 | --- | 62 | 40 | 39 | 1.8 | 279 | 10 | 52 | 0.2 | 2.2 | 490 | 7.4 | Boswell and |
| Minimum | 15 | .1 | --- | 3.3 | .6 | 5.6 | .6 | 4 | 0 | 4.5 | .0 | .0 | 51 | 5.1 | others, 1968 |
| Maximum | 56 | 28 | --- | 248 | 147 | 242 | 7.7 | 911 | 544 | 610 | 1 | 560 | 1850 | 8.2 | |
| Median (64) | 22 | 6.5 | 98 | 120 | 50 | 56 | 1.8 | 650 | 13 | 44 | 0.4 | --- | 671 | 6.9 | Whitfield, |
| Minimum | 2.7 | .26 | .05 | 6.4 | 3.3 | 6.5 | .3 | 36 | 0 | 4.5 | .0 | --- | 100 | 6.2 | 1980 |
| Maximum | 59 | 36 | 5.7 | 480 | 210 | 280 | 6.2 | 1140 | 680 | 450 | 1.8 | --- | 2470 | 7.4 | (Rapides Parish) |
| **Brazos** | | | | | | | | | | | | | | | |
| Median (55) | 21 | 3.5 | --- | 152 | 36 | 119 | 3.0 | 600 | 124 | 111 | 0.3 | .5 | 770 | 7.0 | Cronin and |
| Minimum | 6.2 | .01 | --- | 18 | 3 | 17 | .9 | 230 | .4 | 16 | .1 | .0 | 281 | 6.6 | Wilson, 1967 |
| Maximum | 47 | 13 | --- | 440 | 159 | 380 | 9.3 | 890 | 612 | 890 | .6 | 76 | 2790 | 8.3 | |
| Drinking Water Standards | --- | .3 | .05 | --- | 150 | --- | --- | --- | 250 | 250 | 1.7 | 45 | 500 | --- | US Public Health Service, 1962 |

*Total dissolved solids.

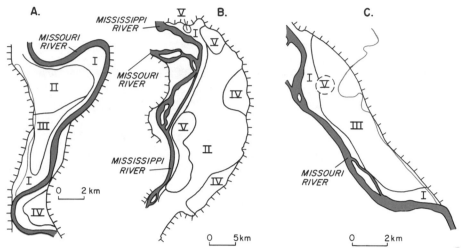

Figure 7. Flood-plain groundwater response zones. A, Missouri River near Glasgow, Missouri (Sharp, 1976; Granneman and Sharp, 1979); B, Mississippi River American Bottoms area (Schicht and Jones, 1962; U.S. Army Corps of Engineers, 1976); and C, Missouri River near McBaine, Missouri, (Nuzman, 1969; Foreman and Sharp, 1981). I, denotes zones of rapidly fluctuating groundwater levels; II, zones of long-term stability/slow groundwater response; III, zones of predominantly down-valley flow; IV, zones of persistent groundwater highs; and V, pumping-induced lows in the potentiometric surface.

and flow over permeable soils. Finally, zones of persistent lows in the potentiometric surface are caused by large pumpage.

### Water quality characteristics

The quality of ground water in the alluvial aquifer is generally suitable for most uses. The water is typically calcium-bicarbonate or calcium-magnesium bicarbonate facies, but occasionally sodium, chloride, and sulfate ions dominate. The water is generally hard; TDS (total dissolved solids) values frequently average around 500 mgl, but can exceed 1,000 mgl. Concentrations of iron and manganese commonly and sulfate occasionally exceed Public Health Service drinking standards. For irrigation use, alluvial waters have low sodium hazards and low to moderate salinity hazards. Median values of ground water from alluvium are plotted in a trilinear diagram (Fig. 8). In most cases, little cementation of the alluvium occurs, despite supersaturation with respect to silica. In a few instances, siderite concretions have been observed (Gorday, 1982).

Deterioration of ground-water quality in the alluvium is caused by three processes. First, where river water is of poor quality, seasonal recharge by river water degrades water quality. Second, point-source pollution by dumping or sewage lagoons, and nonpoint-source pollution by agriculture can be locally significant. Finally, recharge from bedrock aquifers may create high concentrations of sulfate, chloride, and sodium. This phenomenon has been observed in alluvial aquifers in Arkansas (Counts, 1957), Missouri (Foreman and Sharp, 1981), Mississippi (Boswell and others, 1968), and Ohio (Gallaher and Price, 1966).

### TRIBUTARY RIVER ALLUVIAL AQUIFERS

Adjoining the major river-valley aquifers are alluvial systems of the major tributary streams, which are substantial in some regions. These streams develop in a systematic manner (Bhowmik, 1984). Flood-plain width, alluvial depth and cross-sectional area, stream incision, and stream sinuosity all tend to increase toward the confluence with the major stream. Alluvial thickness may decrease abruptly above buried knickpoints. The size of the flood plain size in a tributary river is a poor predictor of hydraulic properties. Gorday (1982) analyzed available aquifer test data from rivers that border on or incise pre-Wisconsin drift. He found that the hydraulic conductivity and transmissivity varied widely, but mean hydraulic conductivities clustered around $10^2$ m/day. Tributary streams are associated with poorer alluvial aquifers than the major rivers because (1) the input of sediment fill from slope processes is proportionately greater, (2) the valleys are smaller, and (3) the streams are smaller. Walton and others (1967) show how cones of depression can extend beneath such tributary streams. This condition is less likely for the major rivers (Fig. 1) because of their greater width and depth, and their generally more permeable alluvium.

### CONCLUSIONS

Because of their common geological histories and similar hydrogeologic properties, a number of general conclusions can be made about the hydrogeology of large river flood-plain aquifers, but site-specific characteristics will alter these generalizations.

1. The major rivers are underfit. Their valleys are filled with

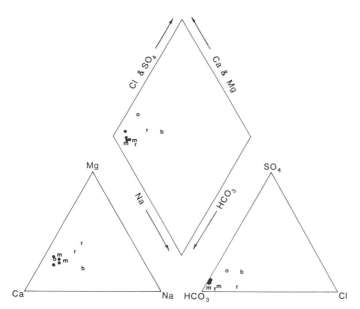

Figure 8. Plot for median values of chemistry of alluvial water for the (●) Missouri River valley; (m) Mississippi River valley; (o) Ohio River valley (Gallaher and Price, 1966); (r) Red River valley; and (b) Brazos River valley.

as much as several hundred feet of alluvial sediments, and the rivers are not fully penetrating. These systems reflect processes generated by Pleistocene glaciation and sea-level fluctuations.

2. Typically, the alluvium can be subdivided into a sand-gravel substratum deposited by vertical accretion, and a more variable topstratum deposited by lateral accretion. The topstratum is further divided into a channel belt, a meander belt, and a flood basin.

3. The substratum is typically a fining-upward sequence. Average hydraulic conductivity is around $10^2$ m/day. Both confined and unconfined conditions are observed. Storativities can exceed 0.20, but under clayey topstrata can be as low as $10^{-4}$.

4. A variety of ground-water response zones are defined. They are zones of rapidly fluctuating ground-water levels, of long-term stability, of slow ground-water response, of predominantly down-valley flow, of persistent ground-water highs, and of pumping-induced lows in the potentiometric surface.

5. Water is generally of good chemical quality and characteristically is a calcium-magnesium-bicarbonate facies. Iron, manganese, and hardness may exceed recommended limits. Recharge from polluted streams and bedrock aquifers or pollution from dumping and by agricultural chemicals can also deteriorate local water quality.

6. Tributary alluvial aquifers tend to be larger near the confluence with the main stem. The yields from these aquifers tend to be smaller because of smaller size, delayed recharge from their streams, and the greater influence of overland flow and slopewash processes that deposit fine-grained, low-permeability sediments. The hydraulic properties of the substratum sands and gravels are

similar to those of the main stream, but site-specific conditions are still paramount.

7. Man has altered the alluvial systems by shortening the rivers, stabilizing banks, dredging, constructing wing dikes, locks and dams, and artificial cutoffs, eliminating islands, urbanizing, and intensive cultivation. The long-term effects of these alterations are not yet known.

## REFERENCES CITED

Bedinger, M. S., 1961, Relation between medium grain size and permeability in the Arkansas River valley, Arkansas: U.S. Geological Survey Professional Paper 424-C, p. 31–32.

Bedinger, M. S., Emmett, L. F., and Jeffery, H. G., 1963, Ground-water potential of the alluvium of the Arkansas River between Little Rock and Fort Smith, Arkansas: U.S. Geological Survey Water-Supply Paper 1669-L, 29 p.

Bhowmik, N. C., 1984, Hydraulic geometry of floodplains: Journal of Hydrology, v. 68, p. 369–401.

Boswell, E. H., Cushing, E. M., and Hosman, R. L., 1968, Quaternary aquifers of the Mississippi Embayment: U.S. Geological Survey Professional Paper 448-E, 29 p.

Cohen, P., 1963, Specific-yield and particle-size relation of Quaternary alluvium, Humboldt River valley, Nevada: U.S. Geological Survey Water-Supply Paper 1669-M, p. M1–M24.

Counts, H. B., 1957, Ground-water resources of parts of Lonoke, Prairie, and White counties, Arkansas: Arkansas Geological and Conservation Commission Water Resources Circular 5, 59 p.

Cronin, J. G., and Wilson, C. A., 1967, Ground water in the flood-plain alluvium of the Brazos River, Whitney Dam to the vicinity of Richmond, Texas: Texas Water Development Board Report 41, 206 p.

Cunningham, A. B., and Sinclair, P. J., 1979, Application and analysis of a coupled surface and groundwater model: Journal of Hydrology, v. 43, p. 129–148.

Dahl, A. R., 1961, Missouri River studies; Alluvial morphology and Quaternary history [Ph.D. thesis]: Ames, Iowa State University, 251 p.

Dalsin, G. J., 1978, The Mississippi River valley alluvial aquifer in Mississippi: U.S. Geological Survey Water-Resources Investigations 78-106.

Davey, J. C., 1982, Hydraulic properties of a major alluvial aquifer; An isotropic, inhomogeneous system; Comments: Journal of Hydrology, v. 57, p. 203–204.

Davis, P. N., Sharp, J. M., Jr., and Falteisek, J. D., 1980, Missouri's instream flow requirements: Missouri Office of Administration Report, 415 p.

Emmett, L. F., and Jeffery, H. G., 1968, Reconnaissance of the ground-water resources of the Missouri River alluvium between St. Charles and Jefferson City, Missouri: U.S. Geological Survey Hydrology Investigations Atlas HA-315, scale 1:125,000.

—— , 1969a, Reconnaissance of ground-water resources of the Missouri River alluvium between Jefferson City and Miami, Missouri: U.S. Geological Hydrology Investigations Atlas HA-340, scale 1:125,000.

—— , 1969b, Reconnaissance of ground-water resources of the Missouri River alluvium between Kansas City, Missouri, and the Iowa border: U.S. Geological Survey Hydrology Investigations Atlas HA-336, scale 1:125,000.

—— , 1970, Reconnaissance of the ground-water resources of the Missouri River alluvium between Miami and Kansas City, Missouri: U.S. Geological Survey Hydrology Investigations Atlas HA-344, scale 1:125,000.

Fishel, V. C., 1948, Ground-water resources of the Kansas City, Kansas, area: Kansas Geological Survey Bulletin 71, 109 p.

Fisk, H. N., 1944, Geological investigation of the alluvial valley of the lower Mississippi River: Mississippi River Commission, U.S. Army Corps of Engineers, 78 p.

Foreman, T. L., and Sharp, J. M., Jr., 1981, Hydraulic properties of a major alluvial aquifer; An isotropic, inhomogeneous system: Journal of Hydrology, v. 53, p. 247–258.

Freeze, R. A., and Cherry, J. A., 1979, Groundwater: Englewood Cliffs, New Jersey, Prentice-Hall, 604 p.

Funk, J. L., and Robinson, J. W., 1974, Changes in the channel of the lower Missouri River and effects on fish and wildlife: Missouri Department of Conservation, Aquatic Series 11, 52 p.

Gallaher, J. T., 1963, Geology and hydrology of alluvial deposits along the Ohio River in the Hawesville and Coverport areas, Kentucky: U.S. Geological Survey Hydrology Investigation Atlas HA-72, scale 1:24,000.

——, 1964, Geology and hydrology of alluvial deposits along the Ohio River in the Henderson area, Kentucky: U.S. Geological Survey Hydrology Investigation Atlas HA-91, scale 1:24,000.

Gallaher, J. T., and Price, W. E., Jr., 1966, Hydrology of the alluvial deposits in the Ohio River valley in Kentucky: U.S. Geological Survey Water-Supply Paper 1818, 80 p.

Gorday, L. L., 1982, The hydrogeology of alluvial aquifers in or bordering the dissected till plains [M.A. thesis]: University of Missouri at Columbia, 106 p.

Grannemann, N. G., and Sharp, J. M., Jr., 1979, Alluvial hydrogeology of the lower Missouri River valley: Journal of Hydrology, v. 40, p. 85–99.

Grubb, H. F., 1974, Simulated drawdown for selected well fields in the Ohio River alluvial aquifer: U.S. Geological Survey Water-Resources Investigation 2-74, 45 p.

Heath, R. C., 1984, Ground-water regions of the United States: U.S. Geological Survey Water-Supply Paper 2242, 78 p.

Johnson, A. I., 1967, Specific yield; Compilation of specific yields for various materials: U.S. Geological Survey Water-Supply Paper 1662-D, 71 p.

Johnson, A. I., Moston, R. P., and Versaw, S. F., 1966, Laboratory study of aquifer properties and well design for an artificial-recharge site: U.S. Geological Survey Water-Supply Paper 1615-H, 42 p.

Kernodle, J. M., 1977, Theoretical drawdown due to simulated pumpage from the Ohio River alluvial aquifer near Siloam, Kentucky: U.S. Geological Survey Water-Resources Investigation 77-24, 46 p.

Krinitzsky, E. L., and Wire, J. C., 1964, Groundwater in alluvium of the lower Mississippi valley (upper and central areas): U.S. Army Waterways Experiment Station Technical Report 3-658, 98 p.

Mansur, C. I., and Kaufman, R. I., 1956, Control of underseepage of Mississippi River levees, St. Louis District: Journal of Soil Mechanics, Foundation Division, Proceedings American Society of Civil Engineers, v. 82 (SM1), p. 864-1–864-31.

Moraes, J.A.P., 1969, Hydrogeology of the McBaine area, central Missouri [M.A. thesis]: University of Missouri at Columbia, 110 p.

Newcome, R., Jr., 1971, Results of aquifer tests in Mississippi: Mississippi Board of Water Commissioners Bulletin 71-2, 44 p.

Newcome, R., Jr., and Page, L. V., 1962, Water resources of Red River Parish, Louisiana: U.S. Geological Survey Water-Supply Paper 1614, 133 p.

Norris, S. E., and Fidler, R. E., 1969, Hydrogeology of the Scioto River valley near Piketon, south-central Ohio: U.S. Geological Survey Water-Supply Paper 1872, 70 p.

Nuzman, C. E., 1969, Ground-water hydrologic study of the Missouri River valley for Columbia Missouri: Kansas City, Missouri, Unpublished report of the Layne-Western Company, Incorporated, 49 p.

Price, W. E., Jr., 1964a, Geology and hydrology of alluvial deposits along the Ohio River between Catlettsburg and South Portsmouth, Kentucky: U.S. Geological Survey Hydrology Investigation Atlas HA-75, scale 1:24,000.

——, 1964b, Geology and hydrology of alluvial deposits along the Ohio River

between Ethridge and the Twelvemile Island, Kentucky: U.S. Geological Survey Hydrology Investigation Atlas HA-97, scale 1:24,000.

——, 1964c, Geology and hydrology of alluvial deposits along the Ohio River between southwestern Louisville and West Point, Kentucky: U.S. Geological Survey Hydrology Atlas HA-98, scale 1:24,000.

——, 1964d, Geology and hydrology of alluvial deposits along the Ohio River between prospect and southwestern Louisville, Kentucky: U.S. Geological Survey Hydrology Investigation Atlas HA-111, scale 1:24,000.

Pryor, W. A., 1973, Permeability-porosity patterns and variations in some Holocene sand bodies: American Association of Petroleum Geologists Bulletin, v. 57, p. 162–189.

Scheurmann, K., 1970, Die Grundwasser bewegung in einer talalluvion, dargestellt am Beispiel der Stadt Landshut: Schrift, Bayersiches Landesstelle fuer Gewaesserkeunde, v. 15, 46 p.

Schicht, R. J., and Jones, E. G., 1962, Ground-water levels and pumpage in East St. Louis area, Illinois: Illinois State Water Survey Report Investigation 44, 39 p.

Schmudde, T. H., 1963, Some aspects of land form of the lower Missouri River floodplain: Annals Geography, v. 53, p. 60–73.

Sharp, J. M., Jr., 1976, Hydrogeologic characteristics of the Missouri River flood plain: Missouri Water Resources Research Center Report, 60 p.

——, 1977, Limitations of bank-storage model assumptions: Journal of Hydrology, v. 35, p. 31–47.

Stewart, C. A., 1982, Interchange of water between the Missouri River and surrounding alluvial aquifer near Huntsdale, Missouri [M.A. thesis]: Columbia, University of Missouri, 85 p.

Trescott, D. C., Pinder, G. F., and Jones, J. F., 1970, Digital model of alluvial aquifer: Journal of Hydraulics, v. HY5, p. 1115–1128.

U.S. Army Corps of Engineers, 1976, American Bottoms area; Preliminary groundwater analysis: St. Louis, Missouri, Corps of Engineers, 30 p.

U.S. Public Health Service, 1962, Public Health Service Drinking Water Publication 956, 61 p.

Walton, W. C., Hills, D. L., and Grundeen, G. M., 1967, Recharge from induced streambed infiltration under varying groundwater level and stream-stage conditions: Minnesota Water Resources Research Center Bulletin 6, 42 p.

Weeks, E. P., and Appel, C. A., 1984, Optimum location of a well near a stream, *in* Rosenshein, Jr., and Bennett, G. D., eds., Ground-water hydraulics: American Geophysical Union Water Resources Monograph 3, p. 4–28.

Whitfield, M. S., Jr., 1980, Chemical character of water in the Red River alluvial aquifer, Louisiana: U.S. Geological Survey Open-File Report 80-1018, 95 p.

Winkley, B. R., 1970, Metamorphosis of a river: Mississippi Water Resources Institute Report, 43 p.

MANUSCRIPT ACCEPTED BY THE SOCIETY MARCH 6, 1987

## ACKNOWLEDGMENTS

I'd like to thank T. W. Anderson, Leo Emmett, Julia Graf, Hayes Grubb, H. G. Jeffery, and O. J. Taylor of the U.S. Geological Survey and Stan Davis of the University of Arizona for their reviews of earlier versions of the manuscript. I also benefitted from numerous conversations with my former students at the University of Missouri-Columbia and other colleagues. Figures were drafted by Jeff Horowitz. Manuscript preparation was provided by the Owen-Coates Fund of the Geology Foundation, The University of Texas.

The Geology of North America
Vol. O-2, Hydrogeology
The Geological Society of America, 1988

## Chapter 34

# Western alluvial valleys and the High Plains

**George H. Davis**
*10408 Insley St., Silver Spring, Maryland 20902*

## INTRODUCTION

The western alluvial valleys and the High Plains are the most extensive and productive ground-water systems in North America. The alluvial basins include the intermontane valleys of the Pacific mountain system (93,000 km$^2$) and the basins of the Basin and Range Province (302,000 km$^2$ in the United States and roughly twice that in Mexico); the High Plains encompasses 458,000 km$^2$ (Fig. 3, Table 2, Heath, this volume). Among the more important subdivisions of the Pacific system are the Puget Sound Lowland, the Willamette Valley, the Central Valley of California, and the South Coastal Plain of California. Most of the arable land and the principal urban areas of the Pacific Coast of the U.S. and Canada are located in these basins. Ample ground-water resources have played a major role in the economic development of the region, supplying water for the irrigation of crops and for industrial and municipal use in the urban centers.

To the east of the Sierra Nevada lies a great expanse of alternating alluvial basins and mountain ranges extending some 700 km to the Wasatch Mountains east of Great Salt Lake. Within this area (132,000 km$^2$), the basins all filled with varying thicknesses of alluvium, make up about one third of the gross area. To the southeast the high Colorado Plateau separates another area of similar geological structure from the northern part of the Basin and Range Province. The southern area includes the desert basins of southeastern California; the southern two thirds of Arizona; the southwestern quadrant of New Mexico; the Rio Grande Depression in Colorado, New Mexico, and Texas; and extends some 1,000 km into Mexico. It is commonly referred to as the Basin and Range Lowland because the ranges are more subdued and make up a smaller proportion of the total area than in the Nevada–Utah sector. Among the more important subdivisions are the Mojave Desert, the Imperial Valley–Gulf of California Trough, the central Arizona valleys, and the Rio Grande Depression. As on the Pacific Coast, most of the farmed land and urban areas lie in the alluvial basins, largely because of locally plentiful ground-water supplies. However, most of the area remains desert because of deficient recharge.

The High Plains is the remnant of a pediment of Tertiary age that once extended eastward from the Rocky Mountain front to beyond the present course of the Missouri River in places. Although much reduced by erosion, this remnant embraces most of Nebraska, about half of Kansas, the panhandles of Oklahoma and Texas, and parts of New Mexico, Colorado, Wyoming, and South Dakota. The present High Plains was the site of alluvial deposition throughout Miocene time, culminating in the deposition in late Miocene and early Pliocene time of a gravelly apron, termed the Ogallala Formation, and related deposits. These gravelly deposits together with overlying Quaternary and underlying Miocene deposits of the Arikaree Group, all of continental origin, make up the High Plains Aquifer (Gutentag and others, 1984).

## THICKNESS OF ALLUVIAL DEPOSITS

The thickness of continental deposits in the western alluvial basins is variable, and in many is unknown. The intermontane basins of the Pacific mountains are generally underlain by marine sediments of Tertiary and Cretaceous age laid down in arms of the sea that formerly occupied these structural depressions. The thickness of the continental deposits generally is less than about 1,000 m except in deep synclines, as at the southern end of the Central Valley of California, where 5,000 m of continental deposits overlie mid-Pliocene marine sediments (de Laveaga, 1952).

Throughout the Basin and Range Province, there was almost no marine deposition following the close of the Laramide revolution in early Miocene time, and great thicknesses of continental deposits, consisting mainly of alluvium with some interbedded volcanics and lacustrine deposits, including evaporites, have accumulated in subsiding and widening basins during Tertiary and Quaternary time. Some of the better-documented areas of great thicknesses of continental deposits are at Tucson, Arizona (3,500 m), and Phoenix, Arizona (5,000 m), Eberly and Stanley, 1978; the Sevier Desert, Utah (4,000 m), Allmendinger and others, 1983; the Carson Desert, Nevada (more than 3,350 m), Anderson and others, 1983; the San Luis Valley, Colorado (9,000 m), West and Broadhurst, 1975; and the Albuquerque Basin, New Mexico (3,360 m), Cape and others, 1983. The water-bearing character of the deeper part of these alluvial

Davis, G. H., 1988, Western alluvial valleys and the High Plains, *in* Back, W., Rosenshein, J. S., and Seaber, P. R., eds., Hydrogeology: Boulder, Colorado, Geological Society of America, The Geology of North America, v. O-2.

fills is not well known because development of water supplies from such great depths is not economical at present. However, irrigation wells as deep as 1,000 m have been operated in the San Joaquin Valley, California. Presumably, reduction of porosity and permeability through compaction and cementation at depths of several thousand meters would physically limit ground-water development, but little specific information on water-bearing properties is available for the depth range below about 1,000 m. Another important limitation on the use of water from deeper alluvial deposits is the common increase in dissolved solids content with depth.

The High Plains aquifer, in contrast, is laterally extensive over an area of 458,000 km$^2$, dips gently eastward about 3 m per km or less and has been little affected by structural deformation. The only significant structural features affecting the High Plains Aquifer other than regional tilting are faults in southeastern Wyoming, where the base of the aquifer has been displaced by 300 m along northwest-trending reverse faults (Gutentag and others, 1984, Fig. 6). Its saturated thickness ranges from 30 m or less in the southern High Plains of west Texas to more than 300 m in western Nebraska (Gutentag and others, 1984, Fig. 9).

## DEPOSITIONAL REGIMES

The material that underlies the western intermontane basins consists almost entirely of material washed down from the flanking mountains in structurally active areas. The basins represent negative structures—synclines, downtilted homoclines, or grabens—while the mountains represent positive structures—anticlines, tilted blocks, or horsts. Commonly, although not everywhere, major faults mark the basin margins. Detritus eroded off the mountains is carried away by streams and laid down as stream-channel deposits or by sheet floods when streams overtop their banks and spread out over lowland areas. The finest fraction of sediment, fine clay and silt, may be carried to lakes that occupy low parts of basins, where it is deposited as broad aprons of lacustrine strata that interfinger upslope with coarser alluvium.

Two principal types of stream deposits can be recognized in most basins: (1) well-sorted gravels, sands, and silts deposited in stream channels and (2) poorly sorted deposits consisting of a clayey matrix enclosing stringers of sand and gravel laid down in torrential runoffs by sheet flows when the streams overflow their banks. Such poorly sorted deposits form the bulk of alluvial fills in the western intermontane basins. Well-sorted sands and gravels predominate only along the courses of major rivers and in the vicinity of major channels leading out from mountain canyons. Poorly sorted deposits also characterize the areas between principal streams, which are inundated only during large floods.

In most alluvial basins the streams shift course frequently as old channels become choked with sediment. Thus, the deposit as a whole typically consists of a poorly sorted fine matrix containing stringers of sand and gravel. The carrying power of a stream is directly related to the velocity of flow, which in turn is controlled by the cross-sectional area of the stream and the channel slope. Accordingly, a stream issuing from a steep mountain drainage basin will deposit much of its load, particularly the coarser materials on the alluvial fan flanking the mountains, owing to reduced carrying power. This reduction in capacity may be due to a flattening of gradient, increase of width, loss of water through infiltration, or commonly, through all three factors (Bull, 1977, p. 261).

In closed drainage basins, all of the sediment load is deposited, but in basins with through-flowing streams, such as the Willamette, the Gila, and the Central Valley of California, a large part of the suspended load and some of the bed load is carried out to sea. In regimes where a large proportion of the fine sediment was carried out of the area, e.g., the Ogallala Formation of the High Plains, the average hydraulic conductivity and specific yield of the alluvial deposits are significantly higher than in closed basins (see Gutentag and others, 1984, Figs. 10 and 11; and later in this chapter).

The sediment regime differs considerably among western alluvial basins depending on the rate of subsidence of the basin with respect to the rate of uplift of the adjoining mountains, and the width of the basin as related to sediment transport. Moreover, in a given basin these factors may vary greatly over geologic time. For example, the typical valleys of the northern Basin and Range Province exhibit fairly steeply sloping alluvial fans extending nearly to the valley axis, where a narrow, level playa lake bed commonly is found. In contrast, the Central Valley of California, which is relatively broad and is traversed by through-flowing rivers, is characterized by elevated channels of the trunk streams contained within natural levees, occupying the topographic axis of the valley. The trunk rivers are flanked on both sides by broad, level flood basins extending out from the rivers as much as 15 km. Historically these flood basins temporarily stored the excess flows of sediment-laden waters that could not be accommodated in the main channels. Drilling has shown that this regime is not merely a transitory feature, but has persisted throughout much of Quaternary time (Olmsted and Davis, 1961, p. 115). Accordingly, the axes of such basins are characterized by thick sequences of fine, laterally extensive clayey silts encasing occasional stringers of well-sorted sand representing ancient channel deposits of the main streams. These flood-plain deposits interfinger with typical alluvial-fan deposits upslope.

It can be readily appreciated that there are many possible combinations of alluvial processes. Furthermore, climatic changes and tectonic movements in the basins and flanking mountains have in the past caused pronounced changes in sedimentary regimes. Among the most common of such changes are: (1) tectonic movement causing changes in drainage patterns and sources of sediment; (2) climatic changes, such as increasing or decreasing precipitation; (3) change from closed-basin drainage to through-flowing drainage, or the reverse; (4) development of subsurface drainage via karstic caverns; and (5) volcanic activity.

The Central Valley of California offers many examples of tectonic control of drainage regimes. While the northern part, the

Sacramento Valley, has been the site of nonmarine deposition since Eocene time, much of the southern part, the San Joaquin Valley, was occupied by an arm of the sea through mid-Pliocene time. Growth of the Southern Coast Ranges beginning toward the end of the Tertiary resulted in the establishment of the present alluvial regime in the San Joaquin Valley. Differential tectonic movements of the Sierra Nevada to the east and the Coast Ranges to the west have caused repeated shifts in the drainage axis of the San Joaquin Valley, and in the provenance and resulting physical character of the basin-fill deposits (Miller and others, 1971, p. E28). Even the present-day drainage system is affected by ongoing tectonic activity. Davis and Green (1962) have shown, for example, that tectonic subsidence in the Tulare Lake area occurs at a rate sufficient to prevent the establishment of a true through-flowing drainage system in the southern part of the San Joaquin Valley, which in historic times has only discharged to the sea during exceptional floods. At Tulare Lake bed and at the similar Buena Vista Lake bed at the southern end of the San Joaquin Valley, tectonic subsidence exceeding sedimentation rates has resulted in semi-permanent sumps, which are the site of lacustrine deposition. Both sumps have served as traps for fine sediment throughout much of late Pleistocene and Holocene time, and both areas are underlain by great thicknesses of fine-grained materials that yield little water to wells.

Perhaps the best-known climatic change affecting sedimentation in the western alluvial basins was that which led to the formation of large lakes in the northern Basin and Range Province in response to increased precipitation and/or reduced evaporation during Pleistocene pluvial episodes. At that time, large lakes occupied much of the lowland area, and many breached low divides to form extensive compound lakes. The largest of these were Lake Bonneville and Lake Lahontan. Lake Bonneville extended from the Wasatch Front to beyond the Utah–Nevada border and from southern Idaho to southern Utah, covering 52,000 km$^2$. At maximum stage the lake was as much as 340 m deep. Lake Lahontan occupied the western lowlands of the Humbolt, Truckee, Carson, and Walker River basins. Other closed basins, including the Imperial Trough, Owens Valley, Death Valley, Mono Basin, and low parts of the Mojave Desert were occupied by lakes as well. During these episodes the former alluvial regime was replaced by extensive lacustrine deposition and deltaic deposits were formed at canyon mouths (Arnow and others, 1970). With the return of arid conditions in Holocene time, the lakes receded to a few scattered relics, the largest being Great Salt Lake. Throughout much of their extent the lacustrine strata act as confining beds and thus exercise an important control over the occurrence and movement of ground water. The deltaic deposits tend to be clean and well-sorted sandy to gravelly materials and locally are important aquifers.

Changes from closed-basin drainage, evidenced by accumulation of salines, to through-flowing drainage (and vice-versa) has been noted in many basins. A modern example is the Imperial Trough of southernmost California, the low point of which is 84 m below sea level. Although the nearby Colorado River had discharged to the Imperial Trough at times in the geologic past, for the past few centuries the Colorado had discharged to the sea in the Gulf of California (Hely and others, 1966, p. C2). Under that flow pattern, the drainage and sediment regime of the Imperial Trough was that of a typical desert basin with an intermittent playa lake temporarily occupying the lowest part of the basin during rainy periods. This regime was terminated abruptly in 1904 when the Colorado River, with some human assistance, broke into the Imperial Trough, and virtually the full flow of the Colorado discharged into the basin for two years to form the huge freshwater lake now known as the Salton Sea. By the time the Colorado was diverted back into its old course, the lake covered 1,300 km$^2$ and was as much as 24 m deep (Hely and others, 1966, p. C20). The lake, now about as salty as sea water because of evaporation, is maintained mainly by irrigation return flows.

The presence of great thicknesses of halite, anhydrite, and gypsum of continental origin in the Phoenix, Pichacho, and Tucson basins of central Arizona provide evidence of closed-basin drainage conditions in that area in late Miocene time (Eberly and Stanley, 1978). As much as 1,800 m of salt, which apparently accumulated over a period of 4.4 m.y., underlies the Phoenix Basin. The initiation of closed drainage coincides with an episode of block faulting, which produced the present horst and graben terrane of the southern Basin and Range Province. Through-flowing drainage was reestablished between 10.5 and 6 Ma, near the end of the Miocene (Eberly and Stanley, 1978, p. 938). During roughly the same time interval, the lower Colorado River Basin, as far north as the southern tip of Nevada, was occupied by an arm of the sea. Saline deposits similar to those of central Arizona no doubt will be found elsewhere in the Basin and Range Province. Several have been reported recently in connection with deep seismic exploration, for example, in the Carson Desert of western Nevada and the Sevier Desert of western Utah (Anderson and others, 1983, p. 1059, 1066). The saline deposits are dense and impermeable, and any water encountered below the uppermost saline deposits would be highly saline and unusuable for most purposes. The saline deposits, of course, represent valuable mineral resources in their own right.

Sea-level changes, particularly during the Pleistocene, which are secondary effects of climate change and continental glaciation, altered the base levels of coastal streams and greatly affected the depositional patterns of many stream systems. For example, Upson and Thomasson (1951, Plate 5) mapped coarse stream deposits extending to 61 m below sea level at the mouth of the Santa Ynez River, near Lompoc, California. During episodes of lowered sea level, the formerly aggrading river carried its sediment load far offshore. With recovery of sea level at the end of the Pleistocene, backfilling occurred, and gravel trains within the valley fill attest to the greater carrying power of the river at that time. Similar relations have been reported for other Pacific Coast streams (Poland and others, 1956, 1959; Olmsted and Davis, 1961; Atwater and others, 1986). Such Pleistocene backfill deposits are generally coarser and better sorted than the enclosing alluvium, and locally are important aquifers.

Development of karstic drainage beneath western alluvial basins is best known from southern Nevada. Most of the intermontane basins in Nevada south of latitude 37°N form a multibasin subsurface drainage system, which discharges via underlying permeable Paleozoic carbonate rocks to large springs near the California border on the west and to springs along Colorado River tributaries to the east (Eakin and others, 1976, Plate 1). Where the vertical connection to the underlying carbonate aquifer system is good, the alluvium of the intermontane basins is drained, resulting in some areas of great thicknesses of unsaturated alluvium and absence of the more-typical playa regime.

Volcanism is another major factor affecting the alluvial regimes in the western basins. Many basins contain several volcanic flows interbedded with alluvium, and most contain abundant reworked volcanic sediments. Most of western North America experienced volcanic activity during the Tertiary, and the volcanoes of the Cascade Range are still intermittently active. Volcanic debris at times furnished most of the sediment load of the streams, and volcanic flows engulfed and filled or dammed many stream valleys. Indeed, basins not affected by Tertiary volcanic activity are the exception in the West. The basins adjoining the Cascade Range have been the most affected. In this range, which consists of a great mass of volcanic rocks extending from the Canadian border to the Sacramento Valley in California, volcanics cover the former terrain to depths of up to several thousand meters, and deposition in nearby basins depends largely on the volcanic regime (Foxworthy, 1979, p. 13). An example of this type of control has been evident since the May 1980 eruption of Mount St. Helens, Washington. The nearby Toutle and Lower Cowlitz rivers were filled with volcanic debris to 4.5 m above their former channels, and a large and continuing sediment load has been delivered to the Columbia River since the eruption (Foxworthy and Hill, 1982). Volcanic deposits may range from highly permeable basaltic flows and rubbly deposits that yield water readily to wells to dense, impermeable tuffs and breccia flows. Reworked materials also have a wide range in permeability. Thus the impact of volcanic activity on the hydrologic regime is highly erratic and subject to great variability over short distances.

To summarize with respect to the principal types of western alluvium terranes: (1) the intermontane basins of the Pacific Coast are mainly synclinal or downtilted monoclinal structures formerly occupied by the sea, containing thick sections of saline-water-bearing marine sediments beneath a few hundred to a thousand meters of alluvium of Tertiary and Quaternary age; (2) the alluvial basins of the Basin and Range Province represent elongate, down-dropped blocks in a region of east–west extension; the basin fills, commonly as much as 3,000 m thick, for the most part recording nonmarine deposition since the end of the Laramide revolution; in many basins, great thicknesses of lacustrine saline deposits and volcanic flows are intercalated with stream-laid materials; and (3) the High Plains are the erosional remnant of a great alluvial apron laid down in Tertiary time east of the Rocky Mountains; except for regional tilting as western North America rose in Pliocene time, the High Plains has been little affected by deformation and erosion is the principal process operating at present.

## PHYSICAL CHARACTER OF ALLUVIAL-FILL DEPOSITS

In general, the physical characteristics of alluvial fills follow systematic patterns, i.e., the coarsest deposits are generally found where streams debouch from high mountains; the average grain size tends to decrease away from these foci, and interfan areas are characterized generally by finer deposits. Deposition of alluvium is caused mainly by decrease in carrying power, which is governed by such factors as stream gradient and width, and loss of flow through infiltration. In the more extensive basins, three distinct types of deposition are observed: alluvial-fan deposits, flood-plain and channel deposits of major perennial streams, and lacustrine sediments.

Alluvial-fan deposits, generally associated with intermittent or ephemeral streams, range in degree of sorting from clean, well-sorted gravels and sands of the larger channels to very poorly sorted mixtures of gravel and sand in a clayey-silt matrix laid down by mudflows. Various intermediate facies, characterized by intermediate degrees of sorting make up most alluvial-fan deposits. For the most part, these represent materials laid down on fan surfaces by overbank flows that deposit their suspended load as water dissipates into the surface materials.

Alluvial deposits of perennial streams are commonly found in basins with through flow, such as those of the Pacific Mountains, the lower Colorado River Basin, and the Rio Grande Basin. Here, perennial streams draining well-watered catchments traverse the alluvial basins and discharge their flow ultimately to the sea. The deposits of these streams range from clean, well-sorted gravels and sands of the channel deposit to clayey and silty materials deposited on flood plains by overflow waters. These latter deposits are almost indistinguishable in physical character from overflow deposits on alluvial-fan surfaces.

Lacustrine deposits found in most alluvial-basin fills, represent deposition under ponded still-water conditions. In closed desert basins not subject to underdrainage, a playa or intermittent lake generally occupies the topographic trough. In basins with through flow, overflows commonly pond in low-lying flood basins flanking the major streams. In both regimes, fine clay and silt in suspension in the flood waters settles out, producing generally well-sorted fine deposits. In deposits of large, perennial lakes, well-sorted sand and gravelly materials of deltas mark former river mouths and beaches. The lacustrine deposits generally are distinguishable from those laid down in moving water by better sorting and greater lateral extent. However, lateral extent is not a reliable criterion for identifying still-water deposits because those of natural-levee backswamps and cut-off meanders commonly are of limited lateral extent.

In most alluvial-basin fills, clay and silt are the predominant grain sizes. For example, Meade (1967, p. C11) reported that among 305 samples studied from three core holes on the west

**TABLE 1. CHARACTERISTICS OF ALLUVIAL-FAN DEPOSITS OF THE WESTERN BORDER OF THE SAN JOAQUIN VALLEY, CALIFORNIA**

| Type of Deposit | Depositional Characteristics | Average Parameters from Grain-size Analysis (%)* | | | |
|---|---|---|---|---|---|
| | | Clay | $S_O$ | $Qd_\phi$ | $\sigma_\phi$ |
| Water-laid sediments | No discernible margins; usually clean sand or silt, cross-bedded, laminated, or massive | 6 | 1.8 | 0.8 | 1.4 |
| Intermediate deposits | No sharply defined margins; clay films around sand grains and lining voids; graded bedding and oriented fragments | 17 | 4.0 | 2.0 | 3.9 |
| Mudflow deposits | Abrupt, well-defined margins, lobate tongues; clay may partially fill integranular voids. May not have graded bedding or particle orientation | 31 | 9.7 | 3.1 | 4.7 |

$$*S_O \text{ (Trask sorting coefficient)} = \frac{Q_{75} \text{ (larger quartile diameter)}}{Q_{25} \text{ (smaller quartile diameter)}}$$

$$Qd_\phi \text{ (phi quartile deviation} = \frac{\phi 25 - \phi 75}{2}$$

$$\sigma_\phi \text{ (phi standard deviation} = \frac{\phi 16 - \phi 84}{2}$$

$\phi = \log_2$ of grain diameter in mm

side of the San Joaquin Valley, nearly two-thirds of the samples were in the silty-clay, clayey-silt, and sand-silt-clay classes. Although more samples fell in the clayey-silt class than any other, coarser sediments were represented also. Figure 1 shows these samples in the form of Shepard diagrams, on which the two principal source areas are indicated. The salient points of this presentation are: (1) particle-size characteristics are similar despite differences in depositional environment and provenance of materials, and (2) pronounced heterogeneity is characteristic of all categories.

Meade (1967, p. 46), on the basis of the samples illustrated plus samples from two core holes on the east side of the San Joaquin Valley and two core holes in the Santa Clara Valley, California, concluded with respect to the alluvial deposits of the areas studied that the most abundant particle-size class was clayey-silt, followed by sand-silt-clay, and silty sand. He further reported that the general degree of sorting was poor in all samples (average phi quartile deviation, $Qd_\phi$, near 2.0) and that skewness measures indicated a disproportionately large mixture of finer particles. The most notable features of these suites of samples were: (1) general fineness of particle size, (2) diverse juxtaposition of coarser and finer deposits, and (3) a high proportion of montmorillonite, which made up 60 to 70 percent of the clay-mineral content in all three areas.

Bull (1964, p. A24), in studies of the surface and near-surface alluvial-fan deposits of the western border of the San Joaquin Valley, was able to subdivide the materials into three categories. The characteristics are shown in Table 1. Typically, water-laid channel deposits and lacustrine sediments are better sorted than overbank deposits, and mud-flow materials are virtually unsorted (Figs. 2 and 3).

Despite the extreme heterogeneity of grain size and sorting characteristics exhibited by alluvial-fill deposits, some other physical parameters are less variable. Porosity, commonly expressed as percentage of total volume occupied by pore spaces, ranges from 20 percent for poorly sorted clayey gravel to 60 percent for well-sorted lacustrine clays and silts. Porosity depends chiefly on: (1) shape and arrangement of particles, (2) degree of sorting, and (3) post-depositional cementation and compaction, which reduce porosity. Well-sorted sands, silts, and clays tend to have high porosity regardless of grain size, while poorly sorted materials have low porosity because small particles occupy the spaces between large grains.

Hydraulic conductivity, the capacity of a material to transmit water, depends upon porosity, grain size, and degree of sorting. In the core samples discussed earlier from the San Joaquin and Santa Clara valleys, vertical hydraulic conductivity ranged over seven orders of magnitude, from $28.5 \times 10^{-7}$ m/day to 26.5 m/day (after Johnson and others, 1968, Table 5). Indeed this range was observed in a single core hole. This is not surprising in view of the extreme heterogeneity of grain size and sorting manifested by alluvial deposits, but it emphasizes the futility of calculating average values for materials of such wide diversity.

Probably the most widely available and systematic data to be found on alluvial fills of western basins is specific yield, the volume of material that will drain under the force of gravity. Specific yield, expressed as percent of total volume, depends upon porosity (which it cannot exceed) and hydraulic conductivity

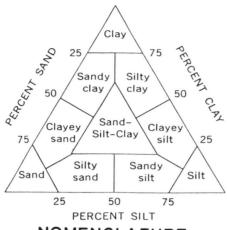

**NOMENCLATURE**

EXPLANATION

MICRONS

Gravel    >2000
Sand      62–2000
Silt      4–62
Clay      <4

SOURCE

•,  Diablo Range
○,  Sierra Nevada
+,  Mixed

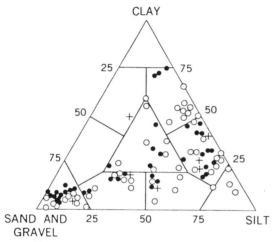

**FLOOD-PLAIN SEDIMENTS**

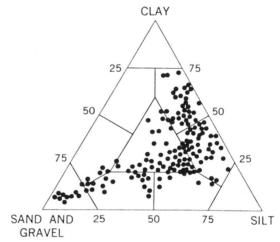

**ALLUVIAL-FAN SEDIMENTS**

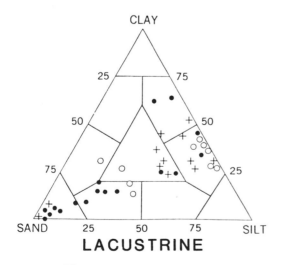

**LACUSTRINE**

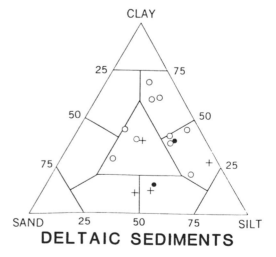

**DELTAIC SEDIMENTS**

Figure 1. Sand-silt-clay percentages of alluvial-fill deposits of west side San Joaquin Valley, California (after Meade, 1967, Fig. 5).

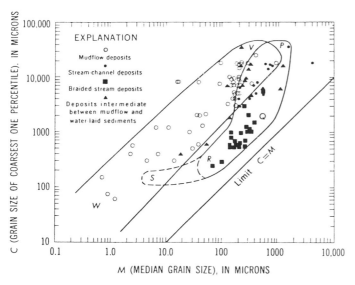

Figure 2. Textural patterns of surficial deposits of alluvial fans of the west border of the San Joaquin Valley, California (after Bull, 1964, Fig. 11).

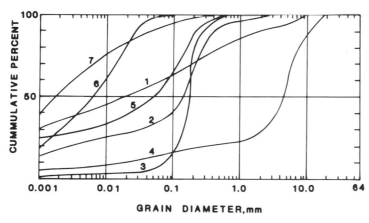

Figure 3. Cumulative curves of grain-size distribution for representative alluvial fills from California. Sources of data: samples 1–4 (Bull, 1964, Fig. 13), samples 5–7 (Johnson and others, 1968, Table 8). Descriptions: (1) mudflow sandy clay, west border San Joaquin Valley; (2) clayey sand, intermediate between mudflow and water-lain condition, Arroyo Ciervo Fan, west border San Joaquin Valley; (3) water-laid sand from Arroyo Ciervo Fan; (4) water-laid silty gravel, west border San Joaquin Valley; (5) diatomaceous clayey silt, lacustrine Corcoran Clay member, Tulare Formation, depth 202 m, core hole 14/13-11D1, San Joaquin Valley; (6) sand-silt-clay from zone of organic lacustrine (?) deposits, depth 596 m, core hole 19/17-22J1, J2, San Joaquin Valley; (7) silty clay alluvial-fan material, depth 231 m, core hole 6S/2W-24C7, Santa Clara Valley.

(which governs the rate and amount of drainage). In unconfined deposits, specific yield is identical to storage coefficient and that term is used for the property in many reports. Generally, specific yield is estimated from drillers' logs, geophysical logs, or other subsurface data by methods described by Eckis and Gross (1934), Piper and others (1939), and Olmsted and Davis (1961). As this parameter is essential in estimating ground-water storage capacity and in estimating water budgets, it has been estimated, albeit roughly, for most of the alluvial basins of the western U.S. Table 2 presents information on specific yield and ground-water storage capacity for the principal regional subunits (Fig. 4), calculated from area and estimated storage capacity of the uppermost 30 m of saturated sediments. The calculated average values range from less than 6 percent for the coastal basins of California to as much as 20 percent for the San Luis Valley, Colorado. In the heavily developed basins, where subsurface records are plentiful, the average specific yield values generally lie close to 10 percent. Specific-yield measurements for various types of alluvial materials range from less than 1 percent for poorly permeable fine-grained deposits to as much as 40 percent for clean, coarse sand (Piper and others, 1939, p. 109–116). However, average values over broad depth intervals show less variability.

Deposition on the High Plains occurred under somewhat different conditions than in the western basins, to the extent that the alluvium was laid down on broad composite flood plains of eastward flowing rivers that drained the Rocky Mountains. The alluvial material of the Ogallala Formation is described as generally unconsolidated, poorly sorted clay, silt, sand, and gravel containing some caliche zones cemented with calcium carbonate (Gutentag and others, 1984, p. 1). This description accords well with those of most of the flood-plain deposits of the western basins.

Gutentag and others (1984, p. 26) observe that the specific yield of the High Plains aquifer within a typical vertical section ranged from less than 5 to 30 percent, depending on the character of the sediments, and averaged 15.1 percent for the entire High Plains aquifer. This value is significantly higher than the averages given in Table 2 for the western alluvial basins, but that is consistent with a scenario of through-flowing streams winnowing the materials and carrying at least parts of the fines beyond the area of alluvial deposition.

## Influence of source area on physical character of alluvium

Several characteristics of source areas influence the amount of sediment transported, grain-size distribution, and sorting characteristics, and thus the storage capacity and transmissivity of the alluvial fills of the western basins. The factors exercising the greatest control are: (1) altitude of catchment areas; (2) amount, areal distribution, and seasonal distribution of precipitation; and (3) petrology of source rocks, particularly initial grain size, mineralogy, degree of consolidation, and solubility. These controls interact in complex ways and are subject to great local variability;

**TABLE 2. SPECIFIC YIELD AND GROUND-WATER STORAGE CAPACITIES OF WESTERN BASINS**

| Unit and Subdivision | Area (km²) | Estimated Average Specific Yield (%) | Approximate Storage Capacity, Upper 30 m of Saturation (km³) |
|---|---|---|---|
| **1. Pacific Mountains** | | | |
| a. Puget Sound Basin | 16,000 | 8.9 | 42 |
| b. Willamette Basin | 9,000 | 9.6 | 26 |
| c. California coastal basins | 25,600 | 5.8 | 49 |
| d. Central Valley of California | 41,900 | 8.0 | 100 |
| **2. Northern Basin and Range** | | | |
| a. Bear River | 4,900 | 7.5 | 11 |
| b. Great Salt Lake | 32,400 | 6.7 | 53 |
| c. Sevier Lake | 10,900 | 9.2 | 30 |
| d. Humboldt | 21,900 | 11.5 | 76 |
| e. Central Lahontan | 8,900 | 11.2 | 30 |
| f. Tonapah | 52,600 | 10.1 | 160 |
| **3. Basin and Range lowland (U.S.)** | | | |
| a. Colorado River | 44,800 | 10.4 | 140 |
| b. Gila River | 40,900 | 11.4 | 140 |
| c. Southeastern California | 60,700 | 9.3 | 170 |
| **4. Rio Grande Depression (U.S.)** | | | |
| a. San Luis Valley | 8,100 | 20.0 | 49 |
| b. Albuquerque | 13,000 | 18.0 | 70 |
| c. El Paso | 2,800 | 11.0 | 9 |

Sources of information and notes:
  1a, 1b. Columbia North Pacific technical staff, 1980. Recomputed and prorated for upper 30 m of saturated zone.
  1c, 1d. Thomas and Phoenix, 1976, Table 3. Recomputed and prorated for upper 30 m of saturated zone.
  2. Eakin and others, 1976, Plate 1. Converted to metric units.
  3a, 3b. Davidson, 1979, Table 1. Converted to metric units.
  3c. Thomas and Phoenix, 1976, Table 1. Recomputed and prorated for upper 30 m of saturated zone.
  4a, 4b, 4c. West and Broadhurst, 1975, Pages D14, D26, and D15. Recomputed and prorated for upper 30 m of
    saturated zone.

thus, it is difficult to relate sediment characteristics quantitatively to source area.

Altitude, for example, affects amounts and distribution of precipitation, rate and type of weathering, vegetative cover, and stream gradients. The amount and distribution of precipitation also clearly affects weathering and vegetative cover. Finally, the petrology of source rocks, in combination with these other factors, exercises strong control over weathering and amount and character of sediment produced.

The structural movements that formed the western basins commonly resulted in an asymmetrical distribution of altitude and slopes on opposite sides of basins. As the structural grain of North America is aligned roughly north–south, and the prevailing moisture-laden winds are westerly, the resulting orographic precipitation is generally heaviest on the western flanks of the mountains, and produces a rain shadow of deficient precipitation on the eastern flanks of the mountains and the western borders of the intermontane basins. Not uncommonly, the geologic history and petrology of the rocks differ greatly on opposite sides. Topographic, geologic, and climatic asymmetry is most pronounced in the Pacific Mountain Province and is less evident in the Basin and Range Province. Where pronounced, as in the San Joaquin Val-

ley, California, the asymmetry accounts for surprising juxtaposition of vastly different alluvial regimes over short distances. In the extreme case, a ground-water quality boundary relating to different sources of sediment and recharge formerly was interpreted as evidence of a major fault (see Davis and others, 1959, p. 81).

## RATES OF DEPOSITION

Rates of deposition are difficult to calculate in alluvial deposits because little datable material survives in most alluvial fills. Organic remains of plants and animals generally are destroyed by oxidation in the prevailing environment of alluvial fills. However, where volcanic-ash markers or lake sediments can be distinguished, they offer opportunities for dating by the potassium-argon and carbon-14 methods, or by paleontology.

Where specific dates are available, it can be seen that depositional rates can be highly variable, even over short distances, and depend largely on rates of tectonic subsidence. Some well-documented sedimentation rates from the Central Valley and Imperial Trough, California, and the Phoenix and Tucson basins, Arizona, are presented in Table 3.

The highest sedimentation rates are found in actively subsid-

Figure 4. Map of western United States showing ground-water storage units and subdivisions of Table 2. Heavy blue lines show borders of U.S. Water Resources Council river basin planning regions. Patterned area is the High Plains.

The rate listed for the western San Joaquin Valley of 0.64 mm/yr was based on denudation of an area of highly erodable sediment in an area of heavy grazing and may be excessive for that reason. The rates shown for the Sacramento–San Joaquin delta and the Santa Clara Valley are about the same as the rate of sea-level rise for the same period (Shepard, 1964, Fig. 2). In both areas the sedimentation rate shown may reflect backfilling of Pleistocene stream trenches, rather than tectonic subsidence. The preceding rates represent maximum values in areas of very rapid local basin filling. More typical rates for large areas of active deposition are approximately one order of magnitude less. The average rate of deposition in the axis of the San Joaquin Valley over the past 615,000 yr is about 0.16 mm/yr. This value is close to that computed for the northern Sacramento Valley of 0.23 mm/yr over the past 3.4 m.y., and for the Tucson Basin of 0.18 mm/yr over 16.1 m.y.

## MODIFICATION OF ORIGINAL MATERIALS

Once deposited, alluvium is immediately exposed to physical processes tending toward compaction and to the chemical effects of air, water, and organic matter. If not covered promptly by new deposits, soil-profile development will begin. Soil moisture containing dissolved oxygen ($O_2$), carbon dioxide ($CO_2$), and other atmospheric gases, augmented by additional $CO_2$ derived from organic decomposition in the soil zone, attacks the constituent rock and mineral fragments. For example, feldspar is readily converted to clay minerals, releasing soluble ions that are transported downward in infiltrating water. Mafic minerals also weather readily, producing clay minerals and poorly soluble oxides of irion (Fe) and aluminum (Al). The clay minerals and oxides thus formed tend to wash downward, producing a low-density surface layer, or A horizon, underlain by a denser B horizon of maximum clay and oxide accumulation. This zone normally grades downward into weathered parent material, termed the C horizon.

Soil development in well-drained alluvial deposits commonly follows these trends with increasing age: (1) increasing thickness of horizons and depth to fresh parent material, (2) redder hues, (3) brighter colors, (4) lower soil pH, (5) sharper definition of horizons and more horizons, and (6) greater development of clayey horizons. Clay and iron compounds first bridge grains, then fill interstitial and tubular pores, then cover grain surfaces, and finally extend to greater depths. In poorly drained soils, hardpans (or duripans) typically accumulate at the base of the B horizon. The typical history of hardpan development is: (1) development of a calcium carbonate–silica hardpan with no overlying B horizon, (2) a fairly strong calcium carbonate–silica hardpan beneath a well-drained clayey B horizon, (3) a weak iron-silica hardpan with calcium carbonate seams, and finally (4) a red, acidic, strongly developed iron–silica hardpan with a heavy clay B horizon above.

The rate of soil-profile development depends to some degree upon precipitation; however, in typical semi-arid climates of the

ing structures. Rapid subsidence in the Imperial Trough can be inferred from the fact that the topographic low point is 84 m below sea level despite the proximity of a large sediment source in the nearby Colorado River. This is confirmed by geodetic measurements (Lofgren, 1974), which indicate present tectonic subsidence of as much as 34 mm/yr near the south end of the Salton Sea. This modern rate exceeds by an order of magnitude the average sedimentation rate of the Imperial Trough of 3 mm/yr over the past 3 m.y. quoted by Meidev and Howard (1979). Rates in the range of 0.4 to 1.7 mm/yr are shown for other areas of known rapid tectonic sinking. These areas include the Phoenix Basin, an intermontane graben filled in an episode of closed-basin deposition, and the sharp synclinal structure at the south end of the San Joaquin Valley, which has accumulated 5,000 m of post mid-Pliocene continental deposits. It is noteworthy that the average rate of subsidence there, calculated over 3.4 m.y., is confirmed by carbon-14 dating of wood from 22 m depth in a well in the same area. Rates of similar magnitude are also observed for accumulation of peaty post-Pleistocene deposits in the subsiding Sacramento–San Joaquin delta, California, as well as the area of maximum tectonic subsidence in the axis of the San Joaquin Valley, Tulare Lake bed (Davis and Green, 1962).

## TABLE 3. THICKNESS AND RATE OF ACCUMULATION OF SELECTED ALLUVIAL FILLS

| Area | Thickness (m) | Time (yrs x 1,000) | Rate of Accumulation (mm/yr) |
|---|---|---|---|
| **California** | | | |
| 1. Northern Sacramento Valley | 700 | 3,300 | 0.21 |
| 2. Sacramento-San Joaquin delta | | | |
| a. Maximum | 17.7 | 10.69 | 1.7 |
| b. Minimum | 10.06 | 6.6 | 1.5 |
| 3. Axis of San Joaquin Valley | | | |
| a. Maximum | 239 | 615 | 0.39 |
| b. Minimum | 91 | 615 | 0.15 |
| 4. Western border San Joaquin Valley | --- | --- | 0.64 |
| 5. Southern border San Joaquin Valley | | | |
| a. Deep syncline | 5,000 | 3,400 | 1.5 |
| b. Wood in well at 22 m depth | 22.4 | 17.13 | 1.3 |
| 6. Imperial Trough | | | |
| a. Modern leveling, maximum | --- | --- | 34 |
| b. Average over 2 m.y. | 6,000 | 2,000 | 3 |
| c. Average over 4 m.y. | 7,500 | 4,000 | 1.9 |
| 7. Santa Clara Valley | 22.26 | 14.35 | 1.6 |
| **Arizona** | | | |
| 8. Phoenix Basin, Miocene salt | 1,815 | 4,400 | 0.41 |
| 9. Tucson Basin, average over 16.1 m.y. | 2,895 | 16,100 | 0.18 |

Sources of information and notes on interpretation:
1. Based on K-Ar age of Nomlaki Tuff (Marchand and Allwart, 1981, p. 19), 213 m above base of Tehama Formation (Olmsted and Davis, 1961, p. 76).
2. Based on C-14 age of peat (Rubin and Alexander, 1960); as reeds that form peat grow only about at sea level, rates appear representative for relative subsidence of delta area as sea level rose at close of Pleistocene.
3. Based on K-Ar age of volcanic ash above Corcoran Clay Member of the Tulare Formation (Marchand and Allwart, 1981, p. 34), map of Corcoran (Davis and others, 1959, Plate 14), and Atwater and others, 1986, Fig. 7.
4. Based on denudation rate calculated for Arroyo Ciervo Basin over period 1955-1965 (Bull, 1964, p. 38); may be affected by modern grazing.
5a. Based on alluviation since end of mid-Pliocene marine deposition (de Laveaga, 1952, p. 120).
5b. Based on ¹⁴C age of wood sample from well in vicinity of deep syncline (Ives and others, 1967).
6a. Based on comparison of precise leveling over a period of a few years, in area of maximum subsidence (Lofgren, 1974).
6b. Average reported by Meidav and Howard (1979, p. 446).
6c. Calculated from Figure 1, Meidav and Howard (1979).
7. Based on ¹⁴C dated wood at 22m (Meade, 1967, p. C14).
8. Based on K-Ar ages of volcanics underlying and overlying saline deposits, and stratigraphic interpretation of Eberly and Stanley (1978, p. 931, 938).
9. Based on K-Ar age of andesite at depth of 2,875 m (Eberly and Stanley, 1978, p. 939); section includes late Tertiary closed-basin deposits and deposits of subsequent through-flowing drainage regime.

western alluvial basins, some profile development may be noted within a few decades. Soils with distinct profiles in the San Joaquin Valley, for example, have been dated as of Holocene age, that is, less than 10,000 yr old, by Marchand and Allwardt (1981, Table 2). Mapping by these authors of alluvial deposits of the northeastern San Joaquin Valley show clear and consistent relationships between the age of soils and various stages of soil development cited above. Some of the oldest soils mapped, of Pliocene age, 3 to 4 m.y. old, have clayey B horizons as much as 4 m thick, and iron–silica hardpans as much as 1 m thick. Generally these ancient soils are characterized by pronounced "mima mound" development as well.

Not uncommonly, renewed alluviation caused by climatic change or tectonic movement results in the burial of well-developed soil profiles. Such buried soil horizons are useful

markers for stratigraphers, as they represent a unique time horizon. Generally the oldest horizons are best exposed near the heads of alluvial fans where they are trenched by the principal stream; farther out on the fan the oldest materials generally are covered by successions of younger deposits.

Alluvium deposited on well-drained fans commonly is subject to continuous aeration and oxidation from the time of deposition. However, materials deposited under swampy or perennially saturated conditions, as on the distal parts of fans or in flood basins, may remain in such a nonoxidizing, or reducing, environment indefinitely. Under these conditions, soil development will not proceed as described above. Instead, the ground water is depleted in oxygen, and organic matter is preserved from oxidation. Iron compounds remain in a reduced state, and the typical yellow and reddish colors of oxidized iron compounds do not develop. The typical colors of reduced sediments are gray sands and greenish, bluish, and dark gray to black silts and clays. In the reducing environment, some solution of minerals may take place, but once the initial $CO_2$ content of the ground water is depleted, little additional solution occurs. Under suitable combinations of organic matter, anaerobic bacteria, and oxygen-depleted ground waters, sulfate and nitrate ions in the ground water may be reduced by bacteria to provide the oxygen required for their life cycle. In such circumstances, methane ($CH_4$), $CO_2$, and hydrogen sulfide ($H_2S$) may go into solution in the ground water. In the presence of such bacterially generated $CO_2$, further solution of mineral matter may take place. Alluvial materials that have once been oxidized, however, commonly retain their characteristic reddish hues even under great depth of burial (Meade, 1967, p. C6), particularly in the absence of organic matter.

A specialized environment of alluvial valleys is that of alkali belts, areas where the water table is sufficiently close to land surface (generally less than 1 m) that evaporation can occur directly from the capillary fringe. Such zones commonly are found around the distal margins of alluvial fans and in low-lying, poorly drained areas such as flood basins. Continuing evaporation leads to the accumulation in the soil of salts grouped under the general heading of alkali. Salts of calcium and magnesium impart a white color to the soil, termed white alkali, while accumulations of sodium carbonate result in a condition termed black alkali. The dark color is attributed to preservation of organic matter that is not subject to oxidation by decay in the sodic environment. These black alkali soils are virtually sterile and support almost no plant life. Soil alkali accumulations may be redissolved in subsequent flooding or infiltration, or they may be buried by later layers of sediment. In the latter event the salts can be dissolved in later infiltration or can go into solution in ground water if the water table rises.

Cementation of alluvial materials is characteristic of soil-profile development under arid conditions as described earlier. It also occurs through deposition of calcium carbonate in stream channels through evaporation of the waning stages of intermittent flows, through precipitation of carbonate and silica compounds in interstitial pores, and as a result of evaporation from intermittent lakes as well as permanent lakes. The latter processes are particularly important in cementation of lacustrine deposits. In stream-laid deposits, precipitation from soil moisture and ground water is the most widespread origin of cemented zones. Precipitation of carbonate, silica, and gypsum in the pore spaces of alluvium is fairly common (Lattman, 1973). If the ground-water solution is close to saturation with respect to any constituent, subtle changes in temperature, pressure, pH, or oxygen content may bring about oversaturation and subsequent precipitation of the less soluble constituents. Deposits of carbonate or silica around spring orifices are well-known examples of this process. Other examples, from alluvial fills, are fault zones that become so cemented through precipitation from percolating ground waters that they act as subsurface dams, preventing lateral movement of ground water.

Probably the most pervasive change affecting western allu-

vial deposits is compaction, or increase in density, the result of various diagenetic changes that begin immediately upon deposition. Loading by overlying deposits is a major cause of compaction, but other processes, including drying, also influence compaction, particularly during its early stages. The amount and rate of compaction are governed by textural properties, such as particle-size distribution and degree of sorting; mineral composition, especially the proportion of mica in sands and differing proportions of clay minerals in the fine fractions; and the chemical properties of interstitial waters, such as pH, dissolved-solids content, and cation distribution. These factors are interrelated in complex ways, which greatly complicates the prediction of compaction due to increase in effective overburden loading. Of special interest to hydrogeologists is artificial compaction, in which aquifer systems are subjected to increased loading through reduction of hydraulic head by pumping (Poland and others, 1984).

The initial porosity of alluvial-fill sediments is largely determined by depositional environment, mean grain size, and degree of sorting. Clays deposited on the floor of Lake Mead, for example, are reported as having initial porosities as high as 88 percent (Sherman, 1953, p. 399). Water is expelled from such materials at very slight loadings.

Clean, nonmicaceous sands, in contrast, generally have initial porosities close to 35 percent and compact only slightly in the load range of 1.2 to 120 bars (1 to 100 kg/cm$^2$), equivalent to up to 915 m of burial. If the mica content of a sand is as much as 10 percent, however, the initial porosity may approach 60 percent, and compressibility will be substantial at loadings of less than 1.2 bars (Gilboy, 1928, cited in Meade, 1968, Fig. 7).

Natural materials of intermediate grain size generally have initial porosities between those of sand and clay, although in poorly sorted mixtures of water-laid sand, silt, and clay, initial porosities commonly are in the neighborhood of 30 percent (Johnson and others, 1968, Table 5). Under special conditions of deposition and dessication, very high initial porosities are possible in dry clayey materials. Bull (1964, Table 17) reported surficial deposits of high montmorillonite content, laid down as mudflows or soupy streamflows on the alluvial fans of the western border of the San Joaquin Valley, with dry density as low as 0.96 gm/cm$^3$, implying porosity on the order of 60 percent. These materials, which have high dry strength, compact readily when thoroughly wetted.

It is well known that sands are compacted to sandstones under heavy loadings. Schmidt and McDonald (1979) concluded from study of quartzose sandstones that, in the early stage of diagenesis, primary porosity is reduced by purely mechanical processes. Uncemented grains slip, rotate, or break to create an arrangement of greater density. At greater depth the remaining porosity is destroyed through pressure solution and cementation. According to Angevine and Turcotte (1983), pressure solution is important under burial depths exceeding 1,500 m, and little porosity remains at depths greater than 3,000 m. Nonetheless, Eberly and Stanley (1978, Appendix I) present core descriptions of al-

**TABLE 4. REPRESENTATIVE CHEMICAL ANALYSES OF PRECIPITATION AND STREAMFLOW IN CENTRAL VALLEY AREA, CALIFORNIA**

| Constituent | 1 11/20/58 | 2 1/25/56 | 3 1/28/53 | 4 1/26/56 | 5 4/21/56 | 6 5/18/70 | 7 5/26/70 | 8 5/13/70 |
|---|---|---|---|---|---|---|---|---|
| Ca (mg/l) | 0.0 | 28 | 209 | 31 | 56 | 5.8 | 0.5 | 25 |
| Mg (mg/l) | 0.2 | 14 | 201 | 14 | 52 | 0.3 | 0.0 | 13 |
| Na (mg/l) | 0.6 | 70 | 554 | 15 | 84 | 1.2 | 0.7 | 55 |
| K (mg/l) | 0.6 | --- | --- | --- | --- | 0.5 | 0.3 | 2.4 |
| $HCO_3$ (mg/l) | 3.0 | 156 | 283 | 139 | 303 | 18 | 3.0 | 100 |
| $SO_4$ (mg/l) | 1.6 | 124 | 1,810 | 36 | 201 | 4.0 | 0.0 | 58 |
| Cl (mg/l) | 0.2 | 12 | 340 | 7 | 44 | 0.4 | 0.8 | 71 |
| TDS (mg/l) | 4.8 | 350 | 3,280 | 195 | 609 | 28 | 7 | 293 |
| pH | 5.6 | --- | --- | --- | --- | 6.6 | 6.1 | 8.1 |

Sources of information and notes:
1. Snow, Spooner Summit, Sierra Nevada, along U.S. Highway 50, Nevada, alt. 2,165 m (Feth and others, 1964).
2. Streamflow, 11,300 l/s, Warthan Creek, T.21S.,R.15E.,Sec.4. High stage of stream-draining area underlain mainly by Tertiary and Quarternary deposits (Davis, 1961, Table 2).
3. Streamflow, 14 l/s, Warthan Creek T.21S.,R.15E.,Sec.7. Base flow (Davis, 1961, Table 2).
4. Streamflow 22,700 l/s, Orestimba Creek, T.7S.,R.7E.,Sec.24. High stage of stream-draining area underlain mainly by Mesozoic marine sediments, (Davis, 1961, Table 2).
5. Streamflow, 16 l/s, Orestimba Creek, T.7S.,R.7E.,Sec.24. Base flow (Davis, 1961, Table 2).
6. Streamflow, 14,800 l/s, East Fork Kaweah River near Three Rivers, California. Snowmelt runoff from Sierra Nevada near San Joaquin Valley (U.S. Geological Survey, 1979).
7. Streamflow, 43,100 l/s, Merced River at Happy Isles Bridge near Yosemite, California. Snowmelt runoff from High Sierra in Yosemite Valley (U.S. Geological Survey, 1979).
8. Streamflow, 70,600 l/s, San Joaquin River near Vernalis, California. High stage outflow from the San Joaquin Valley (U.S. Geological Survey, 1979).

luvial sediments from deeper than 2,500 m in central Arizona, in which "sand" is recorded along with poorly consolidated conglomerate.

In studies of compaction of sediments of areas of land subsidence in central California, Meade (1968) found that increases in effective overburden load in the range between 3.6 and 84 bars had caused an average reduction of 10 to 15 percent of the volume of sediments affected. He notes, however, that the effects of loading are complicated by the effects of other factors on pore volume, particularly grain-size distribution. The factors that directly controlled pore volume in the load range 1.2 to 120 bars included average particle size, degree of sorting, proportion of montmorillonite among clay minerals, and proportion of sodium relative to other cations absorbed by the clay minerals. In general, compaction was inversely related to average particle size and directly related to degree of sorting, proportion of montmorillonite, and proportion of sodium. Thus, the greatest compaction was observed in well-sorted clayey silts with high montmorillonite content, in contrast to well-sorted sand and gravel, which showed very little volume change in the 1.2 to 120 bar loading range. Sands and gravels nonetheless play a major role in compaction phenomena as drains for water expelled from finer sediments as compaction proceeds.

The alluvial deposits of the High Plains aquifer have not been subjected to significantly greater loading than at present. Accordingly, despite their considerable age, they consist generally of unconsolidated clay, silt, sand, and gravel, except where cemented by calcium carbonate.

## CHEMICAL CHARACTER OF GROUND WATERS OF ALLUVIAL FILLS

The principal sources of dissolved substances in the ground waters of the western alluvial valley fills are weathering of mineral matter in the soil and rocks of upland catchments and of the alluvial deposits themselves. Dissolved mineral matter entrained in precipitation plays a minor quantitative role, although near the sea coast, appreciable amounts of sea salts are delivered in rain and fog. The largest sources of recharge to the alluvial basins are infiltration from streams draining well-watered upland areas and infiltration of irrigation water. As a rule of thumb, it can be assumed that where annual precipitation is less than 300 mm in the western basins, little precipitation infiltrates through the soil zone (Blaney, 1933), although recharge can occur through the beds of streams on the valley floors.

When rain falls on the land surface it carries in solution small amounts of the gaseous components of the atmosphere—nitrogen ($N_2$), oxygen ($O_2$), and carbon dioxide $(CO)_2$, as well as other gases including oxides of sulfur and nitrogen. Analysis of snow from the Sierra Nevada, representative of much of the Pacific Coast (Table 4) indicates about 5 mg/l total dissolved solids, mainly bicarbonate ($HCO_3$) and sulfate ($SO_4$) ions. Laney (1967, p. 561) reports that rainwater in the mountains near Tucson, Arizona, contains less than 15 mg/l dissolved solids, of which $SO_4$ made up 90 percent of the anions and calcium was the predominant cation. Presumably, the calcium and sulfate are dissolved from dust particles swept up from the arid lands of the

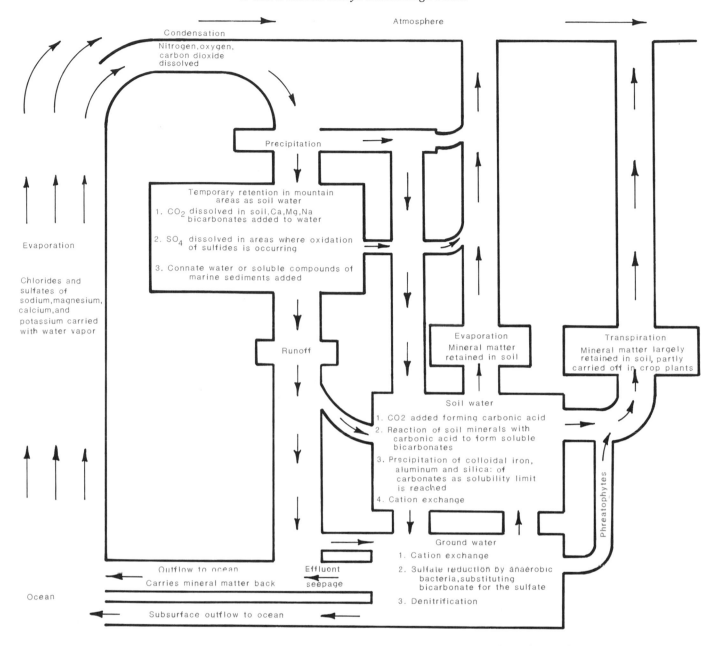

Figure 5. Geochemical cycle of surface and ground waters of western alluvial basins, (after Davis and others, 1959, Fig. 5).

west, although some sulfate no doubt originates from industrial and other artificial sources. Dissolved $CO_2$ dissociates to form the $HCO_3$ ion, which greatly enhances the solvent power of the water. As soon as precipitation enters the soil, it begins to react with mineral and organic matter present. Bacterial respiration normally maintains the soil at 10 times or more the atmospheric $CO_2$ level. Thus, additional $CO_2$ goes into solution in infiltrating water as well as organic acids resulting from decay processes. This weak acid solution is highly effective in dissolving carbonate minerals, and also attacks feldspars and mafic minerals. The latter are converted to aluminum-silicate clay minerals through the

leaching away of calcium (Ca), magnesium (Mg), sodium (Na), and iron (Fe). Potassium (K), which is abundant in orthoclase feldspar, normally is absorbed on clay minerals, and rarely appears in high concentrations in ground water. Sulfate ($SO_4$) and chloride (Cl) present in the soils and rocks are also readily dissolved. These various reactions, as well as those in the saturated zone, are illustrated in Figure 5, which was designed to apply to the San Joaquin Valley; however, the principles apply generally to other western alluvial basins as well. The atmospheric gases $O_2$ and $N_2$ play important roles in plant growth and decay, which maintain the $CO_2$ level of the soil air. The solution that results from reactions in the soil zone generally is a blend of the principal cations Ca, Mg, and Na, and the principal cations $HCO_3$, Cl, and $SO_4$. Iron and aluminum, both sparingly soluble in nonacid waters, generally are retained in the soil zone. Silica ($SiO_2$), the other major component of most rocks, is dissolved at slow rates. Davis (1964) reports that the median silica of ground water is about 17 mg/l and that of stream water is about 14 mg/l.

In the case of the western alluvial basins the cycle of solution and chemical transport commonly takes place in two stages: first in upland catchments; then the ground water discharges to streams, where it is thoroughly reaerated before infiltrating a second time in the alluvial basin.

Rainwater (1962) indicates that in the stream waters that recharge the western basins the prevalent or modal value of total dissolved solids ranges from less than 100 mg/l in the Pacific Mountains to more than 2,500 mg/l in the vicinity of Great Salt Lake and in the Gila–Salt River Basin of central Arizona. With respect to ionic composition, the Pacific mountain streams are of Ca-Mg $HCO_3$ type, while the more concentrated waters are of Na Cl-$SO_4$ type.

Terranes predominating in granitic and metamorphic rocks, carbonate sediments, and marine sandstones or graywackes normally yield Ca-Mg $HCO_3$–type waters. Waters containing a high proportion of $SO_4$ are associated with marine shales, particularly organic shales that contain much sulfide, which when subjected to weathering, oxidizes and dissolves to form $SO_4$ ion in water. Cl in significant amounts nearly always is derived from marine sediments or their secondary products and represents the slow discharge of marine water trapped in sediments at the time of deposition. Even when the rocks exposed at land surface yield $HCO_3$-type waters, springs discharging Cl waters from depth may dominate the chemistry of many streams. The streams of the arid eastern slope of the southern Coast Ranges of California present interesting contrasts within a limited area with respect to range of dissolved-solids content, chemical type, and variability with discharge (Davis, 1961). The ratio $HCO_3/SO_4$ in the low flows of streams was found to be directly proportional to the percentage of the drainage basins underlain by marine rocks of Jurassic and Cretaceous age vis-a-vis Tertiary sediments, which consisted largely of organic shales, and continental deposits derived from them. Some representative analyses of waters recharging the Central Valley of California are given in Table 4.

The mineral matter that makes up alluvial deposits depends of course on the provenance of the materials; however, weathering processes in upland catchments have a homogenizing tendency. Studies of the petrology of the alluvial sediments in the Santa Clara and San Joaquin valleys by Meade (1967) reveal surprising uniformity in many characteristics despite differing provenance. Meade found that sandy sediments of Sierra Nevada provenance (granitic rocks, slates, and schists) cored on the west side of the San Joaquin Valley consisted of about 50 percent quartz and rock fragments, 25+ percent feldspar, 2 to 5 percent biotite, 1 to 2 percent hornblende, and about 15 percent clay minerals, of which 70 percent consisted of montmorillonite.

Sandy sediments on the east side of the valley, closer to the Sierra Nevada, differed mainly in a greater proportion of accessory biotite and hornblende, ranging from 5 to 15 percent. Montmorillonite was the predominate clay mineral there also. Sandy sediments of Coast Range origin consisted of about 25 percent rock fragments (andesite, serpentinite, chert), less than 2 percent mica, and 15 percent clay minerals with montmorillonite predominate; the remaining 60 percent consisted of quartz and feldspar grains.

In alluvial sediments of the Santa Clara Valley, where the source rocks are mainly Mesozoic graywackes, shales, cherts, and mafic and ultramafic rocks, Meade reported that the sands consisted of 30 to 60 percent dark grains (serpentine, red chert, dark mafic rock, lithic sandstone, and shale fragments), less than 1 percent mica, and 10 to 15 percent clay, of which montmorillonite made up 70 percent. The unaccounted-for difference of 20 to 60 percent presumably represented quartz and feldspar. Montmorillonite was the predominate clay mineral in nearly all the sediments studied by Meade regardless of source terrane or depositional environment. The uniform preponderance of montmorillonite in diverse terranes, which included the Sierra Nevada, the Coast Ranges, and the Transverse Ranges, and diverse environments of deposition led Meade (1967, p. C47) to conclude that there was a pervasive but not fully understood influence on the clay mineralogy of the areas studied.

Most of the water infiltrating the land surface of the western alluvial basins typically is in the range of 100 to 1,000 mg/l dissolved solids (Rainwater, 1962). As shown in Figure 5, additional $CO_2$ gas may be taken into solution in the soil zone, and additional solution of Ca, Mg, and Na may occur, mainly through further weathering of feldspars, and highly soluble Cl and $SO_4$ compounds in the soil may be dissolved. Many of these reactions are reversible, and if the soil moisture is concentrated through evaporation, salts may precipitate in the soil zone. Soil moisture in excess of the specific retention can percolate downward to the water table. The chemical character of the shallow ground water corresponds to the long-term average character of water infiltrating through the soil zone.

Figure 6, a transverse geologic and chemical section across the San Joaquin Valley, illustrates a number of the chemical phenomena common to the western alluvial basins. The shallow ground waters in the area of recharge to the east average about 200 mg/l TDS, and Na as percent of total cations averages about 25 percent. This marks a significant increase in TDS as compared with the recharge source, snowmelt runoff typical of the Sierra Nevada from the Kaweah River (Table 4). At greater depth the ground waters are of comparable TDS content but differ notably in proportion of Na. The Na percentage ranges from 70 to 90, evidently due to exchange of Ca and Mg ions in the water for Na ions adsorbed on the abundant montmorillonite of the sediments. In the axis of the valley the shallow ground waters have higher TDS content, about 300 mg/l, and also are high in Na, about 90 percent, reflecting greater solution and cation exchange in ground waters that have migrated over long distances.

On the western end of the section the shallow ground waters

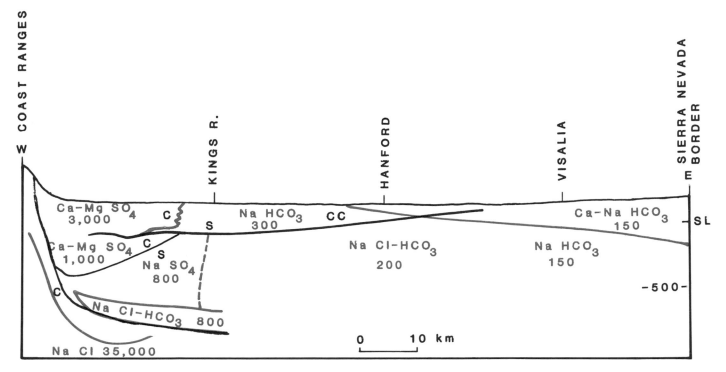

Figure 6. Generalized geochemical section across the San Joaquin Valley, California (modified after Davis and others, 1959, Plate 27, and Miller and others, 1971, Plate 4). Blue lines show boundaries, C = coast range provenance, S = Sierra Nevada provenance, CC = Corcoran Clay member of Tulare Formation. Numbers in blue associated with chemical symbols indicate average total dissolved solids for zone.

are $Ca–MgSO_4$ waters of about 3,000 mg/l TDS content, not unlike the recharging stream (see Warthan Creek, Table 4) that drains a terrane underlain by abundant organic shales. Evidently, little additional mineral matter is dissolved as these waters infiltrate. Two other distinct water types can be seen also on Figure 6. One is a $NaSO_4$ water of about 800 mg/l TDS and about 75 percent Na, which is typical of the principal confined aquifer over broad areas. This water has been ascribed by Davis and Coplen (1984) to mid-Pleistocene recharge to the aquifer. Another distinct water type is a $NaHCO_3–Cl$ water that averages 800 mg/l TDS and 90 percent Na. This water occurs in a zone of organic lacustrine sediments, and its chemical character is ascribed to cation exchange and bacterial reduction of $SO_4$, in which $HCO_3$ was substituted for $SO_4$ ions removed (Davis and Poland, 1955, p. 460).

Similar chemical trends have been observed in the ground waters of other western basins. Laney (1967) described waters of the Tucson Basin, Arizona, and showed a transition from: (1) rainfall of 15 mg/l TDS in which $SO_4$ is the predominant anion, to (2) mountain runoff of less than 100 mg/l TDS with increased proportions of $HCO_3$ and Ca, to (3) shallow ground water in the recharge area of about 350 to 500 mg/l TDS with further increases in the proportions of $HCO_3$ and Ca, to (4) a $Ca–NaHCO_3$ water of about 350 to 500 mg/l TDS, which is characteristic of most of the shallow aquifers of the basin. The transition from (3) to (4) is marked by an increase in Na percentage, ascribed by Laney to cation exchange.

Arnow and others (1970) describe similar transitions in the chemistry of ground waters of the Weber River delta deposits along the eastern border of Great Salt Lake, which were laid down in Pleistocene Lake Bonneville. They recognized three main types of ground water: (1) a widespread $Ca–MgHCO_3$ water containing 150 to 300 mg/l TDS derived by recharge from the Weber River and streams draining the Wasatch Mountain Front, (2) a $NaHCO_3$ water of about 200 to 300 mg/l TDS originating from type 1 but modified by cation exchange, and (3) a $NaCl$ water of 500 to 1,000 mg/l TDS resulting from mixing of type 1 water with residual brines that underlie Great Salt Lake.

To summarize, the chemical history of ground waters of the western alluvial basins typically follows the trends illustrated in Figures 5 and 6. The principal departures occur where readily soluble $SO_4$ and Cl compounds are available for solution in the subsurface environment, in which case waters very high in dissolved solids may evolve. However, even in that event, the development patterns follow similar trends.

The ground waters of the High Plains aquifer, in contrast, are relatively uniform over a wide area in concentration and composition and are mainly of $CaHCO_3$ type. Some 19 percent of the aquifer contains water of less than 250 mg/l, and an additional 62 percent contains water in the range of 200 to 500 mg/l TDS (Gutentag and others, 1984, p. 36). The predominant source of mineralization is solution by infiltrating waters of $CaCO_3$, the cementing material of the sands and gravels. Only 3 percent of the aquifer contains water exceeding 3,000 mg/l TDS, generally where formation waters of underlying Paleozoic and Mesozoic marine deposits have contaminated the High Plains aquifer.

## HYDROLOGIC SIMILARITES AND CONTRASTS

Although separate alluvial basins in the West number in the hundreds (154 basins are identified in California alone; Thomas and Phoenix, 1976, Plate 1), certain trends are common to most of them. Many factors that control alluvial deposition and basin filling operate on a regional or continental scale, such as climatic patterns and their changes over time; isostacy, hence altitude of mountains and depths of basins, which in turn control rates of erosion and deposition; and volcanism, which in much of the West accounts for a major source of the basin fills and to a considerable degree determines the petrology of the alluvial deposits.

The thickness of alluvial fills is fairly consistent in the Pacific mountain system, and in the Basin and Range Province to the extent that it is known. In the Pacific intermontane basins, the fills typically consist of a few hundred to about 1,000 m of late Tertiary and Quaternary alluvium overlying marine Tertiary sediments. The Basin and Range fills occupy grabens initiated at the close of the Laramide Revolution. The well-documented larger basins typically contain 2 to 5 km of continental basin-fill deposits overlying pre-Laramide consolidated rocks. Many of these basin fills include thick sections of saline deposits as well as volcanic flows, tuffs, and breccias.

Depositional regimes are similar throughout the west. All the alluvial basins represent negative structural elements, and have been subjected to similar climatic, tectonic, and volcanic regimes. Just as there is a rough accordance of mountain summit levels, there appears to be a similar accordance of depth of alluvial fills. The main contrasts in regimes depend generally on local factors, such as whether closed-basin drainage or though-flow has prevailed over long time periods. Other such local factors include the prevalence of limestone in the pre-Laramide rocks of southern Nevada—which led to the development of karstic underdrainage of many basins—and very deep, actively subsiding local structural features in the southern San Joaquin Valley and the Imperial Trough.

With respect to physical character, alluvial deposits are noted for pronounced vertical and lateral variability over short distances. However, the local variability is damped by depositional processes, so that on a basin scale it is possible to estimate key hydrogeologic parameters with good reliability. An example of this is specific yield, which varies systematically over broad areas, and can be checked by independent methods. The influence of differing provenance of materials on physical character likewise is smoothed by erosional and depositional processes.

Rates of alluvial deposition, governed mainly by tectonic movements, are surprisingly uniform over wide areas and long time spans. Based on sparse data, there appears to be a bimodal distribution of rates of tectonic subsidence: one applies to local areas of concentrated rapid subsidence with rates ranging from 0.4 to 1.7 mm/yr, and even greater for the Imperial Trough, an active seafloor spreading zone; another applies to broad areas of normal downwarping with rates close to 0.2 mm/yr.

The modification of original materials depends mainly on weathering and compaction and is similar throughout the west. Weathering is governed mainy by climate on a continental scale, and compaction of materials is subject to universal physical laws.

The initial chemical character of ground waters of alluvial basins is subject to considerable variability due to variability in geology of catchment areas, climatic regime, and previous depositional history. Three principal types of water result: $Ca-MgHCO_3$ waters, $SO_4$ waters, and Cl waters. In terranes without deep sources of Cl water or extensive marine shales, or secondary concentrations from these sources, the ground water that develops through normal weathering is of $Ca-MgHCO_3$ type, with less than 500 mg/l TDS content. In most basins, Ca-Mg waters are subject to natural softening through cation exchange, which permits solution of additional mineral matter, and $SO_4$ may be reduced through bacterial action in the presence of organic materials.

The High Plains deposits offer many contrasts with the alluvial fills of the western basins, as well as some similarities. The reason is that the High Plains has been a static environment of weathering and erosion for millions of years since the close of Ogallala deposition in the late Miocene. Other than regional tilting in the Pliocene and faulting on a local scale in a few areas, the High Plains have experienced little tectonic activity, in contrast to the western alluvial basins which have undergone vast and continuing tectonism during the same time period. As a result the alluvial deposits of the High Plains generally are even more uniform and consistent in physical and chemical character than those of the western basins.

# REFERENCES CITED

Allmendinger, R. W., and 7 others, 1983, Cenozoic and Mesozoic structure of the eastern Basin and Range Province, Utah, from COCORP seismic reflection data: Geology, v. 11, p. 532–536.

Anderson, R. E., Zoback, M. L., and Thompson, G. A., 1983, Implications of selected subsurface data on the structural form and evolutions of some basins in the southern Basin and Range Province, Nevada and Utah: Geological Society of America Bulletin, v. 94, p. 1055–1072.

Angevine, C. L., and Turcotte, D. L., 1983, Porosity reduction by pressure solution; A theoretical model for quartz arenites: Geological Society of America Bulletin, v. 94, p. 1129–1134.

Arnow, T., Feth, J. H., and Mower, R. W., 1970, Ground water in the deltas of the Bonneville Basin, the Great Basin, U.S.A.: International Association of Scientific Hydrology, Symposium on the Hydrology of Deltas, v. II, p. 396–407.

Atwater, B. F., and 8 others, 1986, A fan dam for Tulare Lake, California, and its implications for the Wisconsin glacial history of the Sierra Nevada: Geological Society of America Bulletin, v. 97, p. 97–109.

Blaney, H. F., 1933, Rainfall penetration, *in* Ventura County Investigation: California Division of Water Resources Bulletin 46, p. 82–90.

Bull, W. B., 1964, Alluvial fans and near-surface subsidence in western Fresno County, California: U.S. Geological Survey Professional Paper 437-A, 71 p.

—— , 1977, The alluvial fan environment: Progress in Physical Geography, v. 1, p. 222–270.

Cape, C. D., McGeary, S., and Thompson, G. A., 1983, Cenozoic normal faulting and the structure of the Rio Grande Rift near Socorro, New Mexico: Geological Society of America Bulletin, v. 94, p. 3–14.

Columbia–North Pacific Technical Staff, 1970, Water resources, *in* Columbia–North Pacific comprehensive framework study of water and related lands: Vancouver, Washington, Pacific Northwest River Commission, Appendix 5, 1022 p.

Davidson, E. S., 1979, Summary appraisals of the nation's ground-water resources; Lower Colorado region: U.S. Geological Survey Professional Paper 813-R, 23 p.

Davis, G. H., 1961, Geologic control of mineral composition of stream waters of the eastern slope of the southern Coast Ranges, California: U.S. Geological Survey Water-Supply Paper 1535-B, 30 p.

Davis, G. H., and Coplen, T. B., 1984, Pleistocene paleohydrology of the central and western San Joaquin Valley, California: Geological Society of America Abstracts with Programs, v. 16, p. 482.

Davis, G. H., and Green, J. H., 1962, Structural control of interior drainage, southern San Joaquin Valley, California: U.S. Geological Survey Professional Paper 450-D, p. 89–91.

Davis, G. H., and Poland, J. H., 1955, Ground-water conditions in the Mendota–Huron area, Fresno and Kings counties, California: U.S. Geological Survey Water-Supply Paper 1360-G, 180 p.

Davis, G. H., Green, J. H., Olmsted, F. H., and Brown, D. W., 1959, Ground-water conditions and storage capacity in the San Joaquin Valley, California: U.S. Geological Survey Water-Supply Paper 1469, 287 p.

Davis, S. N., 1964, Silica in streams and ground water: American Journal of Science, v. 262, p. 870–891.

de Laveaga, M., 1952, Oil fields of central San Joaquin Valley province: American Association of Petroleum Geologists–Society of Economic Paleontologists and Mineralogists–Society of Exploration Geophysicists guidebook, field trip routes, Joint Annual Meeting, p. 99–103.

Eakin, T. E., Price, D., and Harrill, J. R., 1976, Summary appraisals of the nation's ground-water resources; Great Basin region: U.S. Geological Survey Professional Paper 813-G, 37 p.

Eberly, L. D., and Stanley, T. B., Jr., 1978, Cenozoic stratigraphy and geologic history of southwestern Arizona: Geological Society America Bulletin, v. 89, p. 921–940.

Eckis, R., and Gross, P.L.K., 1934, South Coastal Basin investigation; Geology and ground-water storage capacity of valley fill: California Department of Public Works, Division of Water Resources Bulletin 45, 279 p.

Feth, J. H., Rogers, S. M., and Roberson, C. E., 1964, Chemical composition of snow in the Sierra Nevada and other areas: U.S. Geological Survey Water-Supply Paper 1535-J, 39 p.

Foxworthy, B. L., 1979, Summary appraisals of the nation's ground-water resources; Pacific Northwest region: U.S. Geological Survey Professional Paper 813-S, 39 p.

Foxworthy, B. L., and Hill, M., 1982, Volcanic eruptions of 1980 at Mount St. Helens: U.S. Geological Survey Professional Paper 1249, 125 p.

Gutentag, E. D., Heimes, F. J., Krothe, N. C., Luckey, R. R., and Weeks, J. B., 1984, Geohydrology of the High Plains aquifer in parts of Colorado, Kansas, Nebraska, New Mexico, Oklahoma, South Dakota, Texas, and Wyoming: U.S. Geological Survey Professional Paper 1400-B, 63 p.

Hely, A. G., Hughes, G. W., and Irelan, B., 1966, Hydrologic regime of the Salton Sea, California: U.S. Geological Survey Professional Paper 486-C, 31 p.

Ives, P. C., Levin, B., Oman, C. L., and Rubin, M., 1967, U.S. Geological Survey radiocarbon dates IX: American Journal Science Radiocarbon Supplement, v. 9, p. 505–529.

Johnson, A. I., Moston, R. P., and Morris, D. A., 1968, Physical and hydrologic properties of water-bearing deposits in subsiding areas in central California: U.S. Geological Survey Professional Paper 497-A, 71 p.

Laney, R. L., 1967, Investigation of the geochemistry of water in a semi-arid basin in Arizona, U.S.A.: International Association of Hydrogeologists Memoir, v. VIII, p. 559–565.

Lattman, L. H., 1973, Calcium carbonate cementation of alluvial fans in southern Nevada: Geological Society America Bulletin, v. 84, p. 3013–3028.

Lofgren, B. E., 1974, Measuring ground movement in geothermal areas in Imperial Valley, California: Proceedings, Conference on Research for the Development of Geothermal Energy Resources, Pasadena, California, p. 128–133.

Marchand, D. E., and Allwardt, A., 1981, Late Cenozoic stratigraphic units, northeastern San Joaquin Valley, California: U.S. Geological Survey Bulletin 1470, 70 p.

Meade, R. H., 1967, Petrology of sediments underlying areas of land subsidence in central California: U.S. Geological Survey Professional Paper 497-C, 83 p.

—— , 1968, Compaction of sediments underlying areas of land subsidence in central California: U.S. Geological Survey Professional Paper 497-D, 39 p.

Meidev, T., and Howard, J. H., 1979, An update of tectonics and geothermal resource magnitude of the Salton Sea geothermal resource: Transactions Geothermal Resources Council, Reno, Nevada, September, p. 445–448.

Miller, R. E., Green, J. H., and Davis, G. H., 1971, Geology of the compacting deposits of the Los Banos–Kettleman City subsidence area, California: U.S. Geological Survey Professional Paper 497-E, 46 p.

Olmsted, F. H., and Davis, G. H., 1961, Geologic features and ground-water storage capacity of the Sacramento Valley, California: U.S. Geological Survey Water-Supply Paper 1497, 241 p.

Piper, A. M., Gale, H. S., Thomas, H. E., and Robinson, T. W., 1939, Geology and ground-water hydrology of the Mokelumne area, California: U.S. Geological Survey Water-Supply Paper 780, 230 p.

Poland, J. F., Piper, A. M., and others, 1956, Ground-water geology of the coastal zone, Long Beach–Santa Anna area, California: U.S. Geological Survey Water-Supply Paper 1109, 162 p.

Poland, J. F., Garrett, A. A., and Sinott, A., 1959, Geology, hydrology, and chemical character of ground waters in the Torrance–Santa Monica area, California: U.S. Geological Survey Water-Supply Paper 1461, 425 p.

Poland, J. F., and others, 1984, Guidebook to studies of land subsidence due to ground-water withdrawal: Paris, UNESCO Studies and Reports in Hydrology, no. 40, 305 p.

Rainwater, F. H., 1962, Stream composition of the conterminous United States: U.S. Geological Survey Hydrologic Investigations Atlas HA 61.

Rubin, M., and Alexander, C., 1960, U.S. Geological Survey radiocarbon dates:

American Journal Science Radiocarbon Supplement, v. 2, p. 156.

Schmidt, V., and McDonald, D. A., 1979, The role of secondary porosity in the course of sandstone diagenesis, *in* Scholle, P. A., and Schluger, P. R., eds., Aspects of diagenesis: Society Economic Paleontologists and Mineralogists Special Publication 26, p. 175–207.

Shepard, F. P., 1964, Sea-level changes in the past 6,000 years; Possible archeological significance: Science, v. 143, p. 574–576.

Sherman, I., 1953, Flocculent structure of suspended sediment in Lake Mead: Transactions of the American Geophysical Union, v. 34, p. 394–406.

Thomas, H. E., and Phoenix, D. A., 1976, Summary appraisals of the nation's ground-water resources; California region: U.S. Geological Survey Professional Paper 813-E, 51 p.

U.S. Geological Survey, 1979, Quality of surface waters of the United States; Pt. 11, Pacific Slope basins in California: U.S. Geological Survey Water-Supply Paper 2159.

Upson, J. E., and Thomasson, H. G., Jr., 1951, Geology and water resources of the Santa Ynez River Basin, Santa Barbara County, California: U.S. Geological Survey Water-Supply Paper 1107, 194 p.

West, S. W., and Broadhurst, W. L., 1975, Summary appraisals of the nation's ground-water resources; Rio Grande region: U.S. Geological Survey Professional Paper 813-D, 39 p.

MANUSCRIPT ACCEPTED BY THE SOCIETY APRIL 17, 1987

## Chapter 35

# *Glacial deposits*

**D. A. Stephenson***
*Dames & Moore, 7500 North Dreamy Draw Drive, Phoenix, Arizona 85020*
**A. H. Fleming and D. M. Mickelson**
*Department of Geology and Geophysics, University of Wisconsin at Madison, Madison, Wisconsin 53706*

## INTRODUCTION

Glacial landscapes involve a complex mixture of sediment types that often have different hydrogeologic properties. The single most important factor affecting hydrogeologic characteristics of glacial deposits is the diversity of sediments and the resultant numerous lithologic discontinuities. In this chapter, the genesis, spatial arrangement, and geologic and hydrogeologic properties of these sediment types are discussed.

For the purposes of this chapter, the following terms are defined:

**Diamicton.** An increasingly popular descriptive term for poorly sorted, poorly stratified sediment. Diamicton includes sediments of varying genesis such as till, mudflow, and other mass-movement deposits. Till is a genetic term for sediment deposited directly by glacier ice without significant resorting (Dreimanis, 1982). Because earlier publications typically use "till" as the descriptive term for diamicton, we will continue that usage here.

**Glaciolacustrine deposits.** A term describing all lake deposits associated with glacial activity. These deposits may be sand, silt, or clay, and are mostly well sorted. Stratification may be well developed to absent.

**Glaciofluvial deposits.** A term describing sediments deposited by streams in association with glacier ice. Related terms are: **outwash,** a glaciofluvial sediment deposited by streams flowing away from the glacier; and **ice-contact stratified deposits,** stream sediments deposited in contact with glacier ice. Included in this category are land-forms such as eskers, ice-contact terraces, and deltas. Normally, outwash and ice-contact stratified deposits are composed of sand or of sand and gravel.

Eolian deposits, such as loess and dune sand, are not glacial deposits *per se* but are commonly present in the glacial landscape. Table 1 lists some generalized physical properties of glacial deposits.

Glacial deposits, primarily of Quaternary age, cover about 13 million square kilometers of North America (Flint and Sharp, 1971) and frequently constitute a major ground-water source for domestic, commercial, industrial, and agricultural purposes. Because these deposits are at or near the ground surface, they are used for waste disposal, are easily affected by spills of toxic and hazardous materials, and are subject to agricultural and other nonpoint pollutants. Many studies of glacial deposits have been published, but only a few compilations are available that address the overall character and distribution of deposits in a glacial landscape and the hydrogeologic properties of individual glacial materials. The following discussion focuses on the geologic and geochemical processes of aquifer and aquitard genesis in glacial deposits, with particular reference to midwestern North America.

## SEDIMENTOLOGY OF GLACIAL DEPOSITS

The character of sediments in a glacial landscape is a function of three things: the lithology and geochemical properties of the sediment source, the nature and distance of sediment transport, and the mode of sediment deposition. For example, source terrains underlain by carbonate rocks produce sediments rich in carbonate while source terrains with silica-rich rocks generally produce glacial deposits reflecting that composition.

Texture or grain-size distribution of windblown sediment, of glaciolacustrine sediment, and of glaciofluvial deposits is controlled mostly by the environment of deposition. For glaciofluvial deposits, texture and sorting can change radically over short distances if stream energy during deposition was variable in time or space. For till and of related diamictons, texture is controlled both by source and by mode of deposition. Regional textures are commonly controlled by the position of shale outcrops or lake basins that supplied fine-grained sediment. For example, till of clayey texture is present in the flow-path direction of glacial ice in all the Great Lakes basins and most other lake basins (Mickelson and others, 1983). Sandy till is present mostly in areas where till was derived from Precambrian bedrock or younger sandstones. However, even in areas of clayey till located south of the Great Lakes, till or other diamictons of sandy texture can be found. The reason is that glaciers eroded most of the pre-existing lake sedi-

---

*Present address: Harding Lawson Associates, 2929 N. 44th St., Suite 102, Phoenix, Arizona 85018.

Stephenson, D. A., Fleming, A. H., and Mickelson, D. M., 1988, Glacial deposits, *in* Back, W., Rosenshein, J. S., and Seaber, P. R., eds., Hydrogeology: Boulder, Colorado, Geological Society of America, The Geology of North America, v. O-2.

*D. A. Stephenson and Others*

TABLE 1. GENERALIZED PROPERTIES OF GLACIAL DEPOSITS

| | Diamicton | | Sand and Gravel | | |
|---|---|---|---|---|---|
| Characteristic | Till | Proglacial and Supraglacial Diamicton | Ice-contact Stratified Deposits | Pitted Outwash | Outwash |
| Sorting | Poor | Poor | Moderate | Good | Excellent |
| Stratification | None or poor | None or poor | Locally collapsed | Locally collapsed | Well developed |
| Surface form (all may be fluvially dissected) | Flat, hummocky, or stream-lined | Hummocky | Hummocky or ridges | Gently sloping with depressions | Gently sloping |
| Where deposited | Beneath ice | Ice margin | Ice margin or beneath | In front of or on margin in valley, apron or plain | In front of margin in valley, apron or plain |
| Grain size | Sandy to clayey, uniform | Sandy to clayey, variable | Variable, usually coarse | Variable, gravel near source, sand away | Uniform gravel near source, sand away |
| Rounding of clasts | Angular | Moderate | Variable | Fairly well rounded | Well rounded |
| Compaction | Fairly compact, often over con-solidated | Loose, usually not over consolidated | Loose, granular | Loose, granular | Loose, granular |
| Jointing | Often well developed, vertical | Often closely spaced, fissile | Not common | Not common | Not common |
| Lateral continuity | Well developed | Poorly developed | Poorly developed | Well developed | Well developed |
| Aquifer potential | Poor | Poor to fair | Fair | Good | Excellent |

ment and derived much of their basal load from bedrock composed of sandstone, limestone, and crystalline rocks.

Mode of deposition also affects texture, the presence of fractures, consolidation, and other properties. If debris is carried near the base of the ice and deposited as basal till, little or no change in texture occurs during the process of deposition. If, however, debris is carried high enough in the ice so that it melts out and is deposited from the ice surface, various mass-wasting processes and some washing may affect the material, producing other diamictons that are often more variable than basal till, and perhaps more sandy (Fig. 1).

In the discussion above, the differences in till and related sediments that can occur during a single glacial advance have been described. It is important to recognize, however, that in most glaciated areas, several advances have taken place. The record of glacial advances varies from place to place because of processes within the glacier. For example, valley glaciers commonly downcut their valleys during each advance and erode

away older deposits in the valley bottom. Thus, only the thin deposits are preserved high on valley walls or moraines and terraces that were outside the limit of later advances. The ice sheet that covered the flatter terrain of much of central and eastern North America left a complex record of earlier advances. In general, central areas of the ice sheet experienced severe erosion during each advance. Although isolated valleys contain a stratigraphic record, much of the Canadian Shield has thin, very young till over bedrock. In the other parts of the ice sheet, however, deposition generally prevailed and several stratigraphic units are commonly present.

## DOCUMENTATION OF GLACIAL DEPOSITION

As described above, deposits in a glaciated region are rarely uniform and isotropic. Ground-water studies require information about the vertical and horizontal distribution and hydraulic characteristics of sediments in the landscape. Traditional mapping of

Quaternary glacial deposits, although providing some ground-water resource information, has often fallen short of providing details necessary to make sound interpretations in ground-water studies. Part of the reason is that many studies have concentrated on the distribution of materials or landforms at the surface without providing detailed and widespread subsurface information. Other studies do not provide sufficient information about the sediments and depositional environments necessary to make reasonable interpretations of the continuity and hydrogeologic characteristics of units, nor do these reports contain quantitative geotechnical or hydraulic data. Among approaches to deal with this problem, two appear to be developing along parallel lines; both appear valid at different map scales and different levels of interpretation.

In the central United States, where the stratigraphy is fairly well known, lithostratigraphic units have been defined and described (e.g., in Illinois: Willman and Frye, 1970; in Minnesota and North Dakota: Harris and others, 1974; in Iowa: Hallberg, 1980; in Wisconsin: Mickelson and others, 1984). Normally, units exposed at the surface are mapped and also shown in cross sections. In some cases, a more useful approach is that of stack-mapping—the mapping of units that include an indication of all of the materials in the section present to a certain depth (Kempton and Cartwright, 1984). This approach has been used successfully at the county scale (Berg and others, 1984a) and at the state level as a basis for maps showing susceptibility to contamination (Berg and others, 1984b). Moreover, this approach provides the most useful information for the purposes of ground-water studies, partly because stratigraphic extrapolation is done by the stratigrapher instead of by a map user, as in the case of widely spaced cross sections. However, this method of evaluating glacial deposits is expensive and time-consuming.

Another approach, which is particularly useful for more generalized studies or in situations where glacial deposits are fairly thin, is land-systems mapping. The land-systems approach consists of mapping areas that have similar landforms and sediment types as units that can be further subdivided. For example, a subglacial land system can be subdivided into several categories, including drumlin areas. A drumlin area can then be subdivided into units based on sediment type, slope position, or other features. Intended to be a hierarchical, integrative resource inventory system, land-systems mapping has been used for many years in various parts of the world (Stewart, 1968; Haantjens, 1972). Several reviews of general land-systems approaches are given, for example, in Wright (1972), and Cook and Doornkamp (1974). Eyles (1984a, 1984b), Eyles and Menzies (1984), and Paul (1983) have provided a detailed outline of glacial land systems in low-relief and high-relief areas.

Knowledge of the sediments in the glacial setting is essential to the interpretation of the hydrogeology of glaciated terrains. Figure 2 shows a hypothetical mountain and valley landscape that has been glaciated. Except in valley bottoms, the surficial cover is thin and composed of weathering residuum and isolated patches of till from earlier advances. Alluvial fans and talus cones

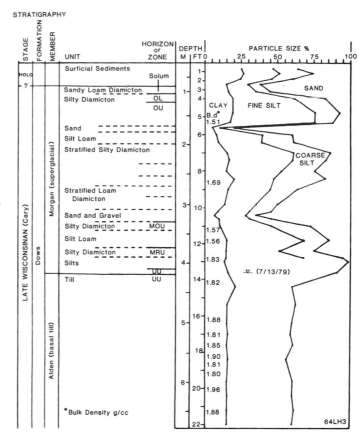

Figure 1. Stratigraphy and grain-size data from a site in northern Iowa where supraglacial deposits overlie basal till. Note uniformity of till in contrast to overlying diamicton. (From Lutenegger and others, 1984).

are common near the base of slopes, and these extend into the basin fill that is typically better-sorted sand or sand and gravel. In some cases, these talus cones are active features, but periglacial conditions during glaciation probably greatly enhanced these processes, producing thick accumulations of mass-movement debris near the valley walls. Lateral moraines along valley sides extend across the valley floor as end moraines, showing in the subsurface as lenses of till and boulders.

In areas of continental glaciation, particularly in areas of clayey or silty till, major ground-water sources are commonly in valleys, some of which are entirely filled with deposits of the youngest glacial advance (Fig. 3). Thus, depth to bedrock and the stratigraphy of deposits above the bedrock greatly influence considerations for water supply and for the disposal of waste that must be isolated from a lower Quaternary aquifer or older bedrock aquifer. Because in many areas lithostratigraphic units can be defined and traced, and ranges in properties can be obtained, a lithostratigraphic approach to defining the hydrogeologic setting is often appropriate.

These lithostratigraphic units should, however, be inter-

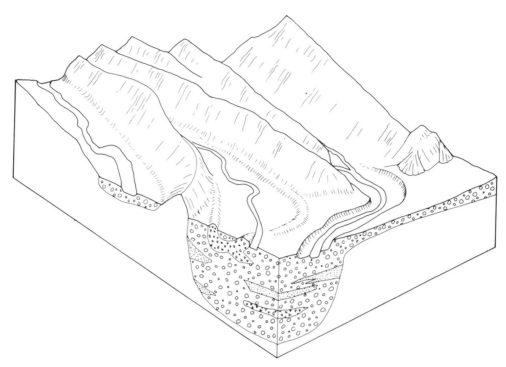

Figure 2. Sketch of a hypothetical mountain landscape showing the distribution of glacial sediments and related deposits and landforms. Triangle symbol represents till. Circles and dots represent fluvial sediments.

preted with genesis in mind. Till and related diamictons have different hydrogeologic properties than do water-laid deposits. Even within deposits traditionally called till, differences in environment of deposition create differences in characteristics of the sediment (Lutenegger and others, 1984). Sediments deposited in the subglacial environment commonly show streamlined forms at the surface (Fig. 3). This drumlin topography normally indicates uniform basal till at the surface (although not necessarily at depth), probably with few sand lenses and similar discontinuities. However, joints might be common, depending on the texture of the till (Grisak and Cherry, 1975; McGown and Radwan, 1975; McGown and Derbyshire, 1977; Connell, 1984). Uniform basal till might also be present as nearly flat plains that have little original relief. In areas that have deposits older than Wisconsinan age, the surface has commonly been fluvially dissected, but glacial deposits in the subsurface still greatly affect hydrologic conditions (Sharp, 1984). In areas where bedrock is near the surface, for example in much of the Canadian Shield, till is commonly so thin that use of the surficial deposits for either waste disposal or water supply is impossible (Fig. 3).

End moraines are ridges that are roughly parallel to the former ice margin, that formed in different ways, and therefore, that are variable in composition. In places where ice evidently had a wet bed during till deposition (e.g., in Illinois, Indiana, Iowa, and Ohio, along the margins of the Great Lakes, and in southern Ontario), the till in the moraines is generally fairly uni-

form, and probably is mostly basal till (Mickelson and others, 1983). In other areas, particularly farther north in parts of Wisconsin, Minnesota, North Dakota, Michigan, and in many parts of New England and Canada, end moraines are much more complex and generally have higher-relief hummocky topography. In these cases, basal till is commonly intermixed with silt, sand, gravel, supraglacially derived diamictons, and sediment melted out in situ from higher in the ice.

Stream deposits vary in character with distance from the ice margin at the time of deposition. Commonly, sediments deposited near the ice margin are poorly rounded, less stratified, and less well sorted than those deposited away from the ice. Ice-contact stratified sediments generally contain collapse of bedding and interbedded lenses or amorphous masses of diamicton. Because of their discontinuity and often limited extent, these deposits provide limited opportunity for high capacity wells, although frequently used for domestic supplies. Sand and gravel deposited by streams flowing away from the glaciated area are generally better stratified, better sorted, and more continuous then that deposited near the ice and, thus, provide greater potential for water supply.

## HYDROGEOLOGY OF GLACIATED TERRAINS

Because many glacial terrains are composed of heterogeneous mixtures of sediments, complex hydrogeologic environments result. Hydraulic properties of the deposits show consid-

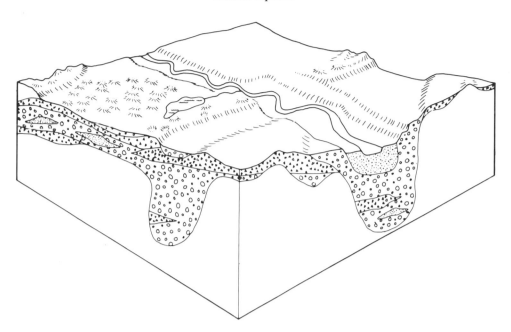

Figure 3. Sketch of a hypothetical midcontinent landscape showing the distribution of glacial sediments and related deposits and landforms in association with other deposits. Triangle symbol represents till. Circles and dots represent fluvial sediment.

erable variability, both between different glacial terrains and for a particular type of deposit. The diversity of these deposits results in numerous lithologic discontinuities and is perhaps the single most important factor affecting the overall hydrogeology of glaciated terrains. In an attempt to examine the hydrogeology of glacial terrains, the remainder of this chapter addresses: (1) The hydraulic and hydrogeochemical properties of specific types of glacial materials; and (2) The flow systems and hydrostratigraphy that characterize different glacial depositional systems as a result of the intercalation of various glacial sediments.

Glacial deposits may be grouped into two general hydraulic categories: (1) materials that transmit water readily, such as outwash and ice-contact stratified deposits and sandy diamicton; and (2) materials that have relatively low hydraulic conductivity, including many tills, loess, and lacustrine sediment (mainly silt and clay). The majority of glacial aquifers consists of the first type, and because these deposits have been widely exploited for water supplies, much is known about their hydraulic characteristics. Hydrogeologic data for the second group are relatively scarce, although their low hydraulic conductivity has recently made these materials attractive for waste-disposal sites. Therefore, much of the available data has come from investigations related to waste disposal. The following discussion of hydraulic properties is directed chiefly at materials of the second category—low hydraulic conductivity—because: (1) the hydraulic properties of outwash are fairly well known and are similar to those of alluvium, which is discussed in detail elsewhere in this volume; (see chapters by Heath, Lennox and others, Cartwright and others,

Cherry and Farvolden, Rosenshein, and Sharp); and (2) of the need to evaluate systematically the hydrogeology of till and related deposits within a wide range of glacial terranes.

*Hydraulic Properties*

Data on the hydraulic properties of glacial deposits are usually provided as follows:

**Hydraulic conductivity,** the most widely reported hydrologic property of glacial deposits, shows much variation among the different types of deposits. Intergranular porosities, where reported, show much less variation and are commonly in the range of 25 to 40 percent for most glacial deposits (see Morris and Johnson, 1967; Grisak and Cherry, 1975).

**Transmissivity** is commonly used to describe the amount of ground water flowing in primarily confined sand and gravel aquifers.

**Values of specific storage** have in some cases been reported for confining layers of till and lacustrine deposits that overlie outwash aquifers (Norris, 1959; Grisak and Cherry, 1975).

Reported values of hydraulic conductivity are frequently calculated by different methods. Field-determined values, from aquifer tests or single-hole methods, are often much greater than laboratory-determined values for the same material. Likewise, different laboratory methods may produce differing values of hydraulic conductivity. Where hydraulic properties are compared in this chapter, an attempt was made to differentiate between the results of different methods. However, some of the scatter in the data is due to the different procedures used.

**TABLE 2. HYDRAULIC CONDUCTIVITY RANGES OF GLACIAL DEPOSITS**

| | Unweathered (m/day) | Weathered (m/day) | Fractured (m/day) |
|---|---|---|---|
| Basal till | $10^{-2}$ to $10^{-6}$ | $10^{-1}$ to $10^{-4}$ | 1.0 to $10^{-4}$ |
| Supraglacial till* | 1.0 to $10^{-4}$ | 1.0 to $10^{-4}$ | 1.0 to $10^{-4}$ |
| Lacustrine silt and clay | $10^{-4}$ to $10^{-8}$ | n.a. | $10^{-3}$ to $10^{-6}$ |
| Loess | 1.0 to $10^{-6}$ | $10^{-2}$ to $10^{-5}$ | n.a. |
| Outwash | 102 to $10^{-2}$ | n.a. | n.a. |

*Includes some poorly stratified sediments and debris flow.

The hydraulic properties of glacial deposits show a systematic variation that can be related to differences between depositional environments and, to some extent, postdepositional modification. Two types of hydraulic conductivity are recognized: (1) Primary hydraulic conductivity, which is a property of the predominant material itself that is in turn the result of its genetic history; and (2) Secondary hydraulic conductivity, which arises through postdepositional modification, such as fracturing and weathering. These secondary features, considered with the intercalation of two (or more) lithologies that have entirely different primary hydraulic conductivities, result in a bulk hydraulic conductivity that may be different from the primary hydraulic conductivity and from the conductivity that ultimately predominates in the landscape.

To separate quantitatively these two kinds of hydraulic conductivity is often problematic, but some attempt has been made to do so in the following discussion.

The values given in Table 2 are compiled from many sources in the literature. Despite the large number of sources and methodologies represented, distinct differences occur among the ranges of hydraulic conductivity for each type of material. These distinctions are especially noticeable in the values for till, both for the genetic categories (basal and supraglacial) as well as for primary (unweathered) and secondary (weathered, fractured) hydraulic conductivities.

In the case of primary hydraulic conductivity, several genetic controls appear to influence the range of values. A particularly strong relationship is evident between laboratory-determined hydraulic conductivity and grain-size distribution, as shown in Figure 4. The major influence here appears to be the amount of clay present. Haefeli (1972) has suggested that as little as 2 percent clay in otherwise clean sand can reduce hydraulic conductivity by two to four orders of magnitude. In Figure 4, a clay content of between 15 and 20 percent appears to mark a threshold above which conductivities are uniformly low. At clay contents below this threshold, a major factor appears to be the sand-silt ratio (higher conductivities associated with a larger ratio).

Also evident is that some conductivity values for till differ by as much as two orders of magnitude, even though the samples reflected similar grain-size distributions, indicating the effect of other factors. One such factor is suggested by a comparison of the conductivity ranges of basal till and supraglacial diamicton, as given in Table 2. Although both types of till show similar spreads of about four orders of magnitude, the values for supraglacial diamicton cluster approximately about two orders above those for basal till. This difference stems directly from contrasts between the subglacial and supraglacial environments. Deposition of till in the subglacial environment often occurs under great confining pressure, due to the weight of the overlying ice. Thus, many basal tills are characterized by a high degree of consolidation and have fine-grained types often referred to as "overconsolidated" (Sladen and Wrigley, 1983; McGown and others, 1975). The product of this process is a dense, compact basal till in which voids are poorly interconnected, resulting in reduced hydraulic conductivity. The degree of consolidation can be influenced by a number of variables, including the ratio of basal shear stress to normal stress, basal porewater pressure, the amount of reworking due to basal meltwater, and the amount of overburden that has been removed by erosion (Boulton and Paul, 1976; Sladen and Wrigley, 1983). Because these factors commonly vary from place to place, not all basal till has the same degree of consolidation; therefore, the effect on conductivity can also be expected to vary.

Supraglacial deposition, on the other hand, occurs at the ice margin, and the deposits are not subject to intense compaction. Thus, pore spaces can be expected to be better connected than those in basal till. Moreover, the supraglacial environment is typified by abundant rainfall and meltwater, which can have an effect that ranges from minor washing of fines from the sediment to complete reworking and sorting (Paul, 1983). Both the lack of consolidation and better sorting tend to increase the hydraulic conductivity of supraglacial diamictons as compared to basal till. However, this sorting process also tends to be irregularly localized in the supraglacial environment, producing a complex mixture of till, sand and gravel, and debris-flow sediment that may have little lithologic continuity. This condition has major implications for

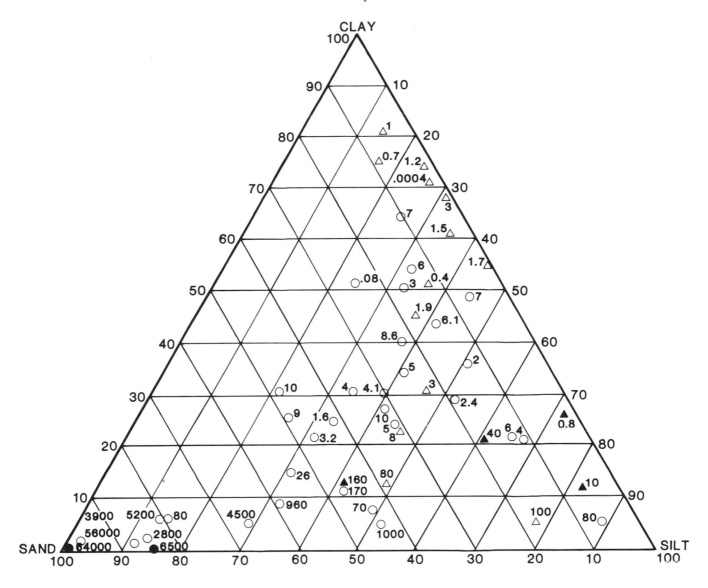

Figure 4. Relationship between hydraulic conductivity (in units of $1 \times 10^{-5}$ m/d) and glacial-deposit type.

| Sources | Value | | | |
|---|---|---|---|---|
| | Till ○ | Loess ▲ | Lacustrine Deposits △ | Outwash ● |
| Gordon and Huebner (1983) | 1.6, 2, 3, 4, 5, 6, 7, 8.6, 10, 26, 70, 170, 1000, 4500 | | 0.7, 1, 1.2, 1.5, 1.7, 1.9, 3, 5, 100 | 6500, 64000 |
| Morris and Johnson (1967) | 0.08, 2.4, 3.2, 4.1, 9, 80, 960, 2800, 3900, 5200, 56000 | 0.8, 40, 160 | 0.0004, 0.4, 8, 80 | |
| Prudic (1982) | 6 | | | |
| Herzog and Morse (1984) | 10 | | | |

**TABLE 3. COMPARISON OF LABORATORY- AND FIELD-DETERMINED VALUES OF HYDRAULIC CONDUCTIVITY**

| Author | Material | Hydraulic Conductivity (m/d) | |
| --- | --- | --- | --- |
| | | Lab | Field |
| Gordon and Huebner (1983) | Clayey till | $3 \times 10^{-5}$ | $1 \times 10^{-2}$ |
| | Clayey till | $3 \times 10^{-5}$ | $3 \times 10^{-3}$ |
| | Lacustrine clay | $7 \times 10^{-6}$ | $2 \times 10^{-4}$ |
| Sharp (1984) | Clayey till | $10^{-5} - 10^{-6}$ | $10^{-1} - 10^{-2}$ |
| Herzog and Morse (1984) | Basal till | $10^{-5} - 10^{-6}$ | $10^{-3} - 10^{-4}$ |
| | Ablation till | $10^{-3} - 10^{-4}$ | $10^{-1} - 10^{-2}$ |
| Grisak and Cherry (1975) | Clay-loam till | $10^{-5} - 10^{-6}$ | $10^{-4}$ |
| | Lacustrine clay | $10^{-7}$ | $10^{-4}$ |
| Hendry (1982) | Sandy-clay till | $10^{-4} - 10^{-5}$ | $10^{-2} - 10^{-4}$ |
| Prudic (1982) | Clayey till | $10^{-5}$ | $10^{-4} - 10^{-5}$ |
| Grisak and others (1976) | Clay-loam till/ lacustrine sediment | $10^{-5} - 10^{-6}$ | $10^{-2} - 10^{-4}$ |

the large-scale hydrogeology of the supraglacial land system, as will be discussed later.

Secondary hydraulic conductivity in glacial deposits is typically difficult to detect using laboratory methods alone, simply because of the larger scale involved than that encompassed by conventional laboratory specimens. Causes of secondary hydraulic conductivity include jointing, weathering, and ground-water erosion. The magnitude of secondary hydraulic conductivity is commonly two or three orders greater than the laboratory-determined primary hydraulic conductivity of a given glacial sediment. Thus, the presence of secondary hydraulic conductivity is frequently identified by comparing laboratory and field tests (Table 3).

For many years, fractures have been considered to influence the movement of ground water in fine-grained glacial till and lacustrine sediment (Williams and Farvolden, 1967). However, the idea that fractures are common to many of these deposits has emerged mostly within the last decade, due in part to problems associated with waste-disposal sites (e.g., Herzog and Morse, 1984; McGown and Radwan, 1975; Grisak and others, 1976). Fractures appear to be particularly prevalent in fine-grained basal till and other deposits that have been overridden by ice (Connell, 1984; Grisak and Cherry, 1975), but fractures are not limited to fine-grained materials. Knutson (1971), for example, reports that a strong fissility is frequently developed in the sandy basal till in the Midwest and New England. Moreover, fracture genesis may not be limited to the subglacial environment. Connell (1984) has summarized several mechanisms that may generate fractures in a variety of glacial environments.

Data in Tables 2 and 3 show that a well-connected fracture network can considerably elevate the hydraulic conductivities of otherwise low permeability materials. Sharp (1984) has described localized fracture-flow systems in pre-Illinoian till from Missouri

where fractures yield sufficient water to meet domestic and some agricultural needs. Similar fracture systems are responsible for the good soil drainage found in some clay-rich tills and lacustrine sediments in parts of Alberta (Hendry, 1982). Fractures can produce strongly anisotropic permeability (Snow, 1969).

Grisak and Cherry (1975) also found fractured tills to be characterized by two specific storage values: intergranular specific storage ranges from $1 \times 10^{-2}$ to $9 \times 10^{-3} \text{m}^{-1}$, whereas specific storage associated with the fracture network is on the order of $1 \times 10^{-5} \text{m}^{-1}$. The distinction between the two types of specific storage became evident during water-level observations in a till confining bed overlying a sand and gravel aquifer (Fig. 5). During pumping, changes in total head are rapidly transmitted through the fractures, producing rapid responses in piezometers that intersect open fractures. On the other hand, piezometers screened in unfractured till blocks respond much more slowly, because water is released by the delayed process of intergranular consolidation (Grisak and Cherry, 1975). Interbedded lenses of sand and gravel probably produce a similar effect on specific storage.

Many glacial terrains, particularly in the north-central United States, exhibit strong weathering profiles that may extend to depths of 10 m or more. Because of the unlithified nature of the parent material, weathering changes and pedological structures can be difficult to recognize at depth (Follmer, 1984). Moreover, glacial sequences sometimes show the effects of more than one pedogenic episode. The effects may be superimposed on one particular stratigraphic level, or may be associated with different horizons, or both. Follmer (1984) emphasized the potential complexity of the weathering profile in glacial terrains, and the variable effect that this can have on ground-water flow and geochemistry, especially where the soil is developed on several glacial deposits.

Weathering results from a complex interaction of chemical

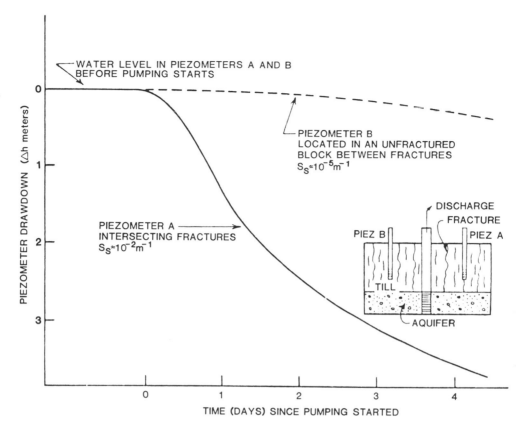

Figure 5. Schematic diagram showing water-level behavior typical of piezometers in fractured and unfractured till overlying a pumped aquifer (from Grisak and others, 1976).

and mechanical processes, the intensity of which varies both spatially and temporally. This interaction is reflected in the wide range of conductivity values associated with weathered glacial deposits (Table 2). Where direct comparisons have been made, however, weathered deposits tend to show greater conductivity than their unweathered counterparts (Herzog and Morse, 1984; Lloyd, 1983). Much of the increased hydraulic conductivity is probably due to the disaggregation of material produced by surficial mechanical weathering and to changes in soil structure due to pedogenesis and fracturing (Bouma, 1973). Some studies suggest that hydraulic conductivity decreases with depth in weathering profiles of till (Lloyd, 1983; Sharp, 1984). In cases where the translocation of clay is an important weathering process, claypans and other argillic horizons can significantly impede vertical conductivity and infiltration (Jamison and Peters, 1967; Quigley and Ogunbadego, 1976). Likewise, the clay-rich B-horizon of the Sangamon paleosol has been reported to behave as a low hydraulic conductivity layer in the glacial deposits of Illinois (Follmer, 1982).

Many glacial sequences are capped by a blanket of loess, derived by wind from rock flour deposited by meltwater in major outwash valleys. Loess consists chiefly of silt-sized grains of quartz and feldspar that typically have a very loose packing arrangement when unsaturated. Thus, porosity of unsaturated loess can be high, reaching 50 or 60 percent in some places. This condition might suggest that recharge rates should be significantly greater in areas mantled by thick loess deposits. However, this is not so. The high porosity of loess will collapse if saturated. Since the porosity of deposits in thick loess can still be measured as high, one could conclude that loess has never been saturated. Recharge is probable in thick loess areas only through prominent joints and root channels (S. N. Davis, personal communication, 1985).

Because loess is frequently at the surface, it is also a principal parent material of soils in glaciated regions. Infiltration tests indicate that weathering and pedogenesis significantly decrease the hydraulic conductivity of loess (Prill, 1977; Van Bavel and others, 1968). The cause of the decrease is probably due to several factors, including compaction, leaching of carbonates, and the filling of pores by clay minerals derived from the weathering of feldspar (Prill, 1977; Morrow, 1964).

Weathering may also affect the specific storage of glacial deposits. Many midwestern U.S. tills contain abundant fine-grained carbonates. Over time, these carbonates are leached from the till, resulting in losses as high as 50 percent of the original volume. Specific storage is probably reduced as volume is lost

TABLE 4. COMPARISON OF AVERAGE LINEAR GROUND-WATER VELOCITIES FOR MATRIX MATERIAL
AND ZONES OF SECONDARY PERMEABILITY IN SOME CLAYEY TILL AND LACUSTRINE SEDIMENT

| Author | Locality | Intergranular Velocity | | Secondary Velocity | |
|---|---|---|---|---|---|
| | | cm/yr | m/10,000 yr | cm/yr | m/10,000 yr |
| Desaulniers and others (1981) | Ontario, Canada | 0.04 to 0.46 | 13 to 26 | n.d. | n.d. |
| Grisak and others (1976) | Interior Plains, Canada | 0.0005 to 0.05 | 0.05 to 5 | 4 to $4 \times 10^3$ | $4 \times 10^2$ to $4 \times 10^{6}$* |
| Gordon and Huebner (1983) | Milwaukee County, Wisconsin | 0.2 to 0.3 | 20 to 30 | $6 \times 10^3$ | $6 \times 10^{5}$† |
| | Winnebago County, Wisconsin | 0.7 to 1.1 | 70 to 110 | 750 to 950 | $7.5 \times 10^4$ to $3 \times 10^{5}$† |
| | Outagamie County, Wisconsin | 0.06 to 1 | 6 to 100 | 20 to 3000 | $2 \times 10^3$ to $3 \times 10^{5}$* |

*Fracture permeability.

†Sand and gravel lenses.

and consolidation occurs. Sharp (1984) has suggested that time-dependent consolidation may be a factor in comparatively low specific storage values for deeply weathered till in Missouri.

Irregularly shaped bodies of sand and gravel are commonly present in most glacial sequences and range in thickness from less than a centimeter to many meters. Where laterally extensive, these bodies can transmit significant quantities of water (Lloyd, 1983). More commonly, they are discontinuous, discrete bodies that act as local high hydraulic conductivity zones amidst materials of lower hydraulic conductivity. Such deposits create a higher bulk hydraulic conductivity because their effects are beyond the scale of that measured in laboratory samples. Like fractures, the presence, size, and frequency of high-conductivity lenses play a critical role in the ability of fine-grained glacial deposits to yield water for small users, as well as in attempts to isolate toxic wastes from the general hydrogeologic environment (Norris, 1961; Engquist and others, 1978; Gordon and Huebner, 1983).

A particularly striking example of secondary hydraulic conductivity comes from southeastern Sweden, where permeable zones have developed through weathering processes in fissile layers in sandy basal till and in thin lenses of outwash (Knutson, 1971). These permeable zones control the flow of water to a majority of springs and domestic wells that were developed in deposits of both the subglacial and supraglacial land systems. The velocities are sufficiently high to cause significant ground-water erosion, marked by cones of sand and gravel at their outlets (Knutson, 1971).

The importance of secondary hydraulic conductivity becomes quite apparent in a comparison of average linear ground-water velocities (Table 4). These data suggest that ground-water

flow in fine-grained glacial deposits is effectively controlled by the secondary hydraulic conductivity. The extremely low average ground-water velocities of the till matrix imply that the intergranular porewater at depth should be very old. Desaulniers and others (1981) obtained radiocarbon dates from lacustrine clays in southern Ontario that indicate ground-water ages in excess of 8 ka. Ground-water ages of 9 to 10 ka have also been found in clayey till from two sites near Superior, Wisconsin (Bradbury and others, 1985). Ground-water in the underlying sandstone bedrock near these sites has been dated at 16 ka. Similar dates are probably typical of porewater at depth from other clay-rich glacial deposits, suggesting that such water may have originated at the time the till was deposited (Grisak and others, 1976). On the other hand, ground water within the upper few meters of these deposits, and within fractures, shows modern oxygen isotope ratios and tritium values, indicative of the much greater velocities in weathered and fractured zones (Grisak and others, 1976; Bradbury and others, 1985; Desaulniers and others, 1981).

## Hydrogeochemical Properties

Because the chemical composition of ground-water in glacial deposits can be quite variable, generalizations are difficult. The hydrogeochemistry at any given site is a function of several related variables, including lithology, depth, hydraulic conductivity, residence time, and weathering. Ground water within basal till typically shows higher total dissolved solids than in other adjacent deposits (Eyles and Sladen, 1981; Lloyd, 1983; Grisak and others, 1976). This hydrogeochemistry is thought to be due to the relatively low permeability that enhances the slow movement of ground water from the matrix into fractures and inhibits oxidation of pyrite. Pyritic shales and carbonates are both important constituents of many tills; thus, many pore waters are of the calcium-sulphate or calcium-bicarbonate-sulphate facies (Spears and Reeves, 1975; Rozkowsky, 1967; Lloyd, 1983; Sharp, 1984). Ground water from till in the midwestern U.S. and in Canada is commonly super-saturated with respect to calcite and dolomite (Grisak and others, 1976). Small, but widespread amounts of chloride in ground water from till of the Interior Plains Region have been ascribed both to localized concentrations of chloride salts in incorporated shale and to the upward movement of brines from bedrock due to ice loading (Cherry, 1972; Grisak and others, 1976). High iron contents have occasionally been reported in ground water from older, deeply weathered till (Drew and Sharp, 1981; Gilkeson and others, 1977).

Grisak and others (1976) suggest that diffusion is the predominant transport process in deposits of low hydraulic conductivity. Where these materials are cut by fractures or other zones of relatively high ground-water velocity, strong ionic gradients exist between the till blocks and the open fractures. The flux across these boundaries can be significant and may produce locally saline conditions in the fractures (Grisak and others, 1976). The ion exchange capacities of many clay-rich glacial deposits can provide an important measure of attenuation for contaminants emanating from waste-disposal sites (Grisak and others, 1976). Finally, hydrogeochemical characteristics have provided significant information concerning recharge and the hydraulic relationships between glacial deposits and underlying bedrock aquifers in some areas (Walton, 1970; Sage and Lloyd, 1978; Lloyd, 1980).

### Hydrostratigraphy

The degree of lithological heterogeneity is the major determinant affecting the hydrostratigraphy of glacial terrains. The most productive aquifers are associated with thick, widespread blankets of outwash that were deposited in front of advancing or retreating ice sheets (Fig. 3). In many places, this outwash filled sizable preglacial river valleys, commonly to depths of over 100 m. These buried valley aquifers tend to yield large volumes of ground water and may constitute the only significant source of ground water in many glaciated regions (Norris and White, 1961). The glacial stratigraphy in these valleys can be quite complex, however, especially where multiple advances or a fluctuating ice margin affected deposition (Norris and Spieker, 1966; Winter, 1973; Stephenson, 1967).

In many areas, much of this proglacial outwash is overlain by extensive sheets of basal till (Fig. 3). Fine-grained types commonly produce confined conditions in the underlying aquifers and can greatly impede recharge (Penman, 1950; Lloyd and others, 1981). Where the till is coarse-grained and relatively homogeneous over a wide area, small quantities of water might be yielded to domestic users and small farms (Olsson, 1974; Engquist and others, 1978). Moreover, abundant basal meltwater beneath temperate glaciers can produce longitudinally extensive deposits of sand and gravel within the till. The above observations suggest that, despite their typically low hydraulic conductivity,

sequences of subglacial deposits have appreciable hydrogeologic continuity. Even where the till is predominantly fine-grained and is considered an aquitard, considerable recharge can occur through time to underlying aquifers by vertical movement (Norris, 1959; Walton, 1960; Lloyd, 1983).

In contrast, deposits of the supraglacial environment commonly exhibit great internal diversity and, except for some large ice-contact stratified deposits, have little lithologic continuity. Although many of the better-sorted deposits have high primary permeabilities, they are of small extent and discontinuous, and, thus, will act only as ground-water reservoirs, not conduits. Therefore the supraglacial deposits might be expected to exhibit relatively low bulk permeability, characterized by many highly localized, poorly connected aquifers.

Little direct evidence is available to test these conclusions. In southern Sweden, the yields of wells and springs situated in drumlin terrane are about one to two orders of magnitude greater than those of wells located in hummocky moraine, which is presumably supraglacial (Knutson, 1971; Engquist and others, 1978). Knutson (1971) has attributed this difference to observable contrasts between the stratigraphic continuity of the two types of terrane, with some of the basal lithologies being traceable for several kilometers. However, many of the supraglacial deposits are rarely greater than a few meters in extent. Tracer experiments conducted by Knutson (1971) in both kinds of terrane also suggest greater hydraulic continuity in the subglacial deposits.

## CONCLUSIONS

Glacial deposits in North America are primarily Quaternary in age, cover about 13 million km$^2$, and constitute both a sediment source for ground water and, increasingly, a sink for waste disposal. The term "glacial deposits" includes:

- Till: sediment deposited directly by glacier ice without significant resorting;
- Glaciofluvial: sediments deposited by streams in association with glacier ice;
- Glaciolacustrine: lake sediments deposited in association with glacial activity; and
- Loess: not strictly a glacial deposit but often present in a glacial landscape.

All glacial deposits comprise silt, sand, or sand and gravel in addition to clay fractions and cobble/boulder fractions.

Hydrogeologic aspects of glacial deposits have not been overly emphasized to date in North American geological research. This oversight has developed because large-volume water supplies frequently have been pumped from bedrock aquifers that have received most study. In more recent years, as practices developed to utilize glacial deposits for waste disposal and increasingly for smaller-scale water supplies, the paucity of information on relationships between glacial geology and ground-water occurrence and movement has been apparent. Programs are being developed to produce quantitative information on sediment, hydraulic, and index properties of glacial deposits.

Continentally glaciated areas are of major concern when

researching relationships between ground water and geology. However, whether in alpine or continental glaciated terrains, the most characteristic geologic feature is the diversity in lithology, both laterally and vertically. Definition of the three-dimensional configuration of materials that have high values of hydraulic conductivity is one of the major challenges facing a glacial-deposit stratigrapher.

The character of glacial sediments is a function of (1) lithology and geochemical properties at the sediment source; (2) the nature and distance of sediment transport; and (3) the mode of sediment deposition. In a given geographic area, differences in till and other glacial sediments may be the result of single or several glacial episodes. Thus, aquifer systems are frequently of more concern than individual aquifers or aquitards.

A knowledge of sedimentary environments in the glacial setting is essential to interpretation of the hydrogeology of glacial terrains. Ground-water studies should include information on vertical and horizontal distribution and hydraulic characteristics of sediments in a glacial landscape. Traditional mapping of glacial deposits has often not included hydrogeologic aspects. New mapping approaches address the issue of heterogeneity and complexity of glacial deposits as related to hydrogeology. These mapping techniques include "stack mapping" and land-systems mapping.

Glacial deposits are grouped into two general hydraulic categories: (1) those that transmit ground water readily, such as outwash and ice-contact stratified deposits; and (2) those that have low hydraulic conductivity, such as till. Complicating transmissive properties, especially of number 2 above, is the presence of fractures. The concept that fractures are common in many unconsolidated glacial deposits has received recent emphasis because of interest in contaminant movement through glacial deposits.

The most widely reported aquifer characteristic of glacial deposits is hydraulic conductivity. The method of evaluating this hydraulic property must be carefully selected. The hydraulic properties of glacial deposits show a systematic variation related to both depositional environment and postdepositional modification.

The chemical composition and age of ground water in glacial deposits are frequently quite variable over small distances. Ground water in fine-grained deposits, such as basal till, generally contains higher total dissolved solids and is older than ground water in coarse-grained materials and fractures that have smaller residence times for the water. The sometimes extreme variability in chemical composition of glacial-deposit ground water makes generalizations exceedingly difficult. This condition is especially true in areas that received multiple glaciations, especially when each advance was from a different source.

Innumerable opportunities for research exist for the person interested in hydrogeology or hydrogeochemistry of glacial deposits. A few of these opportunities include:

1. Development of new generations of vertical variability mapping techniques that emphasize center-of-gravity of high or low hydraulic conductivity zones;

2. Comparison of natural ground-water flow (time-independent) versus aquifer system flow (time-dependent) characteristics;

3. Vadose-zone geologic characteristics in glaciated areas as they affect contaminant distribution and movement or ground-water recharge; and

4. Hydrogeochemical characteristics of naturally occurring ground water in glacial deposits as a mechanism in site selection for hazardous-waste disposal. Brackish ground water in glacial deposits does occur in some portions of the Interior Plains of Canada and the United States. The origin of sulfate as the dominant anion is not known.

The importance of the hydrogeology of glacial deposits in the northern U.S. and in Canada is in contrast to the paucity of data that quantitatively define this geologic environment. This is a paradox in an age where ground-water quality is almost a household concern. Glaciated portions of North America include many of the large population centers, and in many of these population centers, glacial deposits serve as both water supply and waste-disposal medium. Many opportunities exist for increased research, which the authors hope will lead to better understanding of the hydraulic and geochemical characteristics of glacial deposits.

## REFERENCES CITED

Berg, R. C., Kempton, J. P., and Stecyk, A. N., 1984a, Geology for planning in Boone and Winnebago counties: Illinois State Geological Survey Circular 531, p. 69.

Berg, R. C., Kempton, J. P., and Cartwright, K., 1984b, Potential for contamination of shallow aquifers in Illinois: Illinois State Geological Survey Circular 532, p. 30.

Boulton, G. S., and Paul, M. A., 1976, The influence of genetic processes on some geotechnical properties of glacial tills: Quarterly Journal of Engineering Geology, v. 9, no. 3, p. 159–194.

Bouma, J., 1973, Relationships between soil structure characteristics and hydraulic conductivity: Soil Science Society of America Special Publication 5, p. 77–105.

Bradbury, K. R., 1982, Hydrogeologic relationships between Green Bay of Lake Michigan and onshore aquifers in Door County, Wisconsin [Ph.D. thesis]: Madison, University of Wisconsin, 287 p.

Bradbury, K. R., Desaulniers, D. S., Connell, D. E., and Hennings, R. G., 1985, Ground-water movement through clayey till, northwestern Wisconsin, U.S.A.: Proceedings Seventeenth International Congress of the International Association of Hydrogeologists, Tucson, Arizona, v. 17, no. 1, p. 405–416.

Cherry, J. A., 1972, Geochemical processes in shallow ground water flow systems in five areas in southern Manitoba, Canada: Proceedings 24th International Geological Congress, Montreal, Section 11, p. 208–221.

Connell, D. E., 1984, Distribution, characteristics, and genesis of joints in fine-grained till and lacustrine sediment, eastern and northwestern Wisconsin [M.S. thesis]: Madison, University of Wisconsin, 452 p.

Cook, R., and Dooncamp, J. C., 1974, Geomorphology and environmental management: London, Clarendon Press, p. 326–351.

Desaulniers, D. E., Cherry, J. A., and Fritz, P., 1981, Origin, age, and movement of porewater in argillaceous Quaternary deposits at four sites in southwestern Ontario: Journal of Hydrology, v. 50, p. 231–257.

Dreimanis, A. D., 1982, Work Group 1, Genetic classification of tills and criteria for their differentiation; Progress report on activities 1977–1982, and definitions of glacigenic terms, *in* Schluchter, C., ed., Commission on genesis and lithology of Quaternary deposits report on activities 1977–1982: Zurich, International Quaternary Association, p. 70.

Drew, T. A., and Sharp, J. M., 1981, Permeability and hydrology study for Ralls and Monroe counties: Missouri Department of Natural Resources Report, p. 68.

Engquist, P., Olsson, T., and Svensson, T., 1978, Pumping and recovery tests in wells sunk in till: Nordic Hydrology Conference, Hanasaari, Finland, p. 134–142.

Eyles, N., 1984a, Glacial geology; A landsystems approach, *in* Eyles, N., ed., Glacial geology; An introduction for engineers and earth scientists: Oxford, Pergamon Press, p. 1–18.

——, 1984b, The glaciated valley landsystem, *in* Eyles, N., ed., Glacial geology; An introduction for engineers and earth scientists: Oxford, Pergamon Press, p. 91–110.

Eyles, N., and Menzies, J., 1984, The subglacial landsystem, *in* Eyles, N., ed., Glacial geology; An introduction for engineers and earth scientists: Oxford, Pergamon Press, p 19–70.

Eyles, N., and Sladen, J. A., 1981, Stratigraphy and geotechnical properties of weathered lodgement tills in Northumberland, England: Quarterly Journal of Engineering Geology, v. 14, p. 129–141.

Flint, R. F., and Sharp, J. M., 1971, Glacial and Quaternary geology: New York, John Wiley and Sons, p. 892.

Follmer, L. R., 1984, Soil; An uncertain medium for waste disposal: Proceedings of the Seventh Annual Madison Waste Conference, Madison, University of Wisconsin-Madison Extension, p. 296–311.

——, 1982, The geomorphology of the Sangamon surface; Its spatial and temporal attributes, *in* Thorn, C. E., ed., Space and time in geomorphology: London, Allen and Unwin, International Series no. 12, p. 117–146.

Gilkeson, R. H., Cartwright, K., Follmer, L. R., and Johnson, T. M., 1977, Contribution of surficial deposits, bedrock and industrial wastes to certain trace elements in ground water: Proceedings of the Fifteenth Annual Engineering Geology and Soils Engineering Symposium, Pocatello, Idaho State University, p. 17–38.

Gordon, M. E., and Huebner, P. M., 1983, An evaluation of the performance of zone of saturation landfills in Wisconsin: Proceedings of the Sixth Annual Madison Waste Conference, Madison, University of Wisconsin-Madison Extension, p. 23–53.

Grisak, G. E., and Cherry, J. A., 1975, Hydrological characteristics and response of fractured till and clay confining a shallow aquifer: Canadian Geotechnical Journal, v. 12, no. 23, p. 23–43.

Grisak, G. E., Cherry, J. A., Vonhof, J. A., and Bleumle, J. P., 1976, Hydrogeological and hydrochemical properties of fractured tills in the interior plains region, *in* Glacial till: Royal Society of Canada Special Publication 12, p. 304–335.

Haantjens, H. A., 1972, Lands of the Aitape-Ambunti area, Papua New Guinea: Land Research Series 30, p. 248.

Haefeli, C. J., 1972, Ground-water inflow into Lake Ontario from the Canadian side: Ottawa, Department of the Environment, Inland Waters Branch, Science Series no. 9, 101 p.

Hallberg, G. R., 1980, Pleistocene stratigraphy in east-central Iowa: Iowa Geological Survey Technical Information Series no. 10, p. 168.

Harris, K. L., Moran, S. R., and Clayton, L., 1974, Late Quaternary stratigraphic nomenclature, Red River valley, North Dakota and Minnesota: North Dakota Geological Survey Miscellaneous Series 52, p. 47.

Hendry, J. M., 1982, Hydraulic conductivity of a glacial till in Alberta: Ground Water, v. 20, no. 2, p. 162.

Herzog, B. L., and Morse, W. J., 1984, A comparison of laboratory and field determined values of hydraulic conductivity at a waste disposal site: Proceedings of the Seventh Annual Madison Waste Conference, Madison, University of Wisconsin-Madison Extension, p. 30–52.

Hughes, G. M., Schleider, J. A., and Cartwright, K., 1976, Hydrogeology of solid waste disposal sites in northeastern Illinois: Illinois State Geological Survey Environmental Geology Notes no. 80, 24 p.

Jamison, V. C., and Peters, D. B., 1967, Slope length of claypan soil affects runoff: Water Resources Research, v. 3, p. 471–480.

Kempton, J. P., and Cartwright, K., 1984, Three-dimensional geologic mapping; A basis for hydrogeologic and land-use evaluations: Bulletin of the Association of Engineering Geologists, v. 21, no. 3, p. 317–335.

Knutson, G., 1971, Studies of ground-water flow in till soils: Geologiska foreningens, Stockholm, Forhandlingak, Pt. 3, no. 546, p. 533–573.

Lloyd, J. W., 1980, The influence of Pleistocene deposits on the hydrogeology of major British aquifers: Journal of the Institute of Water Engineers, v. 34, p. 346–356.

——, 1983, Hydrogeological investigations in glaciated terrains, *in* Eyles, N., ed., Glacial geology; An introduction for engineers and earth scientists: Oxford, Pergamon Press, p. 349–368.

Lloyd, J. W., Harker D., and Baxendale, R. A., 1981, Recharge mechanisms and ground water flow in the chalk and drift deposits of southern East Anglia: Quarterly Journal of Engineering Geology, v. 14, p. 87–96.

Lutenegger, A. J., Kemmis, T. J., and Hallberg, G. R., 1984, Origin and properties of glacial till and diamictons, *in* Young, R. N., ed., Geological environment and soil properties: American Society of Civil Engineers, p. 310–331.

McGowen, A., and Derbyshire, E., 1977, Genetic influences on the properties of tills: Quarterly Journal of Engineering Geology, v. 10, p. 389–410.

McGowen, A., and Radwan, A. M., 1975, The presence and influence of fissures in the boulder clays of west-central Scotland: Canadian Geotechnical Journal, v. 12, p. 84–97.

McGown, A., Anderson, W. F., and Radwan, A. M., 1975, Geotechnical properties of the tills in west-central Scotland, *in* The Midlands Soil Mechanics and Foundation Engineering Society, The engineering behavior of glacial materials: Norwien, Geoabstracts, p. 81–91.

Mickelson, D. M., Clayton, L., Fullerton, D. S., and Borns, H. W., Jr., 1983, The late Wisconsin glacial record of the Laurentide ice sheet in the United States, *in* Wright, H. E., ed., Late Quaternary environments of the United States, Volume 1, The late Pleistocene: Minneapolis, University of Minnesota Press, p. 3–37.

Mickelson, D. M., Clayton, L., Baker, R. W., Mode, W. N., and Schneider, A. F., 1984, Pleistocene stratigraphic units of Wisconsin: Wisconsin Geological and Natural History Survey Miscellaneous Paper 84-1, 99 p.

Morris, D. A., and Johnson, A. I., 1967, Summary of hydrologic and physical properties of rock and soil materials as analyzed by the hydrologic laboratory of the U.S. Geological Survey, 1948–1960: U.S. Geological Survey Water-Supply Paper 1939-D, p. 39.

Morrow, J. H., 1964, Permeability measurements on loess from the Vicksburg area: Journal of the Mississippi Academy of Science, v. 20, 166 p.

Norris, S. E., 1959, Vertical leakage through till as a source of recharge to a buried-valley aquifer at Dayton, Ohio: Ohio Department of Natural Resources, Division of Water Technical Report 2, 16 p.

——, 1961, Hydrogeology of a spring in a glacial terrain near Ashland, Ohio: U.S. Geological Survey Water-Supply Paper 1619-A, 17 p.

Norris, S. E., and Spieker, A. M., 1966, Ground-water resources of the Dayton area, Ohio: U.S. Geological Survey Water-Supply Paper 1808, 167 p.

Norris, S. E., and White, G. W., 1961, Hydrologic significance of buried valleys in glacial drift: U.S. Geological Survey Professional Paper 424-B, p. 34–35.

Olsson, T., 1974, Ground water in till soils: Striae, v. 4, p. 13–16.

Paul, M. A., 1983, The supraglacial landsystem, *in* Eyles, N., ed., Glacial geology; An introduction for engineers and earth scientists: Oxford, Pergamon Press, p. 71–90.

Penman, H. L., 1950, The water balance of the Stour catchment area: Journal of the Institute of Water Engineers, v. 4, p. 457–469.

Prill, R. C., 1977, Movement of moisture in the unsaturated zone in a loess-mantled area, southwestern Kansas: U.S. Geological Survey Professional Paper 1021, p. 1–21.

Prudic, D. E., 1982, Hydraulic conductivity of a fine-grained till, Cattaraugus County, New York: Ground Water, v. 20, no. 2, p. 194–204.

Quigley, R. M., and Ogunbadego, T. A., 1976, Till geology, mineralogy, and geotechnical behavior, Sarnia, Ontario, *in* Legget, R. F., ed., Glacial till: Royal Society of Canada Special Publication 12, p. 336–345.

Rozkowsky, A., 1967, The origin of hydrochemical patterns in hummocky moraine: Canadian Journal of Earth Science, v. 4, p. 1065–1092.

Sage, R. C., and Lloyd, J. W., 1978, The influence of drift deposits on the Triassic sandstone aquifer as inferred by hydrochemistry: Quarterly Journal of Engineering Geology, v. 11, p. 209–218.

Sharp, J. M., 1984, Hydrogeologic characteristics of shallow glacial drift aquifers in dissected till plains (north-central Missouri): Ground Water, v. 22, no. 6, p. 683–689.

Sladen, J. A., and Wrigley, W., 1983, Geotechnical properties of lodgement till, *in* Eyles, N., ed., Glacial geology; An introduction for engineers and earth scientists: Oxford, Pergamon Press, p. 184–212.

Snow, D. T., 1969, Anisotropic permeability of fractured media: Water Resources Research, v. 5, p. 1273–1289.

Spears, D. A., and Reeves, M. J., 1975, The influence of superficial deposits on ground-water quality in the Vale of York: Quarterly Journal of Engineering Geology, v. 8, p. 255–270.

Stephenson, D. A., 1967, Hydrogeology of glacial deposits of the Mahomet Bedrock Valley in east-central Illinois: Illinois Geological Survey Circular 409, 51 p.

Stewart, G. A., ed., 1968, Land evaluation: Melbourne, Macmillan, p. 392.

Van Bavel, C. H., Brust, K. J., and Stirk, G. B., 1968, Hydraulic properties of a clay-loam soil and the field management of water uptake by roots: Soil Science Society of America Proceedings, v. 32, p. 317–321.

Walton, W. C., 1960, Leaky artesian aquifer conditions in Illinois: Illinois State Water Survey Report of Investigations 39, 27 p.

——, 1970, Ground-water resource evaluation: New York, McGraw-Hill, p. 664.

Williams, R. E., and Farvolden, R. N., 1967, The influence of joints on the movement of ground water through glacial till: Journal of Hydrology, v. 5, p. 163–170.

Willman, H. B., and Frye, J. C., 1970, Pleistocene stratigraphy of Illinois: Illinois State Geological Survey Bulletin 94, 204 p.

Winter, T. C., 1973, Hydrogeology of glacial drift, Mesabi Iron Range, northeastern Minnesota: U.S. Geological Survey Water-Supply Paper 2029-A, 23 p.

Wright, R. L., 1972, Principles in a geomorphological approach to land classification: Zeitschrift für Geomorphologie N. F., v. 16, no. 4, p. 351–371.

Manuscript Accepted by the Society March 10, 1987

The Geology of North America
Vol. O-2, Hydrogeology
The Geological Society of America, 1988

# Chapter 36

# *Coastal Plain deposits*

**James A. Miller**
*U.S. Geological Survey, Federal Office Building, 75 Spring Street, S.W., Suite 772, Atlanta, Georgia 30303*

## INTRODUCTION

Coastal Plain sedimentary rocks underlie an area of about 1.5 million km$^2$, extending from Cape Cod, Massachusetts, southward to the Florida Keys, then westward and southward around the periphery of the Gulf of Mexico and across the Yucatan peninsula where they form part of the western boundary of the Caribbean Sea (Fig. 1). The Coastal Plain varies in width from a few kilometers at its northern end and in parts of east-central Mexico to almost a thousand kilometers in the Mississippi embayment and along the axis of the Florida peninsula. Coastal Plain rocks range in age from Jurassic to Holocene and generally thicken toward the modern shoreline from a featheredge at their updip limit. Onshore, Coastal Plain rocks are more than 6 km thick; offshore, they are at least 7 km thick, and possibly are as thick as 18 km (Jones, 1975). The greatest thicknesses are in southern Louisiana and contiguous offshore areas, where sediments carried by the Mississippi River have been deposited as a massive delta complex over geologic time. Clastic Coastal Plain rocks are generally semiconsolidated to unconsolidated, but where carbonate rocks comprise these strata, they may be well consolidated.

Coastal Plain rocks are generally easily eroded. Where they consist of clastic material, streams are gently to moderately incised, and low, rolling hills are developed. Where carbonate rocks occur near land surface, the topography is usually flat to slightly rolling, streams are widely spaced, and local to subregional sinkhole and other types of karst topography have developed. Near the coast, flat marshes and swamps, some of which are extensive, are found at elevations of only 1 or 2 m above sea level. In upland areas, where relief is highest, elevations range up to 100 to 250 m above sea level. Locally, in northeastern Mexico between latitude 23° and 25° N, mountains that result from domal uplift and igneous intrusions reach elevations of 1,500 m (Murray, 1961).

Extensive, slightly dissected plains are characteristic of large sections of the inland parts of the Coastal Plain. Along the coastline, barrier islands are common. Locally, such islands have sealed off the mouths of streams and are separated from the main body of the continent by extensive water bodies such as the Pamlico and Albermarle Sounds of North Carolina or Galveston and Matagorda Bays in Texas. The bird-foot delta constructed in south Louisiana by the Mississippi River is also a prominent physiographic feature of the coastline. Drowned river systems such as the Chesapeake and Delaware Bays locally cut deeply into, or completely cross, the Coastal Plain.

The sediments underlying the Coastal Plain are all water-laid and were deposited during a series of transgressions and regressions of the sea. They consist primarily of sand, silt, and clay derived from adjacent uplands in the continental interior and transported into the region by coastward-flowing streams. Local gravel beds are present in upland areas. These sediments were deposited in fluvial, marginal marine, and shallow marine environments whose exact locations varied widely, depending on the relative positions of land masses, shoreline, and streams at a given point in geologic time. Complex interbedding and variations in lithology result from these fluctuating depositional conditions. Consequently, the geometry of sand and clay bodies varies greatly. Sands, for example, may occur as fan-shaped deposits if laid down in a fluvial setting; as elongate, sinuous, "shoe-string" bodies if they are part of a delta or if they represent an ancient river channel; as coarse, thick, well-sorted, linear accumulations if they are a dune or beach ridge complex; or as thin, sheetlike deposits if they represent a marine shelf environment. Unique geologic conditions in or adjacent to the basin of deposition are reflected as subregional lithologic variations in clastic rocks. Examples are the extensive lignite that occurs in lower Tertiary rocks from western Alabama through Texas, reflecting deltaic conditions, and the bentonite and tuffaceous material that occurs in many of the fine-grained Coastal Plain units in Mexico, resulting from volcanic activity in the nearby Sierra de los Volcanes.

Carbonate rocks are less widespread than clastic sediments in the Coastal Plain, and generally were deposited in places where clastic materials were not supplied, because of either low relief or great distance to a clastic source. Extensive carbonate platforms underlie the Florida and Yucatan peninsulas, where Tertiary to Quaternary limestone of various original textures has been partly altered to dolomite and locally is interbedded with gypsum and anhydrite. A thick chalk sequence of Late Cretaceous age extends from central Alabama into northeastern Mexico and was laid down in quiet, shallow, open marine waters.

Miller, J. A., 1988, Coastal Plain deposits, *in* Back, W., Rosenshein, J. S., and Seaber, P. R., eds., Hydrogeology: Boulder, Colorado, Geological Society of America, The Geology of North America, v. O-2.

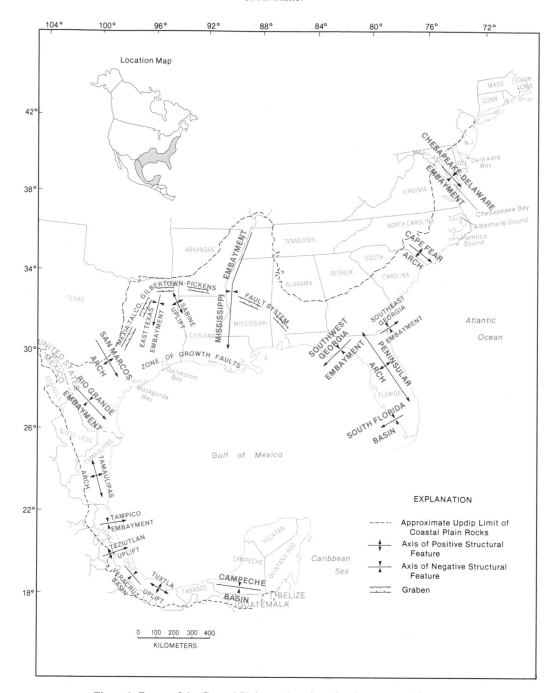

Figure 1. Extent of the Coastal Plain and location of major structural features.

Salt beds of probable early Mesozoic age underlie much of the central part of the U.S. Gulf Coast and part of eastern Vera Cruz and Tabasco, Mexico. Piercement domes and salt ridges, created from salt flowage triggered by uneven sediment loading on the mother salt bed, form important local geologic structures in these areas.

The permeability of clastic Coastal Plain sediments is determined primarily by the texture and degree of sorting of the sediment, which are in turn determined chiefly by the energy conditions in a given depositional environment. Strong currents, whether fluvial or marine, effectively winnow away fine materials and allow only coarse sand or gravel to be laid down. Such coarse sediments, especially if they are well sorted, comprise excellent aquifers. In quiet waters, silt and clay are deposited that form confining units whose effectiveness varies depending on the thickness and continuity of the fine-grained material. The original texture and composition of clastic Coastal Plain rocks are usually preserved because these rocks are, in general, not greatly affected by diagenetic processes. Accordingly, their original hydraulic characteristics are also preserved more or less intact.

By contrast, carbonate rocks in the Coastal Plain may have undergone considerable diagenesis, so that their present texture and mineralogy is quite different from the original. Like clastic rocks, the original texture of limestones is determined largely by the energy conditions that exist where the limestone is deposited. For example, grainstones occur where energy is high, such as in nearshore swash areas or in tidal channels; micrites are deposited in quiet water that may be lagoonal or open marine; and so on. After deposition, however, the hydraulic character of the original limestone may be altered drastically by diagenetic processes, chiefly dolomitization and/or dissolution.

Beds of gypsum, anhydrite, and salt precipitate from solution where features such as barred basins or tidal flats allow hypersaline conditions to exist. These rocks have extremely low permeability and, accordingly, are very effective barriers to ground-water flow. However, they are subject to dissolution in places where a dynamic freshwater flow system exists.

Outcropping Coastal Plain rocks have been studied since the turn of the century. Intensive subsurface investigation began with the discovery of vast reserves of oil and gas in these rocks and continues today, both onshore and offshore. Thousands of papers have been published pertaining to various aspects of Coastal Plain geology. An excellent summary of Coastal Plain geology can be found in Murray (1961), which is by far the best overview volume available. The geology of these rocks, then, is perhaps as well known and documented as that of any geologic province in the world. Coastal Plain hydrology, however, is less well known. For the most part, hydrologic investigations have addressed local problems or have been of a reconnaissance nature. Exceptions are the Florida peninsula, Long Island, New York, and the New Jersey Coastal Plain, where population pressures have created, or have the potential to create, ground-water supply or aquifer contamination problems. Important contributions to the science of hydrology have resulted from early investigations of the Coastal Plain. For example, the storage properties of clastic, confined aquifers were first described quantitatively based on investigations on Long Island by Jacob (1940), and one of the first recognitions of a regional ground-water flow system was that of Stringfield (1936) on the Florida peninsula.

Regional investigations of the Coastal Plain rocks of the United States were undertaken by the U.S. Geological Survey as part of the Regional Aquifer-System Analysis (RASA) program (Sun, 1986). The results of these regional investigations are summarized in chapters 25 through 27 of this volume. In addition, chapters 26 and 28 deal with the hydrogeology of the Coastal Plain of Mexico and the Yucatan peninsula, respectively. The purpose of this chapter is to describe the geologic controls on the occurrence, movement, and chemical quality of ground water throughout the Coastal Plain province of North America.

## DEPOSITIONAL REGIMES

Major rivers carried sediment into the Atlantic and Gulf Coast geosynclines and deposited most of the sediment in a series of constructional deltas along the rims of the basins. Some sediment, however, was carried out into shallow and intermediate marine shelf environments. During major transgressions and regressions of the sea, the positions of the deltas and shelves shifted laterally, creating complex interbedding of fluival to marine rocks. In many parts of the Coastal Plain, consecutively younger strata are found, occurring in offlap relationship, as one approaches the modern shoreline from the inner margin of the Coastal Plain. In other places, younger strata overlie older units in an onlap relationship.

The overall configuration of Coastal Plain rocks is that of a wedge of sediments that dips and thickens seaward. This wedge-shaped mass of sediments forms the flanks of the Atlantic and Gulf Coast geosynclines, both of which have received accumulations of sediments since Early Jurassic time. Superimposed on this wedge are a series of gentle folds that created major headlands and embayments along the coastline during geologic time. The locations of the more important of these positive and negative folds are shown on Figure 1. Most of the embayments have been negative areas since they were first formed and, accordingly, contain thick sections of sediments. By contrast, many of the positive structures, such as the Cape Fear and Peninsular arches, have periodically been covered by transgressive seas and became for brief periods the sites of marine deposition. Other positive structures, such as the Sabine Uplift and the Tamaulipas Arch, have remained as structural highs since their inception. The carbonate platforms of the Yucatan and Florida peninsulas, throughout most of their geologic history, were positive structures that were covered with shallow seas and sank slowly as carbonate deposition kept pace with subsidence. Elsewhere along the rim of the Gulf Coast geosyncline, subsidence has been rapid and continuous, leading to the accumulation of thicknesses of more than 17,000 m of clastic rocks in places such as offshore Louisiana. Along the continental shelf of the Atlantic Coast, total sediment thickness reaches a maximum of about 6,000 m. In general, coarser sediments having higher permeability are in updip areas, at places where sediments are thin or intermediate in thickness, or atop structural highs. This is because higher energy conditions existed in the shallower waters in these general locales, leading to the deposition of coarser materials.

Coastal Plain strata are cut by extensive fault systems that extend from Alabama to northern Mexico. The faults are all of the gravity type and can be grouped into two categories: (1) an inner belt of discontinuous grabens that forms the Mexia-Talco and Pickens-Gilbertown fault systems, and (2) a wide, coast-parallel band of listric or growth faults that are all down-to-the-basin and that primarily cut rocks of late Tertiary age and younger. Some faults disrupt the ground-water flow system because they have downdropped low-permeability strata opposite aquifers (Johnston and Miller, this volume). Small horsts and grabens in the eastern part of the Yucatan peninsula are reflected in the straight linear courses of surface streams and the regular alignment of bays and inlets along the coast (Lesser and Weidie, this volume).

Locally, the movement of salt at great depths has created ridges and sags in the overlying sediments and has culminated in piercement salt domes, particularly throughout Louisiana and in southern Mississippi, east Texas, and the eastern part of Vera Cruz and Tabasco, Mexico. Piercement domes not only locally impede ground-water flow because of the emplacement of a plug of impermeable material, but may also create plumes of high-salinity water down the hydraulic gradient from the salt dome.

Depositional environments in the Coastal Plain generally become more marine as one proceeds seaward from the inner margin of the Coastal Plain. At or near this margin, sediments are largely of fluvial origin, contain large amounts of coarse sand and gravel, and are apt to be thin-bedded and contain many lenticular clay bodies. These fluvial rocks pass into deltaic deposits that are more clayey, medium- to thick-bedded, locally lignitic, and very thick in aggregate. In turn, the deltaic strata pass into marginal marine rocks that are sandy if they represent shoreline deposits and clayey if they were laid down in lagoons or marshes. Finally, open marine environments are encountered; they have strata that are widespread and of relatively uniform bed thickness, and their texture is dependent largely on water depth and the distribution of marine currents. The total thickness of either marginal or fully marine rocks is less than that of the deltaic deposits.

As erosion proceeds and tectonic forces act on both foreland and basin, the general environments described above will shift both parallel and perpendicular to the coastline. The result is a complex stacking of facies and rocks of different textures. Within this intricate mass of strata, however, sufficient interconnection exists among sand bodies such that thick sequences of rock behave hydraulically as a single aquifer.

Coastal Plain carbonate strata were deposited in shallow marine waters that received no clastic material. Like the clastic rocks, carbonate strata vary in original texture, depending on whether they were deposited in a high- or low-energy environment. Coarse-textured limestones generally reflect shallower water where wave and current action winnow out fine material. Deeper water conditions existed where micrites and chalk accumulated. Individual carbonate rock environments of deposition tend to persist for long periods of time, indicating tectonic stability, particularly on carbonate platforms.

## GEOLOGIC FACTORS AFFECTING PERMEABILITY

### Lithology

The texture and mineralogy of clastic Coastal Plain rocks are altered very little after the rocks are deposited. Some compaction may occur, especially if the strata are subsequently deeply buried, but this process affects clay and silt much more than sand or gravel. Accordingly, the final amount of pore space in the rock, and the degree of interconnection between the pores remains much the same as that of the original sediment, particularly in coarse-grained aquifer materials.

Gravel and coarse fluvial sand have the highest permeability and thus compose the best clastic aquifers in Coastal Plain rocks. These materials usually are in updip areas where the streams transporting them first enter the Coastal Plain. Sorting of the gravel and sand varies; it is best where the deposits were laid down in stream channels and poor in overbank or sheetflow deposits. Some of the most prolific aquifer materials are, therefore, lenticular in cross-section and sinuous in plan view. However, where stream channels become clogged and shift course over time, thick anastomosing sheets of coarse sand and gravel result. Where beach and dune sands are preserved, they, too, are coarse, well sorted, and constitute good aquifers, but they are of limited extent because they accumulate in relatively narrow bands along the shoreline.

Deltaic Coastal Plain deposits are widespread, particularly in mid-dip areas, but their associated sands are apt to be poorly sorted and are commonly thin-bedded and clayey. Exceptions are channel and distributary-mouth sands, which constitute important local to subregional aquifers. Overall, deltaic deposits contain large amounts of clay and silt, particularly in interdistributary bay, natural levee, and overbank areas.

Marine sands, although they may cover wide areas, are generally less transmissive aquifers because they are commonly not thick and are finer-grained than the sands described above. Thick marine clays or chalk deposits may be formed in quiet, relatively deep waters, and these fine-grained materials form confining units that effectively retard ground-water movement. The clay content of Coastal Plain rock units generally increases seaward because sand is being lost as the rocks become gradually more marine by facies change.

Unaltered carbonate rocks vary considerably in texture, and to a certain extent, in mineralogy. Carbonate strata reflect environments ranging from sabkha conditions, characterized by micrite, primary dolomite, and gypsum or anhydrite, to quiet-water middle-shelf conditions, evidenced by thick sequences of micrite or chalk. In between, grainstone, coquina, and molluscan-rich limestone result from deposition along beaches, in tidal channels, or on current-swept inner shelves. Very locally, accumulations of rudistid limestone in Florida and Mexico reflect reefal conditions. Grainstone and coquina constitute the best unaltered limestone aquifers. Molluscan and rudistid limestones are less permeable, and chalk and micrite form very effective confining units.

Fresh water circulating through limestone aquifers has a profound effect on their porosity and permeability. Dolomitization occurs largely in freshwater–saltwater mixing zones, and can either increase or decrease the porosity of a limestone, depending on the limestone's original texture (Miller, 1986). A rock that was originally a micrite may be recrystallized into a highly porous, loosely interlocking mosaic of dolomite crystals. Conversely, dolomite crystals may partly or completely fill the pore space in a loosely packed grainstone and create a confining unit out of a rock whose original permeability was high. Fresh water also commonly increases the permeability of limestone by dissolution, developing secondary porosity that ranges in scale from pinpoint vugs to caverns that may be tens of meters across. Evaporites, which form effective confining units, may also be dissolved by vigorously circulating freshwater, and in some cases are reprecipitated elsewhere (Miller, 1986).

### Structure

Ground-water flow is more sluggish in flat-lying Coastal Plain strata than in places where these beds are tilted. The reasons for this are: (1) hydraulic conductivity is greater parallel than perpendicular to bedding, and accordingly, when permeable rocks are tilted, a vertical component of flow is superimposed upon the horizontal component that occurs in flat-lying beds, thus steepening the hydraulic gradient; (2) where aquifers are overlain by confining units, tilted aquifers receive direct recharge at the outcrop, whereas flat-lying aquifers are recharged solely by leakage. In the Coastal Plain, there is an overall gentle flow and tilting of both aquifers and confining units (see Fig. 2 of Grubb and Carrillo, this volume).

Some of the domes and arches shown on Figure 1 have brought permeable rocks to the surface, where they have been partially eroded and exposed to form major recharge areas. Florida's Peninsular arch has exposed permeable rocks that receive high rates of recharge, which has created a regional high on the potentiometric surface of the Upper Floridan aquifer (see Fig. 4 of Johnston and Miller, this volume). The Cape Fear arch is a major recharge area for an extensive Cretaceous sand aquifer (see Fig. 1 of Meisler and others, this volume). Other positive features, such as the San Marcos Arch and the Sabine Uplift, have little or no effect on the flow system.

Growth faults that are instrumental in the creation of geopressured anamolous geothermal conditions in southern Louisiana and Texas (Wallace, 1982) interrupt regional ground-water flow in these areas. Smaller-scale grabens and horsts, while they may have considerable local influence, do not generally affect regional flow. Joints may provide conduits that locally produce anamolous water levels (Lesser and Weidie, this volume) but have little or no effect on regional flow.

### Topography

Topography is a major influence on regional flow in Coastal Plain aquifers (Grubb and Carrillo, this volume). The elevation of recharge areas where aquifers crop out, the degree of incisement of streams of different sizes, and the location and extent of lowland areas largely determine the overall configuration of ground-water flow. Streams and swampy lowland areas are places where ground water discharges either as base flow or diffuse upward leakage. The relative positions and difference in elevation of recharge and discharge areas determine the head gradients and the length of ground-water flow paths (Freeze and Witherspoon, 1967). Major sounds and bays occupy topographic lows in some places and coincide with areas of ground-water discharge (Meisler and others, this volume).

Topography greatly influences the movement of ground water both locally and regionally. This effect is shown schematically on Figure 2, adapted from Toth (1963). Most of the water recharging Coastal Plain aquifers moves quickly along relatively short flow paths, such as those labeled "local" and "intermediate" on Figure 2, and discharges as base flow to streams on the local scale and rivers on the intermediate scale. The component of flow called "regional" on Figure 2 moves relatively slowly and follows long flow paths before discharging to major rivers. A very small component of flow is not intercepted by rivers of any size and percolates slowly into deep, thickly confined parts of the aquifers before finally discharging into shallower aquifers (and ultimately to bays or the ocean) by diffuse upward leakage. Water budget studies in parts of the southeastern U.S. Coastal Plain show that only about 3 percent of total recharge percolates into the deep parts of the aquifers (Barker, 1986).

Karst features make up a special type of topography that greatly influences regional ground-water flow. Little surface drainage is developed in karst plains, and recharge to the carbonate aquifer is largely direct, into sinkholes, cenotes, and other solution pits that are vertical or nearly so. Much ground-water discharge in karst areas, particularly in west-central Florida, is channeled into large springs, some of which form the headwaters of major streams (Rosenau and others, 1977).

### Confining units

The most obvious effect of confining units on ground-water flow is to create confined or artesian conditions in an aquifer. Where ground water becomes fully confined, gradients are not as strongly influenced by topography. Accordingly, flow paths become longer and head gradients more gentle in confined areas. Confining units in the Coastal Plain largely limit the seaward extent of ground-water flow as aquifers lose their permeability by facies change to increasingly finer-grained materials in a coastward direction. Some confining units are of limited areal extent and accordingly do not affect regional flow. Others, such as the thick chalk sequence in central and western Alabama (Meisler and others, this volume) and thick clays of the Midway, Jackson, and Vicksburg Groups in the western Gulf Coastal Plain (Grubb and Carrillo, this volume) constitute extremely effective barriers to the vertical or horizontal movement of ground water. Confining units of regional extent were laid down in shallow shelf environments.

The development of karst topography and resulting high transmissivity in parts of the Upper Floridan aquifer in west-central Florida are directly related to the thickness and continuity of a clay confining unit that overlies the aquifer. Where this upper confining unit is absent or less than about 30 m in thickness, the Upper Floridan aquifer is karstified; as the confining unit thickens, solution features are largely lost, ground-water flow is much less dynamic, and the transmissivity of the aquifer is considerably decreased (Johnston and Miller, this volume).

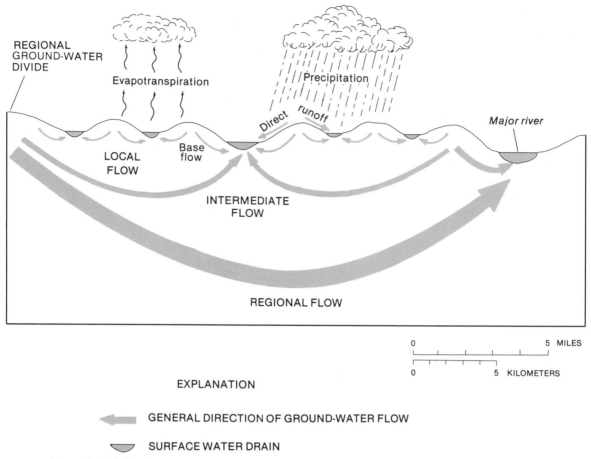

Figure 2. Relation between surface drainage system and scale of ground-water flow system, which discharges into surface drainage (adapted from Toth, 1963).

## Carbonate dissolution

Dissolution activity is most likely to occur in places where ground water is moving rapidly in and out of a carbonate aquifer. This rapid circulation is favored by: (1) a thin or absent confining unit, as discussed above; (2) coarse texture of the carbonate rock; and (3) the presence of joints or fractures that serve as initial conduits for ground water. The latter two factors are important in the Yucatan peninsula, while the first is more important in Florida. Although the effects of dissolution are most easily observed on the modern land surface, paleokarst is developed at several deep subsurface levels in the Florida peninsula. One such ancient solution zone, the "Boulder Zone" of south Florida, is used as a receiving zone for treated municipal wastewater (Johnston and Miller, Ch. 27, this volume). Much younger and shallower solution conduits in the Biscayne aquifer of south Florida, also discussed in chapter 27, are commonly filled with sand, but nevertheless are highly transmissive parts of the aquifer.

## Miscellaneous factors

Fluid pressures higher than the normal hydrostatic pressure

of 107 $g/cm^2/m$ occur over a wide area in offshore and onshore Louisiana and Texas (Wallace, 1982). This abnormal pressure, called geopressure, exists because growth faults have sealed off sands by downdropping them opposite shale beds, thereby preventing the expulsion of pore fluids from the sands. In effect, the fluids in these trapped sands support a large part of the weight of the thick overburden that overlies the sands. No ground-water flow takes place in the geopressured zone, and where the zone is present, it delineates the seaward limit of the active flow system (Grubb and Carrillo, Ch. 26, this volume). The saline water in the geopressured zone has a temperature greater than 100°C and contains dissolved methane in varying quantities. These waters have, accordingly, been studied as a possible energy source (Papadopolous and others, 1975).

Deposition of thick sequences of geologically young sediments in south Louisiana and east Texas has been so rapid that the sediments are undercompacted. Heavy pumping has locally removed large quantities of water and lowered the hydraulic head, allowing the clay beds that separate the pumped aquifers to more completely compact, resulting in subsidence of the land surface (Gabrysch, 1982). This problem has been especially significant at Houston, Texas.

TABLE 1. RELATIVE EFFECT OF GEOLOGIC FACTORS ON COMPONENTS OF THE
GROUND-WATER FLOW SYSTEM IN COASTAL PLAIN ROCKS

| Geologic Factor | Relative Effect On: | | |
|---|---|---|---|
| | Regional Flow | Subregional Flow | Local Flow |
| **Clastic Rocks** | | | |
| Lithology (texture) | **Major** | **Major** | **Major** |
| Structure | Minor | Minor | **Major** |
| Topography | **Major** | **Major** | **Major** |
| Confining Units | **Major** | **Major** | **Major** |
| Clastic Diagenesis* | Minor | Minor | Minor |
| Geopressure* | **Major** | **Major** | Minor |
| Compaction* | Minor | Minor | **Major** |
| Unique Rock Types* | Minor | Minor | **Major** |
| **Carbonate Rocks** | | | |
| Lithology (Texture) | **Major** | **Major** | Minor |
| Structure | Minor | Minor | **Major** |
| Topography | Minor | Minor | Minor |
| Confining Units | **Major** | **Major** | Minor |
| Dolomitization† | **Major** | **Major** | **Major** |
| Dissolution† | Minor | **Major** | Minor |

*Factor does not apply to carbonate rocks.
†Factor does not apply to clastic rocks.

Permeable basalt flows of Pliocene and Pleistocene age cover several thousand square kilometers of the Coastal Plain of Mexico between latitude 18° and 20° north. These basalts provide base flow to the major streams crossing the Coastal Plain and are the source of large springs. Along with thin, narrow bands of Holocene alluvium that parallel the rivers, the basalts are among the best aquifers in Mexico.

## RELATIVE EFFECT OF GEOLOGIC FACTORS

All the factors listed above affect the ground-water flow system of the Coastal Plain. Some of these factors have regional influence and some have only local effects. Table 1 summarizes the relative effect of each geologic factor on different segments of the flow system. Because the flow system is a continuum, it is not possible to exactly quantify its local, subregional, or regional components. In terms of the area affected by each factor, however, a local effect is one that is expressed over tens of square kilometers; a subregional effect can be determined over hundreds of square kilometers; and a regional effect influences conditions over thousands of square kilometers. Table 1 is divided into clastic and carbonate rock segments because individual factors affect these rock types differently.

## GROUND-WATER FLOW SYSTEM

The overall pattern of ground-water flow in major aquifers in Coastal Plain rocks is that of a classic artesian system. The ground water originates as recharge from precipitation in aquifer outcrop areas, moves into confined conditions down a hydraulic gradient that generally parallels the dip of the aquifers, and discharges as base flow to streams and springs, or as diffuse upward leakage to shallower aquifers in downdip areas. Under predevelopment conditions, variations in the direction and rate of flow are determined mostly by variations in hydraulic conductivity and the distribution of natural recharge and discharge.

Most of the flow is perpendicular to the modern coastline. In some aquifers, however, a strong component of flow occurs parallel to the coast (see Fig. 1 of Meisler and others, this volume) that may result from one or several factors, including: (1) considerable relief in recharge areas; (2) higher permeability parallel to the coast (anisotropy); (3) complete overlap of an aquifer by a confining unit; or (4) the downdip damming effect of zones of sluggish circulation containing salt water.

These natural flow patterns become distorted in regional, confined aquifers under large-scale ground-water development. Large areal declines in water levels due to pumping have occurred; in the Upper Floridan aquifer along the Atlantic coast from southeastern South Carolina to northeastern Florida; a large cone of depression has developed in a Cretaceous aquifer straddling the Virginia–North Carolina line; and so on. The changes of hydraulic gradients and flow directions resulting from extensive pumping in regional confined aquifers cause: (1) an increase in recharge to the aquifers that, in turn, causes a reduction in discharge to streams from local to intermediate ground-water flow systems, and (2) a reduction in diffuse upward leakage.

## GROUND-WATER CHEMISTRY

Water in clastic aquifers in the Coastal Plain varies in its chemical composition depending primarily on the residence time of the water in the aquifer, which largely determines the degree of mineral–water interaction. In and just downdip of recharge areas, the water is mostly a calcium bicarbonate type. Downgradient, the predominant water type is sodium bicarbonate, produced largely by sodium-calcium exchange. Such an exchange is greatly facilitated where glauconite is present in the aquifer. Still farther downgradient, mixing of fresh water and salt water produces a sodium-chloride-type water.

Water chemistry in carbonate aquifers varies in a different fashion. In recharge areas, the dominant water type is calcium bicarbonate like that in clastic aquifers, but with a higher dissolved-solids content. However, where dolomite is present, the water changes downgradient to a calcium-magnesium bicarbonate type; or where gypsum or anhydrite occur, the water will chemically evolve to a calcium-magnesium bicarbonate-sulfate or a calcium-magnesium sulfate type. All these changes are due to mineral–water interaction. Farther downgradient, mixing of fresh water with increasing amounts of salt water ultimately results in a sodium-chloride-type water.

Salt water is present in many Coastal Plain aquifers either onshore or offshore. Some of the salt water may be connate, and some of it originates as brine produced by the dissolution of evaporites, particularly halite. The freshwater–saltwater interface, whose exact location depends on hydraulic pressures and gradients, and to a lesser extent the permeability of the aquifer, may be either sharp or diffuse. In general, places where salt water is relatively shallow coincide with areas of ground-water discharge. At some places, however, complex freshwater–saltwater relations appear to have resulted from eustatic sea-level changes in the geologic past (Meisler and others, 1984) or to the cyclic flow of modern seawater (Kohout, 1960). Locally, fresher water may underlie more saline water, especially where the permeability of a deep aquifer is considerably greater than that of a shallower one, thereby allowing more complete flushing of salt water. Where piercement salt domes have brought halite plugs near to the surface, plumes of salt water occur very locally downgradient of the domes as a result of partial dissolution of the salt plug.

## REFERENCES CITED

Barker, R. A., 1986, Preliminary results of a steady-state ground-water flow model of the Southeastern Coastal Plain aquifer system, *in* Proceedings of the Southern Regional Ground Water Conference, September 18–19, 1985, San Antonio, Texas: Association of Ground Water Scientists and Engineers, Division of the National Water Well Association, p. 315–338.

Freeze, R. A., and Witherspoon, P. A., 1967, Theoretical analysis of regional ground-water flow; 2. Effect of water-table configuration and subsurface permeability variations: Water Resources Research, v. 3, no. 2, p. 623–634.

Gabrysch, R. K., 1982, Ground-water withdrawals and land-surface subsidence in the Houston-Galveston region, Texas, 1906–80: U.S. Geological Survey Open-File Report 82-571, 68 p.

Jacob, C. E., 1940, On the flow of water in an elastic artesian aquifer: American Geophysical Union Transactions, part 2, p. 574–586.

Jones, P. H., 1975, Geothermal and hydrothermal regimes in the northern Gulf of Mexico Basin: Proceedings, Second United Nations Symposium on the Development and Use of Geothermal Resources, San Francisco, v. 1, p. 15–89.

Kohout, F. A., 1960, Cyclic flow of saltwater in the Biscayne aquifer of southeastern Florida: Journal of Geophysical Research, v. 65, p. 2133–2141.

Meisler, H., Leahy, P. P., and Knobel, L. L., 1984, The effect of eustatic sea-level changes on saltwater-freshwater relations in the northern Atlantic Coastal Plain: U.S. Geological Survey Water-Supply Paper 2255, 28 p.

Miller, J. A., 1986, Hydrogeologic framework of the Floridan aquifer system in Florida and in parts of Georgia, Alabama, and South Carolina: U.S. Geological Survey Professional Paper 1403-B, 91 p.

Murray, G. E., 1961, Geology of the Atlantic and Gulf Coastal Province of North America: New York, Harper and Brothers, 692 p.

Papadopolous, S. S., Wallace, R. H., Jr., Wessleman, J. B., and Taylor, R. E., 1975, Assessment of onshore geopressured-geothermal resources in the northern Gulf of Mexico Basin, *in* White, D. E., and Williams, D. L., eds., Assessment of geothermal resources in the United States—1975: U.S. Geological Survey Circular 726, p. 125–146.

Rosenau, J. C., Faulkner, G. L., Hendry, C. W., Jr., and Hull, R. W., 1977, Springs of Florida: Florida Bureau of Geology Bulletin 31, 461 p.

Stringfield, V. T., 1936, Artesian water in the Florida peninsula: U.S. Geological Survey Water-Supply Paper 773-C, p. C115–C195.

Sun, R. J., 1986, Regional aquifer-system analysis program of the U.S. Geological Survey—summary of projects, 1978–84: U.S. Geological Survey Circular 1002, 264 p.

Toth, J. A., 1963, A theoretical analysis of ground-water flow in small drainage basins: Journal of Geophysical Research, v. 68, p. 4795–4812.

Wallace, R. M., Jr., 1982, Geopressured-geothermal energy resource appraisal; Hydrogeology and well testing determine producibility: Louisiana Geological Survey Guidebook Series no. 2, 111 p.

MANUSCRIPT ACCEPTED BY THE SOCIETY DECEMBER 4, 1987

The Geology of North America
Vol. O-2, Hydrogeology
The Geological Society of America, 1988

# Chapter 37

# *Sandstones and shales*

**Stanley N. Davis**
*Department of Hydrology and Water Resources, University of Arizona, Tucson, Arizona 85721*

## INTRODUCTION

Early development of artesian wells about a century ago in the central part of the United States stimulated a contemporary interest in the sandstone aquifers that are widespread in the region (Stone, 1886; Chamberlain, 1885; Darton, 1897; Norton, 1912). Many of the early wells were quite deep (some exceeded 300 m) and represented impressive technical achievements of the times. Today, permeable sandstones such as the Dakota, Navajo, Coconino, St. Peter, Mt. Simon, and the Lamotte are widely exploited, but flowing wells, if originally present, are mostly gone, and where overexploited, large cones of depression have formed. For example, some of the sandstone aquifers that originally produced flowing wells in Chicago (Stone, 1886) have present water levels more than 150 m below the land surface.

The importance of sandstone aquifers of regional extent has perhaps been overemphasized in elementary textbooks, owing to the widespread and almost exclusive use of cross sections to depict classical artesian basins with a single sandstone aquifer that is part of a smoothly formed structural basin. The recharge for such aquifers is almost always shown taking place where the sandstone crops out along the margins of the basin. Vast areas in the United States, Canada, Mexico, and other countries are indeed underlain by sandstone aquifers that form the only important local sources of ground water, but their geometry and recharge mechanisms rarely conform to textbook artistry. The major focus of this chapter is the discussion of the geologic variables that give rise to the almost infinite hydrogeologic variations found in sandstone aquifers, but attention will also be given to fine-grained clastic rocks. Although some of their hydrogeologic properties are difficult to measure, an understanding of these properties is essential to a full understanding of water movement in large hydrogeologic systems. Problems of aquifer recharge, variable natural water chemistry, and confinement of artificially injected or emplaced waste are related closely to the hydrogeologic properties of fine-grained clastic rocks. Furthermore, in terms of their total mass, clays and shales are by far the most abundant type of sedimentary rock, accounting for more than 65 percent of the volume of all the sedimentary rocks (Garrels and MacKenzie, 1971).

## CLASSIFICATION OF SANDSTONES AND SHALES

A simple classification of clastic rocks is based primarily on texture. Terms used most commonly are shale, siltstone, sandstone, and conglomerate, which correspond to the lithified-size equivalents of clay, silt, sand, and gravel, respectively. Terms such as lutite, mudstone, arkose, granule conglomerate, and pebbly wacke may be useful in detailed petrologic discussions of sedimentary rocks (Pettijohn and others, 1972) but will be avoided here. Most properly, shale refers to fine-grained rocks having the tendency to split along parallel planes. In actual fact, general usage by geologists extends the term shale to cover all fine-grained sedimentary rocks (lutites) that have been subdivided into claystone and siltstone by sedimentary petrologists.

Some sandstones and siltstones are cemented to the extent that they resemble metamorphic rocks, and many metamorphic rocks retain sedimentary structures of the antecedent sedimentary rocks. The distinction between the two broad rock types, sedimentary and metamorphic, is quite arbitrary and is commonly made on the basis of certain mineral assemblages that can give information concerning past temperature and pressure conditions at the time of lithification. As an example, for many years a large body of rocks along Pacheco Pass in central California was assumed to be normal sandstone by field geologists. Detailed petrographic work, however, discovered that these normal-looking sandstones were actually cemented with jadeite, thus indicating that the rocks were metamorphic (McKee, 1960). Although of profound importance in the reconstruction of the geologic history of the region, this discovery was of only secondary importance in interpreting the hydrogeology of the region because of the hydrogeologic similarity of the metamorphic rocks to thoroughly cemented normal sandstones, which they had been previously assumed to be.

An important generalization can be made at this point. Standard geologic methods and classifications must be used to establish the physical framework of interest to the hydrogeologist. However, as important as standard classifications and methods can be, the hydrogeologist is also concerned with such items as continuity of fractures, the depth of weathering, the presence of traces of gypsum, and a myriad of other seemingly minor items

Davis, Stanley N., 1988, Sandstones and shales, *in* Back, W., Rosenshein, J. S. and Seaber, P. R., eds., Hydrogeology: Boulder, Colorado, Geological Society of America, The Geology of North America, v. O-2.

TABLE 1. GEOMETRY OF CLASTIC SEDIMENTARY ROCKS RELATED TO
IMPORTANT ENVIRONMENTS OF DEPOSITION

| Environment of Deposition | Geometry of Resulting Hydrogeologic Units |
|---|---|
| **Marine** | |
| Shoreline (beach, bars, lagoons, and deltas) | Shale tends to be more or less continuous over areas of 10s to 100s km². Thicknesses are in meters or 10s of meters. Sand beds more elongated unless in prograded sheet. Ancient bar and channel deposits, 100s m wide, 1 to 10s m thick, 10s km long. |
| Shelf and slope | Sand and shale beds continuous over 100s to 1000s km², 10s to few 100s m thick. |
| Floor | Extensive shale over 1000s to 10,000s km², thicknesses 100s m. Cyclic turbidite sands, individual beds, 10s cm to few m thick. Lateral extent measured in km. |
| **Continental** | |
| Fluvial (flood plains, channel deposits, alluvial fans) | Less continuous than marine units. Large rivers produce sands a few km long, 10s m to few km wide and few m to few 10s m thick. Shales more continuous. Ancient alluvial fans produce thick, poorly sorted units with local well-sorted units, few m thick, 10s m wide, and 100s m long. |
| Aeolian (sand dunes, loess deposits) | Extensive sandstones up to a few 100 m thick and 1,000s km² area. |
| Lacustrine (shoreline, deltas, lake bottom) | Lake-bottom shale, extensive, few 1,000 km² to 10,000 km², and up to a few 100 m thick. Deltaic and shoreline sand bodies are elongate and more local than shale. |
| Glacial (till plains, moraines, ice-contact deposits) | Tillite forms continuous sheets 10s of m thick over 1,000s of km². |

that are commonly of overriding importance in understanding the hydrogeologic problems of clastic sedimentary rocks.

## GEOMETRY OF SEDIMENTARY UNITS

The texture, lateral continuity, thickness, and overall shape of sedimentary units are controlled in a large measure by the original environments of deposition of the sediments and secondarily by their postdepositional compaction, cementation, erosion, folding, and faulting. These factors in turn establish the hydrologic and chemical characteristics of the hydrogeologic units.

Rocks of marine origin dominate most sedimentary sequences. These rocks are generally more homogeneous and have a more extensive lateral continuity than rocks of a continental origin. This is particularly true of rocks that originated as deposits some distance from ancient shorelines. Sedimentary processes near the shoreline and within continental settings are much more variable on a local scale, which will be reflected in the rapid lateral variations of some of the sedimentary units. Table 1 presents a very brief summary of the more important depositional environments that govern the geometry of bodies of sandstone and shale.

Structural features, including faults, also affect the geometry and hydrogeologic characteristics of shale and sandstone. Faulting will truncate beds, and if the beds are well indurated, open fractures associated with the faulting may greatly increase local permeability along faults. Quite commonly, faulting produces an opposite effect. Gouge (Chu and others, 1981), rising fluids which deposit minerals along the fault, and low-permeability beds offset against higher permeability beds will produce hydrogeologic barriers along the fault.

Near the surface, vertical or subvertical joints are prominent and ubiquitous in sedimentary rocks. These joints may be spaced from a few centimeters to more than 10 m apart and may increase the overall permeability of the surface rocks by at least an order of magnitude. Joints close rapidly with depth, which is reflected in a rapid reduction of rock permeabilities with depth

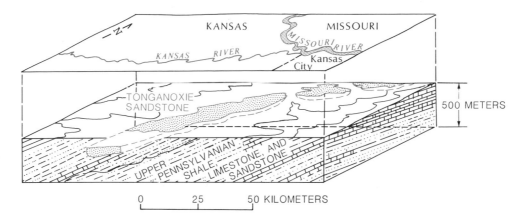

Figure 1. Tonganoxie sandstone, an ancient stream channel deposit within a sequence of Upper Pennsylvanian shales, sandstones, and limestones. Diagram is generalized from data in Lins (1950), Moore (1949), and Homyk and others (1967).

(Smith, 1963; Rats and Chernayshov, 1967). Where fluid pressures are large in the deep subsurface, however, fractures may develop in fine-grained rocks in response to stress-release mechanisms (Palciauskas and Domenico, 1980).

Unconformities of various types play an important role in determining the geometry of sedimentary units. This is particularly striking in the Upper Pennsylvania beds in Nebraska, Iowa, Kansas, Missouri, and Oklahoma (Moore, 1949). Here, ancient stream deposits cut through many of the thin marine deposits to form narrow bodies of sandstone (Lins, 1950), which can be important aquifers on a local scale (Fig. 1). On a much larger scale, regional unconformities can truncate sedimentary units and expose these units to cross-formational flow, which in turn affects recharge to and water quality within these units.

## POROSITY OF SANDSTONE

Porosity of sedimentary materials will vary from less than 0.005 (0.5 percent) to more than 0.60 (60 percent; Davis, 1969). Soon after deposition, sands that are eventually lithified into sandstones generally have original porosities of between 0.25 and 0.45. During the lithification of sands, a large number of natural processes combine to reduce the original porosity of the sand. These processes include the chemical precipitation of minerals including calcite, iron oxides, gypsum, and authigenic clays (Almon and others, 1976; Dapples, 1972; Land and Dutton, 1978; Pettijohn and others, 1972; Waldschmidt, 1941; Wallace, 1976; Wilson and Pittman, 1977); the compaction of the sediments (Athy, 1930; Maxwell, 1964); and pressure solution of minerals, particularly quartz, at contact points (Angevine and Turcotte, 1983; Houseknecht, 1984; Sibley and Blatt, 1976). During the history of a typical sandstone, a number of processes act to counter the loss of porosity. Shrinkage, fracturing, and dissolution in some settings may actually reverse the reduction in porosity (Schmidt and McDonald, 1980). Data on porosity of

sedimentary rocks have been reviewed by Manger (1963), Schoeller (1962), and Davis (1969). Some typical values of porosity for sandstones are given in Table 2. Most porosities of sandstone range from about 10 to 25 percent, although much smaller porosities are found routinely in orthoquartzites such as the late Precambrian Sioux Quartzite of South Dakota.

As a broad generality, the porosity of sandstone decreases with depth (Maxwell, 1964). This decrease, however, is rather gradual and is of direct practical importance primarily in considering the possible production of hydrocarbons from depths in excess of 3,000 m. This decrease is due in part to phase changes and dissolution and precipitation of minerals, processes that are enhanced by higher temperatures at great depths.

Most porosity studies of numerous samples from single aquifers indicate a normal Gaussian distribution of the resulting data (Bennion and Griffiths, 1966; Davis, 1969; Dullien, 1979). However, caution should be used in assuming a normal distribution for extreme values, particularly those at the higher end. Sandstone will not be stable at high porosities, and projected porosity values in excess of about 0.5 would seem unreasonable.

## PERMEABILITY OF SANDSTONE

An attempt has been made to present a few representative values of permeability along with porosity (Table 2). A very large body of information exists, primarily from the work of petroleum engineers and mainly for samples of sandstone. Methods of measurement vary widely and represent various laboratory as well as field methods. Not only do the results reflect unspecified variations of the field settings and effective volumes measured, but for laboratory work, they also reflect different types of fluids used in the permeability tests as well as variations in confining pressures placed on the samples during the tests. These factors, which are discussed in detail by Neuzil (1986), are particularly important for fine-grained rocks. Although sampling and measurement

TABLE 2. REPRESENTATIVE POROSITIES AND PERMEABILITIES OF SANDSTONE AND SHALE
WITH VALUES FOR SALT AND COAL GIVEN FOR COMPARISON

| Lithology | Formation or Geologic Age | Location | Porosity (%) | Permeability (md) | Field (F) or Laboratory (L) | Reference |
|---|---|---|---|---|---|---|
| 1. Sandstone | Paskapoo | Alberta | — | 4285 | F | Garven (1982) |
| 2. Sandstone | Paskapoo | Alberta | — | 2640 | F | Garven (1982) |
| 3. Shale (fractured) | Paskapoo | Alberta | — | 3230 | F | Garven (1982) |
| 4. Shale (fractured) | Paskapoo | Alberta | — | 988 | F | Garven (1982) |
| 5. Mudstone | Cenozoic | North Dakota | — | $4.6 \times 10^{-2}$ | F | Rehm and others (1980) |
| 6. Coal | Cenozoic | North Dakota | — | 334 | F | Rehm and others (1980) |
| 7. Salt* | Permian | Texas | — | $9.61 \times 10^{-5}$ (horizontal) $9.61 \times 10^{-6}$ (vertical) | — — | INTERA (1986) |
| 8. Shale* | Wolfcamp | Texas | — | $9.61 \times 10^{-5}$ (horizontal) $9.61 \times 10^{-6}$ (vertical) | — — | INTERA (1986) |
| 9. Sandstone | —— | —— | — | 0.15   (horizontal) 47.9   (vertical) | L L | Young and others (1964) |
| 10. Siltstone | —— | —— | — | $2.4 \times 10^{-7}$ (vertical) | L | Young and others (1964) |
| 11. Sandstone | Wilcox | —— | — | 4740 | F | Slaughter and others (1983) |
| 12. Sandstone | Navajo | Utah | — | 3.8 | F | Woodward-Clyde (1984) |
| 13. Sandstone | Dakota | Utah | — | 10.9 | F | Woodward-Clyde (1984) |
| 14. Shale (fractured)[†] | Monterey | California | 21 | 2690 | L | Isherwood (1981) |
| 15. Shale (fractured) | Green River | —— | — | 0.32 | F | Isherwood (1981) |
| 16. Shale (fractured, oil shale) | Green River | —— | 8 | 2.07 | F | Isherwood (1981) |
| 17. Shale | Graneros | Kansas | 11.6 | $4.7 \times 10^{-5}$ | L | Isherwood (1981) |
| 18. Sandstone | Bradford | —— | 14.8 | 2.7 | L | Davis and DeWiest (1966) |
| 19. Sandstone | Berea | —— | 19 | 383 | L | Davis and DeWiest (1966) |
| 20. Sandstone | Repeto | —— | 19.1 | 36 | L | Davis and DeWiest (1966) |
| 21. Sandstone[§] | Cambrian | —— | 11.2 | — | L | Manger (1963) |
| 22. Sandstone[‡] | Pennsylvanian | —— | 17.4 | — | L | Manger (1963) |

*Values used in ground-water model.
[†]Average of 56 samples within an interval of 39 m. Porosity range 14% to 29%. Permeability range 89 md to 16,500 md.
[§]Average of 16 samples.
[‡]Average of 587 samples.

techniques may introduce large variations in the values reported in Table 2, these variations are still much smaller than the fundamental variations related to lithology and structure.

Studies of lightly cemented sandstones have shown that, like their nonindurated counterparts, grain size plays a dominant role in determining the permeability of the rock (Johnson and Greenkorn, 1963). Very roughly, the permeability increases as the square of the grain size. This simple relationship tends to break down as cementing and compaction decrease the pore size. For medium-grained sandstones with porosities between 0.15 and 0.25, a very rough correlation also exists between porosity and permeability. This correlation is actually quite good for some data sets, particularly if it is based on measurements of a single, poorly lithified sandstone unit that has a rather uniform grain size. An example of one of the numerous equations relating porosity to permeability is given below:

$$k = D^2 \phi^{5.5/5.6} C$$

in which k is permeability, D is an effective grain size, $\phi$ is porosity, and C is a dimensionless constant. This equation was given by Dullien (1979). Most similar equations also show an exponential relation between porosity and permeability.

As the porosity of a sandstone decreases, particularly below 0.15, the permeability depends more on the presence of through-going secondary fractures rather than on the original porosity within the rock. As porosity decreases below 0.05, permeability is almost entirely dependent on the presence of fractures. Thus, in the case of a quartz sandstone with a brittle cement that is in a highly fractured zone, the present permeability may exceed the original permeability of the antecedent sand (Fig. 2). In general, the hydrogeology of all highly indurated sediments is profoundly affected by fractures, and the fractures account for very large local permeabilities (Huntoon, 1986; Huntoon and Lundy, 1979).

If a large number of permeability measurements are made of a single stratigraphic unit, the resulting values appear to have a log-normal statistical distribution (Fig. 3; Davis, 1969). The ex-

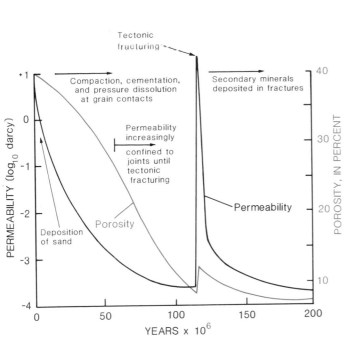

Figure 2. Generalized history of porosity and permeability of a sandstone.

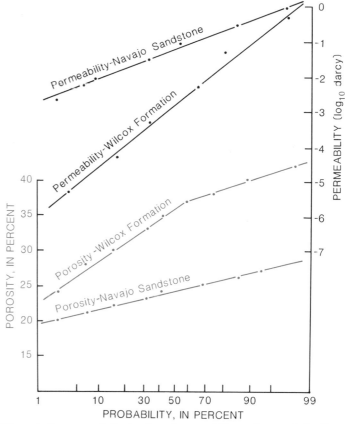

Figure 3. Log-normal distribution of permeability and normal distribution of porosity in two sandstones. Data for the Navajo Sandstone are from Rubenstein (1966); data for the Wilcox Formation are from Slaughter and others (1983).

treme values in a log-normal distribution, however, are not bounded; that is, permeability values approach infinity on the high end and zero on the low end of the distribution. If a medium-grained sandstone is considered, a permeability of 100.0 darcys might be predicted for the 99.99 percent value and 0.2 microdarcy for the 0.01 percent value. On the high end, large voids are needed to account for such high permeabilities. These voids could be produced by fracturing of the sandstone. On the low end, an almost complete cementation of pore space is required to reduce the porosity, as well as permeability, to near zero. Both processes, fracturing and cementation, are different processes than those that govern permeabilities in the central part of the distribution. It is my opinion, therefore, that a single log-normal distribution does not describe the frequency of occurrence of extreme values and, based on my experience, it should not be extended beyond the ninety-fifth percentile on the high end of the scale and lower than the fifth percentile on the low end of the scale. Part of this assumed lack of reliability, if true, may only be apparent and may relate to the random errors introduced by measurement difficulties at extreme values. More likely, most of the departure from the log-normal distribution is due to the fundamental inappropriateness of the log-normal assumption for extreme values.

A number of typical permeability values for sandstones are given in Table 2. Given the nature of most depositional processes where horizontal and near-horizontal stratification results, horizontal permeability should exceed vertical permeability in any

given sample. This is generally true and has been noted by various researchers (Buryakovskiy and Dzhafarov, 1980; Davis, 1969; Muskat, 1946). Many horizontal permeabilities are more than ten times the vertical values, with three-fold differences being about the median. Exceptions, however, are common, particularly where vertical fractures are present in the sandstone. For example, this is probably the explanation for the high value for the vertical component shown for sample 9 of Table 2.

## POROSITY OF SHALE

Compared with our knowledge of sandstone and carbonate rocks, little is known of the hydrogeologic characteristics of shale. Most early interest centered on the porosity of the shale because differential compaction of shale might give rise to structures of interest in petroleum exploration (Athy, 1930; Hedberg, 1936). Also, shale porosity can be related to density, which in turn is of importance in the interpretation of geophysical surveys. More recent interests have included studies of porosity as an indication of the extent to which pore fluids have been extruded into adjacent permeable zones, the assumption being that original porosities of shales were very large, and present low porosities could have only been achieved through the expulsion of pore fluids originally in the shale.

Although the compaction of fine-grained sediments in the deep subsurface is very sensitive to the availability of adjacent permeable zones as avenues for the expulsion of pore water, in broad terms, all deep basin studies show a general decrease in porosity of shale with depth. Porosities of 0.3 to 0.5 are common near the surface. These decrease to values of 0.05 to 0.15 at depths of 5,000 m or more.

Secondary fractures add a small but important porosity, commonly less than 0.01, to shale near the surface. A soft shale with an abundant clay component may not retain significant fracture openings beyond depths of only 30 m. In contrast, brittle shale with large amounts of silica, such as the Monterey Shale in the coastal region of central California, can maintain abundant open frauctures at depths greater than 300 m (Isherwood, 1981). Nevertheless, the total amount of porosity added to the shale by secondary fractures is probably very small, although general studies of this question are evidently lacking.

## PERMEABILITY OF SHALE

The permeability of shale is difficult to measure; therefore, reliable data are scarce in relation to the abundant data on the porosity of shale and the permeability and porosity of sandstone. Nevertheless, enough data are given in Davis (1969), Garven (1982), Young and others (1964), Isherwood (1981), and Neuzil (1986) to draw some useful conclusions. Those interested in a detailed discussion of the topic should consult Neuzil (1986).

The intergranular permeability of shale is very low, as reflected in values given for samples 10 and 17 in Table 2. Most reliable laboratory measurements of the permeability of shale range between $10^{-1}$ and $10^{-4}$ millidarcys. In contrast, field determinations of the permeability of fractured shale will have values comparable with sandstone, or in the range of 10 to $10^4$ millidarcys (samples 3 and 4, Table 2). The very important conclusion to be drawn is the fact that, contrary to conventional ideas, shale can be permeable at shallow depths. Siliceous shales can even maintain high fracture permeabilities at depths in excess of 300 m (Isherwood, 1981; Hunter and Young, 1953; Regan, 1953). Fractured shale can supply water to shallow wells (Upson, 1951), and it can transmit contaminated water away from waste-disposal trenches. Some thick shales at depths of a few hundred meters, such as the Pierre Shale in Colorado, South Dakota, and adjacent states, show evidence of regional upward leakage of water, which is larger than can be accounted for by using Darcy's Law along with laboratory measurements of permeability. The vertical permeability necessary to account of the fluid movement has been attributed to widely spaced fractures (Bredehoeft and Neuzil, 1980).

Despite the evidence of regional flow upward through thick shale units, a number of lines of evidence exist that many shale units, particularly those at depths in excess of 300 m, do form extensive barriers to cross-formational movement of ground water (Warner and others, 1986). Stratigraphic correlations establish the continuity of many of these units over thousands of square kilometers. Contrasts in water chemistry above and below shale units suggest effective separation of waters. And, perhaps most important, large differences in hydraulic head have evidently been maintained across shale units for many thousands if not millions of years (Sharp and Domenico, 1976). No geologic unit is absolutely impermeable, but some shale is about as close to this absolute condition as any geologic material measured within a few kilometers of the Earth's surface.

## CHEMICAL CHARACTERISTICS OF GROUND WATER

The total dissolved solids (TDS) in ground water from sandstone varies from less than 30 mg/l in some shallow aquifers of the Atlantic and Gulf coastal regions (sample 7, Table 3) to brines approaching saturation with respect to halite in deep sandstones within several sedimentary basins in North America (samples 8, 9, 10, Table 3). A number of broad generalities can be made concerning the chemistry of waters in sedimentary rocks (Table 4) with the understanding that numerous exceptions exist to each of the generalizations. For example, although TDS in ground water of a given region commonly increases with depth, one region cannot be compared directly with another. Saline water is found at depths of less than 70 m below valleys in northwestern Missouri, while potable water is recovered from Carrizo Sandstone in Texas at depths greater than 1,200 m.

Although Table 4 is empirically based, explanations exist for most of the statements. As an example, gypsum is a common constituent of sedimentary rocks. Where present, its solubility is high enough to produce water with more than 2,000 mg/l TDS, hence the large amounts of $Ca^+$ associated with $SO_4^{2-}$ in samples 13 and 15 in Table 3 and the resulting qualifications of statements D and E in Table 4. As another example, statement B in Table 4 relates to the fact that most sandstones are rather pure quartz sandstones with clay, quartz, calcite, or iron-oxide cements. The solubilities of both clays and quartz are low, and equilibrium values of less than 12 mg/l $SiO_2$ are to be expected unless opal, volcanic glass, and silicate minerals such as olivine or the feldspars are present. Finally, in very dilute waters that have not undergone extensive mineral-water interactions, the addition of only a small amount of material in solution will alter completely the relative concentrations of dissolved constituents. Thus, in sample 7 of Table 3, a very small amount of NaCl has evidently been dissolved, and even though in concentrations of less than 4.5 mg/l, $Na^+$ and $Cl^-$ dominate the dissolved ions. The relative importance of dissolved constituents to be found in waters with less than 100 mg/l TDS is generally difficult to predict. For this reason, statements D and G in Table 4 have a reference to a lower limit of 100 mg/l TDS. Mixing of waters within aquifers or within wells or springs could, of course, result in numerous exceptions to the generalizations given in Table 4.

TABLE 3. SELECTED CHEMICAL ANALYSES OF GROUND WATER FROM CLASTIC SEDIMENTARY ROCKS

| Lithology | Age | Location | Well Depth (m) | $SiO_2$ | Ca | Mg | Na | K | Fe | $HCO_3$ | $CO_3$ | $SO_4$ | Cl | F | $NO_3$ | Conductivity μ Mho | TDS ppm | pH | References |
|---|---|---|---|---|---|---|---|---|---|---|---|---|---|---|---|---|---|---|---|
| 1. Sandstone | Cretaceous | Wyoming | 1,050 | 20 | 5 | 1 | 473 | 5.2 | 0.01 | 1,220 | — | 2 | 31 | 8 | 4.3 | 1,830 | 1,150 | 8.2 | Littleton and Swenson, 1950 |
| 2. Shale | Triassic | Virginia | 46 | 45 | 14 | 5 | 6 | 1 | — | 29 | — | 6 | 10 | 0.0 | 3.9 | — | 190 | 6.4 | LeGrand, 1960 |
| 3. Sandstone | Jurassic | Utah | — | 10 | 101 | 15 | 343 | 3.4 | — | 200 | — | 806 | 47 | 0.3 | 5.8 | — | — | 7.9 | Woodward-Clyde, 1984 |
| 4. Sandstone | Cambrian | Illinois | 454 | — | 50 | 26 | 1.3 | 2.8 | — | 281 | — | 5.9 | 1.3 | — | — | — | 235 | — | Visocky and others, 1985 |
| 5. Sandstone | Cambrian | Illinois | 951 | — | 1,000 | 930 | 15,000 | 270 | — | 146 | — | 1,400 | 37,000 | 1.0 | 2.6 | — | 55,800 | 8.2 | Visocky and others, 1985 |
| 6. Sandstone | Cretaceous | Wyoming | 221 | 13 | 8.5 | 0.4 | 310 | 5 | 4.1 | 580 | 12 | 2 | 142 | 0.1 | 5.1 | 1,340 | 802 | 8.2 | Rapp and Durum, 1953 |
| 7. Sandstone | Mississippian | Alabama | 76 | 8.2 | 1.6 | 0.9 | 2.6 | 0.7 | 0.2 | 4 | — | 1.4 | 4.2 | 0.2 | 0.0 | 37 | 28 | 5.0 | Robinson and others, 1953 |
| 8. Sandstone | Cretaceous | Alberta | 1,283 | 22 | 5,750 | 1,070 | 31,500 | 585 | 0.0 | 140 | — | 180 | 60,400 | 0.1 | 0.0 | 109,000 | 106,000 | 6.8 | White, 1965 |
| 9. Sandstone | Eocene | Texas | 2,088 | 30 | 6,610 | 593 | 51,800 | 341 | 41 | 225 | — | 735 | 95,200 | 2.1 | 5.1 | 143,000 | 156,131 | — | White, 1965 |
| 10. Sandstone | Cambrian | Missouri | 190 | 17 | 49 | 39 | 5.6 | — | 0.03 | 320 | — | 9.7 | 8.2 | 0.1 | 0.3 | — | 379 | 7.3 | Homyk and others, 1967 |
| 11. Sandstone | Ordovician | Missouri | 157 | 3.8 | 55 | 31 | 16 | — | 0.13 | 307 | — | 19 | 2.5 | 1.2 | — | — | 289 | 7.4 | Homyk and others, 1967 |
| 12. Shale | Paleocene | Alberta | 31 | — | 4 | 1 | 200 | 8.2 | 0.1 | 473 | — | 58 | 1 | 0.5 | 6.5 | — | 513 | — | Garven, 1982 |
| 13. Conglomerate | Jurassic | Wyoming | 13 | 20 | 428 | 149 | 794 | 14 | 0.04 | 334 | 0.0 | 3,000 | 14 | 1.4 | — | 5,190 | 4,590 | 7.6 | Swenson and others, 1951 |
| 14. Shale | Triassic | New Jersey | 92 | 16 | 29 | 16 | 12 | 1.1 | 0.02 | 126 | 0.0 | 22 | 12 | 0.2 | 26 | 340 | 260 | 7.9 | White and others, 1963 |
| 15. Shale | Cretaceous | North Dakota | 11 | 26 | 416 | 143 | 362 | 14 | 3.5 | 104 | 0.0 | 2,170 | 38 | 0.2 | 0.1 | 3,560 | 3,300 | 6.3 | White and others, 1963 |
| 16. Sandstone | Pennsylvanian | Kentucky | 58 | 9.4 | 16 | 6.4 | 13 | 2.2 | 4.2 | 98 | 0.0 | 14 | 3 | 0.1 | 1.0 | 194 | 168 | 7.2 | White and others, 1963 |
| 17. Sandstone | Ordovician | Wisconsin | 587 | 8.7 | 60 | 31 | 12 | 4.0 | 0.4 | 285 | 0.0 | 111 | 12 | 0.5 | 0.8 | 658 | 577 | 7.6 | White and others, 1963 |
| 18. Sandstone | Jurassic | Arizona | 182 | 11 | 0.8 | 0.9 | 144 | 0.8 | 0.8 | 217 | 33 | 64 | 16 | 0.3 | 0.5 | 630 | 489 | 9.2 | White and others, 1963 |
| 19. Sandstone | Miocene | Louisiana | 2,575 | 16 | 9,210 | 1,070 | 63,900 | 869 | 110 | 115 | — | 153 | 124,000 | 1.4 | 0.0 | 158,000 | 200,000 | 6.3 | White and others, 1963 |

TABLE 4. GENERALIZATIONS CONCERNING CHEMISTRY OF WATER FROM SANDSTONE AND SHALE

| Statement | Analyses in Table 3 |
| --- | --- |
| Total dissolved solids (TDS) generally increase with depth. | Samples 8, 9, and 19 are representative of deep ground water. Samples 4 and 5 show effects of increasing depth. |
| Silica in solution is generally lower than 30 mg/l, and the median is close to 10 mg/l for quartz sandstones. | All samples except 2, which probably is high because of feldspars in associated sandstone. |
| Fluoride values are commonly greater than 0.3 mg/l, and may be greater than 5 mg/l in sodium bicarbonate waters. | Sample 1. |
| Ca/Na weight ratios are greater than 1.0 for waters having a TDS of between 100 and 400 mg/l. | Samples 2, 4, 10, 11, 14, and 16. |
| Ca/Na weight ratios are less than 0.1 for waters having a TDS of between 500 and 100,000 mg/l, provided $SO_4$ concentrations are less than 20% of Cl plus $HCO_3$ concentrations. | Samples 1, 6, and 12. |
| Ca/Na weight ratios increase somewhat as TDS exceeds 100,000 mg/l. Values of the ratio are generally between 0.1 and 0.5 for brines. | Samples 8, 9, and 19. |
| $HCO_3^-$ is the dominant anion in waters with TDS concentrations of between 100 and 800 mg/l. | Samples 2, 4, 6, 10, 11, 12, 14, 16, 17, and 18. |
| $Cl^-$ is the dominant anion in waters with TDS concentrations in excess of 50,000 mg/l. | Samples 5, 8, 9, and 19. |

# REFERENCES

Almon, W. R., Fullerton, L. B., and Davies, D. K., 1976, Pore space reduction in Cretaceous sandstones through chemical precipitation of clay minerals: Journal of Sedimentary Petrology, v. 46, p. 89–96.

Angevine, C. L., and Turcotte, D. L., 1983, Porosity reduction by pressure solution, a theoretical model for quartz arenites: Geological Society of America Bulletin, v. 94, p. 1129–1134.

Athy, L. F., 1930, Density, porosity, and compaction of sedimentary rocks: American Association of Petroleum Geologists Bulletin, v. 14, p. 1–24.

Bennion, D. W., and Griffiths, J. C., 1966, A stochastic model for predicting variations in reservoir rock properties: American Institute of Mining and Metallurgical Engineers Transactions, v. 237, p. 9–16.

Bredehoeft, J. D., and Neuzil, C. E., 1980, Regional flow in the Dakota aquifer; A study of the role of the confining layer: EOS American Geophysical Union Transactions, v. 61, p. 952.

Buryakovskiy, L. A., and Dzhafarov, I. S., 1980, Permeability anisotropy in stratified oil- and water-saturated rocks: International Geological Review, v. 22, no. 9, p. 1067–1069.

Chamberlain, T. C., 1885, Requisite and qualifying conditions of artesian flow: U.S. Geological Survey Annual Report, v. 5, p. 131–175.

Chu, C. L., Wang, C. Y., and Lin, W., 1981, Permeability and frictional properties of San Andreas fault gouges: Geophysical Research Letters, v. 8, no. 6, p. 565–568.

Dapples, E. C., 1972, Some concepts of cementation and lithification of sandstones: American Association of Petroleum Geologists Bulletin, v. 56, p. 3–25.

Darton, N. H., 1897, Preliminary report on artesian waters of a portion of the Dakotas: U.S. Geological Survey Annual Report, v. 17, part 2, p. 609–697.

Davis, S. N., 1969, Porosity and permeability of natural materials, *in* De Wiest, R.J.M., ed., Flow through porous media: New York, Academic Press, p. 53–89.

Davis, S. N., and De Wiest, R.J.M., 1966, Hydrogeology: New York, John Wiley and Sons, 463 p.

Dullien, F.A.L., 1979, Porous media, fluid transport, and pore structure: New York, Academic Press, 396 p.

Garrels, R. M., and Mackenzie, F. T., 1971, Evolution of sedimentary rocks: New York, W. W. Norton and Company, 397 p.

Garven, G., 1982, Groundwater hydrology of the Pine Lake research basin, Alberta: Alberta Research Council Earth Science Report 82-2, 153 p.

Hedberg, H. D., 1936, Gravitational compaction of clays and shales: American Journal of Science, 5th series, v. 31, p. 241–287.

Homyk, A., Harvey, E. J., and Jeffery, H. G., 1967, Water resources, *in* Mineral and water resources of Missouri: Missouri Division of Geological Survey and Water Resources, v. 43, 2nd series, p. 253–399.

Houseknecht, D. W., 1984, Influence of grain size and temperature on intergranular pressure solution, quartz cementation, and porosity in a quartzose sandstone: Journal of Sedimentary Petrology, v. 54, p. 348–361.

Hunter, C. D., and Young, D. M., 1953, Relationship of natural gas occurrence and production in eastern Kentucky (Big Sandy gas field) to joints and fractures in Devonian bituminous shale: American Association of Petroleum Geologists Bulletin, v. 37, p. 282–299.

Huntoon, P. W., 1986, Incredible tale of Texasgulf Well 7 and fracture permeability, Paradox basin, Utah: Ground Water, v. 24, p. 643–653.

Huntoon, P. W., and Lundy, D. A., 1979, Fracture-controlled ground-water circulation and well siting in the vicinity of Laramie, Wyoming: Ground Water, v. 17, no. 5, p. 463–469.

INTERA Technologies, Inc., 1986, Second status report on regional groundwater flow modeling for the Palo Duro basin, Texas: Office of Nuclear Waste Isolation, Battelle Memorial Institute, Report ONWR-604, 144 p.

Isherwood, D., 1981, Geoscience data base handbook for modeling a nuclear waste repository: U.S. Nuclear Regulatory Commission, NUREG/CR-0912, vols. 1 and 2, 315 p., and 331 p.

Johnson, C. R., and Greenkorn, R. A., 1963, Correlation of lithology and permeability for a Virgilian sandstone reservoir in central Oklahoma: Geological Society of America Proceedings, v. 73, p. 180–181.

Land, L. S., and Dutton, S. P., 1978, Cementation of a Pennsylvanian deltaic sandstone, isotopic data: Journal of Sedimentary Petrology, v. 48, p. 1167–1176.

LeGrand, H. E., 1960, Geology and ground-water resources of Pittsylvania and Halifax counties: Virginia Division of Mineral Resources Bulletin 75, 86 p.

Lins, T. W., 1950, Origin and environment of the Tonganoxie Sandstone in northeastern Kansas: Kansas Geological Survey Bulletin 86, part 5, p. 105–140.

Littleton, R. T., and Swenson, H. A., 1950, Ground-water conditions in the vicinity of Gillette, Wyoming: U.S. Geological Survey Circular 76, 43 p.

Manger, G. E., 1963, Porosity and bulk density of sedimentary rocks: U.S. Geological Survey Bulletin 1144-E, 55 p.

Maxwell, J. C., 1964, Influence of depth, temperature, and geologic age on porosity of quartzose sandstone: American Association of Petroleum Geologists Bulletin, v. 48, p. 697–709.

McKee, B., 1960, Tectonic significance of phase change; Plagioclase jadeite + quartz: Geological Society of America Bulletin, v. 71, p. 1926–1927.

Moore, R. C., 1949, Divisions of the Pennsylvanian System in Kansas: Kansas Geological Survey Bulletin 83, 203 p.

Muskat, M., 1946, The flow of homogenous fluids through porous media: New York, McGraw-Hill Book Company, 763 p.

Neuzil, C. E., 1986, Groundwater flow in low-permeability environments: Water Resources Research, v. 22, p. 1163–1195.

Norton, W. H., 1912, Underground water resources of Iowa: Iowa Geological Survey, v. 21, Annual Reports, 1910 and 1911, chapters 1–4, p. 31–158.

Palciauskas, V. V., and Domenico, P. A., 1980, Microfracture development in compacting sediments; Relations to hydrocarbon-maturation kinetics: American Association of Petroleum Geologists Bulletin, v. 64, p. 927–937.

Pettijohn, F. J., Potter, P. E., and Siever, R., 1972, Sand and sandstone: New York, Springer-Verlag, 618 p.

Rapp, J. R., and Durum, W. H., 1953, Reconnaissance of the geology and ground-water resources of the La Prele area, Converse County, Wyoming: U.S. Geological Survey Circular 243, 33 p.

Rats, M. V., and Chernayshov, S. N., 1967, Statistical aspect of the problem on the permeability of jointy rocks, *in* Hydrology of fractured rocks: International Association of Scientific Hydrology, v. 1, p. 227–336.

Regan, L. J., 1953, Fractured shale reservoirs of California: American Association of Petroleum Geologists Bulletin, v. 37, p. 201–219.

Rehm, B. W., Groenewold, G. H., and Marin, K. A., 1980, Hydraulic properties of coal and related materials: Ground Water, v. 18, p. 551–591.

Robinson, W. H., Ivey, J. B., and Billingsley, G. A., 1953, Water supply of the Birmingham area, Alabama: U.S. Geological Survey Circular 254, 53 p.

Rubenstein, S. R., 1966, Petrographic and related physical properties of Navajo Sandstone, Arizona: Engineering Geology, Association of Engineering Geologists, v. 3, p. 40–51.

Schmidt, V., and McDonald, D. A., 1980, Secondary reservoir porosity in the course of sandstone diagenesis: American Association of Petroleum Geologists, Continuing Education Course Note Series 12, 125 p.

Schoeller, H., 1962, Les eaux souterraines: Paris, Masson, 642 p.

Sharp, J. M., and Domenico, P. A., 1976, Energy transport in thick sequences of compacting sediment: Geological Society of America Bulletin, v. 87, p. 390–400.

Sibley, D. F., and Blatt, H., 1976, Intergranular pressure solution and cementation of the Tuscarora orthoquartzite: Journal of Sedimentary Petrology, v. 46, p. 881–896.

Slaughter, G. M., White, R. M., and Alger, R. P., 1983, Permeability of selected sediments in the vicinity of five salt domes in the Gulf Interior Region: Office of Nuclear Waste Isolation, Battelle Memorial Institute Report ONWI-356, 71 p.

Smith, B. L., 1963, Geology of the Jersey Central Power and Light Company,

Yard Creek pumped storage project, northern New Jersey [abs.]: Geological Society of America Special Paper 73, p. 246–247.

Stone, L., 1886, Chicago artesian wells: Chicago Academy of Sciences Bulletin, v. 1, no. 8, p. 93–102.

Swenson, F. A., Bach, W. K., and Swenson, H. A., 1951, Ground-water resources of the Paintrock irrigation project, Wyoming: U.S. Geological Survey Circular 96, 45 p.

Upson, J. E., 1951, Geology and ground-water resources of the south-coast basins of Santa Barbara County, California: U.S. Geological Survey Water-Supply Paper 1108, 144 p.

Visocky, A. P., Sherrill, M. G., and Cartwright, K., 1985, Geology, hydrology, and water quality of the Cambrian and Ordovician systems in northern Illinois: Illinois Geological Survey, Cooperative Groundwater Report 10, 136 p.

Waldschmidt, W. A., 1941, Cementing materials in sandstones and their probable influence on migration and accumulation of oil and gas: American Association of Petroleum Geologists Bulletin, v. 25, p. 1839–1879.

Wallace, C. A., 1976, Diagenetic replacement of feldspar by quartz in the Uinta Mountain Group, Utah, and its geochemical implications: Journal of Sedimentary Petrology, v. 46, p. 847–861.

Warner, D. L., Syed, T., and Davis, S. N., 1986, Confining layer study; Supplemental report: Norman, Oklahoma, Engineering Enterprises, Incorporated, Prepared for U.S. EPA Region V, Contract no. 68-01-7011, 168 p.

White, D. E., 1965, Saline waters of sedimentary rocks, *in* Young, A., and Galley, J. E., eds., Fluids in subsurface environments: American Association of Petroleum Geologists Memoir 4, p. 342–366.

White, D. E., Hem, J. D., and Waring, G. A., 1963, Data of geochemistry (sixth edition), Chapter F, Chemical composition of subsurface waters: U.S. Geological Survey Professional Paper 440–F, p. F1–F67.

Wilson, M. D., and Pittman, E. D., 1977, Authigenic clays in sandstones; Recognition and influence of reservoir properties and paleoenvironmental analysis: Journal of Sedimentary Petrology, v. 47, p. 3–31.

Woodward-Clyde Consultants, 1984, Geologic characterization report for the Paradox basin study region, Utah Study areas, v. 6, Salt Valley: Office of Nuclear Waste Isolation, Battelle Memorial Institute Report ONWI-290, 190 p.

Young, A., Low, P. F., and McLatchie, A. S., 1964, Permeability studies of argillaceous rocks: Journal of Geophysical Research, v. 69, p. 4237–4245.

MANUSCRIPT ACCEPTED BY THE SOCIETY FEBRUARY 24, 1988

The Geology of North America
Vol. O-2, Hydrogeology
The Geological Society of America, 1988

# Chapter 38

# *Carbonate rocks*

**J. V. Brahana**
*U.S. Geological Survey, Federal Courthouse, Nashville, Tennessee 37203*
**John Thrailkill**
*Department of Geology, University of Kentucky, Lexington, Kentucky 40506-0059*
**Tom Freeman**
*Department of Geology, University of Missouri, Columbia, Missouri 65211*
**W. C. Ward**
*Department of Earth Sciences, University of New Orleans, New Orleans, Louisiana 70148*

## INTRODUCTION

The hydrogeology of carbonate rocks is more variable than that of any other major rock type. Whereas some carbonate formations are among the most permeable and productive aquifers known, others have such low permeability that they serve as major confining layers and yield virtually no water to wells. In carbonate terranes throughout North America and the world, hydrogeologic variability generally is the rule rather than the exception.

This wide range of hydrogeologic behavior results from the complex interaction of many geologic and hydrologic variables. Most important of these variables is the continually evolving porosity and permeability distribution of the rock mass, which is a result of the chemical interaction of the carbonate rocks and the water moving through them. The flow of water through the rocks is essential for aquifer development, for it is flow that serves as the primary transport mechanism in the dissolution of carbonate rocks.

The purpose of this chapter is to describe the major hydrogeologic controls, principles, and processes that affect the occurrence, movement, storage, and geochemistry of ground water in carbonate rocks. The chapter represents a synthesis of selected concepts and theories currently thought to explain some of the major similarities and differences of carbonate rocks in various settings. It should be emphasized, however, that many ideas in this chapter are not universally accepted as fact. Even among the authors of this chapter there are several major points of disagreement. Generalizations about complex areas have been incorporated to demonstrate major principles. While these simplifications may be valid on a regional basis, they may not accurately define localized areas within a region. For details that concern specific regions, see chapters 3, 19, and 23 through 29 of this volume.

This chapter is limited to concepts selected by the authors as most significant and to examples from the North American con-

tinent. Although isolated references are made to a few studies outside North America, much of the worldwide research is not discussed. Because of space limitations, only a small fraction of the current research in carbonate hydrogeology and the related aspects of karst geomorphology, carbonate petrology, speleology, petroleum geology, civil engineering, and soil science is discussed. Many important works are not referenced, although their conclusions have been distilled and condensed within this chapter.

Reports with extensive bibliographies include: International Association of Scientific Hydrology (1967), Stringfield and LeGrand (1969), LaMoreaux and others (1970), Herak and Stringfield (1972), Burger and Dubertret (1975), Warren and Moore (1975), Dilamarter and Csallany (1977), Tolson and Doyle (1977), Milanovic (1981), Yevjevich (1981), Longman (1982), Back and Freeze (1983), Back and LaMoreaux (1983), James and Choquette (1983), Scholle and others (1983), Castany and others (1984), LaFleur (1984), LaMoreaux and others (1984), Jennings (1985), and LaMoreaux (1986).

## ORIGIN OF CARBONATE SEDIMENTS

Carbonate sediments accumulate in a wide variety of environments, both marine and nonmarine. Depositional environment may have a profound influence on the properties of carbonate aquifers, because depositional setting determines texture, fabric, original porosity and permeability, and, to some extent, thickness of carbonate accumulations. Additional discussion of environments of deposition and diagenesis is in Back (this volume).

The bulk of carbonate sediments are deposited in tropical and subtropical seas where there is slight or no influx of terrigenous detritus from land areas. Marine depositional environments include (1) tidal flat, (2) beach and coastal dune, (3) continental shelf, (4) bank, (5) reef, (6) basin margin and slope, and (7) deeper ocean or basin.

Brahana, J. V., Thrailkill, J., Freeman, T., and Ward, W. C., 1988, Carbonate rocks, *in* Back, W., Rosenshein, J. S., and Seaber, P. R., eds., Hydrogeology: Boulder, Colorado, Geological Society of America, The Geology of North America, v. O-2.

On land the most extensive deposits of calcium carbonate collect in lakes, which may be freshwater, alkaline, or saline. In addition, some calcium-carbonate lake deposits are associated with springs. Carbonate-depositing lakes are not restricted to tropical climates, but are common in temperate latitudes. Other terrestrial carbonate deposits are caliche (soil-zone deposits) and travertine (cave, karst, and hotspring deposits).

Nearly all carbonate sediment in modern oceans is ultimately derived from calcareous algae and $CaCO_3$-producing animals living at or near the site of sediment accumulation. Some marine carbonates and the bulk of nonmarine carbonates are physicochemical precipitates and/or algae- and bacteria-induced precipitates.

Carbonate sand is commonly composed of locally derived skeletal fragments of marine organisms such as molluscs, echinoderms, corals, ostracodes, foraminifers, and red and green algae. In ancient seas, other organisms such as brachiopods, crinozoans, and trilobites were important contributors to coarser skeletal sediments. Mud-sized carbonate sediment is derived directly from certain calcareous algae or is produced by abrasion and bioerosion of larger skeletal particles.

Reef buildups occur where communities of calcareous organisms construct a skeletal framework. Although corals produce the most conspicuous reefs today, many other organisms (algae, sponges, stromatoporoids, and rudistid molluscs) were important reef builders in ancient seas.

Nonskeletal components of carbonate sands are fecal pellets, oolites, and fragments of penecontemporaneously lithified sediment (intraclasts). In some marine and nonmarine environments, carbonate mud is a physicochemical precipitate.

Commonly the quality of carbonate aquifers can be directly related to original mineralogy of the carbonate sediment because the amount of early dissolution and cementation is dependent, in large part, on the stability of the carbonate minerals. Therefore, the original mineralogy is a major constraint on the early-diagenetic processes that alter porosity and permeability. Today, shallow-marine sediments are composed predominantly of "metastable" aragonite and Mg calcite, whereas pelagic sediment is predominantly "stable" low-Mg calcite. Despite ocean-water saturation with respect to magnesite and dolomite, these minerals are unknown as primary products in the marine environment.

Because skeletal grains are irregular in shape, carbonate sediment may have extraordinarily high porosities. Modern carbonate sediment commonly has 40 to 70 percent porosity. Most limestones, however, have no more than a few percent porosity because postdepositional cementation and pressure-solution compaction tend to obliterate original porosity.

## FACTORS THAT AFFECT THE EVOLVING POROSITY AND PERMEABILITY OF CARBONATE ROCKS

The porosity and permeability of carbonate rocks can be affected by more than 60 different processes and controls (Table 1). These have been grouped loosely into seven categories: (1) diagenetic, (2) geochemical, (3) lithologic-stratigraphic, (4) structural-tectonic, (5) hydrologic, (6) weathering-geomorphic, and (7) historical geologic-chronologic. In simplistic terms, these processes and controls can be considered causes in a cause-effect relation. The effects are the resulting distributions of porosity and permeability.

Innovative porosity classification schemes (Choquette and Pray, 1970; Longman, 1982) have been developed on the basis of studies by carbonate petrologists and petroleum geologists who concentrate on the solid phase of the aquifer. Permeability classifications (White, 1969, 1977; LeGrand and Stringfield, 1971), on the other hand, tend to concentrate on the fluid phase or flow characteristics of an aquifer; these classifications characteristically have been the domain of hydrogeologists and karst researchers.

For this chapter, subtle distinctions between types of porosity and permeability are not important. What matters is the presence or absence of porosity and permeability, the degree to which each is developed, and the factors that influenced their development.

Quantitative representation of carbonate aquifers is typically expressed in some terms of storativity, hydraulic conductivity, and yield. Examples of the range and typical values of porosity, hydraulic conductivity, and well yields are given in Table 2; diverse lithologies were chosen to show the wide variation between different types of carbonate rocks. Ranges of values were included to show the huge differences that may occur within the same lithology, a phenomenon particularly noticeable in karst areas.

Figure 1 shows the generalized porosity, pore size, and hydraulic conductivity of several types of carbonate rocks (Smith and others, 1976). Also included is a speculative boundary line suggested by Ford (1980) that differentiates conditions favoring development of karst features (enterable cave systems, intermediate and large surface landforms). The line originates at a rock pore size of about 10 microns, a value reported by Bocker (1969) to be the lower threshold of pore spacing necessary for significant flow and solvent action by water. Figure 1 emphasizes the principle that wherever comparatively high primary porosity occurs, solutional attack is widely diffused. If the solvent attack is not concentrated against restricted zones of permeability, the pores grow more or less equally. Point-centered karst forms such as dolines and linear cave systems cannot be produced, and thus cannot capture additional surface water where solution is diffused at a scale of centimeters or less (Ford, 1980).

### Diagenetic

Diagenetic factors represent an important category of processes and controls that affect porosity and permeability. After deposition, carbonate sediments are subjected to a sequence of changes associated with induration and lithification. An obvious early physical process is compaction, which functions identically to the same porosity reduction mechanism as occurs in terrigen-

**TABLE 1. PROCESSES AND CONTROLS THAT AFFECT POROSITY AND PERMEABILITY OF CARBONATE ROCKS**

| Geologic Factor | Processes | Controls | General Influence |
|---|---|---|---|
| Diagenetic | Compaction<br>Cementation<br>Pressure solution<br>Solution* (includes recrystallization, inversion, micritization) | Original porosity and permeability;<br>Original mineralogy;<br>Grain size/surface area;<br>Proximity to sea level (uplift or burial);<br>Volume and rate of water movement;<br>Fluid chemistry: pH, $pCO_2$, salts in solution; temperature, pressure | Influences initial distribution of porosity and permeability of indurated rock mass. Many of these are geochemical in nature; they occur very early in the history of the rock. |
| Geochemical | Solution* (dissolution)<br>Dolomitization<br>Dedolomitization<br>Precipation<br>Sulfate reduction<br>Redox | Ground-water flux;<br>Original porosity and permeability;<br>Mineralogy;<br>Fluid chemistry: pH, $pCO_2$, salts in solution, temperature, pressure, mineral-water saturation | Influences later development of porosity and permeability; influences water chemistry. |
| Lithologic-Stratigraphic | | Layer thickness;<br>Sequence thickness;<br>Variability in texture (vertical);<br>Variability in permeability (vertical);<br>Original porosity and permeability inherited from diagenesis;<br>Bulk chemical purity;<br>Grain size | Influences anisotropy of rock mass, thereby resulting in zones potentially more permeable if other geologic factors are favorable. |
| Structural-Tectonic | Uplift<br>Tilting<br>Folding<br>Jointing<br>Faulting<br>Metamorphism | Fracture density;<br>Openness of fractures;<br>Layer (permeability) orientation | Influences orientation of permeability zones.<br>Influences integrity of confining layers.<br>In extreme instances (metamorphism), influences existence of permeability zones. |
| Hydrologic | Dynamic ground-water flow* | Climatic—temperature;<br>Climatic—precipitation;<br>Depth of circulation;<br>Location of boundaries;<br>Existence of complete flow systems;<br>Flux;<br>Initial anisotropy-vertical variation;<br>Springs;<br>Surface-water/ground-water relation<br>Recharge;<br>Hydraulic gradient;<br>Size of ground-water basin | Influences existence of flow systems.<br>Influences rate of flow system evolution. |
| Weathering-Geomorphic | Infilling (fluvial and glacial)<br>Unloading | Topography;<br>Relief;<br>Soil development;<br>Cap rock;<br>Degree of karstification;<br>Base level;<br>Surface slope | Influences development of flow systems.<br>Influences destruction of permeability by sedimentation.<br>Influences shallow porosity-permeability development. |
| Historical Geologic-Chronologic | | Sequence of events;<br>Duration of events | Influences stage of development of specific permeability zones. |

*Most important.

ous sediments. The large original porosity of the sediments commonly is reduced by one to two orders of magnitude if they are buried under an accumulating sequence of deposits.

Three processes appear to account for this decrease in porosity: (1) mechanical compaction (grain rearrangement and grain breakage); (2) chemical compaction (pressure dissolution at grain-to-grain contacts and along stylolites); and (3) redeposition of carbonate, as cement, derived through pressure dissolution. Halley and Schmoker (1983) favor chemical compaction as the dominant factor of the three.

It should be noted that chemical compaction can occur under shallow burial. Incipient pressure solution in carbonate

*J. V. Brahana and Others*

## TABLE 2. REPRESENTATIVE VALUES OF POROSITY, HYDRAULIC CONDUCTIVITY, AND WELL YIELD FOR VARIOUS CARBONATE AQUIFERS*

| Lithology | Porosity (%) "Average" | Range | Hydraulic Conductivity (m/day) "Average" | Range | Well Yield (L/s) "Average" | Range |
|---|---|---|---|---|---|---|
| Carbonate mud; nonindurated | 55 | 40 to 70 | $10^{-2\dagger}$ | $10^{-3}$ to $10^{-1\dagger}$ | — | — |
| Dolomite (primary) | 1 | 0.1 to 5 | $10^{-2}$ | $10^{-4}$ to 1 | 0.1 | 0 to 1 |
| Dolomite (secondary) | 3 | 0.1 to 20 | $10^{-2}$ | $10^{-4}$ to 1 | 0.1 | 0 to 2 |
| Tertiary limestone; fine grained, few fractures | 25 | 20 to 35 | $10^{-2}$ | $10^{-4}$ to 1 | 0.1 | 0 to 5 |
| Paleozoic limestone; fine grained, few fractures | 2 | 0.1 to 10 | $10^{-2}$ | $10^{-4}$ to 1 | 0.01 | 0 to 1 |
| Oolitic limestone | 15 | 1 to 25 | $10^{-1}$ | $10^{-2}$ to $10^{-1}$ | 1 | 1 to 10 |
| Holocene coral limestones | 35 | 30 to 50 | $10^{3}$ | $10^{2}$ to $10^{4}$ | $10^{2}$ | 1 to $10^{2}$ |
| Karstified limestone with caverns | 10 | 5 to 50 | $10^{5}$ | $10^{-1}$ to $10^{7}$ | 0.1 | 0 to $10^{3}$ |
| Marble, fractured | 0.7 | 0.1 to 2 | $10^{3}$ | $10^{-3}$ to 1 | 0.1 | 0 to 2 |
| Chalk | 30 | 15 to 45 | $10^{-2}$ | $10^{-3}$ to 1 | 0.01 | 0 to 1 |

Notes:
m/day = meters per day
L/s = liters per second
"Average" commonly reported as a median or average value.

*See also Table 2, Heath, this volume.

†No measurements known; estimated from physical parameters.

Sources:
Clark, 1966
Davis and DeWeist, 1966
Davis, 1969
Smith and others, 1976
Freeze and Cherry, 1979
Matthess, 1982
Scholle, 1979

rocks has been reported from depths of 100 m (Schlanger, 1964) and 120 m (Halley and Schmoker, 1983).

A comprehensive survey of cementation processes and products that affect marine carbonate sediments has recently been provided by James and Choquette (1983). Isotopically light $\delta^{13}$C values suggest that cementation of some limestones occurs during burial diagenesis, with calcium carbonate being provided by pressure solution and dissolution of aragonite skeletons (Hudson, 1975). Tan and Hudson (1974) suggested the cements were emplaced by infiltrating ground water.

Because carbonate sediments commonly include metastable aragonite, substantial changes in porosity and permeability can develop during early diagenesis. Primary aragonite needles and skeletal fragments are preferentially dissolved, creating molds, and the voids later are filled by mold-occluding cement. The process results in a general overall porosity reduction, although zones of greater permeability may be concentrated during this phase.

Dissolution of more stable constituents of carbonate rocks can occur under burial conditions which favor the generation of low pH water. Moore and Druckman (1981) made a case for the enhancement of residual depositional porosity by $CO_2$ and $H_2S$

generated by sulfate-reducing bacteria associated with petroleum migration. Incipient dissolution of this type is typically fabric-selective because of minor differences in fabric-constituent solubility owing to differences in minor-element composition and/or texture, but advance stages of solution typically lead to nonfabric-selective vug development.

Major processes that affect porosity and permeability are shown in Figure 2. The relative importance of each process is plotted as a function of depth. The most important of these processes, from a hydrogeologic standpoint, act in the shallow environment, generally less than 100 m deep.

Controls on diagenesis reflect the chemistry of the reactants and the immediate environment. As previously mentioned, modern marine carbonate sediments are mineralogically composed of aragonite, Mg-calcite, and calcite. These minerals are all stable in shallow sea water but unstable or metastable in fresh water. Under most freshwater conditions, high Mg-calcite is the least stable of these minerals, altering quickly via magnesium expulsion to calcite or by adding Mg ions to form dolomite. Aragonite is next most soluble and may dissolve to produce molds when altered via neomorphism to calcite. Calcite is the most stable; it may be dissolved, recrystallized, or remain unchanged depending

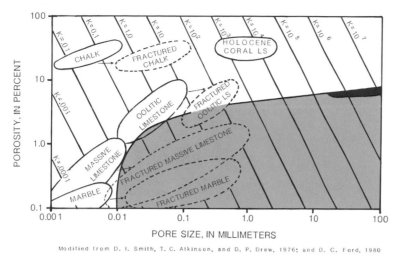

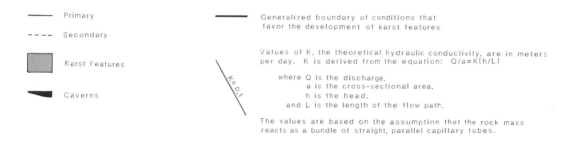

Figure 1. Primary and secondary porosity, pore size, and theoretical hydraulic conductivity of selected carbonate rocks, karst features, and caverns.

on the volume and chemistry of the water. Dolomite may be dissolved, altered to calcite, recrystallized, or remain unchanged depending on the same factors. Organic material and hydrocarbons may modify this "normal" diagenesis by coating grains, inhibiting alteration, or changing water chemistry.

Chalk is an example of a rock that remains relatively unchanged. Chalks have an initial composition that is almost entirely low-Mg calcite, commonly with some admixture of clay. Chalks, despite their fine grain size, have great chemical stability in marine as well as nonmarine settings. Chalks can be exposed to freshwater in near-surface settings for millions of years with little or no alteration (Scholle, 1979). This eliminates much variability in the local diagenetic alteration, which is so characteristic in shallow water limestones because of local, subtle uplift and subaerial exposure.

Original porosity affects secondary porosity by affording voids through which freshwater may move. Original porosity is a very important controlling influence on secondary permeability developed during the early diagenetic phase. Grain size is important in diagenesis. The finer the grain size, the more surface area is available to interact with the interstitial fluids.

If carbonate sediments are exposed to fresh water relatively early after deposition, diagenesis and porosity reduction are initiated early. If the sediments remain buried and the marine pore water is not flushed, diagenetic changes tend to occur much more slowly. Under most conditions, large volumes of fresh water coupled with rapid movement of that water are needed to cause significant dissolution in carbonate rocks. These large volumes over relatively short periods of time create increasing porosity.

Many environmental factors play an important role in dissolution of carbonate rocks. The following list generalizes the most simple of these relations:

| Factor | Relation to dissolution |
|---|---|
| pH of interstitial fluid | Inversely proportional |
| $CO_2$ content of fluid | Directly proportional |
| salts in solution | Directly proportional |
| temperature | Inversely proportional |

The effects of all geologic controls on the early diagenetic processes result in an indurated rock with a unique original porosity and permeability distribution that reflects both composition and environment through the time of lithification of the rock mass. It is often difficult to separate diagenetic factors from geochemical factors, and many researchers do not differentiate between the

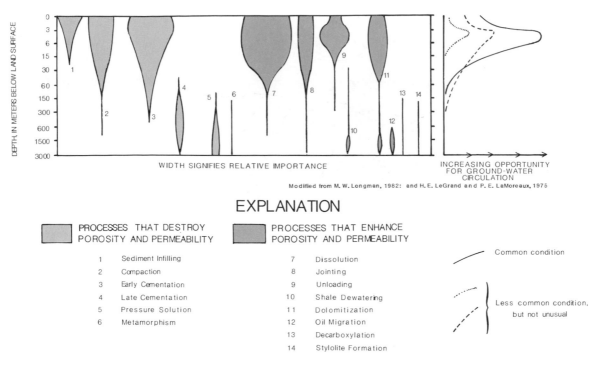

Figure 2. Relation between depth, ground-water circulation, and selected processes that destroy and enhance porosity and permeability in carbonate rocks.

two, but consider them a continuum. The distinction in this chapter is arbitrary; later postlithification processes are defined as geochemical, and early prelithification processes are defined as diagenetic.

### Geochemical

Geochemical processes play a major role in the evolution of carbonate aquifers. Garrels and Christ (1965) provided a detailed description of the reactions and equilibria involved. Abundant evidence has accumulated, both from chemical and geomorphological studies of carbonate aquifers within 100 m of land surface, that dissolution by carbonic acid is the principal process occurring. The opposite process of precipitation also is important, because it results in decreased porosity and permeability. In carbonate aquifers that are deeply buried, dissolution commonly is less active. Carbonic acid is created as $CO_2$ from the soil zone (and to a lesser extent, the atmosphere) and is dissolved in water.

The fundamental reactions for carbonic acid-calcite chemistry are described in Chapter 43 of this volume.

Studies in the central Appalachians (Jacobson and Langmuir, 1970), Mammoth Cave region (Thrailkill, 1972), and elsewhere have shown that water in shallow carbonate aquifers generally is undersaturated with respect to calcite and in equilibrium with a $P_{CO_2}$ somewhat higher than that of the normal atmosphere, as is the water of surface streams flowing on limestone. Although there are wide variations between sites, values of one-half calcite saturation are not atypical. In the Mammoth Cave region, although there was no clear increase in saturation with respect to calcite with distance of flow in the aquifer, Ca did increase substantially. This indicates either an addition of high-Ca water from the vadose zone or an introduction of $CO_2$ to drive calcite solution in the aquifer; some evidence suggests that the latter is true (Thrailkill, 1972). There also is evidence, based on vadose-zone studies, that large flows (with low $P_{CO_2}$) do arrive at the water table still unsaturated with respect to calcite, and that at least some vadose seepage is closed to $CO_2$ and can promote solution in the aquifer (Thrailkill and Robl, 1981).

Aquifers are also found in dolomites, and the pertinent carbonate chemistry of those aquifers deals with the mineral dolomite, and $Mg^{2+}$ as well as $Ca^{2+}$ in solution. In many field settings where both types of aquifers are present in the same area, it appears that dissolution and karstification in dolomites are less than in limestones. Although this appears to be true in the central Appalachians (Parizek and others, 1971), it is less evident in southern Missouri (Imes, 1987) and the southern Appalachians. Because the solubilities of the primary minerals dolomite and calcite are nearly equal, the reason for dolomite aquifers being less soluble than limestone aquifers, where the phenomenon occurs, must be related to kinetic, textural, structural, or other factors.

In addition to solution by carbonic acid, minor solution by sulfuric acid and organic acids has been documented. The basic concepts of the solution processes are the same, although the chemical components and details of the kinetics are different. The effects of organic acids are generally associated with petroleum evolution and migration and may be important in deep settings;

however, these are thought to be limited areally, and not significant hydrogeologically on a regional scale.

Conversion of calcium carbonate sediments to dolomite rocks occurs by means of several processes, which are collectively referred to in this paper as a single process, dolomitization. In general, dolomitization redistributes the porosity and permeability of the rock mass so that small pores become fewer but larger, thereby increasing permeability. Davies (1979) pointed out that dolomitization has a homogenizing effect on carbonate rocks. Intercrystalline porosity in dolomites is pervasive, and permeabilities in dolomites are significantly greater than those within equally porous limestone (Wardlaw, 1974).

Summaries of dolomitization models (Land, 1982; Morrow, 1982) indicate many settings in which the process has occurred, as well as diverse sources of magnesium. There is no unifying principle for these models other than the requirement for magnesium, which can be provided by seawater, Mg-calcite skeletons and cement, blue-green algae, clay minerals, and older dolomite. Environments range from sediment surfaces (evaporative), the shallow subsurface (seepage reflux; marine-fresh mixing), to deep burial settings (dewatering; illitization). All (except illitization in deep burial settings) have been documented for specific dolomites, and probably accurately explain the occurrence, the porosity, and permeability of some of these rocks.

Hanshaw and others (1971) hypothesized that the mixing-zone model accounts for some regionally extensive dolomites in the geologic record. The model calls for mixing fresh and saline ground water in a brackish zone of dispersion, thereby creating undersaturation of the ground-water solution because of nonlinearity of calcite solubility. Increased dissolution of calcite increases $Ca^{2+}$ and $CO_3^{2-}$ in the ground water. This dissolution and magnesium from seawater cause supersaturation with respect to dolomite in the brackish zone. This supersaturation occurring in the mixing zone environment where marine sulfate concentration waters are diluted, causes dolomite to form. In normal seawater, sulfate concentrations can inhibit dolomite formation. Sea-level changes, climatic changes, and uplift or downwarp of the rock mass could affect the position of the brackish zone in the aquifer, which would control dolomitization on a regional scale. A strong argument has been established for this process occurring in the Yucatan (Ward and Halley, 1985), as well as parts of the Floridan and the Edwards aquifers (Hanshaw and Back, 1979).

The process of dedolomitization (Back and others, 1983) involves the mineralogic evolution of a dolomite to a calcite aquifer. It occurs in an aquifer where water with low dissolved solids and high $CO_2$ starts dissolving calcite, dolomite, and gypsum from the mineralogic framework of the aquifer. This dissolution results in an increase in total $CO_2$, $SO_4^{2-}$, and $Ca^{2+}$. Because equilibrium with respect to calcite is obtained first, dolomite and gypsum continue to dissolve, contributing additional $Ca^{2+}$ and $SO_4^{2-}$ to the solution. $Ca^{2+}$ combines with $COCO_2^{2-}$ to precipitate calcite, which causes the water to be undersaturated with respect to dolomite, which continues to dissolve. The process of dedolomitization can have a variable effect on the porosity and permeability and significantly affects water chemistry.

The controls on the geochemical processes are many. The basic geochemical environment, which includes abundance of $CO_2$, pressure, pH, temperature, salinity, and the numerous controls described in the early diagenetic phase, also apply to geochemical processes (Table 1).

It is important to emphasize that although the geochemical processes of solution and precipitation of carbonate minerals have been described as separate entities, isolated from other processes that may occur together naturally, they should not be considered as simple, unrelated processes. Many of these processes and controls commonly occur concurrently within natural hydrochemical systems, and interpretation requires an understanding of all potential interactions.

### Lithologic-stratigraphic

Lithologic and stratigraphic controls play an important role in concentrating flow within specific zones of a carbonate rock, and thus influence subsequent dissolution. Although these factors tend to be passive, the effects of their control are obvious in most carbonate terranes, as shown by localization of permeability zones along bedding planes or within a single facies.

Lithology, texture, and bulk chemical purity affect the original porosity and permeability within a sedimentary sequence. Variations in these parameters define the potential for flow localization and concentration and the potential for dissolution. Properties of the rock that initially favor increased flow favor development of greater permeability.

Within any given stratigraphic sequence driven by a common fluid potential, flow will tend to be concentrated in the zones providing the least resistance. If all layers are fairly homogeneous, flow is not concentrated and the resulting permeability distribution is more evenly distributed.

Because many of the characteristics of shallow carbonate aquifers are controlled by the influx of water and sediment from the surface through conduits, the presence of overlying noncarbonate rocks in which such conduits are not developed is a major factor in inhibiting aquifer development. Many shallow carbonate aquifers in the eastern United States are bounded by such overlying rocks.

Observations in a number of carbonate aquifers show that small areas can be overlain by noncarbonate rocks without a major change in aquifer character. Large conduits can be present in the same aquifer beneath outliers of the overlying rocks as much as 2 km from the nearest contact. A classic example of this is the Mammoth Cave region, where larger aquifer conduits (now in the vadose zone due to lowering base level) are found beneath Upper Mississippian and Pennsylvanian rocks. Similar instances are known in the Inner Bluegrass Karst Region, where the aquifer in Middle Ordovician limestone is developed beneath outliers of Upper Ordovician limestones and shales. This indicates that areally distributed vertical recharge from the surface is not necessary for aquifer evolution and that near-horizontal flow in the aquifer can develop conduits beneath noncarbonate rocks if hydrologic conditions for ground-water circulation are favorable.

Lithologies in which conduits are not developed will, of course, restrict the downward development of a carbonate aquifer and limit the thickness of the permeable zone. For this reason, some shallow aquifers occur above the major drainages. In the Inner Bluegrass Karst Region, the nearly horizontal limestones are thin- to medium-bedded with numerous thin shale partings and occasional beds of argillaceous limestone. Although both shales and the argillaceous limestones consist of more than 50 percent calcite, conduit development is inhibited by the large amount of insoluble residue generated. Conduits conducting water and sediment from the surface must become sufficiently enlarged for these shales and argillaceous limestones to be penetrated before conduit systems can be developed to depth and drainage of an area integrated into a ground-water basin. Conduit enlargement has occurred in less than half the area of the Inner Bluegrass Region; elsewhere the conduit system is apparently still perched on shale interbeds and other lithologies that have inhibited conduit development downward. In these interbasin areas the potentiometric surface is only a few meters beneath the land surface, and the aquifer consists of water bodies of limited extent separated by discontinuities.

The thickness of the individual carbonate beds as well as the thickness of the entire sequence are important, because bed thickness controls relative layer strength and joint spacing, whereas sequence thickness defines the vertical range of the soluble layers. The vertical range sets the limit for potential aquifer development. Generally, thin-bedded carbonate rocks units have closer joint spacing than medium-bedded and massive units. Conduit growth will occur where the flow velocities are the highest, which is a function both of the initial permeability and of the head distribution. Although the most favored site for conduit initiation would seem to be bedding surfaces and other fractures that are parallel to the potential gradient, whether or not this is the case depends on the presolution width of the fracture and the intergranular permeability of the limestone.

### Structural-tectonic

Structural and tectonic factors significantly influence the hydrogeology of carbonate rocks because these processes are responsible for the attitude, the altitude, and to a large degree, the secondary vertical hydraulic conductivity. In extreme cases, intense structural deformation may recrystallize the sedimentary carbonate rocks to metamorphic marble, thereby completely altering the previous porosity and permeability distribution.

Five major processes are classified as structural and tectonic (Table 1). These processes are driven primarily by internal earth forces, and they affect all crustal rocks, not just carbonates.

Uplift vertically raises the rocks above sea level, commonly in combination with other structural results. The effects of increasing the altitude above sea level are (1) to provide water in the aquifer with greater potential energy, and (2) to increase hydraulic gradients.

Tilting and folding reorient the existing porosity and permeability distribution in space. The anisotropy is thus changed, and depending on the hydrologic boundaries, the flow distribution can be profoundly affected. Equally important is the lack of tilting. Flat or nearly flat-lying carbonate rocks are capable of developing extensive zones of secondary permeability if other conditions are favorable.

Depending on local hydrologic conditions, folding and tilting can enhance the solution process. The enhancement is generally due to steeper gradients maintained by uplift, which accompanies the tilting and folding. LaMoreaux and Powell (1963) found that differences in regional dip influenced the direction of ground-water movement in the Huntsville, Alabama, area. The impermeable Chattanooga Shale underlies the area, and the solution of carbonate rocks was most rapid in the trough of a syncline that is parallel to the regional structure.

In many carbonate settings, joints, bedding plane joints, and faults appear to be the preferred site of initial conduit development, and much of the small scale circulation, such as that encountered in most wells, occurs along such features. Fractures may also be important on a larger scale as well. One of the principal tools used for well location in carbonate aquifers is to determine the location of near-vertical fractures, commonly defined by linear topographic features (Lattman and Parizek, 1964).

The size of conduits along fractures is greatly dependent on the amount of flow through the conduits and on the potential gradient in the aquifer. Where a fracture is favorably oriented parallel to and downgradient to large flows in a dendritic system, major development of conduits along fractures may occur. Where a fracture is at right angles to the gradient, or if little recharge is occurring upflow, conduit development and flow along the fracture will be small. This is illustrated in the Inner Bluegrass Karst region, where dye traces from some sinkhole alignments and mapped faults showed flow diverging to widely separated discharge points, and by the lack of correspondence in some areas between mapped fault and joint orientations and flow directions in the aquifer (Thrailkill and others, 1982).

Stringfield and LeGrand (1969) studied the occurrence and movement of water in the Guadalupian reef complex of southeastern New Mexico; they concluded that the greatest amount of solution and highest permeability occur along joints in coarse-textured carbonate rocks rather than in fine-grained carbonate rocks, and in calcareous facies rather than in dolomites. Also, they concluded that lateral movement toward the east is prevented by impermeable anhydrite.

Joints in the flat-lying carbonate rocks of the midcontinent are thought to be caused by tensional forces that are actually initiated by compression. These commonly occur in orthogonal sets in otherwise undeformed rocks. They appear to be ubiquitous in sedimentary rocks in all tectonic settings. In some settings the joints serve as the primary avenue of vertical movement of ground water. Joints that are more open and allow more recharge to infiltrate are dissolved more rapidly, and therefore capture more runoff in a self-perpetuating increase in the evolution of porosity and permeability. Although exceptions exist, there are many well-documented case histories of joint-developed cave passages that formed in the shallow phreatic ground-water system in which flow was concentrated along joints. In many areas, joints are thought to provide the hydraulic connection between otherwise impermeable confining layers in the carbonate rock mass and deeper zones of enhanced permeability.

The hydrogeologic effects of faulting are difficult to generalize for carbonate terranes. Faults may either restrict the lateral movement of water in an aquifer or they may serve as conduits through which the water moves. If zones of high permeability are faulted against relatively impervious beds, flow is restricted. If the fault zone has high permeability, it can provide an avenue of flow through otherwise low-permeability rocks. In some areas, faults provide an essential part of the dynamic flow system for deeply buried carbonate aquifers. Where flow is concentrated along these features, they are subject to dissolution based on the same principles as other parts of the aquifer.

Under metamorphic conditions of extreme temperature and pressure, pressure solution causes carbonate rocks to recrystallize, forming marble. Metamorphism destroys the existing porosity and permeability of the rock. It should be noted that although marbles may initially have extremely low porosity and permeability, they are subjected to the same processes as other carbonate rocks, and with time, in a favorable hydrologic setting, they can evolve into locally important aquifers. Regionally, they are less important than limestones because of their limited areas of occurrence.

### Hydrologic

Hydrologic factors are at the heart of ground-water flow; they are more significant in the development of later permeability in carbonate rocks than any other factor. One process, dynamic freshwater circulation, is considered essential for solution to occur. Without it, the chemical processes cannot act. In order to be vulnerable to solution, the carbonate rocks must (1) receive recharge in an intake area, (2) transmit water, and (3) actively discharge water. This complete circulation allows large volumes of water to come into contact with the aquifer and, if conditions are favorable, to dissolve it. If any part of the flow system is missing, there is little or no freshwater circulation, and little or no dissolutional development of permeability. If anisotropic distribution of permeability concentrates the flow such that the recharge from many square kilometers is focused along a limited flow path, the effects of dissolution are enhanced. As hydrologic factors become dominant, the role of porosity assumes less importance.

The flushing due to recharge of fresh water is a result of dynamic ground-water flow. Flushing initiates the diagenetic processes described previously and can result in mixing, wherein the saline water mixes with fresh water, resulting in an undersaturated solution. This favors additional dissolution, and increased permeability.

Stringfield and LeGrand (1969) note that the methods of discharge are especially important. For water-table conditions, circulation tends to be concentrated in a moderately thin zone just below the water table in the aquifer and to be less at greater depth. Circulation tends to be volumetrically less in recharge than in discharge zones and less in the zone of aeration than in the upper part of the zone of saturation. The concentration of flow lines in the discharge area, together with the greater velocities in that area, will result in the enlargement of the outlet, an increased permeability, and a consequent capture of flow from the deeper flow paths. The shorter, shallower flow paths become larger at the expense of the longer, deeper flow paths. Progressive concentration of flow in the upper part of the zone of saturation produces larger conduits and causes eventual diminution of flow and solution at greater depths.

The relation between ground water and surface water is a complex hydrologic control as well as a hydrologic effect. In an example from the Inner Bluegrass Karst Region of Kentucky, Thraillkill and others (1982) have documented that the major flow of subsurface water occurs in ground-water basins that seldom coincide with surface-water basins. Within most of the ground-water basins of the Inner Bluegrass there is a dendritic system of solution conduits that feed a major spring. Recharge to the basin descends steeply to its floor, and flows to the spring in large (cross-sectional area ranges from 10 to 100 $m^2$), nearly horizontal conduits situated at depths of up to 30 m beneath topographically high areas. The gradient of such conduits is believed to be a hydrologic effect rather than a control, with primary influencing factors being discharge and sediment load. Interbasin areas with no deep solution conduits are characterized by shallow (less than several meters) flow that moves primarily downslope and emerges as high-level springs.

Different geographic areas may show markedly different controls and effects than the Bluegrass Region, and a complete spectrum of surface water–ground water relations exist from examples throughout North America. Hydrologic conditions that favor diversion of surface water to subsurface drainage systems enhance aquifer development.

Hydrologic controls on the development of permeability in carbonate rocks are diverse. Climate is one that has received much study. In addition to the obvious factor of precipitation, an important factor is temperature. Not only does temperature control $CO_2$ content of the ground water, it also affects the viscosity and, in areas of climatic extremes where the average temperature is below 0°C, it can form an impermeable layer of permafrost. As a result, karst and dissolution processes may be greatly retarded in regions where ground water does not exist in the liquid state. However, flow systems developed during different climatic settings can often function under extremely adverse conditions that occur later.

The effect of carbonate dissolution is markedly different in an arid climate than in a humid climate. Recharge is thought to be much more dependent on permanent influent rivers in an arid setting, and much less dependent on precipitation infiltration to closed depressions. This latter source is the major form of recharge for most carbonate terranes in a humid climate.

An extensive study by Smith and Atkinson (1976) on 231 data sets with mean annual carbonate solute concentrations, extrapolated erosion rates, and surface versus underground solutional proportionality, concluded that the traditional latitudinal divisions of climate do not reflect a corresponding division of processes and erosion rates. The principal climatic factor appears to be the mean annual runoff rate.

The steeper the hydraulic gradient, the more water flows through a given aquifer. The greater the flux, the greater the development of porosity and permeability.

### Weathering - geomorphic

Geomorphic factors are defined as processes and controls limited to the interface of the solid earth and the atmosphere. These include weathering processes and near-surface karstification controls that affect subsurface flows. While it is true that many characteristic karst landforms are the passive result of solution of the rock mass, they likewise exert an active control that is intimately related to the evolution of the aquifers. The closed depressions capture an increasing amount of precipitation as they grow, thereby funneling increasing volumes of water into the aquifer.

Two processes are listed as geomorphic. Sediment infilling, caused by clay and sand particles transported in the flowing ground water, fills the voids of the carbonate aquifer and reduces permeability (Fig. 2). This process can literally choke an aquifer by filling its transmissive zones with fine-grained sediment. Alternately, it can favorably enhance the hydrogeology of an aquifer, as the sand within the Biscayne aquifer in southern Florida does. Because the sand in the Biscayne has filled solution cavities and lowered the transmissivity, the aquifer maintains a steeper gradient and a higher water table, thus impeding saltwater encroachment. If the aquifer lacked sand-filled cavities, it would have a much flatter water table, would have a much thinner freshwater lens, and would be more vulnerable to salt-water encroachment.

Conduits in an aquifer in which flow velocities are inadequate to transport the sediment load they receive will be plugged. Although some flow will continue through these filled conduits, much of their former flow will be diverted to, and accelerate the growth of, other conduits. Infilling generally occurs under fluvial conditions, but has been reported from glacial settings as well (Ford, 1984).

A process that has received little mention in the literature, but appears to be very important in shallow aquifer evolution, is unloading. Unloading is a slow process that results from erosion of surface material. It is controlled by erosion rates. Joints, bedding planes, and fractures become wider and thus become more suitable for transmitting ground water due to decreased confining pressure as overlying material is eroded. Nearer the surface there is also greater ease of energy dissipation. Stresses from diverse sources such as thermal heating, earth tides, hydrologic loading, frost wedging, tectonic activity, and the effects of man serve to move the solid blocks relative to one another across joints. These strains, on the order of microns (Brahana and Davis, 1973), are concentrated at the joints and may promote ground-water flow along them. Unloading is thought to be a contributing influence on the depth of development of shallow aquifers in flat-lying carbonate rocks.

Geomorphic controls include the thickness of the soil and regolith. Soil provides a storage zone above the aquifer that releases the water slowly to the aquifer during times of little rain. Generally the thicker a soil, the greater its storage capacity. More importantly, the soil recharges the $CO_2$ in the water, chemically providing additional carbonic acid that fuels the dissolution process.

Relief is another geomorphic control. It affects the difference in head between the input and output of the hydraulic system, defining the gradient.

Subsurface drainage in carbonate rocks can be greatly affected by the surface slope of the land. In the Canadian Rockies, Ford (1978) reports there are many limestones with dip slopes of $20°$ to $50°$, and with similar land surface slopes. In a setting with such steep surface slopes and high energy available to surface process, subsurface solutional processes are not able to compete effectively, and fluvial forms result, despite the existence of geologic factors otherwise favorable to initiation of carbonate aquifer development.

Base level is an important control. Base level defines the altitude of the hydraulic drain or outlet from the flow system. If base level is situated favorably with respect to geologic conditions for a long enough duration, it becomes a major control on the location of permeability zones. In their study of secondary permeability due to solution, Herrick and LeGrand (1964) named four types of base-level control and related them to the resulting hydrogeology. The four types are: (1) sea-level position as it affects the water table or artesian pressure and ground-water discharge in carbonate aquifers in coastal areas, (2) a perennial stream as it affects the water table and ground-water discharge, (3) an impervious formation at the base of the aquifer, and (4) a combination of the above controls.

Karst processes and landforms have been intensively studied, and much of the body of hydrogeologic understanding comes from this discipline. Karstification is a hybrid geomorphological, geochemical, hydrologic process, with the end result being a unique land surface. It occurs from the land surface to the deep phreatic zone of active freshwater circulation. It is most active in the shallow phreatic zone, near the water table.

Although an extended discussion of the development of the karst topography overlying a carbonate aquifer is not within the scope of this chapter (see Ford and others, this volume), it is evident that the evolution of the surface features and the evolution of the aquifer are closely related. Obviously, flow entering the subsurface at the bottom of a large depression must traverse the aquifer in a series of conduits that are sufficiently integrated and have a cross-sectional area great enough to permit the passage not only of the water but usually of substantial amounts of surface-derived sediment as well. Similarly, large conduits are also required in the downstream part of dendritic systems draining a large number of smaller depressions. The development of these larger conduits (or systems of parallel conduits in some cases) will locally alter the direction and magnitude of the potential gradient, and hence the flow directions in the aquifer.

Although cap rock was discussed under the lithologic-stratigraphic aspect of variability in vertical texture and permeability, it is justified to restate its importance in this section. Like soil, the cap rock exerts a significant influence on carbonate aquifer development because it controls the amount and the distribution of recharge that is introduced into an aquifer. Among other parameters, its vertical hydraulic conductivity, its thickness, and its anisotropy are important.

### Historical geologic - chronologic

Historical geologic-chronologic factors refer to the sequence and duration of the processes and controls that are acting to modify the carbonate rock by incorporating the concept of time. In simplistic terms, what comes later is dependent on what went before. The sequence in which specific diagenetic events occurred, the timing of destruction or enhancement of the original permeability, and the duration of a base level being established at an optimum level to a permeable zone are all part of this factor. Time provides the framework for the other factors to operate.

The establishment of dynamic freshwater flow commonly occurs at a later stage in the history of a carbonate aquifer. Geologic factors, which are dominant in the early stages of the history of a rock, essentially describe the potential for permeability in space. Whether or not the potential is realized depends on the interaction of all the processes and controls, including the geologic history. As aquifer development shifts from carbonate sediment toward indurated shallow aquifer, the geologic factors appear to become subordinate to hydrologic factors. The importance of geologic conditions at the earlier time needs to be stressed, and the fact that many carbonate rocks show evidence of interrupted stages of aquifer development suggests that carbonate aquifer evolution is not unidirectional, and the relation between historical geologic-chronologic factors and all other factors is quite complex.

High-permeability zones in the aquifer near Mammoth Cave, Kentucky, occur within several distinct intervals above the zone of active flow. Each permeability zone corresponds to an historic flow zone, established over a relatively long period of base-level stability defined by the Green River. As the river rises in response to modern floods and different permeability zones become saturated, the catchment area, the hydraulic gradient, the flow directions, and the discharge points change. These changes illustrate that the conduit type permeability of shallow carbonate aquifers is highly complex, not only in areal and stratigraphic distribution, but in time as well. Quantification of processes in an aquifer such as Mammoth Cave is extremely difficult on a local

scale, and requires a very complete data base coupled with a very thorough understanding of its hydrogeologic history.

## EVOLUTION OF WATER QUALITY

Hanshaw and Back (1979) observed that the chemical character of water in carbonate aquifers is both a control and a response to the many interrelated factors affecting the hydrogeology of carbonate rocks. The chemistry of the ground water is the result of the aquifer mineralogy, fluid chemistry, and flow regime, because these determine the occurrence, sequence, rates, and progress of reactions. Hanshaw and Back (1979) proposed a conceptual model that depicts the changes in ground-water chemistry through distinct phases of aquifer development. Most of the following discussion is based on their work. The graphic representation of these major evolutionary trends is shown in Figure 3.

With emergence from a shallow marine environment, carbonate sediments are subjected to flushing of the marine pore water with freshwater. This flushing, as mentioned earlier, is thought to be one of the major environmental factors initiating diagenesis. The reaction typically follows path 1 (Fig. 3).

Path 1, from marine to recharge, is characterized by the following changes to the interstitial water: (1) dissolved solids concentrations decrease about two to three orders of magnitude; (2) major ions change from predominantly $Na^+$, $Cl^-$, $Mg^{+2}$, $SO_4^{-2}$ to $Ca^{+2}$ and $HCO_3^-$; and (3) conservative mixing and dissolution are the dominant chemical processes.

Path 2, from recharge to marine, occurs if there is an environmental or hydrodynamic change that favors recharge of marine water to the aquifer. The results are the reverse of path 1—dissolved solids increase, and major ions change. The chemistry of this change is due more to the physical replacement of pore water than to geochemical interaction between aquifer and fluid.

Path 3 represents the evolutionary step from freshwater recharge to confined flow downgradient in the carbonate aquifer. It is commonly characterized by: (1) dissolved solids increase (generally much less than an order of magnitude); (2) major ions change from predominantly $Ca^{+2}$ and $HCO_3^-$ to $Ca^{+2}$, $Mg^{+2}$, $HCO_3^-$ and $SO_4^{-2}$, and (3) dissolution as the dominant chemical process.

Path 4 is caused by subsurface mixing of ocean water that has encroached into the deeper parts of coastal aquifers. Hanshaw and Back (1979) refer to the Floridan as an example of this; dolomitization may occur under conditions along this flow path. This flow path is characterized by: (1) dissolved solids concentrations increase by as much as two orders of magnitude ($\sim 10^5$ mg/L); and (2) $Na^+$, $K^+$, and $Cl^-$ become the dominant ions.

Path 5 may occur where there are extensive accumulations of evaporite minerals within an aquifer. Dissolution of the evaporites may result in ground water with: (1) dissolved solids concentrations increase, commonly to concentrations of brines ($\sim 10^6$ mg/L); (2) continued increases in $Na^+$, $K^+$, and $Cl^-$ as the major ions.

**TABLE 3. COMPARISON OF TYPICAL HYDROGEOLOGIC CHARACTERISTICS OF CARBONATE AND NONCARBONATE ROCKS***

| Rock Type | Primary Porosity and Permiability | Secondary Porosity and Permeability | Flow Regime | Flux |
|---|---|---|---|---|
| | | **CARBONATES** | | |
| Cavernous limestone | Low porosity, low average but high variable permeability | Low porosity anisotropic; huge permeability in conduits, very low permeability in rock mass | Laminar to turbulent, deeply buried, may be stagnant | Regional aquifers >$10^3$ L/s at major springs |
| Chalk | High porosity, low permeability, intergranular | Usually not significant, usually undeveloped regionally, jointing can be important | Usually stagnant, very sluggish, commonly only local if at all | Commonly confining layers-->$10^3$ L/s or less; 0 L/s is common, may be local aquifer if fractured |
| Dolomite | Low porosity, low to moderate permeability | Variable from diagenetic pin-point porosity to conduit flow. Joints and faults can be important | Variable usually laminar, can be turbulent to sluggish | Vary from confining layers to regional aquifers; $10^{-1}$ to $10^2$ L/s |
| Marble | Very low, essentially none | Significant if jointed or fractured, solution conduits can develop if flow system complete | Usually laminar where secondary permeability developed--can be variable, turbulent to sluggish | Typically local aquifers yield $10^{-2}$ to $10^1$ L/s; 0 L/s is common |
| | | **NONCARBONATES** | | |
| Extrusive igneous flows | Cooling features important; moderate to low porosity; high permeability in cinder zones and lava tubes | Joints and fractures can be significant | Turbulent to laminar, highly variable | Regional aquifers >$10^3$ L/s at major springs |
| Pyroclastics | Medium to high porosity; low permeability | Usually not significant | Flow sluggish or stagnant | Commonly serve as confining layer. <$10^{-3}$ L/s; 0 L\s is common |
| Nonindurated granular deposits | Intergranular provides all porosity and permeability | None | Laminar flow | From <$10^{-2}$ to >$10^2$ L/s |
| Sandstone | Intergranular, variable, depends on sorting and degree of cementation | Joints and fractures can be significant, as can dissolution of feldspars, rock fragments, and cements | Laminar flow | From <$10^{-1}$ to $10^2$ L/s |
| Shale | High porosity, low permeability | Usually not significant | Flow sluggish or stagnant | From <$10^{-3}$ to 1 L/s |
| Fractured crystalline | Commonly very low porosity, low permeability | Significant--joints and fractures comprise most porosity and permeability | Laminar | Generally from <$10^{-2}$ to 10 L/s |
| Glacial till | Intergranular, low porosity, low permeability | None | Laminar | Variable from $10^{-2}$ to >1 L/s |

*Table 3 continued on next page.

**TABLE 3. (CONTINUED)**

| Rock Type | QW-dynamic Flow Zone[†] | | | | | Major Hydrologic Characteristics |
| | pH | Conductivity (uS/cm) | Major Constituents | Hardness (mg/L) | Geochemical Processes | |
| --- | --- | --- | --- | --- | --- | --- |
| | | | **CARBONATES** | | | |
| Cavernous limestone | 7-8 | $<10^2$ to $>10^2$ | $Ca^{++}$ $HCO_3^-$ | $<10^2$ to $>10^3$ | Solution, precipitation, possible mixing | Highly anisotropic with flow concentrated in a few favored zones, and most of rock mass unsaturated. Can be important local and regional aquifers; although it commonly is not |
| Chalk | 7-8 | $<10^2$ to $>10^3$ | $Ca^{++}$ $HCO_3^-$ | $<10^2$ to $>10^3$ | Minor solution, minor precipitation | Commonly confining layer, with very poor aquifer properties. Fractures may allow shallow local flow |
| Dolomite | 7-8 | $<10^2$ to $>10^5$ | $Ca^{++}$ $Mg^{++}$ $HCO_3^-$ | $<10^2$ to $>10^3$ | Dedolomitization, solution, precipitation, possible mixing | Shows subdued solutional effects compared to limestone; may be excellent regional aquifer, commonly better than more isotropic limestones |
| Marble | 7-8 | $<10^2$ to $>10^3$ | $Ca^{++}$ and/or $Mg^{++}$ $HCO_3^-$ | $<10^2$ to $>10^3$ | Solution, precipitation, possible mixing | Similar to fractured crystalline non-carbonates, except that marble is more chemically reactive, may be important aquifers, greater yields than carbonates |
| | | | **NONCARBONATES** | | | |
| Extrusive igneous flows | 6-8 | $<10^2$ to $>10^3$ | $Na^+$ $K^2$ $Ca^{+2}$ $Mg^{+2}$ $HCO_3^-$ $SiO_2$ Pollutants | $10^1$ to $10^2$ | Minor solution | Can be major aquifers in volcanic terrains. Highly anisotropic vertically, permeability zones tend to be planar. Some first magnitude springs issue from these rocks |
| Pyroclastics | 6-8 | $<10^2$ to $>10^3$ | $SiO_2$ $Ca^{+2}$ $Mg^{+2}$ $HCO_3^-$ | $10^1$ to $10^2$ | Minor solution | Commonly tight confining layers, unless fractured. Generally not an aquifer |
| Nonindurated granular deposits | 5-8.5 | $<5 \times 10^1$ to $>10^5$ | $Ca^{+2}$ $Na^+$ $HCO_3^-$ | $10^1$ to $10^3$ | Minor solution precipitation ion exchange | Coarse, well-sorted deposits, commonly major regional aquifers. Fine-grained deposits are leaky confining layers |
| Sandstone | 6-8 | $<50$ to $>10^5$ | $Ca^{+2}$ $HCO_3^-$ $SiO_2$ | $10^1$ to $10^3$ | Solution, precipitation | Commonly is regional aquifer if porosity and permeability are favorable. Uniformly anisotropic. Secondary permeability can be important |
| Shale | 5-8.5 | $10^2$ to $10^5$ | $SO_4^=$ | $10^2$ to $10^3$ | Ion exchange | Usually is confining layer, with low vertical hydraulic conductivity; generally does not serve as an aquifer |
| Fractured crystalline | 6-7 | $10^2$ | $Ca^{+2}$ $Mg^{+2}$ $HCO_3^-$ | $10^1$ to $10^2$ | Minor solution | Fracture permeability can be important enough to make these local aquifers. Limited to shallow depths (<100 m); commonly impermeable below this |
| Glacial till | 6-8 | $<10^2$ to $10^4$ | $Na^+$ $Ca^{+2}$ $Mg^{+2}$ $HCO_3^-$ | $10^1$ to $10^2$ | Ion exchange | Unsorted material is poor aquifer, with fines filling space between coarse matrix components. Sandy till forms local aquifers; most till confines flow |

[†]L/s = liters per second; mg/L = milligrams per liter; uS/cm = microsiemens per centimeter at 25° Celsius.

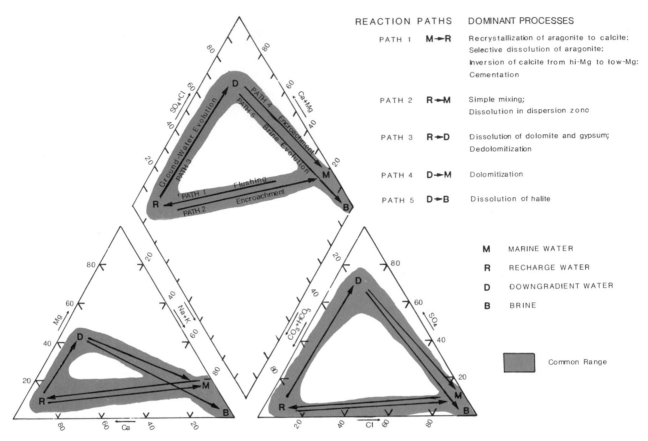

Figure 3. Reaction paths showing evolution of chemistry of ground-water in carbonate rocks.

# HYDROGEOLOGIC SIMILARITIES
# AND CONTRASTS

## *Noncarbonate rocks in comparison to carbonate rocks*

The effects of the previously mentioned processes and controls serve to set carbonate rock apart hydrogeologically from all of the other major rock types. The differences and similarities between seven classes of noncarbonate rocks and four classes of carbonate rocks are summarized in Table 3. The primary differences are: (1) the solubility of carbonate minerals compared to noncarbonate minerals; (2) the development of significant secondary permeability for most carbonate aquifers, which gives rise to dynamic circulation; (3) the anisotropy on a local scale of most carbonate aquifers as compared to granular aquifers (such as nonindurated sands and gravels of coastal plains); (4) the huge range of hydraulic properties of cavernous carbonates as compared to fine-grained, crystalline noncarbonates (such as fractured igneous and metamorphic rocks); and (5) the larger areal extent of carbonate aquifers as contrasted to many other aquifer types, particularly those of volcanic and metamorphic origin.

Similarities between carbonate and noncarbonate rocks may

be observable in hydrologic response, but the cause of the behavior is usually markedly different for the different rock types. As an example, extrusive igneous rocks with lava tubes and very permeable cinder zones commonly show some of the same characteristics of cavernous limestone. The flow in both is highly anisotropic in zones of secondary permeability, with large springs common and ground-water velocities typically variable for naturally occurring hydrologic events; however, the cause of these similar responses is entirely different. The high permeability of the extrusive igneous rocks is due to volume changes associated with cooling magma, whereas the high permeability of the cavernous limestone is due to dissolution of the aquifer along zones of concentrated flow. Especially important is the fact that the permeability of the igneous rocks is much less likely to evolve rapidly because igneous materials typically aren't as reactive as most of the carbonate minerals.

Under conditions where the carbonate rock tends to be less reactive, hydrogeologic behavior tends to be more comparable to noncarbonate behavior, as illustrated by a general comparison of welded tuff and chalk. Both have relatively large porosities and low primary permeabilities, both are fairly homogeneous and

Figure 4. Location of carbonate rock areas used for comparison.

isotropic, and both generally serve as confining layers rather than as productive aquifers. Likewise, unless they are exposed to major jointing and fracturing, both are relatively stable hydrogeologically and experience little if any porosity-permeability evolution after induration-lithification. The processes and controls that formed these hydrogeologic units are distinctly different, yet their behavior is quite similar. Unlike its purer, less stable carbonate relatives, chalk is less likely to develop into an aquifer.

These general observations point out the important need for hydrogeologists not only to describe observable effects, but to conceptualize the processes that acted, the factors that controlled the processes, and the rates at which and the time frame over which the aquifer developed. Without all of this information, many of the subtle interrelations of carbonate hydrogeology will remain hidden behind a seemingly infinite combination of significant factors.

### Carbonate rocks from various terranes

Table 4 represents a simplified summary that compares and contrasts the hydrogeology of 27 different carbonate rock units throughout North America, including the Caribbean. The table lists the major processes, controls, and effects for each carbonate unit on a regional scale. The generalizations are not intended nor should they be used outside the context of coarse regionalization. Most of the area has been interpreted based on published reports, and references are limited to the latest or most significant works. Figure 4 is a map showing the 27 areas studied.

Table 4 graphically demonstrates several important facts:

1. Carbonate rocks serve as significant aquifers throughout North America. They are not limited by location or by age of the formation.

2. Carbonate rocks show a total range of hydraulic conductivity of ten orders of magnitude, from the tightest confining beds to the most prolific aquifers.

3. The hydrogeologic response of carbonates at any time is related to the rock permeability, which is affected most by the processes of dynamic freshwater circulation and solution of the rock.

4. The significant control on dynamic freshwater circulation is the integrity of the hydraulic circuit, which requires that recharge, flow-through, and discharge must be maintained. Without each of these, the system is essentially stagnant, and does not act as a conduit. Its evolution rate approaches that of noncarbonate rocks. Secondary controls on dynamic circulation include

**TABLE 4. GENERAL FEATURES OF 27 CARBONATE TERRANES IN NORTH AMERICA**

| No. | Name | Occurrence | Age | Hydrologic Character | Major Controls | References |
|---|---|---|---|---|---|---|
| 1 | Knox Dolomite | Blue Grass Region, Kentucky | Є-O | Not used as an aquifer because water has dissolved-solids concentrations in the range of 2,000-20,000 mg/L. Freshening of water quality over Jessamine Dome suggests localized and limited flow cells. Permeability zones concentrated in upper part of aquifer. | Dolomite—no active flow system. Deep burial with permeability zones present; >215 m; restricted recharge. No known discharge zone | Hollyday and others, 1981 |
| 2 | Knox Dolomite | Central Basin, Kentucky | Є-O | Minor regional aquifer, thin but ubiquitous permeability zones concentrated in upper part of aquifer, fresh water only in upper zone. Recharge by leakage through 120 m of overlying carbonates. | Dolomite—active flow system, deeply buried. Regional discharge zones along faults upward to rivers that serve as drains. Depth controls flux; much recharge is diverted to shallower formations. | Brahana and Bradley, 1985 |
| 3 | Ozark Aquifer | Ozark Uplift, Missouri | Є-O | Dolomite—active flow system, deeply buried. Regional discharge zones along faults upward to rivers that serve as drains. Depth controls flux; much recharge is diverted to shallower formations. | Dolomite—active dynamic flow system from surface to crystalline basement. High hydraulic gradient. Enhanced anisotropy. Thick soil and regolith. | Imes, 1987 |
| 4 | Selma Chalk | Mississippi K Embayment, Mississippi-Alabama | | Acts as confining layer with very low transmissivity. No evidence of secondary permeability developed. | Chalk—no early permeability developed. No dynamic fresh-water flow system. High porosity diffuses solvent action of recharge. | LeGrand, 1970 |
| 5 | Biscayne aquifer | Southern Florida | Tm-Qp | Prolific aquifer, characterized by isotropic conditions. Hydraulic gradient suggests large transmissivity, with sand infilling large voids thereby maintaining thick lens of fresh water. | Limestone—high initial porosity and permeability (lithology); initial conditions enhanced by sea-level fluctuations and resulting dissolution. Overlying sand cover provides storage. Active dynamic flow. | Johnston and Miller, this volume; Parker and others, 1955 |
| 6 | Ridge and Valley | Appalachian Mountains, Eastern U.S. | Є-M | Many separate, self-contained hydrologic systems. Major stream serves as drain for each. Recharge through dolines; air-filled conduits common, as are deep shafts. | Dolomite and limestone, with interbedded clastics. Lithology, structure, and steep gradients are major controls. Active dynamic flow systems are well developed, but isolated and independent of one another. | Parizek and others, 1971 |
| 7 | Edwards Limestone | Texas | K | Major regional aquifer with 4 first-magnitude springs. Recharge from rivers at higher altitudes; major structural effects as well as late diagenetic effects. Variable water quality defines zones of active and inactive flow. | Limestone, dolomite, evaporite. Faulting created a complete hydraulic circuit. Dolomitization, micritization, and solution occur in preferential lithologies. Excellent example of role of active flow system in evolution. | Maclay and Small, 1983 |
| 8 | Blue Grass | Kentucky | O | Regional aquifer with highly variable character. Surface streams act as drains; surface karst features present, including caves. | Limestone—shallow, nearly flat-lying. Anisotropic development of porosity and permeability as shallow conduit flow system. Major springs are discharge points from ground-water basins. Hydrologic factors are strongly dominant. | Thrailkill and others, 1982 |
| 9 | Mammoth Cave | Kentucky | M | Minor aquifer; highly anisotropic, with dendritic subterranean streams. Topographic escarpments delineate cap rock retreat. Caves and karst features common. | Limestone, dolomite, evaporites—overlying sandstone cap rock; sequence of lithology and mineralogy. Base level of streams and duration of base level are important. Active, dynamic flow system. | Quinlan and others, 1983 |
| 10 | Yucatan | Yucatan | Te-Qp | Major regional aquifer in northern Yucatan. Very low hydraulic gradient; huge permeability, no rivers; infiltration directly from precipitation; maximum freshwater thickness is 70 m. | Limestone, with dolomite in mixing zone. Low hydraulic gradient, enormous secondary permeability, no confining layer, limited soil zone, and an average of 1,050 mm precipitation/year. Dolomite formation occurs in mixing zone. Active, dynamic flow system. | Back and others, 1986; Ward and Halley, 1985 |

*Table 4 continues on next two pages. See end of table for key to Age abbreviations.

**TABLE 4. (CONTINUED)**

| No. | Name | Occurrence | Age* | Hydrologic Character | Major Controls | References |
|---|---|---|---|---|---|---|
| 11 | Floridan aquifer system | Florida, Georgia, South Carolina, Alabama | Te-Tm | Major regional aquifer, with 17 first-magnitude springs. Major rivers, infiltration through sinkholes, maximum freshwater thickness 700 m. Geochemical evolution of water moving down-gradient. | Limestone, with dolomite. Confining layer (Hawthorn Formation). Favorable recharge, steeper hydraulic gradient, very large permeability in paleokarst. Ocean base level and lithology control development of permeability. Active dynamic flow system. | Johnston and Miller, this volume; Stringfield and LeGrand, 1966 |
| 12 | Madison Limestone | Montana, Wyoming, South Dakota, North Dakota, Manitoba | M | Major regional aquifer in north-ern Great Plains; part of inte-grated system. Areally widespread permeability zones developed prior to present geologic setting. High-density brines in Williston Basin (>300,000 mg/L). | Limestone and dolomite, with extensive evaporites. Deep burial and structure exert major control, as do hydrologic boundaries. Active dynamic flow system. Dissolution of evaporites generates brines. | Downey, 1984 |
| 13 | Niagara Dolomite | Illinois, Indiana, Ohio, Wisconsin | S | Major regional bedrock aquifer in upper midwest. Upper zones are reliable water-bearing units, although porosity and perme-bility can show large variations. Glacial drift acts as storage. | Dolomite—nearly horizontal, near surface, covered by glacial drift. Joints and fractures indicate development of system prior to glaciation. Lithology, drift, thick-ness, depth are major controls. Active dynamic flow system. | Zeizel and others, 1962 |
| 14 | Wakefield and Cockeysville Marbles | Maryland | PC | Local water supplies available where secondary permeability developed. Flow systems generally much more localized than in limestone, low porosity. Cone development of large springs and wells. | Marble—intense structural deformation, with joints controlling zone of permea-bility formed after metamorphism; original porosity and permeability destroyed. Wakefield is intimately associated with belt of volcanic rocks. Areally restricted; active, dynamic flow. | Meyer and Beall, 1958 |
| 15 | White Limestone | Jamaica | T | Major regional aquifers are highly variable. Significant karst area has pinnacles (Cockpit karst). Catchment divides range greatly with underground water level stage. Floods affect erosion. | Limestone, chalk, chert, and dolomite. Highly variable original porosity and permeability. Structure important, including joints, folding, and some major faulting. Huge floods input water to karst system at high altitude. Dynamic flow. | Versey, 1972 |
| 16 | North Coast limestone | Puerto Rico | T | Major regional aquifers in northern Puerton Rico are significant karst area, with karst pinnacles, deep dolines, and large caverns. Large variation in permeability; lowland flooding. | Limestones, dolomites, marl, and chalk. Significant uplift (>100 m); monocline during Miocene and continuous subareal exposure. High recharge rate; chronologic sequence; high initial porosity and permea-bility enhanced by dynamic flow. | Giusti, 1978 |
| 17 | Kaibab Plateau | Arizona | C-P | Anisotropic aquifer >1,200 m thick. No surface streams, infil-tration moves primarily verti-cally along faults to impermeable shale; discharges as springs. | Limestone, clastics and evaporites. Lithology, structure, climate, and hydraulic gradient are major controls; climatic effects are limited. Dynamic system. | Huntoon, 1974 |
| 18 | Desert Basins | Nevada | C-D | Regional aquifer in faulted basin and range setting, well-developed secondary permeability; inter-basin movement of ground water, with deep circulation, springs. | Limestones, dolomites, clastics. Regional movement of water controlled by lithology, fractures, and climate. These are dynamic flow systems. | Winograd and Thordarson, 1975; Maxey, 1968 |
| 19 | Helderberg Plateau | New York | S-D | Areally restricted aquifer; ani-sotropic, with extensive cavern developmenmt which is joint con-trolled. Incised streams act as regional drains for tilted car-bonates. Downdip cave passages resurge as gravity springs. | Limestone and dolomite—lithologic and structural controls dominate. Jointing and thin-bedded limestones concentrate flow. Active dynamic flow system. | Kastning, 1984 |
| 20 | Castleguard Alberta | | C | Major portion of aqueous drainage of Columbia ice field. Rapid hydraulic response and recession indicate sparse but well-integra-ted network of conduits. Active karst beneath temperate ice. | Limestone—300-m-thick glaciation has blocked some preexisting conduits. Massive strata and high hydraulic gradients are major controls. Active dynamic flow system. | Smart and Ford, 1983 |

*Table 4 continues on next page; see end of table for key to Age abbreviations.

**TABLE 4. (CONTINUED)**

| No. | Name | Occurrence | Age | Hydrologic Character | Major Controls | References |
|---|---|---|---|---|---|---|
| 21 | Nahanni | Northwest Territories | D | Aquifer that shows minor effect of permafrost in an area where effects should be pronounced. Poljes, dolines, shafts, conduits, and springs show seasonal fluctuation. Accentuated karst landscape. | Limestone, dolomite—at latitude 62-63°N and altitude 800-1400 m is in continuous permafrost condition. Preexisting permeability and steep gradient maintain effective dynamic flow in part of system. | Brook and Ford, 1980 |
| 22 | Goose Arm | Newfoundland | O | Local aquifer formed in incipient secondary permeability zone after glacial injection of fines plugged a large integrated karst aquifer. Constricted and young overspill channels. | Limestone—intensively deformed part of Appalachian orogen. Polygenetic; chronologic control—glacial fines plug previous high permeability zone. Incipient zone during glaciation is now the active flow zone. Lithologic, structural, and chronologic controls predominate. | Karolyi and Ford, 1983 |
| 23 | Winnipeg | Manitoba | D | Aquifer enhanced by glaciation. Presently confined. Subcutaneous karst features developed previously. Recharge at margins. | Limestone dolomite—Preserved and sealed (but not filled) by glacial till and lacustrine clays, thereby protecting it from destructive glacial effects. Low topographic gradient. Frozen ice base with no injection of sediments is responsible for preservation. | Ford, 1983 |
| 24 | Cupido-Aurora | Sierra Madre Occidental, north-northeastern Mexico | K | Aquifers used occur at great depth (200-1,400 m); high transmissivity and storage and high yield (40-240 l/s). Solutional features in outcrop area. | Limestone with shales—thick (>1,000 m); folded into anticlinal and synclinal valleys of eastern Sierra Madres; significant structural control; arid climate; high relief; dynamic flow system. | Lesser Jones, 1967 |
| 25 | Sequatchie Valley | Tennessee-Alabama | M | Aquifers with variable characteristics. Many karst features and conduits. Sequatchie River is major drain for ground-water discharge from Cumberland Plateau. | Limestone—structural control (faulting, folding), also near-horizontal altitude for most of area; breaching of cap rock; base-level control; active dynamic flow. | Crawford, 1984 |
| 26 | Mitchell Plain | Indiana | M | Shallow aquifer, with surface features that funnel recharge into major subterranean streams. Some perennial surface streams; many springs and caves. | Limestone, with evaporites—shallow depth, base level, and lithologic controls dominate. Active anisotropic flow system present. | Quinlan and others, 1983 |
| 27 | Pecos Valley | New Mexico-Texas | P | Major regional aquifer, with huge initial yields (360 l/s). Few surface features, but some well-developed deep conduits. Active flow system west of Pecos River; stagnant east of Pecos. | Limestone, dolomite, and evaporites. Evaporite solution, climate (arid), good dip slope permeability, and effective discharge zone are major controls. Active, dynamic flow system. | Davies and LeGrand, 1972 |

*Key to Age designations:*

| | | |
|---|---|---|
| pϾ = Precambrian | D = Devonian | T = Tertiary |
| Ͼ = Cambrian | M = Mississippian | Te = Tertiary, Eocene |
| O = Ordovician | P = Permian | Tm = Tertiary, Miocene |
| S = Silurian | K = Cretaceous | Qp = Quaternary, Pleistocene |

lithologic, structural, geomorphic, hydrologic, and chronologic aspects (Table 1).

5. The significant controls on the process of solution are: (a) rock solubility and (b) the chemical character of the ground water. Secondary controls on solution include diagenetic, geochemical, and chronologic aspects (Table 1).

6. Ground-water circulation, solution, precipitation, and permeability and porosity evolution are intimately related. Although they are separated in items 4 and 5 above based on physical and chemical aspects, in actual occurrence the interrelation is profound.

## SUMMARY

The broad diversity of the hydrogeology of carbonate rocks is the result of variable combinations of more than 60 processes and controls that act on these rocks. Of the many identifiable processes and controls, two seem to dominate. The first process, dissolution and precipitation, describes geochemically the relative ease with which the fresh water interacts with the solid rock matrix of the aquifer to create varying permeability. The second process, dynamic freshwater flow, is essential if significant permeability evolution in a carbonate aquifer is to occur. The many other processes and controls that occur have varying influence on the two dominant factors, and often the influence can be significant. The variation shown in Table 4, strongly suggests that no factor should be omitted from consideration.

## SELECTED BIBLIOGRAPHY

Back, W., and Freeze, R. A., 1983, Chemical hydrogeology; Benchmark papers in geology: Stroudsburg, Pennsylvania, Hutchinson and Ross, v. 73, 416 p.

Back, W., and LaMoreaux, P. E., eds., 1983, V. T. Stringfield Symposium, Process in karst hydrology: Journal of Hydrology, v. 61, 335 p.

Back, W., and five others, 1983, Process and rate of dedolomitization; Mass transfer and $^{14}C$ dating in a regional carbonate aquifer: Geological Society of America Bulletin, v. 94, p. 1415–1429.

Back, W., Hanshaw, B. B., Herman, J. S., and Van Driel, J. N., 1986, Differential dissolution of a Pleistocene reef in the ground-water mixing zone of coastal Yucatan, Mexico: Geology, v. 14, no. 1, p. 137–140.

Bocker, T., 1969, Karstic water research in Hungary: International Association Scientific Hydrology Bulletin, v. 14, p. 4–12.

Brahana, J. V., and Bradley, M. W., 1985, Delineation and description of the regional aquifers of Tennessee—The Knox aquifer in central and west Tennessee: U.S. Geological Survey Water Resources Investigation 83-4012, 32 p.

Brahana, J. V., and Davis, S. N., 1973, The effect of short-term stress on jointed rock near the land surface: Geological Society of America Abstracts with Programs, v. 5, no. 5, p. 379–380.

Brook, G. A., and Ford, D. C., 1980, Hydrology of the Nahanni karst, northern Canada, and the importance of extreme summer storms: Journal of Hydrology, v. 46, p. 103–121.

Burger, A., and Dubertret, L., eds., 1975, Hydrogeology of karstic terrains: Paris, France, International Association of Hydrogeologists, International Union of Geological Sciences, Series B, no. 3, 190 p.

Castany, G., Groba, E., and Romejn, E., eds., 1984, Hydrogeology of karstic terrains, case histories: International Association of Hydrogeologists, International Contributions to Hydrogeology, v. 1, 264 p.

Choquette, P. W., and Pray, L. C., 1970, Geologic nomenclature and classification of porosity in sedimentary carbonates: American Association of Petroleum Geologists Bulletin, v. 54, p. 207–250.

Clark, S. P., Jr., ed., 1966, Handbook of physical constants: Geological Society of America Memoir 97, Revised Edition, 587 p.

Crawford, N. C., 1984, Karst landform development along the Cumberland Plateau escarpment of Tennessee, *in* LaFleur, R. G., ed., Ground water as a geomorphic agent: Boston, Massachusetts, Allen and Unwin, Incorporated, p. 294–339.

Davies, G., 1979, Dolomite reservoir rocks—processes, controls, porosity development: American Association of Petroleum Geologists Continuing Education Course Note Series #11, Geology of Carbonate Porosity, p. C1–C17.

Davies, W. E., and LeGrand, H. E., 1972, Karst of the United States, *in* Herak, M., and Stringfield, V. T., eds., Karst; Important karst regions of the northern hemisphere: Amsterdam, Elsevier Publishing Company, p. 467–505.

Davis, S. N., 1969, Porosity and permeability of natural materials, *in* DeWiest, R.J.M., ed., Flow through porous media: New York, Academic Press, p. 53–89.

Davis, S. N., and DeWiest, R.J.M., 1966, Hydrogeology: New York, John Wiley and Sons, 463 p.

Dilamarter, R. R., and Csallany, S. C., eds., 1977, Hydrologic problems in karst regions: Bowling Green, Kentucky, Western Kentucky University, 481 p.

Downey, J. S., 1984, Geohydrology of the Madison and associated aquifers in parts of Montana, North Dakota, South Dakota, and Wyoming: U.S. Geological Survey Professional Paper 1273-G, 47 p.

Ford, D. C., 1978, The development of limestone cave systems in the dimensions of length and depth: Canadian Journal of Earth Sources, v. 15, p. 1783–1798.

——, 1980, Thresholds and limit effects in karst geomorphology, *in* Coates, D. R., and Vitek, J. D., eds., Thresholds in geomorphology: Boston, Massachusetts, Allen and Unwin, Incorporated, p. 345–362.

——, 1983, Karstic interpretation of the Winnipeg aquifer (Manitoba, Canada), *in* Back, W., and Lamoreaux, P. E., eds., V. T. Stringfield Symposium, Process in karst hydrology: Journal of Hydrology, v. 61, p. 177–180.

——, 1984, Karst groundwater activity and landform genesis in modern permafrost regions of Canada, *in* LaFleur, R. G., ed., Groundwater as a geomorphic agent, Binghampton Symposium in Geomorphology: Boston, Massachusetts, International Series No. B, Allen and Unwin, Incorporated, p. 340–350.

Freeze, R. A., and Cherry, J. A., 1979, Groundwater: Englewood Cliffs, New Jersey, Prentice-Hall, 604 p.

Garrels, R. M., and Christ, C. L., 1965, Solutions, minerals, and equilibria: New York, Harper and Row, 450 p.

Giusti, E. V., 1978, Hydrogeology of the karst of Puerto Rico: U.S. Geological Survey Professional Paper 1012, 68 p.

Halley, R. B., and Schmoker, J. W., 1983, High-porosity Cenozoic carbonate rocks of south Florida; Progressive loss of porosity with depth: American Association of Petroleum Geologists Bulletin, v. 67, p. 191–200.

Hanshaw, B. B., and Back, W., 1979, Major geochemical processes in the evolution of carbonate aquifer systems: Journal of Hydrology, v. 43, p. 287–312.

Hanshaw, B. B., Back, W., and Deike, R. G., 1971, A geochemical hypothesis for dolomitization by ground water: Economic Geology, v. 66, no. 5, p. 710–724.

Herak, M., and Stringfield, V. T., eds., 1972, Karst; Important karst regions of the northern hemisphere: Amsterdam, Elsevier Publishing Company, 551 p.

Herrick, S. M., and LeGrand, H. E., 1964, Solution subsidence of a limestone terrace in southwest Georgia: International Association of Scientific Hydrology Bulletin, v. 9, no. 2, p. 25–36.

Hollyday, E. F., Brahana, J. V., Harvey, E. J., and Skelton, J., 1981, Comparative hydrogeology of three lower Paleozoic carbonate aquifer systems of Missouri, Kentucky, and Tennessee: Geological Society of America Abstracts with Programs, v. 13, no. 7, p. 475.

Hudson, J. D., 1975, Carbon isotopes and limestone cement: Geology, v. 3, p. 19–22.

Huntoon, P. W., 1974, The karstic ground-water basins of the Kaibab Plateau, Arizona: Water Resources Research, v. 10, no. 3, p. 579–590.

Imes, J. L., 1987, Major geohydrologic units in and adjacent to the Ozark Plateau Province, Missouri, Arkansas, Kansas, Oklahoma; Ozark aquifer: U.S. Geological Survey Hydrologic Atlas 711-E, scale 1:750,000.

International Association of Scientific Hydrology, 1967, Hydrology of fractured rocks, Proceedings Dubovnik Symposium, 1965: Gentbrugge, International Association of Scientific Hydrology, v. 1 and 2, 689 p.

Jacobson, R. L., and Langmuir, D., 1970, The chemical history of some spring waters in carbonate rocks: Ground Water, v. 8, no. 3, p. 5–9.

James, N. P., and Choquette, P. W., 1983, Diagenesis 6, Limestones; The seafloor diagenetic environment: Geoscience Canada, v. 10, p. 162–179.

Jennings, J. N., 1985, Karst geomorphology: Oxford, United Kingdom, Basil Blackwell, Limited, 293 p.

Karolyi, M. S., and Ford, D. C., 1983, The Goose Arm Karst, Newfoundland, Canada, *in* Back, W., and LaMoreaux, P. E., eds., V. T. Stringfield Symposium, Process in karst hydrology: Journal of Hydrology, v. 61, p. 181–185.

Kastning, E. H., Jr., 1984, Hydrogeomorphic evolution of karsted plateaus in response to regional tectonism, *in* LaFleur, R. G., ed., Ground Water as a geomorphic agent: Boston, Massachusetts, Allen and Unwin, Incorporated, 398 p.

LaFleur, R. G., ed., 1984, Ground water as a geomorphic agent, Binghamton Symposium in Geomorphology: Boston, Massachusetts, International Series no. 13, Allen and Unwin, Incorporated, 398 p.

LaMoreaux, P. E., ed., 1986, Hydrology of limestone terranes; Annotated bibliography of carbonate rocks: International Association of Hydrogeologists, v. 2, 341 p.

LaMoreaux, P. E., and Powell, W. J., 1963, Stratigraphic and structural guides to the development of water wells in a limestone terrane: Alabama Geological Survey Reprint Series 6, p. 363–375.

LaMoreaux, P. E., Raymond, D., and Joiner, T. J., 1970, Annotated bibliography of carbonate rocks; Part A, Hydrology of limestone terranes: Alabama Geological Survey Bulletin 94-A, 242 p.

LaMoreaux, P. E., Wilson, B. M., and Memon, B. A., eds., 1984, Guide to the hydrology of carbonate rocks; Studies and reports in hydrogeology: Paris, France, United Nations Educational, Scientific, and Cultural Organization, no. 41, 347 p.

Land, L. S., 1982, Dolomitization: American Association of Petroleum Geologists Education Course Note Series #24, 20 p.

Lattman, L. H., and Parizek, R. R., 1964, Relationship between fracture traces and the occurrence of ground water in carbonate rocks: Journal of Hydrology, v. 2, p. 73–91.

LeGrand, H. E., 1970, Comparative hydrogeology; An example of its use: Geological Society of America Bulletin, v. 81, no. 4, p. 1243–1248.

LeGrand, H. E., and LaMoreaux, P. E., 1975, Hydrogeology and hydrology of karst, *in* Burger, A., and Dubertret, L., eds., Hydrogeology of karstic terrains: Paris, France, International Association of Hydrogeologists, International Union of Geological Sciences, Series B, no. 3, p. 9–19.

LeGrand, H. E., and Stringfield, V. T., 1971, Development and distribution of permeability in carbonate aquifers: Water Resources Research, v. 7, no. 5, p. 1284–1294.

Lesser Jones, H., 1967, Confined fresh water aquifers in limestone, exploited in the north of Mexico with deep wells below sea level, *in* Proceedings Dubrovnik Symposium, 1965, Hydrology of Fractured Rocks: International Association of Scientific Hydrology, v. 2, p. 526–539.

Longman, M. W., 1982, Carbonate diagenesis as a control on stratigraphic traps (with examples from the Williston Basin): American Association Petroleum Geologists Education Course Notes Series no. 21, 159 p.

Maclay, R. W., and Small, T. A., 1983, Hydrostratigraphic subdivision and fault barriers of the Edwards aquifer, south-central Texas, U.S.A., *in* Back, W., and LaMoreaux, P. E., eds., V. T. Stringfield Symposium, Process in karst hydrology: Journal of Hydrology, v. 61, p. 127–146.

Matthess, G., 1982, The properties of ground water: New York, John Wiley and Sons, 406 p.

Maxey, G. B., 1968, Hydrogeology of desert basins: Ground Water, v. 6, no. 5, p. 10–22.

Meyer, G., and Beall, R. M., 1958, Water resources of Carroll and Frederick counties: Maryland Department of Geology, Mines, and Water Resources Bulletin 22, 355 p.

Milanovic, P. T., 1981, Karst hydrogeology: Littleton, Colorado, Water Resources Publications, 434 p.

Moore, C. H., and Druckman, Y., 1981, Burial diagenesis and porosity evolution, Upper Jurassic Smackover, Arkansas and Louisiana: American Association of Petroleum Geologists Bulletin, v. 65, p. 597–628.

Morrow, D. W., 1982, Diagenesis 2, dolomite; Part 2, Dolomitization models and ancient dolostones: Geoscience Canada, v. 9, p. 95–108.

Parizek, R. R., White, W. B., and Langmuir, D., 1971, Hydrogeology and geochemistry of folded and faulted rocks of the central Appalachian type and related land-use problems: Pennsylvania State University Mineral Sciences Experiment Station, Circular 82, 210 p.

Parker, G. G., and others, 1955, Water resources of southeastern Florida, with special reference to the geology and ground water of the Miami area: U.S. Geological Survey Water Supply Paper 1255, 965 p.

Quinlan, J. F., Ewers, R. O., Ray, J. A., Powell, R. L., and Krothe, N. C., 1983, Ground-water hydrology and geomorphology of the Mammoth Cave region, *in* Shaver, R. H., and Sunderman, J. A., eds., Midwestern geology: Geological Society of America and Indiana Geological Survey, v. 2, p. 1–85.

Schlanger, S. O., 1964, Petrology of the limestones of Guam: U.S. Geological Survey Professional Paper 403-D, p. 1–52.

Scholle, P. A., 1979, Porosity prediction in shallow versus deep-water limestones; Primary porosity preservation under burial conditions, *in* Bebout, D., and others, eds., Geology of carbonate porosity: American Association of Petroleum Geologists Continuing Education Course Note Series no. 1, p. D1–D12.

Scholle, P. A., Bebout, D., and Moore, C. W., 1983, Carbonate depositional environments: American Association of Petroleum Geologists Memoir 33, 700 p.

Smart, C. C., and Ford, D. C., 1983, The Castleguard Karst, Main Ranges, Canadian Rocky Mountains, *in* Back, W., and LaMoreaux, P. E., eds., V. T. Stringfield Symposium, Processes in karst hydrology: Journal of Hydrology, v. 61, p. 193–200.

Smith, D. I., and Atkinson, T. C., 1976, Processes, landforms, and climate in limestone regions, *in* Derbyshire, E., ed., Geomorphology and climate: New York, John Wiley and Sons, p. 367–409.

Smith, D. I., Atkinson, T. C., and Drew, D. P., 1976, The hydrology of limestone terranes, *in* Ford, T. D., and Cullingford, C.H.D., eds., The science of speleology: London, England, Academic Press, p. 179–212.

Stringfield, V. T., and LeGrand, H. E., 1966, Hydrology of limestone terraines in the coastal plain of the southeastern United States: Geological Society of America Special Paper 93, 46 p.

—— , 1969, Hydrology of carbonate rock terranes; A review with special reference to the United States: Journal of Hydrology, v. 8, nos. 3–4, p. 349–417.

Tan, F. C., and Hudson, J. D., 1974, Isotopic studies on the paleoecology and diagenesis of the Great Estuarine series (Jurassic) of Scotland: Scottish Journal of Geology, v. 10, p. 91–128.

Thrailkill, J., 1972, Carbonate chemistry of aquifer and stream water in Kentucky: Journal of Hydrology, v. 16, p. 93–104.

Thrailkill, J., and Robl, T. R., 1981, Carbonate geochemistry of vadose water recharging limestone aquifers: Journal of Hydrology, v. 54, p. 195–208.

Thrailkill, J., and five others, 1982, Ground water in the Inner Bluegrass karst region, Kentucky: University of Kentucky Water Resources Research Institute, Research Report 136, 144 p.

Tolson, J., and Doyle, F. L., eds., 1977, Karst hydrogeology: Huntsville, Alabama, University of Alabama, 578 p.

Versey, H. R., 1972, Karst of Jamaica, *in* Herak, M., and Stringfield, V. T., eds., Karst; Important karst regions of the northern hemisphere: Amsterdam, Elsevier Publishing Company, p. 445–466.

Ward, W. C., and Halley, R. B., 1985, Dolomitization in a mixing zone of near-seawater composition, late Pleistocene, northeastern Yucatan Peninsula: Journal of Sedimentary Petrology, v. 55, p. 407–420.

Wardlaw, N. C., 1974, Water above the transition zone in carbonate oil reservoirs: Canadian Petroleum Geology Bulletin, v. 22, p. 305.

Warren, W. M., and Moore, J. D., 1975, Annotated bibliography of carbonate rocks: Alabama Geological Survey Bulletin, Part E, v. 74, p. 31–168.

White, W. B., 1969, Conceptual models for carbonate aquifers: Ground Water, v. 7, no. 3, p. 15–21.

—— , 1977, Conceptual models for carbonate aquifers; Revisited, *in* Dilamarter, R. R., and Csallany, S. C., eds., Hydrologic problems in karst regions: Bowling Green, Kentucky, Western Kentucky University, p. 176–187.

Winograd, I. J., and Thordarson, W., 1975, Hydrogeologic and hydrochemical framework, south-central Great Basin, Nevada-California, with special reference to the Nevada Test Site: U.S. Geological Survey Professional Paper 712-C, 126 p.

Yevjevich, V., 1981, Karst water research needs: Littleton, Colorado, Water Resources Publications, 266 p.

Zeizel, A. J., Walton, W. C., Sasman, R. T., and Prickett, T. A., 1962, Ground-water resources of Dupage County, Illinois: Illinois Geological Survey and Illinois Water Survey Cooperative Ground Water Report 2, 103 p.

MANUSCRIPT ACCEPTED BY THE SOCIETY JUNE 15, 1987

The Geology of North America
Vol. O-2, Hydrogeology
The Geological Society of America, 1988

## Chapter 39

# *Volcanic rocks*

**Warren W. Wood**
*U.S. Geological Survey, 431 National Center, Reston, Virginia 22092*
**Louis A. Fernandez**
*Department of Earth Sciences, University of New Orleans, New Orleans, Lousiana 70148*

## INTRODUCTION

Volcanic rocks are ubiquitous throughout the North American Plate, having formed along convergent plate margins, in rifts, and over hot spots. They range in age from earliest Precambrian to Recent and are exposed or are within easy reach of drilling over a relatively large area of the plate (See Plate 1). Volcanic rocks originate in the earth's upper mantle and lower crust and comprise a large variety of forms. The diversity of forms results in significant diversity of the occurrence, movement, and chemistry of ground water within them. This hydrogeologic variability is caused not only by different geodynamic emplacement and geologic processes but also by differences in hydrologic factors of rainfall, recharge flux, evaporation, topography, soil, vegetation cover, and other factors discussed in Lindholm and Vaccaro (this volume). Geologic controls on the occurrence, movement, and solute composition of ground water in volcanic rocks are largely a function of volume and degree of interconnection of pore space. Primary porosity and permeability of features such as lava tubes, flow breccia, clinkers, flowtop rubble, and shrinkage cracks far exceed those of fractures and joints due to regional tectonic stresses. Other rocks, such as ash-flow tuffs and sheet-flow basalts, may have very low primary permeability and porosity. Thus, processes (chemical, physical, or biologic) that create, enlarge, diminish, or eliminate pore spaces act in concert with the geologic controls and significantly effect the hydraulic characteristics of the rocks.

Ground water and volcanic rocks are intimately related in a variety of complex and contrasting associations. Ground water from volcanic rocks is used for portable water supplies, irrigation, manufacture, mining, geothermal energy, and has produced economic brines and mineral deposits. More than half of the largest springs and some of the most productive wells in the United States issue from volcanic rocks. Yet, the very low ground-water flux in some volcanic rocks has prompted their consideration as potential sites for toxic and radioactive waste repositories. Solutes in water from volcanic rocks range in concentration from less than 10 to greater than 100,000 mg/L.

Chemical composition, mineralogy, volatile content, temperature, and mode of extrusion greatly affect the hydraulic characteristics of volcanic rocks. Some thick flow units of flood basalts have cooled into very dense rock exhibiting very low values of porosity and permeability, while blocky, vesicular, and rubbley interflow zones of lavas have extremely high values of these parameters. Many pyroclastic deposits have large intergranular porosity similar to that found in sedimentary rocks. Fracturing, which increases both porosity and permeability, is a major geologic control on the flux of ground water in both pyroclastics and lava flows, but clearly has a more profound effect on the very dense rocks, which initially have very low values of these parameters. Many volcanic rocks are intercalated with alluvial and lacustrine deposits whose porosity, permeability, and composition may be different than the surrounding rocks and thus affect movement and solute concentrations of ground water moving through the aquifer system. Pyroclastic deposits, by the nature of their origin, tend to exhibit more uniform hydraulic characteristics between regions, but even within a given deposit there may be a significant difference. Thus, the movement of water in volcanic rocks varies significantly from area to area, and although generalizations about the hydrology of volcanic rocks can be useful, local variations are extreme and exceptions to the generalizations are common.

Although volcanic rocks are ubiquitous on the North American Plate, their hydrologic properties have not been defined as extensively as those of other lithologies, probably because their hydrologic variability has discouraged their investigation. Many of the quantitative data available are from the Nuclear Test Site in Nevada, the Hanford Nuclear Facility in Washington, and the National Engineering Laboratory in Idaho. Although the data are limited geographically, they encompass a variety of hydrogeologic environments and are more detailed than those in most hydrologic investigations. Most books on ground-water hydrology give only a brief treatment of water in volcanic rocks. However, Davis and DeWiest (1966), Meinzer (1923), and Stearns, 1942) give excellent introductions to volcanic rocks in relation to water supply, and Custodio (1978) has devoted an entire book to the hydrogeology of volcanic rocks.

Emphasis by ground-water hydrologists has shifted signifi-

Wood, W. W., and Fernandez, L. A., 1988, Volcanic rocks, *in* Back, W., Rosenshein, J. S., and Seaber, P. R., eds., Hydrogeology: Boulder, Colorado, Geological Society of America, The Geology of North America, v. O-2.

cantly over the last 30 years, from a philosophy of exploration to one of management. With that shift, concepts other than those related to water supply have become important. Locations with a low flux of ground water are being considered for toxic and radioactive-waste storage; geothermal energy sources have become important; it is increasingly recognized that many geomorphic features are controlled by ground waters; many brines and mineral deposits are formed by meteoric ground water moving through these rocks, and that low-temperature ground water is important in the diagenesis of volcanic rocks. Thus ground-water hydrology present impinges on traditional geologic problems more than in the past.

The purpose of this chapter is to describe, contrast, and compare the geologic controls on ground-water movement and solute composition in lava flows and pyroclastic deposits. Ground water in intrusive igneous rocks is covered in a companion chapter by Trainer (this volume). Ground water in volcanic rocks is also discussed by Lindholm and Vaccaro (this volume) and Hunt and others (this volume).

## HYDROGEOLOGIC CHARACTERISTICS OF VOLCANIC ROCKS

### *Tuffs*

Pyroclastic deposits include materials with wide ranges of particle sizes, sorting, and fracture density, and therefore their hydraulic properties vary greatly. Bombs, blocks, and lumps of scoria or pumice commonly fall near crater rims and roll downslope, creating locally chaotic deposits of agglomerate or volcanic breccia giving a wide range in values of porosity and permeability. Smaller fragments, including lapilli, pumice, scoria cinders, and ash, are blown further from the volcanic vent to form tuff deposits of more uniform hydrologic character. Mud flows are volumetrically important in younger volcanic rocks such as those of the Cascade Range, but their hydraulic characteristics have not been adequately determined. Most of the hydrologic investigations described in the literature deal with ash-fall and air-fall tuffs, which are volumetrically the most significant pyroclastic deposits.

Ash-flow tuffs, consisting largely of glass fragments and shards, less than 4 mm in diameter, are deposited from a turbulent mixture of gas and pyroclastic material. Ash flows are generally dark in color, unstratified, poorly sorted, and at least partly welded. Deposits from individual volcanic eruptions are usually of small areal extent (to a few hundred square kilometers); however, repeated eruptions create deposits that may cover thousands of square kilometers and may be tens of meters thick (e.g., deposits in the San Juan volcanic field in Colorado and New Mexico). The thickness and areal extent of a deposit depends on the amount of material erupted and the topography over which it was emplaced. Because of their thickness, areal extent and hydrologic characteristics, ash-flow tuffs are hydrologically one of the most important pyroclastic rocks (Winograd, 1971).

Air-fall tuffs are light in color, very fine grained, well sorted, and distinctly bedded. Individual air-fall tuffs generally thin rapidly away from the deposition source, but still may cover thousands of square kilometers. Unlike ash flows, air-fall tuffs are generally thin, are of a relatively uniform thickness, and tend to parallel the surface topography. Because of their limited thickness, air-fall tuffs are usually less important hydrologically than ash-flow tuffs.

Both ash-flow and air-fall tuffs are of a felsic composition generally ranging from rhyolite to dacite. Both may be welded, and the degree of welding directly affects porosity and thus the storage coefficient. Smith (1960) demonstrated an inverse correlation between the degree of welding and porosity. The most densely welded tuffs have a texture like that of obsidian while less-welded zones resemble sandstones. Columnar joints that form on cooling account for a small but important part of pore space in welded tuffs. These joints are characteristic of the degree of welding, and range in spacing from a few centimeters in densely welded material to over a meter in poorly welded tuffs. The joints form perpendicular to cooling surfaces and are thus usually vertical. These features add to the joints and faults formed by regional tectonic stresses. Low-angle foliation and platy fractures are typical of the upper parts of many beds of welded tuff and differ from the high-angle cooling fractures. Low-angle features are believed to be from depositional layering and horizontal movement of the flow as it was degassing and collapsing during cooling. During secondary movement, horizontal zones, less than 1 m thick, containing closely spaced, gently dipping imbricate joints, provide thin zones of high porosity and hydraulic conductivity. High values of hydraulic conductivity in imbricate joints in welded tuff are believed to have contributed to piping and the consequent catastrophic failure of a dam on the Teton River in Idaho in June 1976 (Prostka, 1977).

*Porosity.* The volume of interconnected pore space (effective porosity) controls the volume of water that is free to move in a rock. However, most porosity values reported in the literature are total porosities calculated from rock volume and grain density, and thus do not represent effective (interconnected) porosity. Figure 1 illustrates the range in total porosity for a variety of tuffs quoted in the literature and strongly suggests that nonwelded tuffs have a greater total porosity than welded tuffs. Figure 2 shows the variability of total porosity in single flows of various tuffs and supports the general impression that welding is greater (lower porosity) near the lower middle section than at the top.

Average total porosity of 48 samples of air-fall tuff from the Nevada Test Site was 40 percent (Thordarson, 1965), whereas the total porosity of 48 samples of welded ash-flow tuffs from the Pahute Mesa area on the northern edge of the Nevada Test Site averaged 15.5 percent (Blankennagel and Weir, 1973, p. 9). Total porosity of the Bishop welded ash-flow tuff in California ranged from 3.6 to 37 percent and averaged 14.0 percent (Gilbert, 1938, p. 1843). Thirty-four samples of zeolitized tuffs from the Nevada Test Site ranged in total porosity 19.8 to 48.3 and averaged 37.7 percent (Winograd and Thordarson, 1975, p. 45).

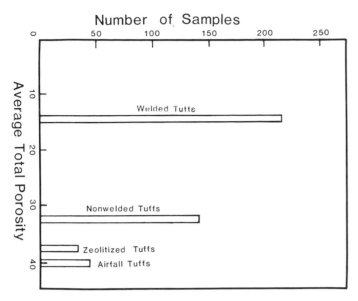

Figure 1. Average total porosity of pyroclastic deposits taken from literature references in text. Values are generally determined in the laboratory on fracture-free samples.

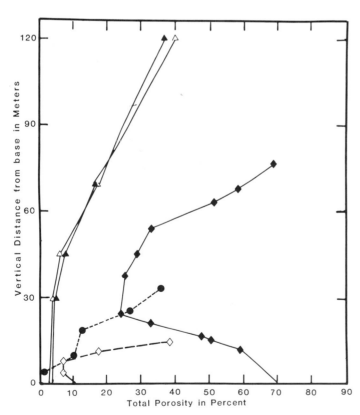

Figure 2. Relationship between porosity and depth for several individual welded tuffs (modified from Smith, 1960, p. 827).

In contrast, the effective porosity of 32 samples of clayey tuff from the Nevada Test Site ranged from 1.8 to 21.6 percent and averaged 11.0 percent (Winograd and Thordarson, 1975, p. 45).

*Hydraulic conductivity.* The ability of primary or secondary pore spaces to transmit fluids depends on numerous factors. In general, the larger the pore spaces and the greater the degree of interconnection, the greater the potential fluid flux. Range in values of hydraulic conductivity from laboratory and field tests reported by Winograd and Thordarson (1975) and Thordarson (1965) from the Nevada Test Site are shown in Figure 3. The pyroclastic rocks shown in Figure 3 consisted of fractured and nonfractured welded tuffs, air-fall tuffs, vitric air-fall tuffs, zeolotized tuffs, and tuffs altered to clay; they range in hdyraulic conductivity over eight orders of magnitude. The ranges in specific capacities of welded tuffs and zeolitized bedded tuffs, as determined from nearly 100 tests at the Nevada Test Site, are given in Figure 4 (modified from Blankennagel and Weir, 1973, p. 8). The specific capacity of 32 clayey and zeolitic tuffs on the Nevada Test Site ranged from $1.7 \times 10^{-2}$ to 5.0 (M$^3$/D)/M (cubic meters per day meter) of drawdown with a mean of 0.8 (M$^3$/D)/M (Winograd and Thordarson, 1975, p. 46). The specific capacities of welded and zeolitized tuffs at the Nuclear Test Site in Nevada appear to be less than those in rhyolites, rhyolite breccias, and vitrophyres (Fig. 4).

The middle to lower middle sections of a given ash flow tend to be the most densely welded and thus have the lowest primary porosity and hydraulic conductivity. However, since these same welded sections also exhibit the greatest number of cooling fractures and thus the highest secondary porosity and hydraulic conductivity, field tests that measure hydraulic conductivity of a flow unit yield much higher values than laboratory tests, which are frequently made on samples that don't include the fractures. Thus, values of porosity and permeability for the middle to lower middle sections of a flow can be contradictory unless the method of determination is clearly stated. The apparent paradox of porosity and hydraulic conductivity distribution in welded tuffs is believed to be due to measurement techniques.

### Lava flows

Chemical composition, the amount of volatiles in the magma, temperature of extrusion, flow thickness and degree of diagenesis, and regional tectonic setting are major controls on the hydrologic characteristics of lava flows. With the exception of massive basalts, tops of most lava flows are surprisingly permeable. Flow tops comprise flow breccias, clinkers, shrinkage cracks, flow-top rubble, and gas vesicles, all of which contribute to their significant porosity and hydraulic conductivity. Zones of fractured and jointed Miocene Columbia River Basalt Group in Washington, Oregon, and western Idaho are given architectural-style names; "colonnade" for the segment dominated by large undulatory columnar joints near the bottom of a flow and "entablature" for the overlying zone of much smaller slender columns of broken vesicular basalt (see Fig. 3, Lindholm and Vaccaro, this

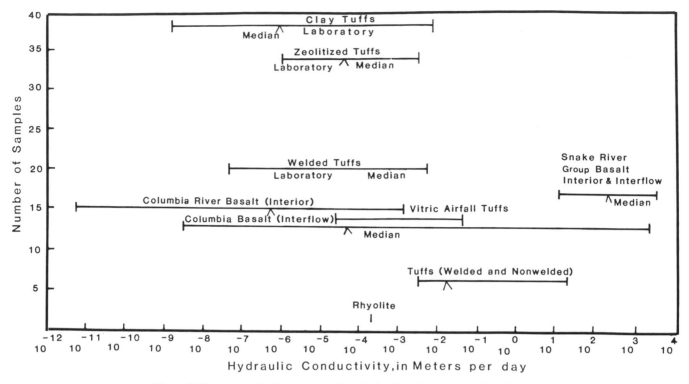

Figure 3. Range in hydraulic conductivity of volcanic rocks (sources given in text).

volume). Porosity and hydraulic conductivity commonly are greater in the upper part of a flow or in the interflow zones between successive flows and lesser in the interim. Most flows in the Columbia River Basalt Group were low-viscosity sheet flows of great fluidity and lateral extent, and they are generally devoid of lava tubes. A significant amount of the ground water in this system moves through the clastic interbeds and in the interflow zones rather than in the central part of the flow.

In contrast to the Columbia River Basalt Group, the lava flows that formed the Quaternary olivine basalts of the Snake River Group in the Snake River Basin of Idaho were quite viscous and smaller in areal extent. They are characterized by rubble, vesicular material, and lava tubes, and because they are geologically younger, usually lack the secondary vesicle and fracture filling common to the Columbia River Basalt Group. Individual flows exhibit significant primary porosity and permeability. The clastic interbeds in the Snake River Group, in general, provide a much smaller ground-water flux relative to the basalts than in the Columbia River basin.

The number and spacing of fractures in lava flows is very important in controlling both porosity and hydraulic conductivity. Figure 5 shows the distance between fractures as a cumulative percentage from ten bore holes in basalt on the Nevada Test Site. The holes ranged in depth from 20 to 72 m below land surface and intercepted 435 fractures in a total linear distance of 354 m. Figure 5 suggests that approximately 50 percent of the fractures in this basalt are less than 0.3 m apart and 95 percent are less than

3 m apart. Carter and others (1967, p. 27) noted that investigations on the Nevada Test Site indicated a decrease in fracture frequency to approximately one-fourth the distance from the top of the flow to its base. At that point, fracture frequency is at a minimum. Below this depth, fracture frequency increases slightly to roughly two-thirds the distance from the top of the flow to its base. Below this depth, fracture frequency increases markedly. Carter and others (1967) reported that of 368 fractures noted on borehole photographs, 71 percent were 0.16 cm or less in width, 22 percent were between 0.16 and .64 cm, and 7 percent were between 0.64 cm and 3.81 cm. They observed that 42 percent of the joints contained no noticeable filling, 44 percent were partially filled, and 14 percent were completely filled. Fillings were usually fine grained and calcareous. Silt from the surface was observed in near-surface fractures.

*Porosity.* The range of total porosity in several hundred samples of basalt is shown on Figure 6. Although the data base for this figure was worldwide, nearly all of the samples are from North America. Total porosity ranges from essentially 0 to more than 75 percent, with values for most samples between 0 and 15 percent, and averaging less than 10 percent. A basalt from the Nevada Test Site (Carter and others, 1967) showed the following relationship between texture and total porosity: dense basalt, 4.8 percent; dense basalt with bands of vesicles, 7.8 percent; vesicular basalt with aligned vesicles, 10.3 percent; and vesicular basalt, 25 percent porosity.

Blankennagel and Weir (1973, p. 10) reported porosities

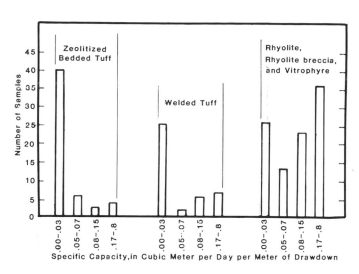

Figure 4. Specific capacity from drill stem tests of major pyroclastic rocks, Nevada Test Site (modified after Blankennagel and Weir, 1973, p. 8).

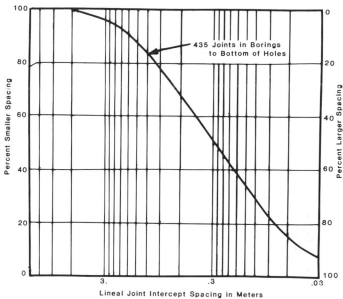

Figure 5. Cumulative frequency of lineal joint intercept spacing in basalts, Nevada Test Site (modified from Carter and others, 1967, p. 24).

determined from density logs of rhyolitic lava at the Nevada Test Site, which ranged from 3 to 15 percent and averaged 12 percent. Total porosities in rhyolite lava determined from neutron logs from an equivalent stratigraphic section ranged from 2 to 16 percent. While rhyolitic lavas appear to have smaller ranges of porosity than basalt lavas, the average values may be similar because flows with lower porosities are controlled by secondary fractures rather than primary fractures. Except for pumice, which is usually pyroclastic, rhyolites typically have fewer vesicles, flow-top breccias, lava tubes, and other of the primary features that give basalts a high porosity. Since fractures in low-porosity rocks of both basalt and rhyolite are a major factor in determining total porosity, similarities are apparent at the lower end of the porosity range.

Effective porosity approximates the storage coefficient of a water-table aquifer more closely than does total porosity. Because of the many; isolated vesicles in basalt, total porosity is usually much greater than effective porosity and averages less than 5 percent for 123 samples collected worldwide (Fig. 7). Garabedian (in review) found that, using a storage coefficient of 0.05, he was able to reproduce observed water levels in a modeling study of basalts in the eastern Snake River Plain. This suggests that the average effective porosity of these rocks is greater than 5 percent, as some water is always retained in the pores. It should be noted that this study includes some rocks other than basalt flows. Although vesicles and interstitial spaces in breccia contribute locally, most effective porosity in basalt is believed to occur along joints and faults that developed during cooling and/or subsequent tectonic activity.

Bulk density, based on fracture-free specimens, is related to rock porosity. The bulk densities of more than 1,000 samples of basalt, collected worldwide, are compared to Columbia River Basalt Group (Fig. 8). These data suggest that Columbia River Basalt Group is fairly typical of the world average and that the average bulk density is approximately 2.8 g/cm$^3$. The bulk densities of breccias range from 2.1 to 2.3 g/cm$^3$; vesicular basalt ranges from 2.3 to 2.5 g/cm$^3$, and dense basalt ranges from 2.7 to 2.9 g/cm$^3$. Based upon a grain density for average basalt of 2.92 g/cm$^3$, the average unfractured total porosity is approximately 4 percent. As the data in Figures 6 and 7 illustrate, the actual average porosity is significantly greater and suggests that fractures play a significant role in controlling porosity in basalts.

*Hydraulic conductivity.* The hydraulic conductivity of basalts ranges over 13 orders of magnitude (Fig. 3). Because hydraulic conductivity is largely controlled by joints and fractures, the interflow zone, the upper part of one flow, and the lower part of the overlying flow (see Figs. 2 to 4, Lindholm and Vaccaro, this volume) generally have hydraulic conductivities greater than those of the more dense interior. In dense basalt, nearly all the permeability in the central parts of individual flows is in fault and joint systems developed during cooling and later regional tectonic activity. Like porosity, hydraulic conductivity in interflow zones is controlled less by fractures and more by primary features such as flow breccias, clinkers, and vesicles.

The distribution of fracturing, and thus permeability, in a given flow unit of a basalt contrasts markedly with that in a flow unit of welded tuff. In most basalts, the greatest density and lowest hydraulic conductivity is in the middle of the flow where

*W. W. Wood and L. A. Fernandez*

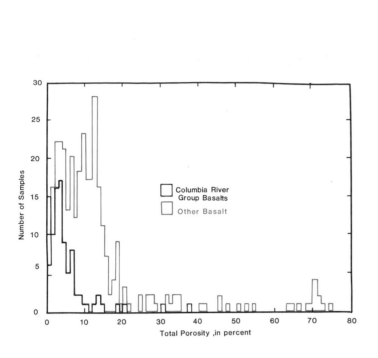

Figure 6. Histogram of total porosity of basalts (modified from Guzowski and others, 1982, E-2).

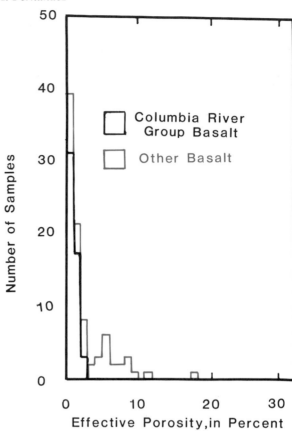

Figure 7. Histogram of effective porosity of basalts (modified from Guzowski and others, 1982, E-3).

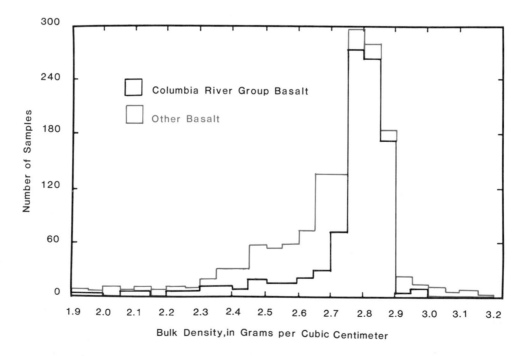

Figure 8. Histogram of bulk density of basalts (modified from Guzowski and others, 1982).

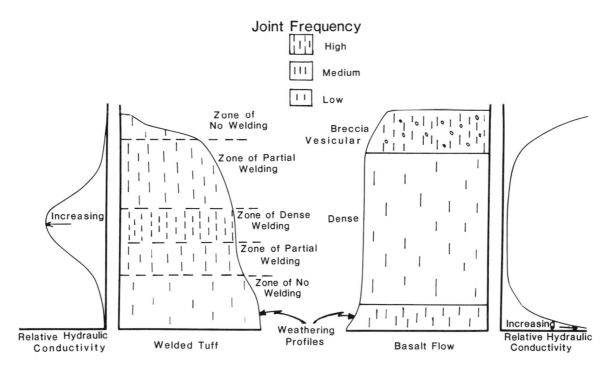

Figure 9. Comparison of the distribution of fractures and hydraulic conductivity in a basalt flow and a welded tuff.

the rock is least affected by cooling fractures. In contrast, the lower midsection of welded tuffs is the most dense but is highly fractured and thus the most permeable. Therefore, interflow zones are generally better aquifers in basalts whereas in welded tuffs the center section with its large number of cooling fractures stores and transmits the greater quantity of water. This concept is illustrated in Figure 9.

### Sills and dikes

Small intrusive bodies like dikes and sills are seldom large enough to be significant sources of water, but their presence may significantly affect ground-water flow. Because they are intrusive features, they frequently exhibit much lower primary porosity and permeability than their extrusive equivalents. As a result they frequently act as a barrier to ground-water flow, and water levels may differ by tens of meters on opposite sides of a dike. Similarly a sill may act as a confining bed for an artesian aquifer system or as a retarding layer in the recharge of a water table aquifer (Hunt and others, this volume).

Dikes and sills exert their most significant hydrologic influence where they intrude rocks of relatively high permeability. Although dikes and sills are common in many volcanic areas, they are not significant in controlling the hydrology of the Columbia River Basalt Group of Washington, the Snake River Group of Idaho, or the pyroclastic deposits at the Nuclear Test Site in Nevada. Large dikes and sills associated with the Mesozoic

basins of eastern North America have been sources of moderate amounts of water where they are fractured or have fractured surrounding rocks during intrusion.

### GEOCHEMISTRY

The geochemistry of ground water in volcanic rocks is important both in terms of the economic usefulness to man and as a means to understanding weathering and diagenetic processes. Solute concentrations and ratios are a function both of the chemistry and mineralogy of the rock matrix and the hydrology of the volcanic-rock aquifer systems. Average chemical analyses of typical rhyolites and olivine basalts, common chemical end members of volcanic rock, are given in Table 1. Although there are large differences in the concentrations of $SiO_2$, $FeO$, $CaO$, $MgO$, and $K_2O$ in these two end members (over 10 percent by weight in some cases), solute concentrations and ratios in water from these two rocks are very similar (Table 2). That is, it is not generally the availability of major ions in volcanic rocks that controls the major concentrations in ground water but rather the thermodynamics of water-rock reactions.

Weathering of the rock by carbonic acid generated by the microbial oxidation of organic carbon is an important process in unconfined aquifers in volcanic rocks, and in recharge zones of confined aquifers, where oxygen is abundant. In confined aquifers, where little or no oxygen penetrates from the atmosphere, and in unconfined aquifers devoid of vegetation cover, little organic carbon is available for the generation of $CO_2$, and silicate hydrolysis plays a significant role in adding solutes to ground water. Volcanic glass, a significant component of the rocks, is a chemically reactive phase in both carbonic-acid and silicate hydrolysis weathering processes, because of large surface areas formed by cooling fractures, structural disorder, and devitrification (Jones, 1966). Cryptocrystalline intergrowths, olivine, pyroxene, and feldspar follow volcanic glass in degree of reactivity. Incongruent dissolution of volcanic glass, coupled with the release of fluid inclusions, is believed to occur in geologically young systems, resulting in relatively high concentrations of $Na^{+1}$ and $Cl^{-1}$ to the ground water. In weathering, major cations are removed with the ground water, leaving a residue of iron oxide and incipient clay minerals. If ground-water flux is high, cation concentrations are likely to be small, and kaolinite tends to form. If ground-water flux is low and concentrations of cations are high, smectite clays tend to form.

Solute chemistry of ground water not only affects the suitability of water for human use directly but also has a significant effect on the hydraulic characteristics of the rock. Fractures in older volcanic rocks generally are filled with mineral precipitates to a greater degree than these in younger rock, and extrusive rocks are more affected by fracture filling than plutonic or metamorphic rocks of the same age. Fracture filling and the resulting loss in hydraulic conductivity is believed to be related to ground-water flux, the large volume of reactive volcanic glass available in extrusive volcanic rocks, and the relatively porous nature of volcanic rocks, which results in a large surface area per unit volume.

### Mechanisms controlling solute chemistry

Weathering of silicate minerals by carbonic acid generally results in a lower pH and a higher bicarbonate concentration in ground water than does weathering by hydrolysis. This is illustrated by the following two equations. Weathering of plagioclase feldspar (labradorite) to calcium smectite by hydrolysis results in generation of cations and hydroxyl ions.

$$(2) Ca_{.6}Na_{.4}Al_{1.6}Si_{2.4}O_{8(s)} + (5.99) H_2O \rightarrow \qquad (1A)$$

$$Ca_{1.7}Al_{2.33}Si_{3.67}O_{10}(OH)_{2(s)} + (1.03)Ca^{+2} + (.8)Na^{+1} + (.87)Al^{+3} + (1.13)SiO°_{4(aq)} + (5.47)OH^-$$

Weathering of labradorite to calcium smectite by carbonic acid results in the additions of bicarbonate ions, rather than hydroxyl ions, as well as cations.

$$2 (Ca_{.6}Na_{.4}Al_{1.6}Si_{2.4}O_{8(s)} + (5.47) H_2CO_3 + (.53)H_2O \rightarrow \qquad (1B)$$

$$Ca_{.17}Al_{2.33}Si_{3.67}O_{10} (OH)_{2(s)} + (1.03)Ca^{+2} + (.8) Na^{+1} + (.87)Al^{+3} + (1.13)H_4SiO°_{4(aq)} + (5.47)HCO_3^{-1}$$

The process of hydrolysis has been documented in the Columbia River Basalt Group by Hearn and others (1984), while that of

### TABLE 1. CHEMICAL COMPOSITION IN PERCENT OXIDE OF TYPICAL VOLCANIC ROCK END MEMBERS*

| | Rhyolite[1] | Olivine Basalt[2] |
|---|---|---|
| $SiO_2$ | 71.27 | 46.42 |
| $Al_2O_3$ | 12.42 | 15.25 |
| $Fe_2O_3$ | 0.49 | 2.07 |
| $FeO$ | 1.38 | 11.49 |
| $MgO$ | 0.36 | 7.20 |
| $CaO$ | 1.36 | 9.88 |
| $Na_2O$ | 2.83 | 2.76 |
| $K_2O$ | 5.26 | 0.65 |
| $TiO$ | 0.47 | 2.76 |
| $P_2O_5$ | 0.09 | 0.77 |
| $MnO$ | 0.05 | 0.20 |
| $H_2O$ | 2.07 | 0.38 |

*All samples from unpublished U.S. Geological Survey files.

[1]Average of 64 samples from the Idavada volcanics, southern Idaho.

[2]Average of 152 samples from the Snake River Group, southern Idaho.

carbonate weathering of the Snake River Group basalts has been verified by Wood and Low (1986).

The addition of hydroxyl ions by hydrolysis to ground water as it moves along a flow path increases the pH to the point where precipitation of $CaCO_3$ occurs. As the concentration of $HCO_3^{-1}$ commonly is much greater than $Ca^{+2}$ in solution, the resultant ground water tends to become a $NaHCO_3$ type, with high pH. The high pH aids in the formation of magnesium clays; thus, magnesium also may be removed from solution.

Cation exchange, a common reaction in weathered and diagenetically altered volcanic rocks, can also yield a $NaHCO_3$ water high in pH and low in $Ca^{+2}$ and $Mg^{+2}$. For example, equation 2 shows that $Ca^{+2}$, generated by either hydrolysis or carbonic-acid reactions, is exchanged for $Na^{+1}$ on sodium saturated clays or zeolites.

$$2NaX + Ca^{+2} \rightleftharpoons CaX + 2Na^{+1} \qquad (2)$$

The removal of $Ca^{+2}$ from solution causes the water to become thermodynamically undersaturated with respect to calcite. Thus, any calcite present will dissolve, releasing $Ca^{+2}$ and $HCO_3^{-1}$ ions in an attempt to regain equilibrium. The $Ca^{+2}$ generated by dissolution of calcite is, in turn, removed by ion exchange, thus further increasing the $Na^{+1}$ and $HCO_3^{-1}$ concentrations. This process also increases the pH by removal of hydrogen ions used in the dissolution of calcite when calcium is removed from solution by either ion exchange or calcite precipitation. Fluoride concentrations can increase dramatically because the thermodynamic saturation of the mineral fluorite, $CaF_2$, is shifted toward dissolution with the removal of calcium. $Mg^{+2}$ is also removed by exchange and may be incorporated into the clay

**TABLE 2. AVERAGE CONCENTRATION OF MAJOR SOLUTES, IN MILLIGRAMS PER LITER**

| Parameter | Snake River Group (olivine basalt) (based on 698 samples) | Columbia River Basalt Group (tholeiitic basalt) (based on 403 samples) | Rhyolite and Tuff | No. of Samples |
|---|---|---|---|---|
| $Ca^{+2}$ | 43 | 27 | 34 | 32 |
| $Mg^{+2}$ | 14 | 12 | 14 | 32 |
| $Na^{+1}$ | 30 | 30 | 33 | 31 |
| $K^{+1}$ | 4.2 | 5.4 | 3.2 | 31 |
| $HCO_3^{-1}$ | 204 | 175 | 175 | 25 |
| $Cl^{-1}$ | 15 | 12 | 74 | 35 |
| $SO_4^{-2}$ | 38 | 23 | 103 | 34 |
| $NO_3^{-1}$ | 1.5 | 2.2 | 1.4 | 6 |
| $SiO_2$ | 36 | 53 | 41 | 29 |
| pH | 7.8 | 8.1 | 7.5 | 38 |
| Dissolved Solids | 285 | 264 | 477 | 30 |

lattice as the pH increases. It is frequently difficult to determine whether a combination of carbonic acid dissolution and ion exchange or silicate hydrolysis is the dominant mechanism controlling the ratios and concentrations of the major cations.

Silica concentrations are commonly much higher in water from volcanic aquifers than in water from other rock types because volcanic glass is more soluble than most silicate minerals. Dissolved silica may eventually precipitate as quartz or chalcedony. Most water in volcanic rocks is supersaturated with respect to quartz and, therefore, some form of quartz would be expected to precipitate. However, the kinetics of this precipitation reaction are slow relative to ground-water movement, and the water can remain supersaturated for long periods of time. Silica in solution also acts as a pH buffer and controls the upper limit of pH of ground water in most volcanic rocks. Both the process of ion exchange and hydrolysis increase pH, and if there were no buffer other than the carbonate system, pH could rise to very high values. However, because of the dissociation of $H_4SiO_4$ to $H_3SiO_4^{-1} + H^{-1}$ and dissociation of $H_3SiO_4^{-1}$ to $H_2SiO_4^{-2} + H^{+1}$, which adds hydrogen to the water, pH values above 9.5 are unusual.

Since volcanic rocks commonly contain small amounts of disseminated pyrite, sulfate concentrations in oxygenated ground water can increase by pyrite dissolution. The following reaction illustrates the release of sulfate and hydrogen ions to solution:

$$FeS_{2(s)} + (3.5)\,H_2O + (3.75)O_2 \rightarrow Fe(OH)_{3(s)} + (4)H^{+1} + (2)SO_4^{-2} \quad (3)$$

Ferrous iron, which is rapidly oxidized to the ferric form in this environment, rapidly combines with hydroxyl ions to precipitate as insoluble ferric hydroxide. The resulting solution is more acidic because of both the addition of hydrogen ions and the removal of hydroxyl ions. It is clear that this reaction can operate only in the presence of dissolved oxygen. In many aquifers devoid of dissolved oxygen and containing organic material, the microbial reduction of sulfate may reverse the reaction, precipitating pyrite and removing iron and sulfur from solution.

Sulfate in ground water also can originate from the dissolution of anhydrite or gypsum, which may have formed in sedimentary interbeds between lava flows in semiarid and arid environments, or from detritus washed in from the surrounding basin. Potassium originates from the dissolution of potassium-rich feldspars, but concentrations seldom exceed 10 mg/l. Potassium is commonly removed from solution by ion exchange with clay minerals and by incorporation within the lattice of the clay minerals.

### Solute concentrations

Average concentrations of major solutes in water from 698 wells completed in the olivine basalt aquifer of the Snake River Group and from 403 wells completed in the semiconfined tholeiitic basalts of the Columbia River Basalt Group are given in Table 2. Concentrations of several constituents are significantly different in the two systems. Water from basalt in the Snake River Group is higher in $Ca^{++}$ and $HCO_3^-$ but lower in silica and pH than is water from Columbia River Basalt Group. These observations are consistent with a carbonic acid weathering of the basalt in the Snake River Group (unconfined aquifer) and silicate hydrolysis of the Columbia River Basalt Group (semiconfined aquifer). Sulfate concentration is higher in water from basalts in the Snake River Group because of oxidation of pyrite in this fully oxygenated aquifer system and the presence of anhydrite washed into the basin from Paleozoic marine sediments, which are present in the eastern end of the basin.

In addition to the major solutes, which constitute approximately 95 percent of the total solute load, there are several ubiquitous minor elements present in concentrations of a few milligrams per liter or less. A very large number of the elements also are present in concentrations of a few micrograms per liter. The average concentrations of selected minor and trace elements from basalt aquifers are given in Table 3. Minor and trace elements in ground water, like those of the major solutes, are derived from the weathering of minerals composing the rock matrix. The

**TABLE 3. SELECTED TRACE ELEMENTS IN MICROGRAMS PER LITER FOR
THREE DIFFERENT LITHOLOGIES OF VOLCANIC ROCKS***

| Parameter | Snake River Group (olivine basalt) | | Columbia River Basalt Group (tholeiitic basalt) | | Rhyolite and Tuff | |
|---|---|---|---|---|---|---|
| | Average Concentration | Number of Samples | Average Concentration | Number of Samples | Average Concentration | Number of Samples |
| Fe | 20 | 18 | 41 | 403 | 32 | 13 |
| Mn | 15 | 15 | 16 | 403 | —— | —— |
| Sr | 230 | 65 | 139 | 337 | —— | —— |
| Li | 40 | 157 | 17 | 337 | 400 | 1 |
| Ba | 134 | 14 | —— | —— | 50 | 1 |
| B | 68 | 54 | 91 | 56 | 47 | 14 |
| F | 840 | 683 | 630 | 403 | —— | —— |

*Sample of rhyolite and tuff taken from WATSTORE files; others from unpublished analyses of the U.S. Geological Survey.

concentration of a particular minor or trace minor element in solution depends upon a variety of factors, many of which are difficult to quantify in natural systems. Concentrations of some elements are controlled by the chemistry of the environment in which they occur, whereas concentrations of others are a function of their availability in the skeletal framework of the rock. In general, those elements that increase in concentration with increasing total dissolved solids or chloride concentration are usually controlled by availability of the element in the skeletal framework. Thus, concentrations of lithium, strontium, and boron generally increase with total dissolved solids and are believed to be controlled largely by their availability. In contrast, the concentration of barium frequently is controlled by the precipitation of barite, while fluoride concentration frequently is controlled by the precipitation of fluorite. Iron concentration appears to be controlled by the formation of iron hydroxide.

Mineral precipitation occurs when water containing dissolved solids becomes thermodynamically supersaturated due to changes in the chemical or physical environment. Often this results from changes in water temperature, loss of gases due to changes in pressure, or mixing of waters having different solute chemistry or different ionic strength.

### Distribution coefficients

The movement of solutes, both natural and anthropogenic, is affected by many chemical mechanisms including mineral dissolution, precipitation, ion exchange, sorption, ion filtration, and others. Because of uncertainties in identifying and quantitatively evaluating the exact mechanism responsible for the removal or addition of solutes as water passes through the rock, an empirically determined coefficient, called the distribution coefficient or more commonly "Kd," is utilized for many sorption-ion exchange reactions. Kd is defined as mass of solute sorbed per unit mass of rock. The distribution coefficient can be very useful in modeling trace amounts of solutes, even in fractured volcanic

rock (Robertson, 1974). Table 4 (modified from Guzowski and others, 1982) gives a range of Kd values for a variety of elements for several different temperature and oxidation conditions in the Columbia River Basalt Group at Hanford, Washington. Values of Kd determined from field studies are generally much lower and more representative than those given in Table 4. For example, Robertson (1974) found that a Kd of 3.0 was appropriate for strontium in fresh olivine basalts of the eastern Snake River basin while the Kd for strontium in the Columbia River Basalt Group (Table 4) ranges from 60 to 480. Thus, values given in the table would yield retention times two orders of magnitude greater than those observed in the field. This difference is not trivial and illustrates the caution that should be exercised when using laboratory-derived Kd values. Nevertheless, Kd values given in Table 4 give some indication of the relative retention of solutes under certain conditions.

### Geothermal waters

Geothermal water is an important resource that exists primarily in the upper few kilometers of the crust along the subduction and rift zones on the western margin of the North American Plate. In the 100 years since the founding of the Geological Society of America, thermal springs from volcanic areas have had constant attention but shifting emphasis from geologists. Because of the general interest in thermal springs, several regions in which geothermal water is present have been designated as U.S. national parks. Throughout the nineteenth and early twentieth centuries, mineralized geothermal water formed an annual multimillion-dollar health-spa and therapeutic-water business that attracted worldwide interest (Moorman, 1867; Fitch, 1927). Although the therapeutic value of mineral hot springs is open to debate, the springs did provide economic incentive for the development of many analytical techniques used for chemical analysis of water. Presently many thermal waters originally associated with health spas are being considered as sources of geothermal energy.

**TABLE 4. LABORATORY VALUE OF DISTRIBUTION COEFFICIENT ($K_d$) FOR THE COLUMBIA RIVER BASALT GROUP***

| Elements | Fresh Basalt, Using Oxygenated water at 60°C | | Fresh Basalt, Using Reduced Water at Various Temperatures | | Altered Basalt, Interbeds and Secondary Minerals Using Oxygenated Water at Various Temperatures | |
|---|---|---|---|---|---|---|
| | $K_d$ Range (ml/g) | Median | $K_d$ Range (ml/g) | Median | $K_d$ Range (ml/g) | Median |
| Cm, Am, Pa, Th, Ac, Po, Bi, Sm, Nb, Zr | 50-200 | 100 | 50-200 | 100 | 190-251,000 | 500 |
| Pu | 5-5000 | 25 | 3-5000 | 100 | 100-5000 | 200 |
| Np | 4-80 | 20 | 0-450 | 30 | 9-180 | 10 |
| U | 3-70 | 5 | 38-52 | 30 | 12-3030 | 20 |
| Ra, Pb | 50-150 | 100 | 90-240 | 100 | 50-2000 | 200 |
| Tc | 0-100 | 0 | 2.5-4500 | 50 | 1-6 | 1 |
| Sr | 60-200 | 100 | 100-400 | 100 | 230-480 | 300 |
| Cs | 100-1000 | 300 | 100-1000 | 300 | 550-225,000 | 2000 |
| I | 0 | 0 | 0 | 0 | 1-4 | 1 |
| Se | 2-10 | 3 | 1-20 | 3 | 0-10 | 1 |

*Modified from Guzowski and others (1982).

Geothermal generating capacity for energy on the North American Plate present is a small percentage of that generated by other means. The economic value of geothermal reservoirs depends largely on the thermal flux in the ground-water system. Power generation with existing technology requires significant quantities of water at high temperatures. Space heating is economical for local use at flow rates as low as 40 liters per minute with a water temperature 15° above average annual air temperature.

One of the major accomplishments in assessing geothermal resources has been the demonstration that most, if not all, geothermal water is of local meteoric origin and not magmatic or juvenile. The use of detailed water budgets, the study of stable isotopes of $^2H$, and $^{18}O$, and the use of radioactive isotopes of $^3H$ and $^{14}C$, has resolved this question for most active systems. The development of chemical geothermometers, which permit inexpensive accurate estimates of reservoir temperature (Brook and others, 1979, p. 21–23) and the use of hydrologic techniques developed for evaluating storage volume and hydraulic conductivity of potable water aquifers, have permitted a quantitative resource assessment within the United States (Muffler, 1979; Reed, 1983) and other areas on the North American Plate.

In general, heat in geothermal waters is acquired by (1) water coming in contact with or in close proximity to recently emplaced igneous bodies, such as the systems at Steamboat Springs, Nevada; on the Aleutian Island, Alaska; at Geyser, California; at Yellowstone National Park, Wyoming; and at Cerro Prieto and Pathe in Mexico; and (2) by deep circulation in an area of normal geothermal gradient such as the hot springs of the Appalachian Mountains and at Hot Springs, Arkansas. Water heated by deep circulation generally is less than 90°C, whereas water heated by recently emplaced igneous rocks frequently is more than 150°C.

## Mineralization and brines

The interaction of hydrothermal fluids and volcanic rocks has long been recognized as playing a vital role in the generation of a variety of important ore deposits; these include uranium and related lithophile elements, precious metals (silver and gold), and base metals (copper-lead-zinc). The flow of hydrothermal solutions occurs when density differences cause hot, low-density water to rise over intrusive igneous bodies, which commonly underlie volcanic terrains. Although the paleohydrology of ore-forming systems has yet to be completely reconstructed, the flow through porous layers in volcanic rocks has become better understood in recent years as a result of studies on modern hot-spring systems. The study of hot-spring chemistry, including precipitates and alteration products, has helped elucidate the processes and mechanisms involves in the formation of hydrothermal ore deposits.

Many hydrothermal solutions are brines, with solute concentration varying from as little as 30,000 mg/l to as much 50,000 mg/l. Solutes may be derived from either a cooling magma or from the rocks through which the water passed; the latter phenomenon appears to be the more important.

Uranium and related lithophile-element deposits (fluorine, beryllium, lithium, etc.) in volcanic rocks are diverse, occurring in a variety of rock types ranging in age from Early Proterozoic to Quaternary. The high uranium contents of felsic, relative to basic, volcanic rocks makes them the most attractive sources. The concentration of uranium in volcanic rocks occurs by direct precipitation from magmatic fluids, precipitation from hydrothermal fluids, and remobilization of secondary uranium by low-temperature ground water. The largest production of uranium from volcanic rocks on the North American Plate (Marysvale, Utah, and the Lakeview District in Oregon) is from fracture-filled veins formed in a shallow, predominantly hydrothermal environment.

Precious metal and base metals were deposited in volcanic rocks by hydrothermal solutions. A good example is the deposits in the San Juan volcanic field, where middle Tertiary rhyolitic and andesitic rocks have been extensively hydrothermally altered in the vicinity of epizonal intrusions emplaced along caldera ring fractures. Although the source of the metals is still debated, studies using oxygen and hydrogen isotopes have shown that the hydrothermal ore fluids vary from predominantly meteoric to mixed magmatic and meteoric waters.

## SPRINGS

In an early report on ground water, Meinzer (1927) noted that more than half of the 65 first-order springs in the United States (greater than 2.8 m$^3$/sec) are in volcanic rocks. All of these springs are in the Tertiary or younger rocks near the western edge of the North American Plate. Olivine basalts of the Snake River Group along a 65-km reach of the Snake River in Idaho are the source rock of 11 of these first-order springs. A significant part of the present spring discharge in this area is surface water recharged from excess irrigation and canal leakages (Mundorff and others, 1964). However, measurements made prior to extensive irrigation development suggest that natural spring discharge in this reach was over 100 m$^3$/sec. The location of springs is determined by topography and structure as well as depth to ground water. Springs from basaslt typically are larger than springs from the more siliceous rhyolites and obsidians, as these rocks tend to be more restricted in areal extent, contain fewer primary openings, and generally yield water from fractures or weathered zones.

Geysers are perhaps the most spectacular form of springs; they flow due to pressure changes brought about by heating of the water rather than by hydraulic head driven by gravity. The famed "Old Faithful" of Yellowstone National Park is perhaps the best known of these.

Ground-water flow through springs is important in changing the landscape in many volcanic terrains, by both erosional and depositional processes. Box canyons, such as those of Blue Lake, Little Canyon, and Box Canyon in the 1,000 springs area of the Snake River in Idaho, are formed and cut headward by spring sapping rather than by overland flow. Piping in altered tuff of the middle Tertiary John Day Formation in eastern Oregon has produced pseudokarst features that have significantly altered the topography of the region (Parker, 1963). Deposition of tufa and travertine by springs and fumaroles issuing from volcanic rocks is well illustrated in Yellowstone National Park. Less spectacular deposits occur in many volcanic systems, particularly those having thermal springs.

## SUMMARY

The occurrences, movement, and solute concentrations of ground water in volcanic rocks are controlled by geologic factors (porosity, intrinsic permeability, topography, mineralogic composition, etc.). These factors reflect different geodynamic emplacement processes and are interrelated with hydrologic factors (recharge flux, concentration of atmospheric gases, climate, vegetation, artesian or water-table condition, etc.). Pyroclastic rocks typical of convergent plates and continental hot spots exhibit a wide range of geohydrologic properties with porosity ranging from less than 10 percent to over 70 percent and hydraulic conductivities that range over nine orders of magnitude from $10^{-8}$ to $10^1$ M/D. Ash-flow tuffs typically exhibit distinct increases in fracture density in the middle to lower part of a flow associated with increased welding. The most densely welded sections, which have the lowest primary porosity and permeability, exhibit the greatest secondary porosity and permeability. Low-angle foliation, associated with collapsing and flowing of tuffs during emplacement, are typical of the upper parts of many beds and differ significantly from the high-angle cooling fractures. These factors commonly produce very high values of porosity and permeability in the upper portions of flows.

In contrast to pyroclastic deposits, flood basalts tend to exhibit less fracturing in the middle to lower section of a given cooling unit. However, the interflow zone comprising the top of one flow and the bottom of the next frequently tends to exhibit high porosity and permeability. Porosities for flood basalts range from less than 1 percent to greater than 75 percent and average close to 5 percent. Hydraulic conductivities occur over 13 orders of magnitude and range from less than $10^{-11}$ to greater than $10^3$ M/D.

Concentrations and ratios of major solutes are controlled more by the type of reactions than by the primary mineralogy of the rocks. Whether an aquifer is confined or unconfined makes a greater difference than does the rock type. Trace-element concentration may be more affected by the rock type than is major-ion concentration, and this phenomenon may lead to economic deposits being associated with specific rock compositions. Low-temperature diagenesis forming calcite, amorphous silica, and clay minerals tends to reduce the hydraulic conductivity. Economic deposits of uranium and other metals are also concentrated by low-temperature leaching, solute transport, and precipitation from ground water.

Geothermal systems are common in many geologically young volcanic systems and have been shown to have been formed largely from the heating of local meteoric water. Solutes in active geothermal systems, which range from less than 50 mg/l to several hundred thousand mg/l, have formed economic mineral deposits in numerous cases. More than half of the major springs in the United States issue from volcanic rocks and have had a major impact on the formation of many box canyons and other spring features. Travertine and tufa deposits form in springs and geysers and thus also significantly change the local topography by the addition of material to the landscape.

# REFERENCES CITED

Blankennagel, R. K., and Weir, J. E., Jr., 1973, Geohydrology of the eastern part of Piahute Mesa, Nevada Test Site, Nye County, Nevada: U.S. Geological Survey Professional Paper 712-B, 35 p.

Brook, C. A., Mariner, R. H., Mabey, D. R., and others, 1979, Hydrothermal convection systems with reservoir temperatures ≥90°C, *in* Muffler, L.J.P., ed., Assessment of geothermal resources of the United States, 1978: U.S. Geological Survey Circular 790, p. 18–85.

Carter, L. D., Bailey, D. M., and Hunt, R. W., 1967, Preshot geological and engineering conditions at the Project Flivver Site, Nevada Test Site: Vicksburg, Mississippi, U.S. Army Corps of Engineers, U.S. Army Engineering Waterways Experimental Station Miscellaneous Paper 3-895, 63 p.

Custodio, E., 1978, Geohidrologia de terrenos e islas volcanicas: Madrid, Instituto de Hidrologia, Centro de Estudios Hidrograficos Publicacion 128, 303 p.

Davis, S. N., and DeWiest, R.J.M., 1966, Hydroecology: New York, John Wiley and Sons, 463 p.

Fitch, W. E., 1927, Mineral water of the United States and American spas: New York, Lea and Feibiger, 799 p.

Gilbert, C. M., 1938, Welded tuff in eastern California: Geological Society of America Bulletin, v. 49, p. 1829–1862.

Guzowski, R. V., Nimick, F. B., and Muller, A. B., 1982, Repository site definition in basalt; Pasco Basin, Washington: U.S. Nuclear Regulatory Commission, Division of Waste Management NUREG/CR-2352, 87 p.

Hearn, P. P., Steinkampf, W. C., Bortleson, G. C., and Drost, B. W., 1984, Geochemical controls on dissolved sodium in basalt aquifers of the Columbia Plateau, Washington: U.S. Geological Survey Water Resources Investigations 84-4304, 38 p.

Jones, B. F., 1966, Geochemical evolution of closed basin water in the western Great Basin, *in* Ray, J. L., ed., Second Symposium on Salt: Northern Ohio Geological Society, v. 1, p. 181–200.

Meinzer, O. E., 1923, The occurrence of ground water in the United States: U.S. Geological Survey Water-Supply Paper 489, 321 p.

—— , 1927, Large springs in the United States: U.S. Geological Survey Water-Supply Paper 557, 94 p.

Moorman, J. J., 1867, The mineral waters of the United States and Canada: Baltimore, Kelley and Peil, 507 p.

Muffler, L.J.P., 1979, Assessment of geothermal resources of the United States, 1978: U.S. Geological Survey Circular 790, 162 p.

Mundorff, M. J., Crosthwaite, E. G., and Kilburn, C., 1964, Ground water for irrigation in the Snake River Basin in Idaho: U.S. Geological Survey Water-Supply Paper 1654, 224 p.

Parker, G. G., 1963, Pipina; A geomorphic agent in landform development of the drylands: International Association of Scientific Hydrologists Publication 65, p. 103–112.

Prostka, H. J., 1977, Joints, fissures, and voids in rhyolite welded ash-flow tuff at Teton Damsite, Idaho: U.S. Geological Survey Open-File Report 77-211, 14 p.

Reed, M. J., 1983, Assessment of low-temperature resources of the United States, 1982: U.S. Geological Survey Circular 892, 73 p.

Robertson, J. B., 1974, Digital modeling of radioactive and chemical-waste transport in the Snake River Plain aquifer at the National Reactor Testing Station, Idaho: U.S. Geological Survey Open-File Report IDO-22054, 41 p.

Smith, R. L., 1960, Ash flows: Geological Society of America Bulletin, v. 71, p. 795–842.

Stearns, H. T., 1942, Hydrology of volcanic terrains, *in* Meinzer, O. E., ed., Physics of the earth; Part 9, Hydrology: New York, McGraw-Hill, p. 678–703.

Thordarson, W., 1965, Perched ground water in zeolitized-bedded tuff, Rainier Mesa and vicinity, Nevada Test Site, Nevada: U.S. Geological Survey Open-File Report TEI-862, 26 p.

Winograd, I. J., 1971, Hydrogeology of ash-flow tuff; A preliminary statement: Water Resources Research, v. 7, no. 4, p. 994–1006.

Winograd, I. J., and Thordarson, W., 1975, Hydrogeologic and hydrochemical framework, south-central Nevada-California, with special reference to the Nevada Test Site: U.S. Geological Survey Professional Paper 712C, 126 p.

Wood, W. W. and Low, W. H., 1986, Aqueous geochemistry and diagenesis in the eastern Snake River aquifer system, Idaho: Geological Society of America Bulletin, v. 97, p. 1456–1466.

MANUSCRIPT ACCEPTED BY THE SOCIETY JUNE 26, 1987

## ACKNOWLEDGMENTS

We wish to extend our sincere thanks to Paul Hearn, Frank Trainer, John Vaccaro, and Gerry Lindholm of the U.S. Geological Survey, and Professors Gary Byerly of Louisiana State University, and Stan Davis of the University of Arizona for their comments and constructive criticism of the manuscript. We wish also to thank Lory Severin for her professional preparation of the manuscript.

## Chapter 40

# *Plutonic and metamorphic rocks*

**Frank W. Trainer***
*U.S. Geological Survey, 345 Middlefield Road, Menlo Park, California 94025*

## INTRODUCTION

Plutonic and metamorphic (crystalline) rocks form the basement in North America, and are exposed over much of the continent (Plate 2). Where near the surface, they commonly yield small supplies of water to wells. Limited permeability at even moderate depth may favor their use locally for waste storage. Both water supply and waste storage present challenging problems because the permeability of the rocks is still poorly understood.

The crystalline rocks differ from unconsolidated deposits in possessing little primary permeability except where weathered. Ground water in crystalline rocks therefore occurs principally in fractures. Other types of consolidated rocks—the volcanic rocks, described in this volume by Wood and Fernandez (this volume) and the consolidated sedimentary rocks, described by S. N. Davis (this volume)—are also fractured, but possess primary permeability that, in many rocks, exceeds that due to fracturing. Some progress has been achieved in describing flow in idealized fracture systems, but relatively little is yet known of the nature of the fractures and of their geologic controls. Generalization and the transfer of experience from one terrane or one region to another are therefore still limited.

## GEOLOGIC SETTING

Understanding of the crystalline rocks has expanded rapidly during the century of the Geological Society of America, and the hydrogeologist, arriving late on the scene, can now take advantage of progress in study of their composition, weathering, and structure. Plutonic rocks and their metamorphic equivalents consist principally of silicate minerals. Other metamorphic rocks reflect the composition of their sedimentary antecedents, and some, like marble or quartzite, approach a monomineralic composition. Rock composition is significant as a chief control on weathering.

The crystalline rocks have formed in mobile crustal belts. In North America they are best known in the Canadian Shield (Pfannkuch and others, this volume) and the Cordilleran and Appalachian regions (see chapters by Mifflin, Riva-Palacio, Gale and others, Seaber and Hollyday, and LeGrand, this volume). In some terranes a complex deformational history is reflected by complex patterns of fractures of different ages; in others, older openings were obliterated by metamorphism, and most of the open fractures date from late stages of structural history. The dominant pattern of fracturing through jointing is modified locally by faulting. Most joints are near the surface, but many faults extend to considerable depth. Hence, the distribution of fractures in three dimensions can be complex.

Fractures are commonly tighter and less abundant with increasing depth beneath the surface—characteristics related to the state of stress in the crust. Two considerations enter the picture. Formation of incipient fractures is due to any of several forces, of which tectonic stress has been considered one of the most widespread and effective. However, the formation of open fractures is related to relief of confining pressure. Uplift and erosion combine to unload the rock, and incipient fractures can open.

Measurements of the stress field near the surface have been made since the 1950s. Hast (1967) described a horizontal compressive-stress field in Scandinavia and Finland that increases with depth. This finding has been confirmed in other regions. Lindner and Halpern (1978), McGarr and Gay (1978), Sbar and Sykes (1973), and Zoback and Zoback (1980) have summarized data from North America.

Of the three principal compressive stresses, two are near the horizontal, and the third, due to load, is near the vertical. The horizontal principal stresses are commonly unequal, and in many areas one or both are greater than the vertical stress. Regional stresses have long been attributed to tectonic forces, and more recently to motion or deformation of lithospheric plates. On the other hand, Voight (1967) suggested that differences between horizontal and vertical stresses may reflect denudation: vertical stress in a given part of the crust decreased significantly with uplift, but the horizontal stresses are still decaying slowly from their earlier magnitudes. The modern stress field appears to be a

———————
*Present address: P.O. Box 1735, Corrales, New Mexico 87048.

Trainer, F. W., 1987, Hydrogeology of the plutonic and metamorphic rocks, *in* Back, W., Rosenshein, J. S., and Seaber, P. R., eds., Hydrogeology: Boulder, Colorado, Geological Society of America, The Geology of North America, v. O-2.

persistent long-term feature, perhaps of Tertiary and Quaternary age in eastern and central North America and of Quaternary age in the West (Zoback and Zoback, 1980).

Detailed consideration of crustal stress is beyond the scope of this chapter, but it is appropriate to note that the stress field controls fracturing of the rocks. For fractures to open, one principal compressive stress must be significantly smaller than the other two. Understanding of the field may therefore serve to explain local and regional patterns of fracturing and to facilitate generalization of the regional hydrogeology of crystalline (and other fractured) rocks. One approach will likely be through study of major patterns of crustal stress. Zoback and Zoback (1980) described stress provinces in the United States and grouped them into two large regions. The subcontinental region west of the Great Plains is one of active deformation characterized by extensional tectonism. The eastern and central United States, in contrast, is characterized largely by compressional tectonism. Extension of such studies to determine whether fractures differ significantly in different stress settings could improve our understanding of the hydrogeology of fractured rocks.

A second approach treats the effect of contemporary compressive stress on the hydraulic conductivity of fractured rocks. Carlsson and Olsson (1979) found a direct relationship between hydraulic conductivity and the orientation of the natural stress field. In the present discussion we can consider examples of joints within a single stress province. Recent study in Georgia (Cressler and others, 1983) has emphasized the hydrogeologic significance of sheeting fractures. Recognition of sheeting in wells has resulted from improvements in down-hole instrumentation, and the question arises whether sheeting fractures are common elsewhere in the Piedmont Province, as data from exposures suggest they may be, even though they have not been widely recognized in wells. Recognition may be dependent on adequate instrumentation; or, on the other hand, abundance of open sheeting fractures may differ from one area to another in response to differences in the stress field.

Joints tend to open most readily normal to the axis of least compressive stress (Hubbert and Willis, 1957). Hence, if any joints are now being formed in eastern North America, or if incipient joints are being opened, sheeting joints should be favored because they would be normal to the least principal stress, which is vertical. With respect to steeply dipping joints, the writer believes, on the basis of an ongoing study of fractured rocks in the northern Piedmont Province, that water flows more readily in some directions than in others, and that the directions of preferred flow bear a systematic relationship to the contemporary stress field.

## FRACTURES

The following discussion, which emphasizes topics of direct interest in hydrology, is based largely on observations made during hydrogeologic investigations. However, it is desirable first to note aspects of rocks and fracturing that strongly influence the response of rocks to stress. The very fact that rocks fracture seems surprising, for the crust is generally under compression, a state in which rocks are far stronger than under tension. Fracture occurs, by shearing or under tension, at places where difference in principal compressive stresses is such that one stress is smaller than the strength of the rock under shear or under tension, respectively. Tiny fractures termed Griffith cracks, chiefly along grain boundaries, are believed to play a major role in tensional fracture (e.g., see Hobbs and others, 1976). Tensional stress accumulates at the tips of the cracks, which grow, and some cracks intersect others. Two aspects of these cracks are significant in hydrogeology: the microcracks can form extensive networks of interconnected openings, as noted in another section of this chapter; and, with other microscopic openings, they can influence the formation of incipient fractures.

Some fractures (joints) form near the land surface from incipient fractures due to earlier deformation. Incipient joints are believed to be discontinuous surfaces or zones of weak material represented by cracked mineral grains, disjointed grain boundaries, tiny faults, and layers of liquid inclusions (Chapman, 1958). Segall and Pollard (1983) concluded that the growth of cracks (as, for example, by integration of such microfeatures) may lead to jointing; and field examples show, by geometric similarity, that joints could have formed from microfractures. Kendorski and Mahtab (1976) tested tensile and compressive strengths of oriented cylinders of quartz monzonite. Many samples failed along preexisting planes of weakness, two of which coincide geometrically with two of three joint sets in the rock. (The third set differs from the others in age and inferred origin.) Harper (1966) correlated joints in outcrops with microfractures in thin sections of Precambrian crystalline rocks in Colorado.

The foregoing examples are of jointed rock. Joints are macrofractures along which dilation has occurred without movement parallel to the fracture surfaces. They are abundant and ubiquitous but relatively small, and they occur near the land surface. Faults, in contrast, are more localized fractures along which movement has occurred parallel to the fracture surfaces. Faults can be much more extensive than joints, and some extend to great depths. Hence, a corollary of the occurrence of zones of weakness—the partitioning of strain in rocks—is of greater hydrogeologic significance in faults than in joints. Recent studies (e.g., Lister and Williams, 1983) have emphasized that deformation tends to be concentrated in localized zones of weakness separated by relatively sound rock. Once formed, such a zone, which can be microscopic or macroscopic, persists as a potential fracture. Repeated deformation tends to perpetuate and extend existing fractures and to open new ones along similar zones of weakness.

Finally, it should be emphasized that the fractured-rock environment is dynamic rather than static. Fractures are deformable, and they reflect temporal change and spatial variation in stress. Hydrogeologists have rarely been concerned with natural stress change, which at most places is probably very slow, but current studies of earthquake mechanisms show how significant it can be in special problems. Spatial variation is of more general concern in hydrogeology, and it deserves increased attention. As an example, consider a single fault zone in heterogeneous rocks. Regardless of any degree of uniformity of the stress field, the rocks in the fault zone can be under effective tension, compression, or shear at different places, horizontally and with depth, with corresponding differences in fracture characteristics and fluid movement.

### Microfractures and pores

Both pores and grain-sized cracks are associated with grains in crystalline rocks, particularly along mineral cleavage and at grain boundaries. It is clear, upon reflection over the presence of well-developed directional structures in rocks and their constituent grains, that there must exist a considerable potential for integration of some microfractures and pores into incipient macrofractures. Microfractures and joints in crystalline rocks have been observed by several investigators (e.g., Harper, 1966; Kendorski and Mahtab, 1976) to possess the same geometry. Harper concluded that the microfractures and joints he studied had developed in the same stress field or fields.

Brace (1972) observed that rocks are electrically conductive at depth whereas the minerals in rock such as granite become conductive only at temperatures greater than 500°C; he concluded that crustal rocks contain a network of electrolyte-filled cavities. The occurrence of saline water in crystalline rocks, noted in a later section, suggests that such networks of cavities may be common and perhaps extensive.

The presence of microfractures and pores in crystalline rock may have been indicated by pumping tests in New Hampshire (Stewart, 1978). A good fit was obtained for drawdown-time curves with type curves for a double-porosity model described by Barenblatt and others (1960). The model assumes the presence of fractures and blocks, each characterized by its own values for transmissivity and storage coefficient. Inasmuch as the blocks of crystalline rock can be assumed to contain negligible porosity except for microfractures and pores, agreement of data curves and type curves implies that interconnected microscopic openings provide the transmissivity of the blocks.

### Joints

Crystalline rocks are jointed virtually everywhere near the land surface, and although most joints are of limited lateral extent and depth, they are so numerous as to form a three-dimensional network of conduits in the near-surface rock. Geologic controls on the character of both joints and network constitute a significant problem that has scarcely been touched.

***Types and origin of joints.*** Four types of joints are representative of plutonic and metamorphic rocks as a group: (1) primary joints related to the intrusion and consolidation of magma, (2) joints related to folding, (3) joints related to regional deformation, and (4) sheeting joints.

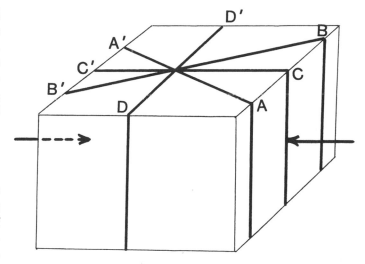

Figure 1. Idealized geometry of representative joint sets (after Billings, 1946, Fig. 86; Hobbs and others, 1976, Fig. 7.31). Arrows indicate direction of inferred compression. A–A′ and B–B′, shear joints; C–C′, extension joint; D–D′, release joint. In folded rocks the shear and extension joints can form in other attitudes as well, producing complex groups of sets (Stearns and Friedman, 1972).

Primary joints, joints related to folding, and some regional joints exhibit the geometry of fractures found in rock samples after failure in laboratory compression tests (Fig. 1). For this reason these joints have commonly been attributed to compressional forces. This common geometry is useful in description of joints, even though it does not indicate common origin. The presence of one joint pattern in three terranes of unlike deformational histories raises the question of whether determination of the origin of the joints is useful in hydrogeology. In the present state of our knowledge, such a determination has limited usefulness because we know so little about the characteristics of these several types of joints. But if significant differences were found, understanding of geologic setting could become an important tool in generalization of the hydrologic character of crystalline rocks.

Kendorski and Mahtab (1976) found similar joint patterns at the surface and at depths greater than 600 m. Raven and Gale (1977) found strong similarity in geometry, at the surface and at depth, in mines whose depths exceed 1,100 m. The same joint sets and systems appear to be present or potentially present throughout the uppermost hundreds of meters of the crust at a given locality. It is not clear how many of the joints found in mines were open prior to local release of stress during excavation, and how many opened during the mining operation. Aside from this problem in interpretation, characteristics of jointed rock, as determined through study of exposures and drill holes, are likely to be representative of enormous masses of rock; the hydrogeologic transfer value of direct observation should prove significant, once relationships have been established between physical features and hydrologic characteristics.

Sheeting is best developed in granite and other massive rocks (e.g., Jahns, 1943; Chapman and Rioux, 1958; Johnson, 1970; Twidale, 1973). The fractures appear to be independent of original macrostructures and fabric of the rock; they range from subparallel to the land surface at shallow depth to horizontal at greater depth. They are also progressively farther apart with increasing depth, and they occur to depths of 100 to 150 m or more. Individual fractures may persist laterally for several hundred meters, beyond which others overlap them en echelon. Steeply dipping joints may intersect them to form a three-dimensional network.

LeGrand (1949) described sheeting in Georgia and noted that inclination toward the valleys produced a pseudosynclinal structure which, with cross joints, facilitates flow from recharge areas on hills to discharge areas in valleys. Despite this early recognition of the hydrologic function of sheeting joints, they have received little attention from hydrogeologists, probably because the subsurface data available until recently have not served to identify this type of fracture.

Near Atlanta, Cressler and others (1983) found productive wells on hilltops where soil cover is thin or absent. Many of these wells derive water from depths of 130 m or more. This combination of topographic situation, thin residual cover, and depth is unfavorable for the success of water wells, according to criteria developed in the Piedmont Province over the past 30 to 40 years. Some of these wells were shown to tap near-horizontal fractures that appear to be sheeting joints. Response of wells to pumping indicated that some fractures may extend laterally for a hundred to several hundred meters.

Sheeting has commonly been attributed to vertical expansion of the rock upon erosional unloading. Johnson (1970) gave a detailed summary and analysis of the unloading hypothesis, calling on enlargement of small cracks present in the rock prior to sheeting. Hast (1967) attributed horizontal fissures in surface exposures and mines in Scandinavia to horizontal overloading of the rock. Twidale (1973) also concluded that lateral compression, rather than unloading, causes sheeting. The unloading and compression hypotheses may turn out to be two faces of the same coin, in light of the interpretation of Voight cited earlier.

***Characteristics of joints.*** Several characteristics of joints control the flow systems in jointed rocks by determining the water-bearing character of individual joints and the nature of the conduit network. The hydraulic characteristics are aperture and wall roughness. The network characteristics include abundance of joints and their extent, depth of occurrence, and geometry. These characteristics are not well documented. Comparison of reported data is difficult because diverse methods of study have been used. Even a brief discussion is valuable chiefly as an indication of research needed.

Aperture in crystalline rocks has been reported on the basis of direct measurements in mine openings (Mahtab and others, 1973) and in surface exposures (Bianchi and Snow, 1969). The measured values probably are not representative of aquifers because both mine walls and surface exposures have been affected by stress release. Snow (1968) used a parallel-plate model and a calculated aperture assumed from permeability determined by packer tests in drill holes. Results did not differ notably with rock type. Computed apertures ranged from $4 \times 10^{-4}$ to $7.5 \times 10^{-5}$m (400 to 75 microns) in the upper 10 m of bedrock; over the depth interval from about 15 to 60 m they decreased from about $1 \times 10^{-4}$m to $5 \times 10^{-5}$m. (According to Snow, only fractures having aperture greater than about $3.5 \times 10^{-5}$m are important water conduits.) The upper limit for the porosity of undisturbed rocks ranged from 0.05 percent near the surface to 0.0005 percent at about 130 m; downward decrease in porosity was commonly at least an order of magnitude in 60 m.

Use of integral sampling may improve determination of fracture aperture in drill holes. Conventional drill core can expand and fracture after removal from the hole, while fractures in the wall may deform, with change in permeability, because of hydraulic pressure applied during injection testing. In integral sampling, the rock is stabilized and sampled by cementing a metal rod into the drill hole and then overcoring the rod and enclosing rock. Such sampling in granite (Rocha and Barroso, 1971) disclosed a downward decrease in average width of open fractures from $4 \times 10^{-4}$m (0.4 mm) in the zone 2.5 to 4.5 m below the bedrock surface to $1 \times 10^{-4}$m at 13.5 to 14.5 m; the greatest average aperture was $7 \times 10^{-4}$m at 7.5 to 8.5 m.

Roughness of fracture walls causes the aperture to vary. The potential flow path along many individual joints is a flat network of tubular and disklike openings, rather than a tabular opening. Use of a parallel-plate model requires assumption of an average aperture in an idealized joint that lacks surface roughness.

Density of fractures can be seen in many exposures to be highly variable. Segall and Pollard (1983) found the perpendicular distance between joints of a single set in exposed granodiorite to range from about 0.2 m to nearly 25 m. Mahtab and others (1973) found the distribution of fracture spacing in quartz monzonite in a mine to be strongly skewed toward the small end of the scale; at a depth of about 650 m, more than 80 percent of the joints in a single set were less than about 0.3 m apart. Wallis and King (1980) found similar spacing in drill holes in granite in Manitoba. The quantitative significance of these direct observations is elusive with respect to conceptual models of aquifers because of likely stress-release effects.

Snow (1968) used injection-test data to estimate spacing and aperture of water-conducting fractures. His assumptions included three mutually orthogonal joint sets and a Poisson distribution of joint spacing. His estimates of average minimum spacing ranged from about 1.3 m at a depth of about 1.6 m to about 4.5 m at 96 m.

Numerous Soviet geologists have determined the density of fractures (coefficient of fracturing) by field or laboratory measurements. Most of this work has been done in sedimentary rocks, but Pisarskiy and Sherman (1967) dealt with crystalline rocks. Their coefficient involves the ratio of the products of length and aperture to the area of exposure. In a phlogopite deposit, this parameter ranged from 1 to 8 units at the surface to 0.15 unit at a depth of 150 m. They found the coefficient to depend on lithology and structure.

Little is known of lateral extent (length) of joints in crystalline rocks. Observations in quarries and natural exposures show that most joints are relatively short—probably in the range from a fraction of a meter to 10 m.

Depth of occurrence involves the vertical extent of individual joints and the vertical thickness of the network of open joints. In many quarries and cliffs, only a small proportion of the joints extend downward as far as tens of meters. With respect to the network, observations in mines and deep tunnels show joints to be present to depths of many hundreds of meters. As already noted, it is difficult to establish the natural abundance of open joints in subsurface workings because of the opening of incipient joints through stress relief, but the fact that apparently unfaulted rock yields water suggests that some of the joints are open at these depths beyond the zone of influence of the excavations. In several mines in the Canadian Shield, evaporation by ventilation removes water that seeps from joints below 300 m (Raven and Gale, 1977); moreover, coatings of carbonate and clay minerals and iron oxide on joints are largely above that level. These observations suggest that circulation along joints there is chiefly above 300 m.

Data on yield of water wells in crystalline rocks commonly suggest that the network of open joints is principally within 100 to 150 m of the surface (e.g., LeGrand, 1954, 1967; Davis and Turk, 1964), although some recent work (Cressler and others, 1983; Daniel, 1985) found significant yields to greater depths. Abundance and aperture of joints decrease with increasing depth, and these changes contribute to downward decrease in porosity and permeability. Snow (1968) concluded that change in aperture is the more important factor.

Geometry of the joints determines their attitude and that of their intersections, while the number and lengths of joints determine the number of intersections. Although the hydrologic significance of intersections of fractures has been emphasized in description of fracture traces and lineaments (e.g., Parizek, 1976), scant attention has been devoted to intersections in hydrogeologic studies. However, Kohut and others (1983) found that the long axes of cones of depression due to pumping from wells in granodiorite were aligned along the strike of the intersection of two prominent joint sets.

Attempts have been made to calculate the permeability tensor in fractured rock at field sites through use of a parallel-plate model with measured or assumed values for orientation, density, and aperture of fractures (Bianchi and Snow, 1969; Parsons, 1972). However, difficulty in obtaining representative values for those parameters and for wall roughness hinders such an approach. Until better data become available, a direct method, using head tests in drill holes with packers (Hsieh and Neuman, 1985) seems a more promising approach.

## Faults

Faults are better understood than joints, with respect to physical characteristics and stress setting, but as a class of hydrogeologic features, faults are poorly known. A few generalizations are widely recognized but there has been little detailed study of the hydrology of faults, particularly in crystalline rocks.

Comparison with joints is an effective approach to the general hydrologic character of faults. Faults and single fault zones range from millimeters to hundreds of kilometers in length and to kilometers in depth. On an areal scale, the larger faults differ from most joints in being major structural elements that can control regional structural and topographic grain as much as do such features as folds and outcrop belts. As compared with joints, which are ubiquitous, faults commonly occur in restricted zones. Within a fault zone, individual fractures can be very numerous, but even in densely faulted terrain there may be extensive tracts of jointed rock not cut by faults. Faults formed under any combination of stress conditions can be either filled with gouge or relatively open.

Joints in the zone of saturation are conduits for water, but faults can be either conduits or barriers. A barrier can be the boundary where a less permeable section of rock offset by the fault abuts a more permeable section and blocks the flow path, or an impermeable segment of the fault itself. Both types can be complex. Raven and Gale (1977) found individual faults to be sealed by gouge at some places and open at others.

Open fractures in a fault zone form a tabular conduit network. Unlike the network of joints, networks in fault zones range from near-horizontal to vertical. Whereas joints facilitate shallow ground-water circulation with flow lines nearly parallel to the land surface, steeply dipping fault zones provide the most effective avenues for deep circulation, as in some hydrothermal-convection systems.

Under natural flow conditions a fault zone that is more permeable than nearby unfaulted rock may have little effect on ground-water flow except locally, or it may have a large effect, depending on its trend relative to the direction of ground-water flow. Where fault and flow line are normal, the fault can cause a local change in transmissivity and hydraulic gradient. Where fault trend and flow direction are similar, the fault may be a hydraulic sink that receives and transmits a large part of the ground water in the vicinity.

Faults and joints can yield substantial quantities of water to an excavation. Observations made in a tunnel in Colorado illustrate differences in fracturing and water flow in different hydrogeologic settings. Wahlstrom and Hornback (1962) found 1,037 faults in 3,400 m of tunnel in a thrust plate, 180 to 300 m beneath land surface. Water entered from fractures along 75 percent of this reach; inflows were as great as 0.2 l/s (liters per second) per meter of tunnel. In a quartz monzonite stock they found 638 faults in 3,825 m of tunnel. Water was entering only 5 percent of this reach, and the few flows measured ranged from 0.001 to 0.17 l/s per meter. In another part of the same tunnel, Warner and Robinson (1967) recorded numbers and ratios of joints and faults in granite and gneiss. They counted 2,951 joints and 1,374 faults in 23,476 m of tunnel, about 450 to 1,350 m below land surface:

one joint in 8 m, one fault in 17 m, and a joint/fault ratio of 2.15. Spatial differences in number of fractures are related to lithology, structure, and depth.

Reflection upon such numbers suggests the hydrologic significance of the nature, abundance, distribution, and interrelationships of fractures in a fault zone—topics for which little information is available. Fractures associated with a fault can be assigned to the stress conditions under which faulting occurred, and the shear fractures among them are miniatures of the fault, but there is no present way of predicting either the width of the fracture network or the number of fractures (Stearns and Friedman, 1972). Experimental evidence (e.g., Tschalenko, 1970) suggests that second-order fractures associated with a fault provide an important part of the porosity and permeability in the fault zone. To compound the problem of understanding the porosity provided by fractures caused by faulting, joints unrelated to the faulting are commonly present. Pohn (1981) used joint spacing as an indicator of concealed faults in slightly deformed sedimentary rocks—an approach that may be useful in crystalline rocks. Wider spacing of the joints of a given set was found near a fault zone than at a distance. Pohn suggested that faulting took up part of the stress that might otherwise have opened incipient joints, or that more joint sets may have formed near the fault than at a distance, although individual joints in a set may be farther apart here than in rock that has not been faulted.

Raven and Gale (1977) found that a typical fault zone in the southern Canadian Shield consists, from the center outward, of (1) gouge; (2) sheared, brecciated, and altered rock, partly resealed; and (3) a 1- to 20-m fractured zone parallel to the main gouge zone. Fault and shear zones as thick as 30 to 60 m, with associated joints, were observed to depths of 2,000 m. Faults that cut similar rocks in near-surface workings and at deeper levels had similar width, orientation, and character. Gouge and secondary alteration differed, through mineralogical control, with rock type; but within a given rock type no significant variation was found with depth.

Raven and Gale (1977) also noted striking differences in discharge from fractures in mine walls. Faults produced the largest seepage inflows, followed in order by shear zones, fractures in dikes and sills, and fractures at contacts of rock masses. Joints, partings along foliation, and microcracks were less effective conduits. However, large differences were found from one fault to another, or from one place to another on a single fault. Only 25 percent of faults and shear zones between the land surface and the 600-m depth seeped continuously, and some faults seeped along only parts of their extent, laterally and with depth.

Some high-yield wells in crystalline rock near Atlanta tap steeply dipping shear zones (Cressler and others, 1983). The more extensive zones are as long as 11 km and as thick as 100 m or more. They consist of fine-grained rock that weathers into fragments bounded by fracture faces. The longer features, which may be series of short, closely spaced parallel zones, form prominent valleys. Raven and Gale (1977) observed shear zones that resemble faults except for the absence of a definitive gouge zone. They

also found thin (1–10 cm) sheared contacts, some with gouge, between many dikes and sills and their host rocks, and shear zones along the foliation in gneiss. About 50 percent of the dikes and sills observed in the subsurface had sheared contacts. In North Carolina, veins and dikes that penetrate gneiss and schist are jointed or have joints along their contacts with the host rock (LeGrand, 1954).

## Linear features

Photogeologic fracture traces and lineaments—linear features on the land surface that include depressions, soil and vegetation patterns, and straight reaches of streams—appear to be fracture controlled. They are of hydrogeologic interest for two reasons: (1) they suggest the presence of localized permeable zones that may be significant in water supply, drainage, and rock stability; and (2) the length and geometry of some features suggest the presence of large blocks bounded by fracture zones. The integrity of the rock, which would therefore be a function of location, is of critical concern in waste-storage applications.

Most hydrogeologic studies of these linear features have been in sedimentary rocks, where wells on fracture traces have statistically higher yields than other wells nearby, and wells at the intersections of two or more linear features have still higher yields (Parizek, 1976). The linear features are commonly believed to reflect the presence of faults, shear zones, or zones of closely spaced joints. Numerous field studies have shown that fracture traces tend to parallel local joint sets, but bedrock exposures in the traces are commonly inadequate to provide good fracture data.

Geologic setting in an area of crystalline rocks (Gross, 1951) permitted distinction of features that were probably fractures from those of other origins, and field checks showed many of these conclusions to be valid.

Raven and Gale (1977) compared lineaments and fracture zones. Lineaments examined on the ground ranged from 0.25 km to more than 20 km long, 1 to 350 m wide, and 1 to 100 m deep. Ratios of depth to width and of amplitude to length were useful in distinguishing faults, shear zones, dikes, and zones of preferential weathering. Good agreement was found between fractures mapped on airphotos and in the subsurface, but no relationship was found between characteristics of photolineaments and ground-water conditions.

In the Piedmont Province, Nutter (1977) and McGreevy and Sloto (1977) found the yields of wells drilled on linear features to be substantially greater than those of wells not on linear features.

Proposals for systems of long lineaments, regional to subcontinental in size, have long attracted strong adherents and dissenters (Wise, 1976). Recent interest in these features, which have been interpreted as regional faults that reach great depth, has been stimulated by the use of aerial photographs and other remote imagery. There is a large Soviet literature on large lineaments and deep faults; recent accessible examples include

Beloussov (1962), Burtman (1980), and Kvet (1983). Three representative studies in crystalline-rock terranes in North America are those of Kisvarsanyi and Kisvarsanyi (1976), Shurr (1981), and Wise (1976). Despite the significant role such features could play in deep circulation of fluids, little consideration has been given them by hydrogeologists.

## HYDROGEOLOGIC TERRANES IN CRYSTALLINE ROCKS

### Permeability as related to depth

Summaries of permeability data (e.g., Davis and DeWiest, 1966; Freeze and Cherry, 1979; Gale, 1982; Schoeller, 1975) indicate a wide range in values for any given type of rock. For plutonic and metamorphic rocks as a group, hydraulic conductivity ranges from about 9 to $9 \times 10^{-9}$ m/day.

Data on the permeability of near-surface crystalline rocks come chiefly from yield tests in water wells and packer tests in geotechnical boreholes. In North America, most of the published well records are from the Piedmont Province (e.g., Cressler and others, 1983; LeGrand, 1954; Meyer and Beall, 1958). Data have been summarized and analyzed by Davis and Turk (1964) and Landers and Turk (1973). Davis and Turk (1964), Johnson (1981), and Snow (1968) summarized and analyzed unpublished packer-test data.

Figure 2 illustrates permeability with increasing depth. Plot A shows a lognormal decline of well yield with increasing depth. Plot B is from water-pressure tests in boreholes; the proportion of packed-off intervals that did not accept water increases logarithmically with depth.

From review of such plots, Davis and Turk (1964) concluded that (1) permeability-depth relationships, determined particularly by weathering and structure, are similar in various crystalline rocks; (2) the proportion of packer-test intervals that do not accept water increases with depth; and (3) frequency distributions of both yield and acceptance are skewed toward the smaller values. We may recall here that frequency distributions of length, spacing, and aperture of joints in crystalline rocks are also skewed, and that Summers (1972) found specific capacity of wells in crystalline rocks in Wisconsin to be lognormally distributed.

Conclusions drawn thus far are very useful. For example, Davis and Turk (1964) concluded that in unweathered rock, 5 to 15 percent of wells drilled area failures, median yield is less than 0.5 l/s, about 10 percent of wells yield 3 l/s or more, and decrease in yield with increasing depth is about 10 fold between 30 and 325 m. Like many other writers, they used such data, with economic considerations, to suggest optimum depths for wells, in this instance less than about 50 to 65 m for domestic wells.

More recent work in the Canadian Shield and in several other regions of crystalline rocks, noted by Farvolden and others (this volume), has shown that the presence of persistent fracture zones complicates the relatively simple relationship of permeability to depth inferred in the earlier studies.

An important aspect of the distribution of permeability is concealed by averaged data such as those represented in Figure 2A. Numerous workers have pointed out that most water wells are drilled for relatively small supplies and that drilling commonly is terminated once adequate yield has been obtained; but where few fractures are found, drilling is often continued in hope of finding water at greater depth. Lateral distribution of fractures affects the form of Figure 2A: the lower part of the plot contains data from all wells in the sample, but the upper part represents mainly sparsely fractured rocks. This affects the form of the plot but does not negate the general conclusion of downward decrease in permeability, which is also shown (Fig. 2B) by packer-test data from dam sites, which are located in areas selected for soundness of rock. Downward decrease in permeability is also evident in many mines (e.g., Raven and Gale, 1977), which commonly are in structurally disturbed areas.

What is known of fracture permeability in crystalline rocks suggests (1) a wide range, (2) large areal diversity, (3) sharp decrease with increasing depth near the surface, and (4) very low values generally at depth (although higher values occur locally). Geologic considerations indicate that flow paths have existed to great depths and that sedimentary and metamorphic rocks have gradually lost water during a process of dewatering by escape of water upward through the crust (Fyfe and others, 1978). Most of the hydrogeologic data available represent the uppermost part of the column, but recent studies have begun to extend the depth range.

Brace (1980) has summarized recent data on permeability at greater depths. Measurements have been made in drill holes at 2 to 3 km. A range in permeability of 4 to 6 orders of magnitude is typical for a single hole. In crystalline rocks in general, conductivity appears to range from about $10^{-6}$ to $10^{-1}$ m/day. Some part of the rock sampled by nearly every well had conductivity in the range from $10^{-3}$ to $10^{-1}$ m/day, and in some wells this was at considerable depth—a test hole in New Mexico penetrated such a zone at 1.8 km, and in Colorado one was found at 3.3 km.

The evidence available indicates some degree of permeability to the maximum depths probed directly, despite enormous overburden pressures. Brace (1980) concluded from indirect evidence that these depths can be extended, perhaps to 10 km. That estimate has been shown to be conservative, for the Soviet super-deep drill hole on the Kola Peninsula, which had reached 11,515 m as of August 1982, encountered gases and mineralized water even near the 11,500-m level (Kozlovsky, 1982).

### Near-surface aquifers

Near-surface fractured rock forms extensive aquifers in many regions of crystalline rocks. The character of such an aquifer results from the combined effects of fracture system, topography, and weathering. Topography commonly exerts a major influence through its effect on both weathering and the opening of joints, while weathering modifies both transmissive and storage characteristics of the rock.

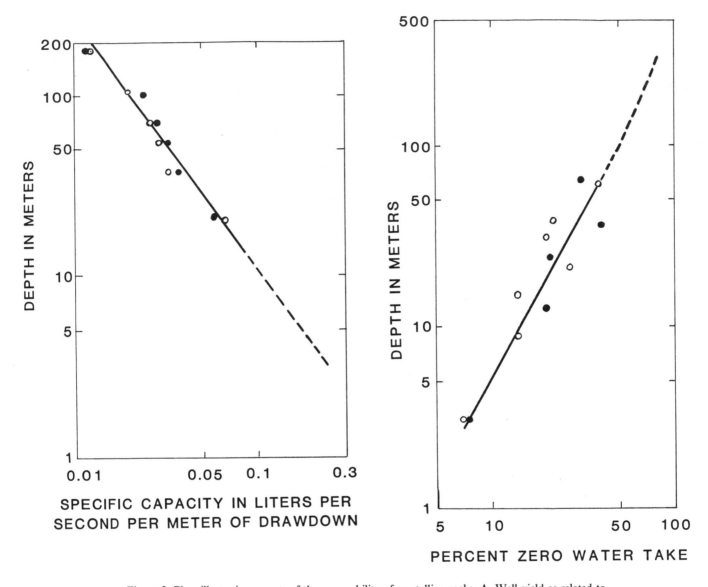

Figure 2. Plots illustrating aspects of the permeability of crystalline rocks. A. Well yield as related to depth of well in granitic rock (open circles) and schist (solid circles) in the eastern United States. B. Percentage of pressure tests in which boreholes did not accept water, as related to depth in granite (solid circles) and amphibolite (open circles) in California. Plots A and B are after Davis and Turk (1964), and are used with permission of the authors and the National Water Well Association.

Weathering can produce a clay-rich regolith through breakdown of silicate minerals. This residual material can attain thicknesses of tens of meters. In North America such a residuum is most extensively developed in the Piedmont Province. In massive rocks such as granite, weathering advances from fracture walls into the intervening blocks, which shrink under continued weathering. The vertical profile passes downward successively through soil, clay-rich residual material, and slightly weathered rock, into fresh rock. Figures 3A and 3B show, by steepening of the curves with depth, that the transition to fresh rock can be fairly sharp. Intensive subsurface investigations at construction sites have re-

vealed complex variants of this ideal profile (Kiersch and Treasher, 1955) in which numerous masses of fresh rock were found to be "boulders" underlain by decomposed rock. Nutter and Otton (1969) noted that residual boulders are typical of massive rocks. In schist and gneiss, openings along the foliation are also important loci of weathering, and fewer unweathered boulders persist.

Not all weathering of granite produces clay-rich residuum. Grus, a granular material formed by disaggregation of the rock with little alteration, is an important weathering product in many temperate regions. Disaggregation suggests that grus is due to

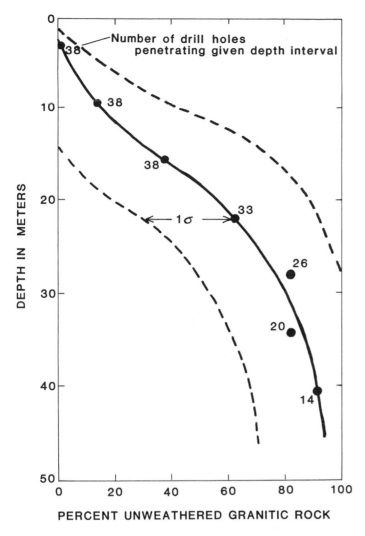

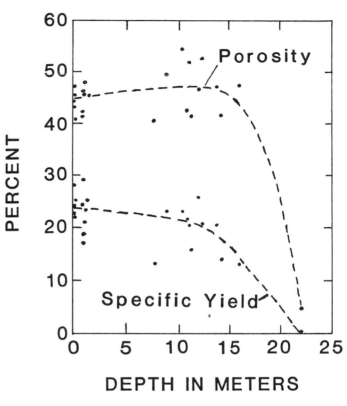

Figure 3. Plots illustrating aspects of the porosity and permeability of crystalline rocks. A. Percentage of unweathered rock as a function of depth at Folsom Dam, California (after Davis, 1981). B. Change in porosity and specific yield with depth, in saprolite on metamorphic rock in Georgia (after Stewart, 1962).

physical weathering, but the depth of the material at many places makes such an origin unlikely. Numerous authors (e.g., Wahrhaftig, 1965) have explained breakdown of the rock by expansion of biotite during initiation of chemical weathering. Unlike clay-rich residuum, grus can yield substantial quantities of water to drilled wells. Landers and Turk (1973) found that median and mean well yields are two to three times greater in grus than in fresh crystalline rock, and that median and mean well depths are considerably smaller. The origin, nature, and distribution of grus pose a significant hydrogeologic problem.

Davis and Turk (1964) concluded that, with the exception of serpentinite, different types of crystalline rocks have similar depth-permeability relationships, but they noted that data plots show considerable scatter. Studies of individual areas (e.g., LeGrand, 1954; Meyer and Beall, 1958; Nutter, 1977) have shown appreciable differences in yield of wells in different plutonic and metamorphic rocks. Lithologic differences in fracturing and in susceptibility to weathering are probably both involved.

Many investigations in the Piedmont Province, based on large samples of wells (e.g., LeGrand, 1954; Meyer and Beall, 1958; Nutter, 1977; Daniel and Sharpless, 1983) have shown that average yield is related to topographic position. Terrain classifications used by different authors differ somewhat; in general terms, average yield is greatest in wells situated in ravines, less on slopes and broad uplands, and least on narrow hilltops. This topographic effect is commonly attributed to greater number and aperture of fractures, and more effective weathering in ravines than on uplands, and to the location of some ravines and valleys on zones of closely spaced fractures. The relative importance of these interrelated factors has not been clearly demonstrated.

In the Piedmont Province the residuum and bedrock form a single aquifer which differs hydrogeologically from unmantled fractured rock (LeGrand, this volume). Typically, the residuum is partly saturated and the water table fluctuates within it. Flow to wells is through the fractured rock. Hence the operative permeability for much of the ground-water system is that of the fractured rock, while the specific yield is that of the residuum. LeGrand (1967) developed a rating system for well-site selection based on topographic situation and depth to fresh rock. This approach, coupled with detailed knowledge of the rocks, provides a good basis for inference of probable hydrogeologic conditions. Inference of fracture zones from photogeologic fracture traces (Parizek, 1976) is a more site-specific approach to site selection, but little has been published on fracture traces in crystalline rocks. In addition, current work in the Piedmont is emphasizing recognition of zones of permeable rock along such features as contacts, shear zones, and folds (e.g., Cressler and others, 1983; Daniel and Sharpless, 1983).

This regolith-bedrock system is but one type of aquifer in crystalline rock covered by unconsolidated material. Other examples, such as bedrock covered by glacial drift or by coastal-plain sediments, seem not to have been studied in detail.

Near-surface flow systems in crystalline roacks have not been investigated intensively in North America. Lawson (1968) constructed a flow net in crystalline rocks in British Columbia and concluded that local, intermediate, and regional flow systems are all present. Local systems in the uppermost 40 to 50 m of rock are the quantitatively significant flow systems. The lower limit of appreciably permeable rock was estimated to be at about 325 m. In the Piedmont Province, effective circulation has extended to at least 325 m, as attested by the flushing of saline water from the crystalline rock to that depth (Feth and others, 1965). Farvolden and others (this volume) cite recent studies of flow systems in the Canadian Shield.

Near-surface ground water in the crystalline rocks has typically been derived from rain and snow. The rock-forming minerals are silicates, and their chief cations are Ca, Mg, Na, and K. Water-rock interaction is at moderate temperatures, within a range of less than 20°C in much of North America. The principal aggressive agent is carbonic acid derived chiefly from $CO_2$ in the soil. Under these conditions the water typically attains a Ca- or N-$HCO_3$ composition (Fig. 4A). Variants within the spectrum reflect major differences in composition of the aquifer rock. Thus, LeGrand (1958) showed that water in igneous and metamorphic rocks in North Carolina can be grouped in two classes by water chemistry. The range in dissolved-solids content in 29 samples of water from "granitic" rocks was 25 to 123 mg/l (milligrams per liter); that in 23 samples of water from "dioritic" rocks was 106 to 696 mg/l.

### Deep waters and flow systems

The combination of chemical and isotopic approaches is leading to new progress in study of connate, metamorphic, and magmatic waters. Recent work in the Pacific tectonic belt in western North America (Barnes, 1970; White and others, 1973) has suggested the widespread occurrence of ground water expelled from marine sediments during low-grade metamorphism. The nature of magmatic water has been inferred from hydrothermal ore deposits and from thermal waters associated with volcanism (White, 1970). Isotopic study of batholithic rocks (Taylor, 1977) and modern hydrothermal waters (White, 1970) shows that water circulating in and near the outer parts of cooling plutons is largely of meteoric origin. Taylor concluded that most of the water in granitic plutons is probably not juvenile but is ultimately derived from dehydration and/or partial melting of the lower crust or of subducted lithosphere. If juvenile water rises from the mantle with basaltic lavas, it seems not to have been identified.

A significant aspect of the hydrology of the plutonic and metamorphic rocks, long known but now the subject of intensified study, is the widespread occurrence of salty water in crystalline-rock terranes (e.g., Canadian Shield; Farvolden and others, this volume; Sweden: Fritz and others, 1979; California: Mack and Ferrell, 1979; and New Mexico: Grigsby and others, 1984). Figure 4B illustrates the principal-ion composition of three examples of these waters.

Similarity in composition of the three waters is striking and may be significant. Fritz and Frape (1982) concluded that the chlorine content of saline water in the Canadian Shield is not a result of intense water-rock interaction at elevated temperatures, and suggested that saline water had moved into the rock and subsequently attained its present composition through water-rock interaction. All three regions represented in Figure 4B have been beneath the sea at some time since crystalline rock was brought near the land surface. The region represented by the Sierra Nevada samples is one of considerable relief, and it is intersected by what seems to be a major fracture zone—possibly an avenue for flow. Interestingly, of the three regions, this is the one in which there is evidence of flow of the saline water under natural conditions (Fig. 4B, diamond diagram).

Salt water in crystalline rocks has important implications for the construction of waste repositories. Its presence suggests near-static conditions under which, with natural gradients, movement of water toward the surface would be so slow as to be negligible. Observation that cavities, once breached, can drain for substantial periods indicates the presence of extensive networks of interconnected openings. Hence the possibility of fluid movement may be ever present, a problem waiting to occur after inadvertent modification of head conditions and flow paths during repository construction.

Contrasting sharply with these porous networks and near-stagnant fluids are localized deep hydrothermal-convection systems in which cumulatively enormous flows of fluids have occurred, commonly with associated water-rock interaction. Numerous systems, of a large range in temperature, are now known; relatively few are hotter than 150°C. The hotter systems are thought to reflect magmatic heat sources; many waters are highly mineralized, and aside from their potential reserves of energy, many of the systems are of interest for their possible implications for the study of hydrothermal-ore deposits. Contrasting with the high-temperature systems are more common low-temperature convection systems (commonly <90°C) in environments of regional conductive heat flow. With much less intense

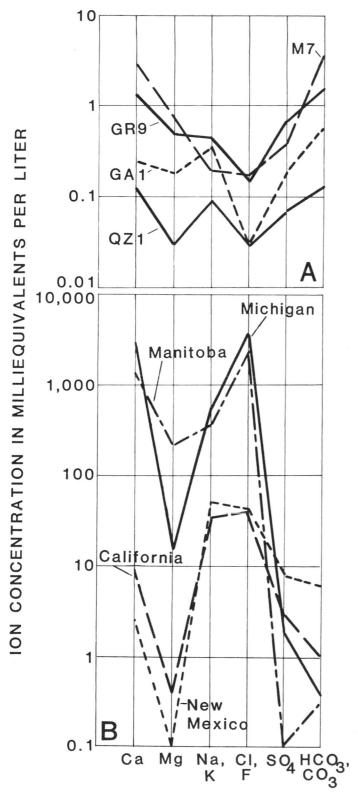

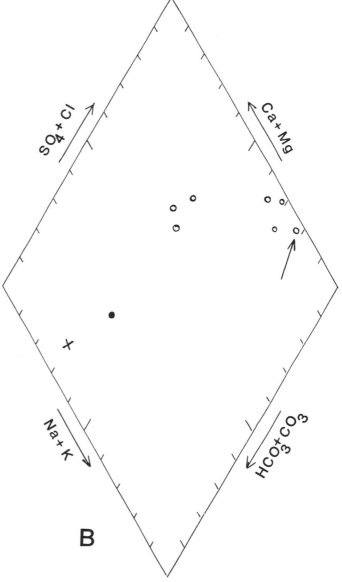

Figure 4. Graphs illustrating water chemistry in crystalline rocks. A. Schoeller (semi-logarithmic) plot of representative analyses of shallow ground water in crystalline rocks. Analyses from White and others (1963). Symbol on plot indicates rock type and analysis number in original table. Rock types represented are: quartzite, QZ; gabbro, GA; granite, GR; and marble, M. B. Schoeller plot for representative saline waters in crystalline rocks, and diamond field of Hill-Piper (trilinear) diagram for one water. Analyses are: Houghton County, Michigan (White and others, 1963); Thompson, Manitoba (Fritz and Frape, 1982); New Mexico, mixture of native and injected (surface) waters (Grigsby and others, 1984); California (Mack and Ferrell, 1979). Diamond diagram shows the California sample (arrow, sample Oakhurst 7S/21E-14C) and similar samples (open circles), representative dilute water (x), and representative potable water (solid circle) believed to be a mixture of the other waters (after Mack and Ferrell, 1979, Fig. 4).

water-rock interaction, waters in these systems are slightly mineralized. Through-flow rates are also much lower than in the magma-related systems.

## SIMILARITIES AND CONTRASTS WITH OTHER ROCKS

The principal and most significant hydrogeologic contrast is between crystalline rocks and those consolidated rocks and unconsolidated deposits whose permeability is largely or entirely primary. Coarse-grained sediments, some sandstones, and some volcanic rocks are moderately to highly permeable. Where strata are thick and laterally extensive, they can form aquifers far thicker and more productive than those at most places in crystalline rocks.

This difference in the nature of permeability leads to a fundamental difference in aquifer analysis. It is now widely recognized that the nonequilibrium (Theis) formula, applied so successfully to relatively homogeneous materials such as granular sediment, is of limited application in fractured-rock aquifers whose properties depart substantially from the assumptions of the Theis solution. Mathematical models have been proposed for double-porosity media consisting of fractures and blocks, for models containing a single fracture that is hydraulically equivalent to a mass of fractured rock, and for various parallel-plate models. Recent reviews include those of Parsons (1972), Witherspoon and Gale (1977), Streltsova-Adams (1978), Gale (1982), and Gringarten (1982).

The strongest similarities are between crystalline rocks and those other consolidated rocks that possess little or no primary permeability. Except for the influence of stratification on fracturing, which may have analogy in some metamorphic terranes, all these consolidated rocks have rather similar fracture permeability except where the rocks are readily soluble at low temperature and pressure. This similarity will eventually facilitate the transfer of experience and knowledge among all types of fractured-rock aquifers.

## REFERENCES

Barnes, I., 1970, Metamorphic waters from the Pacific tectonic belt of the West Coast of the United States: Science, v. 168, p. 973–975.
Barenblatt, G. I., Zheltov, Yu. P., and Kochina, I. N., 1960, Basic concepts in the theory of seepage of homogeneous liquids in fissured rocks (strata): Journal of Applied Mathematics and Mechanics, v. 24, p. 1286–1303 (English translation from original Russian).
Beloussov, V. V., 1962, Basic problems in geotectonics: New York, McGraw-Hill Book Company, Incorporated, 809 p.
Bianchi, L., and Snow, D. T., 1969, Permeability of crystalline rock interpreted from measured orientations and apertures of fractures: Annals of Arid Zone, v. 8, no. 2, p. 231–245.
Billings, M. P., 1946, Structural geology: New York, Prentice-Hall, Incorporated, 473 p.
Brace, W. F., 1972, Pore pressure in geophysics, *in* Heard, H. C., Borg, I. Y., Carter, N. L. and Raleigh, C. B., eds., Flow and fracture of rocks: American Geophysical Union Monograph 16, p. 265–273.
——, 1980, Permeability of crystalline and argillaceous rocks: International Journal of Rock Mechanics and Mining Sciences and Geomechanical Abstracts, v. 17, p. 241–251.
Burtman, V. S., 1980, Faults of middle Asia: American Journal of Science, v. 280, p. 725–744.
Carlsson, A., and Olsson, T., 1979, Hydraulic conductivity and its stress dependence; Workshop on low-flow, low-permeability measurements in largely impermeable rocks: Proceedings, Organization for Economic Cooperation and Development and International Atomic Energy Agency, Paris, 1979, p. 249–259.
Chapman, C. A., 1958, Control of jointing by topography: Journal of Geology, v. 66, p. 552–558.
Chapman, C. A., and Rioux, R. L., 1958, Statistical study of topography, sheeting, and jointing in granite, Acadia National Park, Maine: American Journal of Science, v. 256, p. 111–127.
Cressler, C. W., Thurmond, C. J., and Hester, W. G., 1983, Ground water in the greater Atlanta region, Georgia: Georgia Geologic Survey Information Circular 63, 144 p.
Daniel, C. C., 1985, Statistical analysis of water-well records from the Piedmont and Blue Ridge of North Carolina; Implications for well-site selection and well design: Geological Society of America Abstracts with Programs, v. 17, p. 86–87.

Daniel, C. C., and Sharpless, N. B., 1983, Ground-water supply potential and procedures for well-site selection, upper Cape Fear River basin: North Carolina Department of Natural Resources and Community Development Report, 73 p.
Davis, S. N., 1981, Depth-dependent hydrologic characteristics of dense bedrock; The O- to 500-m interval, *in* Davis, S. N., ed., Workshop on hydrology of crystalline basement rocks: University of California, Los Alamos National Laboratory Report LA-8912-C, Appendix E, p. 52–63.
Davis, S. N., and DeWiest, R.J.M., 1966, Hydrogeology: New York, John Wiley and Sons, Incorporated, 463 p.
Davis, S. N., and Turk, L. J., 1964, Optimum depth of wells in crystalline rocks: Ground Water, v. 2, no. 2, p. 6–11.
Feth, J. H., and others, 1965, Preliminary map of the conterminous United States showing depth to and quality of shallowest ground water containing more than 1,000 parts per million dissolved solids: U.S. Geological Survey Hydrologic Investigations Atlas HA-199, scale 1:3,168,000.
Freeze, R. A., and Cherry, J. A., 1979, Groundwater: Englewood Cliffs, New Jersey, Prentice-Hall, Incorporated, 604 p.
Fritz, P., and Frape, S. K., 1982, Saline groundwaters in the Canadian Shield; A first overview: Chemical Geology, v. 36, p. 179–190.
Fritz, P., Barker, J. F., and Gale, J. E., 1979, Geochemistry and isotope hydrology of groundwater in the Strip granite; Results and preliminary interpretations: University of California, Lawrence Berkeley Laboratory Report LBL-8285, 141 p.
Fyfe, W. F., Price, N. J., and Thompson, A. B., 1978, Fluids in the earth's crust: Amsterdam, Elsevier Scientific Publishing Company, 383 p.
Gale, J. E., 1982, Assessing the permeability characteristics of fractured rock, *in* Narasimhan, T. N., ed., Recent trends in hydrogeology: Geological Society of America Special Paper 189, p. 163–181.
Grigsby, C. O., Goff, F. E., Trujillo, P. E., and Counce, D. A., 1984, Geochemical behavior of a hot dry rock geothermal reservoir: New Mexico Geological Society Guidebook, 35th Field Conference, p. 265–270.
Gringarten, A. C., 1982, Flow-test evaluation of fractured reservoirs, *in* Narasimhan, T. N., ed., Recent trends in hydrogeology: Geological Society of America Special Paper 189, p. 237–263.
Gross, W. H., 1951, A statistical study of topographic linears and bedrock structures: Geological Association of Canada Proceedings, v. 4, p. 77–87.
Harper, M. L., 1966, Joints and microfractures in Glenwood Canyon, Colorado:

The Mountain Geologist, v. 3, no. 4, p. 185–192.

Hast, N., 1967, The state of stresses in the upper part of the earth's crust: Engineering Geology, v. 2, no. 1, p. 5–17.

Hobbs, B. E., Means, W. D., and Williams, P. F., 1976, An outline of structural geology: New York, John Wiley and Sons, Incorporated, 571 p.

Hsieh, P. A., and Neuman, S. P., 1985, Field determination of the three-dimensional hydraulic conductivity tensor of anisotropic media; 1. Theory: Water Resources Research, v. 21, p. 1655–1665.

Hubbert, M. K., and Willis, D. G., 1957, Mechanics of hydraulic fracturing: American Institute of Mining, Metallurgical, and Petroleum Engineers Transactions, v. 210, p. 153–168.

Jahns, R. H., 1943, Sheet structure in granites; Its origin and uses as a measure of glacial erosion: Journal of Geology, v. 51, p. 71–98.

Johnson, A. M., 1970, Physical processes in geology: San Francisco, Freeman, Cooper and Company, 577 p.

Johnson, K. L., 1981, Permeability-depth relationships in crystalline rocks with applications to low-level waste repositories [M.S. thesis]: Tucson, University of Arizona, 112 p.

Kendorski, F. S., and Mahtab, M. A., 1976, Fracture patterns and anisotropy of San Manuel quartz monzonite: Association of Engineering Geologists Bulletin, v. 13, no. 1, p. 23–52.

Kiersch, G. A., and Treasher, R. C., 1955, Investigations, areal and engineering geology; Folsom Dam Project, central California: Economic Geology, v. 50, no. 3, p. 271–310.

Kisvarsanyi, G., and Kisvarsanyi, E. B., 1976, Ortho-polygonal tectonic patterns in the exposed and buried Precambrian basement of southeast Missouri, *in* Hodgson, R. A., Gay, S. P., Jr., and Benjamins, J. Y., eds., First International Conference on the New Basement Tectonics, Proceedings: Utah Geological Association Publication 5, p. 169–182.

Kohut, A. P., Foweraker, J. C., Johanson, D. A., Tradewell, E. H., and Hodge, W. S., 1983, Pumping effects of wells in fractured granitic terrain: Ground Water, v. 21, no. 5, p. 564–572.

Kozlovsky, Y. A., 1982, Kola Super-deep; Interim results and prospects: Episodes, 1982, no. 4, p. 9–11.

Kvet, R., 1983, Global, equidistant fracture systems and their significance (with examples from hydrogeology): International Geology Review, v. 25, no. 2, p. 242–248.

Landers, R. A., and Turk, L. J., 1973, Occurrence and quality of ground water in crystalline rocks of the Llano area, Texas: Ground Water, v. 11, no. 1, p. 5–10.

Lawson, D. W., 1968, Groundwater flow systems in the crystalline rocks of the Okanagan Highlands, British Columbia: Canadian Journal of Earth Sciences, v. 5, p. 813–824.

LeGrand, H. E., 1949, Sheet structure, a major factor in the occurrence of ground water in the granites of Georgia: Economic Geology, v. 44, p. 110–118.

—— , 1954, Geology and ground water in the Statesville area, North Carolina: North Carolina Division of Mineral Resources Bulletin 68, 68 p.

—— , 1958, Chemical character of water in the igneous and metamorphic rocks of North Carolina: Economic Geology, v. 53, p. 178–189.

—— , 1967, Ground water of the Piedmont and Blue Ridge provinces in the southeastern States: U.S. Geological Survey Circular 538, 11 p.

Lindner, E. N., and Halpern, J. A., 1978, *In-situ* stress in North America; A compilation: International Journal of Rock Mechanics and Mining Sciences and Geomechanical Abstracts, v. 15, p. 183–203.

Lister, G. S., and Williams, P. F., 1983, The partitioning of deformation in flowing rock masses: Tectonophysics, v. 92, p. 1–33.

Mack, S., and Ferrell, L. M., 1979, Saline water in the foothill suture zone, Sierra Nevada range, California: Geological Society of America Bulletin, v. 90, pt. 1, p. 666–675.

Mahtab, M. A., Bolstad, D. D., and Kendorski, F. S., 1973, Analysis of the geometry of fractures in San Manuel copper mine, Arizona: U.S. Bureau of Mines Report of Investigations 7715, 24 p.

McGarr, A., and Gay, N. C., 1978, State of stress in the earth's crust: Annual Reviews of Earth and Planetary Science, v. 6, p. 405–436.

Meyer, G., and Beall, R. M., 1958, The water resources of Carroll and Frederick counties: Maryland Department of Geology, Mines and Water Resources Bulletin 22, 355 p.

Nutter, L. J., 1977, Ground-water resources of Harford County, Maryland: Maryland Geological Survey Bulletin 32, 44 p.

Nutter, L. J., and Otton, E. G., 1969, Ground-water occurrence in the Maryland Piedmont: Maryland Geological Survey Report of Investigations 10, 56 p.

Parizek, R. R., 1976, On the nature and significance of fracture traces and lineaments in carbonate and other terranes, *in* Yevjevich, V., ed., Karst hydrology and water resources; Proceedings of U.S.-Yugoslavian Symposium: Fort Collins, Colorado, Water Resources Publications, v. 1, p. 47–108.

Parsons, M. L., 1972, Determination of hydrogeological properties of fissured rocks: International Geological Congress, 24th, Montreal, Proceedings, Section 11, p. 89–99.

Pisarskiy, B. I., and Sherman, S. I., 1967, Parametry treshchinovatosti i ikh znacheniye pri gidrogeologicheskikh issledovaniyakh [Parameters of fracturing and their significance in hydrogeological investigations], *in* Formirovaniye i geokhimiya podzemnykh vod Sibiri i Dal'nego Vostoka: Moscow, Izdatel'stvo Nauka, p. 25–29.

Pohn, H. A., 1981, Joint spacing as a method of locating faults: Geology, v. 9, p. 258–261.

Raven, K. G., and Gale, J. E., 1977, Project 740057, Subsurface containment of solid radioactive waste; A study of the surface and subsurface structural and groundwater conditions at selected underground mines and excavations: Ottawa, Geological Survey of Canada, EMR-GSC-RW Internal Report 1-77, 105 p.

Rocha, M., and Barroso, M., 1971, Some applications of the new integral sampling method in rock masses: International Symposium on Rock Mechanics (Nancy, France), Proceedings, v. 1, Report 1-21, 12 p.

Sbar, M. L., and Sykes, L. R., 1973, Contemporary compressive stress and seismicity in eastern North America; An example of intraplate tectonics: Geological Society of America Bulletin, v. 84, p. 1861–1882.

Schoeller, H., 1975, Analytical and investigational techniques for fissured and fractured rocks, *in* Brown, R. H., Konoplyantsev, A. A., Ineson, J., and Kovalevsky, V. S., eds., Ground water studies; An international guide for research and practice: Paris, UNESCO, supplement 2, ch. 14, 56 p.

Segall, P., and Pollard, D. D., 1983, Joint formation in granitic rock of the Sierra Nevada: Geological Society of America Bulletin, v. 94, p. 563–575.

Shurr, G. W., 1981, Lineaments as basement-block boundaries in western North Dakota, *in* O'Leary, D. W., and Earle, J. L., eds., International Conference on Basement Tectonics, Third, Proceedings: Basement Tectonics Committee Publication 3, p. 177–184.

Snow, D. T., 1968, Rock fracture spacings, openings, and porosities: American Society of Civil Engineers, Soil Mechanics and Foundations Journal, v. 94, SM 1, p. 73–91.

Stearns, D. W., and Friedman, M., 1972, Reservoirs in fractured rock, *in* King, R. E., ed., Stratigraphic oil and gas fields: American Association of Petroleum Geologists Memoir 16, p. 82–106.

Stewart, G. W., 1978, Hydraulic fracturing of drilled water wells in crystalline rocks of New Hampshire: New Hampshire Department of Resources and Economic Development report submitted to U.S. Department of the Interior, Office of Water Research and Technology, 155 p.

Stewart, J. W., 1962, Water-yielding potential of weathered crystalline rocks at the Georgia Nuclear Laboratory: U.S. Geological Survey Professional Paper 450-B, p. B106–B107.

Streltsova-Adams, T. D., 1978, Well hydraulics in heterogeneous aquifer formations, *in* Chow, V. T., ed., Advances in hydrosciences: New York, Academic Press, v. 11, p. 357–423.

Summers, W. K., 1972, Specific capacities of wells in crystalline rocks: Ground Water, v. 10, no. 6, p. 37–47.

Taylor, H. P., Jr., 1977, Water/rock interactions and the origin of $H_2O$ in granitic batholiths: Geological Society of London Journal, v. 133, p. 509–558.

Tschalenko, J. S., 1970, Similarities between shear zones of different magnitudes: Geological Society of America Bulletin, v. 81, p. 41–60.

Twidale, C. R., 1973, On the origin of sheet jointing: Rock Mechanics, v. 5, p. 163–187.

Voight, B., 1967, Correlation of large horizontal stresses in rock masses with tectonics and denudation: Congress of the International Society for Rock Mechanics, First, Proceedings, v. 2, p. 51–56 (in German).

Wahlstrom, E. E., and Hornback, V. Q., 1962, Geology of the Harold D. Roberts Tunnel, Colorado; West portal to station 468+49: Geological Society of America Bulletin, v. 73, p. 1477–1498.

Wahrhaftig, C., 1965, Stepped topography of the southern Sierra Nevada, California: Geological Society of America Bulletin, v. 76, p. 1165–1190.

Wallis, P. F., and King, M. S., 1980, Discontinuity spacings in a crystalline rock: International Journal of Rock Mechanics and Mining Sciences and Geomechanical Abstracts, v. 17, p. 63–66.

Warner, L. A., and Robinson, C. S., 1967, Geology of the Harold D. Roberts Tunnel, Colorado Station 468+49 to east portal: Geological Society of America Bulletin, v. 68, p. 1659–1682.

White, D. E., 1970, Geochemistry applied to the discovery, evaluation, and exploitation of geothermal energy resources: Geothermics, Special Issue 2, v. 1, p. 58–80.

White, D. E., Hem, J. D., and Waring, G. A., 1963, Chemical composition of subsurface waters, *in* Fleischer, M., ed., Data of geochemistry (sixth edition): U.S. Geological Survey Professional Paper 440, Chapter F, 67 p.

White, D. E., Barnes, I., and O'Neil, J. R., 1973, Thermal and mineral waters of nonmeteoric origin, California Coast Ranges: Geological Society of America Bulletin, v. 84, p. 547–560.

Wise, D. U., 1976, Sub-continental sized fracture systems etched into the topography of New England, *in* Hodgson, R. A., Gay, S. P., Jr., and Benjamin, J. Y., eds., International Conference on the New Basement Tectonics, First, Proceedings: Utah Geological Association Publication 5, p. 416–422.

Witherspoon, P. A., and Gale, J. E., 1977, Mechanical and hydraulic properties of rocks related to induced seismicity: Engineering Geology, v. 11, p. 23–55.

Zoback, M. L., and Zoback, M. D., 1980, State of stress in the conterminous United States: Journal of Geophysical Research, v. 85, no. B11, p. 6113–6156.

MANUSCRIPT ACCEPTED BY THE SOCIETY MARCH 14, 1987

## ACKNOWLEDGMENTS

The contributions of the following, through discussion, suggestions for reading, and assistance with obtaining references, are gratefully acknowledged: J. F. Callender, C. W. Cressler, C. C. Daniel, S. N. Davis, R. C. Heath, G. M. Hughes, H. E. LeGrand, Gerald Meyer, J. N. Jones, K. G. Raven, R. E. Reicker, J. E. Robertson, and Trudy Sinnott.

The Geology of North America
Vol. O-2, Hydrogeology
The Geological Society of America, 1988

# Chapter 41

# *Ground water as a geologic agent*

**P. A. Domenico**
*Department of Geology, Texas A & M University, College Station, Texas 77843*

The role of ground water in performing geologic work has long been a topic of interest in the Earth sciences. Among the early studies that we may briefly cite here are the pioneering work of Cox (1911) and Siebenthal (1915) in ground-water mineralization; Munn (1909), and later Hubbert (1953) in hydrocarbon migration and entrapment; Hedberg (1936) in abnormal pressure development due to gravitational compaction; Mead (1925), and later Hubbert and Rubey (1959), on the relationship between abnormal pressures and rock strength and faulting; and numerous investigations of rock–water interactions of various kinds which involve the role of ground water in metamorphism (Yoder, 1955), clastic diagenesis (Hayes, 1979), dolomitization (Vernon, 1969; Hanshaw and others, 1971), solution kinetics (Weyl, 1957), and even landform development including the formation of karst (Stringfield and LeGrand, 1966). This spectrum of diverse happenings, ranging from processes deep within the crust to those that occur at or near land surface, are but a small sample of the role of ground water in geologic processes. Common to all such processes is an understanding that underground fluid does not reside in a passive porous solid, but instead there exists a complex coupling between the moving fluid, entities that might be carried by the fluid, and the solid matrix itself. This coupling, or rock–water interaction, is at the heart of this section, and may be categorized within three broad groups. First, there is the coupling between stress, strain, and pore fluids, most frequently in elevated temperature environments. This coupling is of decisive importance for understanding abnormal pressure development in active depositional basins, earthquake generation and the dissipation of frictional heat, rock strength and faulting, pore-pressure development in response to molten intrusions, and—from an applied science perspective that contains overtones of a purely geologic problem—the fluid and thermo-mechanical response of potential repository rocks to heating. Chapters in this section by Domenico and Palciauskas and by Rojstaczer and Bredehoeft address some of these problems.

The second and third major type of rock-water interactions embedded in this section may be termed simply as mass and energy transport and transfers within saturated porous solids. By transport, the various authors mean the processes that move mass or energy from one point to another in a porous medium. In the case of the former, the transport process is referred to as diffusion, dispersion, or advection of a chemical substance, whereas the latter process may be referred to as conduction, thermo-dispersion, or advection. By transfer, we mean the manner and, more often, the rate at which mass or energy contained in one phase (liquid or solid) may be transferred to the other. Here we have another type of coupling between the transport and transfer in that the physical phenomenon (transport) often intrudes upon the chemical or thermodynamic phenomenon (transfer). Under these categories we include the temperature-dependent liquid-forming phase transformations, such as montmorillonite to illite or kerogen to hydrocarbon; the generation, movement, and entrapment of hydrocarbons; the mobilization and migration of ore-forming fluids and their ultimate precipitation to form ore bodies; and a host of competing diagenetic reactions and transport in clastic and nonclastic environments where ground water is a majority fluid. Chapters in this section by Toth, Sharp and Kyle, and Schwartz and Longstaffe address many of these problems. For purposes of completion, the reader can find chapters on near-surface processes and landform development by Ford, Palmer, and White and by Higgins and others.

In a simple phrase, then, many of the important rock–water interactions involved in geologic processes and discussed in this section can be viewed as problems in mass and energy transport and transfers in deformable bodies, where the deformation can be contemporaneous with deposition, or may occur during uplift, or may be absent entirely. Indeed, as momentum is simply the exertion of a force on a body, it too can be regarded as a transport process, suggesting that the role of ground water in geologic processes can be viewed from the perspective of momentum, mass, and energy transport and transfers within geologic environments. In most cases, these processes occur simultaneously and sometimes they interfere with each other. Although the individual phenomenan are generally separated to facilitate study, the consequences of the coupling are emphasized in several of the chapters.

As the rock-water interactions discussed above take place in a moving medium, namely ground water, the driving forces responsible for water movement assume paramount importance in modern as well as paleohydrologic regimes. Such forces were

Domenico, P. A., 1988, Ground water as a geologic agent, *in* Back, W., Rosenshein, J. S., and Seaber, P. R., eds., Hydrogeology: Boulder, Colorado, Geological Society of America, The Geology of North America, v. O-2.

recognized before the turn of the century by King (1899) who categorized them as gravitational (topographic), thermal, and capillary. For some reason, King (1899) did not mention the importance of tectonic strain as a driving force, although he clearly demonstrated the updip migration of water in response to consolidation of sediments. To these forces we may add yet another responsible for what is commonly referred to as a coupled flow, where ground water moves in response to a gradient of some entity other than hydraulic head, such as temperature or concentration. Virtually all of these forces play an important role in the initiation and maintenance of geologic work in porous media, a point that is well demonstrated throughout the chapters in this section. Interest is focused not only on how such forces can be manifested in certain geologic environments, but what physical and chemical consequences result from these manifestations.

## REFERENCES CITED

Cox, G. H., 1911, Origin of the lead and zinc ores in the upper Mississippi valley district: Economic Geology, v. 6, p. 427–603.

Hanshaw, B., Back, W., and Deike, R., 1971, A geochemical hypothesis for dolomitization by groundwater: Economic Geology, v. 66, p. 710–724.

Hayes, J. B., 1979, Sandstone diagenesis; The hole truth, *in* Scholle, P. A., and Schluger, P. R., eds., Aspects of diagenesis: Society of Economic Paleontologists and Mineralogists Special Publication 26, p. 127–139.

Hedberg, H. D., 1936, Gravitational compaction of clays and shales: American Journal of Science, v. 31, p. 241–287.

Hubbert, M. K., 1953, Entrapment of petroleum under hydrodynamic conditions: American Association of Petroleum Geologists Bulletin, v. 37, p. 1944–2026.

Hubbert, M. K., and Rubey, W. W., 1959, Role of fluid pressures in mechanics of overthrust faulting; Part I, Mechanics of fluid filled porous solids and its application to overthrust faulting: Geological Society of America Bulletin, v. 70, p. 115–166.

King, F. H., 1899, Principles and conditions of the movements of groundwater: U.S. Geological Survey 19th Annual Report, pt. 2, p. 59–294.

Mead, W. J., 1925, The geologic role of dilatancy: Journal of Geology, v. 33, p. 685–698.

Munn, M. J., 1909, The anticlinal and hydraulic theory of oil and gas accumulation: Economic Geology, v. 4, p. 509–529.

Siebenthal, C. E., 1915, Origin of the zinc and lead deposits of the Joplin region, Missouri, Kansas, and Oklahoma: U.S. Geological Survey Bulletin 606, 283 p.

Stringfield, V. T., and LeGrand, H. E., 1966, Hydrology of limestone terranes in the Coastal Plain of southeastern United States: Geological Society of America Special Paper 93, 45 p.

Vernon, R. O., 1969, The geology and hydrology associated with a zone of high permeability (Boulder Zone) in Florida: Society of Mining Engineers, Reprint 69-Ag-12, 24 p. (abstract in Mining Engineers, v. 20, p. 58, 1968).

Weyl, P. K., 1958, The solution kinetics of calcite: Journal of Geology, v. 66, p. 163–176.

Yoder, H. S., 1955, Role of water in metamorphism: Geological Society of America Special Paper 62, p. 505–524.

Manuscript Accepted by the Society March 14, 1987

The Geology of North America
Vol. O-2, Hydrogeology
The Geological Society of America, 1988

# Chapter 42

# *Landform development*

**Charles G. Higgins**
*Department of Geology, University of California, Davis, California 95616*
**Donald R. Coates**
*Department of Geological Sciences, State University of New York, Binghamton, New York 13901*
**Victor R. Baker**
*Department of Geosciences, University of Arizona, Tucson, Arizona 85721*
**William E. Dietrich**
*Department of Geology and Geophysics, University of California, Berkeley, California 94720*
**Thomas Dunne**
*Department of Geological Sciences, University of Washington, Seattle, Washington 98195*
**Edward A. Keller and Robert M. Norris**
*Department of Geological Sciences, University of California, Santa Barbara, California 93106*
**Garald G. Parker, Sr.**
*P.O. Box 270089, Tampa, Florida 33688*
**Milan Pavich**
*U.S. Geological Survey, 928 National Center, Reston, Virginia 22092*
**Troy L. Péwé**
*Department of Geology, Arizona State University, Tempe, Arizona 85287*
**James M. Robb**
*U.S. Geological Survey, Woods Hole, Massachusetts 02543*
**J. David Rogers**
*Rogers/Pacific, 3340 Mt. Diablo Boulevard, Lafayette, California 94549*
**Charles E. Sloan**
*U.S. Geological Survey, WRD, 4230 University Drive, Anchorage, Alaska 99508*

## INTRODUCTION

Subsurface water affects near-surface processes and landforms in a wide variety of ways. Its essential role in weathering, soil development, slope failure, and karst topography has long been acknowledged. However, its importance in other aspects of the development of landforms has just begun to be recognized in the last decade or so. In this chapter we discuss some major aspects of ground-water geomorphology and ways in which subsurface water can shape the Earth's surface. In the next chapter, D. C. Ford, A. N. Palmer, and W. B. White discuss the special conditions of karst landforms; in this chapter we discuss other aspects of the role of underground water in landform development.

This chapter is organized as follows: first, some effects of water in the vadose or unsaturated zone above the water table, with emphasis on weathering and soil development and their influence on landscape (by Pavich), mass wasting and slope failure (Dietrich and Rogers), and hillslope hydrology (Dunne), with

its influence on piping and pseudokarst (Parker) and on development of hillslopes and gully heads (Higgins). Next, the role of water in the saturated zone at or beneath the water table, and its effects on permafrost and pseudokarst (Sloan), land subsidence (Péwé), spring sapping and valley network development (Baker), submarine landforms (Robb), sea cliffs (Norris), scarp retreat (Higgins), and surface stream channels (Keller). Finally, ways are considered in which regional geomorphology can control ground-water behavior and hydrology (Coates).

## WATER ABOVE THE WATER TABLE

### *Weathering and soil development*

Water is essential for weathering, soil development, and life. It is not only the medium for chemical weathering of minerals,

Higgins, C. G., Coates, D. R., Baker, V. R., Dietrich, W. E., Dunne, T., Keller, E. A., Norris, R. M., Parker, G. G., Sr., Pavich, M., Péwé, T. L., Robb, J. M., Rogers, J. D., and Sloan, C. E., Landform development, 1988, *in* Back, W., Rosenshein, J. S., and Seaber, P. R., eds., Hydrogeology: Boulder, Colorado, Geological Society of America, The Geology of North America, v. O-2.

but it is also vital for most mechanical weathering, such as "frost action" or hydration cracking, needle-ice action, salt-crystal wedging, and all types of biological weathering. Weathering is but the first step in soil development, in which these and other water-dependent processes further modify the regolith.

The soil occupies the interface of the atmosphere and lithosphere. Due to its unique textural, mineralogical, and structural properties, the soil forms a special zone of the hydrosphere. This zone has a higher mineral/solution ratio than other hydrospheric bodies such as streams, lakes, or the ocean. Within the soil hydrosphere, many processes occur, including infiltration of fluids into micropores, transmission of fluids along macropores, mineral-solution reactions—such as hydration, hydrolysis, congruent and incongruent dissolution—and dehydration of pores during periods of intense evapotranspiration.

The first-order dependence of soil on climate is a reflection of the importance of rainfall in soil-forming processes. In a general sense, hot-humid environments have higher rates of chemical alteration of rock to soil than do cold-arid environments. However, even in high-latitude cold climates, water is critical in mineral-alteration reactions.

At the temperate latitudes of North America, seasonal variation of rainfall and evapotranspiration related to climate (temperature, wind, solar insolation) is important in determining soil drying, development of macropores, and recharge of ground water to the subsoil through macropores. Under seasonal humid climates, the soil hydrosphere may control the volume and timing of ground-water recharge, so that the underground water not only affects soil development, but the soils themselves influence infiltration rates, thereby controlling further soil formation and landform development.

An example of soil processes as they affect landform in the United States is found in the southeastern Appalachian Piedmont, where one of the major hydrologic functions of clay-rich ultisols beneath relatively flat drainage divides is to partition precipitation. As much as 60 percent of total annual precipitation in Piedmont drainage basins is lost through evapotranspiration. Roughly half of the remaining precipitation runs off, owing to the relatively low infiltration capacity of the soil. The relatively small remainder recharges ground water in the regolith. The soil itself thus limits the volume of water that can infiltrate and reach the regolith, and thereby acts to control the rates of subjacent rock weathering.

On the Piedmont, it appears that the flattened upland landscape is a result of ongoing physical and chemical denudation processes that are in part controlled by soil hydrology. Desilicification in soils and volume reduction during alteration of saprolite or sediment to ultisols are ongoing processes of mass and volume reduction that have lowered land surfaces and upland divides through Quaternary time. In this way, pedogenic processes can produce broad, flat divides of the type classically described as peneplains. These processes are partly responsible for the morphology of large tracts of landscape.

Virtually all erosion is regolith erosion—that is, erosion of soil, saprolite, or weathered rock—because fresh unweathered rock is little affected by most erosional agents. Thus, erosional landscapes must strongly reflect the results of differential weathering and pedogenesis, which in turn are strongly dependent on subsurface water.

Over time, beneath stable geomorphic surfaces where erosion is minimal, depletion of soluble minerals and concentration of insoluble minerals can result in aluminous, siliceous, and/or ferric duricrusts. Although not recognized in the United States, duricrusts on old, cratonic rocks elsewhere in the world exercise control over subsequent landscape evolution. In some areas of Africa and Australia, a change from humid to arid climate over geologic time has left remnants of Tertiary and possibly older duricrusts that retard erosion and serve as caprocks for extensive uplands and plateaus. An extensive discussion of the relationship between soils and landforms can be found in Twidale (1984).

### Mass wasting and slope failure

In recent years, engineering geology and geotechnical studies have advanced our ability to quantify the role water plays in controlling slope stability and evolution. Net movement occurs when shear stress associated with an inclined surface causes the rock or soil to strain slowly, or ultimately to yield, sometimes with catastrophic consequences. Ground water plays a critical role in these mass-wasting processes.

Shear strength of materials generally comprises two components—intrinsic cohesion and frictional resistance between grains of a rock or soil. In rock masses possessing joints, shear strength is derived from friction on opposing sides of the joint planes and from cohesion of the rock itself. The presence of water at less than saturation has been recognized to provide some cohesion through surface tension, which allows uncemented sand and silt to support a low vertical cliff. In argillaceous materials such as shales, however, saturation may cause a decay in pure cohesion and as much as a two-thirds reduction in shear strength (Rogers and Pyles, 1979). A further reduction in frictional resistance results from the buoyancy effect of submergence, which reduces the normal stress between grains or fracture planes. Saturation of the material also increases the downslope driving force. As shown in Figure 1, consequent rotational failure tends to reestablish equilibrium, although inertia involved in such movements often takes the rotation far beyond that which is theoretically predicted. Such rotational slope failures can involve as much as 4 $km^3$ of material.

Failures in rock usually are associated with discontinuities in strength and permeability caused by stratification or jointing. For example, massive slump blocks in the cliffs of the Colorado Plateau are controlled by weak basal shales that yield when saturated. Jointing reduces the overall strength of the rock by diminishing or eliminating its cohesive strength and by providing frictional, but permeable boundaries where high pore pressures can develop.

Shallow soil failures commonly occur during periods of intense rains when high pore pressures are generated from shallow

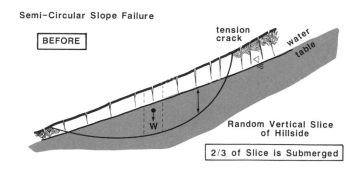

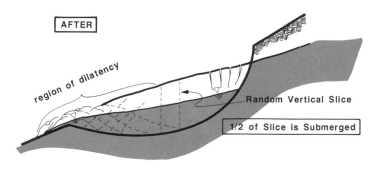

15% INCREASE IN OVERALL STRENGTH

Figure 1. Mechanics of rotational slump failure. Note how the relative proportion of saturated soil in a random vertical slice is lessened by the rotation.

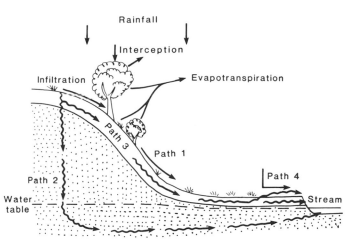

Figure 2. Hillslope hydrologic cycle. See text for discussion of the various flow paths.

subsurface flow perched on the soil-bedrock boundary. Stratification within the underlying bedrock may also bring ground-water flow up into the soil under sufficiently high pressures to cause instability. It is now becoming widely recognized that shallow soil landslides typically occur in small colluvium-mantled bedrock hollows (Dietrich and others, 1986). Hollows can divide hillslopes into small unchannelized basins and focus the drainage of water and the discharge of debris from slopes to channels. Deeper ground-water flow also tends to re-emerge in the lower portions of the larger hollows and contribute to slope instability.

Static liquefaction (Seed, 1983) may be responsible in large part for the remarkable fluidity of debris flows generated in coarse soils, sometimes referred to as "blow outs." Liquefaction begins during the initial slab failure of the debris, when rapid strain occurs under effectively undrained conditions. Excessive pore pressures leading to liquefaction may also occur where deep ground water is forced to the surface or where the B-horizon of the surface soil is much less permeable than the underlying coarse colluvium or weathered bedrock. In sharp contrast to this highly fluid mode of failure, clay-rich material tends to have a complex, time-dependent rheology as it responds to ground-water accumu-

lation by deforming as a slow-moving earth-flow. Substantial recent progress has been made in both the monitoring and modelling of these flows (Keefer and Johnson, 1983). Results from these studies may well help in understanding the mechanics of solifluction.

The least understood but perhaps most pervasive set of physical transport processes influenced by water on hillslopes is that of slow mass movement caused by creep and biogenic transport. Over large parts of soil-mantled landscapes, these processes appear to be the dominant transport mechanism. Some clay-rich soils subjected to strong seasonal drying show a strong time-dependent response to infiltrating water by slowly swelling, flowing downslope several millimeters, and then stopping, despite continued saturation (Fleming and Johnson, 1975).

### Hydrology of subsurface flow processes

The model of runoff processes that until recently has been the most influential in geomorphology is that describing the generation of overland flow when rainfall intensity exceeds the infiltration capacity of the soil (Horton, 1945). Runoff of this kind (Path 1 in Fig 2) has been described thoroughly in other literature. However, over large areas of the Earth's surface, infiltration capacities almost always exceed rainfall intensity, and most runoff travels beneath the soil surface (Hursh, 1944; and others).

When *Horton overland flow* does occur, the soil surface becomes saturated, and if the soil is vertically homogeneous, a wave of saturation extends downward into the soil. If the rainfall intensity is less than the saturated hydraulic conductivity of the soil, the soil surface will remain unsaturated, but its moisture content will increase until its hydraulic conductivity is high enough to pass water downward at the applied rainfall rate. After the end of the rainstorm, soil water may continue to drain downward, reducing the moisture content of the near-surface soil. As water moves to any level in the regolith, it will either be stored in pores, raising the moisture content and therefore the hydraulic conductivity until the flux rate can be accommodated, or if that moisture content already exists, the water will continue its downward path, eventually reaching the water table (Path 2 in Fig. 2). It then follows a relatively long path to a stream channel, where it supplies the base flow of the stream. A simple flow path of this kind usually occurs in deep, permeable, and isotropic materials.

In many landscapes, however, the hydraulic conductivity of the near-surface zone decreases from the topsoil to the undisturbed parent material. This decrease may be gradual or abrupt, and is not necessarily monotonic, though for the sake of simplicity it will be treated as such here. As the infiltrating rainwater encounters a zone of diminishing vertical hydraulic conductivity, the moisture content of overlying layers is raised by storage. At some depth, moisture content will be raised to saturation, and the hydraulic conductivity to its saturated value. At this depth a perched zone of saturation develops over an obvious impeding horizon, or simply at some depth (varying between rainstorms) at which the saturated conductivity is less than the rainfall intensity (Fig. 3).

In the perched saturated zone, water is diverted downslope along the hydraulic gradient. If the soil and rock are sufficiently deep and permeable to conduct all the infiltrated water to the stream channel, the flow path will lie entirely below the ground surface (Dunne and Black, 1970). The pressure distribution in this subsurface flow influences the effective strength of the soil and rock, and may lead to mass failure or to other forms of seepage erosion described elsewhere in this chapter.

Under two circumstances, illustrated in Figure 2, the subsurface water may emerge at the ground surface, where it then has a potential for surface erosion. In one case, the water may emerge at a free face, such as a stream bank, gully head, or cliff (Path 3 in Fig. 2). In the other, the subsurface flux is augmented by infiltration as it travels downslope, until at some point on the hillside it may exceed the conveyance capacity of the soil because of reduced gradient, hydraulic conductivity, or soil thickness. When supply from upslope exceeds drainage downslope, the water content of the soil increases until the perched water table reaches the ground surface. Further excess of supply over drainage generates a pressure gradient that drives water to the surface, in a process referred to as exfiltration. The pressure gradient and associated water exert a stress on the soil particles, which may thus be entrained and carried downslope by the resulting saturation overland flow (Path 4 in Fig. 2). The soil and topographic conditions under which such emergence occurs have been documented by Dunne (1978) and others.

In most soils and sediments, subsurface water flows slowly enough and through pores that are small enough that the flow is dominantly laminar and obeys Darcy's Law. However, in some soils and sediments there are passages large enough to convey

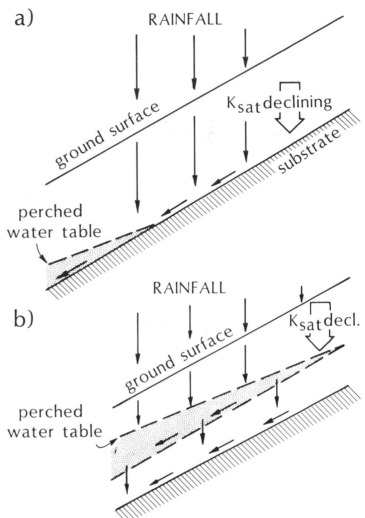

Figure 3. In a soil with a hydraulic conductivity that decreases with increasing depth, the rainfall intensity may exceed the saturated conductivity at a level that depends upon rainfall intensity. In a), a low rainfall intensity can be conveyed to the bedrock surface and then downslope as unsaturated flow. At some distance along the hillside, the increase of discharge caused by downslope flow and vertical percolation may saturate a thin layer (shaded). In b), a higher rainstorm intensity generates a saturated layer at a higher level in the soil, and saturated flow occurs into the subsoil, as well as downslope. With still shallower development of saturation or greater infiltration, the saturated zone may intersect the surface, causing outflow of water as saturation overland flow as described by Dunne (1978).

nonlaminar fluxes of water that are significant from both a hydrologic and geomorphic point of view. Such passages are formed by mechanical and biogenic processes, and include root holes, the burrows of soil fauna, shrinkage cracks, tension cracks between landslide blocks, solution openings, and tectonic joints. The origin and hydraulic characteristics of these "macropores" have recently been described by Jones (1981) and Beven and Germann (1982), who review research on the subject. Macropores provide paths by which water can bypass slow flow through the soil matrix and, especially important, move through horizons of very low conductivity. Even thin, discontinuous zones of low conductivity in the soil may force water into macropores such as root holes and worm holes. Such locally concentrated flow promotes entrainment and transport of soil particles, both underground along the walls of the macropores, and at seepage faces, where saturated flow emerges at the surface.

Subsurface flow can erode mechanically in at least two ways—through the development of a critical seepage force that acts to entrain particles in water emerging from a porous medium, and through the application of a shear stress to the margins of a macropore with consequent removal of particles from the walls. The former process is "seepage erosion," following Hutchinson (1968, p. 691), and the latter is "tunnel erosion" (Buckham and Cockfield, 1950). Either or both of these processes can lead to development or enlargement of pipe-like or tunnel-like openings underground. We use the term "piping" for the effects of the formation and enlargement of such openings. Also, through either of these processes, hillslopes or stream headcuts may be undermined, with consequent failure and retreat of the slope. The term "sapping" traditionally includes the results of such undermining.

***Seepage erosion.*** Seepage erosion is the entrainment of grains by water emerging from a porous medium. Terzaghi (1943) defined the critical conditions required for the simplest case of instantaneous lifting, or "heave," of a relatively large mass of cohesionless material by vertical outflow beneath a horizontal surface. The magnitude of the vertical component of the pore-pressure gradient at the surface must be large enough for the upward drag force on a particle or aggregate at the surface to exceed the immersed weight of the grain.

Zaslavsky and Kassiff (1965) extended the treatment of seepage forces to the case of cohesive materials on a sloping surface. In their analysis, erosion occurs as a tensile failure when the vectorial sum of the fluid drag and the normal component of the soil weight averaged over particle or aggregate exceeds the cohesion of the soil (Fig. 4). These forces in turn depend upon several hydrologic and lithologic factors, as indicated by the following equations.

The fluid drag ($F_d$) on a grain or aggregate is assumed to be proportional to the macroscopic drag force averaged over many particles,

$$F_d = - c_1 \, V \rho_f g \, \nabla \, \phi \cdot \vec{n} = \frac{c_1 V \rho_f g Q \cdot \vec{n}}{K} ,$$

where $c_1$ is a shape factor, V is the volume of the soil element, $\rho_f$ is the fluid density, g is the gravitational acceleration, $\nabla \phi$ is the gradient of hydraulic head in the fluid near the surface, $\vec{n}$ is a unit vector normal to the surface, Q is the specific flux vector of water, and K is the saturated hydraulic conductivity of the porous medium near the surface.

The normal component of the soil's weight is

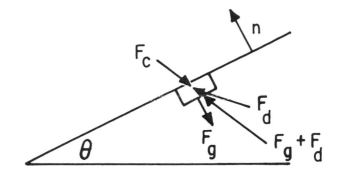

Figure 4. Diagram of forces effective in seepage entrainment of cohesive material on a sloping surface. After Zaslavsky and Kassiff (1965).

$$F_g = -V(\rho_s - \rho_f) \, (1-p)g \cos \theta,$$

where $\rho_s$ is the density of the mineral phase, p is the porosity, and $\theta$ is the slope angle. The negative sign in this case indicates that the force acts downward.

The vectorial sum of $F_d$ and $F_g$ is resisted by the magnitude of the cohesion,

$$F_c = c_2 D^2 \sigma_t ,$$

where $c_2$ is another geometric coefficient relating the particle diameter, D, to the projected area of the grain or aggregate normal to the resultant of the driving forces $F_g$ and $F_d$, and $\sigma_t$ is the tensile strength of the porous medium.

Seepage entrainment occurs when locally

$$F_c = F_d + F_g,$$

and the associated velocity of flow is great enough to transport the grains or aggregates that are mobilized.

Where such a condition occurs along the base of a hillside, seepage erosion saps the slope and the hillside retreats. Where seepage is concentrated at one or a few points, a reentrant forms in the slope (Fig. 5).Such a disturbance of the boundary of the porous medium causes convergence of flow at the head of the reentrant, increasing discharge and therefore the rate of erosion and retreat. Where the soil is sufficiently cohesive, an arch and vertical walls can survive around the site of the initial seepage entrainment. Continued headward erosion may then produce a subterranean conduit or pipe.

***Tunnel erosion.*** Once storm runoff has penetrated into a macropore, the water may then travel underground in sufficient quantity and with sufficient velocity to shear sediment from the walls of the conduit, which may thus be eroded and enlarged. In a general way the fluid shear is proportional to the hydraulic gradient, the diameter or width of the conduit, and its hydraulic roughness. The shear stress can be particularly large beneath a steep hillslope or close to the bank of a stream or gully.

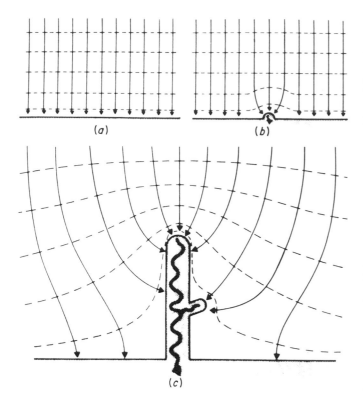

Figure 6. Pseudokarst developed by piping in Triassic Chinle Formation bentonitic and sodic shales near Many Farms, Arizona. From Parker and Jenne (1967).

Figure 5. Plan view of a ground-water flow net during extension of spring heads to form a drainage network. Solid arrows are flow lines, dashes indicate equipotential lines. (a), ground water flows toward the land margin; (b), convergence of ground-water flow at the head of a hillslope embayment produced by seepage entrainment and erosion or by an initial irregularity in the land margin; (c), increased convergence of flow lines around a spring head that has retreated headward from the land margin, extending a valley. A second embayment has been initiated by seepage on one side of the valley and is distorting the flow field in that region. After Dunne (1969).

Where the shear stress is sufficiently large, grains are eroded from the margins of the conduit. The subsequent fate of the eroded grains depends upon the velocity of flow and whether the conduit has an outlet from which water can drain freely. Where there is an open connection to the atmosphere and sufficient head gradient, the velocity may be great enough to transport the eroded sediment as bedload or in suspension.

The shear stress required to erode the margins of a tunnel is radically reduced where the soil or sediment consists of a dispersive clay. All that is then required is a flow velocity sufficient to transport the resulting colloidal suspension. The susceptibility of a soil or sediment to dispersion is enhanced by high concentrations of monovalent cations, particularly sodium, in the pore water and on the clay particles. Water traveling along conduits is drawn into pores and cracks in response to capillary and osmotic potential gradients. There it causes swelling, fracturing into aggregates, and dispersion of particles. The physico-chemical principles governing

the processes of dispersion are summarized in a collection of papers edited by Sherard and Decker (1977).

Under favorable circumstances, dispersion can also reduce the critical pore-pressure gradients required for seepage erosion. For example, where dispersible silty sediments with a conductivity sufficently high to allow significant saturated outflow are connected with a source of ground water, dispersive seepage erosion can be quite rapid. However, in the experience of Sherard and Decker (1977, p. 5), most dispersive clays are so impervious that no significant seepage entrainment occurs, so that in order to initiate erosion it is necessary to provide a path for concentrated leakage. The low conductivity then forces flow from the surface or from a shallow, permeable horizon into cracks and other macropores, which are then rapidly enlarged by dispersion-accelerated shear. Many of the field occurrences of piping described in the papers edited by Bryan and Yair (1982) appear to be of this type.

### Piping and pseudokarst

Although it occurs in many humid areas (e.g., Jones, 1981), piping is a major factor in the erosional process in arid and semiarid lands, where it affects alluvial valley fills, gully walls, hillslopes, and even barren hilltops (Fig. 6).

Piping occurs commonly in unconsolidated silt, clay, and loess. It is also a destructive agent in certain shales, claystones, and chemically altered volcanic ash and tuff. In such rocks the geomorphic effects may simulate solution erosion of calcarous rocks. Where piping is highly developed, erosion of the underground drainage conduits produces a host of karstlike features, such as sinkholes, caves, ragged gullies where pipe roofs have

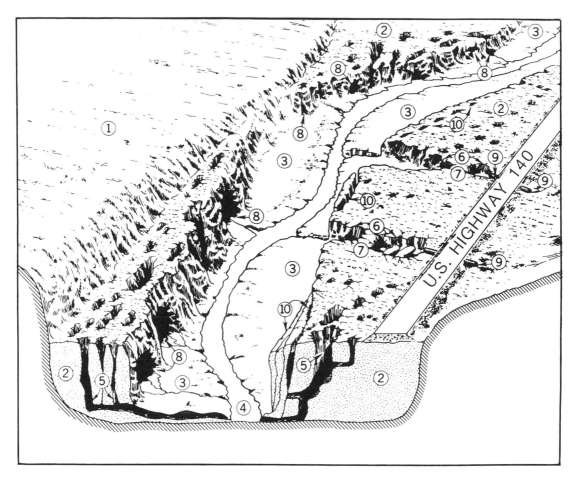

Figure 7. Idealized block diagram of extensively piped valley fill, Aztec Wash, southwestern Colorado. U.S. 140, at right, is being destroyed by effects of piping. 1, Cretaceous Mancos Shale Formation; 2, Quaternary alluvial fill, mainly silt and clay; 3, flood plain of Aztec Wash; 4, channel of Aztec Wash; 5, pipe system; 6, piped gully; 7, natural bridge; 8, cave; 9, culvert; 10, gravity crack system, piped. After Parker and Jenne (1967).

collapsed, and natural bridges. The result is a "pseudokarst" to-pography, complete with ephemeral subsurface streams and a sinkhole-scarred landscape (Fig. 6). However, although the forms produced by piping resemble karst landforms, they are generally not so large nor expansive in area, and their duration as land-forms is relatively short. Whereas a true karst may maintain its physical features for centuries, pseudokarst may develop or change radically as the result of a single heavy rainstorm.

Some of the largest pseudokarst features result from enlarge-ment of soil pipes and can approach karst features in size. Offi-cers' Cave, in the John Day Formation of Oregon, consists of a 200-m-long system of underground drainage tunnels fed by sink-holes. It includes several large cave rooms, and the entire system is drained by a master cave stream. However, despite its appear-ance of permanence, parts of the cave increased in volume about seven times between 1914 and 1962 (Parker and others, 1964), and additional extensive changes have occurred between then

and 1984. Some soil pipes that have developed in irrigated fields recently reclaimed from sagebrush near Phoenix, Arizona, have been found to extend as much as 75 m inland from a nearby arroyo into which they discharge at the base of the arroyo wall. However, such large pseudokarst features are exceptional. Most pipes in valley fill appear to be less than 10 m long and begin in desiccation or stress cracks paralleling the arroyo walls. Such is the situation pictured in Figure 7, where piping in the Quaternary valley fill of Aztec Wash has produced an extensive subsurface drainage system. There runoff locally concentrated by highway culverts has exacerbated the erosional undermining and threat-ened the highway itself.

Pseudokarsts such as these may show marked changes wrought by the erosion of a single heavy rainstorm. The intense summer storms of drylands in the western U.S. may create flash floods in the desert arroyos. The ragged, vertical piped walls of these arroyos, such as those shown in Figure 7, crumble and slump into the arroyos. This forces the thalweg to shift to the opposite wall, and renewed undercutting occurs there. In Aztec Wash, some arroyo channels have doubled in width in only a few hours as a consequence of a single flash flood.

Piping is most common in denuded heavy soils and bedrock with a high proportion of montmorillonite and illite (Parker and Jenne, 1967). Such materials crack widely and deeply when they lose their moisture by evaporation in the hot summer sun and the drying winds. The resulting stress-desiccation cracks are deepest in a strip along each side of a deep arroyo, and their size and depth diminish away from the arroyo. Surface runoff flows into these wide, deep cracks; where they are drainable and the hydraulic gradient is sufficient to cause tunnel erosion, the walls of the cracks erode readily, especially at the base, where a widened floor is created. Eventually, the upper walls become wetted enough to swell shut, but in the meantime a horizontal pipe has developed at the base, bridged by an arch of cohesive soil. This pipe then becomes a locus for subsequent drainage, enlarging with each rainstorm, and in the course of time may become a cave large enough to afford range animals respite from summer heat or winter snow and frost. As the pipe is further enlarged, or as the land surface above it is lowered, it eventually collapses, often in sections, and thus evolves from a subterranean passage to a series of conduits with alternating subaerial reaches, and then to a rapidly growing continuous gully. In a few years, such erosion may destroy a large part of a valley fill or even entire hillsides.

### The role of seepage erosion in hillslope and gully development

Many soils do not form soil pipes. Instead, diffuse or concentrated outflow of soil water from porous granular materials may entrain and carry away particles at sites where saturation overland flow emerges at the surface. Such seepage entrainment and erosion undermine hillslopes and gully heads. Then slope failure causes the oversteepened slope to retreat. Such sapping is similar to the spring sapping produced by outflow of ground water, described below, except that the flows are intermittent—even rare in some cases—and are generally of much smaller discharge. Nevertheless, the effects may rival or even overshadow those of running water in the development of many hillslopes and gully systems. This role of shallow subsurface water in landform development has just begun to be recognized in the last two decades (Leopold and others, 1964; Dunne, 1978; Higgins, 1984; Howard and McLane, in press).

## WATER AT OR BENEATH THE WATER TABLE

The geomorphic effects of ground water in the saturated zone tend to be more pronounced and recognized than those of water table partly because the reservoir is larger, the circulation is more continuous, and it has been more intensively studied. The features that result vary according to many factors, some of which are discussed in the following sections.

### Influence of ground ice on landforms of cold regions

Permafrost is ground that has a temperature colder than 0°C continuously for 2 or more years (Péwé, 1974), but it is not necessarily frozen (Sloan and Van Everdingen, this volume). It forms and is maintained in a climate where the mean annual air temperature is 0° or colder. About 20 percent of the land area of the Earth is underlain by permafrost, including about one-half of Canada and nearly 85 percent of Alaska.

The thermal regime of permafrost is dependent on the quantity of heat affecting the permafrost and the overlying active layer that freezes and thaws annually. Permafrost can built up (aggrade) and thaw (degrade) in numerous ways, all controlled by the thermal regime. Changes in the regime may depend on climatic, geomorphic, and vegetational factors.

Frozen ground water, or ground ice, tends to be concentrated in the upper levels of permafrost. The occurrence of ground water in the permafrost region is discussed by Sloan and van Everdingen in this volume. Ground ice may take many forms, some of the more common being pingo ice, ice lenses, and ice wedges. Ground water influences the geomorphology of permafrost regions in two distinct ways. First, the formation and growth of ground ice produces landforms such as ice-wedge polygons, palsas, and pingos. Second, the thaw of ground ice produces thermokarst features such as thaw lakes, alases, and beaded streams (Fig. 8). The resulting hummocky ground surface resembles karst topography found in limestone areas.

Ice-wedge polygons, the most extensive cold-region landforms created by freezing of ground water, have an ice wedge coincident with their borders. The borders tend to be raised or depressed with respect to the central areas. During the thaw season, low-centered polygons often contain ponds in the central areas, whereas high-centered ones hold water in the bordering depressions. Palsas are mounds or more irregular forms about 1.5 to 6 m high that generally have peat as an important constituent, contain perennial ice lenses, and occur in bogs. Palsas are characteristic of the subarctic, where they commonly occur in areas of discontinuous permafrost. In contrast to palsas, pingos are large perennial ice-cored mounds that tend to be more or less circular in form, 3 to 60 m high, and 15 to 500 m in diameter (Fig. 8). Pingos are necessarily associated with permafrost, and like ice-wedge polygons are key indicators of polar or subpolar environments.

Thermokarst comprises karstlike topographic features produced by the melting of ground ice and the subsequent settling or caving of the ground. One of the most conspicuous kinds of thermokarst is the thaw, or thermokarst lake. The basins of thaw lakes form or are enlarged by the thawing of ice-rich frozen ground.

Beaded streams consist of series of small pools connected by short watercourses. The pools result from the thawing of ice masses, commonly at ice-wedge polygon intersections. The connecting drainage generally lies along thawing ice wedges and therefore tends to comprise short, straight sections separated by

Figure 8. Collapsed pingo on MacKenzie Delta near Tuktoyaktuk, northwestern Canada. Ice-wedge polygons in foreground; oriented thaw lakes in background. Photo 18 August 1954 by Troy L. Péwé.

angular bends. Where ice-wedge polygons are degrading, pronounced intertrough mounds as much as 3 to 15 m in diameter and 0.3 to 3 m high can be left standing in relief.

Permafrost features are reviewed briefly by Black (1976), and a more comprehensive review of periglacial processes is by Washburn (1980). Brown (1973) has discussed ground ice as an initiator of landforms in permafrost regions.

### Land subsidence as a consequence of ground-water withdrawal

Land subsidence is caused by a variety of natural and man-made processes. Surface sinking or collapse may follow thawing of ground ice, as mentioned above. It may also result from subsurface solution of soluble rock masses. Effects owing to solution of carbonate rocks are discussed in Ford and others (this volume). Leaching of salt may produce similar effects. A number of karst-like features of the Texas panhandle can be related to dissolution of subsurface Permian evaporites; local dissolution of as much as 150 m of salt at depths as great as 500 m is largely responsible for regional topographic development there, including the location of the Pecos and Canadian rivers (Gustavson and Finley, 1985).

Subsidence can also be caused by withdrawal of subsurface fluids. At some sites it is caused by pumping of petroleum. This was first noted at the Goose Creek oilfield of Texas in 1925, but the best-known example is the area of Long Beach, California, where as much as 9 m of subsidence has taken place. Extraction of petroleum may also play some part in local subsidence in the Houston-Galveston area, although most of this has been attributed to withdrawal of ground water. Indeed, excessive pumping of ground water, with associated artesian-head decline or water-table lowering, is a major cause of land subsidence in many parts of the world. Subsidence caused by ground-water extraction was recorded as early as 1933 in the Santa Clara Valley, California, and has since been recognized as a serious problem in many areas, notably Venice, Italy; Mexico City, where subsidence now exceeds 7.5 m; and parts of the western San Joaquin Valley, California, where an area of more than 13,000 km$^2$ has subsided as much as 9 m (Poland and Davis, 1969).

A spectacular result of land subsidence in Arizona is the formation of hundreds of earth fissures (Pélé, 1984). These long, narrow, eroded tension cracks occur in unconsolidated sediments, typically near the mountains, along the margins of the basins where ground-water levels have dropped from 60 m to more than 150 m. Some fissures are 1 to 2 km long. They are generally perpendicular to drainage, intercepting surface water, which then erodes hairline-width cracks into gullys as much as 3 m wide and 6 m or more deep. Many, if not most, fissures form as a result of

Figure 9. Distinctive valley morphology associated with spring-sapping processes in massive sandstones of southeastern Utah. Steep valley sides and theaterlike valley heads are created by headward valley growth as sandstone caprocks are undercut by ground-water outflow.

differential subsidence and compaction over buried bedrock hills, ridges, or fault scarps. Some may occur where there are variations in type and thickness of alluvium, and variations in water-level decline. Economic losses caused by this land subsidence and earth-fissure formation are increasing rapidly.

A less pronounced but significant kind of subsidence results from "hydrocompaction," where unconsolidated low-density sediments are saturated for the first time, generally by irrigation water. This has accounted for as much as 4.5 m of settling in part of the San Joaquin Valley, California.

### Spring sapping and valley network development

Where ground water follows a deep route, like that of Path 2 in Figure 2, and where the flow is concentrated it may emerge in a spring. Spring outflow may directly entrain and remove soil or rock particles by seepage erosion, or it may help to concentrate erosion at the site by enhancing rock weathering. In either case the slope locally becomes steepened and undermined so that sapping occurs. This can contribute to the development and headward extension of valleys.

Spring sapping can contribute both to channel and valley development. The relationship to channel networks has been reviewed by Dunne (1980). In a summary of field-experimental and theoretical analyses, he outlined a model of headward channel growth and branching by sapping (illustrated here in Fig. 5) that contrasts with channel network development by overland flow. Models for the latter predict network development by piracy and cross grading (Horton, 1945) or by headward growth through abstraction (Abrahams, 1984). Abrahams finds that networks developed by overland flow can be differentiated from those developed by spring sapping by certain aspects of their morphology.

Unfortunately, geomorphic evidence for channel network processes is rarely discernible in the complex hillslope-channel relationships of valleys. Valley processes involve a considerable component of nonfluvial degradation. Although fluvial incision may drive the hillslope systems of valleys, the picture can be complicated by the action of greatly enhanced past processes that result in a relict valley morphology. One reason that sapping has

been underappreciated as a geomorphic process is that lowered water tables or desiccating climatic conditions have reduced its influence in many Holocene valleys (Higgins, 1984). In other valleys the results of spring-sapping processes may be obscured by modification of valley forms by nonsapping morphogenetic processes.

Excellent examples of valleys formed by sapping have long been known from the Colorado Plateau, where massive sandstone units are eroded by perched water emerging from bedding-plane boundaries (Stetson, 1936; Laity and Malin, 1985). These valleys show numerous distinctive attributes (Fig. 9), including the following: elongate basin shape, low network drainage density, low degree of interfluve dissection, widely spaced and short tributaries to main trunk valleys, theaterlike valley heads, prominent structural control of networks, irregular junction-angle relationships, local examples of long and narrow interfluves between adjacent valley segments that join at unusually acute angles, steep-sided valley walls meeting valley floors at a sharp angle, irregular variation in valley width as a function of valley length, relatively high drainage densities in upstream portions of basins, and local examples of hanging valleys. Many of the attributes of these valleys can be seen in miniature valley systems that can be observed forming entirely by ground-water outflow sapping on some beaches during falling tides (compare Fig. 9 with Fig. 11 or with other photographs in Higgins, 1984).

Permeable volcanic rocks, especially basalts, are also dissected by valleys that may involve spring-sapping processes. The large springs of the Snake River Plain, Idaho, have a probable relation to headward valley growth where seepage-induced weathering concentrates erosion and sapping at the springheads (Johnson 1939). In the Hawaiian Islands, valley development by overland flow is inhibited by the extremely high permeability of the lava flows that comprise the individual shield volcanoes. Nevertheless, less-permeable volcanic ash blankets on the older shield volcanoes (such as Mauna Kea, Kohala, and Haleakala) support the development of long, parallel, V-shaped valleys. Where deep incision encounters the underlying permeable lava flows, the valley morphology is transformed to U-shaped cross section. These steep-walled flat-bottomed valleys are the result of enhanced weathering at the water table (Wentworth, 1928) and development and recession of cirquelike valley heads. Perennial flow is maintained by large springs. Hanging valleys have developed as the deep U-shaped valleys recede headward and capture the more shallowly incised V-shaped valleys.

Perhaps the most exotic valley systems that have been attributed to sapping are the ancient networks that dissect heavily cratered uplands on the planet Mars (Pieri, 1980; Baker, 1982). Nirgal Vallis (Fig. 10) illustrates a Martian valley typical of many that formed early in the history of the planet, when liquid water was an important agent of denudation (Baker, 1985). The network shows strong structural control, theater-headed tributary valleys, and numerous other attributes of sapping morphogenesis.

A variety of small-scale analogs have been invoked to understand the processes involved in drainage system development by sapping. Because of material, time, and scale differences these studies must be considered suggestive of process operation rather than definitive in identifying the precise genesis of complex large-scale valleys. For example, intertidal beach-face channels (Higgins, 1984) provide models of network development as water emerges from sand during falling tides (see Fig. 11).

It is clear that spring sapping has profound implications for morphogenesis on both Earth and Mars. Paleohydrologic, paleoclimatic, and denudational studies of valleys on the two planets must give special consideration to the operation of this fundamental geomorphic process.

### Submarine geomorphic effects of ground-water processes

Processes related to ground-water discharge and pore pressures also modify underwater landscapes. Submarine artesian spring sapping and ground-water-induced mass wasting were proposed many years ago as processes on continental margins by Stetson (1936) and Johnson (1939).

Beneath continental shelves, particularly along passive margins, pressure gradients can be generated within subaerially recharged aquifers that extend to the continental slopes. Pressure gradients may also be caused by eustatic or local relative lowerings of sea level that increase the head. Pore-water overpressures may also result from rapid deposition of sediment or from compression by tectonic forces. Ground water discharging from the sediments can modify submarine terrain by such processes as seepage-face erosion and sapping, by enhancing or triggering of mass wasting, or by solution.

Since large amounts of sediments accumulate along continental margins, where conditions conducive to outflow are most commonly found, submarine erosion, mass wasting, and solution by ground water may help explain extensive submarine unconformities that developed on ancient continental margins during drops in sea level (Vail and others, 1977). A large volume of ground water is stored in sediments within the zone of sea-level variation. Numerical modeling shows that sea-level drops are more significant in creating long-term pressure differentials on continental slopes and fresh ground-water outflow in deep water

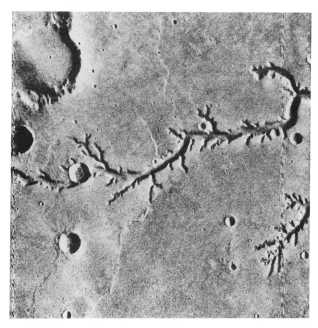

Figure 10. High-resolution image of the upstream reaches of Nirgal Vallis, a probable sapping network on the planet Mars. Specially processed Viking-1 Orbiter orthographic image. This frame shows a scene approximately 70 × 80 km.

than is subaerial recharge of artesian aquifers during high, stable sea levels (Leahy and Meisler, 1982; H. Meisler, personal communication, 1984).

An example of submarine erosion by discharge of ground water may be found on the lower continental slope off New Jersey, where recently acquired sidescan-sonar images and observations from a submersible vessel show valleys in outcrops of truncated coastal plain strata (Robb, 1984). These valleys, located in water depths between 1,500 m and 2,100 m, have theaterlike heads, steep walls, generally flat floors, and basins along their courses. Fragile clastic dikes protruding from cliff faces and excavated trace fossils within tafonilike cavities suggest that erosion in those places took place by some particle-by-particle mechanism such as seepage-face erosion, and not by bottom-current scour or mass wasting. Features resembling kluftkarren, rillenkarren, and solution pits further suggest that solution of chalky Eocene strata may have resulted from discharge of fresh or mixed waters.

Solution and mass wasting related to ground-water discharge may also have contributed to the removal of large volumes of carbonate rocks to form great erosional escarpments like those bordering the east and west sides of the Florida plateau. Active submarine springs, subsurface (Miocene) sinkholes, and surface sinkholes relict from periods of low sea level (Popenoe and others, 1984; Hebert, 1985) are evidence of relatively shallow outflow around the Florida peninsula. Active deep-water outflow of ground water there at depths greater than 3,000 m at the foot of the West Florida escarpment in the Gulf of Mexico was recently discovered by Paull and others (1984).

Another region where groundwater-influenced processes

may modify subocean topography was pointed out by Arthur and others (1980), who suggested that dewatering of sediment accumulations compressed by underthrusting could increase rates of mass wasting in trenches. Complex dendritic canyon systems have been found at some plate boundaries in parts of the oceans that are isolated from the continents, where turbidity-current erosion related to active terrigenous deposition is minimal. Dewatering of sediments, coupled with fracturing, and mass wasting triggered by earthquakes, may play a part in the formation of such networks.

### Sea-cliff erosion by ground-water outflow

Ground water can contribute to sea-cliff retreat by seepage erosion, enchanced weathering, or solution of carbonate minerals (Norris, 1968). Where coastal cliffs are weakly consolidated, emerging ground water readily weakens the rocks by removing intergranular cement, by dislodging grains by seepage pressure, and by encouraging the growth of vegetation near the discharge point. Vegetation adds weight and organic acids to the cliff face and also wedges rocks apart as root growth takes place. In areas where outflow is concentrated, spring sapping may produce alcoves, often with relatively lush patches of vegetation. It is probable that many short, steep-sided canyons cutting the sea cliffs of southern California are due at least in part to intermittent spring sapping.

Along some calcareous coasts, sea-cliff retreat results largely from the development of intertidal nips or notches. Much of this undercutting is caused by the activities of intertidal organisms, but in some places it may be caused largely by solution. Where coastal limestone is porous and permeable, slow effluent seepage of ground water all along the water line may dissolve the rock, especially where fresh ground water has mixed with sea water. Such mixing of waters, even where both are saturated with respect to carbonate minerals, can produce an unsaturated solution, as described by Back and others (1984). Where outflow is locally concentrated in coastal springs, brackish mixed waters floating on sea water may corrode nips along the shore (Higgins, 1980).

The role of ground water in contributing to the development of coastal landforms is more widespread and certainly more important than commonly acknowledged, and should be considered routinely in all studies of coastal erosion.

### Seepage weathering and cliff sapping

Just as concentrated outflow of ground water in springs can initiate the recession of steep-sided alcoves and valley heads, so too can diffuse seepage from the base of a slope promote its wearing away and undermining so that the slope retreats more or less uniformly, maintaining a steep front. As this effect has no formal name, we shall call it *cliff sapping*. Its progression is well illustrated on some beaches where outflow of ground water during falling tides produces miniature landforms that result solely or

mainly from seepage erosion and sapping and that may serve as models of much larger landforms (Higgins, 1984).

In the example illustrated in Figure 11, an extended steep-headed valley is developed in the upper left, where sapping by ground-water outflow was concentrated by a partially buried boulder. Elsewhere in the scene, seepage erosion by relatively nonconcentrated outflow has formed an "escarpment" with shallow alcoves. Characteristic braided outflow channels, terraces, and fallen masses of sand along the miniature cliffs of this example are also typical of larger landscapes that can be attributed to ground-water seepage and cliff sapping. For example, compare the escarpment in Figure 11 with that shown in Figure 12, an aerial photograph of an area at least two orders of magnitude larger. Both of these scenes resemble a view of a much larger embayed escarpment near Cameron, Arizona, reproduced by Stetson (1936, Fig. 2), who proposed that both it and the similar-appearing submarine southern face of Georges Bank were formed by sapping by ground-water seepage at the base. Similar escarpments, still larger by several orders of magnitude, border the "chaotic terrain" of Mars (Sharp, 1973), which is thought to have been formed by liquefaction and outflow following widespread thawing of ground ice.

Simple entrainment and removal of erosible grains suffice to sap low cliffs in loose sediments, but this mechanism cannot, by itself, account for the erosion of consolidated rock in larger escarpments. There, cliff sapping must result from a group of related processes. These may include intensified chemical weathering, leaching, and solution within the seepage zone, coupled with enhanced physical weathering of the periodically damp or saturated rock face by granular disintegration or flaking owing to wetting-drying (Smith, 1978), salt-crystal wedging, root-wedging, rainbeat, and, especially in temperate to cold regions, congelifraction, including needle-ice wedging, which Sharp (1976) found effective even in southern Oklahoma. The combined effect of these processes can be termed seepage weathering. With sufficient hydraulic gradient, seepage outflow may at times be forceful enough to entrain the particles loosened by seepage weathering, but it is unlikely that diffuse seepage can transport the grains very far. Actual removal of the weathered products must then owe to other agents, such as wind, slopewash, and streamwash, or in the case of sea cliffs, to wave action. The combined effect of all these is to sap the base of a cliff so that it retreats by parallel backwasting.

In most places where cliffs have been formed by sapping and seepage weathering, seepage is no longer active and the landforms are now relict. The lack of present activity provides one reason why seepage weathering and erosion have not been generally recognized in the origin of such cliffs and of the extensive landscapes that they front. Another reason is that there have been few studies of this process in the places where it does occur. Preliminary results of one such study—of recession of the Ogallala escarpment of the Southern High Plains—have been published by Osterkamp and Wood (1984). However, more studies are needed of seepage weathering and cliff sapping, which in some places are

 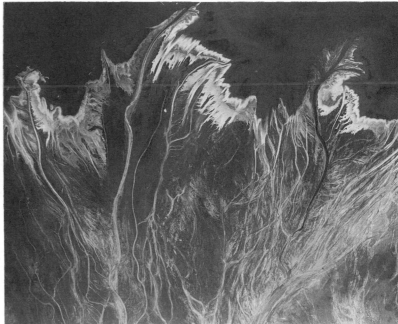

Figure 11. Embayed "escarpment" formed entirely by ground-water outflow and seepage erosion during falling tide on Bermuda Avenue Beach, San Diego, California. Active entrainment and transport of sand grains from the base of the "cliffs" was occurring when this photograph was taken. Pencil shows scale. Width of view about 1.5 m. Compare this scene with Figure 12.

Figure 12. Outflow from springs at the base of an embayed escarpment of the Danakil Basin, Afar Depression, Ethiopia. From Gerster (1976, Fig. 30). Width of view about 300 m. Air photograph copyright by Georg Gerster, reproduced with permission.

among the most important contributors to the denudation of the Earth's surface and the shaping of its landforms.

### Effects of ground water on channel form and process

Discharge of ground water may not only initiate the development of valleys, and even of entire drainage networks, but can influence stream channel form and process in the valleys that result. For example, complex interactions between riparian vegetation, shear strength of channel banks, and channel morphology result from changes in ground-water level.

Redwood Creek, near Orick, California, drains 720 km$^2$ of mixed forest land varying from Douglas fir in the upper part of the watershed to Redwood in the lower. In recent years the watershed has experienced accelerated erosion resulting from high-magnitude winter storms and from timber harvesting. Accelerated erosion has resulted in channel aggradation and subsequent rise in ground-water level adjacent to the main stem of Redwood Creek. In the upper part of the watershed, the local rise in water table has killed Douglas fir trees that had previously stabilized large alluvial terraces. Loss of these trees has contributed to accelerated erosion and widening of the channel (Nolan

and Janda, 1979).

The Carmel River drains 660 km$^2$ in coastal Monterey County, California. Two major episodes of channel widening in the Carmel River have occurred in this century (Kondolf, 1983): one was produced by major flooding, and the second was associated with a lowering of the water table, which killed riparian vegetation stabilizing the channel banks. The 1978 to 1980 erosional episode, which locally increased channel width by more than three times, occurred in response to floods that were of only moderate intensity. This is in marked contrast to the high-magnitude flood of 1911. Detailed studies by Kondolf (1983) clearly suggest that those reaches of the Carmel River that suffered the greatest bank erosion in 1978 to 1980 were the same where the die-off of riparian vegetation due to drawdown of the water table from pumping was also the greatest. These examples suggest that either rising or lowering of the ground-water level may affect the bank vegetation and consequently initiate significant bank erosion, channel widening, and a tendency to braid.

Ground water and surface water are intimately related in the fluvial system, and base flow is often provided by point sources as well as more diffuse ones. Variability of base-flow discharge commonly results from a change in ground-water level, and may produce several effects in stream channels.

Harrison and Clayton (1970) reported that field observations in a small Alaskan stream suggest that seepage of water into or out of the stream bed seems to alter the stream's competence. They reported that gaining reaches, characterized by effluent ground water, had considerably greater competence than losing reaches, characterized by influent ground water. Simons and Richardson (1966) had predicted such an effect by suggesting that effluent ground water should produce a seepage force that would reduce the effective weight of bed material and consequently decrease the stability of the bedload, leading to easier transport. Harrison and Clayton (1970) designed laboratory experiments to test the hypothesis, but the results were conflicting. Later flue studies by Richardson and Richardson (1985) confirm that in sand-bed channels, seepage of water into the channel can increase stream power and sediment transport for relatively low flow events. In bedrock channels of some streams in Indiana, presumed solutional processes create grooves, notches, and large boulders, apparently during low flow. This aspect of the role of base flow discharge in altering channel morphology is largely unstudied.

Effects of ground-water discharge on channel form and process remain a part of fluvial studies that has been little researched. What work has been done is speculative, controversial, and has led to contrasting ideas. In spite of this, future endeavors to explore relationships between ground water and channel process should be fruitful, provided interactions between the ground water and surface water can be measured and evaluated.

## GEOMORPHIC CONTROLS ON GROUND-WATER HYDROLOGY

Just as subsurface water affects surface processes and landforms, so too the landscape and regional geomorphic history affect the behavior of ground water.

### Topography and general terrain conditions

The landscape consists of a series of interconnected hillslopes. The variability of hillslope length, steepness, and shape are reflected in variable water-table configurations, which in turn translate into other aspects of ground-water conditions.

In the crystalline terrane of Statesville, North Carolina, LeGrand (1954) found that topographic slope is the most important control for well yield. The greatest well production there occurs in valleys and broad ravines, with lowest yield where wells are at or near the crests of narrow hills. Also he found that wells drilled in the centers of flat upland areas produce more water than wells farther down on the hillslope.

Water chemistry can also be influenced by topography. Ground water in the lowlands and lower parts of hills commonly contains more dissolved solids than does water upgradient. Because ground-water flows from the upper slopes and discharges at lower elevations, ground water near hilltops and upper slopes is younger and has had a shorter time to dissolve minerals than the waters near the base of hills.

The conditions that determine the position of the water table

in valleys and basins include topographic relief, drainage basin size, amount and composition of the valley fill, permeability distribution, and climate. In all terrains the depth to the water table is greatest at higher elevations and least in valleys and basins. Humid areas also have shallower water tables than do dryland environments. Although effluent systems (with ground water recharging streamflow) tend to occur in humid regions, geomorphic factors are also important. For example, in humid areas it is unusual for second- or third-order basins to have perennial streams, and in fine-grained bedrock it usually requires at least fourth-order basins to have strong base-flow conditions. In arid areas the water table is usually so deep that only the major through-flowing rivers have permanent flow with drainage basin order of at least seven. Topographic slope, water-table position, and ground-water hydrology are also closely related in coastal and small island environments where there is an interface between salt and fresh water (see Hunt and others, this volume; Back, this volume).

### Hydrologic properties of sediments

The geomorphic environment determines the character of unconsolidated sediments during their erosion, transportation, and deposition. When these sediments become aquifers, their history and properties can influence the behavior of the ground water. Thus sediments, which may also include regolith, alluvium, and colluvium, play important roles in the hydrologic realm by determining the recharge-discharge system and the nature and rates of infiltration, transmissivity, throughflow, and underflow. Geomorphic controls are especially significant in controlling ground-water hydrology in fluvial and glacial terranes. Ground water can also occur in significant quantity in loess, dune deposits, and beach deposits.

The caliber and fabric of fluvial sediments are a function of the flow regimes of the depositing river. High-energy streams produce coarse granular materials, whereas quiet waters and lacustrine conditions yield finer grained silts and clay. Although the latter sediments possess high porosity, they form poor aquifers because of their low permeability. Such deposits typically originate as overbank and back-swamp materials of floodplains. Coarse sands and gravels form in point bars and lateral accretion sites. Yields of wells in these sediments are higher because of greater permeability.

In similar fashion, glaciaton has a significant impact on determining ground-water hydrology in glaciated regions. Deposits of till generally have extremely low permeability, in contrast to much more permeable glacial outwash. Thus any search for new ground-water supplies in fluvial or glacial terranes may have to reconstruct the geomorphic history of the sites in order to locate the areas where conditions permitted the greatest amounts of sand and gravel to be deposited.

Bedrock can also be an important consideration in geomorphic and hydrologic studies because it influences both topographic characteristics and drainage density. Carlston (1963) studied

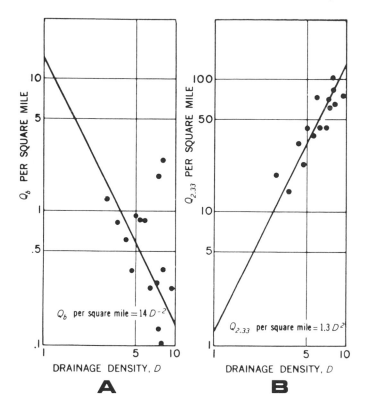

Figure 13. The relation of drainage density to base flow and floods. A, Base flow ($Q_b$); B, Mean annual flood ($Q_{2.33}$). From Carlston (1963).

ments are not restricted to arid regions, they are much more prevalent there, as in the American Southwest. Surficially these landforms seem quite similar, but they differ greatly in the availability of ground water. The chance of developing wells on pediment surfaces is slight because bedrock there occurs at shallow depths, whereas the deep sands and gravels of alluvial fans can yield much larger ground-water supplies.

Buried valleys in glacial terranes can be very important aquifers. The largest buried valley system in the United States is that of the ancestral Teays River. Prior to glaciation this river flowed west through the states of Ohio, Indiana, and Illinois. Norris and Spieker (1966) have provided a detailed analysis of this buried valley near Dayton, Ohio, where it provides one of the principal aquifers for the city. The Binghamton, New York, metropolitan area has several buried preglacial valleys of the Susquehanna and Chenango rivers. The most important of these underlies an 8 km$^2$ area in the urban core, and is one of the highest producing aquifers in the region (Randall, 1977).

Kaye (1976) provides an interesting case history of what happens when there is a failure to identify landforms and to recognize the character of the materials that compose them. Engineers who designed several new buildings in the Beacon Hill area of Boston assumed that the hills there were drumlins and that the sediment that composed them was clay-rich glacial till. It had been interpreted that this till would be largely impermeable and so would not present any serious water-flow problems. Instead, excavations at the sites encountered large ground-water flows that required different and costly construction methods and delayed the projects many months. Instead of drumlins, the Beacon Hill area is part of an end moraine with ice-contact deposits of clay, sand, and gravel that locally contain abundant ground water.

### Slope aspect and orientation

The Appalachian Plateau has often been described as a maturely dissected upland terrain with well-rounded hills interspersed among rivers with floodplains. In the Susquehanna Section, Coates (1971) has called attention to the large-scale asymmetry of the upland bedrock hills, which reflects the thickness of till. On south-facing slopes, till averages about 30 m deep, but the depth to bedrock on the steeper north-facing slopes averages only about 3 m. These hills have been called till shadow hills (Fig. 14). Because till does not yield sufficient water for household wells, water supply in this region must come from bedrock aquifers. In the underlying shale and siltstones it usually requires well penetration into rock of at least 50 m to produce 5 to 10 gallons (17.5 to 35 liters) per minutes. These ground-water conditions have important financial implications. Well yields on north-facing slopes are greater because of increased secondary porosity in the upper bedrock zones and steeper water tables. Thus, homeowners there can save about $2,000 on wells because they do not have to be drilled through so much till or contain so much casing as their counterparts on the south slopes.

Landscape trends can influence hydrologic conditions in

the hydrologic relations of 13 small monolithologic basins in nonglaciated areas of eastern United States. He found that drainage density, surface water discharge, and ground-water movement are all part of a unified physical system. His data (Fig. 13) show that ground-water discharge into streams, as reflected by base-flow conditions, is directly related to drainage density. Furthermore, because the transmissibility of the terrain also controls the proportion of precipitation that infiltrates the ground, as contrasted with that which flows off the surface, surface water or overland discharge is inversely related to drainage density.

In a hydrogeomorphological comparison of basins of the Catskill Mountain Section of the glaciated Appalachian Plateau with those of the Susquehanna Section, Coates (1971) also found drainage density to be an important indicator of base flow in streams. The Catskill basins are underlain primarily by sandstone, whereas the Susquehanna basins are largely of shale and siltstone. Mean base flow is nearly twice as great in the 12 sandstone basins with the lowest drainage density, as compared with the 13 shale-siltstone basins.

### Landforms

Specific landforms can be a predominant control of ground-water hydrologic conditions. Although alluvial fans and pedi-

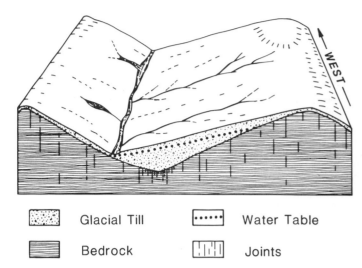

Glacial Till ·········· Water Table

Bedrock Joints

Figure 14. Diagram of till shadow hill. Note the much thicker till on south side of hill than on north side, with the consequent displacement of the original east-west trending thalweg. After Coates (1974).

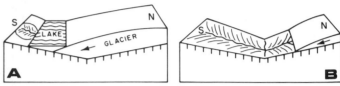

Figure 15. Difference of deglaciation history of north-flowing versus south-flowing streams. A, Sediments in north sloping valley are fine-grained because of ponding in proglacial lakes. B, Sediments in south sloping valley are clean sand and gravel because meltwaters have washed away fines. From Coates (1974).

other ways. For example, in the Northern Hemisphere, southern and southwestern slopes receive greater insolation than northern and northeastern slopes. This leads to differences in soils, vegetation, and water retention and infiltration rates. Such factors create feedback systems that affect the amount of water that enters the ground-water zone.

Directional trends are also produced in glaciated terrains. In a study of the glaciated part of the Susquehanna River basin, Coates (1971) showed that an important relationship exists between the orientation of tributary streams and base-flow discharge. He identified seven stream azimuths that are associated with different levels of base flow. The most dramatic base flow discrepancies occur when south-flowing streams are compared to their north-flowing counterparts. On average, south-flowing streams deliver more than ten times as much ground-water discharge as do north-flowing streams. The reason for this great discrepancy stems from the composition of the valley-fill materials, which can be attributed to the deglaciation history of the

valleys (Fig. 15). During recession of the glaciers, mostly fine-grained sediments were emplaced into proglacial lakes imponded by the ice margin. These sediments settled out on what would later become the north-facing slopes. However, as recession continued and the lakes were drained, the glacial meltwater fines were washed out of the south-facing slopes, leaving lag deposits of sand and gravel. These coarser sediments on south-facing slopes are now much more permeable, provide better aquifers, and produce larger surface and ground-water flow rates than do the finer materials on the north-facing slopes.

These examples give some idea of how the geomorphology of an area can help determine its ground-water hydrology. Topography, surface processes, Earth materials, regional geomorphic history, slope steepness and aspect, and even such features as drainage density all can be important variables that affect ground-water behavior as well as the base flow of surface streams. In such cases it is important to understand the geomorphic history and nature of the landscape in making hydrogeological inferences.

## REFERENCES

Abrahams, A., 1984, Channel networks; A geomorphological perspective: Water Resources Research, v. 20, p. 161–188.

Arthur, M. A., Carson, B., and von Huene, R., 1980, Initial tectonic deformation of hemipelagic sediment at the leading edge of the Japan convergent margin, *in* Marianna, L., ed., Initial reports of the Deep Sea Drilling Project: Washington, D.C., U.S. Government Printing Office, v. 56–57, Part 1, p. 569–613.

Back, W., Hanshaw, B. B., and Van Driel, J. N., 1984, Role of groundwater in shaping the Eastern Coastline of the Yucatan Peninsula, Mexico, *in* LaFluer,

R. G., ed., Groundwater as a geomorphic agent: Boston, Allen and Unwin, p. 281–293.

Baker, V. R., 1982, The channels of Mars: Austin, Texas, University of Texas Press, 198 p.

—— , 1985, Models of fluvial activity on Mars, *in* Woldenberg, M., ed., Models of geomorphology: Winchester, Massachusetts, Allen and Unwin, p. 287–312.

Beven, K., and Germann, D., 1982, Macropores and water flow in soils: Water Resources Research, v. 18, p. 1311–1325.

Black, R. F., 1976, Features indicative of permafrost: Annual Review of Earth and Planetary Sciences, v. 4, p. 75–94.

Brown, R.J.E., 1973, Ground ice as an initiator of landforms in permafrost regions, *in* Gahey, B. D., and Thompson, R. D., eds., Research in Polar and Alpine Geomorphology, 3rd Guelph Symposium in Geomorphology: Norwich, United Kingdom, Geobooks, p. 25–42.

Bryan, R. B., and Yair, A., eds., 1982, Badland geomorphology and piping: Norwich, U.K., GeoBooks, 408 p.

Buckham, A. F., and Cockfield, W. E., 1950, Gullies formed by sinking of the ground: American Journal of Sciences, v. 248, p. 137–141.

Carlston, C. W., 1963, Drainage density and streamflow: U.S. Geological Survey Professional Paper 422-C, 8 p.

Coates, D. R., 1971, Hydrogeomorphology of Susquehanna and Delaware basins, *in* Morisawa, M., ed., Quantitative geomorphology: Binghamton, New York, State University of New York, p. 273–306.

—— , 1974, Reappraisal of the glaciated Appalachian Plateau, *in* Coates, D. R., ed., Glacial geomorphology: Binghamton, New York, State University of New York Publications in Geomorphology, p. 205–243.

Dietrich, W. E., Wilson, C. J., and Reneau, S. L., 1986, Hollows, colluvium, and landslides in soil-mantled landscapes, *in* Abrahams, A., ed., Hillslope processes: Boston, Allen and Unwin, p. 361–388.

Dunne, T., 1969, Runoff production in a humid area: U.S. Department of Agriculture Report ARS 41-160, 108 p.

—— , 1978, Field studies of hillslope flow processes, *in* Kirkby, M. J., ed., Hillslope hydrology: New York, Wiley and Sons, p. 227–293.

—— , 1980, Formation and controls of channel networks: Progress in Physical Geography, v. 4, p. 211–239.

Dunne, T., and Black, R. D., 1970, An experimental investigation of runoff production in permeable soils: Water Resources Research, v. 6, p. 478–490.

Flemming, R. W., and Johnson, A. M., 1975, Rates of seasonal creep of silty clay soil: Quarterly Journal of Engineering Geology, v. 8, p. 1–29.

Gerster, G., 1976, Grand design: New York, Paddington Press, 312 p.

Gustavson, T. C., and Finley, R. J., 1985, Late Cenozoic geomorphic evolution of the Texas Panhandle and northeastern New Mexico—Case studies of structural control of regional drainage development by salt dissolution and subsidence: Austin, University of Texas Bureau of Economic Geology Report of Investigation No. 148, 42 p.

Harrison, S. S., and Clayton, L., 1970, Effects of ground-water seepage on fluvial processes: Geological Society of America Bulletin, v. 81, p. 1217–1226.

Hebert, J. A., 1985, A Miocene karst drainage system; Seismic stratigraphy of the continental shelf west of Florida: Geological Society of America Abstracts with Programs, v. 17, p. 606.

Higgins, C. G., 1980, Nips, notches, and the solution of coastal limestone; An overview of the problem with examples from Greece: Estuarine and Coastal Marine Science, v. 10, p. 15–30.

—— , 1984, Piping and sapping; Development of landforms by groundwater outflow, *in* LaFleur, R. G., ed., Groundwater as a geomorphic agent: Boston, Allen and Unwin, p. 18–58.

Horton, R. E., 1945, Erosional development of streams and their drainage basins: Geological Society of America Bulletin, v. 56, p. 275–370.

Howard, A. D., and McLane, C. F., III, 1987, Erosion of cohesionless sediment by groundwater entrainment: Water Resources Research [in press].

Hursh, C. R., 1944, Report of the sub-committee on subsurface flow: EOS American Geophysical Union Transactions, v. 25, p. 743–746.

Hutchinson, J. N., 1968, Mass Movement, *in* Fairbridge, R. W., ed., Encyclopedia of geomorphology: New York, Reinhold, p. 688–695.

Johnson, D. W., 1939, The origin of submarine canyons; A critical review of hypotheses: New York, Columbia University Press, 126 p.

Jones, J.A.A., 1981, The nature of soil piping; A review, *in* British Geomorphological Research Group Monograph no. 3: Norwich, United Kingdom, GeoBooks, 301 p.

Kaye, C. A., 1976, Beacon Hill end moraine, Boston; New explanation of an important urban feature, *in* Coates, D. R., ed., Urban geomorphology: Geological Society of America Special Paper 174, p. 7–20.

Kondolf, G. M., 1983, Historic channel changes, Carmel River, Monterey County, California, *in* Williams, J., and Kondolf, G. M., eds., Guidebook for the Conference on Channel Stability and Fish Habitat: Monterey Peninsula Water Management District and California Department of Fish and Game, p. 50–76.

Laity, J. E., and Malin, M. C., 1985, Sapping processes and the development of theater-headed valley networks on the Colorado Plateau: Geological Society of America Bulletin, v. 96, p. 203–217.

Leahy, P., and Meisler, H., 1982, An analysis of fresh and saline ground water in the New Jersey Coastal Plain and Continental Shelf: EOS American Geophysical Union Transactions, v. 63, p. 322.

LeGrand, H. E., 1954, Geology and groundwater in the Statesville area, North Carolina: North Carolina Department of Conservation and Development, Division of Mineral Resources Bulletin 68, 68 p.

Leopold, L. B., Wolman, M. G., and Miller, J. P., 1964, Fluvial processes in geomorphology: San Francisco, Freeman, 522 p.

Nolan, K. M., and Janda, R. J., 1979, Recent history of the main channel of Redwood Creek, California, *in* A field trip to observe natural and management related erosion in Franciscan terrane of northern California; a guidebook: San Jose, California, Geological Society of America Cordilleran Section, p. X.–1 to X.–16.

Norris, R. M., 1968, Sea-cliff retreat near Santa Barbara, California: California Division of Mines and Geology, Mineral Information Service, v. 21, p. 87–91.

Norris, S. E., and Spieker, A. M., 1966, Ground water resources of the Dayton area, Ohio: U.S. Geological Survey Water-Supply Paper 1808, 167 p.

Osterkamp, W. R., and Wood, W. W., 1984, Development and escarpment retreat of the Southern High Plains, *in* Proceedings of the Ogallala Aquifer Symposium II: Lubbock, Texas Technical University Water Resources Center, p. 177–193.

Parker, G. G., and Jenne, E., 1967, Structural failure of western highways caused by piping: U.S. Highway Research Board, Highway Research Record no. 203, p. 57–76.

Parker, G. G., Shown, L. M., and Ratzlaff, K. W., 1964, Officers Cave, a pseudo-karst feature in altered tuff and volcanic ash of the John Day Formation in eastern Oregon: Geological Society of America Bulletin, v. 75, p. 393–402.

Paull, C. K., and nine others, 1984, Biological communities at the Florida escarpment resemble hydrothermal vent taxa: Science, v. 226, p. 965–967.

Péwé, T. L., 1974, Permafrost: Encyclopedia Brittanica (15th edition), v. 14, p. 89–95.

—— , 1984, Fissures in Arizona: Earth Science, v. 37, no. 2, p. 19–21.

Pieri, D. C., 1980, Martian valleys; Morphology, distribution, age, and origin: Science, v. 210, p. 895–897.

Poland, J. F., and Davis, G. H., 1969, Land subsidence due to withdrawal of fluids: Geological Society of America Reviews in Engineering Geology, v. 2, p. 187–269.

Popenoe, P., Kohout, F. A., and Manheim, F. T., 1984, Seismic reflection studies of sinkholes and limestone dissolution features on the northeastern Florida shelf, *in* Beck, B. F., ed., Sinkholes; Their geology, engineering, and environmental impact, Orlando, Florida: Proceedings of the First Multidisciplinary Conference on Sinkholes, Boston, A. A. Balkema, p. 43–57.

Randall, A. D., 1977, The Clinton Street–Ballpark aquifer in Binghamton and Johnson City, New York: New York State Department of Environmental Conservation Bulletin 73, 87 p.

Richardson, J. R., and Richardson, E. V., 1985, Inflow seepage influence on straight alluvial channels: Journal of Hydraulic Engineering, v. III, p. 1133–1147.

Robb, J. M., 1984, Spring sapping on the lower continental slope, offshore New Jersey: Geology, v. 12, p. 278–282.

Rogers, J. D., and Pyles, M. R., 1979, Evidence of cataclysmic erosional events in the Grand Canyon of the Colorado River, Arizona, *in* Proceedings of the Second Conference on Research in the National Parks, San Francisco: Washington, D.C., National Park Service, Physical Sciences, v. 5, p. 392–454.

Sharp, R. P., 1973, Mars; Fretted and chaotic terrains: Journal of Geophysical

Research, v. 78, p. 4073–4083.

Sherard, J. L., and Decker, R. S., 1977, Dispersive clays, related piping, and erosion in geotechnical projects: American Society of Testing and Materials Special Technical Paper 623, 486 p.

Simons, D. B., and Richardson, E. V., 1966, Resistance to flow in alluvial channels: U.S. Geological Survey Professional Paper 422-J, 61 p.

Stetson, H. C., 1936, Geology and paleontology of the Georges Bank canyons; Part 1, Geology: Geological Society of America Bulletin, v. 47, p. 339–366.

Terzaghi, K., 1943, Theoretical soil mechanics: New York, Wiley and Sons, 510 p.

Twidale, C. R., 1984, Role of subterranean water in landform development in tropical and subtropical regions, *in* LaFleur, R. G., ed., Groundwater as a geomorphic agent: Boston, Allen and Unwin, p. 91–134.

Vail, P. R., and seven others, 1977, Seismic stratigraphy and global changes of seal level, *in* Payton, C. E., ed., Seismic stratigraphy; Applications to hydrocarbon exploration: American Association of Petroleum Geologists Memoir 26, p. 49–212.

Washburn, A. L., 1980, Geocryololgy; A survey of periglacial processes and environments: New York, Wiley and Sons, 416 p.

Wentworth, C. K., 1928, Principles of stream erosion in Hawaii: Journal of Geology, v. 36, p. 385–410.

Zaslavsky, D., and Kassiff, G., 1965, Theoretical formulation of piping mechanism in cohesive soils: Geotechnique, v. 15, p. 305–310.

MANUSCRIPT ACCEPTED BY THE SOCIETY JUNE 26, 1987

The Geology of North America
Vol. O-2, Hydrogeology
The Geological Society of America, 1988

# Chapter 43

# *Landform development; Karst*

**Derek C. Ford**
*Department of Geography, McMaster University, Hamilton, Ontario L8S 4K1, Canada*
**Arthur N. Palmer**
*Earth Science Department, State University of New York, Oneonta, New York 13820*
**William B. White**
*Materials Research Laboratory and Department of Geosciences, The Pennsylvania State University, University Park,*
*    Pennsylvania 16802*

## INTRODUCTION

Karst landscapes are the foremost examples of ground-water erosion on this planet. The sculpturing and removal of bedrock is predominantly by solution, aided in some cases by soil piping and collapse. Karst landforms develop best in limestones and dolomites, gypsum, and salt. Carbonate rocks crop out over approximately 10 percent of the earth's land area and are found in most nations and all climatic regions. It is estimated that 25 percent of the world's population depends on fresh water in karst aquifers. Karst rocks and their contained minerals, oil, and gas are of considerable importance to the extractive industries, while caves and other karst features have been of great cultural significance.

Karst aquifers are the mavericks of hydrogeology. Closed depressions input a recharge that is intermediate between the classic ideas of infiltration and surface runoff. Integrated conduit systems act as short circuits for the ground-water flow system. Conduits are gross heterogeneities in the permeability distribution, and flow within them does not obey Darcy's Law. The purpose of this chapter is to outline some of the geomorphic features of karst areas and to indicate their relationship to karstic aquifers.

Recent English language books on karst studies include Sweeting (1972), Ford and Cullingford (1976), Bögli (1980), Milanovic' (1981), Jennings (1985), James and Choquette (1988), and White (1988).

### *The karst system*

Karst is a variant of the fluvial geomorphic system. When fully developed, the karst system has three morphological components: input landforms that direct waters underground, subterranean conduit systems, and discharge areas, which include springs, sapping landforms, and erosion residual features such as rock towers.

The input landforms develop on three broad scales. Small forms, on a scale of less than 10 m, include varieties of solution pits, pans, grooves, and runnels in bedrock, collectively known as "karren." The principal intermediate-scale features are closed karst depressions (dolines or sinkholes), which range approximately from 1 to 1,000 m in their greatest dimension. For many researchers, closed depressions are the diagnostic karst landforms. Large-scale forms, normally greater than 1 km in length, include poljes (closed, flatfloored depressions), dry valleys, and gorges. Most karst terrains are dominated by combinations of these input landforms, which may cover thousands of square kilometers.

Systems of connected underground conduits and lesser fissure openings that have been enlarged by solution (cave systems) are fundamental to karst development. They play the role that stream channels of all orders perform in the fluvial system. Sinkholes, poljes, and the deeper karren cannot develop until cave systems are established beneath them at some minimal scale.

Cave systems integrate the drainage from many input points for discharge at a single spring or at a few clustered or aligned springs. Most of the greatest freshwater springs (magnitude defined by mean-annual discharge) are karstic. Landforms at spring points include erosion features such as gorges created by headward sapping of a cave roof or a cliff foot, and constructional features such as travertine terraces, cascades, and dams.

### *Types of karst terrain*

Many different approaches to the classification of karst terrains have been advanced. One that is widely accepted is a division into "holokarst" and "merokarst" (or "fluviokarst"). In the holokarst, all waters are drained underground, including allogenic streams (those draining onto the karst from adjacent nonkarst rocks); there is little or no surface channel flow. In fluviokarst, major rivers remain at the surface because their flow is too large

Ford, D. C., Palmer, A. N., and White, W. B., 1988, Landform development; Karst, *in* Back, W., Rosenshein, J. S., and Seaber, P. R., eds., Hydrogeology: Boulder, Colorado, Geological Society of America, The Geology of North America, v. O-2.

to be adsorbed by the aquifer, or headwater streams survive because the underground channel net has not yet extended to intercept them. "Parakarst" describes terrains that are a mixture of karstic and fluvial, owing to a mingling of karst and nonkarst rocks in outcrop, e.g., limestone and shale sequences. "Covered karst" develops where karst rocks are actively removed beneath a cover of other consolidated rocks such as shale or sandstone. Karst landforms of a collapse type may propagate upward through as much as 1,000 m of overlying rocks. "Mantled karst" refers to deep cover by unconsolidated detritus; this may be local weathering residue, but more often is transported material such as glacial deposits, volcanic ash, and loess. "Paleokarst" refers to karst terrains and cave systems that are buried beneath later strata; they may be exhumed and rejuvenated. "Pseudokarst" describes karst-type landforms and assemblies that are created by processes other than rock dissolution; principal types are thermo-karst, vulcanokarst, and mechanical piping.

### The diversity of karst

A great diversity of forms and combinations occur among the world's karst terrains despite their restriction to the soluble rocks. Major factors contributing to this diversity include: physical and chemical variations in the rocks themselves; geologic structure and tectonic history; relief and regional topography; geomorphic history, including paleokarst in some instances; and past and present climatic conditions. The diversity is demonstrated in the regional accounts provided by Herak and Stringfield (1972) for many countries, by Gams (1974) for Yugoslavia, Jakucs (1977) for Hungary, Gvozdetskij (1981) for the U.S.S.R., and Zhang (1980) for the People's Republic of China.

## CHEMICAL PROCESSES AND KARST DENUDATION

Most minerals are slightly soluble in pure water, and more so if hydrogen ions are present. As a consequence, varieties of small karren (tafoni) are common even on granitic rocks or quartzites if local conditions can reduce the efficiency of competing erosion processes. Specific combinations of topography and geologic structure have permitted some large-scale solutional karst to develop on quartzites. However, such instances are rare. Limestone, dolomite, gypsum, and salt are the principal karst rocks.

Halite and gypsum dissolve by dissociation in the presence of water. For salt,

$$NaCl = Na^+ \, Cl^-.$$

In standard conditions, this reaction proceeds until equilibrium is reached at 368,000 mg/L. Salt is so soluble that its outcrop is confined to the driest places, such as the floor of Death Valley, California, where the rare rains have intensively dissected a salt pan into karren and small dolines. Salt karstification is more significant in covered karst situations such as in Saskatchewan (Quinlan, 1978).

The equilibrium solubility of gypsum is 2,410 mg/L at standard conditions. Gypsum can survive in outcrop in all climates. However, most gypsum karst is in semi-arid and arid regions. The most notable gypsum karsts have thin covers of other rocks.

### Carbonate equilibrium

The greatest extent and diversity of karst features are developed in limestone and secondarily in dolomite. The remainder of this chapter is devoted to karst in these rocks.

Calcite and dolomite solution is primarily the result of reaction with carbonic acid from the atmosphere or the soil. Soil $CO_2$ is generated by organic decay and by the action of microorganisms. $CO_2$ production varies with soil characteristics, plant cover, growing season, temperature, and water availability. Many of the local variations average out over regional scales, and soil $CO_2$ pressure can be modeled by the relation:

$$\log P_{CO_2} \, (atm) = -1.97 + 0.04T \, (°C)$$

where T is mean-annual temperature (Drake, 1983).

The important chemical reactions in karst development are:

$$CO_2 \, (gas) + H_2O = H_2CO_3 \quad K_{CO_2}$$
$$H_2CO_3 = H^+ \, HCO_3^- \quad K_1$$
$$HCO_3^- = H^+ + CO_3^{-2} \quad K_2$$
$$CaCO_3 = Ca^{+2} + CO_3^{-2} \quad K_C$$

The net calcite dissolution reaction appropriate to the near-neutral pH of most karst waters is:

$$CaCO_3 + H_2O + CO_2 = Ca^{+2} + 2 \, HCO_3^-$$

The equivalent reaction for dolomite is:

$$CaMg(CO_3)_2 + 2 \, H_2O + 2CO_2 = Ca^{+2} + Mg^{+2} + 4 \, HCO_3^-$$

These bare equilibria are modified somewhat by the formation of complexes and ion pairs. Precise values for the equilibrium constants are available over the range of temperature encountered in karst systems (Plummer and Busenberg, 1982). Calculations for the chemical equilibria are easily reduced to computer programs, and karst chemistry can be modeled quite precisely.

### Open and closed systems; Solute concentrations at equilibrium

In an ideal open system, carbonic acid consumed by reaction with minerals is replaced by more $CO_2$ drawn from an infinite reservoir of gas in the vadose zone, the soil, or the atmosphere. In an ideal closed system the aqueous solution has an initial concentration of $CO_2$ that reacts with the solid phase. The depleted carbonic acid cannot be replaced, and the amount of solution achieved at equilibrium is much reduced.

Open-system solution occurs where water flows over bare rocks and may occur where abundant carbonate clasts are present in a soil, e.g., in glaciated areas (Ford, 1983). Ideal closed-system conditions prevail where karst conduits are being initiated in bedrock.

As a consequence of variability of $P_{CO_2}$ and temperature, the equilibrium solubility of calcite is found to range between <50 and 350 mg/L as $CaCO_3$ in most natural bicarbonate waters. Beneath alpine glaciers, some waters may be saturated with respect to calcite at approximately 25 mg/L. Where values exceed 350 mg/L, sulfate solution or contaminated water must be suspected.

### Kinetics

Reaction to equilibrium between limestone and carbonic acid requires about ten days. Dye traces and other evidence show that the residence time of water within karst aquifers is on the order of days to months; flow rates through open solution cavities are much higher than in other aquifers. Thus, most karst processes operate out of equilibrium to a considerable extent. They depend on the kinetics of the chemical reactions.

At least three mechanisms have been identified that drive the dissolution rate of calcite, Rc (Plummer and others, 1978):

$$Rc = K_1 a_{H^+} + K_2 a_{H_2CO_3} + k_3 a_{H_2O} - k_4 a_{Ca^{++}} a_{HCO_3^-}$$

The first term describes the dissolution by hydrogen ions, appropriate when there are other sources of acid. The second term is the reaction with carbonic acid, predominant in most limestone solution. The third term is the dissolution in pure water, which has relevance to sculpturing of bare bedrock pavements where $CO_2$ pressures are low. The fourth term describes the back reactions that set in as the system approaches equilibrium.

The dissolution rate for dolomite in highly undersaturated solutions is comparable to that of limestone. However, the rate for dolomite slows dramatically when the water reaches only 1 percent of saturation (Herman and White, 1985). Equilibrium is approached only over times measured in years. Although dolomite aquifers may yield water nearly saturated with dolomite, both caves and surface landforms tend to be more subdued in dolomite terrain.

The rate for dolomite dissolution, $R_D$, far from equilibrium at temperatures below 45°C is (Busenberg and Plummer, 1982):

$$R_D = k_1 a_{H_2CO_3}^{1/2} + k_3 a_{H_2O}^{1/2} - k_4 a_{HCO_3^-}$$

### Denudation

The solutionally transported mass loss from karst basins, expressed as if it were removed uniformly from the land surface, is commonly known as "karst denudation." Many measurements have been made, based mostly on carbonate content of basin runoff, and examined for differences that could be ascribed to the effects of climate. The denudation rates available through 1975 were summarized in three linear regression equations (Smith and Atkinson, 1976):

$$\text{Tropical: } D = 0.063 R + 5.7$$
$$\text{Temperate: } D = 0.055 R + 7.9$$
$$\text{Arctic/Alpine: } D = 0.025 R + 7.4$$

where D is denudation rate in mm ka$^{-1}$ (equivalent to m$^3$km$^{-2}$a$^{-1}$) and R is runoff in mm a$^{-1}$. Denudation rates in most arctic/alpine climates are clearly lower. Rates in temperate and tropical regions could not be statistically distinguished.

From mass balance considerations and dissolution chemistry, the maximum denudation rate, which assumes that the ground water comes into equilibrium with the bedrock, is given by White (1984):

$$D_{max} = \frac{100}{\rho \sqrt[3]{4}} \left( \frac{K_c K_1 K_{CO_2}}{K_2} \right)^{1/3} P_{CO_2}^{1/3} (P - E)$$

The equilibrium constants are as defined previously. $\rho$ is the density of the bedrock. (P–E) is precipitation less evapotranspiration in mm a$^{-1}$, equivalent to runoff if basins are conservative and long-term changes in storage are averaged out. Denudation increases linearly with precipitation (or runoff), but only as the cube root of $P_{CO_2}$. The temperature dependence of denudation rate is contained within that of the various equilibrium constants. The climatic variables of $CO_2$ availability and temperature are muted by the cube root functions, thus explaining the small differences observed in denudation rates. It seems that the marked differences in landform between the doline karsts of many temperate areas and the cone and tower karst typical of many tropical areas must be ascribed to some combination of hydrogeologic setting and differential solution, and not to climate and $CO_2$ production as such. Others have argued that it is the greater runoff intensity at humid tropical sites that explains the differences, but this has not been convincingly demonstrated by field measurements.

Not all water attains chemical equilibrium with bedrock. Actual solute concentrations in streams vary inversely with discharge. The long-term denudation rates are estimated from linear regression of annual values. The quoted range is from 8 to 130 mm ka$^{-1}$ (Jennings, 1985).

## THE SUITE OF INPUT LANDFORMS

### Karren; Surfaces and subcutaneous processes

Falling droplets, sheet and channeled runoff, film flow, and ponded water all create small-scale solution forms (karren) where rain falls onto bare limestone or waves break onto it. The most common forms are approximately circular pits with rounded bottoms and pans with flat bottoms, straight or sinuous channels

Figure 1. A detail of well-developed, but small, clint-and-grike topography on a dolomite pavement. This example is on Ca-rich reefal dolomites in the Bruce Peninsula, Ontario, Canada.

descending slopes, and elongated clefts where joints or dipping bedding planes have been opened. Bögli (1980, p. 50) gives a detailed classification and discussion of processes.

Quaternary glacier scour has left many plains and benchlands of limestone and dolomite that are bare or lightly veneered with drift. Wave action, floods, or wind deflation may achieve that same effect in extraglacial regions, although rarely to the same extent. On such plains, major joints are opened by solution until they terminate at master bedding planes at depths ranging from 0.5 to 25 m. The open joints are "kluftkarren" or "grikes," (Fig. 1). Surfaces between them ("clints") become indented with lesser pit and channel karren draining to them at the surface or via short micro-caves that debouch into the grike walls. Such assemblages are called "limestone pavements." They may cover many square kilometers and are a prominent feature of glaciated karst regions throughout Canada and in the U.S. Rockies.

Beneath soil may be a typical C horizon of weathering rubble if the limestones or dolomites are impure or thin bedded and friable. More commonly, karren develop into which the soil fil-

ters, and plant roots, foci of soil $CO_2$ production, may penetrate. On a broad scale, the pattern of subsoil karren is a clint-and-grike pattern. In detail, the forms developed upon and within the clint blocks to dissect them are perhaps the most diversified morphological family that is to be found anywhere in geomorphology. There is bewildering, uncataloged variety. Contributing factors appear to be small- to micro-scale variations in all rock properties, fracture density and propagation of new fractures as karren are enlarged, spatial and temporal variability of water supply and $P_{CO_2}$, and of open or closed system conditions. The karst plains of Indiana, Kentucky, and Tennessee display excellent examples.

Subsoil karren may be partly or fully exposed as a consequence of soil erosion. Sharp pit and rill karren typical of bare rock are then superimposed upon the rounded subsoil forms. In extreme cases, exposed karren compose forests of rock pinnacles rising 20 m, as in Yunnan, China.

In both pavement and subsoil karren, the frequency of enlarged fissures diminishes rapidly with depth. Usually, permeability is reduced drastically at one or a few key bedding planes, which gather the flow. The karren thus constitute a shallow zone that, in most soil-mantled and/or well-vegetated karst terrains, is the zone of maximum dissolution by autogenic water. Williams (1983) and other geomorphologists have termed this "subcutaneous zone," a hydrologic zone of significant storativity and complex dynamic behavior. It is poised above the conventional vadose zone, which also exists in well-developed karsts. Williams (1983) demonstrated by cross-correlations that response lags of 2 to 14 weeks following rain existed for overhead seepage and flow entering sampled caves in New Mexico and New Zealand. There was no correlation with depth. The extent and variation of the lag was attributed to effects in overlying epikarstic zones. His model for estimating the subcutaneous storage volume is:

$$R_v = [P - (R_s + ET)] - Q_k.$$

$R_v$ is subcutaneous recharge. P is the potential runoff estimated from storm precipitation. $R_s$ is soil moisture recharge. ET is evapotranspiration and $Q_k$ is flood discharge at a karst spring.

## Closed depressions

Closed depressions predominate among surface karst forms of intermediate size. Beck (1984) and Beck and Wilson (1987) present a comprehensive collection of reports that emphasize applied work in the United States.

Four distinct processes, operating alone, may create dolines in karst regions. They are: (1) solution, acting downward; (2) mechanical collapse or stoping, acting upward from a prior solution cavity; (3) subsidence, without rupture, into an interstratal solution cavity; and (4) sapping, suffosion, or seepage erosion of unconsolidated mantle materials, with or without soil piping, into caves, subsoil karren openings, or adjoining dolines. Most often, two or more of these processes operate in conjunction to develop or enlarge a depression. This multiplicity of processes and combinations, coupled with the normal heterogeneity of geomorphologic and geologic variables, results in dolines displaying a wide range of scale, form, and distribution patterns.

Forms range from cylindrical shafts to shallow saucers, but intermediate funnel and bowl shapes are perhaps the most common. In some regions there is an evident progression from initial cylinder or steep funnel to a gentler bowl as the feature ages. Larger, deeper dolines that develop in thick, young limestone may contain secondary stormwater channels, which give them a star-shaped appearance in plan view, such as in a tropical karst "cockpit" (Sweeting, 1972).

Doline density is reported to range from <1.0 to >2,500 per km². The higher densities are rarely encountered, although instances have developed within Holocene time on coastal gypsum in eastern Canada. The range is 1 to 9 per km² in sample areas of the eastern and central United States. Kemmerly (1982) describes doline distribution in the great sinkhole plains of Kentucky and Tennessee.

## Large-scale forms

Dry valleys and gorges are developed by normal fluvial action but lose their water underground due to karstic rocks exposed in the floors. Incipient dry valleys preserve regular surface channels that flow in all wetter periods. In intermediate stages, big sinkholes interrupt the profile, but overflow is sufficiently frequent to maintain recognizable surface channels. Advanced stages show more and larger sinkholes with no channel preserved at the surface. Stream sinkholes may become deeply entrenched, creating blind valleys within the paleovalley. All these stages are well displayed in dry valleys of the central and southern Appalachians (White and White, 1983).

Poljes are large features that may perform both input and output functions. An ideal polje is an elongated, flat-floored, closed depression surrounded by limestone hills that are well karstified. The polje floor may be a corrosion bevel across limestone, or its flatness may be achieved by alluvial filling that buries an antecedant topography. The polje floor receives water from springs around its margin. There is channel flow across the floor to one or more karst sinks. The floor may be seasonally flooded (Fig. 2). The floodwaters undercut the marginal slopes by solution, extending the corrosional plain. In a randpolje the plain extends as a corrosional feature onto adjoining nonkarst rocks that furnish water to it; net water input into the karst aquifer predominates in this case. An open polje is open for the seasonal surface discharge of waters via a breach in its margins; there is net discharge of water from the aquifer (Milanonić, 1981).

Poljes occur in many karst karstlands. They are best known and studied in Yugoslavia, where they are of considerable economic importance. Surprisingly few are recognized in the U.S. karsts. Crawford (1984) defines Grassy Cove, Tennessee, as the greatest American polje; its depression measures 13 by 5 km but is of the irregular type, lacking an extensive flat floor. Smaller, truly ideal, solutional poljes occur in the subarctic Mackenzie Mountains of Canada.

From the open polje there is a progression to the most dramatic of all karst terrains, tower karst. This attains its greatest development in plains and plateaus of Guizhou and Guangxi Provinces, China (Zhang, 1980), but is found in Jamaica, Puerto Rico, Cuba, Mexico, and many other localities. The Chinese form is an alluviated corrosion plain with well-developed epikarst draining to a few, higher-order, allogenic rivers that flow across it. Rising above it are isolated limestone towers or clusters of towers and cones. They are undertrimmed by seasonal floods and by streams sinking into them. Individuals range from a few meters to >500 m in height. They may contain many paleocave fragments attesting to earlier epicycles of erosion. The primary requirements for tower-karst development appear to be a great thickness of pure carbonate coupled with vigorous tectonic uplift during late Mesozoic–Tertiary/Quaternary times. Warm, wet climates are also demanded by many authors, but a small tower karst has developed within the Quaternary in the dry, cold Mackenzie

Figure 2. Third Polje, Nahanni karst. This is an ideal example of a dissolutional polje that was created by coalescence of solutional corridors, accompanied by or succeeded by alluviation. The polje is 1.5 km long and is shown before and after 200 mm of rainfall during an 8-day period in July 1972.

Mountains (latitude 61°N), where vigorous updoming of thick carbonates is occurring (Brook and Ford, 1982).

## THE DEVELOPMENT OF SOLUTIONAL CAVE SYSTEMS

### Karst aquifers and initial conditions

Granular, fracture, and conduit permeability can be recognized as three distinct types. In this section the term "fissure" is used rather than "fracture" because flow in bedding planes is very important in most karsts; geologists do not normally consider bedding planes to be fractures.

In general, karst aquifers originate as fissure aquifers and develop into conduit aquifers with the propagation of cave systems through them (Fig. 3). Most caves form near the center of a broad spectrum of genetic conditions. At one extreme, some karst aquifers function as perfect sponges or ideal granular aquifers; this is true of some young reefs, much chalk, some poorly cemented oolites, or highly dolomitized older reefs. At the other extreme are "perfect pipes,' short, high-gradient, single conduits that neither gain nor lose water underground. Caves that serve as shortcuts through the necks of incised river meanders are examples.

There are two important "threshold" dimensions in karst

water circulation. Flow in larger fissures, or in solution conduits enlarged within them, will be slow and laminar until a width or diameter of about 5 to 15 mm is attained. Turbulent flow may then occur, and suspended load can be carried and deposited. This threshold dimension for conduits is determined by hydraulic gradient and temperature.

### Modeling the initiation and propagation of conduits

Understanding the nature of the extension of the earliest solutional openings in fissures, and the nature of their interconnection, is crucial to understanding much of cave system development. The processes cannot be studied directly in the field. Two different approaches that have been taken are computation from equations believed to be appropriate (White, 1984; Palmer, 1984), and hardware modeling with physical analogs also believed to be appropriate (Ewers, 1982).

The enlargement rate of an initial fissure can be expressed by three equations: a mass-balance relationship, and dynamic equations for laminar flow and chemical mass transfer:

$$\text{Mass balance: } \frac{dw}{dt} = \frac{Q}{b} \frac{dC}{\rho dL}$$

Figure 3. Solutionally enlarged joints in the face of a dimension-stone quarry in Mississippian limestone, Indiana.

Hagen-Poiseuille equation for laminar flow: $Q = \dfrac{b\, w^3\, \gamma\, dh}{12\, \mu\, dL}$

Solution kinetics: $\dfrac{dC}{dt} = \dfrac{2k}{w}(1-C/C_s)^n.$

In any consistent system of units, $dw/dt$ is the rate of solutional widening of the fissure, $Q$ is discharge, $dC/dL$ is change in concentration of dissolved bedrock over length of flow path, $w$ is short dimension of fissure cross section (width), $b$ is long dimension of fissure cross section, $\rho$ is bulk density of bedrock, $\gamma$ is the specific weight of water, $\mu$ is the viscovity of water, $dh/dL$ is hydraulic gradient, $dC/dt$ is change in solute concentration with time, $C$ is concentration, and $C_s$ is saturation concentration of dissolved carbonate, $k$ is empirical rate constant, and $n$ is reaction order. The rate of solutional widening depends primarily on discharge through the fissure and change in solute concentration over its length. The discharge is a function of the cube of the effective fissure width weighted in favor of the narrowest part, but only the first power of other variables. This great sensitivity to width is the primary factor responsible for the selective enlargement of openings. The third equation is a generalized rate equation requiring empirical determination of $k$ and $n$. With distance of flow, the concentration increases toward saturation, and the rate of solution decreases. Experimental data indicate that at low $C/C_s$ values the reaction order ($n$) is initially about 1.5 to 2.2, decreasing with $P_{CO_2}$ and temperature, but the reaction order changes abruptly to about 4 or 5 at higher $C/C_s$. The transition occurs at about 60 to 90 percent saturation, increasing with $P_{CO_2}$ and temperature.

The increase in reaction order represents a sharp decrease in solution rate well before the solutions become saturated. As ground water flows through the initial fissures, most of the solutional potential of the water is dissipated within a few centimeters or meters of the inflow points (Bögli, 1980). Solution rates in this upstream zone are fast, as large as 0.1 cm/yr, but the rate drops rapidly with increasing saturation. As a result, unsaturated ground water is able to penetrate deeply into carbonate rocks, continuing to dissolve them at slow but fairly uniform rates. The discharge through an incipient conduit is strongly limited by the narrow downstream parts. Only when these enlarge sufficiently will the flow velocity be great enough for the entire conduit to be exposed to water with low dissolved solids and high solution rate.

Ewers' (1982) results from a purely hydrologic model are illustrated in Figure 4A. Few initial fissures are of uniform width, and many parts are closed below the 0.01-mm threshold. As a consequence, enlargement begins with many proto-conduits extending wherever openings are greatest. The pattern is distributary, and stochastic in detail. Discharge is governed by the resistance of the downstream parts, which have undergone little or no solutional enlargement.

As is seen, distributaries propagating along the largest openings or (where equal) those most nearly in the direction of the discharge boundary, experience the least resistance. They deform the equipotential field (dotted lines) and steepen its gradient. Other distributaries are, therefore, robbed of part of ther former flow. Eventually one route attains the discharge boundary. Resistance drops, discharge increases to provide a stable maximum solution rate, and the threshold conduit diameter (5 to 15 mm) is relatively quickly attained. A cave system now extends continuously between input and discharge boundaries. The equipotential fields of any adjoining, but more primitive, distributary nets are re-oriented toward the master conduit rather than to the discharge boundary.

The time required for the maximum solution rate to be reached can be estimated by finite-difference analysis using the three equations given. The time requires is typically on the order of 10,000 to 100,000 yr. The variation is determined by initial fissure width, hydraulic gradient, conduit length, $P_{CO_2}$, and temperature.

### Building the plan patterns of common cave systems

Most cave systems are developed by meteoric waters circu-

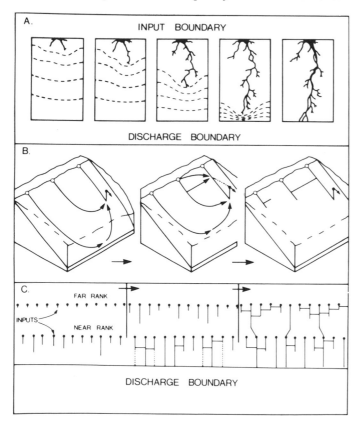

Figure 4. (A) Development of a solutional proto-cave between a single input point and a discharge boundary in a bedding plane. A principal tube makes the connection along a random path, leaving abandoned secondary tubes. (B) Progressive linkage of three separate proto-caves propagating from an input boundary. That on the right establishes a connection with a discharge boundary and the others connect to it. (C) Patterns of growth and connection of principal tubes propagating from two ranks of inputs. The two ranks may be in different bedding planes. (From hardware simulations by Ewers, 1982).

lating in the initial fissure network without major artesian trapping or other distorting effects such as an overlying granular aquifer (Ford and Ewers, 1978).

An interpretation of plans of common cave systems is shown in a highly schematic manner in Figures 4B and C. Here the sinuous and branching form of each connected conduit shown in Figure 4A has been simplified to a single straight line. In Figure 4B, the conduit on the right side of the diagram is taken to be complete. Flow from two adjoining proto-conduits is diverted toward it through favorably located distributaries as the equipotential fields are re-oriented. Eventually these distributaries will link up as conduits. Figure 4C illustrates ranks of inputs near or far from a discharge boundary. Inputs at the near rank will first connect to the boundary in the stochastic manner of Figure 4A because of their greater hydraulic gradient. Once connected, internal linkage (Fig. 4B) proceeds in the near rank; the hydraulic gradient for the far rank is steepened and it proceeds to connect in its turn (Ewers, 1982). The process has obvious affinities to the development of surface river-channel nets. However, the Horton Laws of channel morphometry are rarely obeyed underground because they are overridden by structural and lithologic controls.

### The vertical structure of common cave systems

The nature of cave development in depth below ground was the focus of much controversy among early writers in Europe and North America, leading to neglect of the questions of process, initiation, and plan that are emphasized in this chapter. One school contended that major development was limited to the vadose zone, a second school argued that it occurred at random depth in the phreatic zone, while a third used two distinct lines of argument to propose that development should follow the water table.

The resolution of this problem is given in Figure 5 (Ford and Ewers, 1978). Major caves are found above, below, and along the water table. Most extensive cave systems will contain vadose parts draining to phreatic or water-table parts (Figs. 6 and 7). Systems that have evolved through several developmental stages may contain older phreatic parts and younger water-table passages, in addition to vadose passages.

In many limestones the water table is high above the regional base level before cave development commences because primary and fissure permeability are low (Fig. 5B1). As early caves enlarge, the water table falls until a stable gradient to a spring is achieved (Fig. 5B2). Drawdown vadose caves are developed from the early conduits. They tend to display well-developed phreatic ceilings with vadose entrenchments below them. Invasion vadose caves develop where streams open new routes into rock already drained in earlier stages.

The development of phreatic or water-table caves is chiefly a function of the frequency of penetrable fissures along which conduits may propagate and connect. Low fissure frequency tends to create simple, deeply looping courses (Fig. 5A1). High fissure frequency permits ideally flat, graded, water-table channels (Fig. 5A4). Two distinct intermediate combinations are also

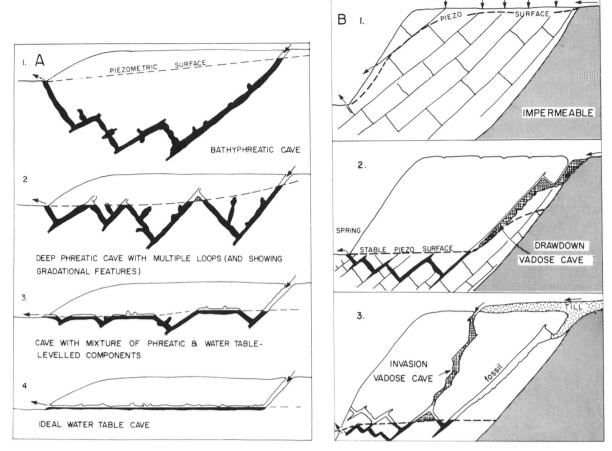

Figure 5. (A) Four types of phreatic and water-table cave systems (from Ford and Ewers, 1978). (B) Distinction between drawdown and invasion types of vadose cave systems.

illustrated. Where strata dip steeply (more than about 5°), deep phreatic caves are more common. Where strata approach the horizontal, perching favors extensive vadose development and only shallow penetration beneath the water table.

### Evolution of cave systems within a single stage

There are changes in the morphology of cave systems that occur within a single stage of development, i.e., they occur without the notable shifts of base level that mark the beginnings of new stages of development. With the enlargement of passages there may be local collapse of roof or walls. Gradational processes tend to flatten the profiles of caves composed of phreatic loops. Their crests may be entrenched. Short loops may become blocked at the base, while flatter bypasses can open across their tops during floods. In long, shallow loops there may be a progressive buildup of insoluble detritus that protects floors from further solution, causing ceilings to dissolve upward above the rising fill.

In place of single bypasses, many vadose and water-table caves contain mazes of floodwater passages along their routes (Palmer, 1975). Some vadose caves can transmit the greatest floods, but more have local points of bedrock construction or debris obstruction behind which there is frequently ponding of floodwaters. Water is forced through many fractures, enlarging them into mazes of diversion conduits. Floodwater mazes are common at the entrances of river caves because the entrance passage is a constriction compared to the flood plain upstream. Flood inundations, accompanied by alluviation, may contribute to the widening of poljes.

### Multi-stage caves

Many caves display the effects of significant downward shifts in their springs, most often caused by entrenchment of outlet valleys. Each time a spring position lowers (usually with an accompanying lateral shift), there is at least a partial reconstruction of the conduit system behind it. New conduits are propagated to tap flow from the preexisting ones. The process proceeds in headward steps and is analogous to the progress of a knickpoint in a surface channel. Some previous passages are abandoned and become relict. They survive above the active passages until consumed by surface processes.

Crossing and re-crossing one another, sequences of active passages below older relict passages compose great multi-stage

Figure 6. An active vadose cave passage in Mammoth Cave, Kentucky. Water flows laterally along the bedding, forming narrow canyon passages; but where it descends vertically along fractures it forms shafts. This stair-step pattern is typical of vadose passages.

Figure 7. Tubular passage forming at or near the water table in the lowest level of Mammoth Cave. Periodic floods fill the entire passage with water. Note banks of silt clay deposited during high flow.

cave systems that are among the most complex of all landforms. Largest and most complex of all is the Mammoth Cave System in Kentucky, with 510 km of interconnected passages mapped in 1986 on at least four distinct levels (Palmer, 1981).

### Some special cases of solutional cave develoment

Approximately 20 percent of accessible solutional caves are created under conditions not considered above. These are formed in at least two ways: by meteoric ground water circulating through inhomogeneous rocks, and by water that acquires its aggressiveness from sources other than surficial carbon dioxide. Such caves lack a close genetic association with surface features and do not simply extend between sink points and springs. Entrances are commonly formed by interception of the caves by surface erosion.

Two-dimensional rectilinear maze caves are formed where highly jointed soluble rock is sandwiched between strata of much lower permeability or where diffuse recharge enters the soluble rock through permeable, but insoluble, rocks such as sandstone (Palmer, 1975). Optimisticeskaja, in the Crimea, is the greatest known maze of this type. It consists of 153 km of integrated passages in gypsum sandwiched between insoluble rocks. Anvil cave, Alabama, is a celebrated U.S. example.

An important, but often overlooked, class of caves is formed by rising thermal waters charged with carbonic acid. Diagnostic features include three-dimensional maze patterns created more or less simultaneously in one stage, rounded cupola-like ceiling pockets believed to be formed by convection currents, and exotic crystalline deposits such as wall coatings of calcite spar. These caves may also display most of the characteristics of normal phreatic caves. We believe the large caves of the Black Hills, South Dakota, to be of thermal origin (Bakalowicz and others, 1987; e.g., 70-km-long Wind Cave in the Mississippian Pahasapa Limestones.

Carbonate rocks are dissolved by carbonic and sulfuric acids during the emplacement of strata-bound sulfide deposits. Many Mississippi Valley–type ore deposits are, thus, syngenetic cave fillings. Basin waters in the process of expulsion and other deep, long-resident ground waters may become acidified by $H_2S$ or iron sulfides. Carlsbad Caverns in the Guadalupe Mountains of New Mexico is believed to have been created by such waters (Hill, 1987).

Mixing of meteoric ground water with brine or salt water can renew aggressiveness, creating or enlarging cavities in the mixing zone. Highly permeable zones in the coastal aquifers of Bermuda, Florida, and the Yucatan are of this type (Back and others, 1984).

# REFERENCES CITED

Back, W., Hanshaw, B. B., and Van Driel, J. N., 1984, Role of groundwater in shaping the eastern coastline of the Yucatan Peninsula, Mexico, *in* LeFleur, R. G., ed., Groundwater as a geomorphic agent: Boston, Allen and Unwin, Inc., p. 281–293.

Bakalowicz, J. M., Ford, D. C., Miller, T. E., Palmer, A. N., and Palmer, M. V., 1987, Thermal gneiss of dissolution caves in the Black Hills, South Dakota: Geological Society of America Bulletin, v. 99, p. 729–738.

Beck, B. F., ed., 1984, Sinkholes; Their geology, engineering, and environmental impact: Rotterdam/Bostom, A. A. Balkema, 429 p.

Beck, B. F., and Wilson, W. L., eds., 1987, Karst hydrogeology; Engineering and environmental applications: Rotterdam, A. A. Balkema, 467 p.

Bögli, A., 1980, Karst hydrology and physical speleology: Berlin, Springer-Verlag, 284 p.

Brook, G. A., and Ford, D. C., 1982, Hydrologic and geologic controls of carbonate water chemistry in the subarctic Nahanni Karst, Canada: Earth Surface Processes and Landforms, v. 7, p. 1–16.

Crawford, N. C., 1984, Karst landform development along the Cumberland Plateau escarpment of Tennessee, *in* LeFleur, R. G., ed., Groundwater as a geomorphic agent: Boston, Allen and Unwin, Inc., p. 294–339.

Drake, J. J., 1983, The effects of geomorphology and seasonality on the chemistry of carbonate groundwaters: Journal of Hydrology, v. 61, p. 223–236.

Ewers, R. O., 1982, Cavern development in the dimensions of length and breadth [Ph.D. thesis]: Hamilton, Ontario, McMaster University, 398 p.

Ford, D. C., 1983, Effects of glaciations upon karst aquifers in Canada: Journal of Hydrology, v. 61, p. 149–158.

Ford, D. C., and Ewers, R. O., 1978, The development of limestone cave systems in length and depth: Canadian Journal of Earth Sciences, v. 15, p. 1783–1798.

Ford, T. D., and Cullingford, C.H.D., eds., 1976, The science of speleology: London, Academic Press, 593 p.

Gams, I., 1974, Kras Slovenska metica: Yugoslavia, Lubljana, 360 p. (in Solovene).

Gvozdetskij, N. A., 1981, Karst: Moscow, Academic Publishing House, 215 p. (in Russian).

Herak, M., and Stringfield, V. T., 1972, Karst; Important karst regions in the northern hemisphere: Holland, Elsevier, 551 p.

Herman, J. S., and White, W. B., 1985, Dissolution kinetics of dolomite; Effects of lithology and fluid flow velocity: Geochimica et Cosmochimica Acta, v. 49, p. 2017–2026.

Hill, C. A., 1987, Geology of Carlsbad Caverns and other caves of the Guadalupe Mountains, New Mexico and Texas: New Mexico Institute of Mining and Technology Bulletin 117, 150 p.

Jakucs, L., 1977, Morphogenetics of karst regions: New York, John Wiley, 284 p.

James, N. P., and Choquette, P. W., 1988: Paleokarst: New York, Springer-Verlag, 416 p.

Jennings, J. N., 1985, Karst: Oxford, B. H. Blackwell, 293 p.

Kemmerly, P. R., 1982, Spatial analysis of a karst depression population; Clues to genesis: Geological Society of America Bulletin, v. 93, p. 1078–1086.

Milanović, P., 1981, Karst hydrogeology: Littleton, Colorado, Water Resources Publications, 434 p.

Palmer, A. N., 1975, The origin of maze caves: National Speleological Society Bulletin, v. 37, p. 56–76.

—— , 1981, A geological guide to Mammoth Cave National Park: Teaneck, New Jersey, Zephyrus Press, 210 p.

—— , 1984, Geomorphic interpretation of karst features, *in* LaFleur, R. G., ed., Groundwater as a geomorphic agent: Boston, Allen and Unwin, Inc., p. 173–209.

Plummer, L. N., and Busenberg, E., 1982, The solubilities of calcite, aragonite, and vaterite in $CO_2$-$H_2O$ solutions between 0 to 90°C and an evaluation of the aqueous model for the system $CaCO_3$-$CO_2$-$H_2O$: Geochimica et Cosmochimica Acta, v. 46, p. 1011–1040.

Plummer, L. N., Wigley, T.M.L., and Parkhurst, D. L., 1978, The kinetics of calcite dissolution in $CO_2$ water systems at 5 to 600° and 0.0 to 1.0 atm $CO_2$: American Journal of Science, v. 278, p. 179–216.

Quinlan, J. F., 1978, Types of karst, with emphasis on cover beds and their classification and development [Ph.D. thesis]: Austin, University of Texas, 325 p.

Smith, D. I., and Atkinson, T. C., 1976, Process, landforms, and climate in limestone regions, *in* Derbyshire, E., ed., Geomorphology and climate: New York, John Wiley, p. 367–409.

Sweeting, M. M., 1972, Karst landforms: London, Macmillan, 362 p.

White, E. L., and White, W. B., 1983, Karst landforms and drainage basin evolution in the Obey River Basin, north-central Tennessee, U.S.A.: Journal of Hydrology, v. 61, p. 69–82.

White, W. B., 1984, Rate processes; Chemical kinetics and karst landform development, *in* LaFleur, R. G., ed., Groundwater as a geomorphic agent: Boston, Allen and Unwin, Inc., p. 227–248.

—— , 1988, Geomorphology and hydrolog of karst terrains: New York, Oxford, 464 p.

Williams, P. W., 1983, The role of the subcutaneous zone in karst hydrology: Journal of Hydrology, v. 61, p. 45–68.

Zhang, Z., 1980, Karst types of China: Geological Journal, v. 4, p. 541–570.

MANUSCRIPT ACCEPTED BY THE SOCIETY MAY 26, 1988

The Geology of North America
Vol. O-2, Hydrogeology
The Geological Society of America, 1988

# Chapter 44

# *Ground water and clastic diagenesis*

**F. W. Schwartz and F. J. Longstaffe**
*Department of Geology, University of Alberta, Edmonton, Alberta T6G 2E1, Canada*

## INTRODUCTION

The term diagenesis refers to changes that occur in a sediment following deposition. These changes are the outcome of a diverse and complex array of physical, chemical, and biological processes. The particular emphasis in this chapter is to examine relationships between the diagenesis of clastic rocks or sediments and mass transport. The possibilities for mechanically deforming clay minerals, chemically altering framework grains, or forming authigenic minerals in pores can produce a bewildering array of diagenetic features. Explaining why these processes occur as they do and what key factors control each process has constituted a major thrust of research in diagenesis, and is our main focus here.

From the perspective of a ground-water volume such as this, one important advance in the study of diagenesis has come from the realization that diagenetic processes can be related to hydrodynamic as well as chemical phenomena. It has become increasingly obvious that in many different settings, considerable quantities of mass must be transported in a flow system to account for observed diagenetic features. For this reason, models of flow in geologic basins are now an integral part of many diagenetic models. Similarly, relationships have been established between local-scale transport in much smaller flow systems and the patterns of diagenesis.

## FUNDAMENTAL PROCESSES—A FRAMEWORK OF CONCEPTS

Diagenesis describes a set of physical, chemical, and biological processes that produce changes in a rock or sediment. The processes of advection, diffusion, and dispersion are responsible for the physical transport of mass from one point to another in the subsurface. The chemical and biological processes include a diverse array of phenomena that constitute the driving force for diagenetic change (Berner, 1980). These processes operate to partition mass among the fluid, mineral, and organic phases and are governed by sets of transport parameters, which themselves may be dependent upon pressure and temperature.

### *Advection*

Advection is a process that moves mass due to the flow of the pore fluid in which it is dissolved. The direction and rate of advective transport are assumed usually to be the same as the fluid velocity. Thus advection is linked closely to the patterns of fluid flow, with flow lines tracing the pathways for mass movement. Advection also refers to solid-phase transport that occurs during compaction in early diagenesis. Although the influence of this process is limited spatially, it can lead to a reduction in pore size, which is one of the important elements of diagenesis.

Patterns of fluid flow are determined by gradients in hydraulic potential, distributions of permeability, and to a lesser extent by fluid density and viscosity. Potential gradients provide the driving forces for flow. They are generated mainly due to (1) gravity, (2) the compaction (or dilation) of a pile of sediments, (3) free convection, and (4) physico-chemical processes, including (a) phase changes such as gypsum to anhydrite, or smectite to illite; (b) the solution or precipitation of minerals; (c) fluid expansion caused by temperature changes; (d) osmotic or membrane phenomena; and (e) metamorphic reactions (Bredehoeft and Hanshaw, 1968).

Ground water moves in a gravitational field because of potential gradients created when recharge is added to a basin in topographically high areas. These gradients result in the flow of ground water from regions of high potential to regions of lower potential. The actual direction of ground-water flow is determined not only by the potential gradient but also by the hydraulic conductivity (permeability) distribution.

Ground-water flow also occurs in response to the compaction of saturated clay-rich sediments. The theory of consolidation describes how the total stress or load borne by a volume of sediment is distributed between the solid matrix and the pore fluid (Domenico and Palciauskas, this volume). In the absence of drainage, incremental loading due to continuing sedimentation is mainly accounted for by a direct increase in pore pressure. This increase in pore pressure provides the potential for fluid motion. It is only when drainage occurs and pore pressures dissipate that a sediment consolidates.

The rate of drainage compared to the rate of sedimentation determines whether water pressures in a sediment will be normal

Schwartz, F. W., and Longstaffe, F. J., 1988, Ground water and clastic diagenesis, *in* Back, W., Rosenshein, J. S., and Seaber, P. R., eds., Hydrogeology: Boulder, Colorado, Geological Society of America, The Geology of North America, v. O-2.

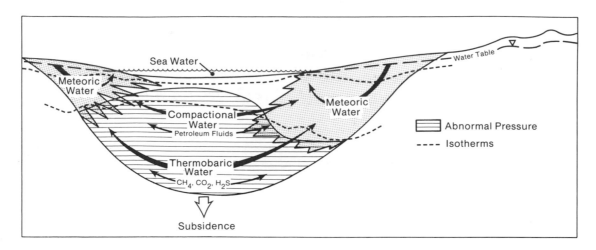

Figure 1. Conceptual diagram illustrating various hydrologic regimes in a large, active sedimentary basin (after Galloway, 1984).

at a given depth or overpressured. Zones of overpressuring such as those observed in sediments of the Gulf of Mexico (Bonham, 1980; Galloway, 1984) provide the hydraulic potential for fluid flow up and out of the deepest parts of the basin.

Fluid flow by convection is another possible mechanism promoting advective mass transfer in some geological basins (Wood and Hewett, 1982; Bodner and others, 1985). Convection cells can develop in thick sandstones or possibly even fractured shale units under normal geothermal gradients. Free convection tends to occur more frequently with increased depth and unit thickness. One of the attractive features of this kind of process is its ability to redistribute relatively large quantities of mass in relatively short times. Convective transport is one of only a few plausible mechanisms to explain the extent of diagenetic changes in some sediments of the Gulf Coast (Blanchard and Sharp, 1985).

Physico-chemical processes become important deep in sedimentary basins. The mineral phase changes can take water from the mineral phase and transfer it to the fluid phase. Potential for fluid flow is generated because the addition of even small quantities of water to a pore results in an increase in pressure. Solution or precipitation of mineral phases, and fluid expansion caused by heating, have much the same effect by altering the volume relationships between a fluid and a pore.

Osmotic or membrane effects contribute to flow in a very different way. Highly compacted sediments, rich in clay minerals, can take on the properties of a semipermeable membrane. When water of differing salinity is in contact across a membrane, the difference in chemical potential causes a flow of water from the fresh-water side through the membrane. Although the ability for some sediments to behave as membranes has been demonstrated in the laboratory, there are no unequivocal examples from the field.

Ground-water flow systems exist at a hierarchy of scales ranging from deep basinal systems, which are the site for burial diagenesis, to localized systems such as in island or sabkha settings, which are often the sites for the early diagenesis of carbonate rocks. The following examples illustrate how these flow systems occur and the generalized patterns of flow that may result.

***Depositionally active sedimentary basins.*** The Gulf of Mexico provides an example of a hydrodynamic system that develops mainly due to sediment compaction, and to a lesser extent, gravity-driven flow and physico-chemical processes (Galloway, 1984). In those parts of the basin that are above sea level, gravity-driven flow dominates. However, deeper portions of the onshore sedimentary sequence in Texas and Louisiana are geopressured. In the shallow offshore (Fig. 1), flow occurs as a consequence of compaction. This 'compaction water' (Galloway, 1984) generally flows upward or toward the basin margins, depending mainly on the permeability distribution, and excess pressure can develop in this environment when free drainage is interrupted. In the abyssal parts of a basin where temperatures and pressures are elevated (thermobaric regime), physicochemical processes operate in addition to compaction to create driving forces for flow.

Galloway (1984) applies this model specifically to the Frio Formation and its updip equivalent the Catahoula Formation. These units are mainly sandstones and mudstones deposited in a major offlap depositional episode. As a consequence of this depositional environment and extensive growth faulting, the thickness of the Frio Formation increases dramatically, especially where it offlaps the older continental platform. Continuing sedimentation in the Gulf of Mexico has resulted in the burial of the Frio Formation under several thousand meters of younger sediments. Landward, however, the Catahoula Formation crops out.

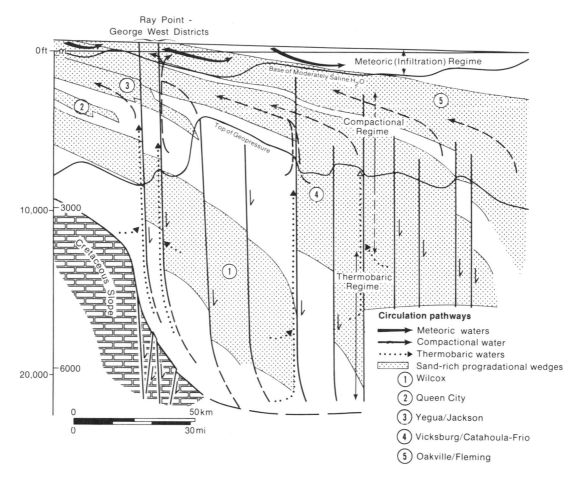

Figure 2. Various hydrologic regimes in the Tertiary fill of the Texas Coastal Plain. Flow is mainly upward along growth faults. There are small meteoric flow systems developed near the surface (after Galloway, 1984).

The relative size and distribution of the meteoric, compactional and thermobaric regimes (Fig. 2) are shown for not only the Catahoula and Frio Formations but also for a deeper section of the Tertiary fill of the Texas Coastal Plain. As Figure 2 illustrates, faults provide pathways for the upward flow of water from the thermobaric zone, which is geopressured, into shallow zones where the compaction water is able to drain normally. Within this latter zone, good lateral continuity in permeability promotes the updip migration of fluids. The meteoric regime (Fig. 2) is shallow, and Galloway (1984) suggests that the depth of this zone is limited by faulting or by the sand-poor character of the updip facies.

In a depositionally active basin, compaction not only contributes to moving water but is itself one of the important diagenetic processes. In muds, the amount of compaction is directly related to loading during burial or tectonism. Initial porosity, which is generally in the range 65 to 85 percent, is reduced to 30 to 40 percent (depending on the clay mineralogy) as the clay particles are squeezed together and free water is expelled. Further loading (and heating) results in removal of adsorbed water and interlayer water from the clay minerals.

Compaction in sands also occurs because of loading, with additional factors being important, such as the mineralogical composition of the framework grains (e.g., quartz, feldspar, rock fragments), grain shape and sorting, and clay content (Blatt, 1982). Well-sorted, quartz-rich sands can have porosities of 40 to 50 percent at deposition. As the overburden load increases, competent grains (e.g., quartz) become more tightly packed, and porosity is reduced to about 30 percent. In less well-sorted sands, smaller grains fill the pore spaces between the larger grains, reducing porosity even further.

The composition of framework grains exerts an important control on the extent of compaction. For example, sediments rich in rock fragments and clays (e.g., graywacke) undergo much more compaction than quartz-rich sands (e.g., quartz arenite) because ductile grains such as mudstone rock fragments become deformed about more resistant grains, and are squeezed into remaining pore spaces (Fig. 3) or are interpenetrated by less ductile, angular grains. Likewise, in sediments rich in detrital feldspar (e.g., arkose), fracturing of these grains can produce greater compaction than observed in very quartz-rich sedimentary materials.

***Depositionally inactive basins.*** Once sedimentation stops and structural uplift produces a topography, the character of flow gradually changes. Gravity-driven flow systems become more important as flow due to compaction ceases and overpressures dissipate. If the basin is particularly deep, the possibility continues for flow due to physico-chemical processes.

Many basins have evolved to the point where gravity-flow predominates. Tóth (1980) describes flow systems of this kind in the Aquitaine region of France, the Great Lowland Artesian Basin of Hungary, the Artesian Basins of the Central Plateau in Iran, and portions of the Western Canada Sedimentary Basin. Other large, gravity-driven flow systems have been described in the East Texas Basin (Fogg and Kreitler, 1982), the carbonate aquifers of Florida (Back and others, 1966; Back and Hanshaw, 1970), the Illinois Basin (Bredehoeft and others, 1963), and the Dakota Sandstone in South Dakota (Bredehoeft and others, 1983).

Hitchon's (1969a, 1969b) study of fluid flow in the Western Canada Sedimentary Basin provides a detailed evaluation of a gravity-driven flow system. Along its western margin, the basin has a maximum depth of approximately 5,000 m and contains a thick sequence of middle and late Paleozoic carbonates and evaporites overlain by clastic rocks of Cretaceous and Tertiary age. The depth of the basin decreases eastward until, along the boundary of the basin in Saskatchewan, rocks of Precambrian age are exposed.

This basin has been explored and exploited for oil and gas for many years. One legacy of this development is a very extensive collection of geophysical and drill-stem test data, which has been used through the years to develop an understanding of fluid flow. Potentiometric data, presented as a series of cross sections and slice maps (based on elevation; Hitchon, 1969a), document the presence of large regional flow systems. Recharge areas coincide with major upland areas, and discharge areas coincide with major lowland areas. The influences of topography are clearly reflected in the potential distributions and provide strong evidence that flow in the basin is gravity driven.

In another paper (Hitchon, 1969b), potentiometric data are presented for individual stratigraphic units, as opposed to slices, in order to evaluate the effects of geology on flow. By reworking the data in this way, it was possible to define in more detail how permeable carbonate rocks of Paleozoic age provide a subsurface drain that channels much of the deeper fluid flow northeastward toward the Athabasca Oil Sands, one of the largest heavy-oil accumulations in the world.

Tóth (1978) presents the results of a detailed investigation of the Red Earth region, which is a small area within the Western Canada Sedimentary Basin. On a much smaller scale, Tóth observed many of the same features of flow observed by Hitchon. A

Figure 3. Compaction of micaceous fragment (M), Upper Cretaceous basal Belly River sandstone, Alberta. Cross-polarized light. Scale bar = 50 $\mu$m.

relationship between topography and fluid flow is shown on an east-west cross section through the study area (Fig. 4). The uplands such as the Buffalo Head Hills and Trout Mountain are areas of recharge, and the lowlands along Peace, Loon, and Wabasca rivers are areas of discharge. Potentiometric maps prepared for the major hydrogeologic units show that the effect of the land surface on the fluid potential is observable down to the top of the Precambrian (Fig. 4). Units D-III and K-I form a high permeability drain, which acts to collect both upward and downward flowing water and channel it to the surface north of the line of section.

Another feature of the cross section (see legends, Fig. 4) is the age of the various flow systems. Tóth demonstrates convincingly that the good correlation between the topography and the potential distributions in the deep carbonate units D-I and D-II should not exist given the low potentials along the permeable drain (D-III and K-I). Apparently, flow in the deeper units has not adjusted yet to the present-day topographic setting, and the distributions of potential are reflecting conditions in the Cretaceous and Tertiary when the drain did not exist. In general terms, changes in the topography of a basin with time may not be reflected in the flow patterns until millions of years have passed. Such a result implies that great care must be exercised in interpreting potentiometric distributions in basins that contain appreciable quantities of low-permeability rocks.

***Small-scale flow systems.*** Not all flow systems that contribute to diagenesis operate on the scale of formations or sedimentary basins. Examples of small-scale flow systems could include the convection cells we discussed previously or systems developed in sabkha settings. These latter flow systems (the so-called reflux models of Land [1983] for shallow environments)

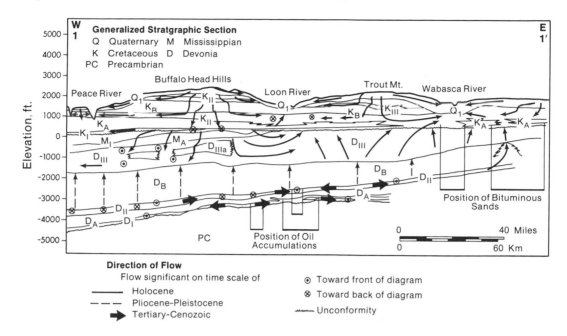

Figure 4. Hydrogeologic cross section 1–1′, Red Earth region, northern Alberta, Canada (after Tóth, 1980).

cycle water back to seas from nearby land areas. They are important because they can transport sufficient quantities of $Mg^{2+}$ to dolomitize carbonate rocks. One variant of this model (summarized by Davies, 1979) starts with seawater trapped in shallow depressions separated from a sea. As a consequence of evaporation, this water becomes saturated with respect to gypsum and sinks downward into underlying carbonate strata, displacing less dense pore fluids.

Land (1983) suggested that, instead of density-driven fluids, a more likely mechanism for moving water is gravity-driven flow between bodies of water of differing elevation that could occur in a sabkha environment. His model proposed that storms drive water up onto a sabkha where it evaporates and returns to the sea in the subsurface, driven by differences in elevation and to a lesser extent by an increase in density. These and other mechanisms, such as evaporative pumping, illustrate that concepts of advective transport that can be difficult to fully describe in sedimentary basins can even be quite complicated in near-surface environments, where the systems are relatively small.

***Limitations in the application of transport concepts.*** The previous examples have demonstrated how a knowledge of the origin of driving forces in a basin provides a general indication of how ground water and dissolved mass should move. For many basins, such generalities may represent all that can really be determined about the dynamics of fluid flow. Except for such areas as the Gulf of Mexico, the flow systems giving rise to diagenesis have long since disappeared, leaving few clues as to their character.

### *Dispersion*

Dispersion is the second important mechanism for physical transport. It refers to processes that spread mass in a region beyond where it could be expected due to advection alone. The effect of dispersion is to mix fluids, displacing one another in a ground-water flow system. Fluid mixing is an important diagenetic mechanism in both clastic and carbonate rocks.

Dispersion is produced by two distinctly different phenomena: molecular diffusion and advection (Bear, 1972; Schwartz, 1985). Diffusion is the flux of ions that result when concentration gradients exist, whether the fluid is moving or not. The process is described in one dimension by Fick's 1st Law:

$$J = -D\,dc/dx \qquad (1)$$

where $J$ = diffusion flux vector; $D$ = diffusion coefficient; $c$ = concentration (mass per unit volume); and $x$ = coordinate direction. In porous media, it is necessary to define an effective dispersion coefficient to account for the diminished effectiveness of free diffusion owing to the tortuosity of the diffusion pathway. A variety of different functional relationships have been established to relate the effective diffusion coefficient to the diffusion coefficient in aqueous solution. All of these relationships involve multiplying the diffusion coefficient by one or more constants with an overall value less than one. For example, Thibodeaux (1979) defines the effective diffusion coefficient ($D'$)

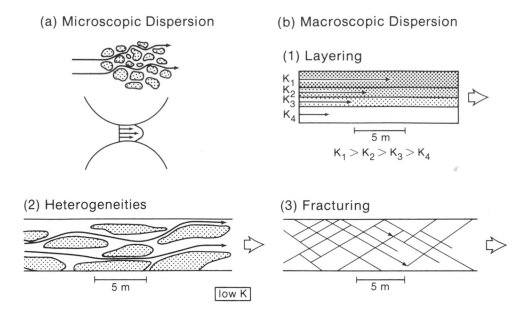

Figure 5. The origin of (a) pore-scale, and (b1 to b3) formation-scale variability in velocity giving rise to dispersion (after Schwartz, 1985).

$$D' = De/\tau \tag{2}$$

where $e$ = porosity of the sediment and $\tau$ = tortuosity factor. Values of the tortuosity factor are usually obtained experimentally and range between one and six (Thibodeaux, 1979).

Dispersion caused by advective mixing is referred to as mechanical dispersion. It results when heterogeneity of the medium at a variety of scales creates variability in the direction and velocity of fluid flow. At the microscopic scale, variability in velocity within or among pores (Fig. 5a) is generated because of frictional drag next to pore walls or subtle changes in the geometry of the pore network. Macroscopic dispersion develops primarily as a consequence of local variability in hydraulic conductivity produced, for example, by layering within a formation or a variable fracture geometry (Fig. 5b). At the megascopic scale, extremely large-scale mixing can result from regional variability in the stratigraphy.

The extent of mechanical dispersion is closely related to the scale of a system. For example, if the most significant dispersion is being generated at the megascopic scale, macro- and micro-scale effects, including diffusion, may be neglected. However, as Schwartz (1985) points out, certain cases exist where small-scale mixing caused by diffusion must be considered. The diffusion of mass from fractures to the porous, nonfractured blocks, for example, can control transport on a macroscopic scale to an important extent.

Dispersion becomes evident when formation waters, which originate in different ways in different parts of a basin and have different compositions, begin to interact. For example, meteoric water recharging a basin typically has a low content of dissolved solids relative to connate water or a highly evolved formation water found in the deepest parts of a basin. A following example illustrates dispersion created through meteoric flushing of a sedimentary basin. For another example of dispersion involving the mixing of fresh water and ocean water in a coastal aquifer, see Back (this volume).

*Mixing of meteoric and formation waters.* Once flow due to the compaction of sediments ceases, and topographic gradients develop, gravity-flow systems expand and begin to flush formation waters out of the basin. The opportunity exists in these cases for a zone of mixing to develop between this meteoric water and the remaining formation water. A good example of this mixing is provided by studies of the Milk River aquifer system located in the southern part of the Western Canada Sedimentary Basin (Schwartz and Muehlenbachs, 1979; Schwartz and others, 1981). This Cretaceous aquifer consists mainly of silty sandstone in its southern part. Northward the aquifer becomes finer grained and disappears in a facies change to shale. The Milk River aquifer is one of several sandstone units developed within a Tertiary-Cretaceous section comprising mainly shale and mudstone. Structurally, rocks in the area have been gently deformed, with beds dipping in a fan-shaped pattern from the Alberta-Montana border northward into the basin. The aquifer crops out in the southern part of the study area. In the northern part of the area, it is overlain by up to approximately 400 m of younger rocks.

The Milk River system is a good example of a classic artesian aquifer. Most of the recharge occurs in the south where the aquifer crops out or is only thinly covered. Flow is generally northward with a significant component of upward leakage through confining beds.

$Na^+$ and $Cl^-$ concentrations are characterized by marked spatial variability. Concentrations are lowest in the south where most of the recharge occurs (Figs. 6a, b). One important feature of the distribution is an obvious northward salient of waters with a lower concentration. Examination of the hydraulic conductivity data for this unit shows that this chemical feature coincides with a well-developed zone of permeability in the aquifer.

The oxygen-18 and deuterium content of water from the aquifer also exhibits marked spatial variability along a south to north trend (Figs. 7a, b). Values range from $-20.5^0/_{00}$ to $-8.69^0/_{00}$ in $\delta^{18}O$ and from $-158.2^0/_{00}$ to $-87.9^0/_{00}$ in $\delta D$. The northward salient, which characterizes the distribution of major ions, is also apparent in the isotopic patterns.

These chemical patterns are interpreted as a broad zone of mixing that forms as meteoric water displaces preexisting formation water. Mixing occurs as mass diffuses from more saline formation water contained in lower permeability units above and below the Milk River aquifer into the less saline water in the aquifer. The isotopic and chemical composition of the end members is characterized by meteoric water recharging in the south and water similar to that in the downdip portion of the aquifer. This pattern of mixing explains why the apparent zone of dispersion is so much larger than might be expected from a simple displacement process. Within the Milk River aquifer, the chemical patterns (Figs. 6, 7) illustrate how localized zones of higher permeability control smaller scale patterns of mixing.

One additional question to be considered is why flushing by meteoric water is not complete in a flow system that has probably operated for several tens of millions of years. As a consequence of a decreasing hydraulic conductivity down the flow system, much of the water recharging the aquifer flows upward through the confining units over a relatively short distance. Mass entering the system by diffusion is effectively lost by advection. As a result, the water chemistry does not change appreciably, producing a quasi-steady-state system with respect to ion concentrations.

### *Chemical and biological processes*

Diagenetic changes occur mainly as a result of a complex set of chemical and (or) biological processes. Reactions describing these processes can be characterized from either an equilibrium or kinetic perspective. For example, an equilibrium description is most appropriate when the reaction rate is fast in relation to other physical transport processes such as advection or dispersion, which also change concentrations (Rubin, 1983). A kinetic description is most appropriate when the reaction is slow in relation to other processes. The effect of an equilibrium description is to produce a system that responds instantly to perturbations in the chemical system.

A compilation of estimates for reaction half times for many of the important reactions listed in Table 1 led Langmuir and

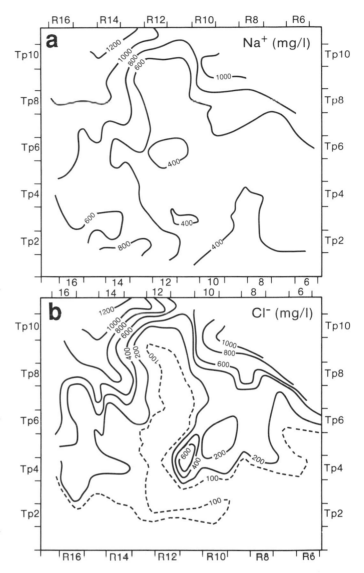

Figure 6. Map showing the concentration of (a) $Na^+$ and (b) $Cl^-$ in waters of the Milk River aquifer (after Schwartz and Muehlenbachs, 1979).

Mahoney (1984) to conclude that physical mass transport in a geologic basin is usually much slower than the rates of the important chemical reactions. Thus, it is reasonable to expect that equilibrium-based approaches will correctly describe most diagenetic reactions (Langmuir and Mahoney, 1984), provided that an adequate source of reactants is available.

The rigorous theoretical framework that equilibrium concepts provide in understanding chemical reactions is of particular importance. Equilibrium expressions can be written for reactions among solids, gases, and solutes or combinations of these.

A general reaction with constituents C and D reacting to products Y and Z can be written:

$$cC + dD = yY + zZ \qquad (3)$$

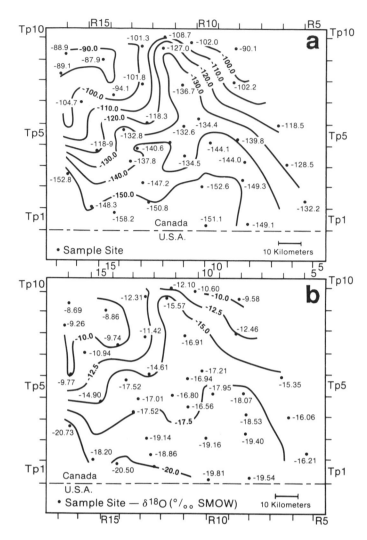

Figure 7. Areal variation in (a) $\delta D$ ($^o/_{oo}$) and (b) $\delta^{18}O$ ($^o/_{oo}$) in waters of the Milk River aquifer (after Schwartz and Muehlenbachs, 1979).

TABLE 1. IMPORTANT TYPES OF REACTIONS, WITH EXAMPLES

1. Complexation

$$Mg^{2+} + CO_3^{2-} = MgCO_3^0$$
$$Cu^{2+} + 2CO_3^{2-} = Cu(CO_3)_2^{2-}$$

2. Acid-Base

$$HCO_3^- = CO_3^{2-} + H^+$$
$$HSiO_3^- = SiO_3^{2-} + H^+$$

3. Dissolution - Precipitation

$$CaCO_{3(s)} = Ca^{2+} + CO_3^{2-}$$
$$FeS_{(S)} = Fe^{2+} + S^{2-}$$

4. Oxidation - Reduction

$$SO_4^{2-} + 2CH_2O + 2H^+ = H_2S + 2H_2O + 2CO_2$$
$$Fe^{2+} + \tfrac{1}{4}O_2(g) + \tfrac{5}{2}H_2O = Fe(OH)_{3(s)} + 2H^+$$

5. Ion Exchange

$$Ca^{2+} + Mg\text{-clay} = Mg^{2+} + Ca\text{-clay}$$
$$Ca^{2+} + 2Na\text{-clay} = 2Na^+ + Ca\text{-clay}$$

6. Mineral Crystallization

$$SiO_2(amorph) = SiO_2 (quartz)$$

7. Gas Solution - Exsolution

$$CO_2(g) = CO_2(aq)$$

where c, d, y, and z are the numbers of moles of the constituents. When the system is at equilibrium in a dilute solution, the relationship between the products and reactants is expressed by the Law of Mass Action as:

$$K = a_Y{}^y a_Z{}^z / a_C{}^c a_D{}^d \qquad (4)$$

where K = equilibrium or dissociation constant and a = activities or effective concentration. There are several commonly used conventions for expressing the Law of Mass Action. In the case of dissolved ion species, activity is related to molal concentration by the following equation:

$$a_i = \gamma_i m_i \qquad (5)$$

where $\gamma_i$ is the activity coefficient, which normally takes on values between zero and one. For very dilute solutions, $\gamma_i$ approaches one. The value usually gets smaller as the system becomes more concentrated, reflecting the effects of electrical interference caused by an increasing population of ions. However, it can increase and become larger than one in solutions with extremely high salt concentrations. Pure solids and liquids such as water are, as a first approximation, considered to have an activity of one. The activities of gases in equilibrium with an aqueous phase are usually taken to be equal to the gas partial pressure in bars.

A variety of theoretical models is available to estimate $\gamma_i$, given information concerning the concentration of various ions in solution. The Debye-Hückel equation, the extended Debye-Hückel equation, and the Davies equations are used commonly in equilibrium calculations. In studies of more concentrated brines, a more sophisticated approach that can accurately predict the large increase in solution activities at high ionic strengths is required. Ion interaction models based on the Pitzer equations (Pitzer and Kim, 1974; Harvie and others, 1984) provide accurate estimates of activity coefficients in solutions up to 20 molal.

Values of the equilibrium constants for many diagenetic reactions can be found in the literature and the data bases of well-known computer codes such as SOLMNEQ (Kharaka and Barnes, 1973), WATEQ (Truesdell and Jones, 1974), or EQ3-EQ6 (Wolery, 1979). Values are usually calculated from free-energy data for the reaction or determined directly from solubility data collected in laboratory and field experiments. For example, expressions like the van't Hoff equation make it possible to determine how equilibrium constants vary as a function of temperature. Not all diagenetic reactions are well characterized in terms of equilibrium constants. Because of the wide variability in the mineralogic composition of clay minerals in particular, progress has been slow in establishing their equilibrium constants.

### Types of reaction

Table 1 summarizes and presents examples of the important types of reactions that contribute to diagenetic change. The complexation reactions (Table 1) occur in the aqueous phase and in some cases can be very important in determining the total mass flux of particular ion species. Stumm and Morgan (1970) define a complex as any combination of cations with molecules or anions containing free pairs of electrons. In natural water systems, mass exists both as simple ions and complexes, which might involve ion pairs (e.g., $MgHCO_3^+$), inorganic complexes of trace elements (e.g., $NiCl^+$), and organic complexes (e.g., copper-glycine). The effects of complexation become more important as the water becomes more saline. In seawater, for example, approximately 11 percent of the total content of magnesium is found in a magnesium-sulfate complex (Garrels and Thompson, 1962). In basinal brines of higher salinity, even more dissolved mass will occur in the form of complexes.

The reason why it is often necessary to characterize the extent of complexation relates to the Law of Mass Action. Ions react according to their free or uncomplexed concentration. Thus, if a considerable proportion of the total concentration is actually distributed among complexes, the net flux of mass through a system might be higher than solubility data alone might indicate. An example of this effect is presented later, and shows how complexation can increase the apparent solubility of aluminum in solution.

Another general group of chemical reactions depicted in Table 1 is the acid-base reactions. These are kinetically very fast and involve the production or consumption of protons ($H^+$). One example of an acid-base reaction is the dissociation of water into hydrogen and hydroxyl ions. Others result from the addition of $CO_2$ gas to water and its speciation among $H_2CO_3$, $HCO_3^-$, and $CO_3^{2-}$. The study of acid-base systems leads naturally to the concept of alkalinity, which describes the relationship between weak acids, such as those of the carbonate system, and strong

bases contributed from the dissolution of carbonate and silicate minerals. These three different kinds of interactions between acids and bases exert a tremendous influence on the chemistry of pore fluids and the nature of diagenesis in rocks.

The dissolution-precipitation reactions are the most important reactions represented in Table 1 because they produce nearly all of the changes to a rock classically considered as diagenesis. They are usually described using equilibrium concepts. Whether a mineral phase dissolves, precipitates, or remains stable depends upon the saturation state of the pore fluid with respect to the mineral. The actual procedure for determining whether a solution is supersaturated, undersaturated, or at equilibrium with a mineral involves comparing ion activity products with theoretical equilibrium constants for the various reactions. These calculations become quite difficult for reactions involving (1) the $CO_2$–water equilibrium, which includes several different reactions, (2) brines, for which it may be difficult to estimate activity coefficients, and (3) complex, multicomponent systems that may be characterized by a large number of free ions, and various complexes. Nordstrom and others (1979) review some common computer codes that have been developed to model ion speciation and the equilibrium status of the solution with respect to a large number of minerals.

A good example in the use of equilibrium concepts to understand mineral reactions is provided by a study by Franks and Forester (1984). This work involved the use of saturation calculations to explain enhanced permeability and secondary porosity in parts of the Wilcox Formation. Calculations of saturation from chemical data for 356 samples indicate a general increase in the solubility index (ion activity product/equilibrium constant) moving toward the Gulf, which coincides with increasing temperature and depth of burial in the Wilcox Formation. A significant variation from this trend is apparent in two areas, where a well-defined zone of undersaturation exists. This zone coincides with a zone of abnormally high permeability and enhanced secondary porosity. Undersaturation develops because of high partial pressures of $CO_2$, especially at temperatures above 100 °C. The $CO_2$ is probably generated as a by-product of the thermal alteration of organic matter in the sediments or upward-moving petroleum and gas.

Oxidation-reduction reactions are distinguished by the transfer of electrons and for this reason involve elements with more than one oxidation state, such as carbon, nitrogen, sulfur, and some metals (e.g., iron and manganese). There is a direct analogy between oxidation-reduction reactions and the acid-base reactions discussed earlier. Instead of a transfer of protons from an acid to a base, there is a transfer of electrons from a reductant (electron donor) to an oxidant (electron acceptor). Unlike nearly all of the other reactions, redox reactions are unique because they are mediated by microorganisms, which use the reaction as a source of energy. The most important oxidation-reduction reactions operate between the liquid and solid phase and result in the precipitation or dissolution of minerals. However, reactions of this kind do not contribute significantly to the flux of elements in geochemical cycles, but are critical in providing the energy necessary to sustain life.

Important oxidation-reduction reactions occur in the fine-grained marine sediments found along continental margins. They involve the depletion and destruction of solid organic compounds, the reduction of $Fe_2O_3$ to $Fe^{2+}$, and the reduction of $SO_4^{2-}$ to $H_2S$ (Berner, 1980). Some of the $H_2S$ gas that forms in this environment may diffuse upward to the sediment-water interface, while the remainder can react with iron to form pyrite. Most of these reactions occur in the uppermost layers of sediment and are examples of early diagenesis.

The ion-exchange reactions (Table 1) describe a complex set of phenomena responsible for exchanging ions in solution with those adsorbed on a solid phase. The most significant sorption occurs with the common groups of clay minerals, including kaolinite, vermiculite, smectite, chlorite, and mica. Metal oxides, principally those of silicon, aluminum, iron, and manganese, and organic matter will also sorb to a variable extent.

It has not clearly been demonstrated that sorption reactions are important to diagenesis. A link should be developed in time demonstrating the ability of sorption reactions to control the cation chemistry of the pore fluid in some circumstances. These reactions may turn out to be important in the zone of bioturbation. Berner (1980) suggests that mass adsorbed on solid particles can be transported very efficiently during bioturbation as solids are moved around.

It is beyond the scope of our review to look in detail at the various equilibrium and nonequilibrium models available to describe ion exchange. The interested reader can refer to a vast literature on the subject, including Sposito (1984) and Langmuir and Mahoney (1984).

Two other types of chemical reactions can contribute to diagenetic change. Mineral crystallization reactions involve the reorganization of amorphous masses into mineral species. Gas solution-exsolution reactions describe the partitioning of gases such as $CO_2$ between gas and liquid phases. Other reactions such as those involving isotopic changes and radioactive decay, do not really contribute to diagenesis but can be used to provide important clues for explaining diagenetic processes.

## KEY DIAGENETIC MINERALS AND REACTIONS

The conceptual framework characterizing the most important diagenetic processes provides a basis for describing real systems. However, going beyond a simple physical description to interpreting observed diagenetic changes is a problem because of the inherent complexity of natural systems. The following section focuses on real systems. In it we examine the most important diagenetic minerals, describing how they occur in clastic sediments, the important reactions involved, and major issues with respect to their origin.

A wide variety of mineral cements can be found in sandstones and conglomerates, including quartz, carbonates, clay minerals, and various oxides, sulfides, and sulfates. Which partic-

ular mineral is favored depends upon the conditions in the system (e.g., temperature, fluid composition, Eh, pH, composition of detrital grains) at that particular stage of diagenesis. Commonly, several diagenetic minerals occur within the pore space. However, careful petrographic observations can be used to determine the relative time of formation of each cement. From such data, changes in the physico-chemical conditions in the sedimentary rock during diagenesis can be deduced.

### Carbonate cements

Minerals such as calcite, dolomite, siderite, and ankerite are among the most common carbonate cements; many sandstones and conglomerates contain more than one of these phases. In many situations, calcite has crystallized as an early phase, cementing grains that have not been greatly compacted (Fig. 8a). Replacement of plagioclase feldspar by calcite is another common mode of occurrence (Fig. 8b). Figure 8c illustrates a paragenetic sequence in which siderite has partly replaced chert grains, and ankerite has partly filled pore space.

Recognition of the relationships among the carbonate cements in sandstones, and their occurrence relative to other cement types, has contributed greatly to understanding clastic diagenesis. For example, one of the great debates in sandstone diagenesis concerns mass transfer (in solution) from shales into sandstones. Using petrographic, chemical, and stable isotope data, Boles (1978, 1981) proposed that late ankerite cement from the Wilcox Group (Fig. 2) formed from preexisting calcite cement at temperatures of about 120° to 160°C. Boles (1978) and Boles and Franks (1979) suggested that the iron and magnesium necessary for the reaction,

$$4CaCO_3 \text{ (calcite)} + Fe^{2+} + Mg^{2+} =$$
$$2CaFe_{0.5}Mg_{0.5} (CO_3)_2 \text{ (ankerite)} + 2Ca^{2+}, \qquad (6)$$

can be derived in sufficient quantities, and at the appropriate temperature, from adjacent shales and mudstones by the transformation of smectite to illite. The change from smectite to illite is a reaction that is typical of shale diagenesis in the Gulf Coast (Hower and others, 1976).

Not all investigators agree that sizeable quantities of magnesium and iron can be transferred from shales into sandstones as a result of the smectite to illite transformation. Hower and others (1976) suggested that shales are closed systems with respect to $Al^{3+}$, $Si^{4+}$, $Mg^{2+}$, and $Fe^{2+}$. Lahann (1980) proposed that illitization of smectite in shales could provide a source of these ions for transport from shales into sandstones. However, the effective distance of transport decreases with higher temperatures of illitization due to the increasing rates of precipitation at higher temperatures relative to rates of diffusion.

The relatively common progression of carbonate cements during diagenesis from Fe-poor calcite to ferroan calcite or ankerite has also been linked to other factors. Brand and Veizer (1980) reported that such a trend in carbonates results from interaction

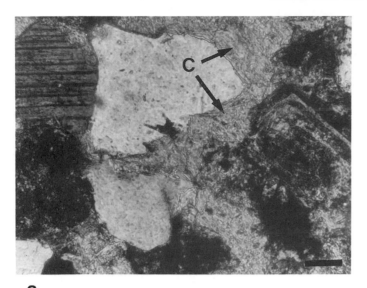

a

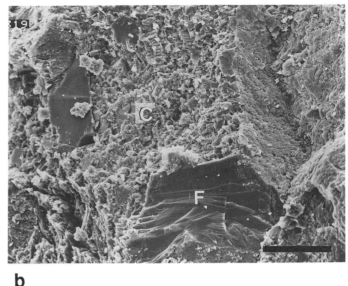

b

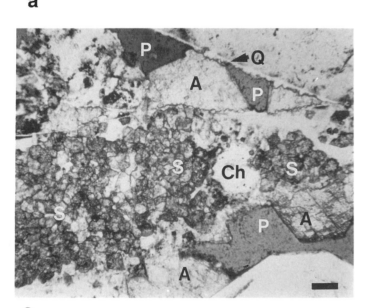

c

Figure 8. Early calcite cement (C), Upper Cretaceous basal Belly River sandstone, Alberta. Cross-polarized light. Scale bar = 50 μm. b. Calcite (C) replacing plagioclase feldspar (F), basal Belly River sandstone, Alberta. Scale bar = 50 μm. c. Siderite (S) replacing chert grain (Ch). Ankerite (A) has developed in the pore space (P). Pore-lining quartz (Q) is also visible. Lower Cretaceous Viking B conglomerate, Alberta. Plane-polarized light. Scale bar = 100 μm.

with meteoric waters. Such fluids were also apparently an important fraction of those responsible for the late-stage, Fe-rich carbonate cements in Tertiary sandstones of the Uinta Basin, Utah (Pitman and others, 1982). The involvement of meteoric water in the ankeritization of calcite has also been reported by Dutton and Land (1985) and Longstaffe (1987). Dutton and Land (1985) proposed that reduction of Fe-compounds during organic maturation may have provided the necessary iron.

Another diagenetic carbonate whose appearance and distribution is of special significance is siderite. It commonly occurs as an early diagenetic cement, especially in fine-grained sediments containing significant amounts of organic matter. Garrels and Christ (1965) and Curtis (1967), among others, have concluded that siderite crystallization requires waters with low dissolved sulfide activity, high bicarbonate activity, and very limited free

oxygen. Because seawater contains abundant sulfate, reducing conditions should lead to the production of $H_2S$ and render siderite unstable in the marine environment (Berner, 1971). In contrast, Postma (1982) has shown that the formation of siderite at or near the sediment-water interface is favored in fresh-water environments, utilizing the following reaction:

$$4FeOOH \text{ (ferric oxyhydroxide)} + CH_2O \text{ (organic matter)} + 3H_2CO_3 = 4FeCO_3$$
$$\text{(siderite)} + 6H_2O. \qquad (7)$$

Gautier (1982) concluded that abundant siderite formation in a marine environment would require not only the reduction of pore-water sulfate, but also the reaction of the resulting aqueous sulfide during early diagenesis. He proposed that such conditions could exist when methane begins to accumulate within a unit.

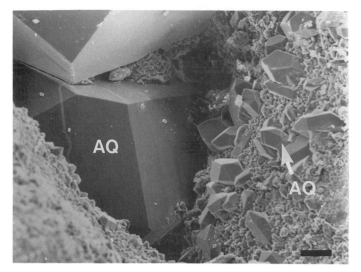

a

b

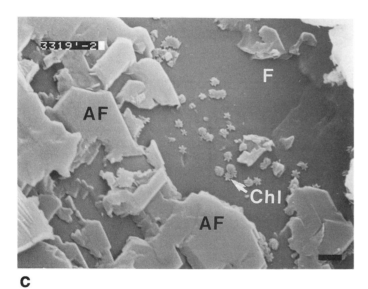

c

Figure 9. a. Quartz overgrowths (AQ) lining and filling pore space, Lower Cretaceous Viking B conglomerate, Alberta. Scale bar = 50 $\mu$m. b. Authigenic quartz (AQ) enveloping kaolinite (K), Lower Cretaceous Viking B conglomerate, Alberta. Scale bar = 50 $\mu$m. c. Authigenic feldspar overgrowth (AF) on detrital feldspar (F), Upper Cretaceous basal Belly River sandstone, Alberta. Authigenic chlorite (Chl) is also present. Scale bar = 1 $\mu$m.

## Quartz

Silica cement is also of great importance to the diagenesis of coarse-grained clastic rocks. If formed early, it can arrest compaction of the unit, and generally resists dissolution during later stages of diagenesis. Quartz is the most common authigenic phase (Figs. 9a, b), but other generally less stable polymorphs such as opal-A, opal-A', and opal-CT are known from relatively deeply buried rocks (Welton, 1984; Williams and Crerar, 1985).

When quartz overgrowths nucleate upon grains, the boundary between authigenic and detrital silica can be indicated by a thin rim of dark-colored material such as clays, organic matter, or iron-oxides. Several generations of overgrowths can be distinguished in some rocks by the presence of such rims (Blatt, 1982; Haszeldene and others, 1984). The use of cathodoluminescence techniques can also reveal the complexity of quartz precipitation during diagenesis.

The diagenetic quartz itself tends to be characterized by a well-developed, euhedral morphology. It can partly or completely coat grains; very large crystals can grow well into the pore space (Fig. 9a). Several such overgrowths in a single pore may completely occlude porosity; in the absence of dust rims about the detrital grains, the resulting texture can be mistaken for pressure solution of quartz grains (Blatt, 1982). Authigenic quartz can also envelop and/or replace other minerals in the pore space (Fig. 9b), provided that a suitable nucleation site is available.

While quartz is the most abundant phase to develop as an overgrowth, other authigenic minerals also develop in similar fashions. Feldspar overgrowths (Fig. 9c), for example, are quite common.

The formation of quartz overgrowths, like other abundant diagenetic cements, requires a significant mass flux into the unit. The origin, quantity, and mode of transport of silica is a matter of fundamental importance to diagenesis on a basinwide scale, yet it remains a subject of some debate.

Several sources of silica have been suggested. Sibley and Blatt (1976), Robin (1978), and Blatt (1979) noted that, in some units, local pressure solution can be an important local source of dissolved silica. However, pressure solution alone cannot account for the abundance of quartz cementation in many sandstones (Blatt, 1979; Land, 1984). Solution of opal, feldspar, mica, and (or) amorphous aluminosilicates can also contribute to the silica content of pore waters (Fisher, 1982; Lewin, 1971; Land, 1984).

Other workers (e.g., Johnson, 1920; Hayes, 1979) have proposed the transport of silica-rich pore waters from shales into adjacent sand units. One frequently cited possibility for increasing the silica (and Mg, Fe, as discussed previously) concentration of shale porewaters involves the transformation of smectite to illite during burial diagenesis (Towe, 1962; Siever, 1962; Hower and others, 1976; Boles and Franks, 1979). The exact amounts of silica released depend upon the nature of the smectite to illite reaction (see summaries by Hower, 1981; Land, 1984). In the reaction proposed by Boles and Franks (1979),

$$\text{smectite} + K^+ = \text{illite} + H_4SiO_4 + \text{cations (Ca}^{2+}, Mg^{2+}, Na^+, Fe^{2+})$$
(from Land, 1984; after Boles and Franks, 1979; and Hower, 1981),   (8)

alumina is conserved between smectite and illite, thus requiring the destruction of some smectite interlayers to provide the alumina necessary for illite formation. An important consequence of this reaction is that about 15 moles of silica are released for each mole of smectite. In the reaction proposed by Hower and others (1976),

$$\text{K-feldspar} + \text{smectite} = \text{illite} + \text{chlorite} + \text{quartz,}   (9)$$

alumina is treated as a mobile component; smectite interlayers are transformed to illite on a mole-for-mole basis. The destruction of the original smectite 2:1 layers is not required. Instead there is a substitution of ions within the structure. The additional $Al^{3+}$ required is supplied by the breakdown of detrital K-feldspar. Silica is still released by this reaction, but in much smaller quantities (3 moles of silica for each mole of smectite).

Attempts to reconcile these two approaches have focused upon the mineralogy and chemistry of thick shale sequences. In general, the results remain equivocal, largely because the detrital mineralogy and chemistry inherited during deposition are rarely constant throughout the lithologic section (Hower, 1981). However, as Land (1984) has pointed out, perhaps of greater importance is not exactly how much silica is produced by the smectite to illite reaction but whether silica derived by any process leaves the shales in sufficient quantities to contribute to the diagenesis of sandstones.

Another problem with the transport of silica (as well as other solutes) from shales into sandstones is the enormous volume of water that is apparently required. Many authors (Blatt, 1979; Bjørlykke, 1979; Land and Dutton, 1978, 1979; Land, 1984) have commented that the volume of water derived from the compaction and diagenesis of basinal shales is quite insufficient (by some orders of magnitude!) to account for the amount of quartz cement present in sandstone units for which such a process has been suggested. Several solutions to this dilemma have been proposed. Descending ground water could provide the additional necessary fluid (Bjørlykke, 1979, 1984). Early precipitation of quartz cements from vertically circulating meteoric water has also been advocated by Sibley and Blatt (1976) and Blatt (1979). The flow of meteoric water deep within certain basins has been demonstrated by Tóth (1980) and Land and Prezbindowski (1981) among others. Moreover, stable isotope studies have demonstrated the importance of meteoric water to the crystallization of diagenetic cements in many basins (e.g., Longstaffe, 1983, 1984, 1986, 1987; Dutton and Land, 1985). The general applicability of this process, however, depends upon the hydrodynamic situation in the basin during various stages of diagenesis.

Another possibility is that the transport of silica is related more to the characteristic distance of diffusion, defined as $(D/P)^{1/2}$, than to the rate of advection (Petrovic, 1976; D is the molecular diffusion coefficient and P is the rate constant for solid precipitation). However, diffusion is not generally accepted to be significant on a scale of greater than a few meters (Wood and Hewett, 1982). Instead, at least for systems not controlled by hydrodynamic flow, Wood and Hewett (1982, 1984) have proposed that most cementation results from mass transport in slowly convecting fluids. A continuous transfer of mass occurs as long as the circulation cells survive. For example, quartz would precipitate as the pore fluid moved from warmer to cooler portions of the system.

Land (1984) has concluded that large-scale convection of fluids appears to be unavoidable if the smectite-illite transformation in shales is of major importance to sandstone diagenesis. Such movement is required to explain why, in some basins, silica (and carbonate) cementation commonly occurs at much higher stratigraphic levels than the smectite-to-illite transformation in underlying shales. Unless the solubility of silica in formation waters is much higher (by orders of magnitude) than presently believed, or unless the supply of dissolved silica within the coarser grained units is much greater than presently understood, some type of large-scale fluid cycling of connate fluids or meteoric waters, linking sandstones and shales within a basin, is essential (Land, 1984).

## Clay cements and the alteration of feldspar

The occurrence of authigenic clay-mineral cements in sandstones and conglomerates has been recognized for some time (Lerbekmo, 1957, 1961; Shelton, 1964; Carrigy and Mellon, 1964; Hayes, 1970), but only recently, as scanning electron microscopy became widely available, have these materials begun to receive much attention (Sarkisyan, 1972; Wilson and Pittman, 1977).

One of the most common diagenetic clay minerals is kaolinite, or its higher temperature polytype, dickite. These minerals normally occur as well-crystallized, face-to-face stacks of pseudohexagonal plates (books) that fill pores (Figs. 10a, b). In some pores, elongated stacks of these plates (vermiform morphology) occur (Welton, 1984; Wilson and Pittman, 1977). In other samples, kaolinite can be engulfed by later authigenic phases (Fig. 9b). A ragged morphology is normally considered to reflect a detrital origin for the clay; authigenic kaolinite, however, can also attain such textures because of solution during later stages of diagenesis (Hurst, 1980).

Authigenic chlorite is another common clay cement that can exhibit a wide variety of morphologies. Its most typical mode of aggregation is as a grain coating or pore lining (Figs. 11a, b). Individual plates are oriented in a face-to-edge manner; this tendency can lead to the formation of a house-of-cards or honeycomb pattern (Fig. 11c) with significant microporosity, or to the growth of rosettes (Figs. 11d, e; Wilson and Pittman, 1977). Small "cabbagehead" growths on grain surfaces (Wilson and Pitman, 1977) are another less common variety of this mineral (Fig. 11f).

Perhaps the most interesting authigenic clay minerals, at least from a morphological point of view, are illite, mixed-layer illite/smectite, and smectite. Because these minerals can have similar appearances, detailed x-ray diffraction and chemical data are needed to identify which phases are actually present. Authigenic illite typically occurs as lathlike or filamentous material that lines and bridges pores (Guven and others, 1980). This so-called "hairy illite" can be virtually indistinguishable from some varieties of illite/smectite (Figs. 12a, b). These delicate textures are not always preserved; for example, matting of the illitic fibres on pore walls commonly occurs as a result of drying (except where special techniques such as critical point drying are employed).

The most common appearance of smectite in sandstones is as a webby to cellular grain coating or pore lining (Wilson and

a

b

Figure 10. a. Booklets and stacks of kaolinite (K) filling pore space. The pore walls are lined by authigenic quartz (Q). Lower Cretaceous Viking A sandstone, Alberta. Scale bar = 10 μm. b. Pseudohexagonal stacks and booklets of pore-filling kaolinite, Lower Cretaceous Viking A sandstone, Alberta. Scale bar = 5 μm.

Figure 11. a. Authigenic chlorite (Chl) coating framework grains. Some feldspar (F) has been partly dissolved. Plane polarized light. Upper Cretaceous basal Belly River sandstone, Alberta (Longstaffe, 1986). Scale bar = 50 μm. b. Authigenic chlorite (Chl) coating grains of the sandstone. Upper Cretaceous basal Belly River sandstone, Alberta (Longstaffe, 1986). Scale bar = 50 μm. c. Authigenic chlorite (Chl) showing house-of-cards texture, Lower Cretaceous Viking B sandstone, Alberta. Scale bar = 10 μm. d. Rosettelike morphology of grain-coating chlorite illustrated in Figures 11a, b. Upper Cretaceous basal Belly River sandstone, Alberta (Longstaffe, 1986). Scale bar = 5 μm. e. Authigenic chlorite rosette (Chl) partly enveloped by authigenic quartz (AQ), Lower Cretaceous Viking A conglomerate, Alberta. Scale bar = 5 μm. f. "Cabbage-head" authigenic chlorite (Chl) on detrital grains, Upper Cretaceous basal Belly River sandstone, Alberta. Scale bar = 1 μm.

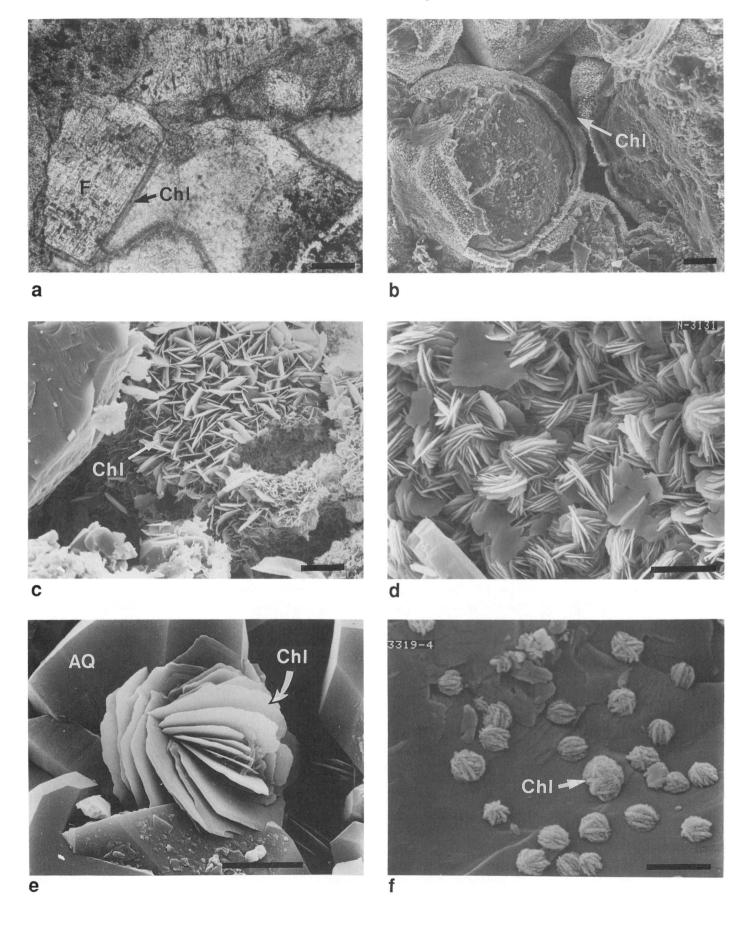

a                                               b

Figure 12. a. Filamentous illite/smectite (I) in pore space, Upper Cretaceous basal Belly River sand-stone, Alberta. Scale bar = 5 $\mu$m. b. Pore-bridging illite (I) and pore-lining authigenic quartz (AQ), basal Belly River sandstone, Alberta (Longstaffe, 1986). Scale bar = 10 $\mu$m.

Pittman, 1977; Welton, 1984; Fig. 13a). The crenulated appearance of this swelling clay mineral probably derives from its dehydration during sample collection and handling. Mixed-layer illite/smectite, especially varieties containing a high proportion of smectite, can also adopt the honeycomblike mode of aggregation; fingerlike projections of clay material from the interlocking network of material are commonly observed (Fig. 13b).

The crystallization of authigenic clay minerals in sandstones and conglomerates during diagenesis is especially dependent upon pore-fluid chemistry. In turn, the pore-fluid composition is controlled by various interrelated factors, the most important of which include depositional environment, dissolution of detrital grains (e.g., feldspar, rock fragments, mica, chert) and cements (e.g., calcite), and the extent of fluid flow and mixing during diagenesis (Bjørlykke and others, 1979; Hurst and Irwin, 1982; Galloway, 1984).

The starting composition of pore waters is largely controlled by depositional environment; fresh water, for example, is acidic and of low ionic strength, whereas seawater is somewhat basic and of moderate ionic strength. Variations in sedimentary facies that result from the depositional environments also define the initial pathways of high permeability and control the transmissivity of the various fluids that may enter or leave the system, at least during the early stages of diagenesis.

One example of how depositional environment, and the associated pore-fluid chemistry, can influence the distribution of diagenetic clay cements is provided by the formation of authigenic Fe-chlorite in the basal Belly River sandstone, Alberta. Within the locality examined, early, grain-coating Fe-chlorite

(Figs. 11a, b) occurs only within sedimentary facies that were very closely associated with the marine to nonmarine transition (e.g., distributary channels; Storey, 1982; Longstaffe, 1986). Oxygen-isotope data showed that this chlorite formed at low temperatures from $^{18}$O-depleted fluids (brackish waters) (Longstaffe, 1986). Such compositions are characteristic of many distributary channel environments. These observations are compatible with those of Hallam (1966) and Porrenga (1965), who proposed that the formation of early diagenetic Fe-chlorite is favored in deltaic environments. Hallam (1966) argued that wave and/or current motion causes the concentric accretion of the chlorite about the sand grains (Fig. 11b), and that the Fe necessary for clay crystallization can be obtained from the fresh water that is fed into the system. While other workers have favored the formation of early diagenetic chlorite from marine pore waters in low-energy environments (Rohrlich and others, 1969), the habit, facies-related distribution, and isotopic compositions of early Fe-chlorite in the basal Belly River sandstone suggest very strongly that mixing of fresh water and seawater have facilitated its crystallization.

Starting pore-fluid compositions are rapidly modified as diagenesis proceeds. In particular, the dissolution and/or alteration of detrital materials such as feldspar, mica, chert, and rock fragments (Figs. 11a, 14) provide an essential supply of species such as $K^+$, $Al^{3+}$, and $Si^{4+}$ that are required for clay mineral formation and/or transformation. One example documented by many workers (e.g., Blanche and Whitaker, 1978; Bjørlykke and others, 1979; Hurst and Irwin, 1982; Longstaffe, 1984, 1986, 1987) is the production of authigenic kaolinite in sandstones from circulat-

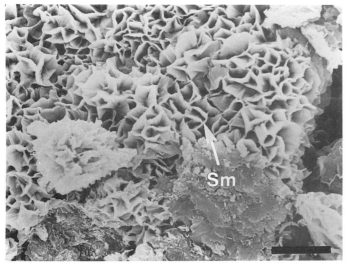

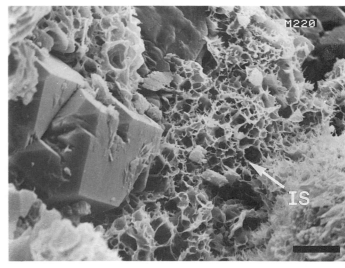

a

b

Figure 13. a. Pore-lining "honeycomb" smectite (Sm). Product of artificial diagenesis of the Clearwater Formation, Alberta; the smectite has been formed by water-rock interaction during tertiary recovery of bitumen (Kirk and others, 1987). Scale bar = 10 $\mu$m. b. Pore-lining "honeycomb" illite/smectite (IS), Lower Cretaceous Viking A sandstone, Alberta. Scale bar = 10 $\mu$m.

ing meteoric water:

$$2KAlSi_3O_8 \text{ (K-feldspar)} + 2H_2CO_3 + 9H_2O = Al_2Si_2O_5(OH)_4$$
$$\text{(kaolinite)} + 4H_4SiO_4 + 2HCO_3^- + 2K^+ \text{ (Bjørlykke, 1984).} \quad (10)$$

The continuous supply of fresh water in an aquifer provides a medium that is undersaturated relative to phases such as K-feldspar (and carbonate), thus permitting dissolution of feldspar and the crystallization of kaolinite and quartz.

Silicate and carbonate dissolution at depth, as manifested by the development of secondary porosity (Schmidt and McDonald, 1979a, 1979b; Hayes, 1979; McBride, 1980), can also be crucial during later (and deeper/hotter) stages of clastic diagenesis. In the Texas Gulf Coast alone, Land and Milliken (1981), Boles (1982), and Land (1984) have shown that dissolution of K-feldspar and albitization of plagioclase have released an enormous volume of material into solution. In the Frio Formation, Texas Gulf Coast, K-feldspar disappears below about 4,000 m; its dissolution probably proceeds in a manner similar to that proposed in Equation 10 (Land, 1984; Land and Milliken, 1981). The potassium produced by this reaction could convert significant volumes of smectite to illite in adjacent shales, provided that there is transport into the shale (Land, 1984). The silica produced could be available to form quartz cement, or alternatively be consumed by albitization reactions (see below).

In a comparable fashion, detrital plagioclase ($Ab_{60}$-$Ab_{80}$) from the Frio Formation has either dissolved or become albitized ($Ab_{98}$) as burial temperatures of about $130° \pm 20°C$ are reached

(Land and Milliken, 1981; Boles, 1982). Both reactions have probably occurred.

The dissolution of the anorthite component of the plagioclase can be described as:

$$CaAl_2Si_2O_8 \text{ (anorthite)} + 3H_2O + 2CO_2 = Al_2Si_2O_5(OH)_4 \text{ (kaolinite)}$$
$$+ Ca^{2+} + 2HCO_3^- \text{ (Land, 1984).} \quad (11)$$

Depending upon pore-fluid chemistry, other clay minerals such as illite might also be formed as a result of this process (Boles, 1982).

The albitization reaction can be described as:

$$CaAl_2Si_2O_8 \text{ (anorthite)} + 2Na^+ + 4H_4SiO_4 = 2NaAlSi_3O_8 \text{ (albite)} +$$
$$Ca^{2+} + 8H_2O \text{ (Land and Milliken, 1981).} \quad (12)$$

With this reaction, a greater volume of albite is produced than the anorthite replaced. The occurrence of diagenetic albite overgrowths may reflect this excess (Land, 1984). Reaction 12 also consumes silica, some of which may have been released by the dissolution of K-feldspar (Equation 10). For the Frio Formation, Land (1984) concluded that a net flux of silica into the sandstone was still required for albitization to proceed to the observed extent.

The albitization reaction also requires a source of sodium. Boles (1982) proposed that interstitial seawater, modified by later diagenetic reactions, such as the smectite to illite transformation in shales, and (or) addition of connate waters from salt diapirs, could provide the necessary mass. He concluded that particularly high sodium concentrations were not necessary for albitization,

Figure 14. Partly dissolved feldspar grain (F), Upper Cretaceous basal Belly River sandstone, Alberta. Scale bar = 10 $\mu$m.

given sufficient flow rates. Land and Milliken (1981) suggested that additional sources of sodium (e.g., dissolution of halite) beyond that of interstitial seawater were required to obtain the observed formation-water chemistry in their study area.

Both reactions involving the alteration of plagioclase (Equations 11 and 12) produce calcium, which may be ultimately consumed in the formation of late carbonate cements (Land, 1984). Alternatively, this calcium may remain in solution in the formation waters; high calcium contents are typical of basinal brines, especially in the Gulf Coast area (Land and Milliken, 1981).

Dissolution of phases formed during early stages of diagenesis may also be of some importance to authigenic clay formation at higher temperatures. In sandstones of the Kootenay Formation, for example, Hutcheon and others (1980) showed a correlation between the corrosion of authigenic dolomite plus the disappearance of authigenic kaolinite and the appearance of authigenic chlorite and calcite:

$$5CaMg\,(CO_3)_2\,\text{(dolomite)} + Al_2Si_2O_5\,(OH)_4\,\text{(kaolinite)} + SiO_2$$
$$\text{(quartz)} + H_2O = Mg_5Al_2Si_3O_{10}\,(OH)_8\,\text{(Mg-chlorite)} + 5CaCO_3$$
$$\text{(calcite)} + 5CO_2. \tag{13}$$

In another example, Bjørlykke (1984) has noted that authigenic kaolinite transforms to illite at about 130° to 150°C in the presence of K⁺:

$$3Al_2Si_2O_5(OH)_4\,\text{(kaolinite)} + 2K^+ = 2KAl_3Si_3O_{10}\,(OH)_2$$
$$\text{(muscovite/illite)} + 2H^+ + H_2O. \tag{14}$$

The dependence of authigenic illite formation in sandstones upon pore-fluid chemistry is clearly indicated by Equation 14, and

explains why, unlike shales, the percent smectite in authigenic illite/smectite in sandstones is not temperature dependent. Reaction 14 may also explain why the formation of authigenic illite or illite/smectite generally postdates that of kaolinite in many sandstones (Hurst and Irwin, 1982).

The lowering of pore-fluid pH caused by the kaolinite to illite reaction (Equation 14) could further facilitate feldspar dissolution and drive the precipitation of additional illite (Bjørlykke, 1984):

$$3Al_2Si_2O_5(OH)_4\,\text{(kaolinite)} + 3KAlSi_3O_8\,\text{(K-feldspar)} = 3KAl_3Si_3O_{10}$$
$$\text{(muscovite/illite)} + 6SiO_2 + 3H_2O. \tag{15}$$

***The role of organic matter.*** Throughout the previous discussion, two problems have emerged repeatedly. As pointed out by Land (1984): (1) an enormous volume of water is required to transport the cations and anions forming cements, and (2) the various reactions that occur require a source of acid.

Given its low ionic strength, meteoric water is unlikely to be the primary source of acidity (and transporter of solute) in many diagenetic settings (Curtis, 1983). One possible solution is that a fundamental relationship may exist among the alteration (maturation) of organic matter in shales and mudstones, the production of solute-rich, acidic pore waters in the fine-grained units, and expulsion of these waters into adjacent sandstones (Schmidt and McDonald, 1979a; Curtis, 1983; Surdam and others, 1984; Franks and Forester, 1984; Moncure and others, 1984; Siebert and others, 1984; Crossey and others, 1984).

One way of producing acidic pore waters is by the thermal decarboxylation of kerogen in organic-rich mudrocks (Tissot and others, 1974; Schmidt and McDonald, 1979a; Al-Shaieb and Shelton, 1981):

$$RCO_2H\,\text{(kerogen)} = RH\,\text{(kerogen)} + CO_2, \tag{16}$$

where R = high molecular weight organic materials (Curtis, 1983).

In this process, $CO_2$ released from the organic matter dissociates in shale pore-waters, forming carbonic acid:

$$CO_2 + H_2O = H_2CO_3 = H^+ + HCO_3^-\,\text{(Curtis, 1983).} \tag{17}$$

Feldspar, mica, and carbonate in the fine-grained sedimentary material are then dissolved by these fluids:

$$CaCO_3\,\text{(calcite)} + H^+ + HCO_3^- = Ca^{2+} + 2HCO_3^-\,\text{(Curtis, 1983).} \tag{18}$$

A variety of processes have been proposed for moving this evolved pore fluid into the sandstone, including compaction (Dickinson, 1953), aquathermal pressuring (Barker, 1972), fluid release resulting from the smectite to illite transformation (Burst, 1969), and the formation of methane (Hedberg, 1974). Upon entering the sandstone units, the acidic solutions continue to dissolve framework grains and cements until, in the absence of

H-donor reactions within the sandstone or continued flux of acid fluids from the shales, the pore waters are neutralized (Curtis, 1983). The observation that authigenic kaolinite commonly fills newly created, secondary pore space (Loucks and others, 1977; Boles, 1982, 1984) may be related to the rising pH of the sandstone pore waters as mineral dissolution proceeds (Curtis, 1983). Over the pH range of 3 to 6, calcite cement continues to dissolve, driving the solution pH to higher values. In contrast, the activity of aluminum drops by several orders of magnitude over the same increase in pH, leading to supersaturation of aluminum and the precipitation of kaolinite.

While carbonic acid is of importance to carbonate dissolution, other investigators have suggested that it is of limited value in the corrosion of aluminosilicate minerals during diagenesis. Instead, organic acids have been proposed both as the solvent for plagioclase (and also carbonates) and the mechanism for increased transport of aluminum (Surdam and others, 1984):

$$CaAl_2Si_2O_8 \text{ (anorthite)} + 2H_2C_2O_4 \text{ (oxalic acid)} + 8H_2O + 4H^+ =$$
$$2H_4SiO_4 + 2 (AlC_2O_4.4H_2O)^+ \text{ (aluminum oxalate complex)} + Ca^{2+}. \quad (19)$$

At temperatures (80° to 200°C) and concentrations of carboxylic acid anions (up to 5,000 ppm) reported by Carothers and Kharaka (1978) for oil field waters, Surdam and others (1984) have shown that carboxylic acids (e.g., acetic acid, oxalic acid) can increase the solubility of Al by up to three orders of magnitude through the formation of organo-Al complexes. One important result of this complexation is a significant increase in mass transport of Al (and perhaps also Si) relative to solely inorganic systems. The concomitant reduction in the amount of water required for mass transport of such cations could alleviate the "volume of fluid" problem (Surdam and others, 1984). Furthermore, changes in pH and carbonate content in the system could subsequently destroy the Al-bearing organic complexes, allowing the crystallization of authigenic kaolinite within both primary and secondary pore space.

Surdam and others (1984) have demonstrated experimentally that significant quantities of organic acids can be produced by the maturation of both algal and humic matter. They further noted that the reduction of mineral oxidants such as $Fe^{3+}$ released by reactions such as the smectite-to-illite transformation can also lead to the oxidation of organic matter, and the release of carboxylic acid groups from kerogen into solution. The timing (and temperature) of the smectite-illite transformation may be such that the organic acids are created and released into sandstone units, creating secondary porosity, just prior to hydrocarbon generation in, and migration from, the adjacent shales or mudstones (Surdam and others, 1984).

The study of natural systems has not yet produced unequivocal support for the ideas of Surdam and his co-workers. Nevertheless, the general observation—that organic and inorganic systems are intimately connected in the dissolution and precipitation of minerals during clastic diagenesis—is of fundamental importance.

## CONCLUDING COMMENTS

We have shown in this chapter how fundamental concepts of mass transport provide a useful framework for interpreting diagenetic processes. It is important to think of diagenesis as more than reactions that fill pores with secondary minerals or dissolve framework grains. Diagenesis is fundamentally a process of mass transport that is dependent on moving ground water to redistribute mass in the system and to determine temperature and pressure. To explain the diagenesis of a given rock sample probably requires an understanding of mass transport and fluid flow on a basinwide scale.

In general, the transport processes are reasonably well understood. However, interpreting the rock record in terms of these processes is a major scientifc obstacle. In looking at the rocks it is difficult to evaluate patterns of fluid migration and the chemistry of the fluids involved. This limitation creates a series of interpretive problems concerning where and how mass that is involved in diagenesis is supplied, what factors might contribute to its transport and mobility, why mass precipitates or dissolves where and when it does, and what specific reactions are actually involved. Nevertheless, in spite of these very obvious problems, good progress has been made in placing the study of diagenesis in a more complete conceptual framework and even using the patterns of diagenetic change as indicators of paleoflow regimes.

## REFERENCES

Al-Shaieb, Z., and Shelton, J. W., 1981, Migration of hydrocarbons and secondary porosity in sandstones: American Association of Petroleum Geologists Bulletin, v. 65, p. 2433–2436.

Back, W., and Hanshaw, B., 1970, Comparison of chemical hydrogeology of the carbonate peninsulas of Florida and Yucatan: Journal of Hydrology, v. 10, p. 330–368.

Back, W., Cherry, R., Hanshaw, B., 1966, Chemical equilibrium between the water and minerals of a carbonate aquifer: National Speleological Society Bulletin, v. 28, p. 119–126.

Barker, C., 1972, Aquathermal pressuring; Role of temperature in development of abnormal-pressure zones: American Association of Petroleum Geologists Bulletin, v. 56, p. 2068–2071.

Bear, J., 1972, Dynamics of fluids in porous media: New York, Elsevier, 764 p.

Berner, R. A., 1971, Principles of chemical sedimentology: New York, McGraw Hill, 240 p.

——, 1980, Early diagenesis; A theoretical approach: Princeton, New Jersey, Princeton University Press, 237 p.

Byørlykke, K., 1979, Discussion, cementation of sandstones: Journal of Sedimentary Petrology, v. 49, p. 1358–1359.

——, 1984, Formation of secondary porosity; How important is it? in McDonald, D. A., and Surdam, R. C., eds., Clastic diagenesis: American Association of Petroleum Geologists Memoir 37, p. 277–286.

Bjørlykke, K., Elverhøi, A., and Malm, A. O., 1979, Diagenesis in Mesozoic sandstones from Spitsbergen and the North Sea; A comparison: Geologische

Rundschau, bd. 68, p. 1152–1171.

Blanchard, P. E., and Sharp, J. M., 1985, Possible free convection in thick Gulf Coast sandstone sequences, *in* McNulty, L., and McPherson, J. G., eds., Transactions of the Southwest Section American Association of Petroleum Geologists: Fort Worth Geological Society, p. 6–12.

Blanche, J. B., and Whitaker, J.H.Mc D., 1978, Diagenesis of part of the Brent Sand Formation (Middle Jurassic) of the northern North Sea Basin: Geological Society of London Journal, v. 135, p. 73–82.

Blatt, H., 1979, Diagenetic processes in sandstones, *in* Scholle, P. A., and Schluger, P. R., eds., Aspects of diagenesis: Society of Economic Paleontologists and Mineralogists Special Publication no. 26, p. 141–157.

—— , 1982, Sedimentary petrology: San Francisco, W. H. Freeman and Company, 564 p.

Bodner, D. P., Blanchard, P. E., and Sharp, J. M., 1985, Variations in Gulf Coast heat flow created by groundwater flow: Gulf Coast Association of Geological Societies Transactions, v. 35, p. 19–28.

Boles, J. R., 1978, Active ankerite cementation in the subsurface Eocene of southwest Texas: Contributions to Mineralogy and Petrology, v. 68, p. 13–22.

—— , 1981, Clay diagenesis and effects on sandstone cementation (case histories from the Gulf Coast Tertiary), *in* Longstaffe, F. J., ed., Short course in clays and the resource geologist: Mineralogical Association of Canada, v. 7, p. 148–168.

—— , 1982, Active albitization of plagioclase, Gulf Coast Tertiary: American Journal of Science, v. 282, p. 165–180.

—— , 1984, Secondary porosity reactions in the Stevens Sandstone, San Joaquin Valley, California, *in* McDonald, D. A., and Surdam, R. C., eds., Clastic diagenesis: American Association of Petroleum Geologists Memoir 37, p. 217–224.

Boles, J. R., and Franks, S. G., 1979, Clay diagenesis in Wilcox sandstones of southwest Texas; Implications of smectite diagenesis on sandstone cementation: Journal of Sedimentary Petrology, v. 49, p. 55–70.

Bonham, L. C., 1980, Migration of hydrocarbons in compacting basins, *in* Roberts, W. H., III, and Cordell, R. J., eds., Problems of petroleum migration: American Association of Petroleum Geologists Studies in Geology no. 10, p. 69–88.

Brand, V., and Veizer, J., 1980, Chemical diagenesis of a multicomponent carbonate system; 1, Trace elements: Journal of Sedimentary Petrology, v. 50, p. 1219–1236.

Bredehoeft, J. D., and Hanshaw, B., 1968, On the maintenance of anomalous pressures; I, Thick sedimentary sequences: Geological Society of America Bulletin, v. 79, p. 1097–1106.

Bredehoeft, J. D., Blyth, C. R., White, W. A., and Maxey, G. B., 1963, Possible mechanism for concentration of brines in subsurface formations: American Association of Petroleum Geologists Bulletin, v. 47, p. 257–269.

Bredehoeft, J. D., Neuzil, C. E., and Milly, P.C.D., 1983, Regional flow in the Dakota aquifer; A study of confining layers: U.S. Geological Survey Water-Supply Paper 2237, 45 p.

Burst, J. F., 1969, Diagenesis of Gulf Coast clayey sediments and its possible relation to petroleum migration: American Association of Petroleum Geologists Bulletin, v. 53, p. 73–93.

Carothers, W. W., and Kharaka, Y. K., 1978, Aliphatic acid anions in oil-field waters; Implications for origin of natural gas: American Association of Petroleum Geologists Bulletin, v. 62, p. 2441–2453.

Carrigy, M. A., and Mellon, G. B., 1964, Authigenic clay mineral cements in Cretaceous and Tertiary sandstones of Alberta: Journal of Sedimentary Petrology, v. 34, p. 461–472.

Crossey, L. J., Frost, B. R., and Surdam, R. C., 1984, Secondary porosity in laumontite-bearing sandstones, *in* McDonald, D. A., and Surdam, R. C., eds., Clastic diagenesis: American Association of Petroleum Geologists Memoir 37, p. 225–237.

Curtis, C. D., 1967, Diagenetic iron minerals in some British Carboniferous sediments: Geochimica et Cosmochimica Acta, v. 31, p. 2109–2123.

—— , 1983, Link between aluminum mobility and destruction of secondary porosity: American Association of Petroleum Geologists Bulletin, v. 67, p. 380–393.

Davies, G. R., 1979, Dolomite reservoir rocks; Processes, controls, porosity development: American Association of Petroleum Geologists Short Course on Carbonate Porosity, Education Course Note Series no. 11, p. c1–c17.

Dickinson, G., 1953, Geological aspect of abnormal reservoir pressures in Gulf Coast Louisiana: American Association of Petroleum Geologists Bulletin, v. 37, p. 420–432.

Dutton, S. P., and Land, L. S., 1985, Meteoric burial diagenesis of Pennsylvanian arkosic sandstones, southwestern Anadarko Basin, Texas: Bulletin of the American Association of Petroleum Geologists, v. 69, p. 22–38.

Fisher, R. S., 1982, Diagenetic history of Eocene Wilcox sandstones and associated formation waters, south-central Texas [Ph.D. thesis]: University of Texas at Austin, 195 p.

Fogg, G. E., and Kreitler, C. W., 1982, Groundwater hydraulics and hydrochemical facies in Eocene aquifers of the East Texas Basin: The University of Texas at Austin, Bureau of Economic Geology Report of Investigations no. 127, 71 p.

Franks, S. G., and Forester, R. W., 1984, Relationships among secondary porosity, pore-fluid chemistry, and carbon dioxide Texas Gulf Coast, *in* McDonald, D. A., and Surdam, R. C., eds., Clastic diagenesis: American Association of Petroleum Geologists Memoir 37, p. 63–79.

Galloway, W. E., 1984, Hydrogeologic regimes of sandstone diagenesis, *in* McDonald, D. A., and Surdam, R. C., eds., Clastic diagenesis: American Association of Petroleum Geologists Memoir 37, p. 3–13.

Garrels, R. M., and Christ, C. L., 1965, Solutions, minerals, and equilibria: San Francisco, Freeman, Cooper and Company, 450 p.

Garrels, R. M., and Thompson, M. E., 1962, A chemical model for seawater at 25°C and one atmosphere total pressure: American Journal of Science, 260, p. 57–66.

Gautier, D. L., 1982, Siderite concretions; Indicators of early diagenesis in the Gammon Shale (Cretaceous): Journal of Sedimentary Petrology, v. 52, p. 859–871.

Guven, N., Hower, W., and Davies, D. K., 1980, Nature of authigenic illites in sandstone reservoirs: Journal of Sedimentary Petrology, v. 50, p. 761–766.

Hallam, A., 1966, Depositional environment of British Liassic ironstones considered in the context of their facies relationships: Nature, v. 209, p. 1306–1309.

Harvie, C. E., Moller, N., and Weare, J. H., 1984, The prediction of mineral solubilities in natural waters: The $Na-K-Mg-Ca-H-Cl-SO_4-OH-HCO_3-CO_3-CO_2-H_2O$ system to high ionic strengths at 25°C: Geochimica et Cosmochimica Acta, v. 48, p. 723–751.

Haszeldene, R. S., Samson, I. M., and Cornford, C., 1984, Quartz diagenesis and convective fluid movement: Beatrice Oilfield, United Kingdom North Sea: Clay Minerals, v. 19, p. 391–402.

Hayes, J. B., 1970, Polytypism of chlorite in sedimentary rocks: Clays and Clay Minerals, v. 18, p. 285–306.

—— , 1979, Sandstone diagenesis; The hole truth, *in* Scholle, P. A., and Schluger, P. R., eds., Aspects of diagenesis: Society of Economic Paleontologists and Mineralogists Special Publication no. 26, p. 127–139.

Hedberg, H. D., 1974, Relation of methane generation to undercompacted shales, shale diapirs, and mud volcanoes: American Association of Petroleum Geologists Bulletin, v. 58, p. 661–673.

Hitchon, B., 1969a, Fluid flow in the Western Canada Sedimentary Basin; 1. Effect of topography: Water Resources Research, v. 5, p. 186–195.

—— , 1969b, Fluid flow in the western Canada basin; 2. Effect of geology: Water Resources Research, v. 5, no. 2, p. 460–469.

Hower, J., 1981, X-ray diffraction identification of mixed-layer clay minerals, *in* Longstaffe, F. J., ed., Short course in clays and the resource geologist: Mineralogical Association of Canada, v. 7, p. 39–59.

Hower, J., Eslinger, E. V., Hower, M. E., and Perry, E. A., 1976, Mechanism of burial metamorphism of argillaceous sediment; 1. Mineralogical and chemical evidence: Geological Society of America Bulletin, v. 87, p. 725–737.

Hurst, A. R., 1980, Occurrence of corroded authigenic kaolinite in a diagenetically modified sandstone: Clays and Clay Minerals, v. 28, p. 393–396.

Hurst, A. R., and Irwin, H., 1982, Geologic modelling of clay diagenesis in sandstones: Clay Minerals, v. 17, p. 5–22.

Hutcheon, I., Oldershaw, A., and Ghent, E. D., 1980, Diagenesis of Cretaceous sandstones of the Kootenay Formation at Elk Valley (southeastern British Columbia) and Mount Allan (southwestern Alberta): Geochimica et Cosmochimica Acta, v. 44, p. 1425–1435.

Johnson, R. H., 1920, The cementation process in sandstones: American Association of Petroleum Geologists Bulletin, v. 4, p. 33–35.

Kharaka, Y. K., and Barnes, I., 1973, SOLMNEQ; Solution-mineral equilibrium computations: Springfield, Virginia, National Technical Information Service Technical Report PB 214-899, 82 p.

Kirk, J. S., Bird, G. W., and Longstaffe, F. J., 1987, Laboratory study of the effects of steam-flooding in the Clearwater Formation: Bulletin of Canadian Petroleum Geology, v. 35, p. 34–47.

Lahann, R. W., 1980, Smectite diagenesis and sandstone cement; The effect of reaction temperature: Journal of Sedimentary Petrology, v. 50, p. 755–760.

Land, L. S., 1983, Dolomitization: American Association of Petroleum Geologists Education Course Note Series no. 24, 20 p.

—— , 1984, Frio sandstone diagenesis, Texas Gulf Coast; A regional isotopic study, *in* McDonald, D. A., and Surdam, R. C., eds., Clastic diagenesis: American Association of Petroleum Geologists Memoir 37, p. 47–62.

Land, L. S., and Dutton, S. P., 1978, Cementation of a Pennsylvanian deltaic sandstone; Isotopic data: Journal of Sedimentary Petrology, v. 48, p. 1167–1176.

—— , 1979, Cementation of sandstone; Reply: Journal of Sedimentary Petrology, v. 49, p. 1359–1361.

Land, L. S., and Milliken, K. L., 1981, Feldspar diagenesis in the Frio Formation, Brazoria County, Texas Gulf Coast: Geology, v. 9, p. 314–318.

Land, L. S., and Prezbindowski, D., 1981, The origin and evolution of saline formation waters; Lower Cretaceous carbonates, south-central Texas, USA: Journal of Hydrology, v. 54, p. 51–74.

Langmuir, D., and Mahoney, J., 1984, Chemical equilibrium and kinetics of geochemical processes in ground water studies, *in* Hitchon, B., and Wallick, E. I., eds., First Canadian Conference on Hydrogeology: National Water Well Association, p. 69–95.

Lerbekmo, J. F., 1957, Authigenic montmorillonoid cement in andesitic sandstones of central California: Journal of Sedimentary Petrology, v. 27, p. 298–305.

—— , 1961, Porosity reduction in Cretaceous sandstones of Alberta: Alberta Society of Petroleum Geologists Journal, v. 9, p. 192–199.

Lewin, J. C., 1971, The dissolution of silica from diatom walls: Geochimica et Cosmochimica Acta, v. 21, p. 182–198.

Longstaffe, F. J., 1983, Stable isotope studies of diagenesis in clastic rocks: Geoscience Canada, v. 10, p. 44–58.

—— , 1984, The role of meteoric water in diagenesis of shallow sandstones; Stable isotope studies of the Milk River aquifer and gas pool, southeastern Alberta, *in* McDonald, D. A., and Surdam, R. C., eds., Clastic diagenesis: American Association of Petroleum Geologists Memoir 37, p. 81–98.

—— , F. J., 1986, Oxygen-isotope studies of diagenesis in the basal Belly River sandstone, Pembina I-Pool, Alberta: Journal of Sedimentary Petrology, v. 56, p. 78–88.

—— , 1987, Mineralogical and oxygen-isotope studies of clastic diagenesis; Implications for fluid flow in sedimentary basins, *in* Hitchon, B., ed., Third Canadian/American Conference on Hydrogeology: National Water Well Association (in press).

Loucks, R. G., Bebout, D. G., and Galloway, W. E., 1977, Relationship of porosity formation and preservation to sandstone consolidation history; Gulf Coast lower Tertiary Frio Formation: Gulf Coast Association of Geological Societies Transactions, v. 27, p. 109–120.

McBride, E. F., 1980, Importance of secondary porosity in sandstones to hydrocarbon exploration: American Association of Petroleum Geologists Bulletin, v. 64, p. 742.

Moncure, G. K., Lahann, R. W., and Siebert, R. M., 1984, Origin of secondary porosity and cement distribution in a sandstone/shale sequence from the Frio Formation (Oligocene), *in* McDonald, D. A., and Surdam, R. C., eds., Clastic diagenesis: American Association of Petroleum Geologists Memoir 37, p. 151–161.

Nordstrom, D. K., and 18 others, 1979, A comparison of computerized chemical models for equilibrium calculations in aqueous systems, *in* Jenne, E. A., ed., Chemical modeling of aqueous systems, speciation, sorption, solubility, and kinetics: American Chemical Society series 93, p. 857–892.

Petrovic, R., 1976, Rate control in feldspar dissolution; II, The effect of precipitates: Geochimica et Cosmochimica Acta, v. 40, p. 1509–1521.

Pitman, J. K., Fouch, T. D., and Goldhaber, M. B., 1982, Depositional setting and diagenetic evolution of some Tertiary unconventional reservoir rocks, Uinta Basin, Utah: American Association of Petroleum Geologists Bulletin, v. 66, p. 1581–1596.

Pitzer, K. S., and Kim, J. J., 1974, Thermodynamics of electrolytes; 4. Activity and osmotic coefficients for mixed electrolytes: American Chemical Society Journal, v. 96, p. 5701–5707.

Porrenga, D. H., 1965, Chamosite in recent sediments of the Niger and Orinoco deltas: Geologie en Mijnbouw, v. 44, p. 400–403.

Postma, D., 1982, Pyrite and siderite formation in brackish and freshwater swamp sediments: American Journal of Science, v. 282, p. 1151–1183.

Robin, P-Y.F., 1978, Pressure solution at grain-to-grain contacts: Geochimica et Cosmochimica Acta, v. 42, p. 1383–1398.

Rohrlich, V., Price, N. B., and Calvert, S. E., 1969, Chamosite in the recent sediments of Loch Etive, Scotland: Journal of Sedimentary Petrology, v. 39, p. 624–631.

Rubin, J., 1983, Transport of reacting solutes in porous media; Relation between mathematical nature of problem formulation and chemical nature of reactions: Water Resources Research, v. 19, p. 1231–1252.

Sarkisyan, S. G., 1972, Origin of authigenic clay minerals and their significance in petroloeum geology: Sedimentary Geology, v. 7, p. 1–22.

Schmidt, V., and McDonald, D. A., 1979a, The role of secondary porosity in the course of sandstone diagenesis, *in* Scholle, P. A., and Schluger, P. R., eds., Aspects of diagenesis: Society of Economic Paleontologists and Mineralogists Special Publication 26, p. 175–207.

—— , 1979b, Texture and recognition of secondary porosity in sandstones, *in* Scholle, P. A., and Schluger, P. R., eds., Aspects of diagenesis: Society of Economic Paleontologists and Mineralogists Special Publication 26, p. 209–225.

Schwartz, F. W., 1985, Modeling of ground-water flow and composition, *in* Hitchon, B., and Wallick, E. I., eds., First Canadian Conference on Hydrogeology: National Water Well Association, p. 178–188.

Schwartz, F. W., and Muehlenbachs, K., 1979, Isotope and ion geochemistry of groundwaters in the Milk River Aquifer, Alberta: Water Resources Research, v. 15, no. 2, p. 259–268.

Schwartz, F. W., Muehlenbachs, K., and Chorley, D. W., 1981, Flow-system controls of the chemical evolution of groundwater: Journal of Hydrology, v. 54, p. 225–243.

Shelton, J. W., 1964, Authigenic kaolinite in sandstones: Journal of Sedimentary Petrology, v. 34, p. 102–111.

Sibley, D. F., and Blatt, H., 1976, Intergranular pressure solution and cementation of Tuscarora orthoquartzite: Journal of Sedimentary Petrology, v. 46, p. 881–896.

Siebert, R. M., Moncure, G. K., and Lahann, R. W., 1984, A theory of framework grain dissolution in sandstones, *in* McDonald, D. A., and Surdam, R. C., eds., Clastic diagenesis: American Association of Petroleum Geologists Memoir 37, p. 163–175.

Siever, R., 1962, Silica solubility, 0–200°C, and the diagenesis of siliceous sediments: Journal of Geology, v. 70, p. 127–150.

Sposito, G., 1984, The surface chemistry of soils: New York, Oxford University Press, 234 p.

Storey, S. R., 1982, Optimum reservoir facies in an immature, shallow-lobate delta system; Basal Belly River Formation (Upper Cretaceous), central Alberta, Canada, *in* Hopkins, J. C., eds., Depositional environments and reservoir facies in some western Canadian oil and gas fields: University of Calgary Core Conference, p. 3–13.

Stumm, W., and Morgan, J. J., 1970, Aquatic chemistry: New York, John Wiley and Sons, 583 p.

Surdam, R. C., Boese, S. W., and Crossey, L. J., 1984, The chemistry of secondary porosity, *in* McDonald, D. A., and Surdam, R. C., eds., Clastic diagenesis: American Association of Petroleum Geologists Memoir 37, p. 127–149.

Thibodeaux, L. J., 1979, Chemodynamics; Environmental movement of chemicals in air, water, and soil: New York, John Wiley and Sons, Incorporated, 501 p.

Tissot, B., Durand, B., Espitalie, J., and Combay, A., 1974, Influence of nature and diagenesis of organic matter in formation of petroleum: Bulletin of the American Association of Petroleum Geologists, v. 58, p. 499–506.

Tóth, J., 1978, Gravity-induced cross-formational flow of formation fluids, Red Earth region, Alberta, Canada; Analysis, patterns, evolution: Water Resources Research, v. 14, p. 805–843.

——— , 1980, Cross-formational gravity-flow of groundwater; A mechanism of the transport and accumulation of petroleum (the generalized hydraulic theory of petroleum migration), *in* Roberts, W. H., III, and Cordell, R. J., eds., Problems of petroleum migration: American Association of Petroleum Geologists Studies in Geology no. 10, p. 121–167.

Towe, K. M., 1962, Clay mineral diagenesis as a possible source of silica cement in sedimentary rocks: Journal of Sedimentary Petrology, v. 32, p. 26–28.

Truesdell, A. H., and Jones, B. F., 1974, WATEQ, a computer program for calculating chemical equilibria of natural waters: U.S. Geological Survey Journal of Research, v. 2, p. 233–248.

Welton, J. E., 1984, SEM petrology atlas: Tulsa, American Association of Petroleum Geologists, 237 p.

Williams, L. A., and Crerar, D. A., 1985, Silica diagenesis; II, General mechanisms: Journal of Sedimentary Petrology, v. 55, p. 312–321.

Wilson, M. D., and Pittman, E. D., 1977, Authigenic clays in sandstones; Recognition and influence on reservoir properties and paleoenvironmental analysis: Journal of Sedimentary Petrology, v. 47, p. 3–31.

Wolery, T. J., 1979, Calculation of chemical equilibrium between aqueous solution and minerals; The EQ3/EQ6 Software package: University of California Lawrence Livermore Laboratory, National Technical Information Service Technical Report UCRL-52658, 41 p.

Wood, J. R., and Hewett, T. A., 1982, Fluid convection and mass transfer in porous sandstones; A theoretical approach: Geochimica et Cosmochimica Acta, v. 46, p. 1707–1713.

——— , 1984, Reservoir diagenesis and convective fluid flow, *in* McDonald, D. A., and Surdam, R. C., eds., Clastic diagenesis: American Association of Petroleum Geologists Memoir 37, p. 81–98.

Manuscript Accepted by the Society July 15, 1987

## ACKNOWLEDGMENTS

We would like to thank Dr. J. Sharp of The University of Texas at Austin for his comments on the manuscript. A. Ayalon, M. Dean, and J. Kirk kindly provided some of the micrographs.

The Geology of North America
Vol. O-2, Hydrogeology
The Geological Society of America, 1988

# Chapter 45

# The generation and dissipation of abnormal fluid pressures in active depositional environments

**P. A. Domenico**
*Department of Geology, Texas A & M University, College Station, Texas 77843*
**V. V. Palciauskas**
*Chevron Research Company, P.O. Box 446, LaHabra, California 90631*

## INTRODUCTION

In any discussion of abnormal fluid pressure in sedimentary environments, the observations of the renowned English geologist Sir Charles Lyell (1871), written over 100 years ago, are still appropriate today:

When sand and mud sink to the bottom of a deep sea, the particles are not pressed down by the enormous weight of the incumbent ocean; for the water which becomes mingled with the sand and mud resists pressure with a force equal to that of the column of fluid above. The same happens in regard to the organic remains which are filled with water under great pressure as they sink, otherwise they would be immediately crushed to pieces and flattened. Nevertheless, if the materials of a stratum remain in a yielding state, and do not set or solidify, they will gradually be squeezed down by the weight of other material successively heaped upon them, just as soft clay or loose sand on which a house is built may give way. By such downward pressure clay, sand, and marl may be packed into a smaller space, and be made to cohere together permanently.

With these statements, Lyell describes the important diagenetic process of sediment compaction in a depositional environment. He notes, correctly, that it is the "weight of the other material" which ultimately provides for porosity reduction associated with the "packing into a smaller space." In modern parlance, we would state that the added sediment load is initially carried by the water contained in the sediments, causing the fluid pressure to rise above its hydrostatic value. With the diffusion of pore water to areas of lower pressure, the sediment load is ultimately transferred to the matrix of skeletal grains, causing the deformation. From the perspective of abnormal pressure environments, we would be concerned mainly with the length of time required for this stress transfer to take place, for this is the time period over which anomalously high fluid pressures may persist.

The subsidence of sedimentary basins is thought to be caused by primary mechanisms in the basement rock, such as thermal contraction or deep metamorphism to the granulite or eclogite facies (Falvey, 1974; Watts and Ryan, 1976). Whatever the cause, basin subsidence assures that the sediments and their contained fluids become subjected to a thermal field. Hence, additional components of anomalous fluid pressures evolve, largely due to the thermal expansion of water and water-release mechanisms associated with phase transformations. As with gravitational loading, fluid flow or other mechanisms must once more be called upon to dissipate the pressures. It thus follows that the study of excess pressure development in depositional environments is the study of competing rates of fluid-pressure production and dissipation.

## EFFECTIVE STRESS CONCEPT

In any porous, water-filled sediment or rock, there are nonuniform stresses within the solid phase due to the points of sediment contact, and there are pressures in the ambient fluid. The former are intergranular in nature and are called intergranular or effective stresses. If the contained fluid is water, the latter are called pore-water pressures or neutral stresses. The total vertical stress acting on a horizontal plane at any depth may be resolved into these neutral and effective components

$$\sigma = \bar{\sigma} + \zeta P \tag{1}$$

where $\sigma$ is total stress, $\bar{\sigma}$ is effective stress, P is neutral stress, and $\zeta$ is a proportionality constant between the pore and the bulk volume changes, described as (Biot and Willis, 1957)

$$\zeta = 1 - K/K_s = 1 - (\beta_s/\beta) \tag{2}$$

where K is the drained bulk modulus, $K_s$ is the bulk modulus of the polycrystalline solids, and $\beta_s$ represents the respective compressibilities $K^{-1}$ and $K_s^{-1}$. If the compressibility of the polycrystal-

Domenico, P. A. and Palciauskas, V. V., 1988, The generation and dissipation of abnormal fluid pressures in active depositional environments, *in* Back, W., Rosenshein, J. S., and Seaber, P. R., eds., Hydrogeology: Boulder, Colorado, Geological Society of America, The Geology of North America, v. O-2.

line grain structure ($\beta_s$) is considerably less than the bulk compressibility $\beta$, which is generally the case for soft clays, then $\zeta \approx 1$. For this condition, equation 1 reduces to the well-known principle of effective stress as first enunciated by Terzaghi (1925).

Although the concept of effective stress was first stated by Terzaghi (1925) in relation to the consolidation of clays, it has found widespread application in studies that encompass a variety of hydrogeologic phenomena. The concept has been used in formulating a theory for landslides (Terzaghi, 1950), for establishing a mechanism for overthrust faulting (Hubbert and Rubey, 1959), and for establishing a theoretical basis for the phenomenon of land subsidence in response to fluid withdrawals (Domenico and Mifflin, 1965). The concept has been invoked by Meinzer (1928) in his classic study of the compressibility of artesian aquifers and further employed by Jacob (1950) in his derivation of the flow equation in elastic aquifers. The ideas of effective stress have been equally useful in establishing a theoretical basis for understanding water-level fluctuations in confined aquifers subject to atmospheric and tidal loading (Hantush, 1964).

Like all fundamental ideas, Terzaghi's concept of effective stress is deceptively simple. The full meaning of the concept is best appreciated from the perspective of stress changes and transfers. If the total stress acting on a plane is increased, say by continuous deposition, this added stress must be reflected on the right-hand side of equation 1. As mentioned for the example of Lyell, this added stress will be borne, in part, by the water, causing a rise in fluid pressure. The fraction of the incremental stress carried by the pore water depends on $\zeta$ and on the compressibility of the water itself. With the diffusion of pore water to regions of low pressure, this fraction of incremental stress is transferred to the matrix composed of the skeletal grains, increasing the effective stress. The time dependence of the response is determined by the material-flow properties and the rate of loading. All measurable effects, such as porosity reduction, distortion, and a change in shearing resistance are due exclusively to such changes in effective stress.

On the other hand, if total stress remains constant, any change in either of the terms on the right-hand side of equation 1 will be reflected by a commensurate increase in the companion right-hand term. For this case,

$$\zeta\,dP = -d\bar{\sigma}, \tag{3}$$

which states that a decrease in fluid pressure results in an increase in effective stress.

For the purposes of this paper, any mechanism that causes a change in $\sigma$, $\bar{\sigma}$, or P itself giving rise to a fluid pressure increase above its hydrostatic value is termed a pressure-producing mechanism. Such fluid pressures are called abnormal fluid pressures, often used synonymously in petroleum literature with excess pressure, or overpressure. At least ten such pressure-producing mechanisms have been cited by Bredehoeft and Hanshaw (1968), virtually all of which can be explained in terms of a direct response to changes in $\sigma$, $\bar{\sigma}$, or the fluid pressure itself. These include

but are not limited to continuous loading and tectonic compression (Dickinson, 1953; Hubbert and Rubey, 1959), the expansion of water due to heating (Barker, 1972; Bradley, 1975), and phase transformations such as gypsum to anhydrite (Heard and Ruby, 1966) and montmorillonite to illite (Powers, 1967).

## CONTINUOUS LOADING AND TECTONIC COMPRESSION

Continuous loading and tectonic compression provide geologic situations where increases in total stress give rise to excess fluid pressure. Our knowledge and appreciation of these pressure-producing mechanisms can be traced to several studies that incorporate observational, theoretical, and experimental aspects of the problem. One of the first observational studies was provided by Watts (1948), who discussed pressure and depth in some oil pools throughout the world, with emphasis on the Ventura Field in California. The first detailed study is attributed to Dickinson (1953) for Louisiana Gulf Coast wells, where a departure from normal pressure was observed at about 2,500 m. Further observational studies are cited by Fertl (1976), indicating that the excess-pressure environment is a common one and undoubtedly has been throughout geologic time. Indeed, it is likely that sections of many basins with an appreciable rate of sedimentation pass through some abnormal pressure period, which, given sufficient geologic time, evolve eventually to gravity-flow systems.

The classic study by Hubbert and Rubey (1959) was one of the first demonstrations of the role of excess pressures in performing geologic work. In their overthrust concept, fault blocks could be moved large distances by relatively small forces, provided the fluid pressures were abnormally high. The special geologic conditions required for the attainment and maintenance of such pressures include (1) the presence of clay rocks, (2) interbedded sandstone, (3) large total thickness, (4) rapid sedimentation. For these conditions, the low-permeability geologic environment is presumed to inhibit fluid flow, thereby maintaining the developed pressures. Indeed, the prevention of fluid flow is a deeply ingrained assumption in all early studies of abnormal pressure development.

### Pressure generation and dissipation

It remained for Bredehoeft and Hanshaw (1968) to demonstrate that fluid flow need not be absent in the generation and maintenance of abnormal pressure basins, but must be slow relative to the rate of sedimentation. The model employed by these authors assumed one-dimensional vertical compression with incompressible fluid and solid grain components. Although several similar papers have since followed (i.e., Smith, 1971), this particular contribution was the first to establish the physical and mathematical basis for the study of excess-pressure generation in combination with pressure dissipation. For this purpose, the effective stress concept provides no clues as to the rate at which the stress transfer takes place from the fluid to the solid phases.

**TABLE 1. COEFFICIENTS OF VERTICAL COMPRESSIBILITY AND SPECIFIC STORAGE
FOR VARIOUS MATERIALS**
(after Domenico and Mifflin, 1965)

| Material | Coefficient of Vertical Compressibility (bar$^{-1}$) | Specific Storage (cm$^{-1}$) |
|---|---|---|
| Plastic clay | $2.12 \times 10^{-1}$ to $2.65 \times 10^{-2}$ | $2.0 \times 10^{-4}$ to $2.6 \times 10^{-5}$ |
| Stiff clay | $2.65 \times 10^{-2}$ to $1.29 \times 10^{-2}$ | $2.6 \times 10^{-5}$ to $1.3 \times 10^{-5}$ |
| Medium clay | $1.29 \times 10^{-2}$ to $7.05 \times 10^{-3}$ | $1.3 \times 10^{-5}$ to $9.2 \times 10^{-6}$ |
| Loose sand | $1.06 \times 10^{-2}$ to $5.3 \times 10^{-3}$ | $1.0 \times 10^{-5}$ to $4.9 \times 10^{-6}$ |
| Dense sand | $2.12 \times 10^{-3}$ to $1.32 \times 10^{-3}$ | $4.9 \times 10^{-6}$ to $1.0 \times 10^{-6}$ |
| Dense sandy gravel | $1.06 \times 10^{-3}$ to $5.3 \times 10^{-4}$ | $1.0 \times 10^{-6}$ to $4.9 \times 10^{-7}$ |
| Rock, fissured | $7.05 \times 10^{-4}$ to $3.24 \times 10^{-5}$ | $7 \times 10^{-7}$ to $3.3 \times 10^{-8}$ |
| Rock, sound | less than $3.24 \times 10^{-5}$ | less than $3.3 \times 10^{-8}$ |

Bredehoeft and Hanshaw (1968) described the time dependence of the response with a fluid flow equation characterized by a hydraulic diffusivity $k/S_s$, where k is the vertical hydraulic conductivity ($Lt^{-1}$) and $S_s$ is the specific storage ($L^{-1}$), the latter reflecting the fluid and matrix compressibility of the porous medium (Table 1). A semiquantitative measure of the time dependence is demonstrated by the dimensionless Fourier Number $N_{FO}$

$$N_{FO} = (k/S_s)/\omega Y \qquad (4)$$

where $\omega$ is the constant rate of loading, i.e., deposition ($Lt^{-1}$), and Y is the accumulated sediment thickness (L) at any time. The smaller the Fourier Number, the longer the time required for the stress transfer.

The semiquantitative nature of equation 4 is seen from the following cases. For a field deformation, which is much slower than the characteristic time for diffusion of pore fluids, the excess fluid pressure adjusts quickly to a value which reflects that the stress transfer takes place more or less at the same rate as the loading. This would be characteristic for high-permeability sediments subjected to low rates of deposition, $k >> \omega$. Conversely, when the load variations are rapid in comparison to the rate of fluid flow, $k << \omega$, the stress transfer takes place slowly compared to the loading rate, and large excess fluid pressures are generated and maintained for long durations. This is the condition alluded to by Hubbert and Rubey (1959), and requires low-permeability sediments undergoing rapid deposition.

The conditions described above are reflected in Bredehoeft and Hanshaw's (1968) study of excess-pressure basins. For Gulf Coast sedimentation, an upper limit of $k/S_s = 3.3 \times 10^{-2}$ cm$^2$ sec$^{-1}$ corresponds to the constant-pressure case where the stress transfer occurs simultaneously with sedimentation (Fig. 1, curve A). Values of $k/S_s = 3.3 \times 10^{-5}$ cm$^2$ sec$^{-1}$) correspond to the approximate constant volume case where the fluid takes on the pressure of the overburden (Fig. 1, curve B). Curve C lies intermediate between these extremes ($K/S_s = 3.3 \times 10^{-4}$ cm$^2$ sec$^{-1}$). The Fourier Number for these three cases is $2 \times 10^1$, $2 \times 10^{-2}$ and $2 \times 10^{-1}$, respectively.

### Constant fluid volume

The maximum pressure generation occurs for the approximate constant fluid volume case of Figure 1 (curve B), where the fluid pressure approaches the overburden pressure. In the mathematical analysis leading to this maximum pressure production situation, both the fluid and the solid grains were assumed to be incompressible. In mathematical terms, $\zeta$ of equation 1 equals one and the specific storage of Table 1 is expressed only in terms of matrix compressibility. For these assumptions, the fluid can take on the total pressure of the overburden. The constant fluid volume case, where both the fluid and the solids are compressible, is more realistically expressed through the effective stress principle in combination with several elastic coefficients defined by Biot (1941):

$$dP/d\sigma = (1/\zeta)\,[1 - (d\bar{\sigma}/d\sigma)] = R/H = \beta_p/[\beta_p + n\,(\beta_f - \beta_s)]. \qquad (5)$$

Here $\beta_p$ is the pore compressibility, taken as $\beta - \beta_s$, n is porosity, $\beta_f$ is the fluid compressibility, $1/R$ is a measure of the change in water content for a given change in fluid pressure, and $1/H$ is a measure of the pore compressibility for conditions of constant fluid pressure. In these developments $1/R = \beta_p + n(\beta_f - \beta_s)$ and $1/H = \beta_p$ (Palciauskas and Domenico, 1982).

A physical interpretation of equation 5 is that $dP/d\sigma$ represents the portion of the incremental stress that is carried by the pore fluid, provided the pore fluid is not permitted to drain. The remainder is carried by the solids. For the Kayenta Sandstone, $dP/d\sigma$ has been calculated to be about 0.67, for limestone 0.25, and for basalt about 0.12 (Palciauskas and Domenico, 1982; Domenico, 1983). Hence, the more rigid the rock, the greater the proportion of the stress carried by the solids. For soft clays, $dP/d\sigma$ will approach 0.99, which agrees with experimental results reported by Lambe and Whitman (1969). Thus, for the constant fluid volume case where both the fluid and the solid grains are compressible, the observed pressures can approach $0.99\sigma$ for highly compressible rocks, but can be considerably less than $\sigma$ as rock rigidity increases. It is noteworthy that for all such rocks, generated pressures caused by gravitational loading cannot equal or exceed the overburden pressures.

## *Pressure generation and dissipation: Horizontal displacements not constrained to zero*

The theoretical developments given as curves A, B, and C in Figure 1 are for one-dimensional vertical compression. The analog field condition is the basin that is constrained laterally; that is, horizontal displacements are not permitted. This means that the basin is not permitted to relax tectonically nor can any form of tectonic compression be initiated. For the laterally constrained case, the effective horizontal stress is determined from the relationship (Landau and Lifshitz, 1959, p. 14)

$$\bar{\sigma}_h = \bar{\sigma}_v \left[ \nu / (1 - \nu) \right] \qquad (6)$$

where $\bar{\sigma}_h$ is the effective horizontal stress, $\bar{\sigma}_v$ is the effective vertical stress, and $\nu$ is Poisson's ratio. For most rocks, Poisson's ratio ranges from 0.25 to 0.33. Hence, the ratio of effective horizontal stress to vertical stress should lie between 0.33 and 0.5. This is contrary to most in situ measurements, which lie between 0.5 and 1.0, with the lower range more characteristic for hard rocks and the higher range more appropriate for soft rocks such as shales and salt (Obert and Duvall, 1967, p. 475). Thus, the horizontally constrained model does not adequately describe the field condition. There are several possible reasons for this. Plastic or viscoelastic deformation of deeply buried salts and some soft clays, for example, could permit partial or complete equalization of the principle effective stresses. In addition, tectonic forces can affect the horizontal stress, but not the vertical stress so that the condition $\sigma_h/\sigma_v > 1$ can be obtained.

Given that horizontal displacements are not constrained to zero in the more realistic cases, the excess pressure analysis takes on a different perspective. First, deformations must be treated within the framework of a mean-stress problem, indicating that fluid-pressure generation will differ somewhat from the one-dimensional case of vertical compression. This is demonstrated by curve D in Figure 1, which was calculated for the same value of parameters used to obtain curve C of Bredehoeft and Hanshaw (1968) where $k/S_s = 3.3 \times 10^{-4}$ cm$^2$sec$^{-1}$, except that the ratio of horizontal to vertical stress was taken as equal to 1:3. Clearly, in this tectonically relaxed state, excess pressure development is not truly excessive, but does increase markedly as the degree of lateral restraint increases. This is demonstrated by curve E, where the stress ratio was taken as 1:2. In addition, as deviatoric stresses are now part of the analysis, the opportunity arises to investigate the possibility of some sort of inelastic behavior. In the one-dimensional analysis, the conditions for failure require that fluid pressures exceed the overburden pressure. In the presence of deviatoric stresses, fluid pressures approaching the overburden pressure appear to be out of the question. For example, according to Hubbert and Willis (1957), the initiation of hydraulic fracture merely requires a fluid pressure in excess of the least compressive stress in a regional stress field. At yet another scale are the pore volume increases which materialize in the form of microfractures, most often at fluid pressures well below the least compressive stress (Handin and others, 1963, p. 753). In either case, if the

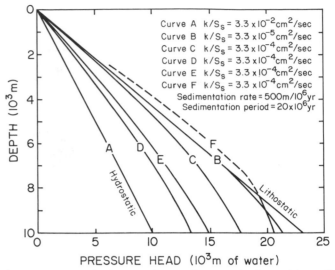

Figure 1. Pressure-depth curves for a sedimentary basin receiving sediments at a constant rate. Curves A, B, and C from Bredehoeft and Hanshaw (1968), curves D, E, and F from Palciauskas and Domenico (1980).

added pore volume due to fracture of some sort is sufficient to accommodate significant expansion of the compressed water, or if the permeability is mechanically increased, an instability may be provided from which there may be no apparent way to further increase fluid pressure above some limiting value. Thus, inelastic behavior must be considered along with fluid flow as an important pressure-dissipative mechanism.

The dilatency mode of failure following Handin and others (1963) was investigated by Domenico and Palciauskas (1979), whereas the problem of failure in shear was addressed by Chia (1980). In the absence of extremely low horizontal stresses (i.e., large deviatoric stresses) during deposition, failure in shear would appear to be relatively uncommon. On the other hand, the dilatency mechanism has some interesting aspects suggesting that the conditions for some sort of inelastic behavior appear to be present for virtually all states of differential stress. For example, for sedimentary rocks containing pore fluids, the lower the fluid pressure with respect to an experimental confining pressure, the larger the deviatoric stress required to initiate dilatency (Handin and others, 1963; Brace and others, 1966). Hence, increasing lateral restraint in the laboratory inhibits dilatency but contributes nothing to the fluid pressure generation. On the other hand, increasing lateral restraint for a field problem that is treated within a mean-stress framework not only inhibits dilatency, but simultaneously acts to increase fluid-pressure generation (curves D and E, Fig. 1). Thus, the conditions for some sort of inelastic behavior appear to be present for virtually all states of differential stress. This is true for both the tectonically relaxed state where the maximum principal stress is the vertical stress, and especially for the case of horizontal compression where the maximum principal stress is the horizontal stress.

## THERMAL EXPANSION OF FLUID

As with gravitational loading, the fluid pressure response in the nonisothermal environment must be viewed from the perspective of competing rates of pressure production and dissipation. Two extreme cases can once more be stated when the pressure-producing mechanism is the thermal expansion of fluid. For a rate of heating much slower than the characteristic time for diffusion of pore fluid, the thermal expansion of fluid can be easily accommodated by the diffusing pore fluid, and the response takes place at constant pressure. Conversely, when the rate of heating is rapid compared to the diffusion time for pore fluid, the response can be approximated as taking place at constant fluid mass. For sediment burial under conditions of a constant geothermal gradient, these relationships are expressed by a special form of the Fourier Number

$$N_{FO} = k/Y\alpha_m\omega G \tag{7}$$

where G is the geothermal gradient ($°CL^{-1}$), and $\alpha_m$ is the thermal expansion coefficient for the fluid and the pores ($°C^{-1}$). This is further described as (Palciauskas and Domenico, 1982)

$$\alpha_m = n(\alpha_f - \alpha_p) \tag{8}$$

where $\alpha_f$ is the thermal expansion coefficient for the fluid, and $\alpha_p$ is the thermal expansion coefficient for the pores.

The coefficient $\alpha_m$ is represented as the differences between two expansivities, one of which pertains to the water and the other to the pores in which the water resides. The coefficient $\alpha_p$ represents all the complicated reversible and irreversible effects arising from the internal pore geometry and stress fields as the temperature is raised. Detailed work by Campanella and Mitchell (1968) suggest that some of these include expansion of the solid grains and volume changes due to structural rearrangement. The pore expansivity is on the order of $10^{-5}°C^{-1}$, whereas the fluid expansivity ranges from $10^{-4}$ to $10^{-3}°C^{-1}$ for the temperature range 10°C to 80°C. As the volumetric expansion of water is greater than the expansivity of the pore volume, $\alpha_m$ will be positive and the fluid will be expelled when the temperature rises in a fully saturated medium. This applies to the drained condition in high permeability rocks. For the undrained condition in rocks of low permeability, the fluid pressure cannot remain constant and rises in accord with whatever boundary conditions are operative. In either case, any thermal expansion of the void volume acts to partially accommodate the fluid expansion.

### Constant fluid mass

The constant fluid mass case arises as the denominator of equation 7 becomes large with respect to the numerator. This assumption of constant fluid mass is at the heart of almost all the research in thermal pressuring of sedimentary basins, including the work of Barker (1972), Bradley (1975), and Magara (1975). Gretener (1981) reviews this work, discussing the experiments in a closed vessel which have led to much of the prevailing theory. Thus, in the current thermal pressuring theory, some formation in

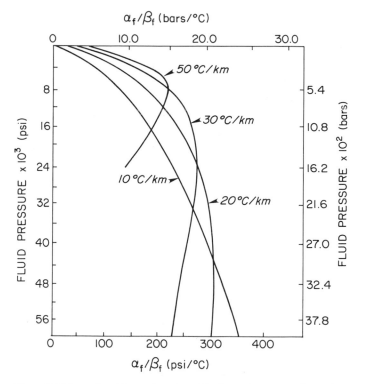

Figure 2. Values of $\alpha_f/\beta_f$ as a function of hydrostatic pressure for various geothermal gradients (modified and extended from data of Knapp and Knight, 1977; Strauss and Schubert, 1977).

a depositional environment must first become "sealed" at some depth and thereafter buried in a thermal field without a loss of water. In addition, from the closed vessel experiments, it is assumed that the thermal expansion of water acts against a rigid, nonyielding matrix that remains at constant pore volume. The pressure change with temperature thus becomes

$$(\partial P/\partial T) = \alpha_f/\beta_f \tag{9}$$

where the parameters of $\alpha_f$ and $\beta_f$ depend on both temperature and pressure (Fig. 2).

It is noted that equation 9 depends only on the fluid properties and is medium independent. Hence, according to this result, the fluid pressure response to a geothermal gradient in a granite under a large confining stress is no different than that in a shale in a tectonically relaxed state. If the medium is accepted as elastic whereby the pores can undergo some volumetric increase in response to thermally induced fluid pressure, the pressure-temperature relationship becomes

$$(\partial P/\partial T)_{m,\sigma} = R\alpha_m = n(\alpha_f - \alpha_p)/[\beta_p + n(\beta_f - \beta_s)] \tag{10}$$

where the properties of both the medium and the fluid are taken into account. Note that for incompressible pores and solids along with no pressure accommodation by expansion of the pore volume ($\beta_p = \beta_s = \alpha_p = 0$), the constant pore volume case of equation 9 is recovered. Equation 10 has been derived for constant stress boundary condition and is referred to as the constant stress case (Palciauskas and Domenico, 1982). Clearly, the fluid pressure development here is considerably less than that described for the constant pore volume case given as equation 9.

At the other extreme is the constant bulk volume boundary condition where the dilation $\epsilon$ of the bulk volume is prohibited. Geologically, this may be a representative situation where a highly expansive porous medium is constrained by a rigid, nonyielding rock. In this case, upon heating, the polycrystalline grains of the porous mixture can expand into the pores due to the higher compressibility of the saturating fluid, thereby decreasing the pore space and increasing the fluid pressure. This is a maximum pressure–producing situation, described by (Palciauskas and Domenico, 1982)

$$(\partial P/\partial T)_{m,\epsilon} = Q\,(\zeta\alpha_b + \alpha_m) \qquad (11)$$

where Q is one of Biot's (1941) constants and $\alpha_b$ is the coefficient of cubical thermal expansion of the porous body, defined as $\alpha_b = (1 - n)\alpha_s + n\alpha_p$, where $\alpha_s$ is the thermal expansivity of the polycrystalline solid grains. The coefficient Q provides a measure of the amount of water that can be forced into a material under pressure while the bulk volume of the material is kept constant. The coefficient Q thus consists of the various compressibilities of fluid, solids, and pores. For nonexpansive and incompressible solids and pores, the constant pore volume case of equation 9 is again recovered.

For realistic field situations, the boundary conditions of constant pore volume, constant stress, or constant volume are not strictly valid. The displacement field and the stress distributions should be derived from the stress-equilibrium equation and the appropriate field-boundary conditions. One simple case, assuming that the displacements are zero far from the heat source, has been solved; the results for the pressure, stress, and volume changes with temperature are shown in Table 2 for a variety of rock types.

Some of the values in Table 2 are measurements, and some are calculations from other measurements, as described by Palciauskas and Domenico (1982) and Domenico (1983). The rock types are arranged from left to right in order of what is expected to be decreasing compressibility. This is demonstrated by the decrease in $\zeta$ from unity to values considerably less than unity, and the values cited for the coefficients 1/H, 1/R, and 1/Q, which are all compressibilities of some sort. These values all decrease in the direction of increasing rock rigidity, except for 1/Q for limestone. The index of rigidity, however, is best expressed by R/H (equation 5), which represents the percentage of the load carried by the pore water, provided the pore water is not free to drain. The calculations and measurements for the drained (1/K) and undrained compressibilities ($1/K_u$) likewise reflect increasing rock rigidity, with the difference between the two becoming imperceptibly small for rigid blocks. The pressure-

temperature relations at constant stress and volume are also noted in Table 2 (equations 9 and 10).

The most interesting calculations are for the changes in fluid pressure, stress, and strain with increasing temperature. These calculations assume that the displacements are zero far from the heat source. In all lithologies, the fluid pressure increases with temperature at a faster rate than increases in stress. The net decrease in effective stress ranges from 0.09 bars °C$^{-1}$ for clays to about 4 bar °C$^{-1}$ for basalts. Further, the elastic strain with temperature is obviously greater for the soft, highly porous rocks than for the rigid rocks, an indication of pressure accommodation by way of elastic volume increases.

The results given above can be compared with calculations presented for the Gulf Coast. Gretener (1981, p. 40) notes that for the prevailing geothermal gradient, the thermal pressure development in the Gulf Coast can be expected to range between 0.255 bars m$^{-1}$ (1.25 psi ft$^{-1}$) and 0.66 bars m$^{-1}$ (3.2 psi ft$^{-1}$), which is considerably above the overburden pressure (0.204 bars m$^{-1}$ or 1.0 psi ft$^{-1}$). Magara (1975) cites a value of 0.285 bars m$^{-1}$ (1.4 psi ft$^{-1}$). Given this situation, the fluid pressure can readily exceed the overburden pressure at given depths. These calculations, however, are based on the constant pore volume case (equation 9), which is medium independent. In reality, an increase in pressure due to fluid expansion is also accompanied by an increase in stress due to the thermal expansion of the solids and the elastic volumetric expansion of the medium. For the mudstone example of Table 2, for example, the fluid pressure increases at the rate of 2.77 bars °C$^{-1}$ whereas the stress increases at the rate of 1.23 bars °C$^{-1}$, which in effect represents a net decrease in effective stress of about 1.54 bars °C$^{-1}$. For a geothermal gradient of 30°C km$^{-1}$, this results in a thermal-pressure development of about 0.0831 bars m$^{-1}$ (0.4 psi ft$^{-1}$), as opposed to the high values cited for the constant pore volume case cited above. In addition, the net decrease in effective stress is about 0.0462 bars m$^{-1}$ (0.22 psi ft$^{-1}$).

It is noted from Table 2 that the greater the rock rigidity, the greater the net decrease in effective stress with increasing depth (temperature). For the sandstone or basalt, for example, the net decrease in effective stress is about 4 bars °C$^{-1}$, compared to 1.54 bars °C$^{-1}$ for the mudstone and 0.09 bars °C$^{-1}$ for the clay. Thus, for soft, highly porous rocks that can accommodate a large component of thermal-pressure development, the decrease in effective stress to the point of failure of some sort is not a straightforward matter, even for the constant-mass case. It is noted, however, that the parameters used in the calculations of Table 2 are for low stress levels typical of the tectonically relaxed state. For basins with a high confining pressure, or for highly expansive substances in a polycrystalline mixture, the grains may readily expand into the pores, resulting in a significant enhancement of fluid-pressure development. Thus, the situations depicted in Table 2 are sample calculations for a particular set of conditions. More exact calculations for a specific area will require site specific data that reflects in situ stress conditions and, of course, consistent laboratory measurements.

## TABLE 2. CALCULATIONS AND MEASUREMENTS FOR VARIOUS ROCK TYPES*

|  | Clay | Mudstone | Sandstone | Limestone | Basalt |
|---|---|---|---|---|---|
| $\zeta = 1 - \beta_s/\beta$ | 1 | 0.95 | 0.756 | 0.69 | 0.23 |
| $1/H = \zeta/K = \beta_p$ (Kbar$^{-1}$) | 1.6 | 0.0446 | 0.0079 | 0.00237 | 0.00052 |
| $1/R = \beta_p + n(\beta_f - \beta_s)$ (Kbar$^{-1}$) | 1.7 | 0.0536 | 0.0119 | 0.00948 | 0.00438 |
| $1/Q = n(\beta_f - \beta_s) + \beta_s\beta_p/\beta$ (Kbar$^{-1}$) | 0.017 | 0.0114 | 0.0059 | 0.00784 | 0.00426 |
| $R/H = (dP/d\sigma)_m = \beta_p/\beta_p + n(\beta_f - \beta_s)$ | 0.99 | 0.83 | 0.67 | 0.25 | 0.12 |
| $1/K = \beta$ (Kbar$^{-1}$) | 1.6 | 0.0468 | 0.0105 | 0.003 | 0.00222 |
| $1/K_u = \beta_u$ (Kbar$^{-1}$) | 0.016 | 0.01 | 0.0052 | 0.00237 | 0.00216 |
| $n$ | 0.35 | 0.20 | 0.20 | 0.178 | 0.08 |
| $(\partial P/\partial T)_{m,\sigma} = R\alpha_m$ (bars °C$^{-1}$) | 0.956 | 1.73 | 5.9 | 9.13 | 8.72 |
| $(\partial P/\partial T)_{m,\varepsilon} = Q(\zeta\alpha_b + \alpha_m)$ (bar °C$^{-1}$) | 11 | 11.23 | 15.5 | 12.58 | 10.13 |
| $dP/dT$ (bar °C$^{-1}$) | 0.18 | 2.77 | 9.1 | 10.39 | 9.34 |
| $d\sigma/dT$ (bar °C$^{-1}$) | 0.09 | 1.23 | 4.7 | 5.51 | 5.4 |
| $d\varepsilon/dT$ (°C$^{-1}$) | $1.88 \times 10^{-4}$ | $9.47 \times 10^{-5}$ | $5 \times 10^{-5}$ | $2.348 \times 10^{-5}$ | $1.5 \times 10^{-5}$ |

*In part after Palciauskas and Domenico, 1982; Domenico, 1983.

### Pressure production and dissipation

The rate of pressure production due to the thermal expansion of fluid during sediment burial can be easily determined for any of the above cited cases. For the constant-stress case expressed in equation 10, the rate of pressure generation with time for a constant geothermal gradient is

$$(\partial P/\partial t)_{noniso} = R\alpha_m\omega G. \qquad (12)$$

The constant volume case of equation 11 has a similar form, as does the more realistic case where neither the stress nor the volume are kept constant. The rate of fluid-pressure dissipation will, of course, be controlled by the rate of fluid flow. In combination with pressure dissipation, the problem can be formulated simply as (Palciauskas and Domenico, 1980)

$$\partial P/\partial t = (k/S_s) \nabla^2 P + (\partial P/\partial t)_{iso} + (\partial P/\partial t)_{noniso} \qquad (13)$$

where the left-hand side describes the change in excess pressure at any given time and the right-hand side contains the various pressure-generating and dissipative terms. The pressure-genera-

tion terms include isothermal gravitational loading and thermal expansion of fluids. The pressure-depth relations given in Figure 1 represent solutions to this differential equation. Curves A, B, and C of Figure 1 are obtained by ignoring the nonisothermal response and considering one-dimensional vertical compression. Curves D and E of Figure 1 are obtained by ignoring the nonisothermal components and considering the pressure change with time to be the result of a mean stress instead of a vertical one. Curve F is obtained for the nonisothermal condition described by the constant-stress case (equation 14) in combination with an isothermal component due to gravitational loading identical to that described in curve C and a geothermal gradient of 30°C km$^{-1}$. The cumulative pressure as predicted in this particular model is seen to be in excess of the overburden pressure, a condition that is not likely to occur because of some sort of inelastic behavior of the rock.

## PHASE TRANSFORMATIONS

Included among the main phase transformations that presumably lead to excess fluid pressures are the gypsum-anhydrite transformation (Hubbert and Rubey, 1959; Hanshaw and Brede-hoeft, 1968) and the montmorillonite-illite transformation (Powers, 1967; Burst, 1969). Both transformations are apparently thermally driven and result in the production of pore fluids. The clay-mineral transformation is perhaps the best studied of the two and will be treated in this section.

Of the naturally occurring clay minerals, the diotahedral clays such as montmorillonite and beidellite are considerably less stable at elevated temperatures. From in situ observations (Gulf Coast) it has been noted that the alteration of montmorillonite to a mixed-layer clay begins at temperatures of the order of 60° to 70°C with a progressively smaller fraction of smectite/illite layers at higher temperatures (Eberl and Hower, 1976). The alteration of smectite to illite layers in a mixed-layer clay results from an increase in the net negative charge on the 2:1 layers through the substitution of Al for Si. When the charge on the layer reaches a critical value (Hower and Mowatt, 1966), water is expelled from the interlayer, converting the expandable smectite structure into that of illite. The expulsion of one interlayer of water decreases the layer spacing by approximately 2.5 Angstroms, resulting in a 16 or 20 percent decrease in the mineral grain volume per layer, depending on whether there was one initial layer of water or two. Thus the potential release of water volume is considerable and can be estimated from the overall clay mineralogy of the medium. At present it is not known if the overall volume change (smectite layer → illite layer + water) is positive, negative, or zero. Even if the net volume change is zero, the increase in the pore-fluid volume content could decrease the grain-to-grain contact area, with a resultant increase in the fluid pressure. Thus the rate of transformation of smectite to illite layers should be proportional to the decrease in the effective stress per unit time interval. This assumes that the clay minerals form a part of the stress-supporting framework in sediments. If, for example, the smectites reside in the pore space of a sand and are not part of the stress-supporting framework, then the clay transformation would have no effect on fluid pressure unless there was an overall volume increase during the transformation.

The fluid-pressure production associated with the smectite-illite transformation is thus considered to be the result of a decrease in effective stress with a commensurate increase in fluid pressure. This differs considerably from the mechanism proposed by Powers (1967). According to Powers (1967), the last few layers of bound water in the montmorillonites have a greater density than free water so that this water must undergo a volumetric increase as it is desorbed. Thus, from Powers' (1967) perspective, the pressure-production mechanism differs very little from that described for the thermal expansion of water.

Figure 3, taken from Powers (1967), is illustrative of the kinetic arguments for the decrease in effective stress discussed above. Since the rate of change in the number of expandable layers depends strongly on temperature, the evolution of the fluid pressure and all other properties depending on the kinetics of this transformation will depend in a sensitive fashion on the thermal evolution of the medium. In this diagram, the contained water

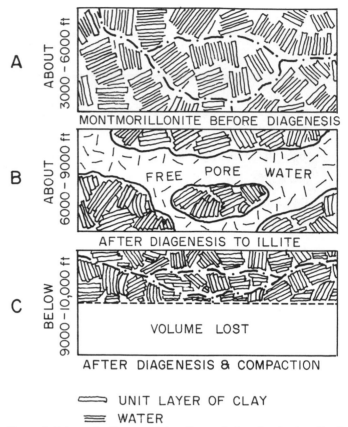

Figure 3. Schematic diagram of clay diagenesis showing A, virtually all bound water with no permeability or porosity; B, the generation of free water at the expense of the solids; and C, free water squeezed out upon completion of compaction (after Powers, 1967).

goes from the interlayered state to a state of free pore water. The pore water produced, assuming it is not free to drain, should immediately take on the pressure of the overburden that the clay minerals bore before the transformation. Thus far, it has not been possible to quantify the rate ($\partial P/\partial t$) at which this pressure is produced in model studies. It is generally assumed that the transformation can be modeled by first order kinetics

$$dN/N = -k^* \, dt \qquad (14)$$

where $k^*$ is the reaction rate parameter which has the Arrhenius form

$$k^* = Ae^{-E/RT} \qquad (15)$$

where N represents the number of expandable layers after time t, A is the frequency factor, E is the activation energy, R is the gas constant, and T is temperature in degrees Kelvin. Although the rate of transformation dN/dt can be quantified in this manner and is expected to be proportional to the decrease in effective stress per unit time, there remains the problem that the coefficient of proportionality cannot be estimated in a rigorous fashion.

Laboratory experiments by Eberl and Hower (1976) on the

kinetics of illite formation yielded an activation energy of 19.6 ± 3.6 kcal/mole for conversion of synthetic beidellite to a mixed layer illite/smectite. This corresponds closely to the energy required to break the chemical bonds in the tetrahedral sheet so that aluminum can be substituted for silicon, thereby building a negative charge on the 2:1 layers. Their experimental result for the frequency factor was $A \simeq 2$ sec$^{-1}$ which is of the same order of magnitude as the recent estimates from Gulf Coast data for this factor ($A \simeq 1 - 10$ sec$^{-1}$; D. Eberl, personal communication, 1980). To arrive at an order of magnitude estimate of the importance of this transformation in freeing water from the interlayers, the half-life $\tau$ (the time required for half of the expanded smectite layers to transform to illite layers) is presented as a function of temperature in Table 3. The data utilized are $E = 20$ kcal/mole, $A = 10$ sec$^{-1}$, $R = 2$ cal/mole deg C, and the results are computed from $\tau = .69/k^*$. If the frequency factor is taken at its upper limit of 10 sec$^{-1}$, the half life as presented in Table 3 decreases by about one order of magnitude.

Table 3 was prepared for a stationary layer, whereas in a sedimentary basin the temperature is a continuously varying function of time. Thus the rate of transformation $dN/dt$ will depend somewhat on the burial history. In particular we would expect the pressure generation through this transformation to be isolated to a rather narrow depth window because of the strong temperature dependence of the kinetics. To illustrate this point, consider the case of a basin growing at a constant rate $\omega$ in a constant geothermal gradient G. Given this time-temperature history, $T = T_s + G\omega t$ where $T_s$ is the surface temperature, equations 14 and 15 can be integrated to determine the rate of transformation $dN/dt$. Even though the constant of proportionality is not known, the decrease in effective stress and increase in fluid pressure must occur in this window if it occurs at all. Curve I of Figure 4 shows the transformation rate versus depth for the smectite-illite transformation. It is easily seen that the depth window is very narrow and is centered on the temperature $T_m$ at which the transformation rate $dN/dt$ is a maximum. This temperature can be determined by the equation

$$E/RT_m = \ln (ART_m^2/G\omega E). \qquad (16)$$

Oddly, for the reported laboratory values for the activation energy, along with assumptions concerning the Arrhenius form, the maximum rate for the transformation occurs, from a geologic perspective, almost instantaneously at near-surface temperatures. This is contrary to observations of smectite-illite occurrences in Gulf Coast sediments. For comparison, the transformation rate of bitumen to oil has also been included in Figure 4 (Curve II). The kinetics of this process has also been described by a reaction rate coefficient of the Arrhenius form with an activation energy on the order of 53 kcals/mole. For these data, it is easily seen that the maximum rate of oil production occurs at a considerably greater depth than the smectite-illite transformation, and indicates that these two processes are not temporally related. Clearly, more work on the clay transformation kinetics and/or values for the

**TABLE 3. THEORETICAL HALF LIFE FOR 50-PERCENT CONVERSION OF SMECTITE TO ILLITE AS A FUNCTION OF TEMPERATURE**

| T (°C) | τ (years) |
|--------|-----------|
| 40 | 1,600,000 |
| 80 | 41,000 |
| 120 | 2,400 |
| 160 | 230 |

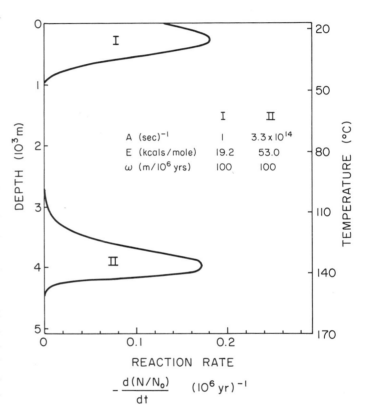

Figure 4. Transformation rate versus depth for smectite-illite (I) and bitumen to oil (II).

activation energy is required before definitive conclusions can be obtained.

Although the rate of release of layered water can be quantified in the manner described above, its significance as a pressure-producing mechanism cannot be tested in model calculations until more details on the physics of how this transformation is expressed as a pressure-production rate are understood. Currently, it is suggested that the free pore water takes on a fraction of the pressure of the overburden, and this magnitude of pressure is continually regenerated in the critical temperature range through which montmorillonite-rich sediments pass in the subsiding basin. Irrespective of the uncertainty regarding the exact location of this critical temperature range, Figure 4 suggests the presence of a rather narrow depth window associated with it.

## CONCLUDING STATEMENT

In this overview, the focus has been on literature that favors the mechanistic-theoretical approach. From this viewpoint, gravitational loading, thermal expansion of water, and the smectite-illite transformation have been singled out for detailed discussion. With rapid deposition of low-permeability sediments, gravitational loading can produce fluid pressures approaching the overburden pressure. The smectite-illite transformation will produce free pore water in a low permeability environment that in some cases can also take on the pressure of the overburden, and this magnitude of pressure can be continually regenerated in the "window" that defines the critical temperature range in a subsiding basin. The pressures produced by either of these mechanisms can be augmented by the thermal expansion of fluids, which itself is most effective in the low-permeability environment. Given that the rate of pressure dissipation by fluid flow is slower than the rate of pressure generation, some sort of inelastic behavior is likely to occur. The point of incipient fracture propagation will occur when the fluid pressure exceeds the least principal stress by an amount equal to the tensile strength of the rock. Although it is difficult to assess the pore volume that may be generated as a result of fracture, an obvious estimate is that it is sufficient to return the fluid to its ambient pressure when the tensile strength of the rock is exceeded. Thus, fracture generation may represent an instability from which there may be no way to further increase fluid pressure. In this sense, fracture generation may serve as the reserve mode of pressure dissipation that comes into play when the rate of pore-water diffusion is inadequately slow compared to the rate of fluid-pressure generation.

## REFERENCES CITED

Barker, C., 1972, Aquathermal pressuring; Role of temperature in development of abnormal pressure zones: American Association of Petroleum Geologists Bulletin, v. 56, p. 2068–2071.

Biot, M. A., 1941, General theory of three-dimensional consolidation: Journal of Applied Physics, v. 12, p. 155–164.

Biot, M. A. and Willis, D. G., 1957, The elastic coefficients of the theory of consolidation: Journal of Applied Mechanics, v. 24, p. 594–601.

Brace, W. F., Paulding, B. W., and Scholz, C., 1966, Dilatancy in the fracture of crystalline rocks: Journal of Geophysics Research, v. 71, p. 3939–3953.

Bradley, J. S., 1975, Abnormal formation pressure: American Association of Petroleum Geologists Bulletin, v. 59, p. 957–973.

Bredehoeft, J., and Hanshaw, B., 1968, On the maintenance of anomalous fluid pressures; I, Thick sedimentary sequences: Geological Society of America Bulletin, v. 79, p. 1097–1106.

Burst, J. F., 1969, Diagenesis of Gulf Coast clayey sediments and its possible relation to petroleum migration: American Association Petroleum Geologists Bulletin, v. 53, p. 73–93.

Campanella, R. G., and Mitchell, J. K., 1968, Influence of temperature variations on soil behavior: Journal Soil Mechanics and Foundations, Division of American Society of Civil Engineers, v. 94, p. 709–734.

Chia, Y. P., 1980, Digital simulation of compaction in sedimentary sequences [Ph.D. thesis]: University of Illinois at Urbana, Geology Department, 99 p.

Dickinson, G., 1953, Geologic aspects of abnormal reservoir pressures in Gulf Coast Louisiana: American Association Petroleum Geologists Bulletin, v. 37, p. 410–432.

Domenico, P. A., 1983, Determination of bulk rock properties from groundwater level fluctuations: American Association of Engineering Geologists Bulletin, v. 20, p. 283–287.

Domenico, P. A., and Mifflin, M., 1965, Water from low permeability sediments and land subsidence: Water Resources Research, v. 1, p. 563–576.

Domenico, P. A., and Palciauskas, V. V., 1979, Thermal expansion of fluids and fracture initiation in compacting sediments: Geological Society of America Bulletin, v. 90, Part I, p. 518–520; Part II, p. 953–979.

Eberl, D., and Hower, J., 1976, Kinetics of illite formation: Geological Society of America Bulletin, v. 87, p. 1326–1330.

Falvey, D. A., 1974, The development of continental margins in plate tectonic theory: Australian Petroleum Exploration Association Journal, v. 14, p. 95–106.

Fertl, W. H., 1976, Abnormal formation pressure: New York, Elsevier Press, 382 p.

Gretener, P. E., 1981, Pore pressure; Fundamentals, general ramifications, and implications for structural geology (revised): American Association of Petroleum Geologists, Educational course note series no. 4, 131 p.

Handin, J., Hager, R. V., Friedman, M., and Feather, J., 1963, Experimental deformation of sedimentary rocks under confining pressure; Pore pressure tests: American Association of Petroleum Geologists Bulletin, v. 47, p. 717–755.

Hanshaw, B., and Bredehoeft, J., 1968, On the maintenance of anomalous fluid pressures; II, Source layer at depth: Geological Society of America Bulletin, v. 79, p. 1107–1122.

Hantush, M., 1964, Hydraulics of wells, *in* Chow, V. T., ed., Advances in hydroscience, v. 1: New York, Academic Press, p. 282–430.

Heard, H. C., and Rubey, W. W., 1966, Tectonic implications of gypsum dehydration: Geological Society of America Bulletin, v. 77, p. 741–760.

Hower, J., and Mowatt, T. C., 1966, Mineralogy of the illite–illite/montmorillonite group: American Mineralogist, v. 51, p. 821–854.

Hubbert, M. K., and Rubey, W. W., 1959, Role of fluid pressure in mechanics of overthrust faulting; I, Mechanics of fluid filled porous solids and its application to overthrust faulting: Geological Society of America Bulletin, v. 70, p. 115–166.

Hubbert, M. K., and Willis, D. B., 1957, Mechanics of hydraulic fracture: American Institute of Mining Engineering Transactions, v. 210, p. 153–168.

Jacob, C. E., 1950, Flow of groundwater, *in* Rouse, H., ed., Engineering hydraulics: New York, J. Wiley and Sons, p. 321–386.

Knapp, R. B., and Knight, J. E., 1977, Differential thermal expansion of pore fluid; Fracture propagation and microearthquake production in hot pluton environments: Journal of Geophysical Research, v. 82, p. 2515–2522.

Lambe, T. W., and Whitman, R. V., 1969, Soil Mechanics: New York, J. Wiley and Sons, 553 p.

Landau, L. D., and Lifshitz, E. M., 1959, Theory of elasticity (translated from Russian by Sykes, J. B. and Reid, W. H.): Reading, Massachusetts, Addison Wesley, 134 p.

Lyell, C., 1871, Student's elements of geology: London, p. 41–42.

Magara, K., 1975, Importance of aquathermal pressuring effect in Gulf Coast: American Association Petroleum Geologists Bulletin, v. 59, p. 2037–2045.

Meinzer, O., 1928, Compressibility and elasticity of artesian aquifers: Economic Geology, v. 23, p. 263–291.

Obert, L., and Duvall, W., 1967, Rock mechanics and the design of structures in rock: New York, J. Wiley and Sons, 650 p.

Palciauskas, V. V., and Domenico, P. A., 1980, On microfracture development in compacting sediments and its relation to hydrocarbon maturation kinetics: American Association of Petroleum Geologists Bulletin, v. 64, p. 927–937.

——— , 1982, Characterization of drained and undrained response of thermally loaded repository rocks: Water Resources Research, v. 18, p. 281–290.

Powers, M. C., 1967, Fluid release mechanisms in compacting marine mudrocks and their importance in oil exploration: American Association of Petroleum

Geologists Bulletin, v. 51, p. 1240–1254.

Smith, J. E., 1971, The dynamics of shale compaction and the evolution of pore fluid pressures: Journal of Mathematical Geology, v. 3, p. 239–263.

Strauss, J. M., and Schubert, G., 1977, Thermal convection of water in a porous medium; Effects of temperature and pressure dependent thermodynamic and transport properties: Journal of Geophysics Research, v. 82, p. 325–333.

Terzaghi, K., 1925, Erdbaumechanik auf Bodenphysikalischer Grundlage: Vienna, Franz Deuticke, 399 p.

Terzaghi, K., 1950, Mechanisms of landslides, *in* Paige, S., and 18 others, Application of geology to engineering practice (Berkey Volume): Geological So-

ciety of America, p. 83–123.

Watts, E. V., 1948, Some aspects of high pressure in the D7 zone of the Ventura Avenue Field: American Institute of Mining Engineering, Petroleum Division, v. 174, p. 191–200.

Watts, A. B., and Ryan, W. B., 1976, Flexure of the lithosphere and continental margin basins, *in* Sedimentary basins of continental margins and cratons: Developments in Geotectonics, v. 12, p. 25–44.

MANUSCRIPT ACCEPTED BY THE SOCIETY MARCH 14, 1987

The Geology of North America
Vol. O-2, Hydrogeology
The Geological Society of America, 1988

## Chapter 46

# *Ground water and fault strength*

**S. A. Rojstaczer**
*U.S. Geological Survey, 345 Middlefield Road, Menlo Park, California 94025, and Department of Applied Earth Sciences, Stanford University, Palo Alto, California 94305*

**J. D. Bredehoeft**
*U.S. Geological Survey, 345 Middlefield Road, Menlo Park, California 94025*

## INTRODUCTION

The influence of pore fluids on crustal failure is a relatively recent field of study. Mead (1925) recognized that the presence of pore water could promote brittle failure in the crust. However, it was not until the classic paper of Hubbert and Rubey (1959) that anyone considered the influence pore water may have on fault motion. According to Hubbert and Rubey, crustal fluids could strongly influence fault strength; they argued that high fluid pressures significantly reduced the shear stress along a fault plane necessary to initiate failure. Hubbert and Rubey were concerned with displacement along overthrust faults, but their analysis has far wider application. Natural mechanisms that raise the pressure of ground water in any fault zone may strongly reduce fault strength.

Direct measurement of high fluid pressures at depth within fault zones is lacking; however, many observations that have been made along active and exhumed ancient fault zones can be readily explained if ground water exists in the mid-to-upper crust at pressures that at least episodically approach lithostatic. Seismic reflection profiles across a creeping portion of the San Andreas fault indicate that at least some portions of this fault possess low-velocity zones at seismogenic depths (Raleigh and Evernden, 1981). Measurements of heat flow along the San Andreas fault suggest that this fault zone is in a state of low deviatoric stress (Brune and others, 1969; Lachenbruch and Sass, 1980). Failure under apparent conditions of low deviatoric stress is not unique to the San Andreas fault. Many exhumed fault zones show no evidence of local melting due to shear heating; this apparent lack of heating suggests that, in these fault zones, fault motion took place under low stress (Sibson, 1973, 1980). The presence of aseismic creep without strain accumulation along the San Andreas fault suggests that the effective normal stress along parts of major fault zones is low enough to promote stable sliding (Thatcher, 1979; Byerlee and Brace, 1972; Dieterich, 1979).

The Hubbert and Rubey theory was invoked to explain fault movement at the Rocky Mountain Arsenal near Denver, Colorado, where deep injection of fluids was postulated to have induced seismicity along historically dormant faults (Evans, 1966). Since this apparent occurrence of pore pressure–induced seismicity, man-made changes in pore pressure have been inferred to have induced fault movement at tens of sites throughout the world (Woodward and Clyde Consultants, 1979). Some of these cases of pore pressure–induced seismicity are well documented and are of great scientific value. They provide considerable insight as to how fault movement can be retarded or enhanced by changes in fluid pressure; they provide partial constraints on the extent to which a process which perturbs pore pressure at depth may influence fault behavior; finally, they provide a partial check on the validity of models which characterize the mechanical interaction of ground water and faults.

In this chapter we use data obtained from induced seismicity studies in an attempt to critically examine current understanding of the influence of perturbations in ground-water pressure on fault strength. The principal topics to be addressed are: (1) the value and limitations of laboratory models that describe the influence of ground water on fault strength; (2) the constraint that a permeable crust places upon the natural generation of pore-pressure perturbations at seismogenic depths.

## INTERACTION BETWEEN GROUND WATER AND FAULT RUPTURE–DEPTH CONSTRAINTS

Most models which attempt to describe the interaction of ground water and fault rupture assume that the crust is fluid saturated at the focus of rupture. The depth to which ground water can be expected to be present provides at least a first-order constraint on the depth to which ground water may influence fault movement.

Geophysical, geochemical, and petrological evidence suggests that ground water is present (at least episodically) to mid-crustal depths in areas of currently active tectonics. In-situ and

Rojstaczer, S. A., and Bredehoeft, J. D., 1988, Ground water and fault strength, *in* Back, W., Rosenshein, J. S., and Seaber, P. R., eds., Hydrogeology: Boulder, Colorado, Geological Society of America, The Geology of North America, v. O-2.

laboratory electrical resistivity measurements suggest that continuous pore spaces can be present and saturated to depths of at least 20 km (Brace, 1971; Nekut and others, 1977). There are ample data from stable isotope and micrographic analyses that indicate ground water of either meteoric of metamorphic origin can be present at least to mid-crustal depths. Stable oxygen isotope studies indicate that meteoric water has penetrated the epizonal perimeters of essentially all continental batholiths (Taylor, 1978; O'Neil and Hanks, 1980) and also indicate that the emplacement of the Skaergaard Intrusion of Greenland induced circulation of ground water of meteoric origin to depths as great as 10 km (Taylor and Forester, 1979). Deep circulation of ground water is not limited to plutons and their surrounding country rocks. Ground water of meteoric origin has been inferred from stable oxygen isotope studies to migrate to depths of 15 km within a Precambrian shear zone (Kerrich and others, 1984). The presence of ground water of metamorphic origin at mid-crustal depths has been inferred from micrographic analyses of syn-metamorphic extension fractures within low-grade metamorphic rocks (Norris and Henley, 1976; Etheridge and others, 1984).

It is fruitful to compare the above data on the depths to which ground water may be found in regions of active tectonism with the depths to which most microseismicity and large shocks ($M_L > 5.5$) occur within the continental crust in regions of elevated heat flow. Sibson (1982) noted that in the parts of the western United States with background heat flow of 1.5 to 2.5 HFU, earthquakes of magnitude greater than 5.5 from 1966 to 1980 were generally confined to a depth range of 8 to 12 km. Figure 1 shows the depth distribution of microearthquake activity during the 1970s in several parts of the western U.S.; microseismic activity, with the exception of the Transverse Ranges of southwest California, appears to be largely confined to the upper 15 km of the crust. It is difficult to state with any degree of confidence the extent to which ground water is present at mid-crustal depths. However, the correspondence between depth constraints on seismicity and ground-water flow point to the possibility that brittle failure within continental crust commonly occurs under fluid-saturated conditions.

## GROUND WATER AND FAULT STRENGTH: THEORY

Frictional slip along pre-existing failure surfaces is generally accepted as the principal mechanism for fault motion in the mid- to-upper crust (Dieterich, 1974). The mechanics of fault motion and the influence that pore fluids have upon frictional sliding can be roughly described with the use of the friction apparatus shown in Figure 2a; a similar apparatus was employed by Hubbert and Rubey (1959). An impermeable concrete rider is placed upon a porous concrete flat surrounded by a water-retaining vessel with a rubber-coated lip (the rubber coating is used to effect a water tight seal between the rider and flat). It should be noted that, unlike this laboratory apparatus, a fault zone in nature will be

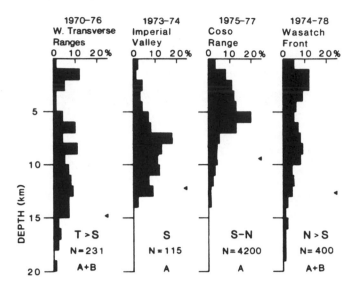

Figure 1. Earthquake depth distributions in the continental crust of the western United States (after Sibson, 1984). Background heat flow in the areas examined is 1.5 to 2.5 HFU. Solid triangles indicate depths above which 90 percent of the activity occurs; the dominant faulting mode (T, thrust; S, strike slip; N, normal), the sample size, and the data quality are listed at the base of each histogram.

composed of and surrounded by rocks of some permeability. In this simple model of fault motion we ignore ground-water flow and concentrate solely on the influence that a pressurized fluid within a fault zone has on fault strength. The effects of ground-water flow are included in a subsequent portion of this section.

In the first experiment, the vessel surrounding the porous flat is simply filled to its top, and the water level in the piezometer is $P_0$. An attached spring of stiffness K and length B-A is used to apply a tangential force to the rider by moving the end of the spring to the right at a velocity V. The force in the spring is plotted as a function of the displacement at the end of the spring in Figure 2b (points along this curve are denoted by the subscript 0). For an explanation of this experiment we quote from Byerlee (1978):

There will be an initial elastic increase in force until the point C where the curve departs from a straight line. This indicates that there is relative displacement between the rider and flat or that the rider or flat is deforming nonelastically. At the point D a maximum is reached and the rider may suddenly slip forward and the force in the spring will suddenly drop to the point E. The force will increase again until sudden slip takes place once more at the point F. This sudden jerky type of movement is known as stick-slip. An alternative mode is stable sliding. In this case the movement between the rider and the flat takes place smoothly and the force displacement curve will be continuous . . .

The two modes of sliding noted in the experiment above, unstable stick slip and stable sliding, are qualitatively analogous to earthquake rupture and stable creep along active faults. Both modes of sliding are caused by a drop in the frictional resistance of the flat

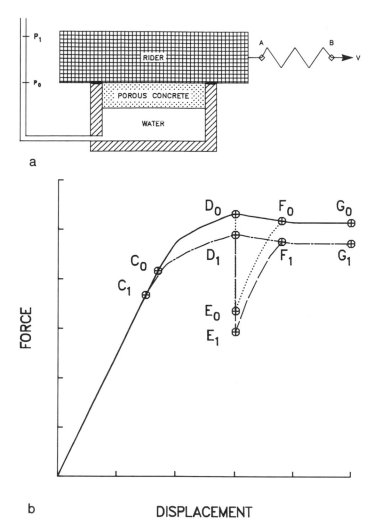

Figure 2. (a) Idealized friction experiment used to analyze motion of simulated wet faults under shear. (b) Qualitative force-displacement curves of the friction experiment. Subscripts 0 and 1 indicate when fluid pressure is $P_0$ and $P_1$ respectively. See text for explanation of curves.

plate once motion occurs. Dieterich (1979, 1981) has shown that the mode of slip is controlled by the rate of decline in frictional resistance with slow rates of decline promoting stable slip. The rate of decline is in turn controlled by the stiffness of the spring, the roughness of the surface, and the effective stress (to be defined below) normal to the failure surface. Stick slip was first postulated as an earthquake mechanism by Brace and Byerlee (1966), who based their hypothesis on the behavior of dry rock samples in ambient humidity under failure induced by compression. Dieterich (1978, 1979, 1981) and Dieterich and Conrad (1984) have persuasively argued that stick slip is due to time-dependent strengthening of contacts by indentation creep along the failure surface. Frictional resistance begins to drop once motion occurs because ". . . the average age of the contacts evolves from the initial hold time to a smaller time of contact characteristic of the sliding velocity" (Dieterich and Conrad, 1984, p. 4197).

In a second experiment, the vessel beneath the porous flat is filled with water to a pressure $P_1$, which is less than the pressure at which water begins to leak out the top of vessel and greater than $P_0$, the hydrostatic pressure of the original experiment. The spring is once again moved at velocity V, and the force in the spring is plotted as a function of displacement at the end of the spring. Comparison of the force displacement curves of the two experiments will typically indicate the results shown in Figure 2b. At the points C, D, E (if the mode of displacement is stick slip), F, and G the curves are displaced by a force $(P_1-P_0)a$, the difference in fluid pressure of the two experiments multiplied by the surface area of the top of the vessel. The increase in fluid pressure may also be sufficient to alter the mode of slip since high fluid pressures promote stable sliding (Byerlee and Brace, 1972; Dieterich, 1981).

The reduction in frictional resistance with increasing pore pressure is not due to any alteration in the frictional characteristics of the sliding surface. Frictional resistance is reduced because the fluid, when pressurized, exerts a buoyant force on the sliding block. Sliding in both experiments is governed by the normal component of the effective stress exerted on the flat by the rider, where the normal component of the effective stress, $\bar{\sigma}_n$, is defined as the difference between the normal stress, $\sigma_n$, and a fraction, $\alpha$, of the fluid pressure P along the flat (Nur and Byerlee, 1971):

$$\bar{\sigma}_n = \sigma_n - \alpha P \qquad (1)$$

It should be noted that compressive stresses are positive. For both experiments, $\alpha$ will be approximately equal to unity. In the first experiment, the fluid pressure on the flat is zero (since the fluid only wets the flat), and the normal component of the effective stress equals the normal stress; in the second experiment, the normal component of the effective stress is less than the normal stress. We quote from Hubbert and Rubey (1959) to explain the impact that this reduction in stress has on sliding: ". . . the reduction of friction is accomplished by reducing the normal component of effective stress which correspondingly diminishes the critical value of the shear stress required to produce sliding."

The phenomena of stick-slip, stable sliding, and the reduction in frictional resistance with increasing fluid pressure have been observed within the laboratory when rocks that contain failure planes are subjected to shear stress. The force per unit area at the points C, D, and G are known as the initial, maximum, and residual friction, respectively. Each is a measure of the frictional strength of the rock. Figure 3 is a plot of the shear stress versus effective normal stress at the point of maximum friction for a variety of rocks under intermediate to high stress (normal stress on the failure plane is greater than 50 bars) and at temperatures up to 400°C. Byerlee (1978) noted that with the exception of certain clay minerals, the relation between shear stress and normal stress appears to be remarkably independent of mineralogy, fracture roughness, and temperature. As shown in the figure, the shear stress at maximum friction is given approximately by the equations:

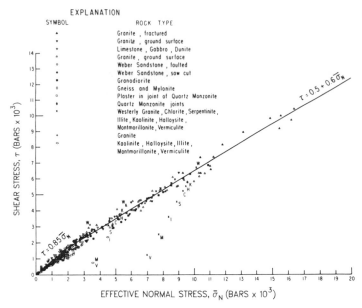

Figure 3. Shear stress along failure plane as a function of effective normal stress at maximum friction for a variety of rock types at normal stresses to 17 kb. Letters denote mineral type for tests where either the entire sample consists of or the failure plane is lined with that mineral (modified from Byerlee, 1978).

$$\tau = 0.85\bar{\sigma}_n \qquad 0.05 < \bar{\sigma}_n < 2\text{kb} \qquad (2a)$$

$$\tau = 0.5 + 0.6\bar{\sigma}_n \qquad 2 < \bar{\sigma}_n < 20\text{kb} \qquad (2b)$$

Equations 2a and 2b have been referred to as Byerlee's Law (Stesky, 1978; Brace and Kohlstedt, 1980) and indicate that the shear stress required to cause sliding can be greatly reduced by the presence of fluids in the failure plane. Zoback and Healy (1984) noted that the coefficient of proportionality given by Byerlee's Law when the stress normal to the failure surface is less than 2 kb (Equation 2a) is roughly a median value and that this coefficient commonly ranges between 0.60 and 1.0. Byerlee's Law is functionally identical to the Coulomb theory of failure, which states that shear fracture takes place across a plane on which the following condition holds:

$$\tau = \tau_0 + \mu\bar{\sigma} \qquad (3)$$

where $\tau_0$ is the cohesive shear strength, and $\mu$ is the coefficient of internal friction. However, Byerlee (1967, 1975) has shown that the maximum friction and fracture strength of rocks are not related in a simple manner.

For purposes of testing the applicability of Byerlee's Law on a crustal scale, it is useful to rewrite it in terms of the maximum and minimum principal effective stresses. Brace and Kohlstedt (1980) have noted that if the failure surface is parallel (or near parallel) to the intermediate principal stress and at an angle with the least principal compressive stress which is not less than 55° and not greater than 75°, then Byerlee's Law can be stated in

terms of the principal effective stresses as:

$$\bar{\sigma}_1 \approx 5\bar{\sigma}_3 \qquad \bar{\sigma}_3 < 1.1\text{kb} \qquad (4a)$$

$$\bar{\sigma}_1 \approx 3.4\bar{\sigma}_3 + 1.8 \qquad \bar{\sigma}_3 > 1.1\text{kb} \qquad (4b)$$

where $\bar{\sigma}_1$ and $\bar{\sigma}_3$ are the greatest and least principal effective stress respectively. The derivation of this transformation can be found in the Appendix.

By assuming that pore pressures are hydrostatic to subhydrostatic, Brace and Kohlstedt (1980) indicate that Byerlee's Law provides an upper bound on the ratio of greatest to least principal stress measured in situ to a depth of 5 km and collected by McGarr and Gay (1978). Zoback and Hickman (1982) present in-situ stress measurements that show that Byerlee's Law provides an upper bound on measured in-situ stress ratios if pore pressures in their study area are hydrostatic. Qualitatively, these observations indicate that the magnitude of the components of the deviatoric stress in the crust can be limited by the laboratory-determined frictional strength of rock and the pressure of pore fluids. They also suggest that fault strength, at least in the upper crust, may be governed by processes similar to those that govern frictional sliding within the laboratory. In the following section of this chapter, we will present some quantitative data that at least partially support the thesis that laboratory observations can be applied on a field scale.

Although laboratory models of frictional sliding have greatly improved understanding of the factors that control fault motion, they are unable to describe the spatial and time-dependent interactions between ground water and faults that can operate within the crust. Within these models, pore pressure and stress are assumed to be homogeneous, and pore pressure is also usually assumed to be time invariant. As a result of these limitations: (1) ground-water flow does not occur; (2) time-dependent changes in stress do not affect fluid pressure; and (3) fluid pressures do not alter the total stresses within and around the zone of sliding. We can incorporate the essential aspects of the temporal and spatial elements of ground water–fault interaction through the use of a simple modification of the poroelastic theory of Biot (1941) if we limit our analysis to the events that occur prior to fault slip. The time and spatial dependence of fluid pressure can be described by a modified diffusion equation:

$$\frac{\partial}{\partial x_i}\left(\frac{k}{\mu_f}\frac{\partial P}{\partial x_i}\right) = \frac{9(\nu_u - \nu)}{2GB^2(1+\nu)(1+\nu_u)}\left(\frac{\partial P}{\partial t} - BA\right) \qquad (5)$$

where k is the permeability, $\mu_f$ is the fluid viscosity, P is the fluid pressure, $\nu$ and $\nu_u$ are the drained and undrained Poisson ratio, respectively, G is the shear modulus, B is one of Skempton's pore-pressure coefficients (Skempton, 1954), and A is a nonspecific source term with dimensions of rate of change of pressure with time. If we take A to equal the effects of the rate of change of mean normal stress with time, then the above equation reduces to the Rice and Cleary (1976) formulation for the diffusive aspects of Biot's theory of poroelasticity. For our purposes, however, A represents the sum of all sources and sinks at a point that affect fluid pressure. These sources and sinks may be due to processes such as pumping from a well, volumetric strain, temperature changes, porosity reduction, and dehydration of rock and sediment.

The relation between deformation (and by extension total stresses) and fluid pressure is described by (Rice and Cleary, 1976):

$$\frac{G}{(1-2\nu)} \frac{\partial \epsilon}{\partial x_i} + G\nabla^2 u_i = \frac{3(\nu_u - \nu)}{B(1+\nu_u)(1-2\nu)} \cdot \frac{\partial P}{\partial x_i} \qquad (6)$$

where G is the shear modulus, $\epsilon$ is the dilatation and $u_i$ are the displacements. This equation is analogous to the displacement equilibrium equation of classical elasticity: the term on the right-hand side plays the part of the body force (Timoshenko and Goodier, 1970, p. 242). Biot (1941) gives the above relations in a slightly different form. In this paper we ignore the effect of pore pressure on total stresses and concentrate on how the diffusive aspects of ground-water flow alter effective stresses within the crust. The impact of shallow fluid pumpage on total stresses has been discussed by Segall (1985).

If we assume that the permeability k is homogeneous, then Equation 5 can be readily placed in dimensionless form. We adopt the following scaling factors for the variables:

$$P = P_c \cdot \hat{P} \qquad (7a)$$
$$t^{-1} = (t_c \cdot \hat{t})^{-1} \qquad (7b)$$
$$x = H_x \cdot \hat{x} \qquad (7c)$$
$$y = H_y \cdot \hat{y} \qquad (7d)$$
$$z = H_z \cdot \hat{z} \qquad (7e)$$

where $P_c$, $t_c$, $H_x$, $H_y$ and $H_z$ are characteristic scalings for fluid pressure, time, and length respectively and $\hat{P}$, $\hat{t}$, $\hat{x}$, $\hat{y}$, and $\hat{z}$ are dimensionless. For the characteristic lengths we take the thickness $H_z$ and widths, $H_x$ and $H_y$ over which the pressure source A operates. For $P_c$ we choose the maximum excess fluid pressure, which is the difference between lithostatic and hydrostatic pressure at depth $H_z$:

$$P_c = (\rho_r - \rho_f)gH_z \qquad (8)$$

where g is the acceleration due to gravity and $\rho_r$ and $\rho_f$ are the density of the rock and the fluid respectively. For $t_c^{-1}$ we pick the sum of the reciprocals of the characteristic diffusion times in each dimension:

$$t_c^{-1} = c(H_x^{-2} + H_y^{-2} + H_z^{-2}) \qquad (9)$$

where c is a diffusion coefficient, which under special conditions is equal to the hydraulic diffusivity:

$$c = \left( \frac{9\mu_f(\nu_u - \nu)}{2kGB^2(1+\nu)(1+\nu_u)} \right)^{-1} \qquad (10)$$

Insertion of the above (Equations 7 through 10) scalings and characteristic parameters into Equation 5 yields:

$$R_x^2 \frac{\partial^2 \hat{P}}{\partial \hat{x}^2} + R_y^2 \frac{\partial^2 \hat{P}}{\partial \hat{y}^2} + \frac{\partial^2 \hat{P}}{\partial \hat{z}^2} + = (R_x^2 + R_y^2 + 1) \cdot \frac{\partial \hat{P}}{\partial t} + D_4 \qquad (11)$$

where $R_x$ and $R_y$ are length ratios $H_z/H_x$ and $H_z/H_y$, respectively, and $D_4$ is analogous to the Damkohler Group IV number, which describes heat generation and transport in chemical reactions (Boucher and Alves, 1959):

$$D_4 = \frac{BAH_z}{(\rho_r - \rho_f)gc} . \qquad (12)$$

Equation 11 allows us to inspect the relative ability of any fluid pressure sourcee to elevate fluid pressure within the crust and by extension reduce the strength of any faults present. Physically, $R_x$ and $R_y$ represent the degree that horizontal flow dissipates fluid pressures in excess of hydrostatic, and $D_4$ is a ratio of the volumetric strength of the pressure source A to the capacity for vertical conductive fluid flow. The ability of any pressure source to significantly raise fluid pressures is clearly enhanced by small values of $R_x$ and $R_y$; under these conditions, fluid pressurization is laterally extensive, and fluid flow is predominately one dimensional. It is also enhanced by large values for $D_4$, which occur when the pressure source operates at a rapid rate over a thick interval in a region of poor permeability. When $D_4$ is much greater than one and $R_x$ and $R_y$ are on the order of one or less, the crust responds to the pressure source in an undrained fashion and fluid flow is ineffective at dissipating pressures in excess of hydrostatic. Conversely, when $D_4$ is much less than one or when $R_x$ or $R_y$ are much greater than one, the pressures produced are rapidly dissipated, and the source does not generate significant excess fluid pressures.

The control on fluid pressures exerted by the dimensionless group $D_4$ and the length ratios $R_x$ and $R_y$ can be clearly shown by solving Equation 11 for the very simple case where pressures are produced by a pressure source that is uniform and operates at a constant rate A over a rectangle with thickness $H_z$ and half-width $H_x$ (the dimension $H_y$ is assumed to be much greater than either $H_z$ or $H_x$); the top and sides of this rectangle are in excellent hydraulic communication with the water table, and the bottom is impermeable. Figure 4a shows the geometry and boundary conditions for this simple model. Figures 4b through 4d show depth profiles of the dimensionless excess pressure as a function of $D_4$ and $R_x$ once steady state has been achieved; the profiles are taken at the mid-line of the rectangle, and dimensionless excess pressure is defined as $P/(\rho_r-\rho_f)gH_z$. If $R_x$ is less than 2.0, excess pore pressure generation is largely limited to the case where $D_4$ is greater than 0.02. The physical meaning of this limitation can be understood if we assign some values to the parameters that make up $D_4$. If we assume that the pressure source A operates at a rate of $2 \times 10^{-12}$ bars $sec^{-1}$ (a rate that approximates the effects of sediment loading along the Gulf Coast on fluid pressure) over a laterally extensive ($H_x$ much greater than $H_z$) thickness of 10 km, B is 0.9, $\rho_f$ is approximately 1.0, and $\rho_r$ is 2.7 gm $cm^{-3}$, then c would have to be less than $5 \cdot 10^{-2}$ $cm^2$ $sec^{-1}$ for there to be any excess pore-pressure generation. A qualitatively similar result was

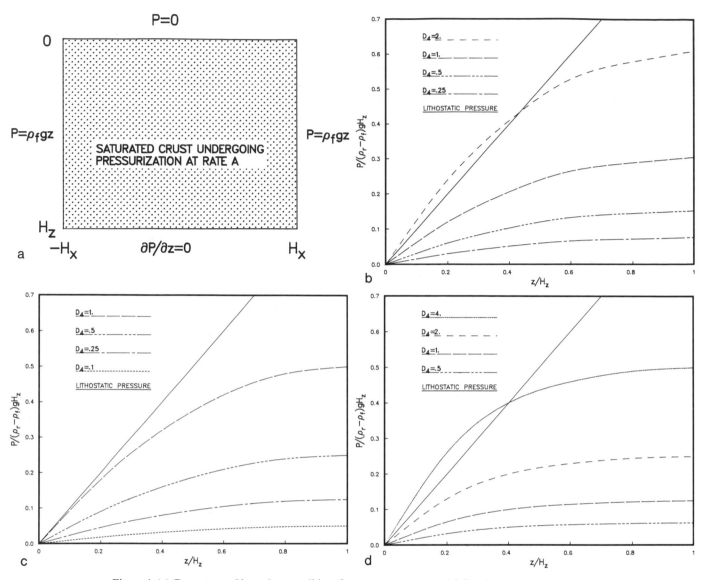

Figure 4. (a) Geometry and boundary conditions for pressure source model. Steady-state excess-pressure profile at x = 0 as a function of $D_4$ for three cases: (b) $R_x = 0$; (c) $R_x = 1$; (d) $R_x = 2$. Initial pressure is assumed to be hydrostatic.

obtained by Bredehoeft and Hanshaw (1968) in their analysis of pore-pressure generation along the Gulf Coast. The generation of excess pore pressures is discussed in Domenico and Palciauskas (this volume), so we will not discuss this topic in any detail. However, the dimensional analysis and the simple model for pore-pressure generation above allow us to examine the effects of natural pressurization mechanisms within the crust on fault strength in a semi-quantitative fashion.

## INJECTION INDUCED SEISMICITY—IMPLICATIONS FOR THE STATE OF EFFECTIVE STRESS AT FRICTIONAL FAILURE

The missing link in the material so far presented is the fluid pressure at which (relative to lithostatic) we can typically expect

ground water to influence crustal failure. We can obtain some information on this topic by comparing in-situ stress measurements with Byerlee's Law. Figure 5 shows the ratio of the least to greatest principal stress at 19 sites throughout the world determined from hydraulic fracturing tests. This ratio varies from 0.3 to 0.8 (values as high as 1.0 are possible), and there appears to be no trend in this ratio with depth. Under the assumption that fault planes are most favorably oriented relative to the principal stresses, we apply Byerlee's Law in terms of effective stresses (Equation 4a) to this data to obtain a family of vertical lines that demarcate the value of this ratio when frictional failure occurs. The lines demarcate the conditions for frictional failure for two different fluid pressures (hydrostatic and ½ lithostatic) and stress orientations. The family of lines indicate that the ratio $\bar{\sigma}_3/\bar{\sigma}_1$ at frictional failure is a strong function of fluid pressure. Comparison of the failure criteria with in situ stress measurements indicates that in many areas, conditions for frictional failure are greatly enhanced within the upper crust if fluid pressures are at least ½ that of the lithostatic load. Since hydrostatic pressure is roughly 2/5 of lithostatic, a fluid pressure rise of about 1/10 of lithostatic at depth will cause failure to occur in many areas if fluid pressures are initially hydrostatic and the least principal stress has a horizontal orientation. If Byerlee's Law is correct, fluid pressure rises of only tens of bars within the upper crust can, when the minimum principal stress is less than 0.6 of the maximum principal stress and fluid pressures are initially hydrostatic, result in frictional movement along appropriately oriented failure surfaces. Partial verification of this result can be obtained from studies of induced seismicity.

Studies of seismicity induced by fluid injection into wells at depth and by reservoir loading can provide a direct means to test the validity of models that describe the influence of pore pressure on fault strength. In areas with induced seismicity where the state of stress prior to failure, the pore pressure under natural conditions, the location and orientation of the fault plane, and the strength of the man-made pressure source are known, models can be tested quantitatively. It is also frequently possible to infer the permeability of the crust on a large scale in these areas because the source(s) of the pore pressure disturbance and the point(s) of failure are often separated by a distance of several kilometers. A review of induced seismicity studies is well beyond the scope of this paper. We will focus on three cases of induced seismicity, all three of which involved fluid injection at depth: (1) seismicity induced by injection of fluid wastes at the Rocky Mountain Arsenal near Denver, Colorado, (2) seismicity deliberately induced by water injection at Rangley, Colorado, and (3) seismicity induced by hydraulic mining in western New York. The data available from these three sites allow us to at least partially test whether laboratory-derived conditions for failure adequately describe how fault strength is affected by fluid pressurization on a field scale. They also yield information on the permeabilities of material near and within fault zones.

### Rocky Mountain Arsenal

Injection-induced seismicity at the RMA (Rocky Mountain Arsenal) near Denver, Colorado, has been the subject of numerous studies (Evans, 1966; Hollister and Weimer, 1968; Healy and others, 1968; Hoover and Dietrich, 1969; van Poollen and Hoover, 1970; Hsieh and Bredehoeft, 1981; Hermann and others, 1981; Zoback and Healy, 1984). The description below summarizes the information contained in these studies pertinent to the subject of this chapter. From 1962 to 1966, contaminated wastewater was injected into the bottom 21 m of a 3,671-m well. Shortly after the initiation of injection, minor earthquakes were detected in the Denver area. More than 1,500 earthquakes (some greater than 4.0 $M_L$) in the Denver area were recorded between 1962 and 1967. A seismic survey conducted by the U.S. Geologi-

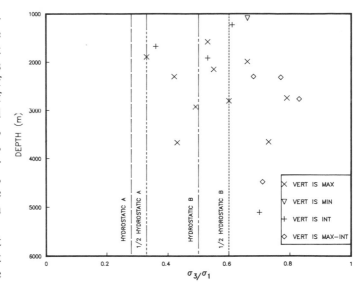

Figure 5. Ratio of least to greatest principal stress at 19 sites throughout the world where depth of measurement is greater than one km. Vertical lines demarcate failure criteria when fluid pressure is hydrostatic and ½ lithostatic; case A is for reverse faulting conditions; case B is for normal faulting conditions. Symbols denote whether the vertical stress is the greatest, least, or intermediate stress; measurements where only the least principal and vertical stress are known are denoted as "MAX-INT." Data from Swolfs (1984) and McGarr and Gay (1978).

cal Survey between 1966 and 1968 at the RMA indicated that earthquake epicenters were confined to an elliptical zone 10 km long and 3 km wide, which also contained the injection well. The trend of the major axis of the seismic zone was approximately N60°W, and the vertical extent of the zone was approximately between the depths of 3.5 and 7.5 km. Recent studies of the surface wave and P-wave first motion of three of the largest earthquakes indicate that seismicity was due to normal faulting; the strike of the nodal planes nearly parallels the epicentral trend. Figure 6a shows a comparison of the volume of fluid injected at the RMA and the frequency of earthquakes.

Table 1 indicates the pertinent parameters for the RMA: principal stresses, shear stress and normal stress on the fault plane, bottom-hole fluid pressure prior to injection, inferred fluid pressure necessary to induce failure, and permeability of the highly fractured Precambrian gneiss into which the wastewater was injected. It should be noted that the techniques utilized to estimate these parameters were not rigorous and that the values indicated here are approximate. The vertical stress was assumed a priori to be a principal stress and was determined simply by multiplying the bottom-hole depth by a lithostatic gradient of .226 bars/m; the minimum stress was inferred from the down-hole fluid pressure at which the injection rate began to rapidly increase (Healy and others, 1968). Water-level measurements after the end of injection were used to infer the bottom-hole fluid pressure prior to injection. The pressure necessary to induce failure and the

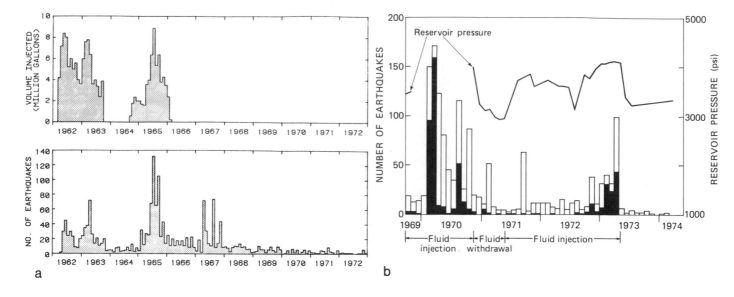

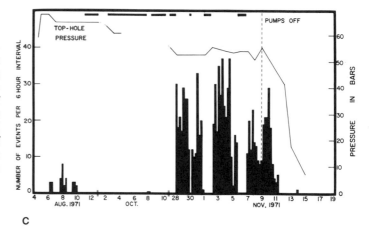

Figure 6. (a) Comparison of fluid injected and the frequency of earthquakes at the RMA. Upper graph shows monthly volume of fluid waste injected in the disposal well. Lower graph shows number of earthquakes per month (after Hsieh and Bredehoeft, 1981). (b) Comparison of pressure in one of the experimental wells (solid line) and frequency of earthquakes at Rangley. Shaded portion of bars indicate earthquakes within 1 km of the experimental wells; clear portion of bars denote all others (modified from Raleigh and others, 1976). (c) Comparison of top-hole pressure of injection well and frequency of earthquakes at Dale; gaps in seismic data indicated by solid bars at top of plot (modified from Fletcher and Sykes, 1977).

permeability of the material within and surrounding the failure zone were determined from a reservoir model of the response of the pore fluid to injection (Hsieh and Bredehoeft, 1981).

The principal stresses and critical values of fluid pressure for the RMA indicate that there is a considerable discrepancy between the actual fluid pressure necessary to induce failure and that predicted by a simple application of Byerlee's Law. Our use of Byerlee's Law indicates that shear stress at RMA is high enough to cause failure if fluid pressures are only 2/3 that of hydrostatic, a condition exceeded by pre-injection fluid pressures. In making the calculation of the theoretical fluid pressure at failure, $P_B$, we assumed that failure occurred along a plane which was favorably oriented relative to the principal stresses; i.e., the failure plane was nearly parallel to the intermediate stress, and the angle between the failure plane and the least principal stress was not less than 55° and not greater than 75°. These assumptions are weakly supported by the character of the seismicity and stress indicators within the region (Hermann and others, 1981; Zoback and Zoback, 1980).

The discrepancy between the actual and theoretical fluid

pressure at failure may or may not be due to any inadequacy inherent in the application of Byerlee's Law. If we attempt to attribute this discrepancy to a deviation from the approximate median coefficient of internal friction given by Byerlee's Law, we find that the coefficient of internal friction for the failure zone at RMA is 1.0; this value is at the upper bound of the common range of values for this coefficient determined experimentally for crystalline rock (Zoback and Healy, 1984). However, other factors not related to Byerlee's Law may contribute to this discrepancy. The minimum principal stress may be greater than has been determined from the injection history of the well; the zone of failure may not be favorably oriented relative to the principal stresses and as a result Equation 4a (Byerlee's Law transformed into a relationship between principal effective stresses) is not applicable; finally, the coefficient $\alpha$ in the effective stress law $\sigma-\alpha P$ may not equal one as was assumed in our calculations. The coefficient $\alpha$ would have to equal to 0.8 for the predicted and actual fluid pressure to correspond, a value which is reasonable for crystalline rock at depth (Rice and Cleary, 1976; Houston and Kasim, 1982).

**TABLE 1. CONDITIONS AT FAILURE FOR AREAS STUDIED\***

| Site | $\sigma_1$ | $\sigma_2$ | $\sigma_3$ | $\tau$ | $\sigma_n$ | $P_o$ | $P_f$ | $P_h$ | $P_B$ | $\mu$ | k |
|------|------|------|------|------|------|------|------|------|------|------|------|
| RMA | 838 | ----- | 362 | 210 | 500 | 269 | 301 | 367 | 240 | 1.0 | 3 |
| Rangely A | 590 | 427 | 314 | 88 | 349 | 170 | 275 | 176 | 245 | 1.19 | 6 |
| Rangely B | 590 | 427 | 314 | 134 | 474 | 170 | 275 | 176 | 316 | 0.67 | 6 |
| Dale | 223 | 133 | 133 | 37 | 174 | 51 | 105 | 51 | 130 | 0.54 | 200 |

\*Symbols are: $\sigma_1$, greatest principal stress; $\sigma_2$, intermediated principal stress; $\sigma_3$, least principal stress; $\tau$, shear stress on fault plane; $\sigma_n$, normal stress on fault plane; $P_o$, fluid pressure prior to injection; $P_f$, actual fluid pressure at failure; $P_h$, hydrostatic pressure; $P_B$, fluid pressure at failure predicted by Byerlee's Law; $\mu$, apparent coefficient os internal friction; k, permeability. Stresses and fluid pressures are in bars; permeability is in $10^{-16}$ $m^2$. Rangley A denotes conditions for the fracture stress field. Rangley B denotes conditions for the composite stress field. Line indicates no data.

### Rangely, Colorado

As noted above, the difficulty in the analysis of failure at the RMA is that the essential physical characteristics that govern fault strength are imprecisely known. In an effort to monitor in detail the influence of fluid pressures on fault strength, an experiment designed to deliberately control seismicity by fluid injection and pumping was performed at Rangely, Colorado (Raleigh and others, 1972, 1976). Between October 1969 and May 1973, fluid pressures within a sandstone oil reservoir were alternately raised and lowered at Rangely, Colorado, an area where earthquakes had been occurring in response to fluid injection at least since 1962. The depth to the reservoir is about 1,700 m, and the reservoir is about 350 m thick. It was found that the frequency of earthquakes near the well field could be dramatically increased when fluid pressures within the reservoir exceeded about 275 bars. The epicenters of the earthquakes were generally along a zone trending N50°E; within this zone two distinct clusters of seismicity were found, one lying in the immediate vicinity of the injection wells and the other 1 to 3 km southwest of the injection wells. The earthquakes close to the wells were at depths between about 2.0 and 2.5 km; the earthquakes southwest of the wells had average focal depths of 3.5 km.

Seismicity was due to strike-slip faulting along a fault which dips 80°NW, and the nodal planes of earthquakes near the wells were near parallel to the epicentral trend. The direction of slip is somewhat in doubt; Zoback and Healy (1984) noted that the direction of slip inferred by Raleigh and others (1972, 1976) was poorly constrained and that the assumed orientation of the stress field conflicts with other nearby stress indicators; by utilizing a stress orientation different from that of Raleigh and others (1972, 1976) and determining slip direction from the direction of the maximum resolved shear stress on the fault plane, they infer that slip occurred along the zone of failure in a right lateral sense with slight reverse motion; alternatively, if it is assumed that the stress orientation of Haimson (1973) used by Raleigh and others (1972, 1976) is correct, then slip occurred in a left lateral sense with

slight normal motion. Figure 6b shows a comparison of the frequency of earthquakes at Rangely with the pressure history of one of the wells used in the experiment.

The pertinent parameters for Rangely, Colorado, are shown in Table 1. Two sets of values are given. One corresponds to the orientation of the stress field determined from a hydraulic fracture experiment at Rangely (Haimson, 1973) and will be referred to as the fracture stress field. The other is consistent with the orientation of the stress field indicated from strain relief techniques and focal plane solutions at Rangely (Raleigh and others, 1972, 1976), the orientation of stress-induced well-bore breakouts at Rangely (Gough and Bell, 1982), and hydraulic fracture experiments at the nearby Piceance Basin in Colorado (Bredehoeft and others, 1976); this will be referred to as the composite stress field. The values given for Rangely in Table 1 are the same as those given by Zoback and Healy (1984). The values of the principal stresses were determined from hydraulic fracturing at a depth of 1,914 m. The shear and normal stress on the fault plane were determined by rotating the principal stresses to a coordinate system relative to the direction of motion within the fault zone. The bottom-hole fluid pressure prior to injection was inferred from pressure measurements made prior to oil field development, which indicated that pressures were hydrostatic. The pressure necessary to induce failure was determined from the bottom-hole pressure of one of the injection wells during times of increased seismicity. The permeability of the sandstone was determined from well field tests.

Table 1 indicates that there is an 11 or 14 percent difference between the critical fluid pressure at Rangely and that predicted by an application of Byerlee's Law, depending on whether the fracture stress or the composite stress field is assumed to be correct. Application of Byerlee's Law indicates that failure will occur when fluid pressures are roughly 3/5 or 3/4 or lithostatic, depending on the orientation of the stress field chosen. The fact that use of the fracture stress field provides a slightly closer match with the actual fluid pressure at failure weakly supports the correctness of this stress field measurement; this observation conflicts with the analysis of Zoback and Healy (1984), who suggest that there is a better match between the laboratory coefficient of internal friction for the reservoir rock (0.85 from Byerlee, 1975) and the apparent field value if the composite stress field is used. If we assume that the discrepancy between the actual and the predicted fluid pressure at failure can be corrected simply by altering the value of the effective stress coefficient $\alpha$, we find that its value would be equal to 0.89 and 1.15 for the fracture stress field and the composite stress field respectively. Although laboratory tests suggest that the value of $\alpha$ for the reservoir rock is 1.0 (Byerlee, 1975), the value of $\alpha$ indicated by the fracture stress field would be reasonable for sandstone at depth. The permeability of the seismically active part of the reservoir is within a factor of two of the permeability at the RMA.

### Western New York

As at the RMA and Rangely, Colorado, seismicity apparently due to fluid pressurization at depth has also taken place in the Attica–Dale region of western New York (Fletcher and Sykes, 1977). A rash of earthquakes of small magnitude were felt in this area during the fall of 1971. The earthquakes were centered near a high-pressure injection well operated for the hydraulic mining of salt and were elongated in a northerly direction about the Clarendon-Linden fault, a major tectonic feature within western New York; the well was hydraulically fractured at the end of July 1971 and high-pressure injection began in August 1971. The depth of the well is about 430 m and is within 50 m of the main segment of the Clarendon-Linden fault. The actual depths of these events were not well determined, but seismicity from 1972 to 1975 had depths typically between 0.5 and 1.0 km. The strike of the nodal planes and the dip of the easterly dipping fault plane of the latter period seismicity agree closely with values determined from subsurface contour data for the Clarendon-Linden fault. Seismicity during the latter period and presumably during the fall of 1971 was primarily due to dip-slip faulting along a plane striking about N15°E and dipping about 50°SE. Figure 6c shows a comparison of earthquakes near Dale with the pressure history of the injection well used for hydraulic mining.

The parameters of importance for Dale are shown in Table 1. As at Rangely and the RMA, there is considerable uncertainty in some of the values given in this chapter. The maximum principal stress was obtained from hydraulic fracturing experiments performed 100 km south of Dale at an average depth of about 510 m (Haimson, 1974); the minimum and intermediate stresses are both assumed to be equal to the overburden pressure at 510 m (using a density gradient of 0.26 bar/m given by Fletcher and Sykes, 1977). The shear stress and normal stress on the fault plane were determined by rotating the principal stresses to a coordinate system relative to the orientation of the fault zone indicated by the first motions of seismic events from 1972 to 1975; since the intermediate and least principal stress are assumed to be equal, fault motion is assumed to be pure dip-slip (Wallace, 1951). The initial fluid pressure is assumed to be hydrostatic, and

the fluid pressure necessary to induce failure was determined by adding the value for hydrostatic pressure at a depth of 510 m to the minimum top-hole pressure of the injection well during periods of increased seismicity. The permeability was determined from the time lag between the initiation of injection and the increase in seismicity, assuming a distance between the point of failure and the injection well of 2 km. It should be noted that the weakest links in the data we use to analyze the critical fluid pressure at Dale are the orientation and magnitude of the principal stresses; the magnitude of the greatest principal stress was essentially determined from a single hydraulic fracture experiment in a well distant from Dale; the orientation of the greatest principal stress was obtained from an average of three hydraulic fracture experiments which individually gave quite dissimilar results. The fracture orientation from each experiment was N93°E, N65°E, and N74°E respectively.

Considerable difference eixsts between the actual fluid pressure necessary to induce failure and that predicted by application of Byerlee's Law. Our use of Byerlee's Law indicates that failure will occur when fluid pressures are roughly lithostatic; failure occurs at Dale at much lower pressures. If we attempt to attribute this difference to a deviation from the approximate median coefficient of internal friction given by Byerlee's Law, we find that the coefficient of internal friction for the fault zone at Dale is 0.54; this value is not within the common range of values for this coefficient determined experimentally for non-argillaceous materials. Fletcher and Sykes (1977) report a similar result. As at the RMA, the discrepancy may not be due to any inadequacy in Byerlee's Law; rather it may indicate that some of the parameter values used to make our calculations are incorrect. If we simply apply Equation 4a to the data, the predicted fluid pressure at failure is within 6 bars of the fluid pressure at failure inferred from the top-hole pressure history of the injection well; however for Equation 4a to be valid at Dale, the orientation of principal stresses and/or the orientation of the fault plane would have to be different than indicated above. The permeability at Dale is roughly two orders of magnitude greater than the permeability at the other sites noted above.

## PERMEABILITY AND THE INFLUENCE OF PORE FLUIDS ON FAULT STRENGTH UNDER NATURAL CONDITIONS

In the above examples, faults are driven to frictional failure by modest rises in fluid pressure. If ground water strongly influences failure within the crust, pore pressures must be elevated under natural conditions at least to the levels indicated by induced seismicity studies. Many natural processes have been suggested as primary causes of high fluid pressure. These processes include compaction due to sediment loading (Dickinson, 1953), porosity reduction by pressure solution (Walder and Nur, 1984), tectonic strain (Davis and others, 1983), phase transformations of minerals that involve dehydration (Heard and Rubey, 1966), and the heating of pore fluids due to burial, shear motion along fault planes, and the emplacement of magma bodies (Barker, 1972; Sibson, 1973; Knapp and Knight, 1977). An examination of the mechanics of many of these processes is presented in Domenico and Palciauskas, (this volume). In this section we focus on the necessary rates at which some of these processes must operate to promote failure within the upper crust.

The degree to which any fluid-pressurization mechanism can promote failure along existing faults depends on a number of hydrologic and geologic factors: (1) the rate, extent, and duration of the pressurization mechanism; (2) the permeability and compressibility of the rock in which it operates; (3) the degree to which the process is isolated from the surface of the earth; (4) the orientation of the fault plane relative to the principal stresses; and (5) the degree of difference between the greatest and least principal stress. Our understanding of these elements at mid-crustal depths is quite poor; it would be difficult if not impossible to assess the degree to which ground water may influence fault strength near the base of the seismogenic zone. However, data from the induced seismicity studies examined above yield some essential information about the conditions under which fluid-pressure-induced failure can occur within the upper crust. It is of interest to examine whether any natural pressurization mechanisms can promote failure within the upper continental crust under the same conditions that are present in the three examples of induced seismicity described in the previous section.

Qualitatively, the hydrologic and geologic conditions under which failure occurs at the RMA, Rangely, and Dale are quite similar. Failure occurs when fluid pressures are less than ½ that of the greatest principal stress, indicating that differences between the least and greatest principal stresses are high and that the fault planes are oriented favorably relative to the principal stresses. The compressibility of the rock at each locality is likely to be within an order of magnitude of $10^{-5}$ bars$^{-1}$, a value typical for rocks (Haas, 1974). The permeability of the crystalline rock fluid reservoir undergoing failure at RMA is about the same as the average permeability for crystalline rocks in the upper crust determined by Brace (1980, 1985), and about four orders of magnitude greater than the average permeability of argillaceous materials; the permeability of the sandstone reservoir at Rangely is similar to that at the RMA, and the permeability of the rocks at Dale is also apparently much higher than that for clays. The fact that permeabilities are high relative to clays at these localities has strong implications on the influence of natural fluid-pressurization mechanisms in these environments. If low-permeability materials such as fault gouge exist within these failure zones, the high inferred-reservoir permeabilities as well as the small time interval between the initiation of injection and the onset of earthquakes suggests that they occupy only a fraction of the reservoir volume. As was noted in a previous section, the impact of a fluid-pressurization mechanism can be inversely proportional to the permeability. The areas of induced seismicity examined above are thus of a mixed character as to their suitability for pore-fluid-induced failure under natural conditions. Although only modest elevations in pore pressure are necessary to promote failure in these areas, mechanisms that elevate fluid pressures must be strong in order to counteract the dissipation of excess pressure by diffusion.

A detailed analysis of the relative influence of proposed fluid-pressurization mechanisms on fault strength is not within the scope of this chapter. We use the diffusion model described by Equation 11 to obtain order of magnitude estimates of the necessary strength three pressure-producing mechanisms must have to strongly promote failure: sedimentation-induced compaction, porosity reduction by pressure solution, and the heating of pore fluids. These processes are idealized as occurring at a constant rate over a thickness and half width of 10 km; the top and sides of the infinitely long rectangular prism are in excellent hydraulic communication with the water table, and the bottom is impermeable. The permeability of the rock is assumed to be homogeneous and constant over time and has a value of $10^{-16}$ m$^2$ (slightly less than the permeabilities at Rangely and at the RMA); the drained compressibility of the rock is assumed to be $10^{-5}$ bars$^{-1}$. The influence the processes have on pore fluid pressure under steady state conditions is determined from the dimensionless curves in Figure 4. Under the above assumptions the fluid pressurization mechanism would have to operate for at least approximately 1,000 yr for the steady state solutions shown in Figure 4 to be applicable. We assume that fluid pressures are initially hydrostatic and frictional failure will occur if fluid pressures attain 0.5 of lithostatic; this latter assumption is poor if failure occurs through reverse faulting (as at Dale) but is approximately correct for the failure conditions at the RMA and at Rangely. Inspection of the curves in Figure 4 indicates that pressures roughly ½ of lithostatic will be exceeded in the center of the rectangular prism throughout its depth if $D_4$ is greater than or equal to 3/5.

Table 2 indicates the rates (when $D_4 = 3/5$) at which the three forementioned pressurization mechanisms would have to operate under these highly idealized conditions in order for fluid pressures to possibly cause failure in the crust. The significance of these rates can be best understood if we examine the effects of these mechanisms after 1,000 yr, the approximate time it takes to achieve steady-state conditions (and for pressures to achieve their maximum value). Sedimentation would create the assumed necessary conditions for failure if the failure zone was buried by 10 km of overlying sediment over the 1,000-yr period; this degree of burial is highly unlikely, indicating that sedimentation-induced compaction is a poor pressure-producing mechanism under the conditions detailed above. Pressure solution could instigate failure if it led to a reduction in porosity of 0.004 during 1,000 yr; since the kinetics of pressure-solution-induced porosity reduction are poorly known (Walder and Nur, 1984), it is difficult to ascertain whether this mechanism can significantly affect fault strength. Temperatures would have to increase 40°C over this time period in order for the effects of thermal expansion to significantly promote failure. Such a temperature increase would not likely occur during burial under a geothermal gradient; however, in the presence of an extensive magmatic heat source, this temperature increase would likely be modest, indicating that in areas of active

TABLE 2. NECESSARY RATES OF THREE FLUID
PRESSURIZATION MECHANISMS*

| | |
|---|---|
| Pressurization due to Compaction | $1 \times 10^{-5}$ cm sec$^{-1}$ |
| Pressurization due to Pressure Solution | $1 \times 10^{-13}$ sec$^{-1}$ |
| Pressurization due to Heating | $3 \times 10^{-10}$ °C sec$^{-1}$ |

*At the rates given, fluid pressures are elevated to roughly 1/2 litho-
static for the idealized model discussed in the text. Compaction rate computed
under the assumption that the strains produced by burial are laterally
constrained. Mechanisms would have to operate for at least 4,000 yr to achieve
required pressure level at the rate shown.

igneous and volcanic activity, thermal pressurization of pore fluids may have a strong influence on frictional failure within the upper crust.

## CONCLUSIONS

In this chapter we have tried to examine how pore pressure may influence fault strength. Our primary focus has been on laboratory observations of frictional failure and their ability to describe quantitatively the state of effective stress at frictional failure on a field scale. Admittedly, the three cases of induced seismicity examined provide neither a complete nor an exhaustive field-scale test of laboratory observations. The greatest difficulty with each case is uncertainty in the value of at least one of the parameters that control frictional failure. The use of these imperfect tests suggests that Byerlee's Law is qualitatively correct in its description of the state of stress at frictional failure. At Rangely, the laboratory-determined coefficient of internal friction is about 30 percent higher or lower than the apparent field coefficient of internal friction, depending on which stress field is chosen. At the other sites, the laboratory values are not known, but the apparent field coefficients of internal friction are at or just outside the range of values typically found in the laboratory. These discrepencies suggest that (as might be expected) the conditions under which frictional failure occurs on a field scale are more complex than those found on a laboratory scale. However, considering the differences in scale and the uncertainties in the field parameters, the qualitative correspondence between laboratory and field conditions under failure also suggests that laboratory tests of fault strength can yield valuable information on the conditions for failure in the crust.

We have also tried to infer whether or not fluid-pressurization mechanisms that are of sufficient strength to significantly influence fault motion within the upper crust exist under natural conditions. Our simple analysis indicates that in areas where low-permeability argillaceous materials are not present in large volume, modest pore-fluid pressurization under natural conditions is limited to processes that operate at very high rates. Pressurization produced by burial under the normal geothermal regime existing in the crust is probably too small to exert a significant influence on frictional failure under the hydrologic and

geologic conditions found in the areas of induced seismicity examined in this paper. Permeabilities in these areas are not high; however, they are large enough to impede the effectiveness of pressurization mechanisms which operate over geologic time at a moderate rate. These mechanisms are apparently only effective within poorly permeable or highly compressible sediment. Pressure-producing mechanisms that operate at a moderate rate within the upper crust will probably only influence fault strength if the faults contain large amounts of poorly permeable and/or highly compressible fault gouge. Mechanisms that operate at a moderate rate may also exert an influence on fault strength near the base of the seismogenic zone; however, the geologic and hydrologic conditions at such depths are too poorly known to test for such a possibility.

### Appendix: Transformation of Byerlee's Law in terms of principal effective stresses

If the strike of the failure plane is aligned parallel to the intermediate principal stress, the shear, $\tau$, and normal effective stresses, $\bar{\sigma}_n$, on the failure plane are related to the greatest, $\bar{\sigma}_1$, and least, $\bar{\sigma}_3$, principal effective stress by the following relations (Jaeger, 1969, p. 8):

$$\rho = -.5\,(\bar{\sigma}_1 + \bar{\sigma}_3)\,\sin 2\phi \qquad (A1)$$
$$\bar{\sigma}_n = .5\,(\bar{\sigma}_1 + \bar{\sigma}_3) + .5\,(\bar{\sigma}_1 - \bar{\sigma}_3)\,\cos 2\phi \qquad (A2)$$

where $\phi$ is the angle between the failure plane and the least principal stress. Substitution of equations A1 and A2 into Byerlee's Law (Equation 2a and 2b) yields the following two relations between $\bar{\sigma}_1$ and $\bar{\sigma}_3$ as a function of $\phi$:

$$\bar{\sigma}_1 = \bar{\sigma}_3\,(.85 - .85\cos 2\phi + \sin 2\phi)/(-.85 - .85\cos 2\phi + \sin 2\phi) \quad (A3a)$$
$$\bar{\sigma}_1 = \bar{\sigma}_3\,(.6 - .6\cos 2\phi + \sin 2\phi)/(-.6 - .6\cos 2\phi + \sin 2\phi) + (-.6 - .6\cos 2\phi + \sin 2\phi)^{-1} \quad (A3b)$$

Figure 7 shows the value of the multiplier following $\bar{\sigma}_3$ as a function of $\phi$ for both equations; when $\phi$ is greater than 55° and less than 75° the multipliers are approximately equal to 5 and 3.4 for equations A3a and A3b respectively. These are the multipliers used in equations 4a and 4b (Byerlee's Law transformed in terms of the principal effective stresses). The offset in Equation A3b is approximately equal to 1.8 over the range of 55 to 75° in the angle $\phi$.

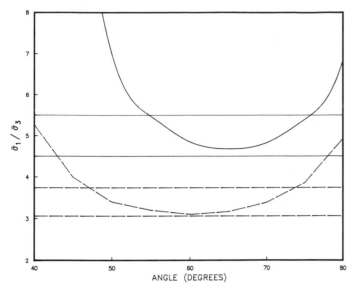

Figure 7. Ratio of greatest, $\bar{\sigma}_1$, to least $\bar{\sigma}_3$, principal effective stress as a function of angle between failure plane and $\bar{\sigma}_3$ when Byerlee's Law is stated in terms of principal effective stresses. Solid curve and dashed curve denote value of ratio for normal effective stresses on failure plane less than 2 kb and greater than 2 kb respectively. Solid and dashed horizontal lines denote bounds where approximate relations given by equations 4a and 4b respectively hold (criterion is 10 percent error).

## REFERENCES CITED

Barker, C., 1972, Aquathermal pressuring; Role of temperature in development of abnormal-pressure zones: American Association of Petroleum Geologists Bulletin, v. 56, p. 2068–2071.

Biot, M. A., 1941, General theory of three dimensional consolidation: Journal of Applied Physics, v. 12, p. 155–164.

Boucher, D. F., and Alves, G. E., 1959, Dimensionless numbers for fluid mechanics, heat transfer, mass transfer, and chemical reaction: Chemical Engineering Progress, v. 55, p. 55–64.

Brace, W. F., 1971, Resistivity of saturated crustal rocks to 40 km based on laboratory measurements: American Geophysical Union Geophysical Monograph Series, v. 14, p. 243–255.

——, 1980, Permeability of crystalline and argillaceous rocks: International Journal of Rock Mechanics, Mining Science and Geomechanics Abstracts, v. 17, p. 241–251.

——, 1985, Permeability of crystalline rocks; New in situ measurements: Journal of Geophysical Research, v. 89, p. 4327–4330.

Brace, W. F. and Byerlee, J. D., 1966, Stick-slip as a mechanism for earthquakes: Science, v. 153, p. 990–992.

Brace, W. F., and Kohlstedt, D. L., 1980, Limits of lithospheric stress imposed by laboratory experiments: Journal of Geophysical Research, v. 85, p. 6248–6252.

Bredehoeft, J. D., and Hanshaw, B. B., 1968, On the maintenance of anomolous fluid pressure; I. Thick sedimentary sequences: Geological Society of America Bulletin, v. 79, p. 1097–1106.

Bredehoeft, J. D., Wolff, R. G., Keys, W. S., and Shuter, E., 1976, Hydraulic fracturing to determine the regional in situ stress field Piceance Basin, Colorado: Geological Society of America Bulletin, v. 87, p. 250–258.

Brune, J. H., Henyey, T. L., and Roy, R. F., 1969, Heat flow, stress, and rate of slip along the San Andreas fault: Journal of Geophysical Research, v. 74, p. 3821–3827.

Byerlee, J. D., 1967, Frictional characteristics of granite under high confining pressure: Journal of Geophysical Research, v. 72, p. 2629–2648.

——, 1975, The fracture strength and frictional strength of Weber Sandstone: International Journal of Rock Mechanics, Mining Science and Geomechanics Abstracts, v. 12, p. 1–4.

——, 1978, Friction of rock: Pure and Applied Geophysics, v. 116, p. 615–626.

Byerlee, J. D., and Brace, W. F., 1972, Fault stability and pore pressure: Seismological Society of America Bulletin, v. 62, p. 657–660.

Davis, D., Suppe, J., and Dahlen, F. A., 1983, Mechanics of fold-and-thrust belts and accretionary wedges: Journal of Geophysical Research, v. 88, p. 1153–1172.

Dickinson, G., 1953, Geologic aspects of abnormal reservoir pressures in Gulf Coast Louisiana: American Association of Petroleum Geologists Bulletin, v. 37, p. 410–432.

Dieterich, J. H., 1974, Earthquake mechanisms and modeling: Annual Review of Earth and Planetary Sciences, v. 2, p. 275–301.

——, 1978, Time dependent friction and the mechanics of rock slip: Pure and Applied Geophysics, v. 116, p. 790–806.

——, 1979, Modeling of rock friction; 1. Experimental results and constitutive equations: Journal of Geophysical Research, v. 84, p. 2161–2168.

——, 1981, Constitutive properties of faults with simulated fault gouge: American Geophysical Union Monograph Series, v. 24, p. 103–120.

Dieterich, J. H., and Conrad, G., 1984, Effect of humidity on time- and velocity-dependent friction in rocks: Journal of Geophysical Research, v. 89, p. 4196–4202.

Etheridge, M. A., Wall, V. J., Cox, S. F., and Vernon, R. H., 1984, High fluid pressures during regional metamorphism and deformation; Implications for mass transport and deformation mechanisms: Journal of Geophysical Research, v. 89, p. 4344–4358.

Evans, D. M., 1966, The Denver area earthquakes and the Rocky Mountain Arsenal disposal well: Mountain Geology, v. 3, p. 23–36.

Fletcher, J. B., and Sykes, L. R., 1977, Earthquakes related to hydraulic mining and natural seismic activity in western New York state: Journal of Geophysical Research, v. 82, p. 3767–3780.

Gough, D. I., and Bell, J. S., 1982, Stress orientation from borehole wall fractures with examples from Colorado, east Texas, and northern Canada: Canadian Journal of Earth Science, v. 19, p. 1358–1370.

Hass, C. J., 1974, Static stress-strain relationships, *in* Touloukian, Y. S., Judd, W. R., and Roy, R. F., eds., Physical properties of rocks and minerals: New York, McGraw-Hill, p. 123–176.

Haimson, B. C., 1973, Earthquake related stresses at Rangely, Colorado, *in* Proceedings, Symposium on Rock Mechanics, 14th: American Society of Civil Engineers, p. 689–708.

——, 1974, A simple method for estimating in-situ stresses at great depths: American Society for Testing and Material Special Technical Publication 554, p. 156–182.

Healy, J. H., Rubey, W. W., Griggs, D. T., and Raleigh, C. B., 1968, The Denver earthquakes: Science, v. 161, p. 1301–1310.

Heard, H. C., and Rubey, W. W., 1966, Tectonic implications for gypsum migration: Geological Society of America Bulletin, v. 77, p. 741–760.

Hermann, R. B., Park, S. K., Wang, C. Y., 1981, The Denver earthquakes of 1967–1968: Seismological Society of America Bulletin, v. 71, p. 731–745.

Hollister, J. C., and Weimer, R. J., eds., 1968, Geophysical and geological studies of the relationship between the Denver earthquakes and the Rocky Mountain Arsenal well: Colorado School of Mines Quarterly, v. 63, p. 1–251.

Hoover, D. B., and Dietrich, J. A., 1969, Seismic activity during the 1968 test pumping at the Rocky Mountain Arsenal disposal well: U.S. Geological Survey Circular 613, 35 p.

Houston, W. N., and Kasim, A. G., 1982, Physical properties of porous geologic materials: Geological Society of America Special Paper 189, p. 143–162.

Hsieh, P. A., and Bredehoeft, J. D., 1981, A reservoir analysis of the Denver earthquakes; A case of reservoir induced seismicity: Journal of Geophysical Research, v. 86, p. 903–920.

Hubbert, M. K., and Rubey, W. W., 1959, Role of fluid pressure in mechanics of overthrust faulting: Geological Society of America Bulletin, v. 70, p. 115–206.

Jaeger, J. C., 1969, Elasticity, fracture, and flow with engineering and geological applications: London, Chapman and Hall, 268 p.

Kerrich, R., La Tour, T. E., and Willmore, L., 1984, Fluid participation in deep fault zones; Evidence from geological, geochemical, and $^{18}O/^{16}O$ relations: Journal of Geophysical Research, v. 89, p. 4331–4343.

Knapp, R. B., and Knight, J. E., 1977, Differential thermal expansion of pore fluids; fracture propogation and microearthquake production in hot pluton environments: Journal of Geophysical Research, v. 82, p. 2515–2522.

Lachenbruch, A. H., and Sass, J. H., 1980, Heat flow and energetics of the San Andreas fault zone: Journal of Geophysical Research, v. 85, p. 6185–6222.

McGarr, A., and Gay, N. C., 1978, State of stress in the earth's crust: Annual Review of Earth and Planetary Sciences, v. 6, p. 405–436.

Mead, W. J., 1925, The geologic role of dilatancy: Journal of Geology, v. 33, p. 685–698.

Nekut, A., Connerney, J.E.F., and Knuckles, A. F., 1977, Deep crustal electrical conductivity; Evidence for water in the lower crust: Geophysical Research Letters, v. 4, p. 239–242.

Norris, R. J., and Henley, R. W., 1976, Dewatering of a metamorphic pile: Geology, v. 4, p. 333–336.

Nur, A., and Byerlee, J. D., 1971, An exact effective stress law for elastic deformation of rock with fluids: Journal of Geophysical Research, v. 76, p. 6414–6419.

O'Neil, J. R., and Hanks, T. C., 1980, Geochemical evidence for water-rock interaction along the San Andreas and Garlock faults: Journal of Geophysical Research, v. 85, p. 6286–6292.

Raleigh, C. B., and Evernden, J., 1981, Case for low deviatoric stress in the lithosphere: American Geophysical Union Geophysical Monograph Series, v. 24, p. 173–186.

Raleigh, C. B., Healy, J. H., and Bredehoeft, J. D., 1972, Faulting and crustal stress at Rangely, Colorado: American Geophysical Union Geophysical Monograph Series, v. 16, p. 275–284.

—— , 1976, An experiment in earthquake control at Rangely, Colorado: Science, v. 191, p. 1230–1236.

Rice, J. R., and Cleary, M. P., 1976, Some basic stress-diffusion solutions for fluid saturated media with compressible constituents: Reviews of Space Physics and Geophysics, v. 14, p. 227–241.

Segall, P., 1985, Stress and subsidence resulting from subsurface fluid withdrawal in the epicentral region of the 1983 Coalinga earthquake: Journal of Geophysical Research, v. 90, p. 6801–6816.

Sibson, R. H., 1973, Interactions between temperature and pore-fluid pressure during earthquake faulting and a mechanism for partial or total stress relief: Nature, v. 243, p. 66–68.

—— , 1980, Power dissipation and stress levels on faults in the upper crust: Journal of Geophysical Research, v. 85, p. 6239–6247.

—— , 1982, Fault zone models, heat flow, and the depth distribution of earthquakes in the continental crust of the United States: Seismological Society of America Bulletin, v. 72, p. 151–163.

—— , 1984, Roughness at the base of the seismogenic zone; Contributing factors: Journal of Geophysical Research, v. 89, p. 5791–5799.

Skempton, A. W., 1954, The pore pressure coefficients A and B: Geotechnique, v. 4, p. 143–147.

Stesky, R. M., 1978, Rock friction; Effect of confining pressure, temperature, and pore pressure: Pure and Applied Geophysics, v. 116, p. 691–704.

Swolfs, H. S., 1984, The triangular stress diagram; A graphical presentation of crustal stress measurements: U.S. Geological Survey Professional Paper 1291, 19 p.

Taylor, H. P., 1978, Oxygen and hydrogen isotope studies of plutonic granitic rocks: Earth and Planetary Science Letters, v. 38, p. 177–210.

Taylor, H. P., and Forester, R. W., 1979, An oxygen and hydrogen study of the Skaergaard intrusion and its country rocks; A description of a 55-m.y.-old fossil hydrothermal system: Journal of Petrology, v. 20, p. 355–419.

Thatcher, W., 1979, Horizontal crustal deformation from historic geodetic measurements in Southern California: Journal of Geophysical Research, v. 84, p. 2351–2370.

Timoshenko, S. P., and Goodier, J. N., 1970, Theory of elasticity: New York, McGraw-Hill, 567 p.

van Poollen, H. K., and Hoover, D. B., 1970, Waste disposal and earthquakes at the Rocky Mountain Arsenal, Derby, Colorado: Journal of Petroleum Technology, v. 22, p. 983–993.

Walder, J., and Nur, A., 1984, Porosity reduction and crustal pore pressure development: Journal of Geophysical Research, v. 89, p. 11539–11548.

Wallace, R. E., 1951, Geometry of shearing stress and relation to faulting: Journal of Geology, v. 57, p. 118–130.

Woodward and Clyde Consultants, 1979, Study of reservoir induced seismicity: Reston, Virginia, U.S. Geological Survey Technical Report Contract 14-08-0001-16809, 222 p.

Zoback, M. D., and Healy, J. H., 1984, Friction, faulting and in-situ stress: Annales Geophycicae, v. 2, p. 689–698.

Zoback, M. D., and Hickman, S., 1982, In situ study of the physical mechanisms controlling seismicity at Monticello Reservoir, South Carolina: Journal of Geophysical Research, v. 87, p. 6959–6974.

Zoback, M. L., and Zoback, M. D., 1980, State of stress in the conterminous United States: Journal of Geophysical Research, v. 85, p. 6113–6156.

MANUSCRIPT ACCEPTED BY THE SOCIETY APRIL 28, 1987

The Geology of North America
Vol. O-2, Hydrogeology
The Geological Society of America, 1988

## Chapter 47

# *The role of ground-water processes in the formation of ore deposits*

**John M. Sharp, Jr., and J. Richard Kyle**

*Department of Geological Sciences, The University of Texas, Austin, Texas 78713*

## INTRODUCTION

Ground water is of paramount importance in a wide variety of ore-forming processes, but the hydrodynamics of these processes have only recently been the subject of quantitative investigations. Fluid flow on both local and regional scales, the mixing of ground waters, and rock-water interactions are research topics of common interest to economic geologists and hydrogeologists. Hydrogeologic models, capable of estimating timing and rates of fluid flow, when coupled with transport and geochemical modeling, should lend insights into the genesis of many kinds of mineral deposits. On the other hand, the analysis of these "fossil" hydrogeologic systems will ultimately increase our understanding of the subsurface portion of the hydrologic cycle by providing paleohydrologic data. We note, however, that most quantitative hydrogeologic analyses have concentrated chiefly on two types of ore deposits: carbonate-hosted lead-zinc and sedimentary uranium. In this paper, we wish to demonstrate the relationship of hydrogeology to economic geology (and vice versa) for a wide range of ore deposits. We review the major factors in relating ground-water processes to ore deposition, classify major ore deposit types hydrogeologically, and discuss selected deposits where the role of various ground-water systems is of documented importance.

The hydrologic cycle represents the circulation of water on the surface, in the atmosphere, and in the crust of the Earth. Some of the minor flow processes, although unimportant in terms of water resources, are of great geologic importance. These include thermally induced free convection near mid-ocean ridges and continental areas of high heat flow and the expulsion of pore fluids during sediment compaction and diagenesis.

The importance of ground water in the geologic cycle (Fig. 1) was emphasized by Hitchon (1976) who stressed the ubiquity of water in processes within the "secondary environment" (weathering, erosion, diagenesis, and metamorphism). Ground water is also of vital importance in the "primary environment." Norton and Knight (1977), for instance, demonstrated that convecting ground water is often the major factor in the cooling of magmatic bodies. Indeed, rocks once considered impermeable have now been shown to have permeabilities that are

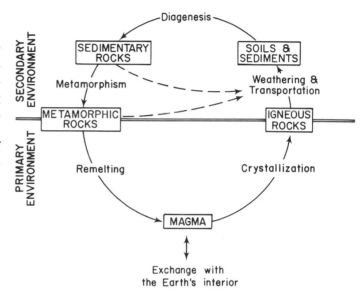

Figure 1. Rock cycle (after Hitchon, 1976). Ground-water processes are ubiquitous in the secondary environment.

significant when considered over a geologic time. Examples include plutonic rocks (Norton and Knapp, 1977), shales (Bredehoeft and others, 1983; Sharp and Domenico, 1976), and evaporites (Bassett and others, 1981; Senger and Fogg, 1983).

Four critical aspects for the generation of an ore deposit by ground water (White, 1968) are: (1) a source for the mineral constituents, (2) dissolution of the minerals in water of whatever origin, (3) migration of the fluids, and (4) precipitation of the minerals in response to physical and chemical changes in the fluid and/or porous medium.

Because many ore-forming processes are outside the common scope of hydrogeology, a few definitions are in order. Ground water encompasses *all* subsurface water. Thus, waters of

Sharp, J. M., Jr., and Kyle, J. R., 1988, The role of ground-water processes in the formation of ore deposits, *in* Back, W., Rosenshein, J. S. and Seaber, P. R., eds., Hydrogeology: Boulder, Colorado, Geological Society of America, The Geology of North America, v. O-2.

meteoric, oceanic, metamorphic, magmatic, and evolved (diagenetic) origin are all included.

Meteoric water is water that has "recently" been a part of the atmospheric portion of the hydrologic cycle. Marine water is sea or ocean water that has recently invaded adjacent rocks and sediments. Evolved or diagenetic water is defined as water that has been in contact with its host strata for a significant period of time, perhaps millions of years. This definition thus includes "connate" waters. Rock-water reactions during this residence time result in chemical compositions that are distinctly different from marine water. Metamorphic and magmatic waters are released from rocks undergoing metamorphism or crystallization from a melt, respectively.

It is also instructive to broaden the common definition of a flow system. Tóth (1963) and Hubbert (1940) described near-surface flow systems as comprising a recharge zone, a zone of lateral flow, and a discharge zone. Our definition of recharge includes all the ways that fluids enter the strata. For example, dehydration and compaction of clays and fluid from crystallizing magmas are considered recharge from internal fluid sources. Thus, recharge is the source of the subsurface waters that form mineral deposits, and discharge is the manner by which those pore fluids leave the subsurface. The zone of "lateral flow" is defined by the processes that circulate water from the recharge to the discharge zone or by the processes that recirculate the water where there is neither recharge or discharge. The processes and water types listed in Table 1 have produced or have been attributed to produce ore deposits.

Most twentieth-century classifications of ore deposits (e.g., Lindgren, 1933) have attempted to incorporate genetic concepts. These classifications suffer from the problems arising from our incomplete understanding of geologic processes, differences in interpretation of geologic relationships, and the deriving of information from "fossils" (i.e., ore deposits, instead of active systems). Recent advances in economic geology include better understanding of mineralization processes based on theoretical and experimental geochemistry, geochronology, hydrogeology, and the recognition of modern ore-forming systems in a number of geologic settings for which there are "fossil" analogs. These advances have resulted in a trend away from rigid classifications toward an appreciation of ore deposits as integral parts of their geologic environments. Major discussions of ore deposit geology in this broad context are provided by, among others, Hutchison (1983) and Sawkins (1984).

The terminology used to describe ore deposits is made complicated by interpretative and semantic differences. Descriptive terms include "stratiform" for strictly bedlike deposits within a sequence of layered rocks and "stratabound" for deposits that are enclosed by definable layers but which may have crosscutting within the layer. A suitable term for deposits that crosscut lithologic layering is not available, although "vein" usually carries this connotation. Genetic (i.e., interpretative) modifiers for ore deposits consist of the end-member terms, syngenetic and epigenetic.

Strictly syngenetic deposits formed at the same time as the

## TABLE 1. ZONES OF GROUND-WATER FLOW

**1. Recharge** (Source of water)
 1.1 Meteoric—by gravity recharge
 1.2 Oceanic—by hydrothermal recharge
 1.3 Diagenetic
  1.3.1 Compaction
  1.3.2 Mineral dehydration
  1.3.3 Metamorphic
 1.4 Magmatic

**2. Mechanisms of Flow**
 2.1 Gravity drive
 2.2 Overpressures
 2.3 Free convection
  2.3.1. Thermally driven
  2.3.2. Concentration-driven
 2.4 Osmotic mechanisms

**3. Discharge** (loss of water)
 3.1 Evapotranspiration
 3.2 To surface waters
 3.3 To oceans
 3.4 Mineral hydration

enclosing rocks and include stratiform mineral concentrations formed from the exhalation of hydrothermal solutions onto the sea floor in volcanic and sedimentary environments, as well as mineral layers that formed during sequential crystallization of layered igneous intrusions. Epigenetic deposits formed at a time subsequent to the lithification of the enclosing rocks. However, these terms are inadequate for the many ore deposit types that form after sedimentation but before the enclosing material is consolidated. For example, several kilometers of burial may be required to convert fine-grained sediments into shale. Ore-forming fluids can evolve within or be introduced into this environment over considerable ranges of time, temperature, and depth. Further complications are provided by mineralogical and textural changes that would accompany compaction and diagenesis in this geologic setting. The term syndiagenetic, including early and late subdivisions, has been used for this type of mineralization. Most types of syndiagenetic and epigenetic ore deposits are clearly related to ground-water processes.

In this chapter we review and evaluate the essentials of ore formation as they relate to ground-water processes, summarize the processes responsible for the various types of mineral deposits, and briefly illustrate these processes for some selected types of orebodies. These include carbonate-hosted lead-zinc deposits (Mississippi Valley and Pine Point as examples); porphyry copper deposits; epithermal deposits associated with shallow hydrothermal systems; stratiform sulfide deposits (both sedimentary exhalative and volcanogenic); uranium roll-front deposits; recharge/weathering zone deposits (lateritic ores of iron, nickel, aluminum, and gold, and supergene sulfide deposits); and metamorphic vein deposits.

**TABLE 2. CRITERIA FOR RECOGNITION OF WATER TYPES**

| Type | Chemical Composition | Isotopic Composition | Original Temperature |
|---|---|---|---|
| Meteoric water | Controlled by surface water and bedrock | Original fractionation along trend line | Near-surface (5-45°C) |
| Marine water | Very similar to sea water; may have more Ca than sea water | Same as or close to sea water | Near-surface (0-35°C) |
| Diagenetic water (chloride type) | Enriched in I, B, $SiO_2$, combined N, Ca, and low in $SO_4$ and Mg relative to sea water | Wide range in $\delta^{18}O$ (meteoric, +10 $\permil$) | Surface to slightly thermal (50-300°C) |
| Metamorphic water | Little known. Tends to be high in combined $CO_2$ and B, $NH_4$, but low in Cl relative to ocean; moderately high in I | Little known $\delta D$ (-60 $\permil$, -20 $\permil$) $\delta^{18}O$ (+4 $\permil$, +25 $\permil$) | Surface to moderately thermal (250-500°C) |
| Magmatic water | Relatively high in Li, F, $SiO_2$, B, S, $CO_2$; low in I, Br, Ca, Mg, combined N(?) | Little known. $\delta D$ (-80 $\permil$, -40 $\permil$) $\delta^{18}O$ (+5 $\permil$, +10 $\permil$) | Strongly thermal (500-800°C) |

Sources: White (1957, 1963), Taylor (1974, 1979), Craig (1961).

## THE HYDROGEOLOGIC FRAMEWORK

Water, because of its ubiquity and physicochemical properties, is the "universal solvent." As such it is integral to most processes operating in the Earth's crust, including ore deposition. The requirements for mineral deposition in a hydrogeologic system are as follows: (1) a source of ore-forming components for transport in solution, or possibly in a colloidal or suspended state, by ground water; (2) water (or other fluids) for transport; (3) existence of a potential field for fluid movement. The most probable driving mechanisms are hydraulic forces, but thermal, chemical-osmotic, or electro-osmotic fields are also possible; (4) sufficient permeability, either primary or secondary, to allow fluid transport within the limits imposed by the driving forces. Secondary permeability may be transient; (5) proper physical/chemical conditions to precipitate the minerals; (6) adequate media dispersivity to account for the observed distribution of minerals, either precipitated or still in solution; and (7) sufficient permeability contrast, to focus the mineralizing fluid through sites of ore formation (Annels, 1979).

### Source of water

Information provided by fluid inclusions (Roedder, 1979, 1984), theoretical calculations (Barnes, 1979), and active ore-forming systems (White, 1981) indicates that water is the overwhelmingly preponderant ore-transporting fluid. Metalliferous waters may have a variety of origins—meteoric, marine, diagenetic, metamorphic, and magmatic. As an example, the involvement of these water types in geologic processes is illustrated by abundant evidence of the widespread deep convective circulation of marine water in modern and ancient oceanic crust. Calculations by Wolery and Sleep (1976) suggest that a volume of marine water equivalent to the hydrosphere undergoes thermal convection through the mid-ocean ridge systems with a time constant of 5 to 30 m.y. The metamorphism of oceanic crust and the generation of metalliferous fluids accompanies the conversion of marine water to diagenetic water to metamorphic water upon heating and rock-water interactions. Consequently, waters may evolve or be present along all the pathways of the rock cycle (Fig. 1).

White (1957, 1963) and Taylor (1974) derived a set of recognition criteria for types of water (Table 2), although the isotopic and chemical composition of magmatic and, especially, metamorphic fluids were not then (and still are not) well known. The overwhelming bulk of the waters found in nonmarine settings has proven to be meteoric. This relationship has been documented by isotopic analyses even for active geothermal systems near magmatic bodies, although in most cases a small percentage of other water types could remain undetected. The boundary between connate waters and other types is very broad, but diagenetic waters would be suspected in stagnant or geopressured systems.

Although metamorphic and magmatic waters are not

abundant, they become increasingly important in deeper crustal environments (Ferry, 1983). Walther and Orville (1982), for instance, calculated that 10 percent of shale weight is released as water during metamorphism to phyllite grade. This could be a significant figure for deep, shale-rich sedimentary basins, such as the Gulf of Mexico. Furthermore, dehydration and decarbonation reactions accompanying burial metamorphism of hydrated rock sequences generate significant fluid volumes over specific pressure-temperature intervals. About 5 wt percent intercrystalline water and volatiles are released from a hydrated rock of mafic composition at the greenschist-amphibolite transition at 450° to 500°C (Fyfe and others, 1978). These processes are favored by high geothermal gradients and likely were particularly important during Archean crustal conditions. The initial water content of magmas with which extensive hydrothermal activity is associated has been estimated to range from about 2.5 to 6.5 wt percent, with a median of about 3.0 wt percent (Burnham, 1979). The significant water content of these melts further controls the development of stockwork fracture zones that are critical to the localization of many pluton-related ore concentrations (Knapp and Norton, 1981).

### Mechanism for fluid flow

The dominant type of ground-water flow system on continents is gravity or topography driven. The gravity (elevation) potential field exists throughout the zone of saturation. Flow of ground water is ubiquitous in almost all strata, even through such low-permeability strata as bedded evaporites (Bassett and Bentley, 1981). Gravity-driven flow systems have been an attributed flow mechanism for a wide variety of deposits, including carbonate-hosted lead-zinc districts at Pine Point (Garven and Freeze, 1984; Garven, 1985) and the Mississippi Valley (Siebenthal, 1915; Bethke, 1986), sedimentary uranium deposits (Adler, 1964; Galloway and others, 1979), sedimentary manganese ores (Force and others, 1986), and deposits formed by near-surface leaching and enrichment—lateritic bauxite, nickel, iron, and gold ores, and supergene copper sulfides (Lelong and others, 1976; Golightly, 1981; Mann, 1984; Anderson, 1955; Brimhall and others, 1985).

Free convection occurs in a wide variety of settings when buoyancy forces, created by temperature and/or salinity variations, overcome viscous resisting forces. Free convection has been studied theoretically (e.g., Lapwood, 1948; Combarnous and Bories, 1975; Straus and Schubert, 1977) and applied to a variety of geologic scenarios—oil and gas fields (Aziz and others, 1973), thick sand sequences (Blanchard and Sharp, 1985), steeply dipping sandstones (Wood and Hewett, 1982), metamorphic terranes (Etheridge and others, 1983), mid-ocean ridges (Wolery and Sleep, 1976; Rona and others, 1983), cooling plutons (Norton and Knight, 1977; Henley and McNabb, 1978), and geothermal systems (Donaldson, 1962; Cathles, 1981, 1983; Norton, 1984; among many studies). Many ore-emplacement processes associated with terranes of high heat flow are possibly the result

of freely convecting ground water. Minerals are selectively dissolved in one part of the system and precipitated elsewhere within the same circulation system. Examples include stratiform sulfide deposits in marine volcanic environments; epithermal vein, replacement, and hot-springs deposits; and porphyritic, intrusion-associated, metallic concentrations.

Overpressured ground-water systems are particularly important in thick sedimentary basins (Sharp and Domenico, 1976; their Fig. 1). Several factors, including sediment loading, rates of thermal expansion, tectonic stresses, mineral dehydration, and formation of gas, are involved in producing overpressures, although sediment loading and delayed compaction are considered most important. While these excess-fluid pressures may be sustained for millions of years, these systems are not steady state. Geochemical and petrographic analyses show that thousands of pore volumes of fluid flux are necessary to account for observed diagenetic features (e.g., Land, 1984). Some of this mass flux may be due to free convection (Blanchard and Sharp, 1985). In addition, the excess pore-fluid pressures are dissipated over time, either gradually (Sharp, 1976) or abruptly (Sharp, 1978; Cathles and Smith, 1983). The upward and outward flow of overpressured, variable-salinity waters into the gravity-driven, predominantly meteoric, or marine-water systems creates opportunities for fluid mixing, cooling, depressurization, and rock/water interactions. The discovery of metalliferous brines in Gulf of Mexico basin sediments (Carpenter and others, 1974) indicates that potential mineralizing fluids are indeed present in sedimentary basins. More recently, Ulrich and others (1984) have suggested that episodic pulses of upward-moving metalliferous brines could be responsible for sulfide concentrations in cap rocks of Gulf Coast salt domes (Price and others, 1983).

While chemical-osmotic and electro-osmotic systems may occur infrequently, some authors (e.g., Graf, 1982) have suggested that reverse osmosis and osmotic solute transport are geologically significant. Electro-osmosis has been demonstrated only on the laboratory scale (e.g., Olsen, 1972). Neither system is yet known to be an important ore-forming process, but they may have importance to forming the brines responsible for metal transport deep within a basin.

### Source of the ore components

The principal metals that form ore deposits can be divided into two groups (Skinner and Barton, 1973): the geochemically abundant metals that have crustal abundances in excess of 0.1 percent, and the geochemically scarce metals that have crustal abundances less than 0.1 percent. The geochemically abundant metals (Fe, Al, Mg, Mn, Cr, and Ti) are characteristically found as essential constituents of relatively common minerals. The geochemically scarce metals may form separate minerals, but commonly occur as trace components within common rock-forming minerals. Many current theories for the genesis of hydrothermal ore deposits involve the leaching of ore components from common minerals in a relatively large volume of source rock and transportation by an aqueous fluid to a depositional site where they are precipitated as stable metallic minerals (e.g., White, 1968; Hutchinson and others, 1980). The active Salton Sea and Red Sea brines are especially spectacular examples.

The composition, lithologic nature, and history of geologic materials are critical to their suitability for elemental sources. Rocks with a high average concentration of certain elements are generally regarded as more likely source rocks for those elements in ore-forming fluids. Thus, mafic and ultramafic rocks are potential sources for copper and nickel, whereas felsic rocks are better sources for lead and barium. Elements that occupy poorly fixed or ephemeral positions in natural materials are likely candidates to be contributed to pore fluids. Possible examples in which the trace components might be released readily include volcanic glass upon devitrification, secondary minerals already present along fractures in igneous rocks, evaporites during dissolution, organic materials upon decomposition, adsorbed metals on clays during compaction, and minerals undergoing alteration. The timing of critical geologic events in a particular region would control the enrichment of a pore fluid with ore components, and the focusing of flow for a significant duration into a physicochemical "trap" would govern the development of a mineral concentration at a particular site.

Ore minerals commonly are metallic sulfides or sulfosalts (more rarely aresenides, selenides, or tellurides), carbonates, oxides, or native elements, and are associated with metallic and nonmetallic minerals including those of the silicate, carbonate, oxide, sulfate, and halide groups. Therefore, consideration of the ore-component sources for a particular mineral concentration is a complex problem involving many elements, each with a specific geochemical behavior. For the residual type of ore deposits, the source of components is generally clear, typically consisting of locally dispersed materials that are concentrated by near-surface processes. Examples include lateritic deposits of nickel (Golightly, 1981) and bauxite (Lelong and others, 1976) where the concentration and mineralogic nature of the ore element is changed by chemical weathering of the source rock, as well as residual concentrations of various minerals (e.g., barite) that are relatively stable in the surficial environment while matrix materials are dissolved and removed by ground water.

Common methods of investigation of the elemental sources for hydrothermal ore deposits include radiometric and stable isotopic studies, constrained by basic geological and geochemical considerations. Even so, the evidence for the sources of components in solution is rarely definitive. Studies of lead-isotopic characteristics of several types of geothermal fluids and ore deposits have identified probable source rocks for lead along the mineralizing fluid flow paths (for a summary, see Doe and Zartman, 1979), but the relationship of this information to the source of other components in particular hydrothermal systems is generally unknown. For example, in a regional lead-isotope study of epithermal ore deposits in the Tertiary San Juan volcanic field in Colorado, Doe and others (1979) concluded that the complex patterns indicate that the major source of ore leads was Precam-

brian basement of two different ages or Phanerozoic clastic sedimentary rocks derived from these sources. However, at least some lead could have been supplied by silicic Tertiary magmas, either directly or during hydrothermal alteration of ore-hosting volcanic rocks. Furthermore, the model for the origin of lead may not apply for the associated zinc, copper, gold, and silver (Doe and others, 1979). The direct supply of ore components to a mineralizing solution may be particularly important for magmatic or metamorphic water systems where some of the chemical species may already be in solution at the time of "recharge."

Strontium isotopic composition and rare-earth-element distribution patterns in carbonate and sulfate minerals have been used to characterize regional paleohydrologic systems in sedimentary terranes (Lange and others, 1983; Graf, 1984). Gradients provided by regional trace element and fluid inclusion data also have been interpreted to reflect paleohydrologic flow regimes (Erickson and others, 1983; Rowan and others, 1984). These types of studies contribute indirectly to the consideration of source materials for the ore-forming components by restricting the materials through which the mineralizing solutions passed.

Stable-isotope investigations of ore deposits have provided much insight into the nature of mineralizing systems. As summarized in Table 2, the oxygen and hydrogen isotopic character of ore and associated minerals, fluid inclusion extracts, and modern geothermal fluids have been used to define the dominant water type in many active and fossil hydrothermal systems (Taylor, 1979). Carbon and sulfur isotopic studies have been widely applied because various compounds of these elements are ubiquitous in many ore deposits (see Ohmoto and Rye, 1979, for a discussion of systematics). These components tend to reflect the overall geologic setting of particular ore deposits with a trend toward nonequilibrium fractionation at lower temperatures, caused at least in part by organic influences. At high temperatures, both of these components may be indigenous to magmas or hydrothermal fluids; interactions with carbon- or sulfur-bearing wall rocks may make significant contributions to the resultant ore deposit. Carbon in low- to moderate-temperature mineralizing systems may be supplied from carbonate wall rocks, or from organic matter resulting in distinctive light isotopic signatures. Controversy exists concerning the importance of sulfur species in metal transport and precipitation in low-temperature formation waters (Anderson, 1975; Barnes, 1979; Sverjensky, 1984). The sulfur reservoir for syngenetic to early diagenetic metallic mineralization from deeply circulating marine waters typically reflects the complex interaction of sulfate supplied locally from marine water, sulfate reduced inorganically at high temperatures, pre-ore sulfides leached by the hydrothermal system, and bacterially generated pre-ore sulfides replaced by ore-stage mineralization (Sangster, 1976; Large, 1981).

### Transport mechanisms

The development of many economic concentrations of elements in the Earth's crust requires some type of focused transport to enrich the chemical element at a particular site. The common primary ore minerals, especially the sulfides, are relatively insoluble in water as simple ions; associated non-ore minerals are only slightly more soluble. It is now generally accepted that the mechanism for attaining solubilities of the magnitude that appear to be required for most major hydrothermal ore deposits is that of forming ion complexes; complexing theory and data as applied to hydrothermal systems are summarized by Helgeson (1969, 1970, 1979) and Barnes (1979). Only a few ligands, generally $Cl^-$, $HS^-$ or $H_2S$, $CO_3^=$, and $OH^-$, are present in sufficient quantities in natural fluids to be important in metal transport. At higher temperatures (above about 200°C), $NH_3$ and $F^-$ may also contribute to complexing, and at lower temperatures, $S_2O_3^=$, $SO_3^=$, $CN_2^-$, $SCN_2^-$, organic ligands, and possibly various aqueous species of Se, Te, As, Sb, $PO_4^=$, and Bi could be of possible importance (Barnes, 1979). The ability of a ligand to form a metal-transporting aqueous complex depends on its activity in solution. This activity is a function of its total concentration, the extent of ion pairing with other aqueous species, and the temperature, ionic strength, acidity, and oxidation state of the solution. These parameters vary considerably for different types of mineralizing systems (Barnes, 1979).

Once in solution, the ore components can be transported by several possible mechanisms. If the ground water is static, then transport will be localized and controlled solely by molecular diffusion, but static systems rarely exist on either regional or geologic time scales. Most systems, even those of very low permeability, have mineralogic, geochemical, isotopic, textural, or structural features that indicate significant amounts of pore fluid have passed through the rocks. In most natural systems, two additional transport mechanisms operate: advection and mechanical dispersion.

Advective transport describes the solute transport via the average velocity of the moving ground water. The relative efficacy of advection versus diffusion can be estimated with the Peclet number (Pe = vL/D) where v is the ground-water velocity, L is a characteristic length of the flow system (such as strata thickness), and D is the diffusion coefficient (for diffusion in water, $10^{-10}$ to $10^{-11}$ m$^2$/sec is an acceptable estimate). A few calculations will demonstrate that the Peclet number will generally exceed 1 for systems larger than several meters or for ground-water velocities commonly found near the earth's surface (centimeters or more per year). Thus, advection commonly dominates over diffusional processes.

Hydrodynamic dispersion (Perkins and Johnstone, 1963; Bear, 1972; Anderson, 1979), although successfully modeled as "hyperdiffusion," is a transport mechanism that must be considered in continuum models of mass transport because we are not able to quantitatively evaluate fluid pathways on the microscopic level. The calculated pore-fluid velocity (v) is actually a fluid velocity averaged over the pore space. Some of the pore fluids will move much more slowly and some much more rapidly than v. Dispersion represents the physical mixing of pore fluids and is most important in heterogeneous media. As shown by Garven

and Freeze (1984), low-dispersivity flow systems may be favorable in ore genesis. Ore fluids thus transport their components by a combination of advection, hydrodynamic dispersion, and diffusion.

## Mechanisms of ore deposition

As a metalliferous fluid migrates along a flow path toward sites of precipitation or discharge, a variety of processes can lead to changes in the physiochemical nature of the fluid and thereby to mineral precipitation. Because these changes are complexly interrelated and may produce similar results, it is seldom possible to specify the exclusive cause of ore deposition. Furthermore, each mineral has a specific geochemical behavior that controls its transport, solubility, and precipitation, and the nature and causes of ore deposition over the commonly estimated $10^4$- to $10^6$-year duration of a "typical" hydrothermal system are complex indeed. Mineral precipitation can be induced by several processes, including changes in temperature and pressure, reaction with wall rocks, intra-aqueous reactions, and solution mixing (Skinner and Barton, 1973).

The solubility of most minerals increases with temperature; therefore, cooling of a solution near saturation commonly causes precipitation. A few minerals, including calcite and anhydrite, have retrograde solubility (Holland and Malinin, 1979) and would not be expected to precipitate upon cooling (ignoring other factors). The solubilities of many minerals involved in hydrolysis equilibria are much more complex; nevertheless, it is apparent that a temperature decrease will cause most sulfide minerals to precipitate (Sverjensky, 1984). The ways and extents to which temperatures of hydrothermal solutions can be expected to vary are dominated by the processes of conduction, convection, fluid mixing, throttling or boiling, and heat exchange with the porous medium. These relationships indicate that abrupt changes in temperature occur only in the relatively small portion of aquifer systems affected by throttling, boiling, or fluid mixing. A change in pressure can cause precipitation, but the isothermal decompression effect is not large. Large pressure changes, however, may play a role in mineral precipitation in deep crustal environments. A large solubility decrease generally occurs when the density of the aqueous phase increases significantly, as in boiling (Drummond and Ohmoto, 1985). Even for those situations where a pressure decrease results in boiling, the pressure effect is often masked because of the greater effects due to temperature and chemical changes, such as increasing concentrations and the loss of acid components to the vapor while the residual solution becomes more alkaline.

Chemical interactions (or "mass transfer") between the hydrothermal fluid and wallrocks are responsible for several features that result in metal precipitation. Henley and others (1984) provide a review of these rock-water interactions. The wallrock may supply a component necessary to form an ore mineral, such as sulfur, that is capable of extracting metals from solution. Also, the hydrolysis reactions resulting in extensive silicate or carbonate alteration zones increase pH, which increases the activity of the sulfide ions and thereby decreases the solubility of sulfide minerals. Interactions of the hydrothermal fluid with the wallrock may alter the oxidation state of the fluid. For example, organic matter causes local deposition of precious metals, uranium, vanadium, and copper by providing a more reduced environment which, in turn, causes precipitation. Chemical reactions between incompatible components, such as methane and sulfate ions, carried metastably in the same fluid may result in sulfide precipitation (Skinner and Barton, 1973). These reactions may be biogenic in nature with bacterial reduction of sulfates being an important process at low temperatures in ground-water discharge zones.

Mixing of the metalliferous solution with other fluids is a common explanation for ore deposition, but is seldom proved conclusively. Mixing may supply a component that is necessary for mineral precipitation, such as reduced sulfur, but which is in low concentrations in the metalliferous fluid (Anderson, 1975). Mixing may also result in a cooler or hotter hybrid solution with attendant precipitation reactions. If metals are carried as complexes in a brine, the ligand concentration will be decreased if the solution mixes with a less-concentrated fluid. This effect will promote precipitation, providing that changes in pH and sulfide concentration do not compensate for the decrease in ligand concentration (Skinner and Barton, 1973). Mixing can also cause precipitation by oxidation in the case of $BaSO_4$ and $MnO_2$ deposits (Force and others, 1986).

It is manifest that the mixing of subsurface waters of different physical and chemical compositions creates a vast range of possible conditions that could lead to economic mineralization. Studies of modern geothermal systems (e.g., Fancelli and Nuti, 1965; Elders, 1981; Smith and Kennedy, 1985) consistently indicate the mixing of several water types. Existing information also suggests that ore concentrations should be strongly localized in zones of boiling of fluid mixing. The combined investigation of hydrology, ore-deposits geology, and geochemistry is obviously a promising area for future research.

## ORE DEPOSITS WITHIN THE HYDROGEOLOGIC FRAMEWORK

While, historically, hydrogeologic theories of ore deposition have had a cyclical acceptance by geologists (Dunham, 1970), we can consider various types of orebodies within the broad scope of hydrogeology. Evaluating the sources of water, the flow mechanism, and the processes of mineral transport and precipitation of various types of ore deposits demonstrates that most are essentially the traces (by-products?) of active or fossil ground-water systems (Table 3). Most ore deposits form in what we might consider undifferentiated discharge/lateral flow zones. Major exceptions include residual ores that form *or* are enriched by shallow ground-water recharge, epithermal deposits associated with the discharge of thermal spring water (as liquid or vapor), and deposits formed by evaporative concentration processes. Deposits formed by ground-water recharge tend to be areally extensive, while those in discharge zones are correspondingly more localized.

Table 3 depicts the mechanisms for precipitation and for driving the flow systems as well as the fluid source. Most ore-forming components are transported in solution by advection and dispersion. The broad classes (1 to 12) of ore deposits are given in the general order of increasing temperatures of formation. Note also that in some cases more than one hypothesis for ore deposit formation by ground-water processes has been proposed. Furthermore, the examples given are from North American deposits except where the only (or best) examples are located elsewhere.

In order to demonstrate the wide variety of ore deposits and the ground-water processes involved in their formation, we review briefly below, in order of generally increasing temperatures, mineral deposits formed in shallow meteoric environments, deep sedimentary environments, marine volcanic environments, shallow continental plutonic environments, and metamorphic environments.

### *Shallow meteoric environments*

Ore deposits in this environment include a variety of concentrations formed by the action of relatively dilute waters at near-surface temperatures. Topography- or gravity-driven flow is paramount. Examples include lateritic concentrations, supergene sulfide enrichment, sedimentary uranium deposits, and sedimentary copper-silver deposits. In addition to ore deposits whose formation is directly related to meteoric water processes, the localization of many mineral concentrations in carbonate strata is greatly influenced by a hydrologic framework that is the result of karstification and attendant near-surface diagenesis such as dolomitization. Although later ore-forming fluids may be of different character, the secondary porosity and permeability trends resulting from meteoric water circulation during earlier periods of subaerial exposure may be critical factors for orebody development (Kyle, 1983).

Lateritic deposits of hydrous aluminum oxides (bauxite) provide excellent examples of the role of ground water in ore concentration. Aluminum is extremely insoluble in most natural waters and forms residual deposits under conditions where other elements are removed. These processes operate most efficiently in humid tropical environments where the topographic and hydrologic framework limits erosion and promotes subsurface drainage (Lelong and others, 1976). Many bauxites in limestone terranes are associated with karst drainage systems, which provide the regional permeability needed for the removal of soluble rock components and the accumulation of aluminum-bearing materials. An example is provided by the bauxite deposits of Jamaica. Although their origin is controversial, they may represent accumulation and alteration of middle to upper Miocene volcanic ash within karst depressions in the middle Eocene to lower Miocene White Limestone (Comer, 1974). Nickel laterites are the result of decomposition and concentration of secondary nickel silicates accompanying removal of the more soluble components of

## TABLE 3. HYDROGEOLOGIC CLASSIFICATION OF ORE DEPOSITS

| Deposit Type (examples) | Dominant Source of Water | | | | | Driving Mechanism | | | | Precipitation Mechanism | | | | | | | | References |
|---|---|---|---|---|---|---|---|---|---|---|---|---|---|---|---|---|---|---|
| | Meteoric | Marine | Diagenetic/Connate | Metamorphic | Magmatic | Gravity | Overpressured | Free Convection | Other | Residual/Syngenetic | Evaporative/Boiling | Cooling | Reduced Pressure | Oxidation/Reduction | pH Changes | Mixing of Fluids | Addition of Metals/Sulfur | |
| **1. Residual Deposits** | | | | | | | | | | | | | | | | | | |
| 1.1 Bauxite (Arkansas; Jamica) | X | | | | | X | | | | X | | | | | | | | Comer, 1974; Lelong and others, 1976 |
| 1.2 Nickel laterites (Cuba; Oregon) | X | | | | | X | | | | X | | | | | | | | Golightly, 1981 |
| 1.3 Precambrian banded iron-formations (Minnesota, Michigan) | X | | | | | X | | | | X | | | | | | | | Marsden, 1968 |
| 1.4 Gold (Alaska; Western Australia) | X | | | | | X | | | | | | | | X | | | | Mann, 1984; Severson and others, 1985 |
| 1.5 Supergene sulfides (Arizona, New Mexico, Montana) | X | | | | | X | | | | | | | | X | X | | | Lowell and Guilbert, 1970; Brimhall and others, 1985 |
| **2. Sedimentary U** | | | | | | | | | | | | | | | | | | |
| 2.1 Roll-front (south Texas, New Mexico, Wyoming) | X | | (X) | (X) | | X | | | | | | | | X | (X) | | | Galloway and others, 1979; Sanford, 1982 |
| 2.2 Unconformity Northwest Territories | X | | (?) | (?) | | X | | | | | | | | X | (X) | | | Hoeve and Sibbald, 1978; Nash and others, 1981 |
| 2.3 Calcrete (Australia) | X | | | | | X | | | | | X | | | | | | | Mann and Deutscher, 1978 |
| **3. Sedimentary Mn (Morocco)** | X | X | | | | X | | | | | | | | X | X | | | Force and others, 1986 |
| **4. Sedimentary Cu-Ag** | | | | | | | | | | | | | | | | | | |
| 4.1 Proterozoic (Michigan, Montana; Northwest Territories) | X | | (X) | (X) | | X | | | | | | | | X | X | | | White, 1971; Renfro, 1974 |
| 4.2 Phanerozoic (Oklahoma, Texas) | X | X | | X | | X | | | | | | X | | X | | | | Smith, 1974; Rose, 1976 |
| **5. Sedimentary Exhalative Pb-Zn-Ag-Ba** | | | | | | | | | | | | | | | | | | |
| 5.1 Lead-Zinc-Silver (British Columbia, Yukon, Alaska) | X | X | (?) | | | X | | | X* | | | X | | | | | | Badham, 1981; Morganti, 1981; Sibson and others, 1975; Gustafson and Williams, 1981; Lydon, 1983 |
| 5.2 Barite (Nevada, Arkansas; Sonora; Yukon) | X | X | | | | X | | | | | | | | | | X | X | Howard and Hanor, 1987 |
| **6. Evaporite-hosted S-Zn-Pb-Ag-Ba** | | | | | | | | | | | | | | | | | | |
| 6.1 Biopigenetic sulfur (Texas, Louisiana; Mexico) | X | (?) | (?) | | | (?) | | | | | | | | X | | X | X | Ruckmick and others, 1979 |
| 6.2 Salt dome cap-rock sulfides (Texas, Louisiana) | (?) | X | X | | | X | (?) | | | | | | | | | X | X | Kyle and Price, 1986 |

Key: X = probable; (X) = possible; (?) = speculative; X*, Sibson and others (1975) suggest "seismic pumping."

## TABLE 3. HYDROGEOLOGIC CLASSIFICATION OF ORE DEPOSITS (continued)

| Deposit Type (examples) | Meteoric | Marine | Diagenetic/Connate | Metamorphic | Magmatic | Gravity | Overpressured | Free Convection | Other | Residual/Syngenetic | Evaporative/Boiling | Cooling | Reduced Pressure | Oxidation/Reduction | pH Changes | Mixing of Fluids | Addition of Metals/Sulfur | References |
|---|---|---|---|---|---|---|---|---|---|---|---|---|---|---|---|---|---|---|
| **7. Sandstone-hosted Pb** (Nova Scotia, Saskatchewan) | X | | | | | | | | | | | | | (?) | | | X | Bjorlyke and Sangster, 1981 |
| (Laisvall, Sweden) | | X | | | | X | | | | | | | | | | X | X | Rickard and others, 1979 |
| **8. Carbonate-hosted Pb-Zn** (Northwest Territories) | X | (X) | | (?) | | | X | | | | X | X | (?) | X | X | X | X | Garven and Freeze, 1984 |
| | (X) | X | | | | | | | | | X | X | | X | X | | X | Jackson and Beales, 1967; Smith and others, 1983 |
| (Missouri, Oklahoma, Wisconsin Tennessee) | X | | | | | X | | | | | X | X | | X | X | X | | Siebenthal, 1915 |
| | X | (X) | | | | X | | | | | X | X | | X | X | X | X | Bethke, 1986 |
| **9. Epithermal deposits** 9.1 Disseminated Au (Nevada, Utah; Dominican Republic) | (X) | X | | (?) | | X | | X | | | X | X | | X | X | X | X | Tooker, 1985 |
| 9.2 Vein Au-Ag-Pb-Zn (Colorado, New Mexico) | (X) | X | (?) | (?) | | X | | X | | X | | X | X | | | | X | White, 1981; Berger and Eimon, 1983 |
| 9.3 Hot Springs Hg-Sb-Au (Nevada, California) | X | (X) | (?) | (?) | | X | | X | | | X | X | | (?) | | | X | Henley and Ellis, 1983 |
| **10. Volcanogenic Massive Sulfides** 10.1 Koroko-type (Ontario, Quebec, Northwest Territories, New Brunswick; Wisconsin, Arizona, Maine, North Carolina, California, Oregon, Alaska) | (?) | | | (?) | | | | X | | | X | (X) | | X | X | X | (X$^{SO_4}$) | Hattori and Sakai, 1979; Franklin and others, 1981; Cathles, 1983 |
| 10.2 Cyprus-type (California, Nevada) | | X | | (?) | | | | X | | | X | (X) | | X | X | X | (X$^{SO_4}$) | Spooner, 1977, Rye and others, 1984 |
| **11. Metamorphic Vein** 11.1 Vein Au (South Dakota, California; Ontario, Quebec, Northwest Territories) | | X | X | | | X | | X | | | X | X | | X | X | | X$^{(S^=)}$ | Henley, 1973; Kerrich and Fryer, 1979 |
| 11.2 Vein Pb-Zn-Ag (Idaho) | | | X | (?) | | X | | X | | | X | (?) | | | X | | | Leach and others, 1985 |
| **12. Intrusion-associated** 12.1 Porphyry Cu (Arizona, New Mexico, Utah, Montana; Sonora; Panama) | (X) | | | | X | (X) | | X | | | X | X | X | X | X | X | X | Norton and Knight, 1977 |
| | X | | | | (X) | | | X | | | X | X | X | X | X | X | X | Beane and Titley, 1981 |
| 12.2 Porphyry Mo-Wo-Sn (Colorado, New Mexico, Nevada, Alaska; British Columbia) | (X) | | | | X | | | X | | | X | X | X | X | X | X | X | White and others, 1981 |
| 12.3 Skarn Cu-Zn (Arizona, New Mexico; Quebec; Sonora, Zacatecas) | | | | | X | | | X | | | | | X | | X | X | X | Shelton, 1983 |
| 12.4 Replacement Pb-Zn-Ag (Colorado, New Mexico; Chihuahua; Honduras) | (?) | | | | X | X | | X | | | | | X | X | X | X | X | Einaudi and others, 1981 |

Key: X = probable; (X) = possible; (?) = speculative

nickel-bearing rocks under tropical weathering conditions (Go-lightly, 1981). Finally, Mann (1984) presented evidence that acidic chloride-rich waters in lateritic profiles dissolve gold and silver. The gold apparently reprecipitates by reduction of $AuCl_4^-$ with $Fe^{+2}$ below the water table. Studies by Severson and others (1985) indicate the potential for widespread remobilization of gold in organic-rich ground water in other climatic regimes.

Supergene sulfide enrichment is an important process that results in economically significant upgrading of many porphyry copper deposits in hydrologically and mineralogically permissive settings. Downward percolating meteoric water becomes acidic through the oxidation of sulfide minerals and leaches copper from primary copper sulfides or earlier formed secondary copper minerals. When this oxidizing, copper-rich water reaches the water table, it encounters more reducing conditions and precipitates secondary copper sulfides by the replacement of primary iron and copper sulfides (Garrels, 1954; Anderson, 1955; Brimhall and others, 1985). These "supergene chalcocite blankets" result in ore-grade enrichment of the "typical" porphyry copper deposit in the southwestern United States by more than 75 percent (Lowell and Guilbert, 1970). Silver is controlled by similar geochemical factors and may also be enriched by supergene processes (May-nard, 1983).

In near-surface environments, lead generally appears as in-soluble sulfate—angesite—and tends to be dispersed in the oxi-dized zone, whereas zinc sulfate is relatively soluble and may be removed by ground-water flow except where locally fixed as smithsonite in carbonate-rich environments (Sangameshwar and Barnes, 1983). Such near-surface concentrations of secondary metallic minerals have been mined in many parts of the world.

Sedimentary uranium deposits may form where ground water becomes more reducing along its flow path, a common situation in most meteoric systems (Baedecker and Back, 1979). Uranium is readily transported as a complexed carbonate, sulfate, or phosphate uranyl ion in oxygen-rich ground water. When the oxygen-rich water encounters a chemically reducing environ-ment, the uranium is reduced to the tetravalent state and precipi-tated (Reynolds and Goldhaber, 1978). Reducing conditions *par excellence* are often associated with the presence of organic mat-ter in porous media where sulfur-reducing anaerobes decompose organics. Roll-type uranium deposits (Fig. 2) indicate ancient or present oxidation-reduction boundaries of a ground-water flow system. The roll-type uranium concentrations generally occur in sedimentary sequences that have considerable permeability con-trasts to influence fluid flow. For example, the uranium deposits in south Texas generally occur within more transmissive fluvial depositional units that form linear permeability trends within finer grained units (Galloway, 1978). The width or thickness of the roll-type orebody is dependent upon the degree of mixing of the oxidizing and reducing waters, which in turn depends upon the aquifer's heterogeneity, the ground-water velocity, and the distribution of organic matter or other reducing agents. The source of the reducing environment may be organic matter, iron sulfides, hydrogen sulfide accompanying oil, gas, coal, or peat

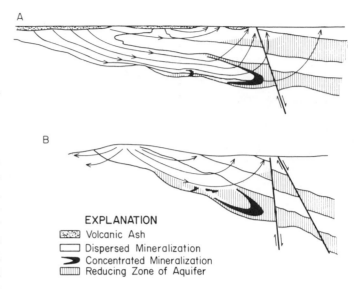

EXPLANATION
- Volcanic Ash
- Dispersed Mineralization
- Concentrated Mineralization
- Reducing Zone of Aquifer

Figure 2. Uranium roll-front deposition (after Galloway, 1978). Concen-trated mineralization occurs where meteoric waters are reduced. Upward movement of hydrocarbons from depths is proposed as one source of reductants. Replacement of detrital organic matter by ore minerals is also commonly observed.

deposits, or highly reducing ground water. The ground-water flow direction could be as shown in Figure 2, a case where mixing is due to longitudinal dispersion. Adler (1964), however, suggested a model where flow is parallel to the roll front, and transverse dispersion controls the width of the roll.

Sedimentary uranium deposits can also occur in evaporative discharge zones as do borates, potash, and vanadium ores. All these deposits are formed because meteoric water (gravity-driven) leaches uranium (or other ore minerals) from volcanic or other source rocks. The solute load is left in the playas upon evapora-tive discharge. Prime examples (Fig. 3) of this environment are the calcrete uranium deposits of western Australia (Mann and Deutscher, 1978).

Strata-bound concentrations of copper and silver sulfides occur in sedimentary strata that accumulated along the shore-ward fringe of a shallow, semi-restricted marine lagoon or inland sea. The metallic concentrations are hosted by reduced siltstones and sandstones or evaporitic carbonate units, which are com-monly underlain by an oxidized continental clastic sequence. The copper deposits typically are contained within a relatively narrow stratigraphic interval but have great lateral extent. The metal concentrations are discordant, both locally and regionally, and contain textural evidence for replacement of pre-existing compo-nents (Brown, 1978). It appears that these strata-bound deposits are the result of early diagenetic mineralization within poorly consolidated sediments. Rose (1976) concluded that these depos-its formed from slightly alkaline, sulfate-rich, chloride brines. The iron-bearing minerals in the associated red beds are commonly

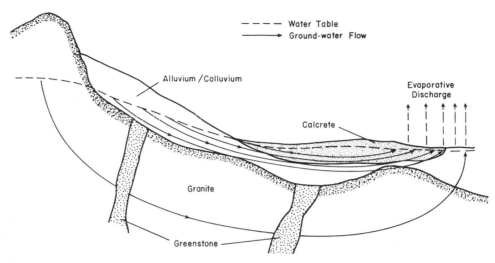

Figure 3. Evaporatively precipitated, uranium calcrete deposits (after Mann and Deutscher, 1978). Ore minerals are leached from crystalline rocks and transported by gravity flow to playa discharge areas.

proposed as the source of metals for the sulfide deposits. Evaporites generally have been regarded as important sources of chlorinity that contribute to metal transport, but D. A. Sverjensky (1987) has proposed that sulfate-rich evaporites are critical to the formation of these deposits by oxidizing basinal brines to an oxidation state that permits transport of metals and sulfate. The associated red beds permit the preservation of the oxidizing brines, and the sulfide- and organic-rich shales serve as precipitants (Sverjensky, 1987).

One genetic mechanism that has a great deal of support for some of these deposits is the sabkha model proposed by Renfro (1974). Because of their unique position, coastal sabkhas are nourished by subsurface flow of landward migrating, low-Eh–high-pH sea water and by seaward-migrating, high-Eh–low-pH meteoric water. Commonly, sabkhas are bordered on the seaward side by intertidal mudflats and lagoons that are carpeted by sediment-binding, blue-green algae. On the landward side, the coastal sabkhas give way to, and initially rest on, sterile, oxygenated desert sediments. Coastal sabkhas and their related evaporite facies prograde seaward across adjacent algal-mat facies. Upon burial, anaerobic bacteria in the algal-mat facies generate hydrogen sulfide. As sabkhas migrate basinward, formation water eventually must flow upward through the buried, strongly reducing algal mat in order to reach the surface of evaporation. This water of low pH and high Eh can mobilize and transport trace amounts of such elements as copper, silver, lead, and zinc. As the water passes through the hydrogen sulfide–charged algal mat, its load of solute metals is precipitated interstitially as sulfides. Resulting metal deposits generally are conformable to the geometry of hydrogen sulfide–bearing host strata. The extent of mineralization is dependent upon quantity of reductant, duration of the sabkha process, and chemistry of the metal-bearing meteoric water (Ren-

fro, 1974). Others (e.g., Gustafson and Williams, 1981; Morganti, 1981; Sverjensky, 1987) have expressed doubt that this mechanism can account for the magnitude of the mineralization and suggested that regional mineralizing fluid flow was needed.

### Deeper sedimentary environments

Deposits within this group include sedimentary exhalative lead-zinc-silver-barite, salt-dome sulfur-lead-zinc-silver-barite, carbonate-hosted lead-zinc, and sandstone-hosted lead. These deposits occur within marine sedimentary sequences and were created by a variety of ground-water processes, generally involving heated, saline formation waters.

***Sedimentary exhalative lead-zinc-silver-barite deposits.*** In contrast to massive sulfide deposits that occur in volcanic rocks and whose origin appears to be directly related to volcanic processes, many stratiform sulfide deposits occur within thick sequences of Proterozoic and Paleozoic siliciclastic sedimentary rocks with only minor associated volcanic strata (Gustafson and Williams, 1981; Sangster, 1983). The orebodies commonly consist of stacked stratiform sulfide lenses separated by weakly mineralized strata. The sulfide ores generally contain zinc and lead (with accessory silver, copper, and barite). Some sulfide-hosting pelitic sediments appear to have been deposited in relatively shallow water, as suggested by associated carbonate (often evaporitic) strata, but restricted euxenic conditions are suggested by the thinly laminated, pyritic, and carbonaceous nature of the immediate sulfide-hosting mudstones. The sulfide deposits are often associated with rapid sedimentary facies changes and/or an abrupt change in thickness of the hosting section. The deposits are generally associated with major regional fracture systems. These relationships and the local conglomerates or breccias and slumped strata that underlie many of the massive sulfide deposits suggest contemporaneous faulting was the major factor controlling the local sedimentation environment, including the discharge of metalliferous fluids onto the sea floor or into permeable sediments (Fig. 4). Some of these "sedimentary exhalative" deposits occur in intracratonic troughs, which developed as the result of tensional rifting and attendant slow subsidence. The role of extensional tectonics in this model is to provide high-permeability zones for metalliferous brines to reach the contemporaneous sea floor (Lydon, 1983).

The origin of the mineralizing fluid for the sedimentary exhalative sulfide and sulfate deposits is controversial. Most authors suggest a brine originating by some mechanism compatible with the tectonostratigraphic setting of the deposits. The debate centers on the extent to which convection is involved in driving the mineralizing system. Russell (1983) suggested deep convective circulation of sea water accompanying crustal extension to form the mineralizing brines. This model has been supported by the thermal balance studies of Strens and others (1987). However, Lydon (1983) maintained that the mineralizing solutions are highly saline formation waters that have evolved by sediment-water interactions accompanying burial. Lydon (1983) modelled his geochemical investigations of the ore-forming system on the data from Gulf Coast oil field brines; the extensional faults with which these deposits are associated are viewed as the features that tap these overpressured reservoirs.

The mechanism by which the mineralizing solution achieved its geochemical character is critical to the metallic composition of individual sedimentary exhalative deposits. Factors for consideration include the nature of the source rock (relative percentages of clastics, carbonates, and evaporites) that would govern the geochemical nature of the formation waters, including metal content and salinity; the sedimentation history of the basin that would control the time-temperature relationships of diagenesis accompanying burial; the tectonic history of the basin that would govern sedimentation and discharge of the ore-forming solutions; and igneous activity that could contribute heat to drive the mineralizing system. It is possible that ore solutions from magmatic activity or metamorphism could be introduced at the base of sedimentary basins. These fluids, as well as formational brines, represent potential mineralizing fluids. The fluids can either remain stagnant or, if a fluid potential gradient is present, migrate.

*Salt-dome sulfur-lead-zinc-barite deposits.* The presence of metal-rich brines in young sedimentary basins raises the possibility that base-metal deposits are currently forming by diagenetic waters escaping from geopressured areas. A possible scenario for the Gulf of Mexico Basin (Fig. 5) is as follows: NaCl brines form by dissolution of basal evaporites (the Jurassic Louann Salt). These brines are expelled upward via cross-formational flow through shales (e.g., the Norphlet Shale) and red beds (e.g., the Hosston Formation). The brines leach (or exchange) cations, picking up lead and zinc, during this process. Rock-water interactions such as albitization of detrital feldspars

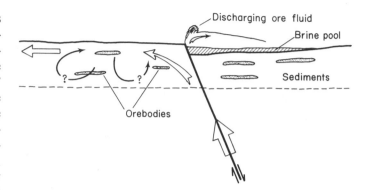

Figure 4. Model for sedimentary exhalative deposits (after Large, 1981). Brines move up along fracture systems to either discharge into the ocean and settle on the ocean floor or flow through existing sediments.

and smectite to illite conversion in deep Gulf Coast sediments also may be effective metal-releasing mechanisms (Land and Prezbindowski, 1981; Light and others, 1987). With continued flow away from geopressured zones, the metal-rich brine cools, depressurizes, and mixes with a variety of fluids (some possibly meteoric), and mineral precipitation occurs. It is common for well scales consisting of a variety of metallic minerals to form in significant amounts in many Gulf Coast oil fields (Carpenter and others, 1974; Hanor, 1979), thereby attesting to the potential of formation waters as ore-forming fluids.

The rate of brine migration may be slow or rapid. The latter is suggested by recent studies of sulfide deposits in the cap rocks of Gulf Coast salt diapirs (Fig. 6). Zones of fracturing, faulting, and brecciation are common along the flanks of diapirs and form zones of, at least sporadically, high permeability. Studies by Price and others (1983) indicate that metal-rich brines are expelled upward along the flanks of the diapirs. Mixing of metalliferous brines with reduced sulfur trapped within the cap rock environment is a likely cause of precipitation, although cooling, depressurization, and fluid mixing are all possible mechanisms of precipitation. Bacterial alteration in a shallow marine environment of initial cap-rock anhydrite is an integral aspect of calcite cap-rock formation and reduced sulfur production; subsequent oxidation of hydrogen sulfide is responsible for extensive elemental sulfur deposits. The salt-dome metallic deposits may represent geologically young examples of the processes responsible for some carbonate-hosted and sedimentary exhalative metallic deposits in older sedimentary environments (Kyle and Price, 1986).

*Sandstone-hosted lead deposits.* Some Mississippi Valley–type districts (e.g., southeast Missouri) contain metallic sulfides in the sandstone units that underlie the main carbonate host (Kyle and Gutierrez, 1988). Thus, these "sandstone-hosted" lead deposits appear to represent metal concentrations that were precipitated in the sandstone aquifer from metalliferous formational brines enroute to the carbonate host. Similar deposits elsewhere (e.g., Laisvall, Sweden) represent large metal accumulations in sandstones without associated carbonate-hosted ores. Rickard and others (1979) provided evidence for the origin of the Laisvall sulfides from concentrated Na-Ca-Cl formation waters at temperatures of about 150°C. They also noted a direct correlation between sediment paleopermeability and ore concentration. Bjørlykke and Sangster (1981), on the other hand, concluded that sandstone-hosted lead deposits are distinctly different from sedimentary copper and carbonate-hosted lead-zinc deposits. Their mineralization model involves a gravity-driven hydrologic system for metal transport in dilute, low-temperature meteoric waters from adjacent basement topographic highs; sulfides would be precipitated in sandstone aquifers in areas of high reduced sulfur. Thus, this model is similar to that commonly proposed for roll-front deposits.

*Carbonate-hosted lead-zinc deposits.* These are particularly well developed in the midcontinent area of the United States and are often referred to as Mississippi Valley–type lead-zinc deposits. Although there are significant differences among districts, there are a number of unifying general characteristics (Ohle, 1959, 1980; Anderson and Macqueen, 1982, Sverjensky, 1986). Most orebodies occur as strata-bound concentrations in shelf carbonate sequences, usually dolostones, on the flanks of major Phanerozoic foreland and intracratonic basins. The deposits generally consist of variable amounts of metal sulfides and associated minerals that formed in zones of secondary porosity in the carbonate strata. Fluid inclusion data indicate that the mineralizing fluids were brines with temperatures generally ranging from 60° to 175°C (Roedder, 1976); these mineralizing fluids are similar to contemporary subsurface brines present in sedimentary basins, such as the Gulf Coast (Carpenter and others, 1974; Land and Prezbindowski, 1981). Summaries on the geochemistry of hydrothermal fluids in deep sedimentary environments are provided by Hanor (1979) and Sverjensky (1984).

For the first half of this century, theories of ore genesis for the Mississippi Valley–type deposits were largely grouped into two major categories: (1) mineralization by meteoric water, and (2) mineralization by low-temperature fluids of magmatic derivation (Bastin and Behre, 1939). Based upon current knowledge of the nature of mineralization, both models, *senso stricto,* are now largely discounted. White (1958), Noble (1963), and Beales and Jackson (1966) proposed generation of ore-forming fluids by expulsion of formation water during compaction of muds enroute to becoming shales accompanying deep burial. During compaction and diagenesis of these water-rich sediments, the metals were released to interstitial fluids as soluble metal chloride and/or organic complexes. The ore-forming fluids moved through permeable zones, and ore deposits such as copper, uranium, vanadium in oxidized sandstones, and lead and zinc sulfides in carbonates were formed along sedimentary or geochemical facies boundaries. Modifications of the basinal evolution model for sulfide mineralization have been applied to explain the origin of many other ore deposits in sedimentary terranes. Although this model had received wide acceptance, no quantitative models of the fluid dynamics were published until that of Sharp (1978).

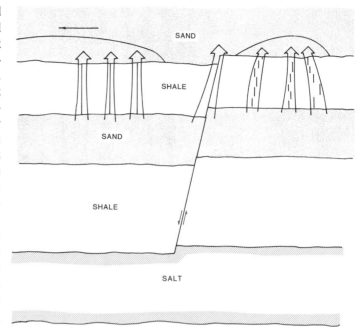

Figure 5. Movement of metalliferous Gulf Coast brines. Water flow from overpressured zones through faults, fractures, or microfractures. The brines can remain concentrated or be dissipated upon entering more active flow systems.

Roedder (1976), using volumetric calculations for typical Mississippi Valley–type ore deposits, fluid inclusions, and crystal growth rates, demonstrated that ore-solution flow rates are in the range of 0.1 to slightly over 100 m/yr. These bracket Sharp's hydrogeological analysis, which estimated peak Darcian flow rates of compaction-driven fluids (migrating from the Ouachita Basin to the Ozark Dome) to range between 1 and 10 m/yr. Cathles and Smith (1983), on the other hand, suggested more rapid pulses of compactional fluids based upon thermal modeling of an Illinois-like basin. Finally, Bethke (1986) suggested a topographically driven flow system for deposits in the Illinois Basin, similar to that developed by Garven and Freeze (1984) for the Western Canada Basin, discussed below. Bethke's study indicates that the overpressured mechanism was not operable in the Illinois Basin, although it may have developed in the Ouachita Basin.

A more complicated scenario is given by Heyl (1969) who, based on geochemical data, suggested several fluid flow patterns for Mississippi Valley–type deposits. Heyl considered modification of the basic compaction concept by several mechanisms, including hydrothermal circulation over local intrusions and mixing with meteoric and "shallowly circulating connate brine" flow systems. These models are summarized in Figure 7.

The Pine Point mining district in the Northwest Territories possesses many common geologic features of the carbonate-hosted deposits of the midcontinent United States. Pine Point's geologic setting was used by Beales and Jackson (1966) to de-

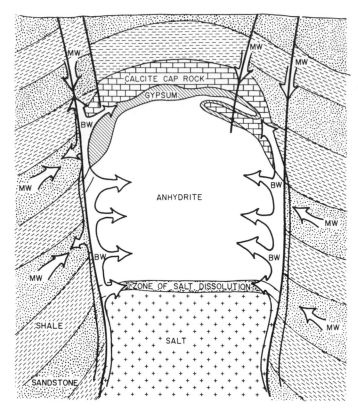

Figure 6. Mineralization of salt-dome cap rock (after Price and others, 1983). Upward flow of metal-rich brines into salt-dome cap rock, mixing with meteoric or marine waters, organic reduction of $SO_4$, and cooling may all be factors controlling mineralization.

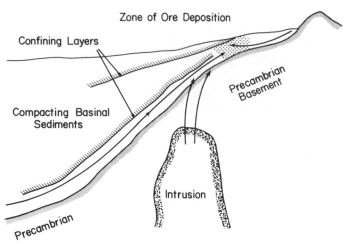

Figure 7. Hydrodynamic models for Mississippi Valley–type deposits. Waters from compacting basinal sediments, magmatic waters, and meteoric waters have all been hypothesized as the ore-bearing fluids, but temperatures and salinities derived from fluid inclusions indicate basinal waters as a dominant source.

velop the "basinal evolution" mineralization model, variations of which have been widely accepted for the formation of carbonate-hosted Pb-Zn deposits. The district consists of about 50 sulfide bodies within a Devonian carbonate complex (Skall, 1975; Kyle, 1981; Rhodes and others, 1984). Orebodies are localized in paleokarstic features that formed during a period of Middle Devonian subaerial exposure of the barrier complex; subsequent modifications are due to the ore-forming solutions. Zn-Pb-Fe zonation apparently reflects local and district-wide fluid movement, including local movement of a metalliferous fluid upward and outward from a deeper aquifer system. The sulfides are concentrated in the local paleosolution structures, which acted as leakage zones between aquifers and loci for possible mixing of fluids of different chemical character.

Fluid inclusion data suggest sulfide precipitation from brines at temperatures between 50° and 100°C. Ore textures, isotopic data, and thermodynamic data indicate rapid precipitation, followed by slower growth (perhaps related to an episodic supply of reduced sulfur). The sulfide could have come from inorganic reduction of sulfate by organic matter (Powell and Macqueen, 1984) or by bacterial reduction of sulfate. The nonsystematic

variation of Pb and S isotopes in galena suggests that metals and reduced sulfur were not contained within the same fluid (Kyle and others, unpublished data). The timing of mineralization has not been uniquely defined, but most estimates suggest either late Paleozoic (Kyle, 1981; Beales and Jackson, 1982) or mid-Tertiary (Garven, 1985).

Two hydrodynamic processes have been postulated as the driving mechanism for the migration of the metal-bearing brines—compaction of Mackenzie Basin shales and gravity or topography-driven flow in the Western Canada Basin. The compaction model was first proposed by Beales and Jackson (1966) and further developed by Smith and others (1983). The compaction theory is based on the large volume of metal-rich shales in the Mackenzie Basin adjacent to the mineralized reef complex; the presence of a permeable subshale carbonate unit, which could have acted as a conduit for compactional fluids; a burial depth for the shales consistent with the geothermometry; a pattern of shale compaction that appears to focus the flow of fluids into linear conduits toward the mineralized strata; and evidence for late-stage shale compaction dewatering downward into the carbonate unit.

Garven (1985) has provided an alternative scenario. Using mathematical models for calculating transport of fluids, heat, and chemical species, he demonstrated that a gravity-driven groundwater flow system in the Western Canada Basin could have

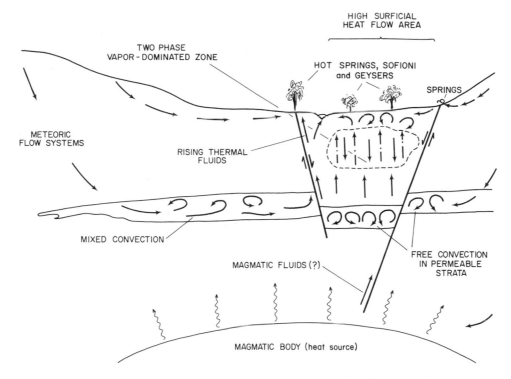

Figure 8. Flow paths for hydrothermal systems. Predominantly meteoric fluids can be driven by gravity, by free convection, or by a combination. Flow systems can be both deep and shallow. Both vapor-dominated and hot-water-dominated systems may exist.

formed the Pine Point deposits in a relatively brief interval of geologic time. His analysis indicates that the Keg River barrier-complex aquifer was a major factor in focusing fluid flow from thick Upper Devonian shales in the western portion of the basin and in discharging metal-rich brines updip toward the northeastern basin margin. Flow rates on the order of 1 m/yr are demonstrated to be sufficient to create fluid temperatures in the 50° to 100°C range. The present fluid-flow and heat-flow patterns in the Western Canada Basin (Tóth, 1978) also support the results of the numerical model simulations.

### Shallow continental plutonic environments

Examples of mineral deposits formed by plutonic-driven hydrothermal circulation include (see Table 3) porphyry Cu and Mo, Cu-Zn skarns, replacement Pb-Zn-Ag, and epithermal Au-Ag, Sb, and Hg deposits. These deposits, at least in part, are the products of deep circulation of meteoric water. Magmatic heat is an efficient force in circulating ground water, and a number of such models have been proposed. White (1968) discussed the circulation of metal-rich brines in the Salton Sea geothermal area. Here, the convecting waters are primarily of meteoric origin. Similarly, silver-gold vein deposits of Neogene age associated with the Kuroko (Japan) district (Hattori and Sakai, 1979)

formed in an environment where meteoric waters presumably leached ore constituents from the country rock. Here, also, fluid circulation is driven by density differences caused by high geothermal gradients. On the other hand, convecting fluids can be predominantly magmatic. White (1968) considered the Providencia ores (Zacatecas, Mexico) as an example of deposits formed by a convecting magmatic fluid.

Epithermal deposits represent paleodischarge areas for deeply circulating fluids. As discussed by White (1981), Henley (1973), and Henley and Ellis (1983), the origins of the mineralizing fluids can be diverse. In most cases, however, a preponderance of meteoric water is indicated by isotopic evidence (Craig, 1961). Epithermal deposits include orebodies of mercury, precious metals, and base metals. A conceptual model for groundwater flow in hydrothermal systems is shown in Figure 8. Note that localized vapor-dominated zones may exist.

Mathematical modeling of modern geothermal systems was reviewed by Garg and Kassoy (1981) who concluded that most of the heat and mass transfer was the result of convection of liquid water and/or steam through fractured media. These convective hydrologic systems result in a wide variety of alteration and mineralization phenomena in modern and ancient geothermal systems as a result of rock-water interactions (White, 1981; Henley and Ellis, 1983; Berger and Bethke, 1985). General fluid flow regimes for fossil geothermal systems, that is, epithermal veins, have been interpreted from integrated structural, paragenetic, and geochemical studies of mineralized areas in the mid-Tertiary volcanic province of the San Juan Mountains, Colorado (Barton and others, 1977; Casedevall and Ohmoto, 1977). Berger and Eimon (1983) proposed that epithermal precious metal deposits could be considered as the result of gradational end-member hydrothermal systems involving stacked-cell convection, closed-cell convection, and hot-springs depositional models.

Mercury deposits form at the lowest temperatures of any epithermal deposits (60° to 150°C). The hydrogeologic scenario presented by White (1981) for these deposits is as follows: deeply circulating waters (T > 200°C) leach Hg from sedimentary rocks and associated organic matter; upon rising toward the surface via advection or convection, the ground-water system possesses a "through-going vapor phase at temperatures below 200°C;" non-condensable gases, such as $CO_2$ (predominantly), $H_2S$, and $CH_4$, are abundant in the vapor phase; elemental Hg is partitioned into the vapor phase and transported to cool, near-surface conditions; and as temperatures decrease, Hg° condenses and HgS precipitates from the liquid phase. This process would be enhanced by the presence of permeable zones for vapor transport and limited vapor discharge to the atmosphere. The vapor-phase transport capability of Hg may account for its typical near-surface deposition in a wide variety of host rocks without economic concentrations of other sulfides.

Precious-metal epithermal deposits are products of high-temperature geothermal systems (Berger and Bethke, 1985). They typically show a depth zoning and selective concentration of a variety of metals, including but not limited to Au, As, Sb, Hg, Tl, B, and Ag. Ag and base-metal deposition apparently occurs at greater depths than the other epithermal metals. This depth zoning was recognized by Nolan (1933) who differentiated them into Au- and Ag-dominated deposits. Base-metal deposits may grade upward into Ag-rich zones and appear to represent the deeper, higher temperature, higher salinity, more acidic portion of the geothermal system. Much of the variation in epithermal deposits is caused by regional geochemical differences (Hedenquist and Henley, 1985), but within a given geothermal setting, White (1981) suggested the following explanation for the zoning: Au, As, Sb, Hg, Tl, and B are deposited at relatively cool, near-surface conditions with Hg transport (and possibly Sb, as per Dessau, 1952) in the vapor phase, but others in the liquid phase, while Ag and base-metal deposition occurs at higher temperatures and greater depths. The dominant complexing agent may change from $Cl^-$ to $HS^-$ as the fluids move upward. Deposition from chloride complexes occurs as the result of dilution by meteoric waters, decreasing temperatures, and increasing pH (caused by $CO_2$ outgassing, boiling, and water-rock interaction).

Other plutonic-driven hydrothermal ore deposits include the stockwork deposits of copper, molybdenum, tin, and tungsten that are associated with fractured multiple intrusions of intermediate of felsic compositions (Lowell and Guilbert, 1970; White and others, 1981; Grant and others, 1977). Where the intrusion wall rocks are reactive carbonate-rich strata, substantial metal concentrations may be localized in envelopes of calc-silicate alteration or skarn (Einaudi and others, 1981). Current genetic models for these deposits involve the interaction of magmatic and meteoric fluids in the vicinity of cooling plutons. Shelton (1983) showed contrasting patterns of fluid flow in porphyry (convection) and skarn (stratigraphically controlled gravity) mineralization. The initial fluids are believed to be magmatic in origin; with time, the pluton is cooled by the convective circulation of ground water that is capable of extracting and transporting metals and is responsible for the dominant meteoric signature of the final alteration suite (Taylor, 1979; Brimhall, 1979; Norton, 1981; Beane and Titley, 1981). As shown by Norton and Knight (1977), convective heat transfer in such systems will dominate conductive transfer whenever permeabilities exceed $10^{-8}m^2$. Typically, only unfractured crystalline rocks, unjointed shales, evaporites, and some clays possess lower permeabilities. Consequently, free convection of ground water in the vicinity of plutons must be expected. Norton and Knight also demonstrated that significant quantities of fluid (either magmatic or meteoric in origin) are involved. A number of general models have been proposed for flow systems near cooling plutons (Fig. 9). In each case, fluids driven by magmatic cooling rise over the pluton. Upon rising, the fluids form either magmatic-vapor or ground-water dominated plumes of dispersion (Henley and McNabb, 1978). The intrusion creates variable patterns of thermal convection dependent on the permeability distribution. Concomitant with circulation is the potential for hydraulic fracturing and an increase in host-rock permeability caused by this fracturing, by cooling cracks, and by leaching of ore components from rocks along flow paths. The permeability increase tends to promote increased influx of meteoric waters. Conversely, precipitation of minerals in fractures and pores in other areas could decrease permeability and create a "self-sealing" system. These processes are considered in the hydrothermal cycle (Fig. 10) proposed by Elders (1981).

### Marine volcanic environments

Ore deposits within this environment include stratiform massive sulfides associated with both mafic and felsic subaqueous extrusive systems (Franklin and others, 1981; Ohmoto and Skinner, 1983). Hydrothermal circulation of marine water near deep-sea rifts is now well documented (Edmonds and Von Damm, 1983; Rona and others, 1983). A heuristic analysis of the hydrothermal system that formed the Miocene Kuroko massive sulfide deposits of Japan is provided by Cathles (1983). He concluded that the Kuroko deposits are the result of multiple small intrusive pulses that cooled by convection of sea water; individual massive sulfide-sulfate lenses were formed from these heated fluids in less than 5,000 years and probably less than 500 years.

Sulfide deposition and alteration of associated volcanics in marine geothermal systems involves circulation of oceanic waters through the volcanic pile to form massive sulfide deposits as shown in Figure 11 (Large, 1977; Reed, 1983). Free convection

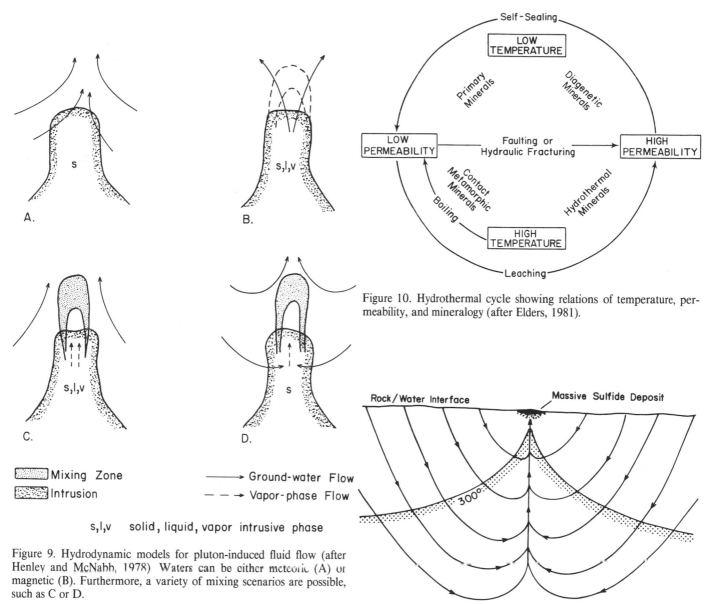

Figure 9. Hydrodynamic models for pluton-induced fluid flow (after Henley and McNabb, 1978) Waters can be either meteoric (A) or magnetic (B). Furthermore, a variety of mixing scenarios are possible, such as C or D.

Figure 10. Hydrothermal cycle showing relations of temperature, permeability, and mineralogy (after Elders, 1981).

Figure 11. Model for marine volcanic sulfide deposition (after Large, 1977). Driven by thermal convection, oceanic waters leach and later deposit metals.

is established over the pile; the downward circulating sea water becomes hotter, more reducing, and more acidic; as metals are leached from the volcanic rocks; mineral-laden fluids rise buoyantly over the pile; and sulfides are precipitated within the volcanic rocks near the rock-water interface, or as stratiform chemical sediments upon issuing from vents and flowing (via density flow) across the sea bottoms as now occurs in the Red Sea (Schoell and Menz, 1976). Discussions of the physical behavior of potential ore-forming fluids of varying temperature and density upon discharge into a subaqueous environment are provided by

Turner and Gustafson (1978), McDougall (1984), and Campbell and others (1984).

Ore deposits in the discharge zone tend to be localized because discharge zones tend to be areally smaller than areas of recharge and lateral flow. Ground water may discharge by flow into lakes and streams by boiling or by evapotranspiration where the rift systems move inland. Ore deposits may be formed by any of the above processes. For example, black smokers (Rona and others, 1983) represent the subaqueous discharge of marine waters hydrothermally circulated by shallow magmatic activity. The

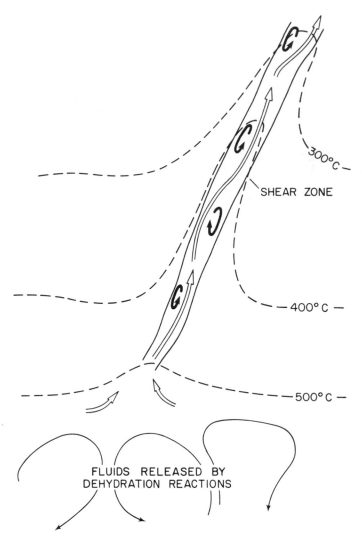

Figure 12. Hydrodynamic model for Archean vein gold deposits (after Henley, 1973). Fluids released during metamorphism convect slowly through the metamorphism zones and rapidly along permeable fracture zones. Net upward flow through the fractures transports ore components.

postulated—a slow, pervasive convection responsible for dissolving the metals, and an intense convection constrained to fracture zones along which ores are deposited. The rapid, upward movement in the fracture zones may be sufficient to induce brecciation.

Many of the lode gold deposits are concentrated along linear zones that transect stratigraphic facies on a regional scale and have been interpreted to be major crustal fault zones; examples include the deposits along the "breaks" of the Superior and Slave provinces of the Canadian Shield and the Mother Lode in California. Some deposits (e.g., the Homestake deposit in South Dakota) may represent gold-rich chemical sediments that formed from discharge of hydrothermal solutions onto the contemporaneous sea floor during quiescence in volcanic activity (i.e., a variation of the volcanogenic massive sulfide model); subsequent deformation and metamorphism is viewed as responsible for remobilization of the metallic concentrations (Ridler, 1976; Rye and Rye, 1974). Other workers (Henley, 1973; Colvine and others, 1984) have emphasized the importance of the regional structural features in gold mineralization and have suggested that the deposits formed as the direct result of late synorogenic to postorogenic discharge of metamorphic fluids along major crustal fracture zones. The origin of the mineralizing fluids in these two hypotheses is a matter of some controversy, although the high $CO_2$ content of fluid inclusions in gold-bearing vein quartz suggests an appreciable contribution of outgassed fluids derived, possibly, from metamorphic reactions associated with the greenschist to amphibolite grade transition (Kerrich and Fryer, 1979).

The origin of the important silver-lead-zinc veins of the Coeur d'Alene district in Idaho has long been debated. These principal veins occur along major structures in metasedimentary rocks of Proterozoic age (Hobbs and Fryklund, 1968). The lead isotopic character of vein galena indicates ages between 1,200 and 1,500 Ma, and it has been suggested that, while the main phase of metal concentration was Proterozoic in age, the ores were remobilized into their present sites during the Cretaceous (Zartman and Stacey, 1971; Harrison and others, 1974). Leach and others (1985) detected appreciable quantities of $CO_2$, $CH_4$, and $N_2$ in fluid inclusions in vein quartz, and have suggested that the ores formed at about 350°C from greenschist metamorphic fluids.

## CONCLUSIONS

We have briefly discussed some of the many ways in which ground-water processes are important in the genesis of mineral deposits. A number of conclusions are evident:

1. The significance of fluid migration is paramount in many models for the genesis of mineral deposits, but in only a few cases have the actual hydrodynamics of the proposed flow system have intensively investigated.

2. Mineral deposits may form where meteoric waters infiltrate into the earth, where ground water discharges, where mineralizing fluids migrate into a zone of different chemical or physical

same flow system is suggested to form possible iron-manganese deposits in the recharge areas, sulfide deposits within the rocks beneath the vents, and chemical sediments in the discharge area.

### Metamorphic environments

Ore deposits within this environment include vein deposits of gold and lead-zinc-silver that appear to originate from fluid generated at least in part by metamorphic dehydration reactions in deep crustal terranes. Flow channelization along deep fracture zones appears to be a fundamental characteristic of these deposits. In the model shown in Figure 12, two convection cells are properties, or where fluids of differing chemical composition or temperature mix.

3. Analytical or numerical hydrogeologic models are capable of predicting times and rates of fluid movement in mineral deposit environments and, when coupled with analysis of mass and energy transport, should lend insight into mineral deposit genesis.

4. The consideration of hydrogeologic concepts by economic geologists may lead to new or revised theories of mineral

deposition. On the other hand, these analyses will ultimately benefit hydrogeology by allowing comparison and calibration of existing models with long-time, large-scale data.

5. There are a number of potentially fruitful lines of research that should benefit both hydrogeology and economic geology. The most promising is the assessment of how the ground-water system changes, both chemically and physically, over geologic time. When we understand this, hydrogeology may prove much more useful in the field of mineral exploration.

## REFERENCES CITED

Adler, H. H., 1964, The conceptual uranium roll and its significance in uranium exploration: Economic Geology, v. 59, p. 46–53.

Anderson, C. A., 1955, Oxidation of copper sulfides and secondary sulfide enrichment: Economic Geology, 50th Anniversary Volume, p. 324–340.

Anderson, G. M., 1975, Precipitation of Mississippi Valley type ores: Economic Geology, v. 70, p. 937–942.

Anderson, G. M., and Macqueen, R. W., 1982, Ore deposit models; 6, Mississippi Valley–type lead-zinc deposits: Geoscience Canada, v. 9, p. 108–117.

Anderson, M. P., 1979, Using models to simulate the movement of contaminants through groundwater flow systems: Chemical Rubber Company (CRC) Critical Reviews in Environmental Control, v. 9, no. 2, p. 97–156.

Annels, A. E., 1979, Mufulira graywackes and their associated sulphides: Institution of Mining and Metallurgy Transactions, v. 88, p. B15–B23.

Aziz, K., Bories, S. A., and Combarnous, M. A., 1973, The influence of natural convection in gas, oil, and water reservoirs: Journal of Canadian Petroleum Technology, v. 12, p. 41–47.

Badham, J.P.N., 1981, Shale-hosted Pb-Zn deposits; Products of exhalation of formation waters?: Institution of Mining and Metallurgy Transactions, v. 90, p. B70–B76.

Baedecker, M. J., and Back, W., 1979, Modern marine sediments as a natural analog to the chemically stressed environment of a landfill: Journal of Hydrology, v. 43, p. 393–414.

Barnes, H. L., 1979, Solubilities of ore minerals, *in* Barnes, H. L., ed., Geochemistry of hydrothermal ore deposits (2nd edition): New York, John Wiley and Sons, p. 404–460.

Barton, B. P., Bethke, P. M., and Roedder, E., 1977, Environment of ore deposition in the Creede mining district, San Juan Mountains, Colorado; Part 3, Progress toward interpretation of the ore-forming fluid for the OH vein: Economic Geology, v. 72, p. 1–24.

Bassett, R. L., Bentley, M. E., and Simpkins, W. W., 1981, Regional ground water flow in the Panhandle of Texas: University of Texas Bureau of Economic Geology Circular 81-3, p. 102–107.

Bastin, E. S., and Behre, C. H., Jr., 1939, Origin of the Mississippi Valley lead and zinc deposits; A critical summary, *in* Bastin, E. S., ed., Contributions to a knowledge of the lead and zinc deposits of the Mississippi Valley region: Geological Society of America Special Paper 24, p. 121–143.

Beales, F. W., and Jackson, S. A., 1966, Precipitation of lead-zinc ores in carbonate reservoirs as illustrated by Pine Point ore field, Canada: Institution of Mining and Metallurgy Transactions, v. 75, p. 278–285.

—— , 1982, Multiple genetic and diagenetic models help the unravelling of Mississippi Valley–type ore genesis 11th International Congress on Sedimentology Abstracts with Programs, p. 17.

Beane, R. E., and Titley, S. R., 1981, Porphyry copper deposits; Part 2, Hydrothermal alteration and mineralization: Economic Geology, 75th Anniversary Volume, p. 235–269.

Bear, J., 1972, Dynamics of fluids in porous media: New York, Elsevier, 764 p.

Berger, B. R., and Eimon, P. L., 1983, Conceptual models of epithermal precious metal deposits, *in* Shanks, W. C., ed., Unconventional mineral deposits (Cameron Volume): New York, American Institute of Mining, Metallurgical, and Petroleum Engineers, p. 191–205.

Bethke, C. M., 1986, Hydrologic constraints on genesis of the upper Mississippi Valley mineral district from Illinois Basin brines: Economic Geology, v. 81, p. 233–249.

Bjørlykke, A., and Sangster, D. F., 1981, An overview of sandstone lead deposits and their relationship to red-bed copper and carbonate-hosted lead-zinc deposits: Economic Geology, 75th Anniversary Volume, p. 179–213.

Blanchard, P. E., and Sharp, J. M., Jr., 1985, Possible free convection in thick Gulf Coast sandstone sequences: American Association of Petroleum Geologists Southwest Section, 1985 Transactions, p. 6–11.

Bredehoeft, J. D., Neuzil, C. E., and Milly, P.C.D., 1983, Regional flow in the Dakota aquifer; A study of the role of confining layers: U.S. Geological Survey Water-Supply Paper 2237, 45 p.

Brimhall, G. H., Jr., 1979, Lithologic determination of mass transfer mechanism of multiple-stage porphyry copper mineralization at Butte, Montana; Vein formation by hypogene leaching and enrichment of potassium-silicate protore: Economic Geology, v. 74, p. 556–589.

Brimhall, G. H., Alpers, C. N., and Cunningham, A. B., 1985, Analysis of supergene ore-forming processes and ground-water solute transport using mass balance principles: Economic Geology, v. 80, p. 1227–1256.

Brown, A. C., 1978, Stratiform copper deposits; Evidence for their post-sedimentary origin: Minerals Science and Engineering, v. 10, p. 172–181.

Burnham, C. W., 1979, Magmas and hydrothermal fluids, *in* Barnes, H. L., ed., Geochemistry of hydrothermal ore deposits: New York, John Wiley and Sons, p. 71–136.

Campbell, I. H., McDougall, T. J., and Turner, J. S., 1984, A note on fluid dynamic processes which can influence the deposition of massive sulfides: Economic Geology, v. 79, p. 1905–1913.

Carpenter, A., Trout, M. L., and Picket, E. E., 1974, Preliminary report on the origin and chemical evolution of lead- and zinc-rich oil field brines in central Mississippi: Economic Geology, v. 69, p. 1191–1206.

Casedevall, T., and Ohmoto, H., 1977, Sunnyside mine, Eureka mining district, San Juan County, Colorado; Geochemistry of gold and base metal ore deposition in a volcanic environment: Economic Geology, v. 72, p. 1285–1320.

Cathles, L. M., 1981, Fluid flow and genesis of hydrothermal ore deposits: Economic Geology, 75th Anniversary Volume, p. 424–457.

—— , 1983, An analysis of the hydrothermal system responsible for massive sulfide deposition in the Hokuroku Basin of Japan, *in* Ohmoto, H., and Skinner, B. J., eds., The Kuroko and related volcanogenic massive sulfide deposits: Economic Geology Monograph 5, p. 439–487.

Cathles, L. M., and Smith, A. T., 1983, Thermal constraints on the formation of Mississippi Valley-type lead-zinc deposits and their implications for episodic basin dewatering and deposit genesis: Economic Geology, v. 78, p. 983–1002.

Colvine, A. C., and others, 1984, An integrated model for the origin of Archean lode gold deposits: Ontario Geological Survey Open-File Report 5524, 98 p.

Combarnous, M. A., and Bories, S. A., 1975, Hydrothermal convection in saturated porous media: Advances in Hydroscience, v. 10, p. 231–307.

Comer, J. B., 1974, Genesis of Jamaican bauxite: Economic Geology, v. 69, p. 1251–1264.

Craig, H., 1961, Isotopic variations in meteoric waters: Science, v. 133, p. 1702–1703.

Dessau, G., 1952, Antimony deposits of Tuscany: Economic Geology, v. 47, p. 397–413.

Doe, B. R., and Zartman, R. E., 1979, Plumbotectonics, the Phanerozoic, *in* Barnes, H. L., ed., Geochemistry of hydrothermal ore deposits (2nd edition): New York, John Wiley and Sons, p. 22–70.

Doe, B. R., Steven, T. A., Delevaux, M. H., Stacey, J. S., Lipman, P. W., and Fisher, F. S., 1979, Genesis of ore deposits in the San Juan volcanic field, southwestern Colorado; Lead isotope evidence: Economic Geology, v. 74, p. 1–26.

Donaldson, I. G., 1962, Temperature gradients in the upper layers of the earth's crust due to convective water flows: Journal of Geophysical Research, v. 67, p. 3449–3459.

Drummond, S. E., and Ohmoto, H., 1985, Chemical evolution and mineral deposition in boiling hydrothermal systems: Economic Geology, v. 80, p. 126–147.

Dunham, K. C., 1970, Mineralization by deep formation waters; A review: Institution of Mining and Metallurgy Transactions, v. 79, p. B127–B136.

Edmonds, J. M., and Von Damm, K., 1983, Hot springs on the ocean floor: Scientific American, v. 248, no. 4, p. 78–93.

Einaudi, L. D., Meinert, L. D., and Newberry, R. J., 1981, Skarn deposits: Economic Geology, 75th Anniversary Volume, p. 317–391.

Elders, W. A., 1981, Applications of petrology and geochemistry to the study of active geothermal systems in the Salton Sea trough of California and Baja California, *in* Hausen, D. M., and Park, W. C., eds., Process mineralogy: American Institute of Mining, Metallurgical, and Petroleum Engineers Transactions, p. 591–606.

Erickson, R. L., Rosier, E. L., Viets, J. G., Odland, S. K., and Erickson, M. S., 1983, Subsurface geochemical exploration in carbonate terrane; Midcontinent, U.S.A., *in* Kisvarsanyl, G., and others, eds., Proceedings, International Conference on Mississippi valley–type Lead-Zinc Deposits: Rolla, University of Missouri, p. 575–583.

Etheridge, M. A., Wall, Y. J., and Vernon, R. H., 1983, The role of the fluid phase during regional metamorphism and deformation: Journal of Metamorphic Geology, v. 1, p. 205–226.

Fancelli, R., and Nuti, S., 1965, Studio sulle acque termali e minerali della parte orientale della Provincia di Siena: Bolletino della Societa Geologica Italiana, v. 94, p. 135–155.

Ferry, J. M., 1983, On the control of temperature, fluid composition, and reaction progress during metamorphism, *in* Greenwood, H. J., ed., Studies in metamorphism and metasomatism (Orville Volume): American Journal of Science, v. 283-A, p. 201–232.

Force, E. R., Back, W., Spiker, E. C., and Knauth, L. P., 1986, A ground-water mixing model for the origin of the Imini manganese deposit (Cretaceous) of Morroco: Economic Geology, v. 81, p. 65–79.

Franklin, J. M., Lydon, J. W., and Sangster, D. F., 1981, Volcanic-associated massive sulfide deposits: Economic Geology, 75th Anniversary Volume, p. 485–627.

Fyfe, W. S., Price, N. J., and Thompson, A. B., 1978, Fluids in the earth's crust: Amsterdam, Elsevier, 384 p.

Galloway, W. E., 1978, Uranium mineralization in a coastal-plain fluvial aquifer system; Catahoula Formation, Texas: Economic Geology, v. 73, p. 1655–1676.

Galloway, W. E., Kreitler, C. W., and McGowen, J. H., 1979, Depositional and ground-water flow systems in the exploration for uranium: University of Texas Bureau of Economic Geology Research Colloquium, 267 p.

Garg, S. K., and Kassoy, D. R., 1981, Convective heat and mass transfer in hydrothermal systems, *in* Raybach, L., and Muffler, L.J.P., eds., Geothermal systems; Principles and case histories: New York, John Wiley and Sons, p. 37–76.

Garrels, R. M., 1954, Mineral species as functions of pH and oxidation-reduction potentials, with special reference to the zone of oxidation and secondary enrichment of sulphide ore deposits: Geochimica et Cosmochimica Acta, v. 5, p. 153–168.

Garven, G., 1985, The role of regional fluid flow in the genesis of the Pine Point deposit, western Canada sedimentary basin: Economic Geology, v. 80, p. 307–324.

Garven, G., and Freeze, R. A., 1984, Theoretical analysis of the role of groundwater flow in the genesis of strata-bound ore deposits; 2, Quantitative results: American Journal of Science, v. 284, p. 1125–1174.

Golightly, J. P., 1981, Nickeliferous laterite deposits: Economic Geology, 75th Anniversary Volume, p. 710–735.

Graf, D. L., 1982, Chemical osmosis, reverse chemical osmosis, and the origin of subsurface brines: Geochimica et Cosmochimica Acta, v. 46, p. 1431–1448.

Graf, J. L., Jr., 1984, Effects of Mississippi Valley–type mineralization on REE patterns of carbonate rocks and minerals, Viburnum Trend, southeast Missouri: Journal of Geology, v. 92, p. 307–324.

Grant, J. N., Halls, C., Avila, W., and Avila, G., 1977, Igneous geology and the evolution of hydrothermal systems in some sub-volcanic tin deposits of Bolivia, *in* Gass, I. G., ed., Volcanic processes in ore genesis: Geological Society of London Special Publication 7, p. 77–116.

Gustafson, L. B., and Williams, N., 1981, Sediment-hosted stratiform deposits of copper, lead, and zinc: Economic Geology, 75th Anniversary Volume, p. 139–178.

Hanor, J. S., 1979, The sedimentary genesis of hydrothermal fluids, *in* Barnes, H. L., ed., Geochemistry of hydrothermal ore deposits: New York, John Wiley and Sons, p. 137–162.

Harrison, J. E., Griggs, A. B., and Wells, J. D., 1974, Tectonic features of the Precambrian Belt Basin and their influence on post-Belt structures: U.S. Geological Survey Professional Paper 886, 15 p.

Hattori, K., and Sakai, H., 1979, D/H ratios, origins, and evolution of the oreforming fluids for the Neogene veins and Kuroko deposits of Japan: Economic Geology, v. 74, p. 535–555.

Hedenquist, J. W., and Henley, R. W., 1985, The importance of $CO_2$ in freezing point measurements of fluid inclusions; Evidence from active geothermal systems and implications for epithermal ore deposition: Economic Geology, v. 80, p. 1379–1406.

Helgeson, H. C., 1969, Thermodynamics of hydrothermal systems at elevated temperatures and pressures: American Journal of Science, v. 267, p. 729–804.

—— , 1970, A chemical and thermodynamic model of ore deposition in hydrothermal systems, *in* Morgan, B. A., ed., 50th Anniversary Symposium: Mineralogical Society of America Special Paper 3, p. 155–186.

—— , 1979, Mass transfer among minerals and hydrothermal solutions, *in* Barnes, H. L., ed., Geochemistry of hydrothermal ore deposits, 2nd ed.: New York, Wiley, p. 568–610.

Henley, R. W., 1973, Some fluid dynamics and ore genesis: Institution of Mining and Metallurgy Transactions, v. 82, p. B81–B89.

Henley, R. W., and Ellis, A. J., 1983, Geothermal systems ancient and modern; A geochemical review: Earth-Science Reviews, v. 19, p. 1–50.

Henley, R. W., and McNabb, A., 1978, Magmatic vapor plumes and groundwater interaction in prophyry copper emplacement: Economic Geology, v. 73, p. 1–20.

Henley, R. W., Truesdell, A. H., and Barton, P. B., Jr., 1984, Fluid-mineral equilibria in hydrothermal systems: Society of Economic Geologists, Reviews in Economic Geology, v. 1, 267 p.

Heyl, A. V., 1969, Some aspects of genesis of zinc-lead-barite-fluorite deposits in the Mississippi Valley, U.S.A.: Institution of Mining and Metallurgy Transactions, v. 78, p. B148–B160.

Hitchon, B., 1976, Hydrogeochemical aspects of mineral deposits in sedimentary rocks, *in* Wolf, K. H., ed., Handbook of strata-bound and stratiform ore deposits, v. 2: Amsterdam Elsevier, p. 53–66.

Hobbs, S. W., and Fryklund, V. C., Jr., 1968, The Coeur d'Alene district, Idaho, *in* Ridge, J. D., ed., Ore deposits of the United States, 1933–1967: American Institute of Mining, Metallurgical, and Petroleum Engineers, p. 1417–1435.

Hoeve, J., and Sibbald, T.I.I., 1978, On the genesis of Rabbit Lake and other unconformity-type uranium deposits in northern Saskatchewan, Canada:

Economic Geology, v. 73, p. 1450–1473.

Holland, H. D., and Malinin, S. D., 1979, The solubility and occurrence of non-ore minerals, *in* Barnes, H. L., ed., Geochemistry of hydrothermal ore deposits (2nd edition): New York, John Wiley and Sons, p. 461–503.

Howard, K. W., and Hanor, J. S., 1987, Compositional zoning in the Fancy Hill stratiform barite deposit, Ouachita Mountains, Arkansas, and evidence for the lack of associated massive sulfides: Economic Geology, v. 82, p. 1377–1385.

Hubbert, M. K., 1940, The theory of ground-water motion: Journal of Geology, v. 48, p. 785–944.

Hutchinson, R. W., Fyfe, W. S., and Kerrich, R., 1980, Deep fluid penetration and ore deposition: Minerals Science and Engineering, v. 12, p. 107–120.

Hutchison, C. S., 1983, Economic deposits and their tectonic setting: New York, Wiley-Interscience, 365 p.

Jackson, S. A., and Beales, F. W., 1967, An aspect of sedimentary basin evolution; The concentration of Mississippi Valley–type ores during late stages of diagenesis: Canadian Petroleum Geologists Bulletin, v. 15, p. 383–433.

Kerrich, R., and Fryer, B. J., 1979, Archaean precious-metal hydrothermal systems, Dome mine, Abitibi greenstone belt; 2, REE and oxygen isotope relations: Canadian Journal of Earth Science, v. 16, p. 440–456.

Knapp, R. B., and Norton, D., 1981, Preliminary numerical analysis of processes related to magma crystallization and stress evolution in cooling pluton environments: American Journal of Science, v. 281, p. 35–68.

Kyle, J. R., 1981, Geology of the Pine Point lead-zinc district, *in* Wolf, K. H., ed., Handbook of strata-bound and stratiform ore deposits, v. 9: Amsterdam, Elsevier, p. 643–741.

—— , 1983, Economic significance of subaerial exposure of carbonate strata, *in* Scholle, P. A., and others, eds., Carbonate depositional environments: American Association of Petroleum Geologists Memoir 33, p. 73–92.

Kyle, J. R., and Gutierrez, G. N., 1988, Origin of sandstone-hosted lead deposits, Indian Creek district, southeast Missouri, U.S.A., *in* Proceedings, Seventh Quadrennial Symposium of the International Association on the Genesis of Ore Deposits, Lulea, Sweden: Stuttgart, Germany, E. Schweizerbach, (in press).

Kyle, J. R., and Price, P. E., 1986, Metallic sulphide mineralizaton in salt dome cap rocks, Gulf Coast, U.S.A.: Institution of Mining and Metallurgy Transactions, v. 95, p. B6–B16.

Land, L. S., 1984, Frio sandstone diagenesis, Texas Gulf Coast; A regional isotopic study, *in* McDonald, D. A., and Surdan, R. C., eds., Clastic diagenesis: American Association of Petroleum Geologists Memoir 37, p. 47–62.

Land, L. S., and Prezbindowski, D. R., 1981, The origin and evolution of saline tormation water, Lower Cretaceous carbonates, south-central Texas, U.S.A.: Journal of Hydrology, v. 54, p. 51–74.

Lange, S., Chaudhuri, S., and Claues, N., 1983, Strontium isotopic evidence for the origin of barites and sulfides from the Mississippi Valley–type ore deposits in southeast Missouri: Economic Geology, v. 78, p. 1255–1261.

Lapwood, E. R., 1948, Convection of fluid in a porous medium: Cambridge Philosophical Society Proceedings, v. 44, p. 508–521.

Large, D. E., 1981, Sediment-hosted submarine exhalative lead-zinc deposits; A review of their geological characteristics and genesis, *in* Wolf, K. H., ed., Handbook of strata-bound and stratiform ore deposits: Amsterdam, Elsevier, v. 9, p. 469–507.

Large, R. R., 1977, Chemical evolution and zonation of massive sulfide deposit in volcanic terrains: Economic Geology, v. 72, p. 549–572.

Leach, D. L., Landis, G. P., and Hofstra, A. H., 1985, Precious and base metal mineralization in the Coeur d'Alene district, Idaho, during Proterozoic metamorphism of the Belt Basin: U.S. Geological Survey Circular 949, p. 28–29.

Lelong, F., Tardy, Y., Grandin, G., Trescases, J. J., and Boulange, B., 1976, Pedogenesis, chemical weathering, and processes of formation of supergene ore deposits, *in* Wolf, K. H., ed., Handbook of strata-bound and stratiform ore deposits: Amsterdam, Elsevier, v. 3, p. 93–173.

Light, M.P.R., Posey, H. H., Kyle, J. R., and Price, P. E., 1987, Model for the origins of geopressured brines, hydrocarbons, cap rocks, and metallic mineral deposits, Gulf Coast, U.S.A., *in* Lerche, I., and O'Brien, J., eds., Dynamical geology of salt and related structures: Orlando, Florida, Academic Press, p. 787–830.

Lindgren, W., 1933, Mineral deposits (4th edition): New York, McGraw-Hill Book Company, Incorporated, 930 p.

Lowell, J. D., and Guilbert, J. M., 1970, Lateral and vertical alteration-mineralization zoning in porphyry ore deposits: Economic Geology, v. 65, p. 373–408.

Lydon, J. W., 1983, Chemical parameters controlling the origin and deposition of sediment-hosted lead-zinc deposits, *in* Sangster, D. F., ed., Short course in sediment-hosted stratiform lead-zinc deposits: Mineralogical Association of Canada, p. 175–250.

Mann, A. W., 1984, Mobility of gold and silver in laterite weathering profiles; Some observations from western Australia: Economic Geology, v. 79, p. 38–49.

Mann, A. W., and Deutscher, R. L., 1978, Genesis principles for the precipitation of carnotite in calcrete drainages in western Australia: Economic Geology, v. 73, p. 1724–1737.

Marsden, R. W., 1968, Geology of the iron ores of the Lake Superior region in the United States, *in* Ridge, J. W., ed., Ore deposits of the United States, 1933–1967: American Institute of Mining, Metallurgical Petroleum Engineers, p. 489–506.

Maynard, J. B., 1983, Geochemistry of sedimentary ore deposits: New York, Springer-Verlag, 305 p.

McDougall, T. J., 1984, Fluid dynamic implications for massive sulfide deposits of hot saline fluid flowing into a submarine depression from below: Deep-Sea Research, v. 31, p. 145–170.

Morganti, J. M., 1981, Ore deposit models; 4, Sedimentary-type stratiform ore deposits; Some models and new classification: Geoscience Canada, v. 8, p. 65–75.

Nash, J. T., Granger, H. C., and Adams, S. S., 1981, Geology and concepts of important types of uranium deposits: Economic Geology, 75th Anniversary Volume, p. 63–116.

Noble, E. A., 1963, Formation of ore deposits by water of compaction: Economic Geology, v. 58, p. 1145–1156.

Nolan, T. B., 1933, Epithermal precious-metal deposits, *in* Ore deposits of the western States (Lindgren Volume): American Institute of Mining and Metallurgical Engineers, p. 623–640.

Norton, D. L., 1981, Fluid and heat transport phenomena typical of copper-bearing pluton environments, *in* Titley, S. R., ed., Advances in geology of the prophyry copper deposits, southwestern North America: Tucson, University of Arizona Press, p. 59–72.

—— , 1984, Theory of hydrothermal systems: Annual Review of Earth and Planetary Sciences, v. 12, p. 155–177.

Norton, D., and Knapp, R. B., 1977, Transport phenomena in hydrothermal systems; The nature of porosity: American Journal of Science, v. 277, p. 913–936.

Norton, D., and Knight, J., 1977, Transport phenomena in hydrothermal systems; Cooling plutons: American Journal of Science, v. 277, p. 937–981.

Ohle, E. L., 1959, Some considerations in determining the origin of ore deposits of the Mississippi Valley type, Part 1: Economic Geology, v. 54, p. 769–789.

—— , 1980, Some considerations in determining the origin of ore deposits of the Mississippi Valley type, Part 2: Economic Geology, v. 75, p. 161–172.

Ohmoto, H., and Rye, R. O., 1979, Isotopes of sulfur and carbon, *in* Barnes, H. L., ed., Geochemistry of hydrothermal ore deposits (2nd edition): New York, John Wiley and Sons, p. 509–567.

Ohmoto, H., and Skinner, B. J., eds., 1983, The Kuroko and related volcanogenic massive sulfide deposits: Economic Geology Monograph 5, 604 p.

Olsen, H. W., 1972, Liquid movement through kaolinite under hydraulic, electric, and osmotic gradients: American Association of Petroleum Geologists Bulletin, v. 56, p. 2022–2028.

Perkins, T. K., and Johnstone, O. C., 1963, A review of diffusion and dispersion in porous media: Society of Petroleum Engineers Journal, v. 19, p. 70–84.

Powell, T. G., and Macqueen, R. W., 1984, Precipitation of sulfide ores and organic matter: Sulfate reactions at Pine Point, Canada: Science, v. 224,

p. 63–66.

Price, P. E., Kyle, J. R., and Wessel, G. R., 1983, Salt dome related zinc-lead deposits, *in* Kisvarsanyi, G., and others, eds., Proceedings of the International Conference on Mississippi Valley–type Lead-Zinc Deposits: Rolla, University of Missouri, p. 558–571.

Reed, M. H., 1983, Seawater-basalt reaction and the origin of greenstones and related ore deposits: Economic Geology, v. 78, p. 466–485.

Renfro, A. R., 1974, Genesis of evaporite-associated stratiform metalliferous deposits: A sabhka process: Economic Geology, v. 69, p. 33–45; Discussion: Economic Geology, v. 70, p. 407–409.

Reynolds, R. L., and Goldhaber, M. B., 1978, Origin of a south Texas roll-type uranium deposit; I, Alteration of iron-titanium oxide minerals: Economic Geology, v. 73, p. 1677–1689.

Rhodes, D., Lantos, E. A., Lantos, J. A., Webb, R. J., and Owens, D. C., 1984, Pine Point ore bodies and their relationship to the stratigraphy, structure, dolomitization, and karstification of the Middle Devonian barrier complex: Economic Geology, v. 79, p. 991–1055.

Rickard, D. T., Willden, M. Y., Maringer, N. E., and Donnelly, J. H., 1979, Studies on the genesis of the Laisvall sandstone lead-zinc deposits, Sweden: Economic Geology, v. 74, p. 1255–1285.

Ridler, R. H., 1976, Stratigraphic keys to the gold metallogeny of the Abitibi Belt: Canadian Mining Journal, v. 97, p. 81–87.

Roedder, E., 1976, FLuid inclusion evidence on the genesis of ores in sedimentary and volcanic rocks, *in* Wolf, K. H., ed., Handbook of strata-bound and stratiform ore deposits: Amsterdam, Elsevier, v. 2, p. 67–110.

——, 1979, Fluid inclusions as samples of ore fluids, *in* Barnes, H. L., ed., Geochemistry of hydrothermal ore deposits (2nd edition): New York, John Wiley and Sons, p. 684–737.

——, 1984, Fluid inclusions: Mineralogical Society of America Reviews in Mineralogy, v. 12, 644 p.

Rona, P. A., Bostrom, K., Laubier, L., and Smith, K. L., Jr., 1983, Hydrothermal processes at seafloor spreading centers: New York, Plenum Press, 796 p.

Rose, A. W., 1976, The effect of cuprous chloride complexes in the origin of red-bed copper and related deposits: Economic Geology, v. 71, p. 1036–1048.

Rowan, L., Leach, D. L., and Viets, J. G., 1984, Evidence for a Late Pennsylvanian-Early Permian regional thermal event in Missouri, Kansas, Arkansas, and Oklahoma: Geological Society of America Abstracts with Programs, v. 16, p. 640.

Ruckmick, J. C., Wimberly, B. H., and Edwards, A. F., 1979, Classification and genesis of biogenic sulfur deposits: Economic Geology, v. 74, p. 469–474.

Russell, M. H., 1983, Major sediment-hosted exhalative zinc and lead deposits; Formation from hydrothermal convection cells that deepen during crustal extension, *in* Sangster, D. F., ed., Short course in sediment-hosted stratiform lead-zinc deposits: Mineralogical Association of Canada, p. 251–282.

Rye, D. M., and Rye, R. O., 1974, Homestake gold mine, South Dakota; 1, Stable isotope studies: Economic Geology, v. 69, p. 293–317.

Rye, R. O., Roberts, P. J., Snyder, W. S., Lahusen, G. L., and Montica, J. E., 1984, Textural and stable isotopes studies of the Big Mike cupriferous volcanogenic massive sulfide deposits, Pershing County, Nevada: Economic Geology, v. 79, p. 124–140.

Sanford, R. F., 1982, Preliminary model of regional Mesozoic groundwater flow and uranium deposition in the Colorado Plateau: Geology, v. 10, p. 348–352.

Sangameshwar, S. R., and Barnes, H. L., 1983, Supergene processes in zinc-lead-silver sulfide ores in carbonates: Economic Geology, v. 78, p. 1379–1397.

Sangster, D. F., 1976, Sulphur and lead isotopes in strata-bound deposits, *in* Wolf, K. H., ed., Handbook of strata-bound and stratiform ore deposits: Amsterdam, Elsevier, v. 2, p. 219–266.

——, ed., 1983, Short course in sediment-hosted stratiform lead-zinc deposits: Mineralogical Association of Canada Short Course Handbook 9, 304 p.

Sawkins, F. J., 1984, Metal deposits in relation to plate tectonics: New York, Springer-Verlag, 325 p.

Schoell, M., and Menz, D., 1976, Chemische Untersuchungen an den salzreichen Tiefenwaessern der Hydrothermalsysteme des Roten Meeres (Chemical studies of the saline bottom waters of the hydrothermal system in the Red Sea): Gesellschaft Deutscher Metallhuetten- und Bergleute, Schriften, no. 28, p. 69–85.

Senger, R. K., and Fogg, G. E., 1983, Regional modeling of ground-water flow in the Palo Duro Basin, Texas Panhandle, *in* Geology and geohydrology of the Palo Duro Basin, Texas Panhandle: University of Texas Bureau of Economic Geology Circular 83-4, p. 109–115.

Severson, R. C., Crock, J. G., and McConnell, B. M., 1986, Processes in the formation of crystalline gold in placers, *in* Dean, W. E., ed., Proceedings, Symposium on Organics and Ore Deposits: Denver Region Society of Exploration Geologists Society, p. 69–80.

Sharp, J. M., Jr., 1976, Momentum and energy balance equations for compacting sediments: Mathematical Geology, v. 8, p. 305–322.

——, 1978, Energy and momentum transport model of the Ouachita basin and its possible impact on formation of economic mineral deposits: Economic Geology, v. 73, p. 1057–1068.

Sharp, J. M., Jr., and Domenico, P. A., 1976, Energy transport in thick sequences of compacting sediment: Geological Society of America Bulletin, v. 87, p. 390–400.

Shelton, K. L., 1983, Composition and origin of ore-forming fluids in a carbonate-hosted porphyry copper and skarn deposit; A fluid inclusion and stable isotopic study of Mines Gaspe, Quebec: Economic Geology, v. 78, p. 387–421.

Sibson, R. H., Moore, J. M., and Rankin, A. H., 1975, Seismic pumping; A hydrothermal fluid transport mechanism: Geological Society of London Quarterly Journal, v. 130, p. 163–177.

Siebenthal, C. E., 1915, Origin of the zinc and lead deposits of the Joplin region; Missouri, Kansas, and Oklahoma: U.S. Geological Survey Bulletin 606, 283 p.

Skall, H., 1975, The paleoenvironment of the Pine Point lead-zinc district: Economic Geology, v. 70, p. 22–45.

Skinner, B. J., and Barton, P. B., Jr., 1973, Genesis of mineral deposits: Annual Reviews of Earth and Planetary Science, v. 1, p. 183–211.

Smith, G. E., 1974, Depositional systems, San Angelo Formation (Permian), north Texas—facies control of red-bed copper mineralization: University of Texas Bureau of Economic Geology Report of Investigations 80, 74 p.

Smith, S. P., and Kennedy, B. M., 1985, Noble gas evidence for two fluids in the Baca (Valles Caldera) geothermal reservoir: Geochimica et Cosmochimica Acta, v. 49, p. 893–902.

Smith, N. G., Kyle, J. R., and Magara, K., 1983, Geophysical log documentation of fluid migration from compacting shales; A mineralization model from the Devonian strata of the Pine Point area, Canada: Economic Geology, v. 78, p. 1364–1374.

Spooner, E.T.C., 1977, Hydrodynamic model for the origin of the ophiolitic cupriferous pyrite ore deposits of Cyprus, *in* Gass, I. G., ed., Volcanic processes in ore genesis: Geological Society of London Special Publication 7, p. 58–71.

Straus, J. M., and Schubert, G., 1977, Thermal convection of water in a porous medium; Effects of temperature- and pressure-dependent thermodynamic and transport properties: Journal of Geophysical Research, v. 82, p. 325–333.

Strens, M. R., Cann, D. L., and Cann, J. R., 1987, A thermal balance model of the formation of sedimentary-exhalative lead-zinc deposits: Economic Geology, v. 82, p. 1192–1203.

Sverjensky, D. A., 1984, Oil field brines as ore-forming solutions: Economic Geology, v. 79, p. 23–37.

——, 1986, Genesis of Mississippi Valley-type lead-zinc deposits: Annual Reviews of Earth and Planetary Sciences, v. 14, p. 177–199.

——, 1987, The role of migrating oil-field brines in the formation of sediment-hosted Cu-rich deposits: Economic Geology, v. 82, p. 1130–1141.

Taylor, H. P., Jr., 1974, The application of oxygen and hydrogen isotope studies to problems of hydrothermal alteration and ore deposition: Economic Geology, v. 69, p. 843–883.

——, 1979, Oxygen and hydrogen isotope relations in hydrothermal mineral deposits, *in* Barnes, H. L., ed., Geochemistry of hydrothermal ore deposits: New York, John Wiley and Sons, p. 236–277.

Tooker, E. W., 1985, Geologic characteristics of sediment- and volcanic-hosted disseminated gold deposits; Search for an occurrence model: U.S. Geological Survey Bulletin 1646, 150 p.

Toth, J., 1963, A theoretical analysis of ground-water flow in small drainage basins: Journal of Geophysical Research, v. 68, p. 4795–4812.

—— , 1978, Gravity-induced cross-formational flow of formation fluids, Red Earth region, Alberta, Canada; Analysis, patterns, and evolution: Water Resources Research, v. 14, p. 805–843.

Turner, J. S., and Gustafson, L. B., 1978, The flow of hot saline solutions from vents in the sea floor; Some implications for exhalative massive sulfide and other ore deposits: Economic Geology, v. 73, p. 1082–1100.

Ulrich, M. R., Kyle, J. R., and Price, P. E., 1984, Metallic sulfide deposits in the Winnfield salt dome, Louisiana; evidence for episodic introduction of metalliferous brines during cap rock formation, *in* White, B. R., ed., Transactions Gulf Coast Association of Geological Societies, v. 34, p. 435–442.

Walther, J. V., and Orville, P. M., 1982, Volatile production and transport in regional metamorphism: Contributions to Mineralogy and Petrology, v. 73, p. 252–257.

White, D. E., 1957, Magmatic, connate, and metamorphic waters: Geological Society of America Bulletin, v. 68, p. 1659–1682.

—— , 1958, Liquid inclusions in sulfides from Tri-State (Missouri, Kansas, Oklahoma) is probably connate in origin: Geological Society of America Bulletin, v. 69, p. 1660–1661.

—— , 1963, Fumarole, hot springs, and hydrothermal alteration: EOS American Geophysical Union Transactions, v. 44, p. 508–511.

—— , 1968, Environments of generation of some base-metal ore deposits: Economic Geology, v. 63, p. 301–335.

—— , 1981, Active geothermal systems and hydrothermal ore deposits: Economic Geology, 75th Anniversary Volume, p. 392–423.

White, W. H., and 6 others, 1981, Character and origin of Climax-type molybdenum deposits: Economic Geology, 75th Anniversary Volume, p. 270–316.

White, W. S., 1971, A paleohydrologic model for mineralization of the White Pine copper deposit: Economic Geology, v. 66, p. 1–13.

Wolery, T. J., and Sleep, N. D., 1976, Hydrothermal circulation and geochemical flux at mid-ocean ridges: Journal of Geology, v. 84, p. 249–275.

Wood, J. R., and Hewett, T. A., 1982, Fluid convection and mass transfer in porous sandstones; A theoretical model: Geochimica et Cosmochimica Acta, v. 46, p. 1707–1713.

Zartman, R. E., and Stacey, J. S., 1971, Lead isotopes and mineralization ages in Belt Supergroup rocks, northwestern Montana and northern Idaho: Economic Geology, v. 66, p. 849–860.

MANUSCRIPT ACCEPTED BY THE SOCIETY JULY 16, 1987

## ACKNOWLEDGMENTS

We would like to thank Grant Garven of Johns Hopkins University, Jon Price of the Texas Bureau of Economc Geology, Kevin Shelton of the University of Missouri-Columbia, Don White of the U.S. Geological Survey, and one anonymous reviewer for their reviews of this manuscript. Impetus for this chapter was provided by a 1982 Penrose Conference: Hydrodynamic and Geochemical Controls of Ore Deposition in Sedimentary Environments. The support of the Alexander von Humboldt Foundation, the American Chemical Society, the Geological Society of America, the National Science Foundation, and The University of Texas is greatly appreciated. Figures were drafted by Jeff Horowitz. Manuscript preparation was provided by the Owen-Coates Fund of The University of Texas Geology Foundation.

The Geology of North America
Vol. O-2, Hydrogeology
The Geological Society of America, 1988

## Chapter 48

# *Ground water and hydrocarbon migration*

**J. Tóth**

*Department of Geology, University of Alberta, Edmonton, Alberta T6G 2E3, Canada*

## INTRODUCTION

Migration of hydrocarbons is axiomatically assumed by most geologists to occur during certain evolutionary stages of sedimentary basins. No other mechanism is believed to be able to produce the huge petroleum accumulations known around the world. However, migration is hidden and occurs presumably on the geologic time scale. Its nature and controls are, therefore, largely matters of conjecture. Hydrocarbons are thought to move in the subsurface either as a separate phase or along with water. Although the latter is considered to be more probable, separate phase migration would have to occur also through rocks that are mostly water wet. It may thus be expected that in either of its likely modes the movement of petroleum is affected by the dynamic conditions of the ambient ground water.

Based on this premise, it appears that an understanding of the role of ground water in the migration and accumulation of petroleum may aid in exploration. Also, attemps to unravel the complex and hidden processes that link the flows of ground water and petroleum, and result in different natural phenomena, pose exciting challenges to the human intellect and may contribute large amounts of new knowledge to science.

Factual knowledge concerning petroleum migration is woefully deficient. Uncertainties exist regarding almost every major aspect of the process including the forms of transport, the impelling forces, the mechanisms of entrapment, and even the relevant details of a basin's geologic history. This state of the art was aptly summarized by Roberts and Cordell (1980, p. vi).

We know there is little we can prove, and we do not excuse ourselves for assumptions or conclusions which may later prove to be erroneous. . . our ideas and our arguments are based on the *interpretation* of factual observations, sponsored and enhanced by imagination. . . we have devised concepts and models to represent our interpretation of certain evidence as we see it, fixed in place. We cannot actually prove when and whence a particular show of oil or gas came or how long it will remain where we see it. To be completely honest, we can only conjecture about oil and gas movement.

This chapter presents: (1) an overview of the current principal concepts and models of petroleum migration; (2) petroleum migration as an example for the geologic agent of ground-water flow; and (3) the potential ramifications of this agent with respect to petroleum exploration.

The objectives are accomplished by: (1) reviewing the process, factors, and mechanisms that control the transport, concentration, and entrapment of hydrocarbons; (2) reviewing representative models that attempt to fit the above controls into the context of the basins' evolutionary history; (3) relating petroleum migration to other hydrogeologic processes and phenomena through one of the models presented, namely the hydraulic theory of petroleum migration; and (4) by outlining a hydrogeological approach to petroleum exploration and thereby demonstrating a utilitarian value of understanding the role of ground-water flow in geologic processes.

## THE PROCESS OF PETROLEUM MIGRATION

The topic of petroleum migration may be introduced appropriately by a review of some basic aspects of the process itself, which include: (1) the modes in which petroleum hydrocarbons move in the subsurface; (2) the forces inducing and resisting migration; and (3) the geometry of the flow paths.

### *The modes of migration*

A great deal of uncertainty exists as to the physical forms and chemical composition of petroleum or its precursors during migration (Levorsen, 1958; Tissot and Welte, 1978; Hunt, 1979; Hodgson, 1980; McAuliffe, 1980; Kinghorn, 1983). The main reason is that any conceivable migration mode may be excluded by some physical, chemical, or dynamic constraints under certain situations. Most researchers agree, however, that several forms of petroleum hydrocarbons exist that may facilitate migration under different circumstances. It is even possible that certain modes prevail at a given place and time during the generation-migration history, while other forms are dominant at other times or other places.

The chief forms in which hydrocarbons are generally believed capable of migrating are: (1) as oil droplets and gas bubbles; (2) in a continuous oil or gas phase; (3) as colloidal or micellar solutions; (4) as true molecular solutions; (5) as petroleum in a framework of organic matter; and (6) as soluble organic matter.

Tóth, J., 1988, Ground water and hydrocarbon migration, *in* Back, W., Rosenshein, J. S., and Seaber, P. R., eds., Hydrogeology: Boulder, Colorado, Geological Society of America, The Geology of North America, v. O-2.

## Forces and patterns of migration

Migration is effected by impelling forces. The forces are proportional to differences in the mechanical energy levels of fluid masses and are represented by the hydraulic gradient vector. Hydraulic conductivity of the rock framework is, in general, heterogeneous and anisotropic and affects the intensity and direction of flow. The relation between the gradient dh/dl, hydraulic conductivity K, and flux $\vec{q}$, is expressed by Darcy's Law:

$$\vec{q} = -K \frac{dh}{dl} \tag{1}$$

In equation 1, the hydraulic head h in any point P is related to the mechanical energy per unit mass of fluid, or fluid potential $\phi$, by

$$h = \frac{\phi}{g} = z + \frac{p}{\rho g} \tag{2}$$

(Hubbert, 1940), where z = elevation of point P; p = pore pressure; $\rho$ = mass density of fluid; and g = acceleration due to gravity.

Various energy sources may give rise to forces causing migration. They are not mutually exclusive and may act singly or in combination, and their relative importance may change with place and time in an evolving basin. The sources of energy generally include: (1) compaction and elastic rebound of the rocks; (2) buoyancy; (3) gravitation in either confined carrier beds or aquifers, or in regionally unconfined rock frameworks; (4) gas expansion; (5) thermal expansion of liquids; (6) molecular diffusion of hydrocarbons; and (7) osmosis.

Compaction of the rock framework will result from loading due, for instance, to sedimentation. The process involves a reduction in pore volumes, which is greater in the more compressible rocks, such as shales, than in the harder ones, like sandstones. Any pressure disequilibrium thus created tends to move formation fluids from the shales to the sandstones.

Elastic rebound of the rock framework might result, for instance, from erosional unloading (Neuzil and Pollock, 1983; Tóth and Corbet, 1986) and may cause pore volumes to expand differentially in shales and sandstones. Consequently, shales become relatively underpressured and act as sinks for fluids moving out of the sandy carrier beds.

Buoyancy is the tendency of relatively light fluids to rise in an ambient fluid of greater density. It enables hydrocarbons to move in an aqueous environment without water flow and was the first type of force considered to be responsible for petroleum accumulation. Buoyancy is also operative in a static environment, with the buoyant force acting vertically upward. These conditions can be illustrated on force vector diagrams (Hubbert, 1953).

Gravitational flow is induced by gravity acting directly upon the fluid particles. It has been considerd by many researchers since Munn (1909) to be the chief mechanism of petroleum migration and accumulation in nondeforming sedimentary basins. Flow in such environments may take place in either regionally confined or regionally unconfined systems. In both instances, water flows toward regions of low fluid potentials. In the first

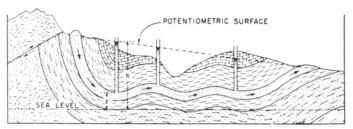

Figure 1. Regional gravity flow of water through confined sand aquifer (after Hubbert, 1953, Fig. 11).

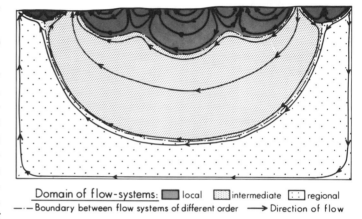

Domain of flow-systems: ■ local ▦ intermediate ⬚ regional
—·— Boundary between flow systems of different order ⟶ Direction of flow

Figure 2. Regional gravity flow of water through unconfined homogeneous drainage basin (modified from Tóth, 1963, Fig. 3).

case (Fig. 1), however, elevation differences of the water table are considered to have no effect on fluid potentials across the confining strata. Flow is, therefore, oriented in the aquifer in one direction only, namely from the high to low outcrops. In the case of regionally unconfined flow (Fig. 2), on the other hand, the relief of the water table does have an influence on the flow at depth. This influence depends on the degree of hydraulic communication across the aquitards and may result in: complex flow patterns, comprising cross-formational flow systems of different orders; opposing flow directions with zero lateral force components, where systems meet or part, such as quasi-stagnant conditions; and ascending flows and fluid discharges in local and regional depressions (Fig. 2). Formation fluids may thus be exchanged between aquifers. Unlike flow by deformation of the rock framework, gravitational flow can be depicted in distributed systems and it can be evaluated from field measurements of pore pressures and topographic elevation, or calculated from the water-table configuration and appropriate boundary conditions (Tóth, 1963, 1978; Freeze and Witherspoon, 1967; Blair and others, 1985).

When methane or other gases are generated from thermally maturing organic matter (Hedberg, 1980) or are heated, their volume expands, resulting in pore-pressure increases. If the increase is not uniform, the pressure differentials will tend to drive

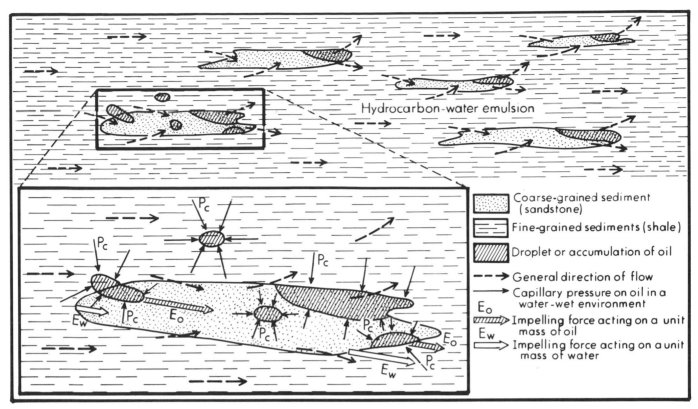

Figure 3. Interplay of capillary and impelling forces acting upon volume elements of hydrocarbons at sand-shale interfaces and possibly resulting in entrapment at the "downstream" end of sandstone lenses (after Tóth, 1970, Fig. 9).

the fluids (both liquids and gases) toward regions of lower pressures.

Pore pressures increase and fluid flow may result upon heating of saturated rocks, because the fluids' coefficients of thermal expansion are greater than those of minerals (Magara, 1978).

Hydrocarbons have the tendency to migrate by the process of diffusion from sites where their thermodynamic activity, or escaping tendency, is high to regions where it is low (Hunt, 1979). Diffusion may be most effective in moving hydrocarbons out of fine-grained source rocks into carrier beds, and it may be enhanced or opposed by other forces.

Osmosis is the tendency of water molecules to move through a semipermeable membrane from a region of low to one of high salinity. Argillaceous strata separating aquifers that contain formation waters of differing salt concentrations have been considered to act as semipermeable membranes and induce osmotic flow (Berry, 1959). A subsequent study indicated, however, that this flow may be insignificant in some natural basin conditions (Bradley, 1975).

## MECHANISMS OF PETROLEUM ENTRAPMENT

Petroleum accumulation constitutes a direct manifestation of hydrocarbon migration. It results from the segregation and retention of migrating volume elements of hydrocarbons from

formation waters by a suitable combination of physical, chemical, and fluid dynamic conditions. The process of retention and accumulation of petroleum within limited rock volumes is termed entrapment.

### Physical and chemical factors of entrapment

The main factors recognized to cause or enhance entrapment are reductions in pore size, temperature, and pressure, and an increase in formation-water salinity.

A decrease in rock-pore (i.e., grain) size in the direction of flow may result in mechanical screening, also called micropore filtration, of the migrating fluids. The effective molecular diameter of water is smaller than those of the most common hydrocarbon components. Many lithologic situations exist in which only water or water and light-molecular-weight hydrocarbons can travel, those particles with large diameters being retained.

Another effect of pore-size reduction is the creation of capillary barriers (Levorsen, 1958). The capillary barrier associated with a grain-size boundary equals the difference in displacement pressures in the two rock types. A volume element of petroleum straddling such a boundary in a static, water-wet environment will have the tendency therefore to migrate by capillary forces from the fine into the coarse-grained rock and to resist movement in the opposite direction (Fig. 3). These tendencies are enhanced

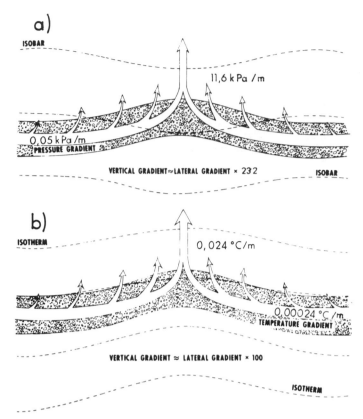

Figure 4. Changes in physical conditions imposed on convergent, upward-turning formation waters and their contents: a, pressure; b, temperature (after Roberts, 1980, Figs. 4 and 5, respectively).

and reduced, respectvely, by a flow of formation water oriented from fine- into coarse-grained rocks.

Because of the effect of temperature and pressure on the solubility of hydrocarbons, migrating hydrocarbons tend to come out of solution in situations where they encounter reduced temperatures and/or pressures, thereby reducing their ability to continue migration. Such situations may exist typically in regions where the lateral direction of moving formation fluids changes upward (Fig. 4a and b). The vertical gradients of temperature and pressure are respectively 100 and 232 times greater than the corresponding lateral gradients, and the resulting relatively rapid cooling and depressurization of the upward-turned fluids will enhance the exsolution of oil and gas.

Salinities of formation waters are known to increase in many regions of converging formation fluid flow (Roberts, 1980). Because solubility is reduced by increases in water salinity, hydrocarbons will tend to come out of solutions in areas of increased salinity, and especially in regions of upward-changing flow direction.

## Hydrodynamic conditions of petroleum entrapment

The basic dynamic requirement of entrapment is the existence of a fluid-potential minimum with respect to the fluid in question. Minima of fluid potential may develop in three essentially different situations, namely: (1) no water flow, that is, hydrostatic conditions; (2) unidirectional water flow; and (3) opposite-directed water flows.

In the absence of water flow the sole force that impells hydrocarbons is buoyancy. However, at the upper boundary of a coarse-grained bed a potential minimum occurs with respect to a buoyant petroleum particle. There, additional energy would be required to force the hydrocarbon into the fine-grained caprock against the resistance of micropore filtration and capillary barrier. This situation is actually a special case of the various versions of dynamic entrapment that may be created by certain lithological and structural configurations in a field of unidirectional water flow (Hubbert, 1953). In regions of oppositely directed flow fields, on the other hand, minimum fluid potentials with respect to hydrocarbons can develop without particular lithological or structural requirements; a horizontal permeability barrier suffices. Fluid-potential minima developed between opposing flow fields were termed hydraulic traps and the conditions of their formations discussed by Tóth (1980).

A demonstration of the concept of fluid-potential minima as a mechanism of hydrodynamic entrapment in a unidirectional flow field, and a method of application, can best be based on Equation 2. In a system of two immiscible fluids, the hydraulic head (or the fluid potential) of either fluid can be expressed in terms of the hydraulic head of the other fluid and the fluid-density ratios. Because the pressure at a given point in the system is the same for both fluids, it can be eliminated from Equation 2 when written for oil and water. By equating the two forms of Equation 2 through p, the following expression for an oil-water system is obtained:

$$h_o = \frac{\rho_w}{\rho_o} h_w - \frac{\rho_w - \rho_o}{\rho_o} z \qquad (3)$$

where the subscripts denote the two fluids, respectively. Dividing Equation (3) by $\rho_w - \rho_o / \rho_o$ results in

$$\frac{\rho_o}{\rho_w - \rho_o} h_o = \frac{\rho_w}{\rho_w - \rho_o} h_w - z \qquad (3a)$$

which, for short, has been written as (Hubbert, 1953):

$$u = v - z. \qquad (4)$$

In equation 4, v is a function of the hydraulic head of water $h_w$, differing from it only by a constant ratio of fluid densities. v may be determined and its contours mapped if hydraulic heads of water and the fluid densities are known. The contours of v are parallel to those of the water's potentiometric surface. z denotes the vertical distance from a datum plane, commonly sea level. The function u is directly proportional to the fluid potential for oil and may be determined even in the absence of hydraulic head data in the oil phase by contouring the difference between v and z, as illustrated in Figure 5. In a hydrodynamic situation the minima of u and v are separated in space resulting in a separation of the foci of convergence of oil and water flow. Consequently, also the

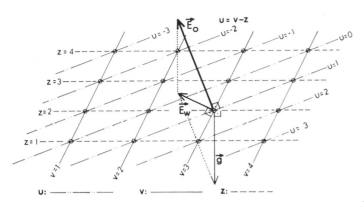

Figure 5. Relations of the function u and oil-impelling force $\vec{E}_o$ to the function v, the water-impelling $E_w$, the function z (elevation), and the acceleration due to gravity $\vec{g}$.

impelling forces $\vec{E}_o$ and $\vec{E}_w$, which are normal to contours of u and v, respectively, have different directions. Figures 6a and 6b show oil-trapping conditions in an anticlinal structure in a static and a dynamic situation, respectively. Owing to the orthogonality of the oil-driving forces and u surfaces if a water-oil interface exists it will be tilted in the direction of water flow (Hubbert, 1953). Figure 7 illustrates how optimum positions of entrapment for oil and gas are separated due to water flow through an anti-

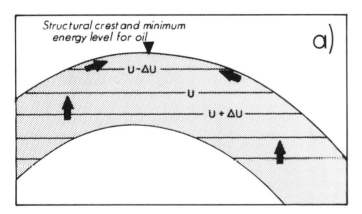

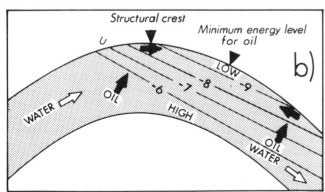

Figure 6. Trapping conditions in an anticline under static (a) and dynamic (b) water environments where u is a function directly proportional to the fluid potential for oil after Dahlberg, 1982, Figs. 7-3 and 7-7, respectively).

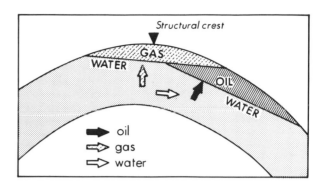

Figure 7. Separation of optimum trapping positions for oil and gas due to water flow (after Dahlberg, 1982, Fig. 7-10).

cline, while Figure 8 shows a hydrodynamic trap caused by a change in permeability of a monoclinal carrier bed.

In ground-water discharge areas, flow is radially converging, the potentiometric contours form closed minima, and lateral escape of transported hydrocarbons is not possible (Figs. 9 and 10). Upward flow of water is possible, if the rocks are water wet, despite reductions in pore size across boundaries of different strata. Micropore filtration and capillarity, however, will present an energy barrier for hydrocarbons at these boundaries, thus completing a three-directional sink, or trap, for petroleum. In regions where descending flows part, namely beneath recharge areas, fluid potentials decrease radially outward. However, the lateral gradients are so slight, if they exist at all, in the centers of these potentiometric mounds that descending hydrocarbons may be unable to overcome the resistance of facies changes and other barriers along the bedding and may accumulate between the diverging flow-system branches. Nevertheless, these accumulations are unstable and relatively sensitive to shifts in the flow fields. They can thus be expected to be less common in nature than those associated with converging flows.

The hydraulic traps discussed above have been attributed to cross-formational gravity-flow systems induced by topographic relief. The nature of the energy source, however, is not an essential aspect of the formation of these traps: fluid-potential fields of different sources even may contribute to a single situation. For instance, hydraulic traps may possibly develop in offshore regions where continental gravity-flow systems oppose landward-oriented compaction flows (Fig. 11).

Potentiometric minima associated with different flow configurations as summarized by Kudryakov (1974) are shown in Figure 12.

## BASIN EVOLUTION AND MIGRATION MODELS

Basinal migration constitutes a hydraulic link between the regions of origin and accumulation of petroleum. It may be expressed in terms of pathways, rates, and volumes of migrating fluids, which depend on the various modes of transport and the different types of driving forces. In turn, these factors themselves

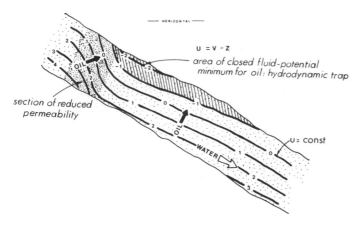

Figure 8. Hydrodynamic trap in monoclinal carrier bed caused in a field of water flow by a reduction of permeability (after Dahlberg, 1982, Fig. 7-15).

depend on the changing environmental conditions during the evolutionary history of a sedimentary basin. Basinal migration thus becomes a sensitive response function, indeed a component, of an area's geological evolutionary history.

Attempts to conceptualize the functional relation between petroleum migration and basin evolution are commonly expressed by means of migration models. The models are generated mostly with either or both of two purposes in mind, namely (a) to portray an invisible natural process, and (b) to assist in exploration. Basinal migration, as perceived by various researchers to be related to basin evolution, is discussed through selected models. The models are arranged in four groups according to the dominant impelling forces: (1) compaction; (2) compaction-heat; (3) compaction-gravitation; and (4) gravitation.

### Compaction drive models

*Model of Jacquin and Poulet.* Jacquin and Poulet (1970, 1973) attempt to reproduce ". . . the hydrodynamic conditions existing in a sedimentary basin during its evolution. . ." (Jacquin and Poulet, 1970, p. 1) by calculating the flux, pressure, and vertical sense of flow as functions of time. Their model's principal assumptions regarding the shape of the basin, geology, and history of sedimentation and subsidence are summarized in Figure 13a, b. In addition, it is assumed that the sand-to-clay permeability ratio is $0.3 \times 10^8$; flow is vertical in the clay, parallel to the clay-sand boundary in the sand and obeys Darcy's Law.

An imaginary surface, $\Sigma$, can be defined in the clay (Fig. 13a) that separates the depth zones of downward- and upward-moving waters. It coincides with the top of the sand at the time the first clay particles are deposited on a horizontal seabottom, then migrates upward with progressive basin subsidence maintaining a distance from the top of the sand of approximately one-fifth of the clay's total thickness.

The main conclusions of the model are summarized in Figure 14 and as follows. The vertical sense of flow in most points in the clay is complex over the time span of the basin's evolution.

Due to the gradually increasing depth of $\Sigma$, flow reverses at some point in time, being oriented upward at most places initially and turning downward later. It remains uniformly descending or ascending only in the depth zones that are proximal to the sand or to the ultimate basin surface, respectively. The reservoir sand receives water only from a thin lower portion of the clay; 50 percent of the total recharge into the sand is derived from a basal band of clay approximately 50 m thick (Figs. 14a, b). A relatively small (4 percent) portion of water is produced by compaction after subsuidence stops and most of this water is supplied by areas relatively distant from the basin's center. Possible migration of petroleum into the reservoir will take place essentially during the period of sedimentation/subsidence. Flow velocities in the sand decrease as depth of burial increases, but a water particle starting at the basin's center will complete the travel to its edge 300 km away in approximately $150 \times 10^6$ years. Pressures in the reservoir remain near hydrostatic but will be superhydrostatic in the clays during and long after subsidence.

*Model of Bonham.* In an attempt to conceptualize the evolution of migration conditions that lead to the building of petroleum deposits in a subsiding basin, Bonham (1980) differentiates a developing, or first, phase and an equilibrium, or second, phase. Because porosity decreases progressively with depth due to compaction, water is squeezed during stages A, B, and C of phase 1, upward relative to stratification into younger layers from sediment units deposited earlier (Fig. 15). Simultaneously, additional water is accommodated by the gradually increasing basin volume resulting in net downward water movement with respect to sea level. In Stage D of phase 2, sediment unit 1 reaches the depth of maximum compaction, the minimum porosity. Continuing subsidence in stage E will not increase the total pore volume but will keep reducing the porosity of the subsiding layers, thus maintaining an upward flow relative to stratification. Because of the now-constant pore volume, however, no additional water enters the basin and thus no flow occurs with respect to sea level: water rises across the strata at the same rate as the basin subsides.

Based on the postulated flow conditions, Bonham (1980, Fig. 3) concludes that exsolution of petroleum can occur only if, for any reason, subsurface isotherms move downward. Even in this case, however, a cooling of formation waters can take place only if either the basin has reached the equilibrium phase or if the rate of isotherm migration is greater than the rate of water descent with respect to sea level.

### Compaction cum heat drive models

*Model of Welte and Yükler.* The objective of the Welte and Yükler (1980) model is to predict expectable distribution of source-rock maturity levels, that is, prospects for hydrocarbon generation in sedimentary basins from calculated vitrinite reflectance values. Vitrinite reflectance is a function of basin temperature that, in turn, depends on conductive heat flow and forced convection by formation-fluid flow (Fig. 16).

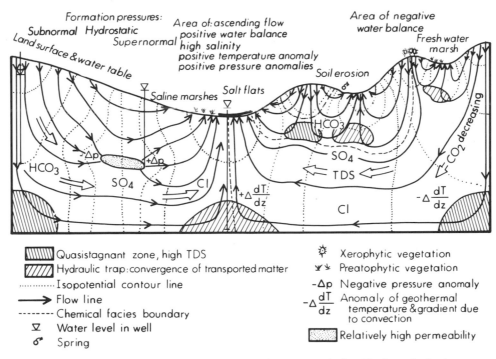

Figure 9. Regionally unconfined gravity flow of formation water and related hydrogeologic phenomena (after Tóth, 1980, Fig. 10).

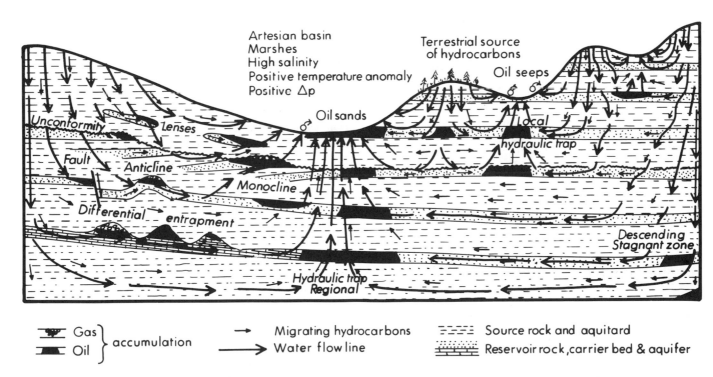

Figure 10. Hydraulic and hydrodynamically aided geologic traps in regionally unconfined fields of gravity flow of formation water (after Tóth, 1980, Fig. 44).

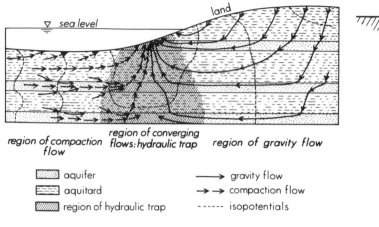

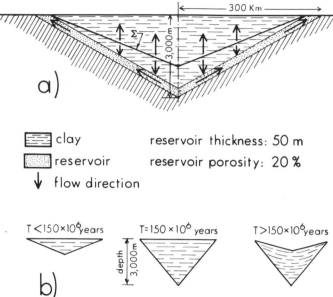

Figure 11. Hydraulic trap created by convergence of continental gravity flow and marine compaction flow of formation waters (modified from Tóth and Corbet, 1983).

Figure 13. Model of subsiding basin: a, structure; b, evolutionary history (after Jacquin and Poulet, 1973, Figs. 1 and 2, respectively).

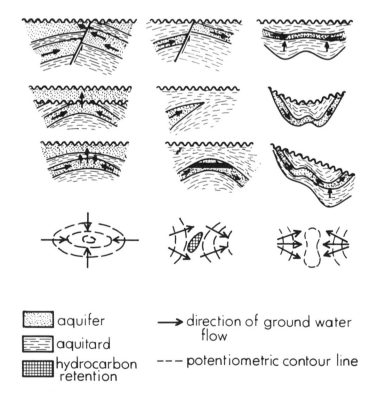

Figure 12. Flow configurations resulting in potentiometric minima or saddles indicative of potential trapping conditions (after Kudryakov, 1974, Fig. 1).

Migration is considered by the model to be initiated by volumetric expansion of thermally maturing kerogen and modified further by compaction and buoyancy. The prevailing nature, direction, and intensity of the driving forces are calculated from assumed conditions of grain-size distribution, rates, and amounts of successive stages of sedimentation and their effects on the distributions of pressure and temperature as calculated by a one-dimensional deterministic model. Figure 16 is the model's conceptualized scheme of basin development from initial stage $T_0$ through present stage $T_3$, while Figure 17 shows the vertical distribution of pore pressures and temperatures associated with the successive stages $T_i$. The vitrinite reflectance values calculated from the temperature distribution are compared with bore-hole data and suitably adjusted. Subsequent iteration between calculated and observed reflectance values is thus supposed to yield temporal and spatial distributions of basinal fluid flow.

***Model of Magara.*** The basis of Magara's model of "aquathermal" migration is the observation that the specific volume of water increases with increasing temperature and thus with, and during, increasing depth of burial (see Fig. 18; Magara, 1978, Fig. 14-6). Noting that the model is valid in relatively open systems, such as those developed in the shallow to intermediate depth ranges, Magara summarizes its main conclusions as follows (Magara, 1978, p. 282):

The directions of fluid migration due to the aquathermal effect are from a hot place to a cold, from a deep section to a shallow, and from a basin's centre to its edges. These directions are essentially the same as those of fluid movement caused by sediment compaction. Therefore, the significance of the aquathermal effect in the subsurface may simply be to increase the effectiveness of compaction fluid flow at deep burial.

***Model of Bethke.*** Although independently developed, the numerical flow model of Bethke (1985) may be considered as a sophisticated and advanced version of the Jacquin and Poulet (1970, 1973) model and leads to more detailed but similar general conclusions. It is formulated in two dimensions for a heterogeneous, anisotropic, and gradually subsiding and aggrading rock framework, with heat transfer by conduction and advection and the resulting aquathermal pressuring taken explicitly into account.

The modelled basin is 400 km wide (Fig. 19) and characterized by a set of physical parameters (Bethke, 1985, Table 1). The basin's edge is kept at hydrostatic potential and constant temperature gradient. Pressure and temperature at the sedimentation surface are held constant at 0.1 MPa (1 atm) and 25°C. Subsidence rate is in equilibrium with sedimentation so that the basin's surface remains at constant elevation. A relatively highly permeable, 500 m thick sand stratum is present at the beginning of the subsidence, followed by the continuous deposition of 5,000 m of uncompacted silty shales at the center of the basin over the course of the subsequent $100 \times 10^6$ years, being the approximate equivalent of a sedimentation rate of $5 \times 10^{-5}$ m/yr and subsidence rate of $3 \times 10^{-5}$ m/yr. The overburden sediments are assumed to consist of vertically interlayered shales and sands, compacting differentially and having anisotropic permeabilities.

Time-dependent results of a sample calculation are illustrated in Figure 19 (Bethke, 1985, Fig. 5). It appears that there is a tendency for the water to flow laterally, due largely to the assumed point anisotropy and layered heterogeneity of the strata. Vertical flow becomes dominant where the ratio of vertical to lateral permeabilities is greater than the ratio of vertical to lateral pathway lengths to the surface, that is, within the upper kilometer of sediments in general. The basal aquifer causes the potential gradients to reverse with depth thus forming a wedge of high fluid potentials. Fluids above this wedge move toward the surface, while below it they tend to migrate obliquely across time lines into stratigraphically lower units.

Fluid fluxes increase toward the basin's margins owing to the additive nature of compaction driven flow. Nevertheless, true flow velocities do not exceed 2 mm/yr (Fig. 19). Excess potentials remain under approximately 0.01 MPa (0.1 atm) during compaction. Although this value would increase with decreasing permeabilities, high potential gradients such as those observed in the U.S. Gulf Coast would be difficult to maintain. Owing to the low flow rates, the dominant mode of heat transfer is conduction, and isotherms remain horizontal (Bethke, 1985, Fig. 6). The same results were obtained with the same amount of sedimentation occurring over only $10 \times 10^6$ years, suggesting that thermal effects due to sedimentation and compaction-drive flow in intracratonic basins are negligible. Variational studies showed that changes in permeabilities affect the magnitudes and duration of excess fluid potentials but not the fluids' flow velocities.

Bethke (1985) summarizes compaction-driven fluid flow in intracratonic sedimentary basins as being characterized by flow velocities of only millimeters per year, very small excess fluid

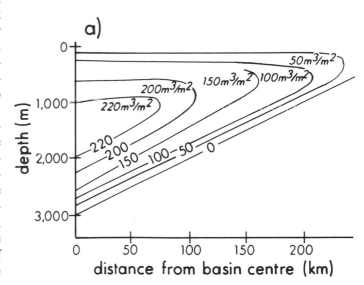

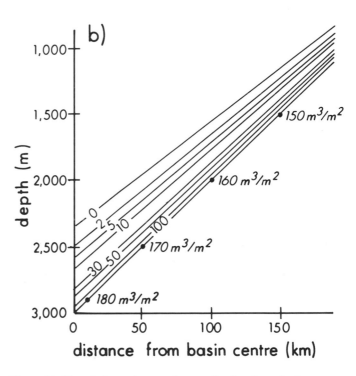

Figure 14. Cumulative volumes of water flowing through clay source rock per square meter during entire period of deposition ($150 \times 10^6$ years) as function of distance r from basin centre: a. upward flow, b. downward flow (after Jacquin and Poulet, 1970, Figs. 9 and 10, respectively).

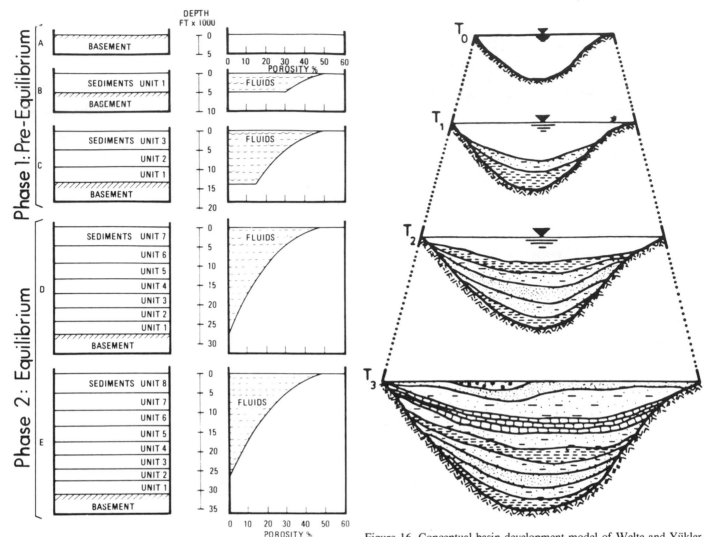

Figure 15. Conceptual basin development model of Bonham (after Bonham, 1980, Fig. 4).

Figure 16. Conceptual basin development model of Welte and Yükler through successive time steps (after Welte and Yükler, 1980, Fig. 1).

potentials, and temperature distribution basically undisturbed by flow. Based primarily on the above flow considerations, he then expresses doubt as to the possible effectiveness of compaction-drive basinal fluid flow as a mechanism for: long-range lateral migration of petroleum, formation of subsurface brines by reverse osmosis, generation of Mississippi Valley–type ore deposits, and certain other basinal-flow-related geologic processes.

### Compaction cum gravity drive: Confined flow models

***Model of Coustau and others.*** Coustau and others, (1975) and Coustau (1977) propose a three-stage model of basin evolution. Each stage is characterized by different basinal flow conditions resulting in different prospects for petroleum occurrence. During the juvenile stage (Fig. 20), compaction induces vertical flow of marine waters out of the shales, and lateral flow in the sands toward the basin's edges. This is the flow credited

with the creation of hydrocarbon deposits. Subsequently, during the mature stage, meteoric waters enter the carrier beds at the uplifted outcrops and discharge cross-formationally in the center of the basin. These waters are considered to destroy previously generated accumulations by mechanical or chemical effects, except in centrally located petroleum "shelters." The third or senile stage is characterized by hydrostatic conditions.

***Model of Kartsev and Vagin.*** Basin history and migration patterns as envisaged by Kartsev and Vagin (1964) are closely similar to those proposed by Coustau and others. However, the Russian model considers repeated occurrences of marine sedimentation and contintental erosion resulting in alternating cycles of compaction flow and gravitational flow, respectively (Fig. 21). Assuming that marine compaction flows create petroleum deposits and meteoric waters destroy them, Kartsev and Vagin propose to express the prospects for petroleum occurrence in a basin by the water-exchange number N. This number is calculated as the ratio of the volume of water that passed through a reservoir bed during a preceding stage of sedimentation or erosion to the volume of water (i.e., the pore volume) contained in that bed at a given instant. In other words, N equals the number of times the amount of fluid initially present in the reservoir bed has been replaced by fluids derived either from the shales or from the the land surface. The water-exchange numbers are estimated from the lengths of the various marine and continental stages, thickness of deposits, degree and rates of compaction, and the hydraulic gradients as assumed for the entire history of the basin. The prospects for petroleum in a basin are perceived to improve with increasing values of N for the sedimentation stage, $N_{sed}$, and to decrease with increasing values for the continental water-exchange number, $N_{cont}$.

### Gravity-drive cross-formational flow model

**Model of Tóth: the hydraulic theory of petroleum migration.** The model of Tóth (1980, Fig. 44), Figure 10, may be considered as the quantitatively defined distributed version of the continental stages of the models of Kartsev and Vagin, and Coustau and others. It is based on the earlier-developed theory of regional ground-water flow (Fig. 2; Tóth, 1963). According to this model, the migration, accumulation, and entrapment of hydrocarbons may be controlled by topography-induced cross-formational gravity flow of formation waters. Hydrocarbons or their protoforms are moved along well-defined migration paths from source or carrier beds toward discharge foci of converging flow systems. They may accumulate en route in hydraulic or hydrodynamic traps, possibly aided by the various trapping factors discussed before. Hydrocarbons do not have to cross through aquitards but may be mobilized and driven into the carrier beds by compaction, diffusion, gravity flow, or some other forces. Changes in the migration patterns can result from epeirogenic movements, orogenic deformations, and erosional modifications of a basin's topographic surface. Normally, these changes in flow paths are delayed with respect to their inductive causes due to the time required for pressures to propagate through compressible and slightly permeable rocks (Tóth and Millar, 1983). Figure 22 illustrates the staged adjustment of a shallow and deep flow system to a change in topography and the resulting, and further delayed, remigration of oil deposits. The time lag between cause and effect may well result in incongruent associations between actual positions of hydrocarbon accumulations and favorable hydrodynamic trapping conditions.

## HYDROGEOLOGICAL RAMIFICATIONS OF THE HYDRAULIC THEORY OF PETROLEUM MIGRATION

The hydraulic theory of petroleum migration is predicated on the premise that formation waters in sedimentary basins move in distributed systems with identifiable regions of recharge, mid-line, and discharge. Owing to its ability to interact with its environment, water in these systems can dissolve, mobilize, transport,

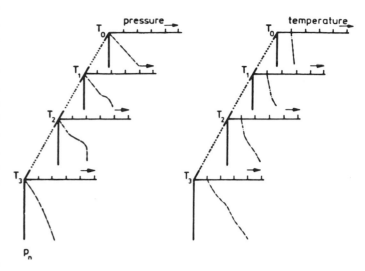

Figure 17. Hypothetical examples of changes in pressure and temperature with respect to space and time in developing basins (after Welte and Yükler, 1980, Fig. 3).

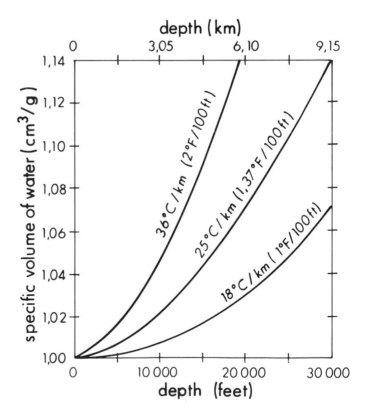

Figure 18. Specific volume of water as function of depth for specified geothermal gradients (after Magara, 1978, Fig. 14-6).

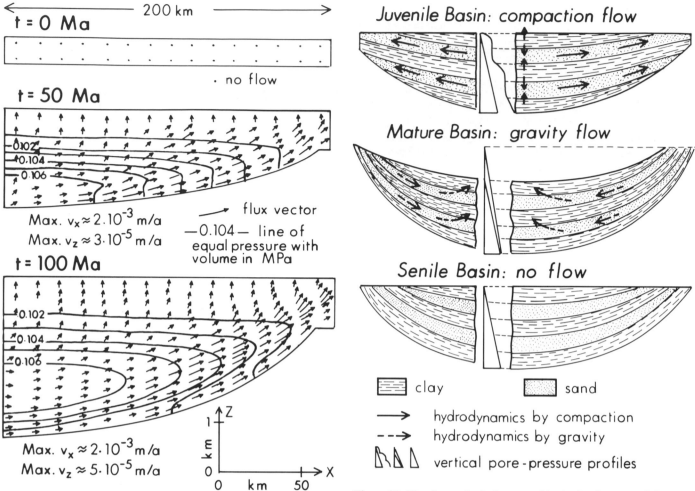

Figure 19. Compaction-driven flow calculation within a basin cross section. Equipotentials are at 0.001 MPa (0.01 atm) intervals. $v_x$, $v_z$: flow velocities relative to subsiding medium (modified from Bethke, 1985).

Figure 20. The three principal stages of basin development of Coustau and others (after Coustau, 1977, Fig. 2).

and deposit organic and inorganic matter, transport heat, and modify pore pressures. Systematically distributed secondary effects give rise to a great diversity of natural flow manifestations related to hydrology, hydrochemistry, mineralogy, botany, soil and rock mechanics, geomorphology, and transport and accumulation of mineral matter and heat; collectively they may be designated as hydrogeologic phenomena. In the context of this theory, these phenomena are linked genetically to migration patterns and certain types of petroleum deposits; indeed, such deposits themselves may be considered as manifestations of regional ground-water flow. The theory provides therefore a unifying framework for the use of hydrogeological phenomena as indicators of potential sites or actual presence of hydrocarbon deposits, and thus constitutes a natural basis for the development of a hydrogeological approach to petroleum exploration.

### Gravity-flow related hydrogeologic phenomena

Figure 9 presents a diagrammatic summary of hydrogeologic phenomena related genetically to petroleum migration and accumulation through the agency of regionally unconfined gravity flow of ground water. These phenomena have been discussed extensively in the literature (e.g., Tóth, 1971, 1980, 1984; Roberts, 1980; Jones and Drozd, 1983; Meyer and McGee, 1985; Zielinski and others, 1985). In this chapter, only a cursory review is given in order to highlight the genetic kinship between hydrocarbon accumulation and other hydrogeologic manifestations of regional ground-water flow.

Because surface moisture is reduced in recharge areas and augmented in discharge areas by ground-water flow, systematic differences in the local water balances and contrasts in soil mois-

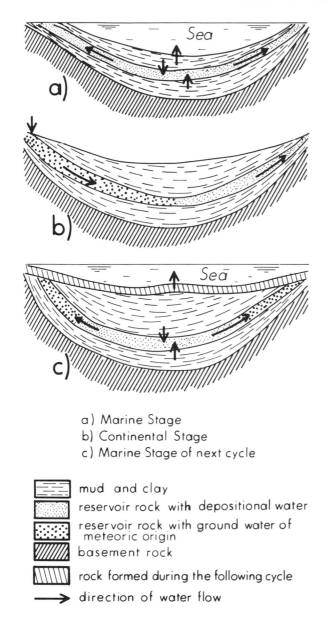

a) Marine Stage
b) Continental Stage
c) Marine Stage of next cycle

mud and clay
reservoir rock with depositional water
reservoir rock with ground water of
 meteoric origin
basement rock
rock formed during the following cycle
direction of water flow

Figure 21. Conceptual model of cyclic basin development of Kartsev and Vagin (after Kartsev and Vagin, 1964, Fig. 1).

ture may develop between the two regions. Relatively moisture deficient conditions are to be expected in inflow areas, and wet, possibly marshy, conditions in outflow areas.

As a result of ionic species transport in ground-water flow systems, characteristic spatial variations in the amounts and kinds of chemical constituents of ground water may be produced. Commonly, water salinity increases in the direction of flow and with depth, reaching maxima in stagnation zones of hydraulic traps. Anions are often distributed in those directions according to a $HCO_3^- \to SO_4^{2-} \to Cl^-$ pattern. Relatively high $O_2$ and $CO_2$ contents render recharging waters oxidizing and acidic as opposed to the reducing and basic nature of water farther along the systems and in discharge areas. Brought to discharge areas, chemical constituents may precipitate out of the water, possibly giving rise to

salt accumulations of different types and intensities, such as salt-affected soils, playas, sebkas, carbonaceous tufas, limonite, and so on. Mixing of relatively fresh discharging ground waters with sea water in, for instance, littoral environments may led to dolomitization of limestone. Precipitation of carbonates and silica can form secondary cements and, thus, reduce porosity and permeability.

Distinct differences in vegetal associations may develop between recharge and discharge areas as a result of differences in moisture and salinity conditions. Recharge-area vegetation tends to be dominated by poorly salt-tolerant xerophytes or mesophytes. Ecosystems in discharge areas, on the other hand, often include halophytes and phreatophytes.

Due to the ascending direction of ground-water flow, pore pressures normally are superhydrostatic in discharge areas. The corollary reduction in the effective stress in the near-surface rocks and soils, and hydraulic discharge of ground water, may lead to soil erosion, reduced slope stability, quicksands, quick clays, frost mounds, pingos, ice fields, geysers, mud volcanoes, assymetric river valleys, karst springs, and so on.

The ability of ground water to transport heat may result in reduced geothermal gradients and negative temperature anomalies in areas of descending cool waters, and in increased gradients and positive anomalies in discharge areas (Fig. 9). Certain metals, such as Pb, Zn, Cu, and U, may be mobilized in the oxidizing recharge environment and precipitated in reducing discharge conditions to form, for instance, Mississippi Valley–type sulfide ore deposits (Garven and Freeze, 1984; Garven, 1985) or roll-front–type sedimentary uranium accumulations (Galloway, 1978). However, should chemical conditions change, either because of a change in ground-water flow or independently from it, previously formed accumulations may reverse their role and become eroding sources that supply metallic ions to other areas and possibly to newly developing deposits. Similarly, hydrocarbon particles may also be transported toward places of accumulation, or washed off from existing deposits and possibly brought to or near the land surface to appear as dissolved gases in water, microseeps of gaseous or liquid hydrocarbons, oil springs, tar pits, and so on.

## Hydrogeologic phenomena as indicators of petroleum migration and accumulation

It is possible to classify the above-outlined hydrogeologic phenomena as indirect or direct indicators of petroleum. Indirect petroleum indicators are those hydrogeologic phenomena created by processes that may also generate petroleum deposits. Phenomena, however, that are produced by regional ground-water flow with a contribution from actual deposits are considered as direct indicators, or signatures, of petroleum.

According to this classification, indirect petroleum indicators include: properties of the flow, fluid potential and pressure fields that characterize discharge areas and hydraulic and hydrodynamic traps (e.g., fluid potential minima, superhydrostatic pore

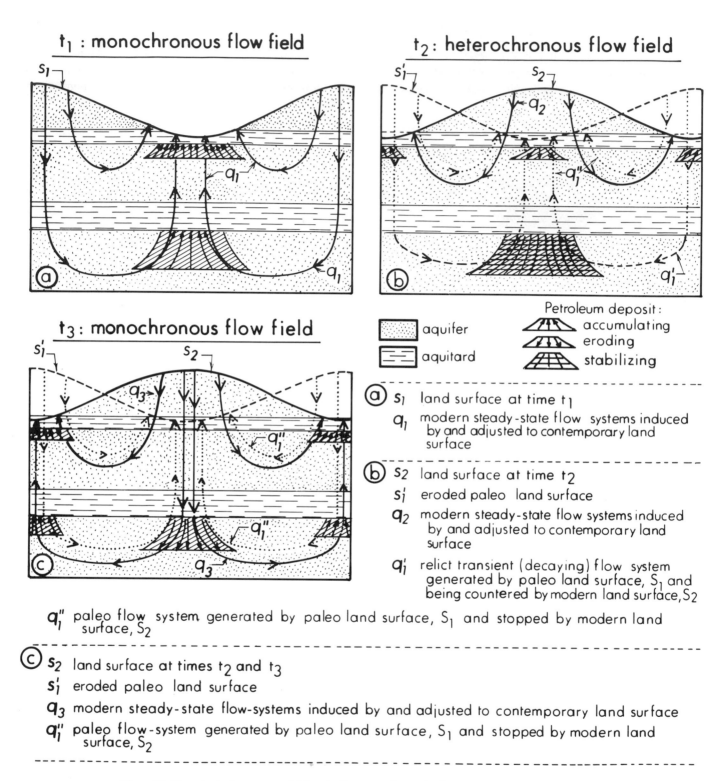

**t₁ : monochronous flow field**

**t₂ : heterochronous flow field**

**t₃ : monochronous flow field**

Petroleum deposit:

aquifer    accumulating

aquitard    eroding

   stabilizing

(a) $s_1$   land surface at time $t_1$

$q_1$   modern steady-state flow systems induced by and adjusted to contemporary land surface

(b) $s_2$   land surface at time $t_2$

$s_1'$   eroded paleo land surface

$q_2$   modern steady-state flow systems induced by and adjusted to contemporary land surface

$q_1'$   relict transient (decaying) flow system generated by paleo land surface, $S_1$ and being countered by modern land surface, $S_2$

$q_1''$   paleo flow system generated by paleo land surface, $S_1$ and stopped by modern land surface, $S_2$

(c) $s_2$   land surface at times $t_2$ and $t_3$

$s_1'$   eroded paleo land surface

$q_3$   modern steady-state flow-systems induced by and adjusted to contemporary land surface

$q_1''$   paleo flow-system generated by paleo land surface, $S_1$ and stopped by modern land surface, $S_2$

Figure 22. Hypothesized sequence of delayed adjustments in gravity flow patterns of ground water, and subsequently petroleum deposits, to a change in boundary conditions (i.e., the topography; modified from Tóth and Corbet, 1983, Fig. 8).

pressures, flowing artesian wells, slight or zero lateral potential gradients, slighter theoretically calculated tilt of oil-water interfaces than the dip of caprocks); positive geothermal anomalies; marshes, soil salinization, and salt-water discharge; increases in formation water salinity; reducing near-surface conditions; altered soils; $H_2$; sulfide ores; radiometric and electrotelluric anomalies; and so on (Roberts, 1980; Tóth, 1980).

Direct indicators, on the other hand, comprise oil springs and seeps, hydrocarbon gases in spring waters, soil wax, and anomalous growth of hydrocarbon-metabolizing bacteria.

## HYDROGEOLOGICAL APPROACH TO PETROLEUM EXPLORATION

Hydrocarbon deposits are envisaged by the hydraulic theory of petroleum migration (Tóth, 1980) to be related to ground-water movement in two different ways. First, ground-water flow may generate deposits by transporting hydrocarbon particles toward sites of fluid-dynamically conducive conditions for accumulation and entrapment. Second, ground-water flow may reveal the presence of deposits by transmitting indirect or direct signatures from accumulations to locations where they can be detected. As a version of the second case, flow may also mask deposits by obliterating signatures that would exist otherwise. These two relations may be used as bases for the development of indirect and, respectively, direct methods of exploration; together constituting a hydrogeological approach to petroleum exploration. Indirect methods may be used to search for subsurface regions where hydraulic and hydrodynamic conditions are favorable for the concentration and entrapment of hydrocarbons. They are based on the indirect indicators outlined earlier. Direct methods, on the other hand, rely on direct indicators and attempt to detect and interpret water-flow–related signatures of actual deposits.

The possibility of genetic relations existing between some of the indirect and most direct indicators, on the one hand, and petroleum deposits, on the other, has been recognized and attributed to vertical migration of petroleum since the early 1930s (Laubmeyer, 1933; Horvitz, 1939, 1980; Davidson, 1982; Klusman, 1985). The exploration technologies developed on the basis of these relations have been called collectively surface exploration or geochemical exploration, and include: soil-gas analysis (Siegel, 1974; Stahl and others, 1981; Jones and Drozd, 1983), geothermal methods (Rose and others, 1979), Landsat imagery, geothermics (Zielinski and Bruchhausen, 1983; Land and others, 1985a, 1985b; Meyer and McGee, 1985; Zielinski and others, 1985), radiometry (Thomeer, 1965), electrotelluric potential measurements (Pirson, 1971), and so on. However, results obtained by these methods are notoriously inconsistent. Signatures, presumed to indicate vertical migration, are often found without corresponding petroleum deposits, or, conversely, deposits may exist without associated signatures. In other instances, signatures and deposits apparently related by isotopic similarity, for instance, may not be positioned along a common vertical. A key to resolve these inconsistencies seems to be an understanding of, and

an ability to evaluate—site specifically—the role of ground-water flow in creating, modifying, or obliterating signatures of petroleum deposits.

Provided that such an understanding can be developed, a hydrogeological approach might be applied to four types of problems in petroleum exploration, namely: (1) areal reconnaissance, that is, location of extensive regions with hydraulically favorable conditions for petroleum accumulation (Tóth, 1980); (2) volume estimates of potentially migrated hydrocarbons (Kartsev and Vagin, 1964; Kortsenshteyn, 1964; Jacquin and Poulet, 1973; Bethke, 1985); (3) location of hydraulic and hydrodynamic traps, extension of known accumulations, and analysis of communication between deposits (Hubbert, 1953; Dahlberg, 1982; Tóth, 1980); (4) direct search for petroleum deposits by survey and interpretation of direct hydrogeological indicators.

Thanks to the previously outlined and many-faceted relations between ground-water flow and hydrocarbon migration, some hydrogeologically based exploration technique should exist for virtually any scale of exploration target, from general basin analysis to single fields or even pools. When properly integrated with the traditional methodologies, namely geology, geophysics, and organic chemistry, the hydrogeological approach may prove to be a very powerful addition to the arsenal of the explorationist.

## SUMMARY AND CONCLUSIONS

Migration is generally assumed to be the only possible process by which petroleum hydrocarbons can be concentrated in sufficient quantities to form large deposits. The modes, forces, mechanisms, and patterns of migration and entrapment are, however, poorly understood and still largely speculative.

Most theories recognize that water is essential in some aspects, or at least during some phases, of migration. For example: most of the proposed modes of migration depend on aqueous solutions of petroleum; water generates or transmits the forces that impel petroleum particles to migrate; and the migration paths of petroleum are considered to be coupled to flow patterns of formation waters.

The modes of transport believed by most theories to be effective in one or other phases of migration include: oil globules or gas bubbles; colloidal, micellar, or molecular solutions; water-free migration through an oleophilic solid network; water-soluble prematuration organic matter. The principal driving forces thought to play a role in the subsurface transport of petroleum are attributed to one or any combination of the following energy sources: (1) compaction and/or elastic rebound of sediments, (2) buoyancy, (3) gravitation, (4) volumetric expansion of gas by heating or by conversion from solids, (5) thermal expansion of liquids, (6) molecular diffusion of hydrocarbons, and (7) osmosis. Factors and mechanisms possibly effecting accumulation and entrapment are thought to include: (1) physical screening or micropore filtration by pore-size reduction; (2) capillarity; (3) exsolution owing to decreases in temperature and pressure, and an increase in water salinity along the migration path; and (4) minimum local values of fluid potentials in two or three dimensions.

The modes, forces, and patterns of migration and their temporal and spatial variations depend directly on the evolutionary history of a sedimentary basin. If, therefore, migration is to be understood it must be analyzed in the context of basin evolution. Numerous models have been proposed to explain and depict basinal migration of petroleum and to provide rational working hypotheses in aid of exploration.

Most of the basinal migration models fall in the category of lumped models. These models portray migration in terms of fluid volumes and flow direction, but neither attempt nor are able to specify the geometry of migration routes beyond general statements about highly permeable carrier beds, fault zones, or rock fractures being preferred pathways; impelling forces are derived usually from compaction and fluid expansion. Models in this category consider movement only of water originally deposited with the sediments, and are exemplified by those of Jacquin and Poulet (1970), Magara (1978), Bonham (1980), Welte and Yükler (1980), and Bethke (1985).

A second category of models considers the subsequent or alternating effects of compaction-driven marine waters and topography-induced gravity flow on the basinal migration and distribution of petroleum. Specified flow paths are confined to regionally extensive carrier beds conveying saline compaction waters updip to, or fresh meteoric waters downslope from, elevated outcrops of these aquifers located at the basins' rims. These models attempt to relate prospects for petroleum occurrence in the basins to the degree of physical and chemical sheltering of previously formed deposits from invading meteoric waters (Coustau and others, 1975; Coustau, 1977) or to the intensity and frequency of replacement of petroleum carrying feedstock water by, or for, fresh meteoric waters (Kartsev and Vagin, 1964). The model of Hubbert (1953) presents the process of, and criteria for, the entrapment of petroleum under hydrodynamic conditions in extended and confined carrier beds.

A third category of models portrays migration only during the continental stage of basin evolution and is exemplified by the hydraulic theory of petroleum migration (Munn, 1909; Tóth, 1980). In these models, basinal flow is allowed to occur across formations, and the flow field's upper boundary is the erosion-sculptured land surface. Consequently, the pattern of the hydrocarbon-transporting water flow is distributed into systems of well-defined geometry. Hydrocarbons migrating in these systems toward discharge depressions tend to be entrapped in focal hydraulic traps or in hydrodynamic traps (Tóth, 1980), according to the principles formulated by Hubbert (1953).

In addition to proposing a distributed type of migration pattern, and based upon this characteristic, the hydraulic theory postulates a genetic relationship between a variety of hydrogeologic phenomena, on the one hand, and regional ground-water flow, on the other. These phenomena result from the interaction between the rock framework and ground water flowing in gravitational systems and include: fluid-potential minima, superhydrostatic pore pressures, flowing artesian wells, marshes maintained by discharge of deep and possibly saline ground water, high formation-water salinities, particular ionic distribution patterns, positive geothermal anomalies, reducing chemical conditions at depth and near the land surface, saline soils, playas, salt deposits, diagenetic changes in rocks and minerals, salt-tolerant and water-tolerant vegetation, mud volcanoes, assymetric river valleys, gas and oil seeps and springs, radiometric and electrotelluric anomalies, colonies of hydrocarbon-oxidizing bacteria, deposits of sulfide ores, uranium, hydrocarbons, and so on. With respect to their relations to petroleum deposits, the other manifestations of ground-water flow may be classified as indirect and direct indicators of petroleum depending on whether they reflect potential conditions for accumulation and entrapment or the actual presence of deposits.

The presence, and probably the movement also, of ground water appears to be a prerequisite for hydrocarbon migration and accumulation. Furthermore, petroleum migration is linked genetically to distributed systems of ground-water flow by some of the basin-evolution models, and the relations are considered to be manifest by hydrogeologic phenomena. Based on these relations and phenomena, that is, applying hydrogeological principles and techniques, the conventional arsenal of geology, geophysics, and organic geochemistry of the explorationist may be broadened by a hydrogeological approach to basin analysis and petroleum exploration to be called, perhaps, Petroleum Hydrogeology.

# REFERENCES CITED

Bair, E. S., O'Donnell, T. P., and Picking, L. W., 1985, Potentiometric mapping from incomplete drill-stem test data; Palo Duro Basin area, Texas and New Mexico: Ground Water, v. 23, no. 2, p. 198–211.

Berry, F.A.F., 1959, Hydrodynamics and geochemistry of the Jurassic and Cretaceous systems in the San Juan Basin, northwestern New Mexico and southwestern Colorado [Ph.D. thesis]: Stanford, California, Stanford University, 213 p.

Bethke, C. M., 1985, A numerical model of compaction-driven groundwater flow and heat transfer and its application to the paleohydrology of intracratonic sedimentary basins: Journal of Geophysical Research, v. 90(B7), p. 6817–6828.

Bonham, L. C., 1980, Migration of hydrocarbons in compacting basins, *in* Roberts, W. H., and Cordell, R. J., eds., Problems of petroleum migration: American Association of Petroleum Geologists Studies in Geology no. 10, Tulsa, Oklahoma, p. 69–89.

Bradley, J. S., 1975, Abnormal formation pressure: American Association of Petroleum Geologists Bulletin, v. 59, no. 6, p. 957–973.

Coustau, H., 1977, Formation waters and hydrodynamics: Journal of Geochemical Exploration, v. 7, p. 213–241.

Coustau, H., Chiarelli, A., Rumeau, J. L., Sourisse, C., and Tison, J., 1975, Classification hydrodynamique des bassins sédimentaires; Utilisation combinée avec d'autres méthodes pour rationaliser l'exploration dans des bassins non productifs: Proceedings, 9th World Petroleum Congress, v. 2, no. 4, p. 105–119.

Dahlberg, E. C., 1982, Applied hydrodynamics in petroleum exploration: New York, Springer-Verlag, 161 p.

Davidson, M. J., 1982, Toward a general theory of vertical migration: Oil and Gas Journal, June 21, p. 288–300.

Freeze, R. A., and Witherspoon, P. A., 1967, Theoretical analysis of regional groundwater flow; 2. Effect of water table configuration and subsurface permeability variation: Water Resources Research, v. 3, no. 2, p. 623–634.

Galloway, W. E., 1978, Uranium mineralization in a coastal-plain fluvial aquifer system: Catahoula Formation, Texas: Economic Geology, v. 73, no. 8, p. 1655–1676.

Garven, G., 1985, The role of regional fluid flow in the genesis of the Pine Point deposit, Western Canada Sedimentary Basin: Economic Geology, v. 80, no. 2, p. 307–324.

Garven, G., and Freeze, R. A., 1984, Theoretical analysis of the role of groundwater flow in the genesis of strata-bound ore deposits: American Journal of Science, v. 284, p. 1085–1124.

Hedberg, H. D., 1980, Methane generation and petroleum migration, *in* Roberts, W. H., and Cordell, R. J., eds., Problems of petroleum migration: American Association of Petroleum Geologists Studies in Geology no 10, p. 179–206.

Hodgson, G. W., 1980, Origin of petroleum; In-transit conversion of organic compounds in water, *in* Roberts, W. H., and Cordell, R. J., eds., Problems of petroleum migration: American Association of Petroleum Geologists Studies in Geology no. 10, p. 169–178.

Horvitz, L., 1939, On geochemical prospecting: Geophysics, v. 14, no. 3, p. 210–228.

—— , 1980, Near-surface evidence of hydrocarbon movement from depth, *in* Roberts, W. H., and Cordell, R. J., eds., Problems of petroleum migration: American Association of Petroleum Geologists Studies in Geology no. 10, p. 241–269.

Hubbert, M. K., 1940, The theory of groundwater motion: Journal of Geology, v. 48, no. 8, p. 785–944.

—— , 1953, Entrapment of petroleum under hydrodynamic conditions: American Association of Petroleum Geologists Bulletin, v. 37, no. 8, p. 1954–2026.

Hunt, J. M., 1979, Petroleum geochemistry and geology: San Francisco, W. H. Freeman and Company, 617 p.

Jacquin, C., and Poulet, M., 1970, Study of the hydrodynamic pattern in a sedimentary basin subject to subsidence: Society of Petroleum Engineers of American Institute of Mining, Metallurgical, and Petroleum Engineers, Incorporated, Paper no. SPE2988, 10 p. (preprint).

—— , 1973, Essai de restitution des conditions hydrodynamiques régnant dans un bassin sédimentaire au cours de son évolution: Révue de l'Institut Français du Pétrole, v. 28, no. 3, p. 269–297.

Jones, V. T., and Drozd, R. J., 1983, Predictions of oil or gas potential by near-surface geochemistry: American Association of Petroleum Geologists Bulletin, v. 67, no. 6, p. 932–952.

Kartsev, A. A., and Vagin, S. B., 1964, Paleohydrogeological studies of the origin and dissipation of oil and gas accumulations in the instance of Cis-Caucasian Mesozoic deposits: International Geology Review, v. 6, no. 4, p. 644–655.

Kinghorn, R.R.F., 1983, An Introduction to the physics and chemistry of petroleum: New York: John Wiley and Sons, 240 p.

Klusman, R., coordinator, 1985, Surface and near-surface geochemical methods in petroleum exploration: Association of Petroleum Geochemical Explorationists Special Publication no. 1, 285 p.

Kortsenshteyn, V. N., 1964, Predicting oil and gas by means of groundwater investigations at depth and estimating the potential reserves of oil and gas: Doklady Akademii Nauk SSSR, Earth Science Section, v. 58, no. 4, p. 856–859.

Kudryakov, V. A., 1974, Piezometric minima and their role in the formation and distribution of hydrocarbon accumulations: Doklady Akademii Nauk SSSR, v. 207, no. 6, p. 1424–1426.

Lang, H. R., Curtis, J. B., and Kovacs, J. S., 1985a, Lost River, West Virginia, Petroleum test site report, *in* Paley, H. N., ed., The Joint NASA/Geosat Test Case Project, Final Report, pt. 2, v. II: American Association of Petroleum Geologists, p. 12-i–12-96.

Lang, H. R., Nicolais, S. M., and Hopkins, H. R., 1985b, Coyanosa, Texas, Petroleum test site report, *in* Paley, H. N., ed., The Joint NASA/Geosat Test Case Project, Final Report, pt. 2, v. II: American Association of Petroleum Geologists, p. 13-i–13-81.

Laubmeyer, G., 1933, A new geophysical prospecting method, especially for deposits of hydrocarbons: Petroleum, v. 29, no. 18, p. 1–4.

Levorsen, A. I., 1958, Geology of petroleum: San Francisco, W. H. Freeman and Company, 703 p.

Magara, K., 1978, Compaction and fluid migration; Practical petroleum geology: Amsterdam, Elsevier Scientific Publishing Company, 319 p.

McAuliffe, C. K., 1980, Oil and gas migration; Chemical and physical constraints, *in* Roberts, W. H., and Cordell, R. J., eds., Problems of petroleum migration: American Association of Petroleum Geologists Studies in Geology no. 10, p. 89–107.

Meyer, H. J., and McGee, H. W., 1985, Oil and gas fields accompanied by geothermal anomalies in Rocky Mountain Region: American Association of Petroleum Geologists Bulletin, v. 69, no. 6, p. 933–945.

Munn, M. J., 1909, The anticlinal and hydraulic theories of oil and gas accumulation: Economic Geology, v. 4, no. 6, p. 509–529.

Neuzil, C. E., and Pollock, D. W., 1983, Erosional unloading and fluid pressures in hydraulically "tight" rocks: Journal of Geology, v. 91, no. 2, p. 179–193.

Pirson, S. J. 1971, New electric technique can locate gas and oil, part 1 and 2: World Oil, v. 172, no. 5, p. 69–74.

Roberts, W. H., 1980, Design and function of oil and gas traps, *in* Roberts, W. H., and Cordell, R. J., eds., Problems of petroleum migration: American Association of Petroleum Geologists Studies in Geology no. 10, p. 217–240.

Roberts, W. H., and Cordell, P. J., eds., 1980, Problems of petroleum migration: American Association of Petroleum Geologists Studies in Geology no. 10, 273 p.

Rose, A. W., Hawkes, H. E., and Webb, J. S., 1979, Geochemistry in mineral exploration (2nd edition): London, Academic Press, Incorporated, 657 p.

Siegel, F. R., 1974, Geochemical prospecting for hydrocarbons, *in* Applied geochemistry: New York, Wiley-Interscience, p. 228–255.

Stahl, W., Faber, E., Carey, B. D., and Kirksey, D. L., 1981, Near-surface evidence of migration of natural-gas from deep reservoirs and source rocks:

American Association of Petroleum Geologists Bulletin, v. 65, no. 9, p. 1543–1550.

Thomeer, H.H.M.A., 1965, Exploration for oil and gas by emanometric methods: Geologie en Mijnbouw, v. 44, no. 9, p. 301–306.

Tissot, B. P., and Welte, D. H., 1978, Petroleum formation and occurrence: New York, Springer-Verlag, 538 p.

Tóth, J., 1963, A theoretical analysis of groundwater flow in small drainage basins: Journal of Geophysical Research, v. 68, no. 16, p. 4795–4812. Also *in* Freeze, R. A., and Back, W., eds., 1983, Physical hydrogeology; Benchmark Papers in Geology, v. 72: Stroudsburg, Pennsylvania, Hutchinson Ross Publishing Company, p. 328–345.

—— , 1971, Groundwater discharge; A common generator of diverse geologic and morphologic phenomena: International Association of Scientific Hydrology Bulletin, v. 16, no. 1–3, p. 7–24.

—— , 1978, Gravity-induced cross-formational flow of formation fluids, Red Earth region, Alberta, Canada; Analysis, patterns, and evolution: Water Resources Research, v. 14, no. 5, p. 805–843.

—— , 1980, Cross-formational gravity-flow of groundwater; A mechanism of the transport and accumulation of petroleum (The generalized hydraulic theory of petroleum migration), *in* Roberts, W. H., and Cordell, R. J., eds., Problems of petroleum migration: American Association of Petroleum Geologists Studies in Geology no. 10, p. 121–167.

—— , 1984, The role of regional gravity flow in the chemical and thermal evolution of groundwater, *in* Hitchon, B., and Wallick, E. I., eds., Proceedings, First Canadian/American Conference on Hydrogeology; Practical Applications of Groundwater Geochemistry, Banff, Alberta, Canada, June 22–26, 1984: Worthington, Ohio, National Water Well Association,

p. 3–39.

Tóth, J., and Corbet, T., 1983, Investigations of the relations between formation fluid dynamics and petroleum occurrences, Taber area, Alberta: Second "Twelve Month" Report, Energy Resources Research Fund Contract no. U81-3R, Alberta Department of Energy and Natural Resources, 125 p.

—— , 1986, Post-paleocene evolution of regional groundwater flow-systems and their relation to petroleum accumulations, Taber area, southern Alberta, Canada: Canadian Petroleum Geology Bulletin, v. 34, no. 3, p. 339–363.

Tóth, J., and Millar, R. F., 1983, Possible effects of erosional changes of the topographic relief on pore pressures at depth: Water Resources Research, v. 19, no. 6, p. 1585–1597.

Welte, D. H., and Yükler, A., 1980, Evolution of sedimentary basins from the standpoint of petroleum origin and accumulation; An approach for a quantitative basin study; Organic Geochemistry, v. 2: Great Britain, Pergamon Press, Limited, p. 1–8.

Zielinski, G. W., and Bruchhausen, P. M., 1983, Shallow temperatures and thermal regime in the hydrocarbon province of Tierra del Fuego: American Association of Petroleum Geologists Bulletin, v. 67, no. 1, p. 166–177.

Zielinski, G. W., Drahovzal, J. A., Decoursey, G. M., and Rupterto, J. M., 1985, Hydrothermics in the Wyoming Overthrust belt: American Association of Petroleum Geologists Bulletin, v. 69, no. 5, p. 699–709.

MANUSCRIPT ACCEPTED BY THE SOCIETY MAY 15, 1987

The Geology of North America
Vol. O-2, Hydrogeology
The Geological Society of America, 1988

# Chapter 49

# *Scientific problems*

**Leonard F. Konikow**
*U.S. Geological Survey, 431 National Center, Reston, Virginia 22092*
**Stavros S. Papadopulos**
*S. S. Papadopulos and Associates, Inc., Suite 290, 12250 Rockville Pike, Rockville, Maryland 20852*

## INTRODUCTION

Hydrogeology is a science that has great practical value. Because of this, it has traditionally been a field in which geologic aspects have melded into engineering aspects, and a field in which geologists and engineers with common interests and goals have often crossed disciplinary lines. In the past, the main driving force for hydrogeologic studies has been the need to assess the water-supply potential of aquifers. However, during the last decade, the emphasis has shifted from water supply to water quality, driven by the need to predict the movement of contaminants through the subsurface environment. One consequence of the change in emphasis has been a shift in perceived priorities for scientific research. Formerly, the focus was on developing methods to assess and measure the water-yielding properties of high-permeability aquifers. The focus is now on transport and dispersion processes, retardation and degradation of chemical contaminants, and the ability of low-permeability materials to contain wastes. In this chapter, we will attempt to outline some of the significant scientific problems facing hydrogeologists today.

The chapter in the sections of this volume on Hydrogeologic Regions and Comparative Hydrogeology contain a wealth of information demonstrating geologic controls on occurrence, movement, and chemical character of water in different regions within North America. These chapters collectively represent many thousands of man-years of hydrogeologic research in the field. In these times of computer models and simulations, it should be reemphasized that the real world is complex, three-dimensional, heterogeneous, and commonly anisotropic. This reality can only be described accurately through careful hydrogeologic research in the field. Although computer models are certainly one of the most valuable tools available for analyzing ground-water systems, they cannot collect and interpret field data, nor can they replace sound hydrogeologic judgment.

During the last decade, the geologic community has become more aware of the role of ground water in geologic processes. The chapters in the section of this volume on Ground Water and Geologic Processes address these issues, and we will not attempt to summarize these detailed discussions of the role of ground water in landform development, diagenesis of rocks, earthquakes, origin of and exploration for ore deposits, and migration and entrapment of hydrocarbons. Research on these geologic problems has not yet resolved all issues, and additional work is warranted.

## SCIENTIFIC PROBLEMS IN HYDROGEOLOGY

The processes affecting contaminant migration in porous media have been reviewed by Gillham and Cherry (1982, p. 32). They state, "Although considerable research on contaminant migration in ground-water flow systems has been conducted during recent decades, this field of endeavor is still in its infancy. Many definitive laboratory and field tests remain to be accomplished to provide a basis for development of mathematical concepts that can be founded on knowledge of the transport processes that exist at the field scale."

As the focus of hydrogeology has shifted from flow to transport, so have the needs for data and prediction shifted from the regional scale to a more local or site-specific scale. This is a natural outcome of the physics of these processes. Responses to hydraulic stresses may spread rapidly as a pressure wave through an aquifer system, but effects of transport processes on ground-water quality depend on the actual movement of the fluid, which is much slower. In a recent review of hydrogeologic work in Canada, Jackson (1987, p. 35) observed, "While Canadian hydrogeologists in the 1960s made their mark by investigating the large groundwater flow systems of the Prairie Provinces, the present generation will probably be remembered for their detailed studies of test sites whose dimensions are measured in meters rather than kilometers." This generalization is broader than just the Canadian experience, and it is a trend that will likely persist for at least the next decade.

The transport of nonreactive solutes in saturated porous media is related to advective and dispersive processes. One limitation in applying transport models to field problems is the observed scale dependence of dispersivity, which implies that dispersivity is not simply a physical constant of the system of interest. In practice, uncertainty about the true ground-water ve-

Konikow, L. F., and Papadopulos, S. S., 1988, Scientific problems, *in* Back, W., Rosenshein, J. S., and Seaber, P. R., eds., Hydrogeology: Boulder, Colorado, Geological Society of America, The Geology of North America, v. O-2.

locity and contaminant concentration distribution are often compensated for by adjusting the dispersion coefficient during model calibration. Also, if transient changes in the flow field and contaminant source are not explicitly accounted for by advection, then their effect will be to induce an apparently greater dispersion than actually exists. Because of these factors, the relatively large dispersivities indicated for many field-scale problems are, at least in part, an artifact of our inability to measure accurately and precisely the velocity and concentration distributions in three dimensions. The more accurately and precisely we can define spatial and temporal variations in ground-water velocity and solute concentration, the lower will be the apparent magnitude of dispersivity. In this sense, the advective, dispersive, and source terms in the solute-transport equation are interdependent, and their linkage is through the uncertainty and variance in the fluid velocity. Therefore, improvement of hydrogeologic and geophysical field methods are needed to obtain direct or indirect measurements of fluid seepage velocity and its variability.

Conceptual and mathematical models of transport processes need to account for scale-dependent dispersion. Although much recently published theoretical work on solute transport in ground water has focused on the nature of dispersion phenomena in ground-water systems, a scientific consensus has not yet been reached, and this will probably continue to be a fruitful area of research in the near future. Gelhar and Axness (1983) and Neuman and others (1987) present general three-dimensional stochastic analyses of macrodispersion (or field-scale dispersion) in anisotropic media. Their theories should be tested in controlled laboratory and field experiments.

Smith and Schwartz (1980) conclude that macroscopic dispersion results from large-scale spatial variations in hydraulic conductivity, and that the use of relatively large values of dispersivity with uniform hydraulic conductivity fields is an inappropriate basis for describing transport in geologic systems. Thus, it becomes particularly important to make maximum use of geological information to characterize the variability in both hydraulic conductivity and effective porosity so that the velocity can be defined as accurately and precisely as possible. However, heterogeneities are not easy to measure, and their effects are not easy to quantify. In the future, hydrogeologists will be expected to determine not only the mean hydraulic properties of a system or spatial trends in these properties, but also the variance and spatial correlation statistics for these properties.

More attention needs to be given to the relation between field measurement techniques and the conceptual models that are considered. For example, in modeling three-dimensional solute movement, we commonly visualize concentration, hydraulic conductivity, and effective porosity as continuous, point-wide properties. In practice, these quantities are measured over some finite scale, which may be different for each of the properties. The use of such data and its effect on our perception of the physics of the system have not been adequately addressed in the literature. It also is important to differentiate the influence of different measurement techniques. For example, measuring concentration of a constituent in the discharge from a pumped well can yield a different value than measuring it in an observation well with an ion-specific probe set at a specific depth. In one instance, the concentration is a weighted average associated with the volume of fluid entering different intervals in the well, whereas in the other instance, the concentration is indicative of the mass in a specific volume of the porous medium about the measuring probe. The need for measurement of various properties at a consistent scale also must be analyzed. The scarceness of such discussions in the literature may be a consequence of a gap between theoreticians and practitioners; if such a gap exists, it should be minimized.

There are several comprehensive and long-term field tracer tests in progress or recently completed that aim to measure and analyze dispersion in controlled field-scale experiments. These are clearly needed to provide control and feedback to the theoretical investigations. Of necessity, these field tests have tended to be located in relatively homogeneous unconsolidated deposits (see, e.g., Garabedian and LeBlanc, 1987; Molz and others, 1987; Mackay and others, 1986). The results of these studies will probably lead to a future need to conduct similar tests in more heterogeneous and anisotropic systems, including consolidated rocks that have secondary permeability.

Because there will always be some uncertainty in even the most controlled field experiment, there also is a need for, and an increasing trend toward, large sand-box experiments in the laboratory on materials of known heterogeneity and/or anisotropy as a means to test new theories on scale-dependent dispersion (see, e.g., Refsgaard, 1986; Silliman and Simpson, 1987; Silliman and others, 1987). Although these laboratory tests have the advantage of minimizing uncertainty in media properties, they have the disadvantage of being limited in the range of length (or correlation) scales over which the experiments can be conducted. This may lead to the use of even larger sand boxes in future experiments.

There are many aspects of solute transport and dispersion in anisotropic porous media that are still poorly understood. The dispersivity of an isotropic porous medium can be defined by two constants—the longitudinal dispersivity of the medium, $\alpha_L$, and the transverse dispersivity of the medium, $\alpha_T$. Most applications of transport models to ground-water contamination problems documented to date have been based on this conventional formulation. The consideration of solute transport in an anisotropic porous medium may require estimating more than two parameters. In practice, it is rare that field values for even the two constants $\alpha_L$ and $\alpha_T$ can be defined uniquely. Thus, although anisotropy in hydraulic conductivity (a second-order tensor) is recognized and accounted for in ground-water flow simulation, it is commonly assumed out of convenience that the same system is isotropic with respect to dispersion. Based on sand-box experiments performed on anisotropic porous media, Silliman and others (1987) concluded that longitudinal dispersion depends on the direction of flow in anisotropic porous media, and that the relation between hydraulic and dispersive anisotropies is not always consistent. That is, higher longitudinal dispersivity is not always aligned with the same axis of the hydraulic conductivity tensor, but instead depends on the nature and scale of the geologic factors causing the hydraulic anisotropy.

There are additional complications and practical problems when the solutes being transported are reactive. Traditionally, such problems have been approached either from a geochemical perspective that the reactions are the primary control or from the hydraulic perspective that flow and transport are the primary controls. However, during the last few years, there has been a notably increased effort at developing integrated approaches and models. Such an integrated approach is more realistic and is expected to be a priority research need in the foreseeable future. Difficult numerical problems arise when reaction terms are highly nonlinear, or if the concentration of the solute of interest is strongly dependent on the concentration of numerous other chemical constituents or if it reacts with the solid phase. The commonly assumed linear relation between the concentration of a chemical species in solution, and the mass of that species adsorbed on the porous matrix, may not always be applicable. Similarly, the assumption that chemical equilibrium is achieved very rapidly in comparison to fluid travel time may not always be valid. If the transported solute is strongly affected by precipitation-dissolution reactions, a relatively difficult moving-boundary problem may arise as a necessary descriptor of the movement of a sharp front between different geochemical zones (see Willis and Rubin, 1987; Dria and others, 1987). Furthermore, these types of reactions may significantly alter the hydraulic properties of the porous media, creating a feedback between solute transport and fluid flow. For field problems in which reactions are significantly affecting solute concentrations, simulation accuracy is less limited by mathematical constraints than by data constraints. That is, the types and rates of reactions for the specific solutes and minerals in the particular ground-water system of interest are rarely known and require an extensive amount of data to assess accurately. Mineralogic variability may be very significant (e.g., see Mackay and others, 1986), and the effects of such variability on transport needs to be explored further. This is not to say that mathematical constraints also do not exist. Rubin (1983) discusses and classifies the chemical nature of reactions and their relation to the mathematical problem formulation. Continued progress is certainly needed toward the development of accurate and efficient multiple-constitutent transport models that incorporate complex reactions.

Further complexities arise when ground-water contaminants are either immiscible or partly miscible, as is the case for many organic chemicals. In such cases, additional processes and parameters may significantly affect the fate and movement of the contaminant. A multiphase modeling approach may be required to represent phase composition, interphase mass transfer, and capillarity (see Pinder and Abriola, 1986). This would concurrently impose more severe data requirements to describe additional parameters, nonlinear processes, and more complex geochemical and biological reactions. Faust (1985) states, "Unfortunately, data such as relative permeabilities and capillary pressures for the types of fluids and porous materials present in hazardous waste sites are not readily available." Well-documented, multidimensional, and efficient multiphase models applicable to contamination of ground water by immiscible organic chemicals are not yet generally available.

Studies of organic chemicals in ground water have highlighted the importance of microbial processes in transforming and degrading these contaminants (see, e.g., Wilson and McNabb, 1983). Populations of viable anaerobic bacteria also have been identified in uncontaminated hydrogeologic environments, such as deep coastal-plain sediments in Maryland (Chapelle and others, 1987). Chapelle and others (1987) demonstrate the presence of methanogenic and sulfate-reducing bacteria and propose that bacterial metabolism is a valid mechanism for $CO_2$ generation in ground water. In hydrogeologic environments near some oilfields, $H_2S$ generated by sulfate-reducing bacteria can have disastrous effects (Cord-Ruwisch and others, 1987). This increased awareness of the role of microbes will lead to more interactions between hydrogeologists and microbiologists in the future as new research efforts focus on defining the hydrogeologic and hydrogeochemical controls and constraints on microbial activity in ground water. Such efforts not only will lead to increased conceptual understanding, but may lead to practical biologically based approaches for in situ clean up of some contaminated aquifers.

Aquifer cleanup is being implemented with increasing frequency. If the contaminants involved include high-density, nonaqueous-phase liquids, then the hydrogeologist is faced with a relatively new set of exploration challenges that will likely create a significant demand for future research. In a manner analogous to petroleum exploration, the hydrogeologist may be required to locate the stratigraphic, structural, and hydrodynamic traps for these liquids, which may be slowly but steadily releasing organic contaminants into solution in flowing ground water by interphase transfer. Cleanup obviously will be most efficient if the separate phase can be located and removed or treated. Of course, because the target nonaqueous-phase liquid has a higher density than water, the trap environments are analogous to inverted petroleum traps. The hydrogeologist may have to draw heavily on scientific advances and exploration technology used by petroleum geologists, including geochemical and geophysical exploration.

The use of surface and borehole geophysical techniques to locate and/or map organic contaminants in the shallow subsurface environment is not well developed. This capability would represent a valuable scientific and technological advance. Existing geophysical methods may have to be refined or new approaches developed to accomplish this goal. Most geophysical methods cannot readily map plumes of dissolved organic chemicals where there is not a separate phase or sufficient ion activity of associated inorganic solutes to affect the fluid resistivity. However, the results of recent research in frequency-domain electromagnetics, ground-penetrating radar, and complex resistivity methods indicate that, under certain field conditions, these methods can be used to map organic contaminants (Valentine and Kwader, 1985;

Olhoeft, 1986). Additional theoretical and field studies are needed on these methods to eliminate present constraints.

The past emphasis within hydrogeology on water-supply issues has led to an abundance of collected data and experience in high-permeability systems. On the whole, we have relatively little experience and insight into ground-water flow and solute transport within low-permeability environments, such as clay, shale, or salt formations, yet these are exactly the types of settings we commonly want to use for waste emplacement. Long-term and regional responses in these environments are difficult to predict on the basis of presently available small-scale laboratory and field measurements.

Many subsurface environments are dominated by the effects of secondary-permeability development, such as fractures and solution openings. In spite of many years of research on flow and transport in secondary-permeability environments, such environments still remain an enigma, and research on these topics will undoubtedly remain a priority issue for years. The difficulties are both conceptual and practical. Newman (1987, p. 551) notes that measurements of hydraulic properties of fractured rocks tend to be erratic and sensitive to the volume of rock sampled by the test. He further notes that this erratic behavior may not necessarily be overcome because ". . .there is generally no guarantee that an REV [Representative Elementary Volume] can be defined for a given rock mass."

In the analysis of regional flow and transport, fractured formations are sometimes represented as an equivalent porous media. For flow problems, this approach often leads to satisfactory results. However, for solute-transport problems, immiscible fluid problems, or where the emphasis of a problem is on a local scale, other concepts are necessary. In recent years, conceptual models for flow and transport in a single fracture or in a network of fractures have been developed (see Gale, 1982). These concepts are based on assumptions of flow between parallel plates and (or) a well-defined distribution of fractures. However, because of geologic complexities, fracture walls are not necessarily smooth or parallel, and an accurate and comprehensive definition of fracture spacing and orientation may not be possible in a natural geologic environment. The understanding of thermomechanical processes and their influences on fluid and solute movement in fractured rocks may have to be improved, particularly where temperature variability is significant. Another approach has been to assume that the fractured medium can be described as a double-porosity medium consisting of porous blocks separated by fractures. However, this concept requires a measure of the degree of hydraulic and diffusive connection between the fractures and the porous blocks. Regardless of the conceptual model selected, it is very difficult to measure the relevant properties of the discrete secondary features, such as dependence of fracture permeability on the stress tensor, fracture geometry, distribution of fracture apertures, and degree of fracture interconnection. In fact, the very existence of fractures is not always known. Thus, although the conceptual models mentioned above show some

promise, the nature of ground-water flow and transport in fractured media is not yet completely understood.

During the energy crisis of the early 1970s, there was considerable emphasis on hydrogeological research associated with energy development. The hydrogeology of geothermal systems (including low-temperature systems, such as the Gulf Coast geopressured system), hydrogeologic effects of oil-shale development, aquifer thermal-energy storage systems, in situ coal gasification, in situ uranium mining, and hydrologic effects of coal mining, are all examples of areas of investigation that the present energy "glut" had removed from the limelight by the early 1980s. Although it is difficult to predict when the next energy crisis will occur, we will be better prepared to face it if research on hydrogeological problems in these areas were to continue, even at a minimal level.

The evolution of ground-water flow systems over geologic time may influence the modern distributions of hydraulic head and solute in regional aquifers. Most ground-water studies focus on problems associated with recent effects of human activities on present-day hydrogeologic conditions, and these studies may involve the prediction of hydrologic changes for a few years or decades into the future. Ground-water flow and solute-transport models commonly are calibrated with historical data collected during a limited time period of only a few years or decades. For purposes of model calibration, it generally is assumed that the aquifer system is in a natural steady-state condition prior to human influence. However, ground-water flow systems dynamically change over geologic time in response to changing boundary conditions or aquifer properties induced by long-term climatic changes and geologic processes, such as sea-level changes, glacial loading or unloading, mineral dissolution or precipitation, sediment deposition or erosion, and tectonic forces. In some cases, modern systems may still be undergoing slow transient responses to such stresses, especially in low-permeability environments. Such responses may be evident in anomalous hydraulic heads or by inconsistencies between heads and distributions of solutes or isotopes. These anomalies or inconsistencies can have important implications for understanding many geological problems; for example, the position of a freshwater-saltwater interface, diagenesis, ore deposition, porosity and permeability enhancement, origin of brines, and petroleum migration. Conversely, a knowledge and understanding of the dynamics of ground-watter flow systems may help in explaining or interpreting certain geologic features observed in rocks. Although such research may seem at first glance to be fairly esoteric, it may have real practical value as an analog for risk assessment at high-level radioactive-waste repositories, where regulators require predictions of ground-water flow and transport for 10,000 years into the future.

As computer technology continues to evolve, there will likely be advances in the development of more efficient three-dimensional simulation models. It is a three-dimensional world, and so, in spite of data deficiencies, there are some distinct advantages to three-dimensional simulations. Winkler and others

(1987) state, "Only in three spatial dimensions do we have a chance to capture all the essential physics of fluid flow." However, they also note that the requirements for the visualization of three-dimensional time-dependent problems are considerably greater than for two dimensions. The next decade will likely yield significant new developments in the coupling of ground-water simulation models with interactive graphics capabilities for post-simulation processing and visualization of model results.

## CONCLUSIONS

Hydrogeology is a growing multidisciplinary science, and much of the recent growth has arisen from concern about the hazards associated with ground-water contamination. Significant advances in the field of hydrogeology have been made during the last 50 years. However, there are still numerous unsolved or poorly understood scientific problems that hydrogeologists face today and are likely to continue to face in the future. We have tried to outline some of these.

Particularly difficult problems arise when or where the hydrogeologist interfaces with other scientific disciplines or sub-disciplines, such as geophysics, microbiology, geochemistry, organic chemistry, soil physics, and environmental engineering. Advances may require that the hydrogeologist serve as part of a multidisciplinary team of scientists. Other major challenges within the field of hydrogeology include, but are not limited to, problems associated with the evaluation of ground-water flow and contaminant transport in low-permeability environments, in systems having significant secondary permeability development (such as fractures, solution openings, and lava tubes), and in heterogeneous systems. Small-scale and local heterogeneities are much more important for predicting transport phenomena than for analyzing flow problems. Improved methods are needed to define these heterogeneities explicitly and to characterize the variability in aquifer properties statistically. The scale problem, reflecting the transition from pore-scale to local-scale to regional-scale phenomena, is a difficult one both in terms of measurement of properties (and their variability) as well as in the prediction of responses.

The recognition and use of quantitative hydrogeology as an approach to solving other classical geological problems, such as ore deposition, earthquake prediction, and petroleum migration, have not yet reached their full potential. These also are examples of nonlinear types of problems, which, in general, include those situations in which fluid or hydraulic properties are affected by transport or reactions. Very few of these complex types of problems have been intensively studied and described in the field or simulated adequately in transient, three-dimensional models.

This range of problems will continue to challenge hydrogeologists and maintain hydrogeology as a vibrant and important field of study for decades to come. The generation of meaningful solutions to these scientific problems in hydrogeology will likely require that theoretical and laboratory investigations be attuned to and integrated with the reality and complexity of hydrogeological field problems.

## REFERENCES CITED

Chapelle, F. H., Zelibor, J. L., Jr., Grimes, D. J., and Knobel, L. L., 1987, Bacteria in deep Coastal Plain sediments of Maryland; A possible source of $CO_2$ to groundwater: Water Resources Research, v. 23, no. 8, p. 1625–1632.

Cord-Ruwisch, R., Kleinitz, W., and Widdel, F., 1987, Sulfate-reducing bacteria and their activities in oil production: Journal of Petroleum Technology, v. 39, no. 1, p. 97–106.

Dria, M. A., Bryant, S. L., Schechter, R. S., and Lake, L. W., 1987, Interacting precipitation/dissolution waves; The movement of inorganic contaminants in groundwater: Water Resources Research, v. 23, no. 11, p. 2076–2090.

Faust, C. R., 1985, Transport of immiscible fluids within and below the unsaturated zone; A numerical model: Water Resources Research, v. 21, no. 4, p. 587–596.

Gale, J. E., 1982, Assessing the permeability characteristics of fractured rock: *in* Narasimhan, T. N., ed., Recent trends in hydrogeology, Geological Society of America Special Paper 189, p. 163–181.

Garabedian, S. P., and LeBlanc, D. R., 1987, Results of spatial moments analysis for a natural-gradient tracer test in sand and gravel, Cape Cod, Massachusetts [abs.]: EOS Transactions of the American Geophysical Union, v. 68, no. 16, p. 322–323.

Gelhar, L. W., and Axness, C. L., 1983, Three-dimensional stochastic analysis of macrodispersion in aquifers: Water Resources Research, v. 19, no. 1, p. 161–180.

Gillham, R. W., and Cherry, J.A., 1982, Contaminant migration in saturated unconsolidated geologic deposits, *in* Narasimhan, T. N., ed., Recent trends in hydrogeology: Geological Society of America Special Paper 189, p. 31–61.

Jackson, R. E., 1987, A survey of contaminant hydrogeology in Canada: EOS Transactions of the American Geophysical Union, v. 68, no. 3, p. 35–36.

Mackay, D. M., Ball, W. P., and Durant, M. G., 1986, Variability of aquifer sorption properties in a field experiment on groundwater transport of organic solutes; Methods and preliminary results: Journal of Contaminant Hydrology, v. 1, p. 119–132.

Mackay, D. M., Freyberg, D. L., Roberts, P. V., and Cherry, J. A., 1986, A natural gradient experiment on solute transport in a sand aquifer; 1, Approach and overview of plume movement: Water Resources Research, v. 22, no. 13, p. 2017–2029.

Molz, F. J., Güven, O., Melville, J. G., and Keely, J. F., 1987, Performance and analysis of aquifer tracer tests with implications for contaminant transport modeling; A project summary: Ground Water, v. 25, no. 3, p. 337–341.

Neuman, S. P., 1987, Stochastic continuum representation of fractured rock permeability as an alternative to the REV and fracture network concepts: Proceedings of the 28th U.S. Symposium on Rock Mechanics, Tucson, p. 533–561.

Neuman, S. P., Winter, C. L., and Newman, C. M., 1987, Stochastic theory of field-scale Fickian dispersion in anisotropic porous media: Water Resources Research, v. 23, no. 3, p. 453–466.

Olhoeft, G. R., 1986, Direct detection of hydrocarbon and organic chemicals with ground penetrating radar and complex resistivity, *in* NWWA/API Conference on Petroleum Hydrocarbons and Organic Chemicals in Ground Water: Worthington, Ohio, National Water Well Association, p. 284–303.

Pinder, G. F., and Abriola, L. M., 1986, On the simulation of nonaqueous phase organic compounds in the subsurface: Water Resources Research, v. 22, no. 9, p. 109S–119S.

Refsgaard, A., 1986, Laboratory experiments on solute transport in non-homogeneous porous media: Nordic Hydrology, v. 17, no. 4/5, p. 305–314.

Rubin, J., 1983, Transport of reacting solutes in porous media; Relation between mathematical nature of problem formulation and chemical nature of reactions: Water Resources Research, v. 19, no. 5, p. 1231–1252.

Silliman, S. E., and Simpson, E. S., 1987, Laboratory evidence of the scale effect in solute transport: Water Resources Research, v. 23, no. 8, p. 1667–1673.

Silliman, S. E., Konikow, L. F., and Voss, C. I., 1987, Laboratory investigation of longitudinal dispersion in anisotropic porous media: Water Resources Research, v. 23, no. 11, p. 2145–2151.

Smith, L., and Schwartz, F. W., 1980, Mass transport; 1, A stochastic analysis of macroscopic dispersion: Water Resources Research, v. 16, no. 2, p. 303–313.

Valentine, R. M., and Kwader, T., 1985, Terrain conductivity as a tool for delineating hydrocarbon plumes in a shallow aquifer; A case study, *in* NWWA Conference on Surface and Borehole Geophysical Methods in Ground Water Investigations: Worthington, Ohio, National Water Well Association, p. 52–63.

Willis, C., and Rubin, J., 1987, Transport of reacting solutes subject to a moving dissolution boundary; Numerical methods and solutions: Water Resources Research, v. 23, no. 8, p. 1561–1574.

Wilson, J. T., and McNabb, J. F., 1983, Biological transformation of organic pollutants in groundwater: EOS Transactions of the American Geophysical Union, v. 64, no. 33, p. 505.

Winkler, K-H. A., Chalmers, J. W., Hodson, S. W., Woodward, P. R., and Zabusky, N. J., 1987, A numerical laboratory: Physics Today, v. 40, no. 10, p. 28–37.

Manuscript Accepted by the Society March 15, 1988

## ACKNOWLEDGMENTS

The authors appreciate helpful discussions with and comments from Allen Shapiro, Peter Haeni, Gordon Bennett, and David Aronson.

The Geology of North America
Vol. O-2, Hydrogeology
The Geological Society of America, 1988

# Chapter 50—Epilogue

# *Societal problems*

**J. S. Rosenshein**
*U.S. Geological Survey, 414 National Center, Reston, Virginia 22092*
**William Back**
*U.S. Geological Survey, 431 National Center, Reston, Virginia 22092*

Hydrogeology has played an integral role in the cultural evolution of civilization throughout ancient and modern history, often in association with aspects of water supply and management. For example, in North and South America, the cultural diversity of the various Indian tribes and nations can be attributed, to a large extent, to their hydrologic beliefs and use of water (Back, 1981). A more recent analogy is the economic expansion of much of the United States, which has been based on advances in well drilling, pumping equipment, and an understanding of the basic principles of ground water (Rosenshein and others, 1986). Societal and cultural responses to this development, in turn, spurred intense activity in the water-related aspects of geology that laid the foundations for the science of hydrogeology.

Societal and technological pressures that resulted from World War II also had an impact on hydrogeology. Demands for ground water in support of the war effort caused by establishment of military bases and industrial sites increased markedly. This resulted in the origination and application of quantitative ground-water techniques. These demands brought to the forefront such hydrogeologic problems as well-field design, stream-flow depletion and interactions, and salt-water encroachment. Investigations associated with the resolution of these problems were of fundamental importance in elucidating much of the basic understanding of the science of hydrogeology.

Many of the societal problems concerning hydrogeology are related either to man's conflict with man or man's conflict with nature. In most parts of North America, competition exists between use of water for public supply, agriculture, and industry. Pressures that result from this competition have led to the establishment of water-rights doctrines and to many legal battles between individuals, between individuals and states, and between states and other public governmental bodies. Recognition of the relation of aquifers to streams has led to a broadening of concerns of legality relative to ground- and surface-water rights. The competition for water has, in turn, resulted in pressures for more intensive and systematic studies of water-resources development and management that have broadened the demand for research applications to hydrogeology. In North America these societal pressures may, in the future, result in establishing multi-state,

multi-province, or bi-national compacts similar to those for streams to govern water use from regional aquifer systems.

Use of ground water in the more arid parts of North America, where withdrawal exceeds recharge, has led to "mining" of ground water. This "mining," in conjunction with the drought of the mid-1970s in the United States, led to the systematic evaluation of regional aquifer systems by the U.S. Geological Survey. Many chapters in the section on Regional Hydrogeology of this volume have benefited from the hydrogeologic understanding that resulted from these regional investigations.

Extensive use of ground water has caused a number of problems of great economic significance. Among the more notable problems are those related to land subsidence and salt-water encroachment. Gradual subsidence of land as a result of large ground-water withdrawals has occurred, for example, in parts of California and Texas. A more catastrophic type of subsidence—sinkhole collapse—such as occurs in Alabama and Florida, is also attributed to extensive withdrawals of ground water. Much of our understanding of processes associated with subsidence has come from pioneering work done by ground-water scientists in the mid-1950s to the mid-1970s.

Ground-water quality and related hydrogeological problems have been a long-term societal concern. Prior to the late 1960s, this concern was primarily with the natural occurrence of undesirable chemical constituents in ground water, the water's bacterial quality, and to some extent the presence of viruses. The effect of human activities on the quality of ground water was slow to be recognized. Initially this recognition usually was by detection of isolated concentrations of metals or, more commonly, detection of salt-water contamination of ground water resulting from industrial activities or ground-water pumpage. In the U.S., the pressures of the 1970s to store toxic fluids in the deep subsurface, and to dispose of sewage-treatment plant effluent and land-fill wastes at the surface and in the subsurface, gradually led to a heightened public concern about underground waste management (Braunstien, 1973). These concerns have markedly intensified as the hazards of synthetic organic compounds to ground-water quality have become recognized. Although many of the contamination problems associated with these substances are site specific, the sites commonly are in urban areas, and the resultant contamination can affect large numbers of people. More pervasive is the contamination from modern synthetic organic agricultural compounds and their degradation products. These concerns about effects of fertilizers, herbicides, and pesticides on ground-water quality are being addressed. Apprehensions about ground-water contamination have led the U.S. Congress to pass legislation such as Superfund, and various clean water and safe drinking water acts. This legislation, which deals with control of disposal, storage, and cleanup of hazardous substances, has greatly increased demand for trained hydrogeologists (Moore, 1987). The role of hydrogeology and the responsibility of the hydrogeologist in addressing modern-day societal concerns about ground-water contamination problems has grown markedly in the past few decades and will continue to grow in the future.

## REFERENCES CITED

Back, W., 1981, Hydromythology and ethnohydrology in the New World: Water Resources Research, v. 17, p. 257–87.

Braunstein, J., ed., 1973, Underground waste management and artificial recharge: American Association of Petroleum Geologists, v. 1–2, 927 p.

Moore, J. E., 1987, Job outlook is good for hydrogeologists: Geotimes, May 1987, p. 15.

Rosenshein, J. S., Moore, J. E., Lohman, S. W., and Chase, E. B., eds., 1986, Two hundred years of hydrogeology in the United States, U.S. Geological Survey Open-File Report 86–480, 110 p.

Manuscript Accepted by the Society September 9, 1987

# *Index*

[Italic page numbers indicate major references]

Typeset by WESType Publishers Services, Inc., Boulder, Colorado
Printed in U.S.A. by Malloy Lithographing, Inc., Ann Arbor, Michigan